Plant Life of a Desert Archipelago

The Southwest Center Series

Joseph C. Wilder, Editor

Plant Life of a Desert Archipelago

Flora of the Sonoran Islands in the Gulf of California

Richard Stephen Felger

Benjamin Theodore Wilder

in collaboration with
Humberto Romero-Morales

Foreword by Exequiel Ezcurra

THE UNIVERSITY OF
ARIZONA PRESS
TUCSON

The University of Arizona Press
www.uapress.arizona.edu

Printed in the United States of America
21   20   19   18   17   16   7   6   5   4   3   2

ISBN-13: 978-0-8165-0243-1 (cloth)
ISBN-13: 978-0-8165-3430-2 (paper)

Cover designed by Leigh McDonald
Cover photo by Benjamin Wilder. Isla Cholludo. Cardón (*Pachycereus pringler*) and a few
organ pipes (*Stenocereus thurberi*). Isla Tiburón in the background to the north.

Publication of this book is made possible in part by a grant from the
Southwest Center of the University of Arizona, with the support of
the National Commission for the Knowledge and Use of Biodiversity
(CONABIO), Mexico, and by The David and Lucile Packard Foundation.

Library of Congress Cataloging-in-Publication Data
Felger, Richard Stephen.
  Plant life of a desert archipelago : flora of the Sonoran islands in the Gulf of California /
Richard Stephen Felger and Benjamin Theodore Wilder in collaboration with Humberto
Romero-Morales ; foreword by Exequiel Ezcurra.
      p.   cm. — (Southwest Center series)
  Includes bibliographical references and index.
  ISBN 978-0-8165-0243-1 (cloth : alk. paper)
  1. Island plants—Mexico—California, Gulf of. 2. Island plants—Sonoran Desert. I.
Wilder, Benjamin Theodore. II. Romero-Morales, Humberto. III. Title. IV. Series:
Southwest Center series.
  QK211.F387 2012
  582.13—dc23
                              2012015047

♾ This paper meets the requirements of ANSI/NISO Z39.48-1992 (Permanence of Paper).

For the Comcaac

&

Cathy Moser Marlett
and Stephen Alan Marlett

For my parents, Rebecca and Janos Wilder
&
my late grandfather, Harvey Bracker,
whose steps I follow in the borderlands

—Ben

For Silke Schneider
and Mary Beck Moser and the late Ed Moser

—Richard

Para mi esposa Herminia Morales Comito
y mi hijo Adán Humberto Romero Morales

—Humberto

Contents

Tables

Cactus scrub, Isla Cholludo. Cardón (*Pachycereus pringlei*) and a few organ pipes (*Stenocereus thurberi*). Isla Tiburón in the background to the north. BTW, 2 September 2008.

Foreword

Exequiel Ezcurra

The archipelago is a little world within itself, or rather a satellite attached to America, whence it has derived a few stray colonists, and has received the general character of its indigenous productions. Seeing every height crowned with its crater, and the boundaries of most of the lava-streams still distinct, we are led to believe that within a period geologically recent the unbroken ocean was here spread out. Hence, both in space and time, we seem to be brought somewhat near to that great fact—that mystery of mysteries—the first appearance of new beings on this earth.

—Charles Darwin, *The Voyage of the Beagle* (1860)

In 1967 Robert MacArthur and Edward O. Wilson brought a breeze of excitement to the scientific community with the publication of their book *The Theory of Island Biogeography*, in which they proposed a model identifying and explaining the factors that drive the species richness of natural communities. Although the theory was developed to explain the biological richness and species distribution patterns in oceanic islands, it was obvious for most scientists that the implications were potentially immense. In a world increasingly fragmented by human activities, montane forests surrounded by drylands, continental lakes, tropical forest fragments surrounded by induced pastures, or protected natural areas surrounded by agriculture, can all be considered, in a way, ecological islands. A theory explaining the dynamics of species immigration, extinction, survival, and evolution in these enclosed microcosms promised a completely new approach for better understanding and managing the natural world. A scientific revolution was under way.

As a result, myriad researchers set out to study island ecosystems, trying to understand the patterns observed and to test the theory's main tenets. Island research became the driving force expanding the conceptual frontier of ecological sciences. Many American researchers were quick to realize that one of the most unspoiled archipelagos in the world is found in the Gulf of California, and many expeditions were organized to inventory and survey the biological richness of the Gulf Islands. Michael Soulé, the founder of the science of conservation biology, started working in 1970 with the genetics and population biology of island lizards. The exciting results of his fieldwork, and the sheer inspiration brought by the experience of working in the Gulf Islands, led him eventually to apply his knowledge into a new discipline. The emerging science of conservation biology was launched with the publication of an article in *Bioscience* (Soulé 1985) that opened his new ideas to the scientific world. Island biogeography in general, and the Gulf Islands in particular, were at the root of the development of conservation as a science. Many other researchers followed the same track. The California Academy of Sciences, together with the University of California, Stanford University, and the San Diego Natural History Museum, regularly organized research trips to the Gulf Islands, and many seminal publications were derived from these expeditions.

At that same time, Richard S. Felger was also working in the Gulf of California for his PhD at the University of Arizona, studying the flora of the Sonoran Midriff Islands. His approach, however, was different. Most researchers analyzed the biota of the islands mostly as a result of natural causes—immigration, emigration, extinction, and distance to the continental source—and believed, following the theory that over time, the countervailing forces of extinction and immigration would result in an equilibrium level of species richness. Felger, however, realized that human activities had been going on for a long time in the gulf, and that these anthropogenic factors were potentially very important. With strong human influences moving species around, it was difficult to conceive of the islands as in biological equilibrium.

Thus, and following the ethnobotanical tradition of researchers such as Arturo Gómez-Pompa and Andrea Kaus (1992), Richard Felger approached the island biogeography problem by "taming the wilderness myth." Instead of ignoring human influences in the gulf's Midriff ecosystem, he worked with the Comcaac (Seri people) on the coast of Sonora and made a serious effort to understand their profound interaction with the island and coastal environments. In 1976, at the same time he and Charles Lowe published their immensely important report "The Island and Coastal Vegetation and Flora of the Northern Part of Gulf of California, Mexico," he also published a paper describing his own ethnobotanical perspective on the natural world (Felger 1976). He clearly showed that areas, which to many researchers seemed remote and pristine, had been used by humans from very early times, and that understanding these uses and the associated lore was fundamental to understanding the natural world. Faithful to his own approach, in 1985 with Mary Beck Moser he published *People of the Desert and Sea*, without any doubt one of the most comprehensive ethnobotanical treatises in the world.

In 2006, thirty years after the publication of his first analysis of the vegetation and flora of the gulf coast, Richard Felger returned to the Midriff Islands with a young botanist, Benjamin Wilder, and both worked in collaboration with a Seri ecologist and friend of both, Humberto Romero. They revisited Tiburón, San Esteban, San Pedro Mártir, San Pedro Nolasco, Dátil, Cholludo, Rasa, and all of the islets that surround the larger Midriff Islands. They inventoried, surveyed, collected, and pressed thousands of plants, ran life-threatening adventures, and published jointly some truly exciting papers (including Felger et al. 2011; Wilder & Felger 2010; Wilder et al. 2008a, 2008b).

This book is the result of their work. It is many things at the same time. On the one hand, it gives the most precise species account of the islands that has ever been made and forms an invaluable database for ecologists and biogeographers that want to test the ideas of island biogeography models in this archipelago. The authors climbed every hill, walked every slope, and collected with a meticulous care as it had never been done before. They also reviewed all the existing herbarium records, and checked all the previous collections, often correcting misidentifications and erroneous records. Although a flora is never complete, if a plant species is on an island, rest assured it has been recorded in their study.

On the other hand, the flora establishes a benchmark for future studies. At the accelerated rate that environmental changes are happening in the Gulf, a serious database describing with rigor what grows there now is likely to be of immense value in the future. The flora is an invaluable snapshot of current environmental conditions in the Gulf of California's Midriff and a witness of current conditions against which we may compare future change.

Third, the flora is a fundamental tool for any researcher working in the region. Conservation biologists, reserve managers, students, and researchers in general will find in this book an aid in their fieldwork that has not existed before. Mexico needs desperately floras and field guides in order to better manage and preserve its natural resources, and this flora is extraordinarily detailed and painstakingly crafted for field biologists.

Lastly, the publication of this flora opens the door for many more investigations. Darwin's "mystery of mysteries"—the appearance of life in geologically recent environments—is still as fascinating as it was when the *Beagle* anchored for the first time in the Galapagos. The evolutionary history of Mexico's northwestern deserts is recorded in the flora of these islands; it is written in the Pleistocene relics of their mountainous peaks, in the plant litter accumulated in packrat middens, in the DNA of many of their plants, in the age distribution of their populations of giant cacti. A detailed and taxonomically rigorous flora is a basic step toward this purpose, as it establishes a unique and powerful tool for the unraveling of these enigmas.

Paraphrasing Darwin's own observations in the Galapagos, this book seems to be able to bring us somewhat near to that great fact, that mystery of mysteries: the fascinating evolutionary history of the Midriff Islands and of the Sonoran Desert.

References

Felger, R. S. 1976. Gulf of California—an ethno-ecological perspective. *Natural Resources Journal* 16(3):451–464.

Felger, R. S., & C. H. Lowe. 1976. The island and coastal vegetation and flora of the northern part of Gulf of California. *Natural History Museum of Los Angeles County, Contributions in Science* 285:1–59.

Felger, R. S., & M. B. Moser. 1985. *People of the Desert and Sea: Ethnobotany of the Seri Indians.* University of Arizona Press, Tucson.

Felger, R. S., B. T. Wilder, & J. P. Gallo-Reynoso. 2011. Floristic diversity and long-term vegetation dynamics of Isla San Pedro Nolasco, Gulf of California, Mexico. *Proceedings of the San Diego Society of Natural History* 43:1–42.

Gómez-Pompa, A., & A. Kaus. 1992. Taming the wilderness myth. *BioScience* 42:271–279.

MacArthur, R., & E. O. Wilson. 1967. *The Theory of Island Biogeography.* Princeton University Press, Princeton, NJ.

Soulé, M. E. 1985. What is conservation biology. *Bioscience* 35(11):727–734.

Wilder, B. T., & R. S. Felger. 2010. Dwarf giants, guano, and isolation: vegetation and floristic diversity of Isla San Pedro Mártir, Gulf of California, Mexico. *Proceedings of the San Diego Society of Natural History* 42:1–24.

Wilder, B. T., R. S. Felger, & H. Romero-Morales. 2008a. Succulent plant diversity of the Sonoran Islands, Gulf of California, Mexico. *Haseltonia* 14:127–160.

Wilder, B. T., R. S. Felger, T. R. Van Devender, & H. Romero-Morales. 2008b. *Canotia holacantha* on Isla Tiburón, Gulf of California, Mexico. *Canotia* 4(1):1–7.

Southern foothills of the Sierra Kunkaak, Isla Tiburón.
PRG, October 2001.

Preface

There is something about this region that captures the imagination, curiosity, and sense of wonder of many, and certainly this is the case for us.

Richard's perspective: I was about eight years old when I started growing cactus and succulents, and because we lived by the sea it was sea life that first caught my imagination. So when I was a freshman at the University of Arizona and first encountered the cactus-rich islands in the Gulf of California, there was no doubt that I wanted to see and learn more. It did not take much to ditch a few classes on Fridays or Mondays so I could make it to Bahía Kino and go with fishermen in their small, open pangas. They went fishing and I explored the islands. Back in Tucson I would spend hours and days in the herbarium and library identifying newly discovered plants and searching for more information. Eventually this interest led to work on my dissertation. But before that pseudo-magico-initiation process, I had the good fortune to meet Jean Russell and her husband, Alexander "Ike" Russell, who took me on trips to the Sonoran region. Just as I was fascinated with plants, Ike loved to fly his airplane. It was a different and simpler world—we landed at a number of places on Isla Tiburón as well as the mainland, and one day Ike introduced me to Mary Beck and Ed Moser at the Seri village of El Desemboque. Through the Mosers I began meeting elderly Seris, and our common interest and knowledge of the sea and the desert opened my universe. Twenty-five years later Becky Moser and I finished our book on the ethnobotany of the Seris. Over the years I had botanized on all the Sonoran Islands, spent more than 175 days on Isla Tiburón, made extensive botanical collections, studied herbarium specimens of my predecessors and colleagues, campaigned for conservation including the heart-wrenching situation of sea turtles, managed to publish assorted papers and books, and botanized in distant parts of the world. Yet the completion of my studies of the flora of these magnificent desert islands eluded completion until Ben drew me back into an earlier and now expanded vision.

Ben's perspective: Upon my first visit to the region in 2004, desert islands, their flora, and questions of how insular floras differed from one another and the mainland filled my head. From Punta Chueca, the view of Isla Tiburón and the island's expansive and uninterrupted bajadas filled me with awe, and the region firmly nestled into my heart. Upon returning to Tucson, and following the suggestions of ecologists at the Desert Laboratory on Tumamoc Hill and at the University of Arizona Herbarium, I met Richard. I asked, "Is there a flora of Isla Tiburón?" Richard responded that a flora of the island had not been produced, but that such a project represented something he had long wanted to do. He continued by saying, "Why don't you do the flora?" In shock, I explained that my botanical skills were largely undeveloped and I would not know where to begin. "This is how you learn," Richard said. I thought very seriously about this offer and soon realized I had the opportunity of a lifetime in front of me, which I had to follow.

Over the New Year holiday of 2005–2006, accompanied by good friend Edward Gilbert, I made my first collecting trip to the region. As we neared Bahía Kino, and the extensive Sierra Kunkaak on Tiburón rose out of the Gulf, the enormity of cataloguing and studying the plants of Mexico's largest island, as well as seven others off the coast of Sonora, began to set in. We went to Punta Chueca, and with the guidance of Steve Marlett and Cathy Moser Marlett met Humberto Romero-Morales and explained the goal of the project. Within hours, Humberto, Ed, and I were exploring the sheltered canyons of the Sierra Kunkaak and the gaining and sharing of knowledge had begun. Suddenly this overwhelming botanical task seemed probable.

After so many years and dozens of field trips to the islands by Ben, Richard, and Humberto, in October 2007, at the top of Tiburón, when the massive body of land finally appeared as an island, we realized that we knew enough to get this work into your hands.

Acknowledgments

Our work has been made possible by the wonderful community of family, friends, and colleagues we are fortunate to have in our lives. Our efforts were enhanced by funding provided by several institutions, but principally that of the David and Lucile Packard Foundation and the Comisión Nacional para el Conocimiento y Uso de la Biodiversidad (CONABIO). Additional generous support was received by Ben from Dora and Barry Bursey; the University of Arizona Mycological Herbarium; Arizona-Nevada Academy of Science undergraduate grant-in-aid; University of Arizona Honors College alumni legacy grants; Department of Ecology and Evolutionary Biology, University of Arizona, Leslie N. Goodding Memorial Scholarship; and the Research Committee of the Cactus and Succulent Society of America; and this work is in part based upon Ben's work supported under a National Science Foundation graduate research fellowship. Richard received support from the Wallace Research Foundation for portions of the earlier work as well as the most recent efforts.

We also received support from the World Wildlife Fund in collaboration with El Área de Protección de Flora y Fauna "Islas del Golfo de California" en Sonora, de la Comisión Nacional de Áreas Naturales Protegidas (CONANP), de la Secretaría de Medio Ambiente y Recursos Naturales (SEMARNAT); T&E Inc. foundation of New Mexico; and Global Green Grants.

A great deal of advice and general assistance were provided by many individuals, especially A. Elizabeth (Betsy) Arnold, Julio L. Betancourt, Thomas Bowen, Janice Emily Bowers, Alberto Búrquez-Montijo, Mark A. Dimmitt, Exequiel Ezcurra, Ana Luisa Rosa Figueroa-Carranza, James Henrickson, Cathy Moser Marlett, Stephen Alan Marlett, the late Paul S. Martin, Angelina Martínez-Yrízar, Francisco Molina-Freaner, Ana Lilia Reina-Guerrero, Peter Michael Sherman, Raymond Marriner Turner, and Thomas R. Van Devender.

Our knowledge of the Gulf of California was greatly aided through rewarding communication with the following friends and colleagues: Xavier Basurto, Louis Bourillón-Moreno, Diane E. Boyer, Georgiana Boyer, David E. Brown, Richard C. Brusca, Ana Luisa Rosa Figueroa-Carranza, Juan Pablo Gallo-Reynoso, Powell B. (Gil) Gillenwater III, R. James Hills, Paul Knight, José Luis León de la Luz, Rodrigo A. Medellín, Lorayne R. Meltzer, Gary Paul Nabhan, Michael E. Oskin, Tad Pfister, Adrián Quijada-Mascareñas, Patricio Robles Gil, Robert (Bob) Russell, Araceli Samaniego-Herrera, Norman Scott, Jorge Torre-Cosío, Enriqueta Velarde, and Richard S. White.

We were privileged to be accompanied in the field by many friends who shared in the adventure, including Erik Salvador Barnett, Juan Alfredo Barnett-Díaz, Raymundo Barnett-Morales, Ivan Eliseo Barnett-Romero, C. David Bertelsen, Bradley (Brad) Lorne Boyle, Exequiel Ezcurra, Pedro Ezcurra, Abram B. Fleishman, Andrea Galindo-Escamilla, Juan Pablo Gallo-Reynoso, Edward Erik Gilbert, Mikhal Gold, Servando López-Monroy, Maximiliano Damián López-Romero, Brigitte Marazzi, Raúl Eduardo Molina-Martínez, Nestor Cristóbal Montaño-Herrera, Gloria Guadalupe (Lupita) Morales-Figueroa, Michael E. Oskin, Ana Lilia Reina-Guerrero, Adán Humberto Romero-Morales, Jesús Sánchez-Escalante, Alexander Swanson, José Ramón Torres, Seth Marriner Turner, and Jesús Ventura-Trejo.

Our efforts in the field were supported by many friends and family: Cosme Damián Becerra, Bruce E. Bracker, Nellie J. Bracker, Bill Broyles, Florencio Cota-Moreno, Lloyd Findley, Ernesto Inojosa, Lawrence A. Johnson, Lorayne R. Meltzer, Ernesto Molina, Erminia Morales-Comito, Tad Pfister, William (Bill) J. Risner, the late Alexander Russell, Jean Russell, Barbara Straub, and Jesús Ventura-Trejo. We are also appreciative to several individuals for additional assistance: Travis M. Bean, Carrie A. Bracker, Roger Howard Hamstra, Linda Klasky, Bete Jones Pfister, Jeffory Kent Pralle, and Michael Francis Wilson.

We are indebted to numerous colleagues who shared their knowledge of the flora of the Sonoran Desert: Daniel F. Austin (Convolvulaceae), J. Travis Columbus (Poaceae), Mihai Costea (*Cuscuta*), Thomas F. Daniel (Acanthaceae), Mark Fishbein (Apocynaceae), James Henrickson (many taxa), Philip D. Jenkins (general knowledge of Sonoran Desert flora), José Luis León de la Luz (general knowledge of Sonoran Desert flora), Michelle (Shelley) McMahon (problem legume specimens), Jon P. Rebman (general knowledge of Sonoran Desert flora, especially Baja California plants), the late Charlotte and John Reeder (Poaceae), Andrew Michael Salywon (Brassicaceae), Jesús Sánchez-Escalante (general knowledge of Sonoran Desert flora), Richard W. Spellenberg (Nyctaginaceae), Victor W. Steinmann (Euphorbiaceae), Gordon C. Tucker (Cyperaceae), Thomas R. Van Devender (general knowledge of Sonoran Desert flora), and George Yatskievych (Pteridaceae and systematics).

The staff of several herbaria were incredibly helpful and tolerant of our multiple requests, and their support was invaluable: Bradley (Brad) Lorne Boyle, Benjamin Daniel Brandt, W. Eugene Hall, Sarah Hunkins, Philip D. Jenkins, and Michelle (Shelley) McMahon made ARIZ a most wonderful home institution; Judy Ann Gibson at SD provided innumerable search results and other data; Jesús Sánchez-Escalante, USON; Wendy Caye Hodgson and Andrew Michael Salywon, DES; Sula Elizabeth Vanderplank, RSA; Andrew S. Doran, UC; Debra Trock, CAS; Andrew C. Sanders, University of California, Riverside; Carolyn Beans, GH; Christopher K. Frazier and Jane E. Mygatt, UNM; and Rusty Russell, US.

The work of several professionals has substantially added to the quality of our work, and we are greatly appreciative of their efforts: Jennifer Bain-Probst (translation of manuscript to Spanish, to be available in a forthcoming product), Tiernen Erickson (assistance with the formation of the specimen database), Kyle Hartfield (production of the plant distribution maps), Michelle Kostuk (scanning of line drawings and photos), and Nancy Jill Arora (production editor, University of Arizona Press), Norman Earl Tuttle (book designer and typesetter), Patricia J. Watson (copy editor), and Joseph Carlton Wilder (visionary director of the Southwest Center who encouraged this book at its every step).

The following institutions were of significant assistance: the University of Arizona Herbarium (ARIZ); Comisión Nacional de Áreas Naturales Protegidas (CONANP); the Desert Laboratory at Tumamoc Hill; Kino Bay organization "Rescue 1" for the safety net provided to people venturing to the Midriff Islands; Prescott College Center for Cultural and Ecological Studies, Bahía de Kino Bay; San Diego Natural History Museum and their herbarium (SD); the Herbarium of the University of Sonora (USON); and the herbarium of Centro de Investigaciones Biológicas del Noroeste (HCIB).

Financial management assistance was given by the following individuals: Richard Cudney-Bueno, program officer, and Jeanne McGinnis, program associate, the David and Lucile Packard Foundation Gulf of California Program; Susana Aguilar-Romero, Fondo Acción Solídaría, A.C. (FASOL) project manager, Global Green Grants (GGG), and René Córdova, GGG supporting member; Sandra Martínez, Red Fronteriza de Salud y Ambiente (non-profit for GGG); Julio César Mora-Sandoval, Red Fronteriza de Salud y Ambiente officer, local contact; and Erin Harm and Peggy Shorenstein, management of accounts at San Diego Natural History Museum.

Botanical collections were made under Mexican Federal Collecting permit NOM-126-SEMARNAT-2000 with the generous assistance of Exequiel Ezcurra and under a collection permit for the state of Sonora graciously provided by Jesús Sánchez-Escalante.

Abbreviations

Illustrators and Photographers

AE	Amy Eisenberg
AR	Alexander "Ike" Russell
BA	Bobbi Angell
BB	Brad Boyle
BM	Brigitte Marazzi
BTW	Benjamin T. Wilder
CDB	C. David Bertelsen
CMM	Cathy Moser Marlett
EHD	Edward H. Davis
ETN	E. Tad Nichols
EWM	Edward "Ed" W. Moser
FB	Felicia Bond
FR	Frances Runyan
GEL	George Edmund Lindsay
HRM	Humberto Romero-Morales
JRT	José Ramón Torres
KH	Kyle Hartfield
LBH	Lucretia Brezeale Hamilton
MBJ	Matthew B. Johnson
MBM	Mary Beck "Becky" Moser
NLN	Nancy L. Nicholson
NSB	Naomi Sara Blinick
PB	Phyllis Brick
PRG	Patricio Robles Gil
RJH	R. James Hills
RSF	Richard Stephen Felger
SMT	Seth Marriner Turner
WNS	William Neil Smith

Archives, Herbaria, and Institutions

ARIZ	University of Arizona Herbarium, Tucson
ASU	Arizona State University Herbarium, Tempe
BRIT	Botanical Research Institute of Texas Herbarium, Fort Worth
CAS	California Academy of Sciences Herbarium, San Francisco (includes Dudley Herbarium of Stanford University [DS])
CONANP	Comisión Nacional de Áreas Naturales Protegidas, Secretaría de Medio Ambiente y Recursos Naturales (SEMARNAT), Mexican federal government
DES	Herbarium, Desert Botanical Garden, Phoenix
DS	Dudley Herbarium of Stanford University; see CAS
GECI	Grupo de Ecología y Conservación de Islas, A.C., Ensenada, Baja California
GH	Gray Herbarium, Harvard University, Cambridge
HCIB	Herbarium of Centro de Investigaciones Biológicas del Noroeste, La Paz, Baja California Sur
MEXU	Herbario Nacional, Universidad Nacional Autónoma de México, Mexico City
MO	Missouri Botanical Garden Herbarium, St. Louis
NY	New York Botanical Garden Herbarium, Bronx, NY

POM   Pomona College Herbarium; see RSA

RSA   Herbarium, Rancho Santa Ana Botanic Garden, Claremont (includes Pomona College Herbarium [POM])

SARH   Secretaría de Agricultura y Recursos Hidráulicos, Mexico City

SD   San Diego Natural History Museum Herbarium, San Diego

SEMARNAT   Secretaría de Medio Ambiente y Recursos Naturales, Mexico City

SRSC   Sul Ross State University Herbarium, Alpine, Texas

TEX   University of Texas Herbarium, University of Texas at Austin

UC   University of California Herbarium, Berkeley

UNAM   Universidad Nacional Antónoma de México, Mexico City

UNM   University of New Mexico Herbarium, Albuquerque

US   U.S. National Herbarium, Smithsonian Institution, Washington, DC

USON   Herbario de la Universidad de Sonora, Hermosillo

Plants not native to flora area are marked with an asterisk (*).

Older spellings of most Seri language words have been standardized with those in use in Moser and Marlett (2010).

Plant Life of
a Desert Archipelago

PART I

The Islands and Their Vegetation

Introduction

Like rocks, from large boulders to small pebbles, lying in the middle of a flowing stream, the Midriff Islands of the Gulf of California stand as a partial bridge between the Baja California Peninsula and the Sonora mainland (fig. 1.1A,B). These desert islands, in the central part of the Gulf of California, between latitudes 28°20' and 29°40', display a continuum of adaptation by life to a harshly arid environment in an isolated setting. The high level of endemism found in the mammals and reptiles of the region (Grismer 2002; Lawler et al. 2002) is largely not mirrored in the flora; however, the vegetation and floristic makeup of each island is unique. From the cardón forests of the tiniest island, Cholludo, to the agave-dominated slopes of San Esteban, topographic, climatic, and biological forces have sculpted a set of unparalleled desert worlds. The Comcaac (Seri people) and their ancestors or predecessors established an existence from the eastern Midriff Islands and the Sonora mainland for thousands of years, based on the diverse terrestrial and marine life found there. Remarkably, these islands remain much as they were centuries ago, when the Comcaac were the only human presence in the region. Keep in mind, however, that there have been long and continuing human influences on these islands (Bowen 2009).

The present work focuses on the majority of the Sonoran Islands: Tiburón and its five neighbors, Alcatraz, Cholludo, Dátil, Patos, and San Esteban; San Pedro Mártir, in the middle of the Gulf; and San Pedro Nolasco to the northwest of Guaymas (fig. 1.2). We have built on the local knowledge of the Comcaac and insights gained from more than a century of scientific focus to better understand the ecology and botanical diversity of the world's best-preserved archipelago.

GEOLOGIC HISTORY AND ISLAND PHYSIOGRAPHY, WITH MICHAEL OSKIN

The Gulf of California opened between 15 and 5.5 million years ago (mya), with most evidence indicating two stages of evolution, a proto-Gulf (middle to late Miocene, 15–6 mya) and a modern Gulf (late Miocene and younger,

3

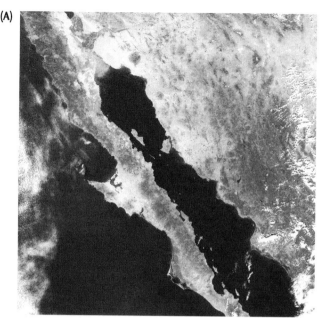

Figure 1.1. Midriff Region, Gulf of California. (A) Satellite image of the Gulf of California and Baja California Peninsula, with the Midriff Islands at center. NASA Rapid Fire image, *Aqua* satellite. (B) Looking from the northwest over the Baja California Peninsula (foreground), the Midriff Islands (middle), and Sonora to the southeast (background). NASA STS-004 shuttle mission, 4 July 1982.

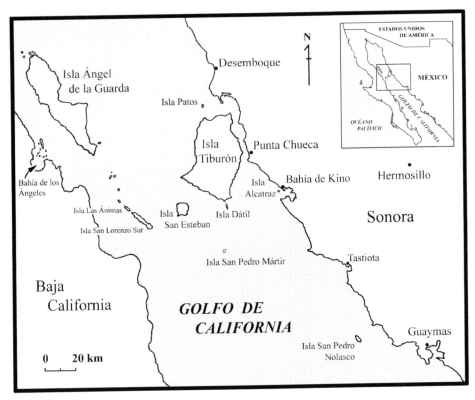

Figure 1.2. Midriff Region, Gulf of California. CMM.

ca. 6 mya to present; Helenes & Carreño 1999). Outcrops of marine rocks on the margins of the Gulf of California support that marine waters advanced as far as Santa Rosalia, midway up the Baja California Peninsula, by 8 mya, and into southeastern California by 6.5 mya (Oskin & Stock 2003a). Marine sedimentation prior to 11 mya has been suggested from core-hole samples drilled from the northern Gulf of California (Helenes et al. 2009). The Sonoran Islands are a mix of land-bridge islands (part of the mainland during the Pleistocene) and oceanic islands (in deep water, isolated throughout the ice ages). Tiburón and the small islands of Patos to the north, Dátil and Cholludo to the south, and Alcatraz in Bahía Kino are land-bridge islands and were connected to the Sonora mainland as recently as 11,000 years ago. San Esteban and San Pedro Mártir are oceanic islands, and San Pedro Nolasco is a geologically isolated island but was once part of the mainland.

Tiburón is part of the transition from the obliquely rifted Gulf of California to the Sonoran Basin and Range Province, marked by mountain ridges with a generally north-south orientation and interspersed with wide, alluvium-filled, low-lying valleys. The Canal del Infiernillo separates Tiburón from the Sonora mainland. This narrow and shallow Canal is a sea-filled valley between two large mountain ranges—the Sierra Seri on the mainland and the Sierra Kunkaak on the island. Farther west on Isla Tiburón, the large Central Valley separates the Sierra Kunkaak from the Sierra Menor. Off the western side of Tiburón there are deep, sediment-filled ocean basins that plummet hundreds of meters into the Gulf of California. Tiburón is composed of Mesozoic and older (pre-65 mya) intrusive rocks and adjacent metamorphosed sedimentary rocks. These are overlain by volcanic rocks extruded, intruded, or deposited either as part of the magmatic arc that existed from 24 to 16 mya in the area of the present Gulf (Gastil et al. 1999; Hausback 1984) or later in relation to rifting that led to the opening of the Gulf of California (Oskin & Stock 2003b).

Tiburón is the largest island in North America south of Canada and encompasses 1223 km^2 (472 mi^2; all island sizes and elevations are from Murphy et al. 2002, unless indicated otherwise). This massive extension of the mainland is composed of a multitude of unique regions, which are discussed in detail in the section on island diversity. Among the principal physiographic features is the Sierra Kunkaak, the extensive range along the eastern part of the island. This range runs the majority of the island's length and culminates in multiple high peaks, deep canyons, and an expansive upper ridge near the midpoint of the island. The peak elevation has been reported to be 1218 m (3995 ft; Murphy et al. 2002); however, an October 2007 trip to the summit and the use of a barometric altimeter and a GPS showed the highest elevation to be 885 m (2904 ft). The middle of the island is an alluvium-filled valley as high as 200 m elevation (656 ft) that separates the Sierra Kunkaak from the Sierra Menor, the rugged but lower lying western mountains. The Sierra Menor likewise runs the majority of the island's length but lacks distinct high peaks. Top elevation is approximately 670 m (2198 ft) at the southern part of the range (Felger 1966). Most of the southern part of the island consists of relatively lower but extremely rugged mountains. Two large arroyo systems drain the central parts of the island, Arroyo Agua Dulce at the north and Arroyo Sauzal on the south.

Islas Dátil and Cholludo are part of a geologic complex on the south end of Tiburón that has a distinct history from the rest of the island. As the San Andreas Fault opened the Gulf and separated Baja California from mainland Mexico, this newly liberated landmass drifted to the northwest. As it passed by the land that now forms Tiburón, a sliver of the Baja California Peninsula was transferred back to the south part of Tiburón (Oskin & Stock 2003a). The La Cruz Fault separates this sliver from the rest of the island. On aerial images of Tiburón, the fault is visible as a southeast-to-northwest line on the southern part of the island. The cover of volcanic rocks is much thinner on the southern block of Tiburón and is absent from Islas Cholludo and Dátil. These smaller islands are made up entirely of metamorphosed sedimentary rocks. Cholludo, at 0.2 km², is the smallest island in the region and reaches 61 m elevation (200 ft). Dátil is about 3 km long from north to south but not much more than 0.5 km wide, encompassing 1.25 km². The island is oriented north to south, with six steep and jagged peaks (the tallest is roughly 180 m, 590 ft). The peaks are interspaced by pronounced canyons that run east to west, making this small island extremely rugged and providing a multitude of niches for the island's relatively diverse flora.

The other two land-bridge islands are Patos and Alcatraz. Patos is a guano-covered island 0.45 km², in open water 8 km north of Tiburón. A single conical peak, topped by a defunct lighthouse, rises to approximately 70 m (230 ft) from a broad and relatively level plain about 4 m in elevation. Alcatraz is the island closest to the Sonora mainland and sits directly in front of the coastal town of Bahía de Kino. The island is 1.44 km² and consists of a large flat on the east and a mountain on the west that reaches approximately 165 m (541 ft).

Islas San Esteban, San Pedro Mártir, and San Pedro Nolasco have been separated from the mainland either throughout their existence or, like Nolasco, for several million years, and each has a markedly different geologic history. San Esteban is a calc-alkaline volcanic island that erupted between 2.5 and 2.9 mya, making it the youngest landmass in the region. Its eruption was about 9.5 million years after subduction beneath the continental margin ceased and at least 4.5 million years after extension began to open the Gulf (Desonie 1992). The island is separated from Tiburón by a deep channel (400 m; Carreño & Helenes 2002). San Esteban is the second largest island in the region, at 41 km², and has multiple distinct high mountains, the tallest being approximately 540 m (1771 ft). The island is drained by the large Arroyo Limantour, which is formed from the drainages of numerous rugged canyons across the island, coalescing in the center and then winding to a broad terminus at the eastern shore.

Isla San Pedro Mártir is roughly equal distance from Sonora and Baja California (ca. 50 km) or from the nearest islands (San Esteban, ca. 40 km, and Dátil, ca. 35 km). Mártir is volcanic in origin and has never been connected to any other landmass (Carreño & Helenes 2002). The island is roughly triangular in shape, girdled by steep sea cliffs and then rising in a dome-like fashion, with a peak elevation of 300 m (984 ft). The island encompasses an area of 2.67 km² (calculated by Grupo de Ecología y Conservación de Islas, A.C. [GECI] in 2008 based on a 60-cm-resolution *QuickBird* satellite image). Above the high sea cliffs the topography shows gradual relief and consists roughly of two main plateaus, or mesas, one at about mid-elevation and the other at the top and occupying the western portion of the island.

San Pedro Nolasco, 3.45 km², is situated in deep water about 30 km northwest of Guaymas, Sonora, and 14.6 km west of Bahía San Pedro, the closest mainland location. The channel between Nolasco and the Sonora mainland is 244 m deep (Murphy et al. 2002).

"Nolasco is mainly composed of granodiorite, apparently of the same petrographic type as that cropping out near Guaymas, and may be related to the granitic rocks of Cretaceous age in Baja California" (Carreño & Helenes 2002:26–27). The island is the outcrop of an uplifted block termed the Pedro Nolasco High bounded by the East Pedro Nolasco and West Pedro Nolasco Faults, which constitute the southernmost segments of the Tiburón Fault system. Seismostratigraphic analysis by Aragón-Arreola et al. (2005) shows that the Pedro Nolasco High is not covered by the oldest marine sediments in the adjacent Yaqui Basin. Marine rocks draping the Pedro Nolasco High date to when Yaqui Fault activity and associated basin subsidence was waning. This waning subsidence is dated to the late Pliocene, at which time there was a migration of deformation from the east to the west in the northern Gulf of California (Aragón-Arreola & Martín-Barajas 2007). The correlation of the sequences that bound the Pedro Nolasco High with the termination of active faults in the eastern-central Gulf of California and migration of activity to the western Gulf yields a best estimate of separation of San Pedro Nolasco Island from the Sonora mainland between 2 and 3 mya. For this reason we are considering Nolasco to be a "geologically isolated" island.

VEGETATION COMMUNITIES

The Sonoran Islands occur in the Central Gulf Coast subdivision of the Sonoran Desert, which is characterized by low and uncertain rainfall. The distinctive morphology of the vegetation of the region has led to it being termed a "sarcocaulescent desert" due to its many succulents and semi-succulents with exaggerated stem (trunk and limbs) diameters (Shreve 1951).

Due to the complex topography and correspondingly complex array of microenvironments, the study region houses diverse biotic communities. The majority of the vegetation communities defined here initially were established by Felger and Lowe (1976). We are presenting a simplified classification of the vegetation, and in reality the distribution of species and vegetation is mostly a broad continuum. Mosaic patterns do occur where environmental gradients are sharp and correspondingly discontinuous. The major communities among these islands are as follows:

I. Seagrass meadow
II. Sea cliffs
III. Guano areas
IV. Beach dunes
V. Littoral scrub
    V-A. Mangrove scrub
    V-B. Salt scrub
VI. Desertscrub
    VI-A. Coast scrub
    VI-B. Creosotebush scrub
    VI-C. Mixed desertscrub
    VI-D. Cactus scrub
    VI-E. Xeroriparian desertscrub
    VI-F. Riparian tinajas
VII. Desertscrub-thornscrub ecotone
    VII-A. Canyons
    VII-B. Exposed slopes and ridges

I. Seagrass meadow

Three seagrasses are found in the protected waters of the Gulf of California that have muddy-sandy substrates in benthic zones about 0.5–3 m below the low tide level and are not exposed to the desert air. The distinctive vascular plant species of these seagrass meadows are *Halodule wrightii*, *Ruppia maritima*, and *Zostera marina* var. *atam*. *Halodule wrightii* and *R. maritima* are not the primary members of this community and are seen only during the warm summer months. *Zostera marina* (eelgrass) is the dominant seagrass of the region, distinctive as an annual in the Gulf of California (as opposed to its perennial form throughout the rest of its range). The seeds were an important food resource for the Comcaac. In the Gulf of California, this community is best developed in the Canal del Infiernillo from the north of Tiburón, corresponding to the northern-most limits of mangroves on the shores, to the south end of the Canal. Seagrasses also occur in Bahía Kino but in reduced quantity. Eelgrass meadows extend sporadically southward along the coast to the vicinity of Altata in the state of Sinaloa. These seagrasses provide a major food source for wintering black brant (*Branta bernicla nigricans*) and the once-abundant green sea turtles (*Chelonia*) in the region (Felger et al. 2005; Felger & Moser 1985; Torre-Cosío 2002).

II. Sea cliffs

This community is seen on vertical rocky faces that are relatively free of bird guano on Tiburón, San Esteban, Dátil, Alcatraz, San Pedro Mártir, and San Pedro Nolasco. At these sites a group of rock-holding plants make up the sparse vegetation. Prominent species in this community include *Eucnide rupestris*, *Ficus petiolaris* subsp. *palmeri*, *Hofmeisteria crassifolia*, *H. fasciculata*, and *Pleurocoronis laphamioides*. The sea-cliff community is a natural extension from the tropical and subtropical canyon rock-cliff habitats.

III. Guano areas (fig. 1.3)

Many of the smaller islands in Midriff Region (Patos, San Pedro Mártir, Rasa, Partida Norte, Alcatraz, and certain islands in Bahía de Los Ángeles) support large breeding colonies of seabirds. The associated excessive amounts of nutrients from the guano, principally nitrogen and phosphorus, drastically limit the diversity of plant species that can exist in these environments. Those that tolerate and in some cases thrive in guano-rich soils include certain cacti (cactus diversity is relatively greater on smaller guano islands than on large non-guano islands; see Wilder et al. 2008a) and members of the amaranth

family. Patos and the mountain on Alcatraz are the only significant guano-dominated areas in the vicinity of Tiburón. Much of the southern part of Nolasco and the island's sea cliffs are white with guano. San Pedro Mártir is one of the most important seabird nesting sites in Mexico and is especially rich in guano. Principal species of this community include *Amaranthus watsonii*, *Atriplex barclayana*, *Cylindropuntia cholla/fulgida*, *Lophocereus schottii*, *Nicotiana obtusifolia*, *Pachycereus pringlei*, *Perityle emoryi*, and *Viscainoa geniculata*.

IV. Beach dunes (fig. 1.4)

Coastal dunes are present along much of the Gulf Coast mainland of Sonora and limited areas on Tiburón. Dunes on Tiburón reflect the prevailing winds from the north. Thus, significant dunes are encountered in only a few places, such as on the north side of the island and the southwest corner north of Punta Willard. Lesser dunes occur in some places along the Infiernillo coast, such as at Punta Tormenta (Palo Fierro).

Many dune-inhabiting plants have deep roots, and there is a preponderance of silvery- or gray-pubescent-leaved species. Characteristic species include dune species (or sometimes extending into adjacent sandy habitats) such as *Euphorbia leucophylla*, *Frankenia palmeri*,

Figure 1.3. Guano area vegetation, vicinity of guano-workers' village, Isla San Pedro Mártir: *Pachycereus pringlei* and small plants of *Sphaeralcea hainesii*. BTW, 11 April 2007.

Figure 1.4. Dune vegetation with *Frankenia palmeri*, northeast side of Isla Tiburón. BTW, 6 April 2008.

Helianthus niveus, and *Psorothamnus emoryi*; sandy-soil species that are also found on sandy non-dune habitats, such as *Abronia maritima*, *Aristida californica* var. *californica*, and *Croton californicus*; and species that occur on a conspicuously wider spectrum of substrata, such as *Atriplex barclayana*, *A. linearis*, *Lycium brevipes*, and *Palafoxia arida*.

V. Littoral scrub

This vegetation, defined by low scrub of halophytic character, is tidally inundated by seawater, either daily or occasionally. The vegetation is almost entirely evergreen and mostly perennial, with a predominance of succulents, semi-succulents, and saltgrasses. There are two major types, mangrove scrub and salt scrub.

V-A. Mangrove scrub. (fig. 1.5) The northern-most mangroves in North America are found on the northeastern shore of Tiburón and the opposite mainland at Estero Sargento, plus a small and extremely endangered northern outlier population of *Avicennia* farther north at Puerto Lobos, Sonora. Mangroves grow in quiet bays and lagoons (*esteros*) southward through the Gulf of California and on into the tropics (e.g., Felger et al. 2001; Turner et al. 1995). Mangroves occur on both sides of the Canal del Infiernillo, including three well-developed mangrove esteros on the east side of Tiburón: San Miguel, Punta Tormenta, and Punta Perla. Three mangrove species make up the Tiburón and Gulf of California *manglares*: *Avicennia germinans*, *Laguncularia racemosa*, and *Rhizophora mangle*.

Occasional winter freezing weather defines the northward distribution of the mangroves—those at Estero Sargento periodically suffer severe freeze damage. Within the mangroves there is overlapping zonation, with red mangrove (*Rhizophora*) extending into the deepest water, black mangrove (*Avicennia*) in the shallowest water, and white mangrove (*Laguncularia*) reaching maximum density between the peak zones of the other two (Felger et al. 2001). *Rhizophora*, *Laguncularia*, and *Avicennia* are among the most widespread mangroves in the world and share many characteristics of ecological adaptation, growth form, and reproductive biology.

V-B. Salt scrub. (fig. 1.6) This community occurs in coastal areas on salt flats, beaches, and mangrove margins and is composed mainly of saltgrasses, halophytic shrubs (salt bushes), succulent perennials, and a few succulent annuals. The saltgrasses are *Distichlis littoralis*, *D. spicata*, and *Sporobolus virginicus*. The halophytic shrubs and subshrubs include *Atriplex barclayana* and

Figure 1.5. Mangrove vegetation at Estero Santa Rosa on the Sonora mainland opposite Isla Tiburón. NSB, 20 March 2008.

Figure 1.6. Salt scrub vegetation, near the shore at Arroyo Sauzal, Isla Tiburón: *Batis maritima* just above the water surrounded by *Distichlis spicata*. BTW, 29 January 2008.

Maytenus phyllanthoides. Succulent herbs and shrubs are represented by *Allenrolfea occidentalis, Batis maritima, Salicornia bigelovii, S. subterminalis, Sesuvium portulacastrum, Suaeda esteroa,* and *S. nigra.*

Salt scrub communities are well developed on Tiburón and Alcatraz, the two islands in the region with the largest area of flat to gradually sloping land abutting the coast, and abbreviated salt scrub can be found on San Esteban and is even more reduced at the shores of other smaller islands.

VI. Desertscrub

The majority of the flora area supports desertscrub vegetation composed of several easily recognized communities, often in a complex patchwork with indistinct boundaries (see Búrquez et al. 1999; Martínez-Yrízar et al. 2010). In addition, during times of favorable rains there are seasonally rich developments of ephemerals (short-lived annuals)—each season has a different suite of species. These seasonal communities are primarily differentiated into winter-spring (cool season) and summer-fall sets, as well as various non-seasonal species that may appear with rains at almost any time with sufficient soil moisture. Six main desertscrub types can be recognized on the islands:

VI-A. Coast scrub. (fig. 1.7) The coast scrub occupies an often well-defined zone near the shore on Tiburón, between salt scrub and the more extensive and inland desert shrub communities. Coast scrub consists of low, monotonous vegetation with few species, dominated by *Frankenia palmeri.* This species is so common in this community that this region can be called the *Frankenia* zone. Coast scrub is best developed along the eastern shore of Tiburón (mirroring that of the adjacent mainland), where it extends inland to about 1 km, but also occurs on other shores of the island.

VI-B. Creosotebush scrub. (fig. 1.8) This community of relatively low diversity is dominated by *Larrea divaricata,* which extends across much of the northern Sonoran Desert. In the mid-Gulf region, creosotebush scrub is found on Tiburón and the adjacent Sonoran coast, where it occurs in limited areas at lower elevations. On Tiburón *Larrea* is dominant in much of the Central Valley and extends to about 350 m on particularly dry slopes (e.g., Sierra Menor). *Larrea* reaches its southern limit in Sonora in the vicinity of Guaymas, where it is restricted to exposed, low-elevation rolling hills and ridges. Ephemeral (annual) species comprise the highest percentage of plant diversity found in this community.

VI-C. Mixed desertscrub. (fig. 1.9) This is the most widespread and characteristic vegetation on Tiburón. It also reaches notable complexity on Dátil and San Pedro Nolasco, with reduced complexity on San Esteban. This vegetation is highly varied, with numerous small-leaved, drought-deciduous small trees, shrubs, and subshrubs. Dominance is shared primarily by desert shrubs and small

Figure 1.7. Coast scrub vegetation, vicinity of Punta Tormenta, Isla Tiburón, looking northwest toward Sierra Kunkaak: *Frankenia palmeri* (foreground) and *Maytenus phyllanthoides* in a small drainageway (middle). BTW, 1 January 2006.

Figure 1.8. Creosotebush scrub, *Larrea divaricata* subsp. *tridentata*, Central Valley, Isla Tiburón. The pole is 1.52 m. RSF, 18 February 1965.

Figure 1.9. Mixed desertscrub, mainland bajada on west side of Sierra Seri, mirroring the vegetation of the opposite bajada on Isla Tiburón. Major elements include *Pachycereus pringlei*, *Bursera microphylla*, and *Simmondsia chinensis*. ETN, March 1951.

trees, which comprise many distinctive, usually localized communities. Representative species include *Bursera microphylla*, *Colubrina viridis*, *Desmanthus fruticosus*, *Jatropha cuneata*, *Lippia palmeri*, *Olneya tesota*, *Parkinsonia microphylla*, *Ruellia californica*, and *Viscainoa geniculata*.

VI-D. Cactus scrub. (fig. 1.10) Cactus scrub, or cactus "forest," is dominated by columnar cacti and may include other succulents and spinescent shrubs. This unique plant community is the dominant vegetation on

Isla Cholludo, the upper elevations of San Pedro Mártir, and the more exposed west- and south-facing slopes of San Pedro Nolasco. Cactus scrub on Tiburón is largely limited in distribution to west-facing bajada slopes of the Central Valley (especially along the northeastern portion of the Central Valley). The columnar cacti are *Carnegiea gigantea*, *Lophocereus schottii*, *Pachycereus pringlei*, *Stenocereus gummosus*, and *S. thurberi*. Other succulents variously occurring in cactus scrub on the islands include *Cylindropuntia* and *Agave*.

Figure 1.10. Cactus forest on Isla Cholludo. Juan Alfredo Barnett-Díaz is on the left, and Ivan Eliseo Barnett-Romero on right. BTW, 2 September 2008.

VI-E. Xeroriparian desertscrub. (fig. 1.11) These communities are characterized by desert trees and shrubs restricted to the usually dry watercourses of desert arroyos and their floodplains. Xeroriparian desertscrub is well developed on Tiburón, less so on San Esteban, and is present to a limited extent on Dátil. Trees and shrubs form an irregular canopy, overtopping the adjacent desertscrub vegetation. Characteristic species include *Acacia greggii* and *A. willardiana*, *Hyptis albida*, *Olneya tesota*, *Parkinsonia florida*, *Sideroxylon occidentale*, and *Vallesia glabra*. Also included are mixed desertscrub species that often attain larger-than-usual stature along these dry watercourses.

VI-F. Riparian tinajas. (fig. 1.12) Waterholes, locally called *tinajas*, in this markedly arid region support rare freshwater wetland communities and in the flora area occur only on Tiburón. The one exception is a tiny water seep on Nolasco, called Agua Amarga, which supports a precariously small colony of *Cyperus elegans*. The Tiburón wetland sites are highly localized at, or near, spring sources or water-retaining bedrock depressions along

canyons and arroyos. Common spring-supported species include *Cyperus elegans*, *Eleocharis geniculata*, *Ficus petiolaris* subsp. *palmeri*, *Phragmites australis*, *Salix exigua*, *Stemodia durantifolia*, the invasive *Tamarix chinensis*, and *Typha domingensis*. Although small in area, these perennial water sources were absolutely essential to the survival of the Comcaac in traditional times.

VII. Desertscrub-thornscrub ecotone

The higher elevations and sheltered habitats of the deep canyons of the Sierra Kunkaak on Tiburón support diverse and dense vegetation best classified as ecotone between mixed desertscrub and thornscrub. Thornscrub has been distinguished using structural criteria, having dense cover with few open spaces, in which dominance is usually shared by many small trees and large shrubs, many with tropical affinities (Brown 1982; Búrquez et al. 1999; Felger & Lowe 1976; Gentry 1942; Martínez-Yrízar et al. 2010). Thornscrub in mainland northwestern Mexico occurs primarily between the southern edge of the Sonoran Desert as defined by Felger and Lowe (1976)

Figure 1.11. Xeroriparian vegetation, northern portion of the Sierra Kunkaak. BTW, 29 December 2005.

and the tropical deciduous forest (Búrquez et al. 1999; Felger et al. 2001; Martin et al. 1998).

Certain more favorable (more mesic) habitats within the desert, such as mountain canyons, north-facing slopes, and higher elevations support a wealth of species that are disjunct from their primary source populations. As Búrquez et al. (1999:45) point out, "Within the desert, tropical 'islands' develop where areas are protected from extreme temperature oscillations, furnished with shade and humidity, and provided with more available water. These mountain oases are like closed microcosms that maintain relict and disjunct populations of organisms." The Sierra Kunkaak on Tiburón in the center of the Sonoran Desert exemplifies this description. Numerous

species with biogeographical connections to the entirety of the Sonoran Desert, including the Baja California Peninsula and the northern and southern edges of the desert in Sonora, are found in the Sierra. The vegetation of the Sierra Kunkaak is unique in supporting not only disjunct tropically evolved species but also highly isolated populations of species from above the desert in northeastern Sonora. Two habitats, separated by elevation and exposure, are found in the Sierra Kunkaak: canyons and exposed slopes and ridges.

VII-A. Canyons. (fig. 1.13) A dense community composed of more mesic-adapted desertscrub species, many of which range into the northern extent of the tropics in

Figure 1.12. Riparian (tinaja) vegetation, Pazj Hax waterhole, Sierra Kunkaak: reedgrass (*Phragmites australis* subsp. *berlandieri*) in center and *mauto* (*Lysiloma divaricatum*) at middle and left center. BTW, 1 January 2006.

southern Sonora, occurs in the deep interior canyons of the Sierra Kunkaak of Tiburón. We classify these canyon communities as ecotone between desertscrub and thornscrub because the species are principally affiliated with desert habitats but the vegetation is as dense as thornscrub, and certainly much denser than usual desert vegetation. This community is best developed on the north side of the Sierra but also occurs to a more limited extent wherever a sheltered habitat is present in the mountains. The canyon vegetation is characterized by a relatively well-developed understory with an overstory comprising trees often to about 5 m tall. Representative species include *Acacia willardiana, Ambrosia carduacea, Celtis pallida, C. reticulata, Croton magdalenae, Ficus petiolaris* subsp. *palmeri, Justicia candicans, Lysiloma divaricatum, Plumbago zeylanica,* and *Tetramerium fruticosum.*

VII-B. Exposed slopes and ridges. (fig. 1.14) At the highest elevations and on exposed south-facing slopes,

an arid-adapted sparse vegetation is seen, marking a sharp transition from what is encountered in the canyons below. Scattered low shrubs provide partial ground cover, including *Dalea bicolor* var. *orcuttiana, Encelia farinosa,* and *Dodonaea viscosa.* The vegetation is generally of short stature, with the taller components reaching about 2–3 m, for example, *Fouquieria splendens,* an occasional *Acacia willardiana,* and *Lysiloma divaricatum.* As one ascends from the sheltered canyons to the exposed highest elevations of the island, a striking transition is apparent. The dense taller communities with many mesic-adapted species are quickly left behind, and a number of desertscrub species from hundreds of meters below reappear intermixed with a group of outstandingly disjunct species confined to the top of the island. An example of the reappearance of desertscrub species is *Bursera microphylla,* the common *Bursera* of the desert lowlands on the island. It is replaced by the more "tropical" *Bursera fagaroides* in the canyons above about 400 m, but *B. microphylla* is again common on the exposed ridges and peaks above about 800 m. In

Figure 1.13. Dense canyon vegetation, interior of Sierra Kunkaak. *Lysiloma divaricatum* (right center) and *Ficus palmeri* (left center) are common members of these canyon communities. BTW, 25 October 2007.

Figure 1.14. Desertscrub vegetation on the exposed upper ridges of Isla Tiburón: *Agave chrysoglossa, Fouquieria splendens,* and *Stenocereus thurberi* in the foreground. Peak at left background is the island summit. BTW, 26 October 2007.

addition to the composition of desertscrub species, the vegetation has the open feel of the desert lowlands. The most striking aspect of the vegetation on the upper ridges and peaks of Tiburón is the occurrence of ice-age relicts, species isolated by hundreds of kilometers from their nearest populations and best understood as remnants of historical vegetation thought to have once been present on the island (e.g., *Canotia holacantha, Dasylirion gentryi,* and *Forestiera phillyreoides*).

HISTORICAL HUMAN USE AND INFLUENCE

The islands, especially Tiburón and San Esteban, have long histories of human use (Bowen 1976, 1983, 2009). Normally this association results in significant changes to an ecosystem, often to the obvious detriment of the environment. However, this is not the case for most of the Sonoran Islands. Tiburón and its neighboring islands, including San Esteban, are in the historical and current homeland of the Comcaac. (The name Comcaac is the term used by the Seri people for themselves. It is a plural noun, of which the singular form is Cmiique.) They and their ancestors or predecessors have lived in the region for thousands of years (fig. 1.15). In traditional times their population probably did not exceed several thousand. The number of people able to live in any given locality, such as on Tiburón or San Esteban, was strongly influenced by the meager supply of freshwater. Historically, the various groups of Comcaac ranged over an expanse of desert lands, from the vicinity of Guaymas northward along the Sonoran coast to the vicinity of Puerto Libertad and the islands of Tiburón and San Esteban. There is also substantial evidence within Seri oral history that cross-Gulf voyages were made to the Baja California Peninsula and the adjacent islands (Bowen 2009). The Comcaac recognized six major traditional groups speaking three different dialects (Moser 1963). During historical times Isla Tiburón and adjacent mainland areas were a primary center and the focus, at least in part, for three major groups—one group lived

primarily in the interior of Tiburón, another occupied the eastern part of the island and the adjacent mainland, and the third group lived on San Esteban and the nearby south side of Tiburón, where the largest waterhole (Xapij or Sauzal) supplied them with permanent water and *carrizo* (reedgrass, *Phragmites australis*) (Bowen 2000; Felger & Moser 1985; Moser 1963).

The different groups of Comcaac primarily lived a hunter-gatherer and seafaring existence. Agriculture and sedentary settlements generally were not feasible due to the extreme aridity of the region and scarcity of freshwater (Felger & Moser 1985). They developed diverse uses and numerous food resources from the local flora. Their knowledge of the flora and fauna of the islands is profound and remains strong, although known to a lesser extent by fewer people than during earlier, traditional times. A substantial portion of this traditional information has been documented and preserved by Felger and Moser (1985), Moser and Marlett (2005, 2010), and Nabhan (2003), among others.

There was a long and tragic history of violence during Spanish colonial times that continued into the early twentieth century. The Comcaac as a whole were targets of several centuries of Spanish colonial and then Mexican military violence (e.g., Sheridan 1999). For much of this time Isla Tiburón served as a refuge for the Comcaac—the isolation and scarcity of freshwater made military pursuit difficult, especially with horses. For nearly 200 years the Comcaac suffered population losses, and their ability to live an autonomous existence diminished. In a horrible act, the distinctive San Esteban people were apparently eliminated by a punitive military expedition in the late nineteenth century (Bowen 2000). An amalgamated population of Comcaac persisted, largely due to the aridity of the land and the associated hardships placed on military expeditions and to the resiliency of the people themselves.

By the middle of the twentieth century the Comcaac had largely moved away from Isla Tiburón and settled in villages on the mainland. Modern ways of life entered the region, and the majority of the people settled in the villages of El Desemboque del Río San Ignacio to the north (fig. 1.16) and later Punta Chueca to the south (fig. 1.17). From a devastatingly low point in the early twentieth century, their population has steadily increased, and by and large the people "were accepting what they could of the new culture and rejecting that part they did not want" (Felger & Moser 1985:19). The Comcaac continue to inhabit a core area of their historical

Figure 1.15. The famous Chico Romero (ca. 1888–1974), who often represented the Comcaac to outsiders, including serving as a consultant to Alfred L. Kroeber (1931) and subsequent authors. EHD, 1924. Heye Foundation (*HF-23817*).

Figure 1.16. El Desemboque. Edward and Mary Beck Moser's home is the white house in the center surrounded by trees (*Tamarix aphylla* and *Parkinsonia florida*). RJH, May 1977.

Figure 1.17. Punta Chueca. RSF, 11 August 1964.

lands, speak their native language, and maintain a keen interest in the plants and natural world of their region.

Environmental changes wrought by the Comcaac have been subtle. One manifestation of their light touch on the land is exhibited by the very low frequency of non-native plant species on the islands (Felger & Moser 1985; West & Nabhan 2002). Non-native species are generally best established in areas of significant disturbance due to human activity and movement between areas. Furthermore, there were reduced sources for foreign non-native species in the region until the last half of the twentieth century. However, this is not to say that the Comcaac did not interact with or transport native plants and animals, either purposely or accidentally. People lived in and visited even remote areas in the region for thousands of years (Bowen 2009).

Plants were harvested almost daily, with people transporting one species or another locally and sometimes regionally. A few examples indicate some of the ways in which plants were moved about. The Comcaac had a cultural practice of burying the placenta from a newborn next to a large cactus such as cardón (*Pachycereus pringlei*), saguaro (*Carnegiea gigantea*), or sinita (*Lophocereus schottii*; Felger & Moser 1985; Moser 1970; Nabhan 2000, 2002; Ben & Humberto, personal observations, 2006). Nabhan (2000, 2002) reports associated succulents sometimes were transplanted to the placenta burial sites, and he discusses distributions by the Comcaac of some cacti and other species. There is also evidence that the Comcaac planted chollas (*Cylindropuntia fulgida*) in association with game circles (Bowen et al. 2004). Oral histories document the Comcaac transplanting and planting certain species. *Agave subsimplex* and two prickly pears (*Opuntia engelmannii* and *O. gosseliniana*) "were said to have been planted near the base of Punta Sargento many years ago by the people of the Sargento Region" (Felger & Moser 1985:225).

Human activities should be addressed when examining unusual plant distributions. However, when the multitude of disjunct plant populations in the flora is analyzed, the explanation of transport by people can be documented in very few cases. This is in part due to lack of information for events that occurred long ago, but for many of the species with anomalous distributions there is no known current or historical cultural use. Instead, paleo-climatic factors may be more plausible to explain disjunct occurrences of more than 150 km, such as *Canotia holacantha* at the summit of Sierra Kunkaak (Wilder et al. 2007b, 2008b).

NON-NATIVE AND INVASIVE SPECIES

The islands of the Gulf of California contain some of the most undisturbed regions of the Sonoran Desert. The historical and current low number of non-native species is a testament to this fact. However, the nearby mainland of Sonora, as well as the Baja California Peninsula, faces an onslaught of invasive species (e.g., Tellman 2002). On the islands there are only a few exotic species of major concern—ones that have the ability to drastically alter habitats and ecosystem processes. It is important to understand that the islands of the Gulf of California are in a globally unique circumstance, in which certain non-native, invasive plant species are newly becoming established and eradication is not only achievable but also relatively inexpensive. In this section we provide a summary of the non-native flora of the Sonoran Islands as of 2011. (See part III, "Species Accounts," for more detailed information.)

No non-native plant species poses a greater threat to the island ecosystems than *Cenchrus ciliaris* (oot iconee, zacate buffel, buffelgrass), native to the Old World. The massive conversion of native Sonoran desertscrub and thornscrub to dominance by this highly invasive grassland species, originally introduced for cattle forage, has changed vast areas into a landscape resembling an African grassland (e.g., Franklin et al. 2006).

Seeds of this perennial grass are enclosed in bur-like fascicles that can lodge into fur and feathers. People also unwittingly become dispersal vectors, with the burs finding their way into cars, clothing, shoes, gear, and so forth. Buffelgrass is able to reproduce in a single season, functioning as a facultative annual, in addition to sustaining as a hardy perennial that quickly responds to moisture, often more rapidly than native species. Once a buffelgrass plant has established, it may persist through more arid years until favorable rain events, at which point seeds in the seed bank germinate and the population can greatly expand. Due to the ability to out-compete native species and a substantial seed bank, there is a potential for exponential population expansion. Once large monoculture paddocks are established, a matrix of tinder fuel for wildland fire develops within the once essentially fireproof desert. Most Sonoran Desert plant species are not adapted to fire and are subject to buffelgrass replacement following fires, which then becomes a positive feedback cycle due to the ability of buffelgrass to thrive after a fire. Buffelgrass has become an invasive species of great concern in many arid lands across the world, for example, Sonora and elsewhere in Mexico (Tellman 2002) and Central Australia (Clarke et al. 2005).

Buffelgrass is abundant along the Sonoran coast, including Bahía Kino to Desemboque. It was reported to be on Alcatraz and subsequently eradicated (see "Species Accounts"). It was found at three places on the east side of Tiburón in 2007, associated with roads and the Caracol research station. Seri youths, working with Ben and Humberto, began active eradication of buffelgrass on Tiburón in 2007 and all known plants were manually removed in 2007, 2008, and 2009. Such small populations can be controlled, thus avoiding unwanted ecological transformation by one of the world's most notorious invasive species. However, continued monitoring is needed.

Tamarix chinensis (*pino salado*, salt cedar), another invasive species of concern, is well established at a few sites on several Midriff Islands. This species is an aggressive invader in riparian habitats and a potentially serious threat to the riparian tinajas on Tiburón and some alkaline-rich coastal habitats. There are two long-established populations on the Midriff Islands. A sizable population has been present at the Xapij (Sauzal) waterhole in the south part of Tiburón since at least 1958. The lack of increase of *Tamarix chinensis* at Xapij, and the fact that an eradication effort would be extremely damaging to the spring, suggests that the best course of conservation action is continued monitoring. *Tamarix* also has been established on Isla Ángel de la Guarda since at least 1983, when first collected by Ray Turner (Turner et al. 1995:384), and was discovered on San Lorenzo by Larry Johnson in 2004 (Tom Bowen, personal communication, 2009). A few individuals were found on Alcatraz in a wet area of the saline flat, and a small population was found near the shore at the terminus of Arroyo Sauzal (see "Species Accounts"). Significant populations of salt cedar occur throughout the Bahía Kino area, which is likely a major seed source for the region. The minute seeds are thought be dispersed primarily by wind (Carpenter 2003), but birds visiting wetland habitats are also likely vectors of seeds. When control of this species is not detrimental to ecologically sensitive habitats, and is logistically feasible, eradication is recommended.

The giant reed *Arundo donax* (xapijáacöl, *carrizo*) has been established at the Sopc Hax waterhole on Tiburón since at least the middle of the twentieth century. This Old World riparian species has been widely naturalized and cultivated in the New World, including the Sonoran Desert, probably since early historical times. It is similar in appearance to the native reedgrass, *Phragmites australis* subsp. *berlandieri* (xapij, *carrizo*), which occurs at other waterholes on Tiburón, including Pazj Hax, within several kilometers of Sopc Hax. *Arundo* has not spread to the other waterholes on the island, and we recommend against attempted eradication for cultural reasons and because, as in the case of *Tamarix* at Sauzal, attempted eradication could cause more harm than benefit. Both *Arundo* and *Phragmites* have been culturally important for the Seris (see "Species Accounts" and Felger & Moser 1985).

A large population of goathead (*Tribulus terrestris*) has been on Alcatraz since at least 1966, and a few individuals were found and removed from Punta Tormenta on Tiburón in 2006. Due to the fruits' sharp spines and their proficiency at becoming embedded in shoes, tires, and other artifacts, as well as the difficulty of control, this species is expected to spread. The best prevention is to check your shoes and equipment.

Other exotic species in the flora do not appear to be of major concern. The spiny bur of the exotic grass *Cenchrus echinatus* was collected in 1993 at Ensenada del Perro on Tiburón but has not been recorded there since that time. This is a common weedy species in Kino and will likely make its way to the islands via human traffic in the future.

A few annuals in the flora are reported as native to the Old World and exotic in the New World: *Chenopodium murale*, *Mollugo cerviana*, and *Physalis pubescens*. In the flora area they have limited distributions and appear to play a minor role in ecosystem processes, and in fact have distributions like those of native species. Three species of shade trees have been planted on Tiburón: two in association with the Mexican marine outposts, *Tamarix aphylla* (athel tree, *pino*) at Tecomate and Punta Tormenta, and *Eucalyptus camaldulensis* (Murray red gum, *eucalipto*) at Tecomate and Ensenada de la Cruz. The athel trees are thriving, but the eucalyptus trees are faring poorly, and neither species is reproducing. *Pithecellobium dulce* (*guamúchil*) was clandestinely planted at the Sopc Hax waterhole, where one juvenile tree was found in 2007.

A number of additional exotic species occur in Sonora and the Baja California Peninsula in disturbed habitats such as coastal communities and abandoned agricultural fields but have not been detected on the Sonoran Islands. These are the source populations from which colonization of island habitats are most likely. The critical step to successful control of invasive species is early identification and removal of pioneer populations. It is important to avoid confusing invasives with natives—for example, on Mártir the locally small population of the native perennial grass *Digitaria californica* could be confused with buffelgrass, especially when not reproductive. (Elsewhere we have witnessed well-intentioned removal of native plants confused with exotics.) West and Nabhan (2002) provide a baseline treatment of invasive species in the Midriff Island region, both on the islands and coastal area (other sources include Felger 2000a and Van Devender et al. 2009). Here we highlight a few of the aggressive invasive species of Old World origin that we believe are significant threats to the Sonoran Islands.

Mesembryanthemum crystallinum (crystal iceplant, *hielitos*) is an aggressive invasive of coastal ecosystems of California and Baja California Norte (California Invasive Plant Inventory 2009). A single specimen is documented from Dátil, and it is a potential threat for expansion to the Midriff Region. This highly succulent cool-season annual has established in the Gulf of California region, including the Sonora coast. In Baja California Norte large populations occur at Bahía de Los Ángeles and adjacent islands (West & Nabhan 2002; they also state that large populations occur on Isla Rasa, but there are no records for it on the island, e.g., Enriqueta Velarde, personal communication, 2008). Other invasive species of concern include Sahara mustard (*Brassica tournefortii*), Bermuda grass (*Cynodon dactylon*), London rocket (*Sisymbrium irio*), sow thistle (*Sonchus oleraceus*), and especially fountain grass (*Cenchrus setaceus*).

Vagrant animals (principally birds and people) that visit islands for brief periods of time are potentially significant in bringing invasive plant species to the islands (Rose & Polis 2000). Invasive species on islands typically face fewer challenges to establishment and rapid expansion than in mainland systems, often causing irreversible changes to ecosystems and subsequent local extinctions (Quammen 1996). Additional non-native species on the islands are inevitable, especially given development of small roadways on Tiburón as part of officially sanctioned hunting activities and increased island visitation by people (Bowen 2009). Early detection and eradication of non-native species, both plants and animals, are needed to preserve these unique insular desert environments.

Honeybees have been in the Midriff Region since about 1900 (Felger & Moser 1985). Since at least the late twentieth century, however, honeybees have become abundant on most of the Gulf Islands and are undoubtedly affecting native pollinator bees (Bowen et al. 2006).

Different non-native and potentially invasive animals come and go on the islands with human activities (e.g., Carabillas Lillo et al. 2000; Lawler et al. 2002; Mellink 2002; Velarde & Anderson 1994). The Seris had dogs on Tiburón (Felger & Moser 1985) and most likely on San Esteban, but none have survived. Lawler et al. (2002:349) state that "domestic species are absent on . . . islands inhabited by coyotes." (Tiburón is the only Sonoran Island with coyotes.) The house mouse (*Mus musculus*) and black rat (*Rattus rattus*) were on San Pedro Mártir but have been eradicated, and the brown rat (*Rattus norvegicus*) is reported for San Esteban. Their impact on the native plant life is unknown. Domestic cats may be present on San Pedro Nolasco (Felger et al. 2011). New introductions of non-native animals can have devastating effects on island ecosystems.

ADMINISTRATION AND CONSERVATION

Mexico's environmental legislation, the Ley General del Equilibrio Ecológico y la Protección al Ambiente, recognizes six categories of natural protected areas that can be established by the federal authority: (1) biosphere reserves (*reservas de la biosfera*), (2) national parks (*parques nacionales*, including both terrestrial and marine parks), (3) natural monuments (*monumentos naturales*), (4) areas for the protection of natural resources (*areas de protección de recursos naturales*), (5) areas for the protection of plants and wildlife (*areas de protección de flora y fauna*), and (6) natural sanctuaries (*santuarios*).

The islands of the Gulf of California were declared an Area de Protección de Flora y Fauna in 1978. In 1995 all the Gulf of California Islands were made a Biosphere Reserve registered in the Man and Biosphere Program of UNESCO. As a wildlife protection area, under Mexican law the islands of the Gulf of California do not have the same strict restrictions that are imposed on biosphere reserves. The reasons to designate the islands within Mexican legislation with a different category than the one they hold internationally are possibly related to the large size and the spatial complexity of the whole archipelago, and the difficulties involved in strict law enforcement within the whole protected area. In spite of their less restrictive status under Mexican law, the islands are in practice managed as a large reserve, with substantial efforts devoted to their protection (Ezcurra et al. 2002). In July 2005 the Gulf of California and its islands were granted World Heritage status by UNESCO. This listing represents international acknowledgment of the unique biological diversity in the waters and on the islands of the Gulf of California and adds an important layer of international conservation pressure over the region. The Sonoran Islands are managed by the Guaymas office of the Comisión Nacional de Áreas Naturales Protegidas (CONANP), of the Secretaría de Medio Ambiente y Recursos Naturales (SEMARNAT) of the Mexican federal government. CONANP also works with the nongovernmental conservation organization Comunidad y Biodiversidad (COBI), World Wildlife Fund, Prescott College, and Grupo de Ecología y Conservación de Islas, A.C. (GECI) to jointly address issues that face management and conservation.

Among the eight Sonoran Islands, San Pedro Mártir and Tiburón have additional levels of federal protection. In 2002 the Mexican federal government established San Pedro Mártir and the surrounding waters as a Mexican Biosphere Reserve encompassing 30,165 ha (Diario Oficial 2002). A comprehensive management and conservation plan serves to formalize the commitment of the Mexican federal government for the conservation of the island (CONANP 2007; Diario Oficial 2011).

Tiburón was the first island in the Gulf of California to receive official status as a protected area. The island was declared a conservation area in 1963, when it was made a Wildlife Refuge and Nature Reserve by Mexican president Adolfo López Mateos. This action was due in part to an initiative started by Enrique Beltrán, an eminent Mexican conservationist and then director of the Direction of Forestry and Wildlife in the Mexican federal government. The primary motivation for this action was to create a refuge for wildlife such as mule deer, a species suffering from extensive poaching on the mainland. Conservation of the island was overseen by the Secretaría de Agricultura y Recursos Hidráulicos (SARH). At that time SARH built basic facilities on the island, including the structures at Tecomate, Punta Tormenta, Ensenada de la Cruz, and Caracol at the northern part of the Sierra Kunkaak, established 130 km of dirt roads, maintained two airstrips (Tecomate and Punta Tormenta), and built some water reservoirs to "improve" habitat quality for game species (Ezcurra et al. 2002). The presence of the Comcaac and their hunting activities was seen as a hindrance to the protection of the island, and they were given virtually no consideration by the government despite the fact that Tiburón had always been a part of their traditional

territory. Only in 1975 was jurisdiction of Isla Tiburón officially placed in the hands of the Comcaac. In that year President Luis Echeverría decreed Isla Tiburón the communal property of the Comcaac and declared the coastal waters of the island for the exclusive use of the Comcaac. Isla Tiburón officially remains the land of the Comcaac. Since the 1970s Mexican marines have had a presence on the island and are stationed on the east side of the island at Punta Tormenta—their primary role has been to thwart drug smuggling. They also had small stations in the north at Tecomate and in the south near Ensenada de la Cruz. Exotic ornamental shade trees (*Tamarix aphylla* and *Eucalyptus camaldulensis*) were planted at the different marine stations.

Since 1975, Isla Tiburón has been used as a breeding ground for the Sonoran Desert subspecies of desert bighorn sheep (*Ovis canadensis mexicana*). Sixteen females and four males were introduced as part of a Mexican federal program to protect the Sonoran Desert bighorn (Medellín et al. 2005). The animals were captured in the Sierra Seri and Sierra Bacha on the mainland across from the island by staff from the New Mexico Department of Game and Fish (Montoya & Gates 1975). There is no historical record of bighorn sheep occurring on Tiburón prior to this introduction. Due to the absence of a top predator on the island and a favorable habitat, the bighorn population grew dramatically, reaching between 480 and 967 individuals as evaluated through an aerial census in 1993 (Lee & López-Saavedra 1994), and by 2001 conservatively estimated to be fluctuating between 360 and 500 animals (Medellín et al. 2005). In the late 1990s permits for the hunting of bighorn sheep on Tiburón were sold on the international market to provide revenue for the Comcaac as well as to decrease the bighorn sheep population. The initial two permits offered in 1998 at auction in Reno, Nevada, garnered U.S. $395,000 total (Navarro 1999). In the early 2000s individual permits sold for around U.S. $100,000, but prices were decreasing by 2006 to about U.S. $75,000 due to the development of private game ranches throughout Sonora. Between 2005 and 2009 five permits were sold annually.

These introductions represent novel attempts at conservation of Sonoran Desert mammals, and from a number of perspectives this approach has been successful for the desert bighorn. But how are the bighorn and the associated hunting activities affecting the island's native flora? No quantitative studies have yet addressed this question;

however, there are a number of obvious and significant impacts. The bighorn diet is based on a wide variety of plants. More than 30 species have been observed by Humberto to be a part of the bighorn diet, the most apparent component being succulent species. Vast numbers of the Tiburón barrel cactus (*Ferocactus tiburonensis*), *amole* (*Agave chrysoglossa*), saguaro (*Carnegiea gigantea*), and other succulents are seen bashed open or otherwise damaged throughout the Sierra. Most concerning is the lack of baseline knowledge for the unique vegetation at the upper elevations of the Sierra, the primary habitat of the bighorn. In addition to physical damage caused by the bighorn, the minor network of roads on Tiburón has been expanded by the Comcaac to aid the movement of hunters around the island. The disturbance caused by these roads and their associated vehicular traffic is a significant concern in terms of the increase in suitable habitat for the establishment of non-native invasive plant species.

Despite these issues, we believe that Tiburón has been maintained as one of the most undisturbed places in the Sonoran Desert in part because of the bighorn program (Wilder et al. 2007a). In addition, from the perspective of bighorn sheep conservation, the contribution these sheep make to mainland populations through reintroduction efforts is significant. The revenue for the Comcaac community generated through the sale of highly priced hunting permits initiated in 1998 is one of the driving forces in their economy in addition to the sale of seafood. The economic incentive for the Comcaac community to maintain Isla Tiburón in an undisturbed state for bighorn conservation is a significant factor that has helped keep the island well preserved in a time of widespread habitat destruction on the Sonora mainland.

This does not mean the bighorn hunting program is a long-term conservation solution. There is evidence of a dangerously low level of genetic diversity within the population and continued extraction of about a hundred sheep every two years, in combination with increased severity of droughts as anticipated due to climate change, could lead to the extinction of the bighorn population (Colchero et al. 2009). The creation and use of roads will only expand over time, contributing to habitat degradation, and the continued bighorn impact upon the vegetation is a serious concern. In the late 2000s sheep extracted from the island, in addition to being used for repopulation efforts, were being purchased by private game ranches in northern Mexico (fig. 1.18). These hunting ranches are direct competition to the hunting experience offered by the Comcaac.

While the mystique of hunting a big game animal on a desert island will remain, the high sums of money previously paid for this experience will likely decline.

The Sonoran coastal highway from Puerto Peñasco to Guaymas has the potential to significantly change development realities in the region. Development projects offering quick financial gains in the Comcaac territory have included the ill-conceived idea of seawater farming of frost-sensitive mangroves in this desert region and a shrimp farm in Estero Sargento. Development pressures are likely to increase as infrastructure expands and opportunities to profit from the natural wealth become economically attractive. As this reality takes hold, it will be useful to consider the model of the Tiburón bighorn hunting program as a starting point for providing an economic incentive for ecological conservation that takes into account the realities of the modern world.

Figure 1.18. Bighorn sheep moved from Isla Tiburón to the mainland at Punta Santa Rosa. HRM, November 2008.

Island Diversity

The flora of all the islands under consideration includes 388 taxa (species, subspecies, and varieties) in 251 genera and 78 families (appendix A). The majority of these are found on Isla Tiburón. A statistical breakdown of the island flora is presented in table 1.1, in the order that the islands are described in this chapter. Asteraceae, Fabaceae, and Poaceae are the largest families on the islands. It is also of note that Cactaceae is the fifth most species-rich family in the total flora of the islands. These findings compare well with those for the flora of the Sonoran Desert as a whole (Felger 2000a; Van Devender et al. 2010; Wiggins 1964).

Each island in the Midriff Region is a unique desert microcosm. The geologic histories and varied topographies of the islands set the stage for their botanical diversity, and a suite of biotic and abiotic factors shapes the community of plants present on the islands. The principal factors that control the floristic diversity on the Gulf Islands are topography and island size, isolation (or lack thereof), soil conditions, aridity, and the presence or absence of herbivores.

Distance and isolation among Gulf of California islands are not nearly as extreme as in oceanic archipelagos like Hawaii and the Galapagos; however, the intermediate level of isolation has allowed radiation to occur in certain taxa of flora and fauna. An expected relationship of increased diversity of plant species is seen with increasing island area and topographic heterogeneity (including elevation). Tiburón, the largest island in Mexico, is correspondingly the most diverse island in the Gulf, and the rugged Sierra Kunkaak is the most diverse area of the island.

To assess the area–diversity relationship for the Sonoran Islands, we used the typical species–area model employed in island biogeography theory: $s = kA^z$, where s is the number of species, A is island area, z is the exponent, and k is a scale coefficient (MacArthur & Wilson 1967). An estimation of z was made by calculating the slope of the log–log function when island area and species diversity are graphed (fig. 1.19). By regressing the species richness of each island against the exponent-transformed area (A^z), k was estimated. Knowing k and z, the expected number of species for each island was calculated. Using Pearson residual analysis (Duffy 1990), we were able to identify

Table 1.1 Statistical Summary of the Flora of the Sonoran Islands

Island	Number of species (percentage of flora)	Three largest families	
		Family	Number of species (percentage of island flora)
Tiburón	340 (87.6%)	Fabaceae	34 (10.0%)
		Poaceae	33 (9.7%)
		Asteraceae	32 (9.4%)
San Esteban	114 (29.4%)	Poaceae	15 (13.2%)
		Asteraceae	13 (11.4%)
		Fabaceae	11 (9.6%)
Dátil	101 (26.0%)	Asteraceae	11 (10.9%)
		Fabaceae	10 (9.9%)
		Euphorbiaceae	10 (10.0%)
Cholludo	30 (7.7%)	Cactaceae	7 (23.3%)
		Poaceae	5 (16.7%)
		Asteraceae	3 (10.0%)
Alcatraz	54 (13.9%)	Cactaceae	9 (16.7%)
		Amaranthaceae	7 (13.0%)
		Asteraceae	6 (11.1%)
Patos	14 (3.6%)	Cactaceae	5 (35.7%)
		Amaranthaceae	4 (28.6%)
		Asteraceae	3 (21.4%)
Mártir	24 (6.2%)	Asteraceae	5 (20.8%)
		Cactaceae	3 (12.5%)
		Poaceae	3 (12.5%)
Nolasco	58 (14.9%)	Poaceae	14 (24.1%)
		Asteraceae	8 (13.8%)
		Cactaceae	7 (12.1%)
All islands	388	Asteraceae	40 (10.3%)
		Poaceae	37 (9.6%)
		Fabaceae	35 (9.0%)

the islands that have significantly more or significantly fewer species than predicted by McArthur and Wilson's model of island biogeography (table 1.2).

The islands identified as having depressed levels of diversity—San Pedro Mártir and Patos—are guano islands with high levels of nitrogen and phosphorus, which undoubtedly exerts strong chemical control on the flora (Felger & Lowe 1976; Wilder & Felger 2010). Dátil, which is remarkably rich in species, is a topographically diverse land-bridge island closely associated with Tiburón. Comparing the similar-sized islands of Nolasco and Dátil, which have skewed diversities (58 and 101 species, respectively), shows the importance of the geologic legacy of an island. While Dátil is a land-bridge island, Nolasco has been geologically isolated for several million years, and its diversity is consistent with the expected number of species predicted from the above model (for more detail, see Felger et al. 2011). It is also interesting to note that the flora of Cholludo, both a guano and land-bridge island, while relatively rich in species, also does not significantly differ from the prediction. It seems that the close proximity and past connection of Cholludo to Tiburón outweigh the filtering of species imposed by the guano soil conditions. This may provide evidence that the geologic history of an island is the most important factor controlling its observed diversity.

Inter-island distance is not sufficient to be a primary controlling factor on the flora of the islands due to the short distances (certainly the case for Tiburón) and vagrant species, such as seabirds, which create connections between the islands and the mainland (Rose & Polis

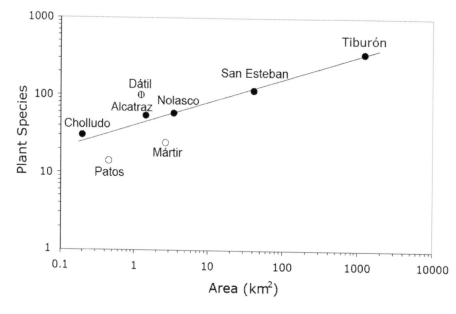

Figure 1.19. Species–area relationships of vascular plants on the Sonoran Islands on log-transformed axes. The slope of the line (z) is 0.304 and $r^2 = 0.71$. The majority of the Sonoran Islands fall very close to the regression line, but with three outliers. Two islands have significantly fewer species than expected (Patos, and Mártir; open circles) and one island has significantly more species than expected (Dátil; line filled circle). Graph by Exequiel Ezcurra.

Table 1.2 Pearson Residual Analysis for Species–Area Relationships of the Flora of the Sonoran Islands

Island	Area (km²)	Number of species	Expected number of species	Pearson residual	Probability
Tiburón	1223	340	340	−0.02	0.4906
San Esteban	41	114	121	−0.67	0.2516
Nolasco	3.45	58	57	0.10	0.4594
Mártir	2.7	24	53	**−4.00**	**0.0000**
Alcatraz	1.44	54	44	1.53	0.0635
Dátil	1.25	101	42	**9.09**	**0.0000**
Patos	0.45	14	31	**−3.03**	**0.0012**
Cholludo	0.2	30	24	1.20	0.1146

Note: Pearson-residual scores and probabilities for islands with significantly fewer species than expected (Mártir and Patos) and those with significantly more than expected (Dátil) are shown in boldface.

2000). The general aridity of the Midriff Region defines the vegetation of the region, and a gradation from arid to extremely arid is seen from the east to the west. The west side of Tiburón is much harsher than the east side, and San Esteban is one of the most arid of the Sonoran Islands, with those on the Baja California side of the Midriff, especially Ángel de la Guarda and the islets in Bahía de los Ángeles, being among the most arid sites in the entire Gulf of California region.

Soil conditions often determine the species that can establish, especially on guano-rich bird islands (also see discussions of Alcatraz, Cholludo, Patos, and San Pedro Mártir). Another significant factor that shapes the plant communities on the Sonoran Islands is the presence or absence of herbivores (the principal ones are rodents, chuckwallas, iguanas, rabbits, mule deer, and bighorn sheep). A major factor for the dense vegetation on some of the smaller islands is the absence of native rodents. San Pedro Mártir and Cholludo have remarkably dense cardón forests largely because there are no rodents to prey on the seedlings (it is believed that the introduced black rat on Mártir had a negligible effect on the cardón population). Similarly, the dense vegetation of Dátil is likely due at least in part to the absence of larger herbivores. On Nolasco, San Esteban, and Tiburón, substantial browsing is carried out by an array of animals. The endemic iguana on San Pedro Nolasco (*Ctenosaura nolascensis*) is the primary herbivore on the island and has been observed to eat the fresh, young leaves of the *palo blanco* (*Acacia willardiana*). The full extent of the pressures by this species and a red land crab (*Gecarcinus* [*Johngarthia*] cf. *planatus*) on Nolasco is not known.

On San Esteban the endemic chuckwalla (*Sauromalus varius*) is a prolific herbivore. Its diverse diet includes cholla flowers, buds, fruits, and young stems (*Cylindropuntia alcahes*), leaves of ironwood (*Olneya tesota*), and fruits of *pitahaya agria* (*Stenocereus gummosus*; e.g., Sylber 1988). Tiburón has a full array of herbivores nearly equivalent to the mainland. Significant herbivores include antelope jackrabbit (*Lepus alleni tiburonensis*) and mule deer (*Odocoileus hemionis sheldoni*) in the lowlands and bighorn sheep (*Ovis canadensis mexicana*) in the mountains.

It is important to assess the current diversity of the Sonoran Islands through the lens of historical conditions, especially the period from the waxing of the last ice age (the end of the Pleistocene, 12,000 yr BP) to the present. During the last ice age, sea levels were much lower than at present due to water storage in the world's ice sheets. Global sea levels during the Last Glacial Maximum, about 21,000–19,000 years ago, were about 120–135 m below the present level (Clark & Mix 2002; Peltier 2002). As global temperatures rose and ice sheets melted, sea levels rose to about 40 m below the current level about 10,000 years ago. Modern sea-level was attained about 6000 years ago (Bowen 2009; Davis 2006; Lambeck & Chappell 2001). Tiburón and its satellites (Cholludo, Dátil, and Patos), as well as Alcatraz, were connected to the mainland, while San Esteban, San Pedro Mártir, and San Pedro Nolasco maintained their existences as islands (fig. 1.20). The connection and separation of the Tiburón complex to the mainland occurred multiple times during the changing sea levels of the ice ages since the Gulf of California opened. Desert climates and vegetation

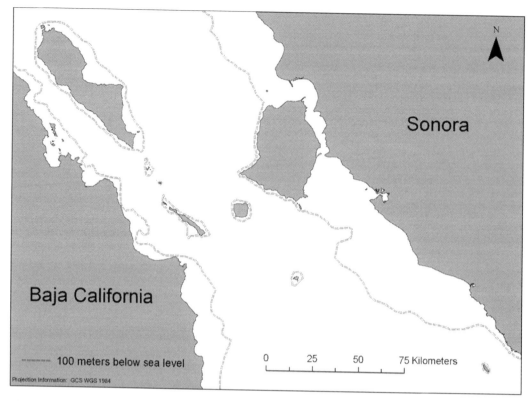

Figure 1.20. Ice-age and present-day coastline of the Midriff Region, Gulf of California. Areas within the 100 m bathometric contour (dashed lines) indicate approximate coastlines during the Last Glacial Maximum. KH.

similar to today did not develop until about 8900 yr BP. Modern desertscrub in Sonora along the Gulf of California first appeared in the middle Holocene (between 9720 and 5340 yr BP; Van Devender et al. 1994).

Pleistocene climatic conditions and the floristic composition of the islands in the Gulf of California and the mainland on either side of the Gulf are not known. However, knowledge of conditions for this period from much of the Sonoran Desert and distribution patterns of current species on the islands allow for the formation of hypotheses of past environmental conditions in the Midriff Region. Reconstruction of Pleistocene vegetation through much of what is now the Sonoran Desert via the analysis of fossil packrat middens confirms the presence of a relatively mesic woodland vegetation and flora in low desert regions in the late Wisconsin, the last glacial period, prior to 11,000 yr BP. An important baseline to consider is that desert communities were present for only about 5 percent of the 2.4 million years of the Pleistocene, while ice-age woodlands in the current desert lowlands persisted for about 90 percent of this period (Betancourt et al. 1990). Ice-age dominants included *Pinus monophylla* Torrey

& Frémont (singleleaf pinyon; Pinaceae), *Juniperus osteosperma* (Torrey) Little (Utah juniper; Cupressaceae), *Quercus turbinella* Greene (shrub live oak; Fagaceae), and *Yucca brevifolia* Engelmann (Joshua tree; Asparagaceae). The early Holocene from 11,000 to about 9000 yr BP was a transitional period, with some mesic species, including *Juniperus californica* Carriére (California juniper) and Joshua tree, in the Tinajas Altas Mountains in southwestern Arizona (Van Devender 1990). The only published packrat-midden records from Sonora are 10,000-year Holocene sequences from the Hornaday Mountains in the Gran Desierto in northwestern Sonora and the Sierra Bacha on the coast of the Gulf of California (Van Devender et al. 1990a, 1994). Ice-age climates with greater winter rainfall and cooler summers favored the expansion of northern and higher elevation species southward and into what are now desert lowlands.

Desert vegetation was constrained to relatively small refugial areas during the Pleistocene pluvial periods (Cody et al. 1983). Recent molecular phytogeographical analyses combined with paleo-vegetation distribution data from fossil packrat middens indicate three probable locations for the Pleistocene desert refugia in the Sonoran Desert region: (1) Lower Colorado Valley, (2) eastern Plains of Sonora, and (3) Cape Region of Baja California Sur (Fehlberg & Ranker 2009; Nason et al. 2002; Van Devender 1990).

Distributions of present-day plant species on Midriff Islands provide evidence for two major biotic responses to climatic changes during the late Wisconsin glacial period. First, a number of species have distributions indicative of migration from the Baja California Peninsula to the Sonora mainland via the Midriff Islands. This connection, discussed by Cody et al. (1983), highlights the sizable number of species with distributions that range over much of the Peninsula, from there onto Midriff Islands, and then in relatively limited coastal areas of Sonora. A phylogenetic study of *Stenocereus gummosus* (*pitaya agria*), a columnar cactus with a "Baja migrant" distribution, established preliminary support for island and mainland Sonora populations originating on the Baja California Peninsula (Clark-Tapia & Molina-Freaner 2003).

The presence of species with outstandingly disjunct distributions at higher elevations on Tiburón indicates paleo-flora (and vegetation) much different than the current suite of desert species (Wilder et al. 2008b). About a half-dozen species confined to higher elevations on the island (above 500 m) are remarkably disjunct from their conspecifics—isolated by more than 200 km of intervening Sonoran Desert (table 1.3). This distributional pattern is defined by the occurrence of relatively mesic-adapted perennial species on the insular mountains that can be likened to a "sky island within a desert island." The presence of these species on Tiburón is best explained in terms of Pleistocene relicts (Turner et al. 1995; Wilder et al. 2008b).

The crucifixion thorn *Canotia holacantha* is one of the best examples of historic, or ice-age, vegetation patterns on Tiburón. This species has a distribution that is transitional into higher elevations and more northern vegetation types, with its primary range in central and northern Arizona. The limited population of *C. holacantha* at the highest elevations on Tiburón is the southernmost locality for the species and extends its known range 230 km to the southwest. The isolated populations in northern Sonora in the vicinity of Magdalena de Kino suggest that during the Pleistocene glacial periods, characterized by higher winter rainfall, *Canotia holacantha* and other more cold-tolerant chaparral and woodland species were widespread across the present-day Arizona Upland subdivision of the Sonoran Desert and likely farther south in Sonora.

Pleistocene climatic conditions in the Midriff Region show evidence of increased winter rainfall and cooler summers with diminished or non-existent monsoonal activity due to persistent westerly winds at the Last Glacial Maximum. This climate facilitated the occurrence of relatively more mesic-adapted plant species in lowland regions throughout the area now referred to as the Sonoran Desert. While a woodland dominated in central Arizona, mixed desert and mesic-adapted communities occurred on both sides of the Gulf of California, and winter-rainfall–adapted Baja California species were prevalent in mainland Sonora opposite the Midriff Islands. Species with a Baja California center of distribution potentially crossed the Gulf of California to the Mexican mainland using the Midriff Islands as stepping-stones. Likewise, northerly derived species from the mainland likely occurred in central Sonora and on Isla Tiburón at least on the higher elevations. The onset of the Mexican monsoon in the early Holocene led to the establishment of the modern desert after a several-thousand-year lag, and the distribution of mesic-adapted taxa contracted to the present localities in sky island mountains or more northern latitudes.

In the following discussion we provide descriptions of the vegetation and ecology for the eight Sonoran Islands as they are today.

Table 1.3 Disjunct Species on Sierra Kunkaak, Isla Tiburón

Family	Species	Nearest population	Distance from Tiburón (km)
Asparagaceae	*Dasylirion gentryi*	W Sonora	155
Celastraceae	*Canotia holacantha*	NE Sonora	230
Cannabaceae	*Celtis reticulata*	W Sonora	155
Oleaceae	*Forestiera phillyreoides*	N Sonora	220
Oleaceae	*Fraxinus gooddingii*	N, NE Sonora	220
Rhamnaceae	*Sageretia wrightii*	NE Sonora	235

ISLA TIBURÓN

The enormous size of Tiburón is one of the first things that strike someone when thinking about the island and the region. Less obvious are the depth of the interactions between the Comcaac and the entire island and the discoveries yet to be made.

Isla Tiburón has the highest number of plant species of all the islands in the Gulf of California (Felger & Lowe 1976; Rebman et al. 2002). The size of the island, topographic complexity, and close proximity to the mainland underlie the relatively high diversity. We document 340 vascular plant species on Tiburón (appendix A), compared to the 298 species reported by Case et al. (2002) and the 286 species reported by Felger and Lowe (1976).

The large landmass of Tiburón is closely associated with the mainland. The waxing and waning of the insularity of Tiburón, combined with close proximity to the mainland, has led to a general uniformity of the island's flora and fauna with those of the mainland. Yet, due to the size and topographical diversity of the island, there are differences, some of them profound. While there are no endemic plants on the island, much of the island's fauna is endemic, at least at a subspecific level.

One of the original questions at the heart of this project was just how different Tiburón is from the Sonora mainland. A fairly consistent ecological region can be defined as the area in the vicinity of Bahía Kino to the vicinity of El Desemboque, and inland to Pozo Coyote, Sierra Seri, and Playa San Bartolo, an area we term the Seri Region. At least 110 plant species in this region (18 of them not native) are not known to occur on Tiburón (appendix B). This large pool of species includes many taxa that are surprisingly absent from the island.

The mountains and complex topography of Tiburón are extensive enough to produce several distinctive climatic zones that affect the distribution of plant species. The rather massive, complex, and high Sierra Kunkaak (peak elevation, 885 m), mirrored only by the Sierra Seri on the mainland, underlies two prominent floristic patterns on the island. (1) The higher elevations and canyons of the Sierra Kunkaak support desertscrub-thornscrub ecotone, which includes species and communities that are northern outliers, disjunct from subtropical regions farther south and inland in Sonora (e.g., *Lantana velutina*, *Lysiloma divaricatum*, and *Rhynchosia precatoria*). Some of these populations are separated from their nearest conspecific populations by the intervening Sonoran Desert. (2) There exists a strong east-west climatic gradient, but apparently a

lesser gradient from north to south. Various plant species on the western (Gulf) side of the island, in the summer rain shadow of the Sierra Kunkaak, are conspicuously more xeromorphic than those of the eastern (Sonora) side closer to the Canal del Infiernillo. Vegetation on the western margin of the island tends to be sparser than elsewhere on Tiburón and includes species with geographic affinities to highly arid regions of the Lower Colorado Desert (e.g., *Ambrosia dumosa*, *A. ilicifolia*, and *Senna confinis*).

Due to the large size of Tiburón, dividing the island into biologically significant regions facilitates discussion of its floristic diversity. In the following discussion we describe the ecological underpinnings of the floristic communities and general features of the flora.

Coastal communities

The coastal communities of Isla Tiburón are distinctive, reflecting the physical conditions of the shore regions. For example, along the edge of the island, littoral (mangrove and salt scrub), sea cliff, and beach dune communities are found at different locations. A well-developed littoral scrub reaches maximum development on the east shore, bordering the protected Infiernillo coast, which has three main lagoons (esteros) and expansive beaches. The Canal del Infiernillo also provides habitat for a large seagrass community (with *Halodule wrightii* and *Zostera marina*, and perhaps *Ruppia maritima*). This seagrass community is the most expansive in the Gulf and is critical to the marine ecosystems (e.g., food for sea turtles and migrating brant).

The three major esteros along the Infiernillo coast of Tiburón support locally extensive mangrove scrub: Punta Perla at the north, Punta Tormenta (Palo Fierro) at about the middle of the island, and San Miguel at the south (fig. 1.21). These are the only places where mangroves occur on the island. The three dominant Gulf of California mangroves (*Avicennia germinans*, *Laguncularia racemosa*, and *Rhizophora mangle*) are locally common in the intertidal zone on muddy substrate, inundated and exposed daily by tides. There is the usual pattern of overlapping zonation among mangroves in Mexico, from the deepest intertidal limit to the water's edge at the shore (Felger 2004; Felger & Moser 1985; Lopez-Portillo & Ezcurra 1989). *Rhizophora mangle*, with its large stilt roots, occurs in the deepest water. *Avicennia germinans* reaches maximum density on the landward side of mangrove areas in the shallowest water. *Laguncularia racemosa* reaches maximum development

Figure 1.21. Estero at Punta San Miguel, Isla Tiburón, looking eastward from Hast Cacöla, with Estero Santa Rosa on the mainland in the background. BTW, 7 April 2008.

between the deeper water *Rhizophora* and the shallower water *Avicennia*.

The Tiburón mangroves are generally 2–3 m tall (sometimes reaching 4+ m) and form dense and often impenetrable (100 percent) cover. Like in other Gulf of California mangrove areas, the water becomes hypersaline since freshwater does not regularly flow into the esteros, and these areas have been called "negative estuaries." These desert-fringe mangroves are truly dwarfed compared to the 20-m-tall mangrove forests growing in brackish water in tropical regions. Occasional frost events, especially at Punta Perla, may contribute to the dwarfing, although aridity is probably the overriding factor.

Isolated patches of salt scrub occur here and there along the shores of the island. Salt scrub is absent in most places where sea cliffs and steep rock slopes and bluffs reach down into the sea, such as along portions of the southern and western shores. There are low beach dunes in some areas of the island, the largest being at the southwestern part north of Punta Willard as well as at the northeast corner of the island. Some characteristic dune

and upper beach species on the island include *Aristida californica* var. *californica, Cenchrus palmeri, Euphorbia leucophylla* subsp. *comcaacorum, Helianthus niveus, Psorothamnus emoryi,* and *Triteleiopsis palmeri.*

The eastern bajada

An expansive bajada marks the eastern flank of the island, between the Sierra Kunkaak and the Canal del Infiernillo (fig. 1.22). This bajada is the western counterpart of the mainland bajada on the opposite shore of the Infiernillo. The absence of cattle grazing on the island is especially noticeable here, resulting in a relatively uniform distribution pattern of many major perennials (not aggregated as they are in cattle-grazed areas on the mainland). Additionally, the increased density and sizes of many woody perennials are notable.

Plant communities on the eastern bajada form bands or zones running parallel to the Infiernillo, and we arbitrarily divide this continuum into lower, middle, and upper sections for ease of discussion. Closest to the water are the mangroves and salt scrub. The *Frankenia* zone

Figure 1.22. East-facing bajada of Isla Tiburón. Front range of Hast Cacöla on the left, mainland in background. BTW, 15 September 2007.

extends inland to about 1 km and often occurs on pavement-like transported felsite. Throughout its range on Tiburón *Frankenia palmeri* occurs only at sites directly facing the sea. Associated species richness is relatively low, coverage and stature of plants are greatly reduced, and many of the species found in this zone are noticeably xeromorphic. This vegetation reflects the lack of protection from the full force of winds and the role of saline soil in an arid environment. Associated species include *Atriplex barclayana, A. linearis, Fouquieria splendens, Maytenus phyllanthoides,* and *Peniocereus striatus.*

Within the *Frankenia* zone are pockets of mixed desertscrub with higher species richness (fig. 1.23). These

populations are associated with swales or depressions that are the termini of dendritic drainages running perpendicular to the shore. Here *Frankenia* plays only a minor role and the major plant cover consists of drought-deciduous desert shrubs and cacti. Species found here include *Atamisquea emarginata, Bursera microphylla, Chromolaena sagittata, Cordia parvifolia, Hibiscus denudatus, Jatropha cinerea, Larrea divaricata, Lycium brevipes, Maytenus phyllanthoides, Stenocereus gummosus,* and the geophytes *Talinum paniculatum* and *Tumamoca macdougalii.* Similar communities occur on the mainland side at roughly equivalent distances inland.

Figure 1.23. Mixed desertscrub on the eastern bajada of Isla Tiburón, vicinity of Zozni Cmiipla. *Encelia farinosa* (foreground) and *Jatropha cinerea* (middle-center and left) are common in this area. BTW, 24 November 2006.

The region we refer to as the middle bajada commences on the inland margin of the *Frankenia* zone, and the vegetation consists of mixed desertscrub species. Low, flat-topped ridges merge into the main body of the expansive east-shore bajada, and the vegetation and flora gradually become richer and more complex with increasing elevation westward toward the Sierra. Immediately inland from the *Frankenia* zone are various xerophytic shrubs that are nearly leafless or have only reduced leaves for much of the year and, like *Frankenia palmeri*, are presumed to be long-lived, for example, *Castela polyandra, Condalia globosa, Krameria bicolor,* and *Phaulothamnus spinescens. Castela polyandra* is known only from this confined zone on the eastern bajada of Tiburón and an equivalent zone on bajadas along the Gulf Coast of the Baja California Peninsula. It does not occur on the Sonora mainland. Several cacti are common in the lower portion of the middle bajada: *Cylindropuntia fulgida, Ferocactus tiburonensis, Lophocereus schottii, Pachycereus pringlei, Stenocereus gummosus,* and *S. thurberi. Ferocactus tiburonensis* and *Lophocereus schottii* are scattered through much of the remaining middle bajada, but other cacti are generally scarce farther inland on the bajada.

Unlike the western-sloping bajada on the mainland side of the Infiernillo, much of the eastern-sloping bajada of Tiburón does not house extensive cardón (*Pachycereus pringlei*) stands (except at Valle de Águila), although cactus scrub does develop on the western side of the Sierra Kunkaak (*see* discussion of the Central Valley, below). The absence of *P. pringlei* might be due in part to the shadow cast by the Sierra Kunkaak onto the eastern bajada, which decreases insolation relative to the west-facing bajada of the mainland. This subtle difference in shadow and slope exposure seems to affect conditions sufficiently to produce floristic and vegetational differences between the bajadas on opposite sides of the Canal del Infiernillo.

As elevation rises closer to the base of the Sierra Kunkaak, species found only along usually dry streamways (arroyos) in the lower bajada begin to occur away from stream courses. Yet, at any point along the bajada, plants along drainages tend to be larger in stature than those of the adjacent desert flats. The inter-streamway flats or uplands are characterized by xerophytic desert shrubs that form the dominant cover, such as *Cordia parvifolia, Encelia farinosa, Larrea divaricata,* and *Ruellia californica*. Dry watercourse (xeroriparian) communities of the middle bajada, adjacent to the desert

flats, are contrastingly represented by large shrubs and small desert trees, including *Bursera microphylla, Colubrina viridis, Olneya tesota,* and *Parkinsonia microphylla*. These communities represent the densest and richest desertscrub of the bajada, having greater diversity of perennial and ephemeral species than any other region of the eastern bajada (Felger 1966).

The upper bajada, starting at about 75 m elevation, is rocky and rough as a result of outwash from the east slopes of the Sierra Kunkaak. Arroyo channels are usually deeper than in the lower or middle bajada. Shrubs and small desert trees largely dominate vegetation of the upper bajada. Cacti and many xerophytic desert shrubs (e.g., *Larrea divaricata*), common on the lower portions of the bajada, are largely absent or are present in reduced numbers in the upper bajada. However, those individuals that are present attain significantly larger stature here. Some of the species present on the upper bajada but absent at lower elevations show affinities with subtropical scrub regions to the south or inland from the desert in Sonora and appear as outliers from the Sierra Kunkaak, for example, *Guaiacum coulteri* and *Galphimia angustifolia*.

Sierra Kunkaak

The Sierra Kunkaak is a large and extremely rugged range along the majority of the eastern side of the island. The Sierra has a north-south alignment and rises rapidly above the bajadas to 885 m at the peak. A large front range, Hast Cacöla—the visible portion of the Sierra from Punta Chueca on the mainland—is followed on its western side by the Sierra Kunkaak Mayor, where the highest peaks occur. The western side of the main portion of the Sierra falls dramatically into the Central Valley. These mountains contain networks of arroyos and canyons providing sheltered niches and microhabitats that support a diverse array of xeroriparian desertscrub and thornscrub-like vegetation (fig. 1.24). Fifty species, or roughly 13 percent, of the total flora of Sonoran Islands occur only in the Sierra Kunkaak—besides the disjunct relict species, examples include *Bursera fagaroides, Justicia candicans, Lycium berlandieri,* and *Zapoteca formosa*.

Extensive areas of vegetation resembling Sonoran foothills thornscrub occur at higher elevations and in canyons, particularly on east- and north-facing slopes. In contrast, south- and west-facing slopes of the Sierra Kunkaak and the exposed upper ridges and peaks are harsher and more exposed, supporting mixed

Figure 1.24. Interior canyon of Sierra Kunkaak. BTW, 25 November 2006.

desertscrub communities, exhibiting the importance of slope exposure in the distribution of vegetation (Shreve 1915). For a more detailed discussion of the vegetation of the Sierra, see the discussion of desertscrub-thornscrub ecotone in the Vegetation Communities section.

Tinajas, or canyon waterholes, are widely scattered in the Sierra Kunkaak and host small, highly localized communities of wetland plants and lush vegetation. Historically, the tinajas were critical to the Comcaac, providing their only reliable sources of water (Felger & Moser 1985). Archaeological artifacts are copious around tinajas and associated trails. Water pools and seepages are marked by locally dense colonies of the reed grasses *Arundo donax* and *Phragmites australis*. The stems of these grasses were employed by the Comcaac to build ocean-going balsas or reed boats (Felger & Moser 1985; McGee 1898). On the nearby mainland and at one waterhole on Tiburón, *Phragmites* has been replaced by the Old World *Arundo donax*. Thus, the localities where *P. australis* is present on Tiburón serve

as refugia for desert populations of this cosmopolitan wetland species. The tinaja plant communities of the Sierra are especially botanically diverse due to their occurrence in xeroriparian thornscrub habitats as well as the opportunities for growth afforded by perennial freshwater.

The Central Valley

Between the Sierra Kunkaak and the Sierra Menor lies an expansive valley that runs much of the length of the island, often referred to as the Agua Dulce Valley due to the waterhole in the vicinity of Tecomate to the north; here it is termed the Central Valley (fig. 1.25). Historically, a group of Comcaac occupied the interior region of the island for at least part of each year (Moser 1963). However, the middens and potsherds that are so abundant around the periphery of the island are not readily seen in the Central Valley (Moser 1963). Only a relatively small number of Comcaac appear to have frequented

Figure 1.25. Central Valley, Isla Tiburón, looking northwest from the top of Cójocam Iti Yaii, Sierra Kunkaak. The middle and northern areas of the Central Valley are just beyond the western foothills of the Sierra Kunkaak (middle), with the Sierra Menor in background. BTW, 26 October 2007.

the area, which has been uninhabited for about one hundred years. The vegetation consists wholly of desertscrub, even to peak elevations of the adjacent Sierra Menor range along the western margin. Along dry drainage margins there is rich and varied arborescent and semi-arborescent growth, while non-dissected flats consist largely of extensive stands of creosotebush (*Larrea divaricata*). This mosaic involving creosotebush scrub, mixed desertscrub, and xeroriparian desertscrub follows a dendritic-reticulate pattern, with the overall drainage coursing northward through the Central Valley and issuing into the sea near Tecomate at the northeastern shore.

The xeroriparian desertscrub of the Central Valley is especially well developed and courses through many kilometers. Especially noteworthy are species in the upper reaches of the interior valley, with individuals of enormous size relative to their usual maximum size elsewhere in the Sonoran Desert. Examples include individuals of *Sideroxylon occidentale* at 11 m tall (fig. 1.26), *Condalia globosa* at 5.5 m, and *Koeberlinia spinosa* at 5.5 m. *Olneya*

tesota (*palo fierro* or ironwood) is the most conspicuous element in the community and generally exceeds any other single species in coverage.

Cactus scrub, with conspicuous "forests" of columnar cacti, occurs primarily on the bajada that comes off the Sierra Kunkaak along the east side of the Central Valley. The valley floor slopes gradually westward and supports a relatively large population of saguaros (*Carnegiea gigantea*), small desert trees (e.g., *Parkinsonia microphylla*), and an array of desert shrubs. Slope exposure compares well with that of the counterpart bajada of the mainland side of the Infiernillo, where extensive stands of the more maritime cardón (*Pachycereus pringlei*) occur rather than a prevalence of saguaros.

Sierra Menor and the west coast

The western margin of the island is extremely rugged and is delimited by the Sierra Menor, which runs the length of the west coast. The east-west divide of the range is close

Figure 1.26. A large *Sideroxylon occidentale*, Central Valley, Isla Tiburón. Robin Thomas holds a 25-foot telescoping pole. RSF, 18 February 1965.

to the shore and in some places drops away into sheer sea cliffs. Elsewhere the western, or seaward, side rises steeply with several discernible terraces. The entire west coast is more arid than the rest of the island. Vegetation is sparse and mostly dominated by small to medium-sized desert shrubs with relatively small leaves (fig. 1.27). Some small desert trees are present, but larger trees are absent. On the east side of the Sierra Menor there are diverse communities that include mixed desert shrubs, small desert trees, and columnar cacti.

Dunes are found near Punta Willard at the southwest corner of the island. Otherwise, where beaches exist instead of sheer sea cliffs, they are of cobblestone-like rock and preclude development of littoral scrub, although a few halophytic species frequently occur along the inland edge of the rock beaches (e.g., *Atriplex polycarpa* and *Suaeda nigra*). The gravelly/stony pediments between the beach and the first hills support sparse stands of *Frankenia palmeri*. The coastal pediment here is narrow and seldom exceeds 100–300 m in width.

Figure 1.27. Ensenada Blanca (Vaporeta) on the west side of Isla Tiburón. RSF, March 1965.

Mesa-like ridge crests of desert pavement are partially barren with only a sparse cover of xerophytic desertscrub. Characteristic species include *Atriplex barclayana*, *Errazurizia megacarpa*, *Fouquieria splendens*, and *Jatropha cuneata*. Winds are persistent, and their desiccating effects, combined with the lack of runoff from adjacent areas, underlie the paucity of vegetation. Significant areas are nearly devoid of perennial plants, and maximum ground coverage reaches only 5.5 percent (Felger 1966). Many perennials aggregate in the shelter of large rocks or in slight depressions where water may briefly pool following rainfall. Species occurring here generally are smaller than in adjacent or nearby sites and appear to be stragglers on the mesa tops.

Canyons support richer desertscrub than adjacent slopes; however, desertscrub here is quite xerophytic compared with other regions of the island. The largest such canyon is at Pooj Iime, where the drainage system winds back into the Sierra Menor with large side canyons extending nearly to the ridge crest of the Sierra (fig. 1.28). Prominent species of this area include *Acacia willardiana*, *Ambrosia ilicifolia*, and *Pachycereus pringlei*. The north and south slopes above the canyon floors support a thin cover of xerophytic shrubs. The opposing slopes show little physiognomic differences due to the aridity of the region.

The North (Tecomate)

Located on the north coast is an area that was once a principal Seri village, Hajháx or Tecomate, at the large Bahía Agua Dulce (fig. 1.29). The region is seldom used nowadays except by fishermen who sometimes camp along the shore. This area also served as a Mexican marine station in the latter decades of the twentieth century but has since been abandoned. Local development included a small pier, since washed away by storms, and the planting of exotic trees—tamarisk (*Tamarix aphylla*) and Murray River eucalyptus (*Eucalyptus camaldulensis*). A small but reliable waterhole sustained the village and gave the bay and valley their names. Localized, halophytic littoral growth occurs in saline soils at the west end of Bahía Agua Dulce.

Figure 1.28. Pooj Iime, west side of Isla Tiburón. *Bursera microphylla*, *Pachycereus pringlei*, and *Stenocereus thurberi* (foreground) and western flank of the Sierra Menor (background). BTW, 29 September 2007.

Figure 1.29. North shore of Isla Tiburón: the Seri camp at Tecomate. WNS, May 1951; Special Collections, University of Arizona Library.

However, at least in part due to the absence of tidal inlets or esteros and the seasonal occurrence of frost, mangroves are absent from the north shores of the island. Large, dense mats or mounds of succulent species, including *Atriplex barclayana*, *Batis maritima*, *Maytenus phyllanthoides*, and *Suaeda nigra*, occur along the shore. The coast has a discernable population of *Frankenia palmeri* along a narrow coastal strip on rocky hills, coastal dunes, and level terrain. The *Frankenia* zone is not as clearly delimited, or as extensive, as on the eastern shore of the island and the opposite mainland.

The occurrence of beach dunes is the most distinctive aspect of the coastal region on the north end of the island. Plant cover, which is thin and highly variable, is richest on the leeward (south) side and on the relatively more stable dunes. Common dune species here include *Abronia maritima*, *Amaranthus fimbriatus*, *Croton californicus*, *Palafoxia arida*, and *Psorothamnus emoryi*. At the east end of the bay, low, stabilized dunes extend inland for several kilometers, and sand has piled against the windward side of the rock hills bordering the east side of the Central Valley.

The non-dissected valley plains so prevalent in the Central Valley reach to the north and terminate at the

coast in the vicinity of Tecomate. Plant coverage of the northern inland valley is much like that of the Central Valley, with *Larrea divaricata* dominant. In addition, the dendritic arroyo system of the Central Valley terminates in the Tecomate region, where it converges into a floodplain at Bahía Agua Dulce. The meandering shallow channels have left several successively higher levels of old floodplain surfaces, which today appear as islets or benches near the present channels. Small desert trees (*Bursera microphylla*, *Olneya tesota*, *Prosopis glandulosa*), mixed desert shrubs (*Colubrina viridis*, *Simmondsia chinensis*, *Viscainoa geniculata*), various smaller shrubs or subshrubs (e.g., *Ditaxis lanceolata*, *Encelia farinosa*, *Errazurizia megacarpa*), and several vines (e.g., *Cardiospermum corindum*) are common. Cacti and succulents are relatively scarce. The desert trees in the arroyos are dominant on the landscape, and local dense stands of the shrubs *Ambrosia salsola* and *Hyptis albida* occur along the intermediate arroyo margins and extend into sandy arroyo beds.

The northern extremity of the Sierra Menor is approximately 1 km west of the old village site of Tecomate. Its hillslopes support mixed desert shrubs, a few small desert trees, and columnar cacti. The vegetation generally becomes richer on east- and north-facing slopes and toward higher elevations. *Agave subsimplex* and *Asclepias albicans* become common above 200 m, and the fern *Notholaena californica* is found on the north slopes among volcanic rocks. On the rocky, south-facing slopes xeromorphic species are prevalent, including *Carnegiea gigantea, Jatropha cuneata, Pachycereus pringlei, Parkinsonia microphylla,* and *Stenocereus thurberi.* This is one of the few localities on the island where *C. gigantea* and *P. pringlei* occur together in substantial numbers. However, *P. pringlei* shows greater development on lower slopes with eastern exposure, while *C. gigantea* reaches maximum density and size on the more arid, south-facing, higher, and steeper slopes.

The South (Sauzal)

The south-central sector of the island contains the large Arroyo Sauzal system, which drains the southern portion of the Sierra Kunkaak and one of the major intermountain valleys (fig. 1.30). The broad channel winds through rugged mountains and hills in a generally southward direction and produces a complex topography. Arroyo Sauzal principally drains the western side of the Sierra Kunkaak and the eastern side of the Sierra Menor, and it is the counterpart to the dendritic arroyo system reaching the vicinity of Tecomate in the northern part of the island. Desertscrub in this part of the island appears somewhat more xeromorphic than on the eastern side of the island but is significantly richer than on the western shore. Hillsides, bajadas, and pediments generally support sparse desert shrub communities. Floodplains and arroyos are characterized by large desert shrubs and small trees.

Low-relief hills and plains away from drainageways have a relatively sparse coverage of xerophytic desert shrub communities. Larger shrubs are widely scattered and are generally aggregated. Smaller shrubs (e.g., *Encelia farinosa*) far outnumber larger ones in total coverage and quantity of individuals. *Encelia* is common throughout the area, and the populations appear to consist of the same cohort, presumably as the result of a particularly fortuitous season. Summer-fall ephemerals sometimes reach 100 percent coverage, with *Tidestromia lanuginosa* playing a major role. The substrate includes loose rock and rough, yellowish gravel in a relatively shallow

Figure 1.30. Arroyo Sauzal, south end of Isla Tiburón. RSF, 29 January 1965.

profile, often extensively tunneled by rodents, and their trails form conspicuous patterns. The sparseness of plant cover, and particularly the absence of cacti, may be due in part to the activities of large rodent populations. Additionally, a large population of mule deer (*Odocoileus hemionis sheldoni*) is present in the area, and many shrubs (e.g., *Bursera hindsiana*) are seriously browsed.

As elsewhere, streamways and floodplains harbor rich stands of desert shrubs and trees. The composition of plant communities at the valley floor varies greatly with the substratum and elevation above the arroyo channel or streamway. Thickets of *Hyptis albida* occur along gravelly channels and reach 3–4 m in height. The vegetation is markedly richer and more varied along the valley floor or canyon bottom at the margins of these shifting streambeds or washes. These communities represent the richest vegetation found in the area outside of the Sauzal waterhole. Coverage by perennials is around 70 percent,

and the structure of the vegetation is relatively complex. *Acacia willardiana* forms a well-developed trunk, and there are some unusually large specimens of *Stenocereus thurberi* (8 m tall). Other species adjacent to the washes or watercourses are *Bursera microphylla, Caesalpinia intricata, Cardiospermum corindum, Colubrina viridis, Jacquemontia abutiloides, Melochia tomentosa, Ruellia californica, Stegnosperma halimifolium,* and *Tephrosia palmeri.* The relative scarcity of *Encelia farinosa* in this community, compared to its distribution on low-relief hills and plains, is notable.

The Sauzal waterhole (from *sauce,* for willow or *Salix*), known by the Seris as Xapij (the name for *Phragmites* or *carrizo*), about 5 km inland from the coast, is one of the better known watering places on the island and played a vital role in the life of the local Comcaac (fig. 1.31A,B). The San Esteban people came to this waterhole when water supplies on Isla San Esteban were insufficient (Bowen

(A)

(B)

Figure 1.31. Xapij waterhole, Arroyo Sauzal, Isla Tiburón. (A) Looking northwest, dense colonies of reedgrass (*Phragmites australis*) with tamarisk shrubs (*Tamarix chinensis*) at the periphery. BTW, 29 January 2008. (B) José Torres at Xapij. EHD, 1936. Heye Foundation (*HF-23927*).

2000). The Comcaac also utilized the extensive stand of *xapij* or *carrizo* (*Phragmites australis*) at the waterhole to build their balsas (Felger & Moser 1985). There are two main *aguajes* (waterholes) about 1 km apart, each marked by a separate thicket of *Phragmites*. A solid rock substratum and dykes preventing rapid percolation of water into the streambed form the aguajes in this small valley plain. Perennial pools of water throughout the island lack fish, but in late spring large numbers of tadpoles of the red-spotted toad (*Anaxyrus punctatus*, = *Bufo punctatus*) and Couch's spadefoot toad (*Scaphiopus couchii*) are seen in the warm, shallow water. The water is highly alkaline, and rotting *Phragmites* and *Salix* produces a sulfuric odor. Alkaline surface encrustations are evident, and characteristic alkaline-adapted species are common, including *Distichlis spicata*, *Heliotropium curassavicum*, and *Tamarix chinensis*.

The perennial water supports a local community of wetland species. This vegetation here is essentially evergreen. The phenology of the riparian community is keyed more to temperature than to the precipitation pattern, which so severely limits the majority of the Gulf Coast vegetation. Birds, due to their high use of the waterhole,

might have introduced the population of the non-native, invasive riparian species *Tamarix chinensis* at the waterhole. The *Tamarix* population appears to be stable, without appreciable change during the last half century, and the many small, young plants probably seldom become established.

ISLA SAN ESTEBAN

Isla San Esteban (41 km²) is about 12 km south of the southwest shore of Isla Tiburón. It is the second largest of the Sonoran Islands and has the second highest floristic diversity (114 species; appendix A). Most of the species are common to both shores of the Gulf. The island is quite arid, and the general appearance of the vegetation resembles that of the western Midriff Islands and the Gulf Coast of the northern Baja California Peninsula (fig. 1.32). The distributional affinities of San Esteban's flora with that of Tiburón are generally more with Tiburón's arid southwestern portion than with the southern coast. Thus, the relationship is correlated more with climatic similarity than spatial distance. One

Figure 1.32. Isla San Esteban, looking westward from near the shore at Arroyo Limantour. *Agave cerrulata* subsp. *dentiens* in the foreground. Note the thin band of fog near the summit. RSF, 5 November 1967.

endemic is known only from Isla San Esteban (*Agave cerulata* subsp. *dentiens*). Two cacti occur only on Islas San Esteban and San Lorenzo (*Echinocereus grandis* and *Mammillaria estebanensis*). *Lyrocarpa linearifolia* is endemic to Islas Ángel de la Guarda, San Esteban, and the Gulf Coast of Baja California Norte (Moran 1983a).

San Esteban is of volcanic origin (Desonie 1992) and includes high sea cliffs interrupted by widely spaced canyons. On the east side of the island the sea cliffs lessen, and a broad valley-like canyon known as Arroyo Limantour meets the sea and is the place where most visitors and biologists have gone ashore. A rugged, horseshoe-shaped series of steep mountains surrounds the central valley and rises to 550 m. The floodplain is strewn with transported rock and gravel heaped into uneven ridges that are traversed by shallow washes. In the interior the valley narrows into a number of small, short, and rugged canyons. The island is extremely arid, receiving about 100 mm of precipitation annually (Bowen 2000; Felger & Moser 1985). Summer precipitation in some years is drastically reduced or even lacking. The central valley is sheltered by the surrounding mountains and contains a "heat sink" due to intense insolation and re-radiation from the partially barren rock slopes. Local heating may contribute to the observed aridity.

Historically, San Esteban was inhabited by a small group of seafaring Comcaac that spoke a distinct dialect (Bowen 2000; Moser 1963). They could not spend the entire year on San Esteban due to the lack of perennial freshwater. Their historical range also incorporated the Arroyo Sauzal region of Isla Tiburón (Bowen 2000; Felger & Moser 1985). The endemic *Agave* forms extensive colonies on the coarse rocky slopes and is one of the major landscape elements of the island. It provided significant and reliable food resources. Unlike other agaves, it was harvestable year-round, and large roasting pits are evident on the landscape. Members of the Comcaac community vividly recalled expeditions to San Esteban to collect agave (Felger & Moser 1985). Most of the San Esteban people were executed by the military in the 1860s (Bowen 2000).

The vegetation consists largely of shrubby plants and succulents. Few species develop an arborescent habit, for example, *Ficus petiolaris* subsp. *palmeri*, *Olneya tesota*, and *Pachycereus pringlei*. Both *P. pringlei* and *O. tesota* reach maximum development and abundance along the valley floodplain. *Ficus petiolaris* subsp. *palmeri* is widely scattered on San Esteban and is generally restricted to steep cliffs, where it usually remains a facultatively

dwarfed shrub and only rarely becomes a small tree. *Sideroxylon leucophyllum*, occurring occasionally in canyons and at the top of the island, may become a small tree. *Bursera microphylla* has a well-developed trunk but seldom becomes tree sized. *Stenocereus thurberi* was once considered rare on the island (Felger & Moser 1974), but fieldwork by Tom Bowen (2003) has shown it to be locally common on the island.

Succulents are abundant and diverse. Large succulents (columnar cacti and chollas) reach greatest height and density along the valley floodplain, while the smaller succulents such as *Echinocereus grandis*, *Agave cerulata* subsp. *dentiens*, and *Mammillaria estebanensis* are most abundant on rocky slopes. Maximum succulent growth generally develops on east-facing slopes, diminishes with south and west exposure, and is least developed on north slopes. The higher elevations of the island are often shrouded in sea fog, and this added moisture likely contributes to the development of succulent communities and may underlie certain morphological distinctions as well. The San Esteban agaves feature grayer coloration and stouter marginal spines on plants at lower elevation, and are distinguishable from greener-leaved plants with smaller and few marginal spines usually occurring at higher elevations (Felger 1966; Felger & Moser 1970).

Sea cliffs and short, rocky beaches preclude all but a rarified shore vegetation of semi-littoral species (e.g., *Atriplex barclayana*, *Suaeda nigra*, and *Vaseyanthus insularis*). These plants occur along rock beaches mostly in front of arroyos and at the base of cliffs, but such habitats are few and of limited scope. Characteristic sea-cliff species include *Eucnide rupestris*, *Ficus petiolaris* subsp. *palmeri*, *Hofmeisteria fasciculata*, and *Pleurocoronis laphamioides*.

Desertscrub along the main valley floodplain is richer than on adjacent slopes but is relatively xerophytic. Common elements in the valley floodplain include desert shrubs, columnar cacti, cholla, and various smaller succulents. Large succulents (cacti) reach maximum development along the floodplain and together with *Jatropha cuneata* form dominant elements.

Desert shrubs and succulents form a relatively thin cover over the rugged and rocky mountain slopes. The density, coverage, and number of species generally increase toward higher elevations and on north- and east-facing slopes, and diminish on south- and west-facing slopes and at lower elevations. Plant cover thins on steeper slopes and is largely absent from the steep rockslides, which occur throughout the island. Rocky mesa-like ridge crests likewise may be partially barren. Less xerophytic species begin to appear toward the top and north side of the island, yet desert shrubs and succulents occur to peak elevation. A number of species are generally confined to higher elevations, for example, *Acalypha californica, Argemone subintegrifolia, Bursera hindsiana, Ephedra aspera, Euphorbia misera, Gambelia juncea, Lyrocarpa linearifolia,* and *Sideroxylon leucophyllum.* Several species are common at higher elevations and occur again along the valley floor but are generally absent from the intervening arid slopes (e.g., *Cryptantha fastigiata, Enteropogon chlorideus,* and *Nicotiana obtusifolia*).

ISLA DÁTIL

This island is a narrow, north-south-oriented, remnant mountain lying at the south end of a partially submerged peninsula extending south from the southeast shore of Tiburón (fig. 1.33). This submerged peninsula is roughly 5 km long, with its emergent portions consisting of Dátil at the south and Cholludo between Dátil and Tiburón. Reefs connect Islas Dátil and Cholludo, and both islands appear to represent rather recent fragmentations from Tiburón. The island is locally called Isla Dátil; map names Isla Turners or Turners Island are not used by local fishermen. Dátil is the local Spanish word for the date palm (*Phoenix dactylifera* Linnaeus), and in northern Mexico it is also employed for various yuccas (*Yucca* spp.). Here it probably refers to the small century plant (*Agave subsimplex*), which is abundant on the island.

The rugged topography of Dátil provides a wide range of microhabitats. This small island (1.25 km²), with a peak elevation of 183 m, supports a remarkable diversity of 101 species (appendix A). Dátil and Cholludo may receive somewhat more summer-fall precipitation than the adjacent mainland, reflecting the orographic effects of the Sierra Kunkaak in the southwestern part of Tiburón and the large size of Tiburón itself. The ameliorating effect of the surrounding sea is a likely factor in the high species richness.

There are a few small rock beaches where salt scrub species are present, but these do not form distinctive communities. Most of the shore consists of high sea cliffs. The island is rocky and steep throughout, and the soil is generally shallow. There are several major canyons that fall steeply and are rock strewn without sandy watercourses or washes.

Figure 1.33. Isla Dátil, looking southward from Isla Cholludo.
BTW, 2 September 2008.

The vegetation consists of mixed desertscrub and some xeroriparian species. The vegetation is noticeably denser and shrubbier than on the adjacent south shore of Tiburón and is likewise distinctive from other Gulf Islands. In fact, the plants on Dátil attain such density that the vegetation has the feel of a thornscrub community, but the species composition is entirely of desertscrub affinity. East- and north-facing canyon slopes and canyon floors support dense desertscrub that may be termed xeroriparian, though the distinction is subtle. The communities with maximum species richness occur on north-facing slopes near the summit on the east side of the island. Elevation gradients in the vegetation are weakly developed.

Large herbivores such as mule deer and rabbits are absent from Dátil, although three rodents occur on the island: a rock pocket mouse (*Chaetodipus intermedius minimus*), white-throated woodrat (*Neotoma albigula varia*), and cactus mouse (*Peromyscus eremicus insularis*).

Most of the species on Dátil are also found on neighboring landmasses. For the most part the flora shows affinity with the south side of Tiburón, but there are some interesting extra-limital populations. *Calliandra californica* and *Cnidoscolus palmeri* are common on protected north- and east-facing slopes and canyons on the east side of the island, and are of limited distribution on Tiburón. *Tiquilia canescens* is a xerophytic species typical of the northern part of the Sonoran Desert, such that its location on Dátil and the southern portion of Tiburón is remarkably disjunct.

Succulents are common but markedly less prevalent than on adjacent Cholludo. Succulents on Dátil include *Agave subsimplex, Asclepias albicans, Hofmeisteria fasciculata*, and 10 species of cacti. The presence or absence of herbivores and differences in soils and slope exposure contribute to the marked vegetational differences between these neighboring islands. The rich flora on Dátil seems to result from the locally favorable maritime conditions; the highly varied island topography, which provides a relatively wide range of microenvironments; and the relative proximity and past connection of the island to the larger flora of Tiburón.

ISLA CHOLLUDO

Cholludo (0.2 km²) is a tiny, rocky isle between Dátil and Tiburón along a sunken peninsula that connects the three islands. Older maps name the island Roca Foca or Seal Island, but Cholludo is the locally used name. The south side falls away abruptly into high cliffs, where numerous seabirds roost. From the south ridge crest, 75–100 m high, the surface of the island quickly slopes to the sea at the north shore. This north-facing slope encompasses most of the island's surface. There are no arroyos, and although the slope is steep and rocky, its surface is continuous. The island lacks a beach or littoral scrub.

Cholludo supports a flora of 30 species (appendix A) and has the highest number of species in proportion to island size of any of the Sonoran Islands. On Cholludo, as on Dátil, summer-fall rains are possibly more significant than in adjacent areas. The surrounding sea contributes to a relatively high humidity and amelioration of temperature extremes. These factors, combined with the north aspect of the island, the absence of rodent populations, and close proximity to a larger flora and landmasses, are major factors contributing to the comparative floristic richness. All the species on Cholludo occur on the neighboring islands of Tiburón and Dátil except for *Carnegiea gigantea*, which is rare on Cholludo and absent from Dátil. Floristic similarity with neighboring landmasses reflects their close proximity.

Cholludo, Spanish for "many chollas," is aptly named—it supports a formidable and scarcely penetrable forest of cactus scrub that is unique relative to its neighboring islands and the mainland. Dominant elements in this succulent landscape are first and foremost *Pachycereus pringlei*, as well as *Agave subsimplex, Cylindropuntia fulgida, Stenocereus gummosus*, and *S. thurberi*. *Agave subsimplex* is abundant over the eastern half of the island, which faces north-northeast, but it is absent from the western part (with a north-northwest exposure). Plant cover over most of the island is 70–80 percent (Felger 1966). Non-succulent leaf-bearing plants play a decidedly minor role, and large-leaved species (*Bursera hindsiana* and *Ficus petiolaris* subsp. *palmeri*) are rare. Non-succulent trees are absent. *Perityle emoryi* seasonally produces a carpet of green among the spiny perennial plant cover.

The locally unique appearance of the vegetation is due to the relatively different biotic and environmental conditions of the island. Similar cactus-covered islands of volcanic origin occur in Guaymas harbor, where maritime conditions, calm weather, and an absence of rodents likewise prevail (Turner et al. 2003). Over the past century those islands have experienced significant increases in cactus density, principally *Pachycereus pringlei*. Comparisons of photos from 1958 and 2007 of Cholludo show a similar situation (fig. 1.34A,B).

(A)

(B)

Figure 1.34. Isla Cholludo. (A) 30 October 1958, RSF. (B) 16 March 2007, BTW.

ISLA ALCATRAZ

Isla Alcatraz (Isla Pelícano, Isla Tassne, Pelican Island) is situated in Bahía Kino, 1.4 km west of the adjacent mainland shore. The island probably has been isolated for a relatively brief period of time. The flora consists of 54 species (appendix A), and all are found on the adjacent mainland. The island has two distinctive topographic areas, each comprising roughly one-half of the 1.44 km² area of the island. The northeastern part is relatively flat and not more than 1 or 2 m above mean high tide level. The southwestern portion rises steeply into a rocky

mountain mass to approximately 130 m. Vegetation, flora, and white guano from extensive seabird rookeries in the mountain are sharply distributed according to these major topographic features. Due to the close proximity of the island to human settlement, Alcatraz has the highest level of disturbance and the highest ratio of non-native to native species of any of the Sonoran Islands—four non-native species: *Cenchrus ciliaris* (removed in the early 2000s), *Chenopodium murale*, *Tamarix chinensis* (removed in 2007), and *Tribulus terrestris*.

The extensive, low flat on the northeast side of the island supports salt scrub and meager, isolated desertscrub.

Much of the interior of this flat consists of a low-lying playa dominated by *Allenrolfea occidentalis*. In this area, the plant community shows little seasonal variation, reflecting a perennial source of groundwater that is absent in the adjacent, higher ground. Sparse stands of herbaceous salt scrub species are strewn along the upper beach. Very open mixed desertscrub of low stature occurs on low sandy rises and flats between the beaches and the small playa (an interior-draining desert flat).

The eastern slopes of the mountainous region support sparse mixed desertscrub. Differences in slope exposure, steepness, and drainage patterns produce moderately diverse but simplified communities of localized populations. Common species are *Amaranthus watsonii*, *Cylindropuntia fulgida*, *Nicotiana obtusifolia*, *Pachycereus pringlei*, and *Viscainoa geniculata*. Following seasons of heavy rains, *Amaranthus watsonii* forms extensive, dense stands throughout the island that seasonally color the island green or brown, giving the island a grassy appearance when seen from the mainland. Near the recess at the base of the mountain there is a localized area in which *Pachycereus pringlei* and *Cylindropuntia fulgida* are so dense that other plants are virtually excluded. Fallen joints of *C. fulgida* cover the ground and readily take root. The higher elevations on the east side of the mountain support plants not found elsewhere on the island, for example, *Justicia californica*, *Olneya tesota*, and *Pleurocoronis laphamioides*. These species generally occur on northeast exposures. Small populations of *Ferocactus emoryi* and *Opuntia engelmannii* appear to be unique among the Sonoran Islands.

Aridity and extensive bird guano deposits render the southwestern and western portion of the island inhospitable to most vegetation. Principal plants found in these guano deposits are a few widely scattered *Pachycereus pringlei* and seasonally localized and relatively sparse stands of *Amaranthus watsonii* and *Nicotiana obtusifolia*. The extent of the non-vegetated ground due to guano deposits is surpassed only on Islas San Pedro Mártir and Patos.

In many ways the flora of Isla Alcatraz is the most dynamic of the islands in this region. For example, *Bursera microphylla* and *Prosopis glandulosa*, documented on the island in the 1960s and 1970s, were not found there in 2007, and *Avicennia germinans*, *Ferocactus emoryi*, *Mammillaria grahamii*, and *Maytenus phyllanthoides*, found on the island in 2007, were not found there in previous decades. As a result of the relatively high level of human activity and close proximity to the mainland, non-native

plants have established at higher rates than on the other islands, principally due to the increased opportunities for dispersal of seed by humans, birds, and wind.

ISLA PATOS

Isla Patos is a small (0.45 km^2) guano-covered island with a peak elevation of 70 m (fig. 1.35). The island lies in open water several kilometers north of Tiburón. In 1946 the major perennials, mostly cacti, were deliberately destroyed and the island cleared to facilitate guano mining (Gentry 1949). The commercial guano-mining ventures were short-lived, and the island has otherwise remained uninhabited (Bowen 2000, 2009). There is no freshwater.

Photographs of the island prior to its clearing in 1946 reveal extremely sparse vegetation. *Atriplex barclayana* can seasonally produce nearly 100 percent plant cover. Longer-lived perennials, however, produce only extremely sparse cover that is the lowest perennial coverage of any of the Sonoran Islands. Twelve species are known to have been present, reflecting work by Johnston (1924), Gentry (1949), and photos by B. F. Osorio-Tafall (February 1946) from the Allan Hancock Foundation at the University of Southern California.

Eight species were not found on the island in the 1960–70s but were documented earlier. Small *Lophocereus schottii* and *S. gummosus* plants from re-colonization, however, were found on the island in 2007, and *Perityle emoryi* and *Trianthema portulacastrum* were again documented. The 1946 photos reveal that the vegetation was essentially similar to that of the 1960–70s and 2007. In 2007 we found two plants previously not recorded from the island: *Chenopodium murale* and *Suaeda nigra*. Although extreme aridity and a large quantity of guano place severe limitations on the species that can establish on the island, the flora appears dynamic. Table 1.4 lists the 14 species known from Isla Patos.

ISLA SAN PEDRO MÁRTIR

Situated in deep water in the middle of the Gulf, this is the most isolated island in the Gulf of California. For its size and elevation, the flora is the most impoverished of any island in the Gulf of California, comprising only 24 species. Of these, the most striking is the cardón (*Pachycereus pringlei*), which forms a massive forest that covers the upper elevations of the island (fig. 1.36). The shore rises

Figure 1.35. Isla Patos, looking northeast, with *Atriplex barclayana* in foreground; lighthouse on summit. BTW, 30 September 2007.

abruptly to high sea cliffs, which are claimed by thousands of noisy birds throughout the year. Mártir is one of the most important seabird nesting sites in Mexico, for which 85 species of birds have been recorded, including eight species of breeding seabirds (Tershy & Breese 1997). The colonies of brown boobies (*Sula leucogaster*) and blue-footed boobies (*S. nebouxii*) are among the world's largest, and the colonies of brown pelicans (*Pelecanus occidentalis*) and red-billed tropicbirds (*Phaethon*

aethereus) are among the largest in Mexico (Tershy et al. 1997). There are no beaches, alluvial deposits, or outwashes, so there is no development of shore vegetation.

Rainfall is undoubtedly unpredictable, the largest amounts coming in the form of tropical storms and hurricanes, which develop between July and October and occasionally strike the island. Often in the early mornings during cooler seasons all but the lower-most portion of the island is enveloped by fog, which remains until several hours after dawn and sometimes nearly all day (fig. 1.37). The vegetation receives significant moisture from this fog—the plants at such times are drenched with dew.

Riparian or semi-riparian canyons and arroyos are absent. There is no source of freshwater (Bowen 2009). Most of the surface of the island is exposed and offers little or no shelter from wind. The island is volcanic in origin, and its surface is covered with rocks of varying size. Soil is poorly developed and shallow. At its west end a rock headland about 75–100 m high juts out into the sea and is connected to the rest of the island by a low saddle. The principal landfall, although a poor one and occupied by sea lions, is at the southeast corner of the island. Just above the southeast landing are the ruins of a village of rock-walled houses and shelters built by guano workers (fig. 1.38).

Starting in 1885 and lasting until 1891, the Mexican Phosphate and Sulphur Company, under two different

Table 1.4 Collection history for the 14 species known from Isla Patos

Species	1921	1946	1960/70s	2007
Amaranthus fimbriatus	X		X	
Atriplex barclayana	X		X	X
Bouteloua barbata var. *barbata*	X			
Carnegiea gigantea		X		
Chenopodium murale				X
Cylindropuntia fulgida		X	X	X
Encelia farinosa var. *farinosa*	X			
Lophocereus schottii var. *schottii*		X		X
Pachycereus pringlei		X	X	X
Perityle emoryi	X			X
Stenocereus gummosus		X		X
Suaeda nigra				X
Trianthema portulacastrum	X			X
Trixis californica var. *californica*	X			

*Non-native species.

Figure 1.36. Isla San Pedro Mártir, cardón forest at summit. BTW,
5 December 2007.

Figure 1.37. Isla San Pedro Mártir, morning fog enveloping dwarf cardons
(*Pachycereus pringlei*). BTW, 6 December 2007.

Figure 1.38. Isla San Pedro Mártir, looking to the southeast at the ruins of the guano-workers' village. BTW, 11 April 2007.

ownerships (1885–1888 as the U.S.-based International Company, and 1888–1891 under British ownership), intensely mined guano on the island (Bowen 2000). One of the few observations of this period comes from ornithologist Nathaniel Goss, who visited the island 15–28 March 1888 by means of a small steamer from Guaymas used to bring supplies to the workers on the island (Bowen 2000; Goss 1888). Goss observed a workforce of Yaqui Indians collecting guano, 135 on the payroll and many with their families. He reported that most of the valuable guano had been removed. This is the only report stating that the workers were Yaquis. Conditions must have been severe for the miners and their families, but the American and British companies would have paid wages, contrary to popular belief that they were prisoners or slaves. As difficult as conditions may have been, the personal safety was likely preferable to the pogroms and deportations to slave-condition henequen plantations in southern Mexico being inflicted on the Yaquis at the time (see Spicer 1980). Bowen (2000:134–136) provides the best overall discussion of the guano operation.

The guano mining activity had significant impacts on the island. Most breeding seabirds temporarily abandoned the island (Slevin 1923), failed to reproduce, or perished, and there was likely decreased plant cover and increased erosion (Tershy et al. 1992). The most significant long-term impact of the guano operation was the introduction of the ubiquitous Old World black rat (*Rattus rattus*). The rats preyed on bird eggs and chicks, especially the boobies, and are reported to have used some of the island plants as an important food source when the seabirds were not reproducing (Tershy et al. 1992). The effects on the plant life, however, are difficult to determine. The area that was mined and where the workers lived is mostly below the major vegetated parts of the island, and there is little direct evidence of major human-influenced disturbance at the higher elevations. Yet so much activity on this small island would have had impacts throughout.

The island has been uninhabited since the end of mining operations. Human use, however, increased in the late twentieth and early twenty-first centuries from commercial and sport fishing, private and commercial tourism, researchers, and other uses (Tershy et al. 1999).

Terrestrial vertebrate fauna includes a rattlesnake (*Crotalus atrox*), a kingsnake (*Lampropeltis getula*), and two small endemic lizards, *Aspidoscelis martyris* (*Cnemidophorus martyris*) and *Uta palmeri* (Grismer 1999; Murphy & Aguierre-Léon 2002). There are no native land mammals.

The island has an unusually small flora in relation to its size and elevation, with 24 species in 23 genera and 13 families, which has been documented by botanists since 1887. Despite the isolation, there are no endemic plants (unlike the reptile fauna), and all species can disperse long distances or over water. Is the depauperate Mártir flora due to its isolated location? Those species not adapted to long-distance dispersal are probably effectively restricted from reaching Mártir, greatly reducing the potential source pool of species. This lack of endemism indicates genetic connectivity for the plant species on the island with other island and/or mainland sources, reducing the isolated nature of the island from the perspective of a wind- or water-dispersed plant. The role of vagrant birds (visitors and unsuccessful colonizers), which often carry seeds between areas and may serve as pollinators for various plants, is likely significant in creating a link to other genetic sources (Rose & Polis 2000).

There are also intrinsic factors to Mártir that limit species numbers. Due to the large number of seabirds, there are high concentrations of guano and associated nutrients, principally nitrogen and phosphorus. The large amount of guano on the island certainly limits the plant species that are able to establish. However, the effects of guano are difficult to separate from the extreme aridity. Certain species tolerate the soil conditions of guano islands and are commonly found on such islands in the Gulf—six of these species occur on Mártir: *Cylindropuntia alcahes*, *C. cholla*, *Nicotiana obtusifolia*, *Pachycereus pringlei*, *Perityle emoryi*, and *Viscainoa geniculata*. On Mártir, *Sphaeralcea hainesii* also extends into guano areas, but no *Sphaeralcea* is found associated with guano habitats on other Sonoran Islands. After each significant non-summer rainfall, a thick blanket of the cucurbitaceous vine *Vaseyanthus insularis* covers much of the vegetated part of the island, forming 100 percent coverage in many areas, draping over the cardons and shrubs, painting the island bright green, smothering and inhibiting the establishment of other species (fig. 1.39). After the vines dry, they impart a golden brown hue to the landscape. Thus, isolation certainly limits those species that can establish, but in addition, high concentrations of guano, intense aridity, *Vaseyanthus insularis*, and lack of topographic complexity add to the impediments that species face on the island, all resulting in one of the most depauperate floras in the Gulf of California.

Permanent vegetation monitoring sites were established in 2007 to serve as a baseline to study potential

Figure 1.39.
Isla San Pedro Mártir, cardons with a ground cover of *Vaseyanthus insularis* after Hurricane John of September 2006. J. A. Soriano/GECI archive.

change, including the effects of eradication of the introduced *Rattus* population, accomplished that year. Analysis of vegetation structure shows seasonally dense ephemerals (annuals), scattered shrubs, and an overstory of cardons. A cardón plot, 0.1 ha in size, contained 209 individuals, 63 percent of which were less than 1 m in height, making this the youngest cardón population recorded. For a detailed discussion of the Mártir cardón population, see Wilder and Felger (2010). No non-native species occur on the island, but this situation has the potential to rapidly change, and vigilant monitoring is needed.

Two principal phytogeographic patterns are evident among the flora: (1) species common to both sides of the Gulf, and (2) plants occurring otherwise only on the Baja California side of the Gulf. Most of the species show the former pattern (20 species, 83 percent of the flora), and for these colonization could have originated from either side of the Gulf or from other islands. The second pattern is shown by only four species (16 percent of the flora): *Cylindropuntia alcahes, C. cholla, Pelucha trifida,* and *Sphaeralcea hainesii. Pelucha trifida* and *S. hainesii* have distributions primarily in the arid northern part of the Gulf. *Cylindropuntia alcahes* and *C. cholla* are found throughout most of the Baja California Peninsula, including on the Pacific side and commonly in Baja California Sur (Rebman 1995).

Significantly, the plants on Mártir are small seeded or with disseminules adapted to over-water colonization, both of which are features of long-distance dispersal. The fruits of *Vaseyanthus insularis* are lightweight structures with air-filled chambers readily suited to ocean dissemination (Gentry 1950). The sticky fruits of *Mentzelia adhaerens* and the aril-surrounded seeds of *Viscainoa geniculata* might readily be carried to the island by birds, while others such as *Ficus petiolaris* subsp. *palmeri, Pachycereus pringlei,* and *Stegnosperma halimifolium* have edible seeds and fruits and may also have been introduced by birds. Lightweight cypselas ("achenes") of the composites might be wind transported from neighboring landmasses and/or become entangled in bird feathers. It is interesting to note the relatively large representation of Asteraceae on the island: six species or nearly 25 percent of the flora.

Various species and growth forms are strangely absent from Mártir. There are no legumes, present on nearly every other island in the Gulf. *Amaranthus watsonii,* with small edible seeds, so prevalent on most islands, even guano islands such as Alcatraz at Bahía Kino, and

Boerhavia, with its small sticky fruits, would seem likely long-distance colonizers. Chenopods are also absent. *Chenopodium murale,* also with small edible seeds, is certainly tolerant of high guano concentrations and is the only plant on the guano-covered Isla San Jorge southeast of Puerto Peñasco. *Cenchrus palmeri* with its nasty, clinging burs also seems a likely colonizer, and it is surprising that it evaded transportation during the guano mining years. Smaller cacti such as *Mammillaria* are conspicuous in their absence, as is *Agave.*

The isolation, current absence of major human influences, small flora, and long collection history make Mártir an ideal site to monitor and document the effects of climate change. Based on historical collections and observations, there are several examples that provide evidence of change on the island in the twentieth century. Five species have not been seen on the island since they were collected at the latest in 1962. *Cyperus squarrosus* was collected only once on the island, by Palmer in 1887. It is a short-lived plant that very well could have a metapopulation dynamic on the island. *Petalonyx linearis* has only been collected by Palmer in 1887 and Johnston in 1921. *Stegnosperma halimifolium* was collected by Palmer in 1887 and again in 1962 by Moran, who noted seeing only two, one a massive individual 4 m tall. All five of these species could still exist as rare populations on the island, yet the possibility of their local extirpation is more likely and intriguing.

The collection history of two species shows evidence of being recent immigrants. *Viscainoa geniculata* was not collected by Palmer in 1887 and 1890 or by Johnston in 1921, and Johnston (1924) specifically listed it as absent from Mártir. It was first encountered by Moran in 1962, who noted seeing only one plant, and the following year Richard noted that it was rare on the island. In 2007 and 2008 we found it scattered at lower elevations and common at higher elevations. In December 2008 we found *Digitaria californica* at the top of the island, the first confirmed presence of this species on Mártir. Moran's (1983b) checklist shows it for Mártir, but we have not located a confirming specimen. Table 1.5 lists the collection history of the 24 species on Isla San Pedro Mártir.

ISLA SAN PEDRO NOLASCO

San Pedro Nolasco is a rugged and precipitous island situated in deep water about 30 km northwest of Guaymas, Sonora, and 14.6 km west of Bahía San Pedro, the

Table 1.5 The Collection History of the 24 Species on Isla San Pedro Mártir

Species	1887–1890	1921	1950s	1960s	1970s	2007–2008
Abutilon palmeri	X	X		X	X	X
Aristida adscensionis				X		X
Baccharis sarothroides	X	X	X	X	X	X
Chylismia cardiophylla subsp. cardiophylla	X	X		X		
Cylindropuntia alcahes var. alcahes		X	X	X		X
Cylindropuntia cholla	X			X		X
Cyperus squarrosus	X					
Digitaria californica						X
Euphorbia petrina	X	X		X		
Ficus petiolaris subsp. palmeri	X	X		X	X	X
Lycium brevipes var. brevipes		X		X		X
Mentzelia adhaerens	X	X		X		X
Muhlenbergia microsperma	X	X		X		X
Nicotiana obtusifolia	X	X		X		X
Pachycereus pringlei	X	X		X		X
Pelucha trifida	X	X	X	X	X	X
Perityle emoryi	X	X		X	X	X
Petalonyx linearis	X	X				
Pleurocoronis laphamioides	X	X		X	X	X
Sphaeralcea hainesii	X	X	X	X	X	X
Stegnosperma halimifolium	X	X				
Trixis californica var. californica	X	X		X	X	X
Vaseyanthus insularis	X	X		X	X	X
Viscainoa geniculata				X	X	X

closest mainland location (fig. 1.40). The channel between Nolasco and the Sonora mainland is 244 m deep (Murphy et al. 2002). The island consists of a narrow north-south-oriented mountain 3.5 km long and not more than 1.5 km wide that embraces about 3.45 km². A prominent crest, reaching a peak of 315 m, runs along most of the length of the island (fig. 1.41). On either side of the ridge crest the terrain quickly falls away into the sea. There are many short, steep canyons but no alluvial developments or fans. The shore is abrupt, and the only beach consists of cobble rocks at Cala Güina, a tiny cove at the southeast side of the island. Sheer cliffs and steep slopes of barren rock reach directly down to the water, and thus halophytic shore plants are absent. Along the entire east side of the island there are only two landfalls: the above-mentioned cove and the other at approximately one-third the distance from the north end of the island. The northeastern landfall, recently named Cañón el Farito, provides the best access to the island, and most collections have been made in this area. There is also limited access from the sea to canyons on the west-central side of the island.

The 58 species of vascular plants documented for the island are classified into 52 genera and 27 families, and only five genera have more than one species (appendix A). Four species are endemic to the island: three small cacti, *Echinocereus websterianus*, *Mammillaria multidigitata*, and *M. tayloriorum* (for discussion of the radiation of small cacti on Gulf Islands, see Wilder et al. 2008a), and one composite, *Coreocarpus sanpedroensis*. Nolasco and Ángel de la Guarda share the title for highest single-island plant endemism for Gulf Islands (Rebman 2002). The vegetation on Nolasco is generally dense and shows sharp differences in vegetation structure and species distributions, which correlate with different slope exposures and soil characteristics.

A strong relationship is seen between the flora and vegetation of Nolasco and Isla Tiburón. Thirty-eight species (65 percent) of the flora of Nolasco are also found on Tiburón, 32 of which occur on the Sierra Kunkaak of Tiburón (see Wilder et al. 2007b). The higher elevations of Nolasco have dense populations of *Agave chrysoglossa* and *Acacia willardiana*, which are also abundant at the higher elevations of Sierra Kunkaak. Both of these species, in addition to *Aristida ternipes* var. *ternipes* and *Cyperus elegans*, occur on no Gulf Islands other than Nolasco and Tiburón, and *Notholaena lemmonii* also occurs on Tiburón and Cerralvo. The *Notholaena* fern

Figure 1.40. East side of Isla San Pedro Nolasco. BTW, 2 February 2008.

on Nolasco is a rare genotype with its distinctive golden farina, different from nearby mainland and other island populations.

The 58 species on Nolasco can be grouped into six generalized biogeographical groups. (1) Four species are endemic to the island, and (2) two species are endemic to the Guaymas region—*Hofmeisteria crassifolia* and *Porophyllum pausodynum*. Taken together, 10 percent of the flora is endemic to the Guaymas region and Nolasco. (3) The largest biogeographical group, 29 species (50 percent

of the flora), is composed of species with distributions including both sides of the Gulf of California in Sonora and the Baja California Peninsula. (4) Other connections to the Peninsula are *Euphorbia magdalenae* and *Bahiopsis triangularis*, which are common on the Peninsula but are not known to occur in mainland Mexico. (5) Three species occur in mainland Mexico and not the Baja California Peninsula. (6) The final biogeographical group is composed of 18 species (31 percent) that are widespread in the Sonoran Desert and regions beyond,

Figure 1.41. Ridge crest of Isla San Pedro Nolasco, looking southward from above Cañón de Mellink to the summit. BTW, 29 September 2008.

often New World or cosmopolitan or nearly so, for example, *Aristida adscensionis* and *Nicotiana obtusifolia*.

A number of taxa are conspicuously absent from Nolasco, such as the family Acanthaceae, which is represented by at least 16 species in the Guaymas region (Felger, unpublished data) and 9 species on Tiburón. This is especially surprising given the presence of sheltered habitats on Nolasco and the otherwise strong floristic affinity with Tiburón, as described above. There are no prostrate *Euphorbia* (*Chamaesyce*) species, although all the Sonoran Midriff Islands, except Patos, have at least one. Besides *Acacia willardiana* there are no legumes (Fabaceae) on Nolasco, while they are abundant and diverse on the opposite mainland (Felger 1966, 1999; Ray Turner, observation, see part II, Botanical Explorations).

The chances for colonization of Nolasco by mainland species have been limited by 14.6 km of intervening water present for several million years, which has promoted single-island endemism for three species of cactus, one composite, one mammal, and four reptiles. The opportunities for plants to reach and gain a presence on the island via rafting events are likely limited by the steep sea cliffs that surround the island and absence of colonizable beaches. The island episodically experiences periods of severe drought that could potentially act as a barrier to establishment by species that are mesic/tropically adapted, as is much of the mainland source pool of species (e.g., Felger 1999). The rich vegetation and unique sheltered habitats on Nolasco would logically support a larger flora, yet this is not seen. Nolasco presents a fascinating study system in which the various factors that affect insular species diversity can be investigated. At present we conclude that the island's isolation, steep perimeter, and periods of extreme drought are major factors controlling species numbers, resulting in a relatively depauperate flora compared to the floristically rich adjacent mainland.

Another unique aspect of the flora is the large number of grass species, all of which are native. Nolasco has the highest relative percentage of the family Poaceae of any island in the Gulf of California (24 percent compared to an average of 8 percent ± 2.4 percent for the islands listed by Rebman et al. 2002). Dense fields of grasses are seen on the eastern side of Nolasco on north-facing slopes, principally *Setaria macrostachya*, on sites where the soil is deep and rich with humus. In January 2008 we found charred nubs of grasses and wood on a north-facing slope. The continued presence of cacti and the

absence of substantially burned shrubs suggest it was a low-intensity fire that spread across only a portion of the island. The biomass and density of grasses on these north-facing slopes are certainly enough to carry a fire, especially during dry seasons. The source of this fire is unknown (natural or anthropogenic), yet it is possible that low-intensity fires have been a part of the island's history. This is the only known evidence for a wildland fire on a Gulf Island and one of the few examples of a fire having been carried by a native perennial grass in a Sonoran Desert habitat (*Hilaria rigida* was observed to have burned in the Pinacate region of northwestern Sonora, where it grows in dense stands; Turner 2007). Fire in the Sonoran Desert has been historically rare and limited to dry seasons following above-average winter rainfall that produces an abnormal buildup of winter annuals (Felger et al. 2007a; Humphrey 1974; McLaughlin & Bowers 1982; Turner et al. 2003). This situation is changing as invasive Old World annual and perennial grasses expand their ranges and increase fuel loads (e.g., Felger et al. 2007b). More information is needed about the role of fire on the island, but it might be worthwhile to consider fire ecology when attempting to understand the ecological dynamics on Nolasco.

The vegetation consists of desertscrub that is sharply delimited into unique communities segregated according to slope exposure (Felger 1966; Felger & Lowe 1976). Corresponding with abrupt topographic changes, the vegetation generally shows a mosaic-like pattern, and vegetational contrasts on different exposure gradients are among the most striking encountered anywhere in the Gulf Coast of Sonora. At higher elevations on both sides of the island there is a general trend toward relatively less xerophytic vegetation and a substantial increase in number of species. A number of species occur only at higher elevations on the east and west sides of the island.

The east side of the island supports three distinct communities: (1) The grassy meadows of north-facing slopes, which are probably the most productive community and show evidence of being stable and long established. The deep soil, rich in humus, is consistently built up by the death and decay of the dense grasses, which in turn promote the recruitment of diverse grasses and slower erosion rates. Data taken in 2008 show an increase in shrub coverage on north-facing slopes. It is possible that occasional low-intensity fires reduce the presence of shrubs, maintaining the dense grass stands, a situation that was seen across the grasslands of the Southwest prior to the control of fire on the landscape (Turner et al. 2003). (2) South-facing slopes support dense populations of cacti and shrubs, but with considerably reduced vegetation cover, and ephemerals play a minor role. Although there is little soil buildup, the rock surfaces are sufficiently weathered to allow ample root penetration. In fact, this fractured rock seems ideal for many succulents and other xeric-inhabiting species, with populations of the endemic *Mammillaria multidigitata* occurring in such large concentrations that we term these areas "Mammopolis" (fig. 1.42). (3) East-facing slopes have the most diverse communities encountered on the island, and vegetation here often has four distinct strata; herbaceous cover, xerophytic succulent species, shrubs, and one tree. Erosion seems to be a significant force on these slopes, and within a specified area changes in species composition may thusly be seen on decadal time scales while the general vegetation structure is maintained (Felger et al. 2011). The west side of the island is

greatly dissected by steep canyons. Steep slopes support communities of cacti and shrubs and, except for large canyons, generally show comparatively reduced vegetation cover and reduced species richness comparable to that of south-facing slopes on the east side of the island.

We recognize six major growth forms on the island: annual (ephemeral) species (18), shrubs (15), herbaceous perennials (12), xerophytic succulents (10), vines (2), and a single tree, *Acacia willardiana* (*Pachycereus pringlei*, a xerophytic succulent, attains tree size, and *Ficus petiolaris* and *Fouquieria diguetii* are large shrubs that approach tree sizes). The 18 species of ephemerals represent 32 percent of the flora (*Vaseyanthus insularis*, both a vine and an ephemeral, is counted as a vine). The annual flora is split between hot-season (summer-fall; 8) and non-seasonal (9) species, and there are only two strictly cool-season (winter-spring) ephemerals, *Perityle californica* and *Parietaria hespera*.

The majority of all species on the island have small, simple leaves, and some are aphyllous. Only two species

Figure 1.42. Isla San Pedro Nolasco, *Mammillaria multidigitata* and *Echinocereus websterianus*, midelevation of Cañón el Farito. BTW, 3 February 2008.

have compound leaves: *Bursera microphylla* and *Acacia willardiana*. Most of the leaf-bearing species are drought deciduous. *Agave chrysoglossa*, the only truly evergreen-leafed species present, has thick, succulent leaves. Several species may be evergreen during years of high rainfall but are eventually drought deciduous in extended dry periods, such as *Ficus petiolaris*, *Hofmeisteria crassifolia*, and *Simmondsia chinensis*.

Non-native plant species continue to be absent on Nolasco, which is somewhat surprising in view of the numerous times people have visited and camped on the island. This system seems ideal to support the establishment and expansion of non-native grass species such as buffelgrass (*Cenchrus ciliaris*) and natal grass (*Melinis repens* (Willdenow) Zizka) that could drastically increase the intensity of fires on the island, rapidly transforming the island's ecosystem. Both of these grasses are abundant on the opposite Sonora mainland. *Melinis repens* has been found in substantial quantities even in the pristine, never-grazed areas opposite from Nolasco (e.g., Cañón las Barajitas, vicinity 28°03'03.6"N, 111°11'01.7"W, 100 percent cover on many sand bars in the canyon bottom, 18 February 1995, *Felger 95-188*). In addition, the island's perennial and ephemeral water sources should be monitored for potential arrival of wetland or quasi-wetland invasives such as *Tamarix chinensis*. A few tamarisk plants could devastate the tiny spring at Agua Amarga with its isolated *Cyperus* population. Early detection and removal of any invasives will be critical although potentially difficult.

PART II

Botanical Explorations on the Sonoran Islands

Collectors, Associates, and Selected Personalities

Botanical explorations focusing on the flora and vegetation of the lands and islands associated with the Gulf of California began with Edward Palmer's collections in 1887 and 1890 (Vasey & Rose 1890; Watson 1889). The most exhaustive works on the plants of the Gulf Islands are Johnston's classic work, produced from three months of "collections, field observations, and subsequent herbarium studies" (1924:951); Gentry's (1949) evaluation and interpretations, based largely on collections by Rempel in 1937 and Dawson in 1940; Felger's (1966) dissertation on the islands and Gulf Coast of Sonora and Felger and Lowe's (1976) description and interpretation of the vegetation and flora; Moran's (1983a, 1983b) island checklists and treatment of the flora of Isla Ángel de la Guarda; Case and colleagues' *Island Biogeography of the Sea of Cortés* (Case & Cody 1983; Case et al. 2002); and the revision of Moran's checklists by Rebman et al. (2002), which provides the most recent and comprehensive listings for the flora of the Gulf Islands. Other principal works that cover vascular plants of the Sonoran Islands, include Felger and Moser (1985), Turner et al. (1995), and most recently our work (e.g., Felger et al. 2011; Wilder & Felger 2010; Wilder et al. 2007b, 2008a, 2008b).

The majority of collections of vascular plants from the Sonoran Islands have been made by Ivan Johnston, Richard Felger, Reid Moran, Joseph Rose, and Benjamin Wilder. Many others, however, drawn to these magnificent islands, have made collections that have added importantly to the known flora. Here we present a summary of these individuals and their contributions, as well as some noted associated observers and researchers.

Several multiple-island expeditions to the Gulf of California are especially noteworthy. The first was the 1911 expedition of the U.S. Fish Commission aboard the steamer *Albatross* (Townsend 1916; see entry for Joseph Rose). The California Academy of Sciences expedition in 1921 was the pivotal scientific effort in the region (Johnston 1924; also see entry for Ivan Johnston). Other significant expeditions include those of the Allan Hancock Foundation in the 1930s and 1940s (Gentry 1949; see entry for Howard Scott Gentry), the Sefton Foundation–Stanford University Expedition in 1952 (see entries for George Lindsay and Reid Moran), the Belvedere expedition of the San Diego Natural

History Museum in 1962 (Lindsay 1955a; see entries for George Lindsay, Reid Moran, and Ira Wiggins), and between 1985 and 1987 the Universidad Nacional Antónoma de México (UNAM) Instituto de Biología organized eight expeditions to the Gulf Islands aboard Mexican Coast Guard ships of the Armada de México, led by Enriqueta Velarde. Four expeditions went to the northern islands, and four to the southern islands. Both Johnston (1924) and Gentry (1949) provide excellent summaries of earlier botanical explorations in the Gulf, and Lindsay's obscure 1955 publication on the botanical collectors of Baja California is especially significant. A "History of Scientific Explorations in the Sea of Cortés" is provided by Lindsay and Engstrand (2002), but no mention is made of the 1921 California Academy of Sciences expedition or the works of the various University of Arizona researchers.

In addition to the large formal-survey expeditions, researchers, largely from the University of Arizona, the Arizona-Sonora Desert Museum, and others, mainly from Arizona, New Mexico, and Sonora, have contributed hugely to the herbarium collections from the Sonoran Islands. These researchers were mostly engaged in low-energy consumption and low-key field trips, often botanizing on a shoestring budget or supported by individual research grants (Bowen 2000).

The general format in the following discussions is as follows: the person's name, with life dates if deceased and known, island(s) visited, and brief biographical and pertinent information when available. Information for island(s) visited and date(s) is presented chronologically. Specific collection localities and herbarium collection numbers are listed if available.

Juan Alfredo Barnett-Díaz.
San Esteban, 2008.

A member of the Comcaac community at Punta Chueca, Barnett-Diaz collected specimens on 3 September with other members of the Comcaac plant team (see entry for Benjamin Wilder, 3 September 2008).

Arroyo: Limantour, *08-01* to *08-05*.

Raymundo Barnett-Morales.
Tiburón & San Esteban, 2008.

A member of the Comcaac community at Punta Chueca, Barnett-Morales collected specimens on 3 September with other members of the Comcaac plant team (see entry for Benjamin Wilder, 3 September 2008).

San Esteban: Arroyo Limantour; *08-01* to *08-04*.
Tiburón: Coralitos, *08-05* & *08-06*.

Dennis L. Bostic (1937–1975).
San Esteban, 1965.

An intrepid field biologist who had a strong interest and knowledge of Baja California ecology (e.g., Bostic 1971), Bostic was a research associate in the herpetology department of the San Diego Natural History Museum and an instructor at Palomar College in Southern California. His natural history book on Baja California (Bostic 1975) contains his original accounts, as well as those by Brian Cooper, Steve Leatherwood, Frank Rokop, Janet Sprague, and Margie Stinson. This splendid book deserves to be reprinted. On 21 June he collected cactus specimens in the vicinity of Arroyo Limantour.

Luis Bourillón-Moreno. Tiburón, 1993.

Bourillón-Moreno's doctoral studies focused on the human uses and fisheries in the Seri region and Midriff Islands (Bourillón-Moreno 2002). He is active in ocean conservation programs at the non-governmental conservation organization Comunidad y Biodiversidad in Guaymas and in Yucatan. He visited many sites on Tiburón, including the majority of the fishing camps, and collected the only Gulf Island record of the non-native grass *Cenchrus echinatus* at Ensenada del Perro.

Thomas Bowen.

Anthropologist and author of *Unknown Island* (2000), the first and most comprehensive ethnohistory of the Gulf of California, and other landmark publications of the region, including *The Record of Native People on Gulf of California Islands* (Bowen 2009), Bowen has conducted archaeological and historical research on the Midriff Islands (Bowen 1976) and has visited most of the larger islands elsewhere in the Gulf. He was ashore on San Pedro Mártir in May 1984 and has conducted extensive fieldwork on San Esteban and Tiburón. On numerous occasions he has provided us with insightful information, reviews, and photos.

James E. Canright (1920–2008).
Dátil, 1998.

Botanist and professor at Arizona State University, Canright collected the only specimen of *Mesembryanthemum crystallinum* from a Sonoran Island.

Fred Cooper. San Pedro Nolasco, 1964.

Cooper and his wife, Joy, were students at the University of Arizona and friends of Richard, Ike Russell, and Jean Russell. The Coopers live in Portland, Oregon.

On 11 August, Fred collected 10 numbers above Cala Güin, the cobble-beach cove at the southeast side of Nolasco. The specimens were given to Richard to process and were cataloged as *Felger 10399* to *10408*.

E. Yale Dawson (1918–1966).
Tiburón, Dátil, San Esteban, &
San Pedro Nolasco, 1940.

Dawson collected specimens and information for his dissertation as a member of Captain Allan Hancock's expedition on the *Velero III* to the Gulf of California. Yale's dissertation was published as *The Marine Algae of the Gulf of California* (Dawson 1944). "Dawson was a dynamo of energy in the field and at the typewriter. He finished graduate school in a year and a half," being the last formal graduate student of the noted phycologist William A. Setchell (Readdie et al. 2006:10; also see entry for Ivan Johnston). Dawson "published 165 books and papers, 96 of which were related to marine algae, the others mostly to cacti and succulents" (Smithsonian Institution Archives 2011). In 1945 he co-authored a paper on Seris, which unfortunately repeated some of the unpleasant misconceptions of the times (Davis & Dawson 1945). Among his prestigious positions, he was director of the San Diego Natural History Museum in 1963 and 1964, and in 1965 he became curator of cryptogamic botany at the herbarium of the National Museum of Natural History at the Smithsonian Institution. Dawson drowned on 22 June 1966 while diving for marine algae in the Red Sea off Egypt.

Tiburón: 25 January. Gentry (1949:99) reports that Dawson collected 21 taxa on the island. He apparently collected at the south shore of the island (e.g., *Cylindropuntia leptocaulis*).

Dátil: 25 (probably) January, three species collected, including *Cylindropuntia bigelovii*.

San Esteban: 1 February, probably at Arroyo Limantour, six species, including *Echinocereus grandis*.

San Pedro Nolasco: 6 February. Six specimens collected at the place now known as Cañón el Farito. This important historical collection includes *Cyperus elegans*, *Echinocereus websterianus*, *Mammillaria tayloriorum*, and the holotype of *Coreocarpus sanpedroensis*.

Exequiel Ezcurra.
San Pedro Mártir, Tiburón, & Dátil, 2007.

Noted ecologist, conservationist, and author, originally from Argentina, Ezcurra served in the administrations of two Mexican presidencies, was at the San Diego Natural History Museum and later the University of California, Riverside as the director of the University of California Institute for Mexico and the United States. Exequiel accompanied Ben and Richard on two field trips to the Sonoran Islands (see their entries for the dates listed here).

San Pedro Mártir: 11 April (fig. 2.1).

Tiburón: 12 April, Ensenada del Perro.

Dátil: 12 April, canyon at the northeast side.

Tiburón: 23–26 October, Sierra Kunkaak.

Figure 2.1. Exequiel Ezcurra, San Pedro Mártir, 11 April 2007. BTW.

Richard Felger.

Tiburón, 1954, 1958, 1963, 1964, 1965, 1966, 1968, 1973, 1974, 1976, 1983, 2007, 2008. **San Esteban,** 1954, 1958, 1963, 1965, 1966, 1967, 1968, 1975. **Dátil,** 1954, 1958, 1963, 1965, 1966, 1968, 2007. **Patos,** 1963, 1964, 1973, 2007. **Alcatraz,** 1965, 1966, 2007, 2008. **Cholludo,** 1954, 1958, 1963, 1965, 1968. **San Pedro Mártir,** 1963, 2007. **San Pedro Nolasco,** 1963, 1964, 1965, 2000, 2006, 2008.

I went to the University of Arizona as a freshman because Tucson was close to Sonora, where intriguing plants, animals, places, and people drew my interest. I soon discovered the islands in the Gulf of California and went there at every possible opportunity—going with fishermen in their small, open pangas. It did not take much to cut classes at both ends of a weekend to go botanizing in Sonora and on the islands.

I am reminded of the early trips I made with Porfirio "Pilo" García from Bahía Kino—then a village with no electric wires and no paved streets—the small wooden pangas of low draft, rickety and smelling of fish, with a chugging 15 hp outboard motor, crossing from San Esteban or Tiburón back to Kino, sometimes in horrific weather. Too often it would be calm in the shelter of the south side of Tiburón, but as soon as we left the protection of the big island we would be in open water, leaving late in the day as dark descended and sometimes the horizon gone because the waves were so high—always on Saturday because they would have to get back to town for a dance, which I soon realized was a euphemism. They would have money from their catches and women were at the dozen bars because the fishermen had money. Jesús Lizarraga, a student at the University of Sonora, went with me on the first trip to San Esteban, and when he got back to Hermosillo he wrote an indignant article for the local newspaper, "13 bars and no schools."

Porfirio "Pilo" García, a tough fisherman from Kino, took me to new island places (fig. 2.2). His poverty and that of his family weighed on me. His stepfather, Don Nacho, struggled with poor health. The front of their house was converted into a small restaurant run by Doña Gavina, Pilo's mother. Hard-packed earthen floors always freshly watered down and swept clean, and good simple food was served. Ordinary things like shoes and books were enormous financial concerns. I realized how privileged I was. Some years later, as fishing declined with crashing marine resources, Pilo was a cab driver in

Figure 2.2 Left to right: Don Nacho García, Doña Gavina García, Porfirio "Pilo" García, and Jean Russell. Bahía Kino. AR, 1959, courtesy of Bob Russell.

Hermosillo, and late in life he was pushing an ice cream cart (or something like that) on the streets in Kino—a brilliant and vibrant life held back by hardship. Midday on the islands, when not working, the fishermen would lie around reading comic books, but Pilo had a dog-eared paperback of García Lorca's works. He would ask me about Spanish and Latin American authors.

Alexander "Ike" Russell and Jean Russell lived at Kino for part of a year (before I met them) and were friends of Doña Gavina and her family. Later I made innumerable trips with Ike in one of his planes or his boat, often together with his wife, Jean, and other friends. Ike would land just about anywhere—it all seemed so ordinary then (see the entry for Alexander Russell; see also Felger 2000b, 2002). I met Ben Wilder in 2005, and we embarked on studies of the floristic diversity of the Sonoran Islands.

The primary set of my collections are at ARIZ, with many duplicates sent to CAS, MEXU, UC, USON, and especially SD, as well as other herbaria, including ASU, HCIB, GH, MO, RSA, UC, and US. Some collections at SD are not at ARIZ. Gaps in some of my early collection numbers generally represent zoological specimens. Two large gaps in my early field numbers are related to an unscrupulous professor absconding, twice, with my field notebooks. When I finally recovered the notebooks, I had too many "temporary" numbers and did not attempt to correct the unfortunate gaps in collection numbers. On field trips with Ben we traded off numbering, so that some are *Felger & Wilder* and others are *Wilder & Felger*.

San Esteban, Tiburón, Cholludo, & Dátil, 28–30 May 1954, with Porfirio "Pilo" García and Jesús Lizarraga.

Left from Bahía Kino in a small, open, wooden skiff (*panga*). Plank seats and no deck. Pilo is the boatman in charge, with another fisherman. Jesús Lizarraga is a friend and student at the Universidad de Sonora in Hermosillo. We went to Arroyo Limantour on the east side of San Esteban and later to the northeast side of the island. Pilo and his helper fished while Jesús and I hiked about the island and I collected plants. This was my first trip to the Midriff Islands, and it opened up a wonderful universe except for the *jejenes* (Ceratopogonidae, or biting midges, also called "no-see-ums" in the United States). The first night on the island Pilo urged us to sleep on the boat anchored offshore, but the uncrowded beach looked better. That was a poor decision—I got covered with painful jejene bites. The next night I slept doubled up on a seat in the fishy-smelling panga.

San Esteban: 28 May, Arroyo Limantour, *466* to *470*.

29 May. Inland in Arroyo Limantour, near the center of the island, *471* to *477*. Northeast side of the island, *478*.

Tiburón: 30 May, near shore opposite Cholludo, *483*.

Cholludo: 30 May, *485* to *489*.

Dátil: 30 May. After a brief stop on Cholludo we went to the northeast side of Dátil, *492 & 493*. The usual calm weather at this time of year made for an uneventful trip back to Bahía Kino in the afternoon. Giant manta rays seemed to be everywhere, jumping into the air and splashing back into the sea.

Dátil, Tiburón, & San Esteban, 6–9 April 1958, with Pilo García and C. H. Lowe (1920–2002).

From Bahía Kino with University of Arizona ecologist C. H. Lowe; Pilo, boatman in charge; and Leopoldo "Polo" Blanco (mate), in the *SC. Tobari*, a fishing panga owned by Don Nacho, Pilo's stepfather. The boat was 16 feet long with a 15 hp outboard.

Dátil: 6 April. Northeast side, in the northeast branch of an arroyo within 100 feet of the ocean, *2541* to *2550*. My journal notes (abridged) read:

> Left Kino at 7:55 a.m. At 9:10 we are in trouble . . . taking in lots of water, getting soaked, bailing out water and bouncing and slapping hard on the swells and waves. Many whitecaps. At 9:20 we turn back and head for the mainland. 9:50 we hit shore about 4 miles north of Punta Cerro Prieta. . . . Porfirio took a gas can and headed for Kino on foot for more gas and oil . . . 22 km . . . barefoot. Pilo said it would be calmer by about 2 p.m. and by then he would be back and we could go to the islands. He left at 10:05 a.m. and returned by another boat at 2:30 p.m. At 2:37 we boarded, loaded, and off—straight for Isla Turners [Dátil]. No whitecaps . . . wind essentially dead. . . . Landed at Turners at 5:30 p.m. . . . east side near north end. . . . Spent several hours collecting plants and then pressing them. 8:20 p.m., *Centruroides sculpturoides* crawling near sleeping bag. Mexican crew took off to sleep in boat. 9:10 p.m. they are singing happily, or madly? . . . and still at 10 p.m. [They were not sharing their tequila.]

7 April. North end of island, collected plants on ridges, steep slopes, canyons, and peaks. Pilo and Polo scrambled about loose rocky slopes with chollas, etc, in their bare feet. In the afternoon we went around Dátil by boat. I counted 132 *Ficus petiolaris* on sea cliffs. Whales were blowing loudly about 0.5 km offshore. Great numbers of

seabirds, especially blue-footed boobies, brown pelicans, cormorants, and gulls, and sea lions with pups clustered on rocks under cliffs. I saw 11 pairs of ospreys with nests: two at the south end of the island, three on the west side, and six on the east side.

At about 5 p.m. we went to the next canyon south of our camp, which we called Rattlesnake Canyon. A short walk along the shore at low tide, but it is high tide so we went by boat. Lowe started looking for snakes at 5:15. Apparently Leopoldo did not get the spirit of herpetology and probably got cold (according to Pilo), so he built a fire. Lowe and I were above in the narrow canyon, which Leopoldo was about to burn up. We managed to put out the fire, probably saving our lives and the island as there is much dry brush. No snakes. Pilo said they don't come out until May. Back at camp at 8 p.m. A strong northwest wind blew all night. Pilo is concerned since we plan to go to San Esteban in the morning. I slept on the boat because of the jejenes.

Northwest part of island, about 150–200 feet elevation, steep and rocky west-facing slope, *2551* to *2553*. Northeast side, rocky narrow arroyo back in from beach, *2554* to *2559*. Southeast side, rocky steep hills forming cliffs, 25–30 feet elevation, *2560*. "Rattlesnake Canyon," southeast part of island, rocky steep slopes of canyon and canyon bottom, 30–150 feet elevation, *2561* to *2564*. Sea cliffs at the northeast shore, about 20 feet elevation, *2565* & *2565B*.

Tiburón: 8 April. Collections from Ensenada de la Cruz, *2568* to *2609*. Abridged from my journal:

Left Dátil at 7:30 a.m. for San Esteban in very choppy water, big swells. Wind still blowing. Pilo is concerned. About halfway to San Esteban we found it too rough. The boat going up and slapping down hard and taking in water. We turned back toward Tiburón. Made it to Playa (Ensenada) de la Cruz, 9:40 a.m. I collected plants the rest of the day. Wind blew all day. About 1 mile inland there is a stand of giant *Ficus petiolaris* trees.

At 8 p.m. I went fishing with Pilo and Leopoldo as we are about out of food. A lantern was fastened to the bow and a metal shield, made from a 5-gallon tin can, was tied over it to reflect light into the water. We rowed over submerged tide pools and rock, the water clear, and it was easy to spot fish. Fish and lobsters were speared with a long pole with two straight metal spears. Got two small lobsters, one striped rock bass (*mojarrón*), two trigger fish (*cochi*), and a few others. A small manta ray 1½ foot across speared but fell off. I saw fish-eating bats dipping down to the ocean. Got back at 10:50. We ate barbecued lobsters, triggerfish, and bass. Crawled in sleeping bag by midnight.

San Esteban to Bahía Kino: 9 April. Left Ensenada de la Cruz at 6:45 a.m. Re-gassed at sea. Made it to San Esteban at 8:30, against a terrific current and some wind and went to Arroyo Limantour. I walked up the south fork of Arroyo Limantour until it ends at a steep mountain slope. Chuckwallas (*Sauromalus varius*) are common and easily approached, but sometimes scurry under rocks. About 10 a.m., a chuckwalla with *Lycium andersonii* (*2618*) leaves in the corner of its mouth. Saw more chucks under chollas (*Cylindropuntia alcahes*) eating the joints. Back to the boat at noon, Pilo working on the outboard motor. Left at 1 p.m. Arroyo Limantour, *2610* to *2626*.

We were out from San Esteban about half an hour when the water got choppy as hell and wind steadily increasing. By 3:30 p.m. we were passing the north side of Dátil. Pilo wanted to camp on Dátil, but Leopoldo wanted to go to Kino. Wind and whitecaps in this protected area had reduced, but the swells still large. Whales blowing off the southwest part of Dátil. 3:50 p.m. I saw a manta ray, about 8 feet across, rolling about, white belly up. (The only large manta seen on entire trip. In May 1954, we saw hundreds of them jumping out of the water.) 4:10 p.m., about ¼ mile east of El Monumento at the southeast corner of Tiburón, stopped to re-gas. No wind but big swells. Pilo looked out on the horizon and reluctantly gave in to Leopoldo, and we headed for Kino. (Next time bring life jackets.)

About halfway to Kino we got into trouble with choppy water and huge, deep swells. Still not so big that we couldn't go straight into them without swamping. We took in a terrific amount of water. About 5 p.m., when waves too big to go into anymore, Pilo had gained the distance needed, and he headed due east for Cerro Prieto at the north end of Bahía Kino, riding the crests of the waves and swells all the way. More than once we almost went over. Fortunately the wind was blowing north to south, and the waves and swells going south, so we rode on the crests into the shore at Cerro Prieto. One more hour at sea would have done us, we were taking in water badly and were in doubt about making it the last half hour. Hit the beach at 6:15, drenched, the load soaked, shivering and teeth chattering, a boatload of seawater. A group of Mexicans and an American family anxiously watching us from the beach. Sun setting, the wind really furious. Got out some miraculously dry clothes from under the tarp and changed in a shack on the beach. The American family gave us coffee and later drove me to Kino village (ca. 6 miles to the south) so I could bring my car back from Kino.

Pilo was pretty shook up and we were happy to be on terra firma. After drying out, Pilo and Leopoldo took the panga close to shore, home to Kino. We met them at Pilo's home at 8:30. Don Nacho and Doña Gavina were very worried as the wind had blown every day since we left four days ago. We gave papa $50 (150 pesos per day, about $12 U.S.; we had already paid for the gas and oil) and tips for Pilo and Leopoldo.

Cholludo, Tiburón, & Dátil, 30 October to 1 November 1958, with Pilo García.

30 October. Yesterday at the Garcías' home in Bahía Kino I made arrangements for a trip to the islands. Don Nacho, Pilo's stepfather, had been sick in bed for a month and was forced to sell his boat, so I have to rent one from someone else and hire two boatmen, one of course being Pilo. The other turned out to be Leopoldo Blanco, again. Charge for boat, Pilo, Leopoldo, food, water, gas and oil is 200 pesos per day (about $15 U.S.). Left Kino at 11:13 a.m. Clear weather, slightly choppy water.

Cholludo: Mid-afternoon on 30 October I spent two hours on the island. Got very windy. Cholludo and Turner are green from recent rains. Collected several specimens, *2721* to *2726*.

Tiburón: Left Cholludo in the late afternoon in very choppy water to the opposite shore at Ensenada de la Cruz. Beached the panga and camped. Tiburón is green, but not as green as Cholludo. Pilo tells me about a place where there is water all year and many large, green plants. Says it is 6 km inland and a 3–4 hour walk. Slept on the beach, wind strong all night.

31 October. Left camp by boat at dawn for Arroyo Sauzal. Just behind the beach Pilo hid his shoes under a *Jatropha cuneata* shrub and walked barefoot—didn't want to wear out his shoes. We left Leopoldo in charge of the boat. About 9 a.m., already hot and the sky clear. I was glad we brought the canteen. *Hyptis albida* (*H. emoryi*) is dominant along the arroyo bottom. Pilo was right—Sauzal is an amazing oasis. Substantial bedrock pools with carrizo (*Phragmites*), some cattails (*Typha*), and much *pino*, the pink-flowered shrubby *Tamarix chinensis*. In the late afternoon we found some toads (*Bufo punctatus*, later renamed *Anaxyrus punctatus*). Arroyo Sauzal and the Sauzal waterhole, *2738* to *2757* and *2764* to *2768* (gaps are zoological specimens).

On the way back, along the gravelly arroyo bed at about dusk, I found a coral snake (*Micruroides euryxanthus*) swallowing a banded sand snake (*Chilomeniscus*)—looked like one snake with a tail at each end and each half a different color pattern. Encountered two diamondback rattlers (*Crotalus atrox*). Got back to the beach after dark and returned to our camp at Ensenada de la Cruz because the shore is too rocky to leave the boat overnight at Arroyo Sauzal. Rather windy.

1 November. Windy all night except the wind stopped just before dawn when I got bitten by a jillion insects. Up early and pressed the rest of the plants in plastic bags from yesterday.

Dátil: From Ensenada de la Cruz we had planned to go to San Esteban, but it was too windy and decided to head back early to Kino. We stopped at Dátil about 10 a.m., and I hurriedly collected specimens and put them in plastic bags to press later. Small annoying flies, called bobos, were everywhere, crawling in nose, ears, and eyes but they do not bite. Northwest part of Dátil, *2770* to *2781*.

Left the island at 11 a.m. Water was rough, but Pilo said it was o.k. Later we couldn't have done it. He headed the boat north into the wind and waves and then turned southeast right to Cerro Prieto going with the swells. Arrived at Kino wet at about 3 p.m. I spent a few hours pressing plants on hand in plastic bags and left for Tucson.

Tiburón & Sonora, January 1963, with Alexander "Ike" Russell.

Although I had made several trips to Sonora with Ike in his plane, my first trip to Tiburón with Ike was in the early 1960s. From Felger (2000b:522):

> On one trip we landed at Palo Fierro on the east side of the island and hiked up to the mountains. I got a lot of new plant records for the island. I carried a small pair of pruning shears or clippers in my pocket and put the samples I collected in plastic bags to keep them fresh until I got back to camp. At the end of the day I would work on my notes and press the specimens—often a rather lengthy affair. This time when we got back to the plane, Ike said we should go to the Seri Indian village of El Desemboque. He seemed surprised that I hadn't been there and didn't know Ed and Becky Moser.
>
> We flew over to Desemboque.... Becky claims that the first thing I said when I walked into their house was, "Where can I press my plants." I suppose I was rather single minded in those days. Soon after, on another trip to Desemboque, Becky showed me some plants and asked if I knew what they were. She had Seri names and some uses for about 50 species of plants. I thought it would be interesting if I added scientific names and other pertinent

botanical information, and we could write it up for publication in about three weeks. Twenty years later we finished our book on the ethnobotany of the Seri although we published various papers along the way. (See fig. 2.3.)

Tiburón & San Pedro Mártir, 22–26 January 1963, with Ike Russell.

Tiburón: 22 January. After several days' fieldwork in western Sonora in Ike's Cessna 180, we went to Tecomate, landing at the fine airstrip and explored the valley floor and bajada, walking inland for several kilometers, *6225 to 6238.* Later in the day we climbed the small steep rocky peak to the west of the airstrip, *6239 to 6285.* Left Tecomate 5:27 p.m. and flew southward to Ike's small airstrip at the place he called Ensenada Blanca, landing at 5:40 p.m., *6287 to 6312.* Left at 6 p.m. and went to Kino, landing at 6:24, nearly dark.

San Pedro Mártir: 24 January. I wanted to go San Pedro Mártir, and Ike thought we might hitch a ride on a shrimp boat from Guaymas since they often spent the day anchored offshore from the island. At the Guaymas harbor we made inquiries for boat to take us to Mártir. We were directed to some Yaquis who took us to their boat, the *Cd. Obregón,* to meet the captain. He was sitting on the stern with a roll of toilet paper in his lap. In spite of his predicament, we shook hands and exchanged "*¿Buenos tardes, como esta usted?*" The crew was having fun at the captain's expense, and he quickly agreed to our request. Left Guaymas harbor at 5 p.m. The operation was part of a cooperative of 300 men from Guaymas. The *Cd. Obregón* was a tall, rusty, all-metal shrimp boat that pitched and rolled day and night. I didn't get seasick but I skipped dinner. They trawled for shrimp all night between Kino and San Pedro Mártir.

The nighttime trawls yielded enormous bulging nets of marine life scoured from the seafloor. The whole wet, writhing mass was dumped onto the deck, and only a small fraction consisted of large, succulent shrimps, which were picked out by hand, put in large baskets, and stored on ice below deck. An occasional edible fish was saved, but the rest of the bycatch, now crushed, mangled, and dead or dying, was shoveled overboard.

25 January. Shortly after dawn the *Cd. Obregón* was anchored near the landing below the guano workers' village site. Ike and I were taken to the island in a dinghy. Getting ashore involved avoiding slippery sea lion shit on the rough, rocky landing. We spent most of the day

Figure 2.3. Richard with Becky and Ed Moser, vicinity of Desemboque. Richard is holding a cross section of a fallen cardón stem. AR, ca. 1963, courtesy of Bob Russell.

exploring for plants, from the village site to the summit. The island was green from recent rains. I was surprised that even at this time of year the island was rather hot, and of course humid. I collected 20 numbers, *6346* to *6366*. Got back to Guaymas in the late afternoon.

Tiburón: 26 January. We left Guaymas early, made several landings along the coast and around noon went to Palo Fierro (Punta Tormenta). I botanized along the bajada within about 1 km of the west end of the airstrip, *6392* to *6429*.

Tiburón, 9 March 1963, with David, Jean, and Ike Russell.

On a trip to Sonora and Baja California in Ike's Cessna 180, we landed at Ike's Sauzal landing place and walked up Arroyo Sauzal to the waterhole, *6435* to *6483*. Jean declared she would not land again at this place—there were too many rodent holes, making for somewhat dicey takeoffs and landings.

Tiburón, 20–23 March 1963, with Ike Russell.

20 March. Left Tucson from Ryan Airfield in Ike's Cessna 180, and after a number of landings in Sonora we went to the Tecomate airfield and camped for the night.

21 March. Tecomate, *6812* to *6866*. We left Tecomate about midday, inspected Patos from the air at low elevation and I recorded the perennials, landed on the Sonora mainland, and then went to Palo Fierro on Tiburón, where we camped.

22 March. My collections are from the vicinity of the Palo Fierro airfield to the foothills and the first canyons on the east side of Sierra Kunkaak, *6947* to *7010*.

It was on this trip that I learned the importance of carrying plenty of water, a flashlight, and a jacket or at least a warm shirt. Ike and I planned to go to a waterhole [Pazj Hax] with carrizo, which we had previously located from the air. We left Palo Fierro at 7:45 a.m. and headed for the waterhole, but sighting a place in a mountain canyon from the air is easier than finding it on the ground. We hiked to the general area, but failed to find the waterhole. I stopped to collect plants and got separated from Ike. Later in the day he left a note on a stick next to a packrat midden that he set on fire as signal, "I'll meet you at the big *tescalama* [*Ficus petiolaris*] that blocks the canyon." I waited for Ike at a big tescalama, but unknowingly we were in different canyons waiting by different big tescalamas. I waited quite a while and realized it was getting late in the afternoon and headed back

to camp at the airplane. I had very little water. Moonless night descended, and I tried to make my way but was halted by impenetrable mangroves and then a forest of chollas (*Cylindropuntia fulgida*). No choice but to wait for dawn. I fell asleep among sharp rocks and scratchy *Cryptantha*, but with only a T-shirt I woke shivering to see a coyote breathing in my face. At first light, making my way back, I was desperately out of water. Popping *Bursera microphylla* fruits in my mouth to make saliva flow, as I had learned from the Seris, helped at first, but I soon experienced a thick tongue, cottonmouth, and hallucinations. I fought the urge to head for the shining water of the sea and tried to keep on course. I climbed into a cardón to look for the plane where I thought it should be (he couldn't have left) but didn't see it, not even later when I stumbled right onto the plane. Ike gave me the last tiny bit of water he was saving for me. Without a word we flew over to Kino Viejo, landed on the highway, and taxied across town to Doña Gavina's little restaurant-home, walked in, upended pitchers sitting on the tables and gulped water, slopping half of it on us and the floor.

Tiburón, San Esteban, & Patos, 4–7 April 1963, including C. H. Lowe and Ike Russell.

This bizarre field trip was orchestrated in part by Dr. Charles H. Lowe, a University of Arizona professor, in order to rendezvous with Dr. Ken Norris on San Esteban. The trip resulted in my obtaining substantial collections.

Tiburón: 4 April. Left Ryan Airfield, Tucson, at 3:30 p.m. in Ike's Cessna 180. Landed at Nogales, Sonora, for visas. Leave 4:40 p.m. and flew southward to San Esteban where UCLA professor Ken Norris and party of about six people were camped near the shore at Arroyo Limantour. We dropped three parachute messages. The first two were lost, but they got the third one, written by Lowe, "Ken—If you get this, we will land on Tiburon & hike to the beach to make a fire (about 25° . . .) from you. Can you come over & pick us up tomorrow?" We landed at Ike's Sauzal landing place and hiked to the beach in the early evening, full moon.

5 April. We later learned the Norris party did not have a boat on San Esteban, having made arrangements to be picked up later by Kino Bay fishermen. On the shore at Arroyo Sauzal, we hailed American tourists in a speedboat. Ike walked back to the airplane and flew on to Guaymas. Arroyo Sauzal and vicinity, *7014* to *7035c*.

San Esteban: 5 April. The Americans took Lowe and me to San Esteban, where we joined the UCLA party. After the tourists left, we found out that Kino Bay fishermen were supposed to return to San Esteban for the Norris group. We had sufficient water for only a few more days. The Kino Bay boat failed to show up, so in the afternoon I used my signal mirror and American tourists in their small fishing boat came to our rescue. They agreed to take Lowe to Kino to arrange a boat to pick up the UCLA party. I stayed on at San Esteban. The American tourists were extremely kind and helpful, but at one point one of them said, "That big fellow sure doesn't know how to say thank you." Ken mentioned it to Lowe, and typical to form, Lowe began thanking them so profusely and offensively that they were sorry for mentioning it.

6 April. I collected and pressed plants in the morning, and hiked inland in Arroyo Limantour to hills at the southwest part of the island. Collections, 5 and 6 April, *7036* to *7094*.

In the afternoon a huge (250 ton) shrimp boat appeared on the eastern horizon and steamed straight for us—everyone just knew that "Uncle Charlie" would be standing on the bow in his red shirt like a big white god—and he was. The shrimp boat crew took us to Ensenada Blanca [Vaporeta], where we met Ike Russell waiting, as arranged, with his airplane. Lowe and I stayed with Ike and the UCLA party returned to Kino. I pressed plants that night inside the airplane to avoid the wind.

Tiburón: 7 April. In the morning I collected and pressed plants in the vicinity of the Ensenada Blanca landing strip, the nearby rocky hills, and an arroyo and rocky slopes near the ocean, *7100* to *7156*. A newly developed small gully at one end of the landing strip had to be avoided—making for shorter takeoffs and landings.

Patos: 7 April. Later in the morning we flew low over the Central Valley of Tiburón and landed on Patos where I collected *Atriplex barclayana*, *7157*. We returned to Tucson via Nogales.

Tiburón, 16 September 1963, with Ike Russell.

By plane with Ike Russell at Tecomate, *8861* to *8913*. Later in the day we landed on Patos, but finding nothing new, I did not collect plants. We then went to the landing field at Palo Fierro, *8915* to *8943*.

Tiburón, San Esteban, Dátil, & Cholludo, 20–23 October 1963, with Pilo García.

19 October. I drove from Tucson to Kino and made arrangements to go to the islands, spending the night at the Garcias' house. Survey crews and workers are setting out

streets, running water, street lights, and lots. The Municipio is selling lots for about $500 U.S. It would be nice to have a little beach house in Kino. I accompanied Pilo, with another fisherman, on a several-day fishing trip in his wooden panga, the *Sin Tiki*. Pilo said, "No es *con* tiki, entonces es *sin*," in deference to Thor Heyerdahl's *Kon Tiki*.

Tiburón: 20 October. From Bahía Kino to Ensenada de la Cruz. Vicinity of the bay, rocky hills, arroyos and slopes, *9066* to *9075*.

Tiburón, Dátil, & Cholludo: 21 October. In the early morning, from Ensenada de la Cruz across the short channel to Dátil. Spent the morning at Dátil, and then to Cholludo. In the afternoon we returned to our campsite at Ensenada de la Cruz. Dátil, northeast side of the island, *9079* to *9133*. Tiburón, Ensenada de la Cruz and vicinity, *9134* to *9151c*. Cholludo, *9152–9167B*.

San Esteban: 22 October. Pressed plants in the early morning and then went to San Esteban. Engine trouble. Arrived midday at El Monumento, but the island was extremely dry and saw no annuals or plants in flower. It was a very windy day and we stayed only a few hours at El Monumento because Pilo was afraid to loose the boat, *9168* to *9187*. We then went to the northeast corner of the island, *9192* to *9199*. Left in the afternoon for Ensenada de la Cruz because of wind, arrived at dark.

Tiburón: 23 October. Ensenada de la Cruz: "This camp is a mess. The fishermen just toss their trash, cans, etc." There is a rude hut for shelter, made of ocotillo and *Acacia willardiana* sticks. Blotters and dryers from my plant press served for additional shade. Collections *9200* to *9243b*. We tried to leave in the early morning, but Pilo announced *"pinche motór no sirve."* Finally leave after Pilo fixed the engine. Wind rough for a while, then calmed. Arrived at Kino in the afternoon.

Tiburón & Punta Chueca, Sonora, 24–27 October 1963, with Jesús Morales.

Punta Chueca: 24 October. Yesterday, in Desemboque, I made arrangements to drive Jesús Morales (1904–1975; see Felger & Moser 1985:413) to Punta Chueca and then go to a major waterhole on Tiburón. He was a tough little guy—leathery, wrinkled, and skinny, and wore oversized dark glasses. His Comcaac knowledge was vast (e.g., Bowen 2000). We got to Punta Chueca before noon, and I expected we would leave soon for the island. When I asked Jesús when we were leaving, he calmly said *"prontito."* I went away and returned repeatedly only to get the same answer, and again the next day and the following morning.

Tiburón: 26 October. Reluctant patience paid off. Early in the morning we loaded our gear and climbed into a Seri panga. From Zozni Cmiipla on the shore at Tiburón, we walked across the long bajada on the east side of the island and into canyons at the base of the mountains. I had a full backpack, and Jesús and a Mexican helper each had a *palanca* (carrying yoke) with water in white plastic detergent bottles, food, and blankets suspended from each end, and the ubiquitous battered, blue and white-speckled enamel coffee pot. We went to Sopc Hax and slept on a rock ledge near the waterhole.

27 October. Jesús cut cane, or *carrizo* (*Arundo donax*), and carried back two large bundles to make model balsas (reed boats) to sell (fig. 2.4; a photo from this trip in Felger and Moser 1985, p. 307, is mistakenly labeled November 1969). He provided plant names and extensive information, making it one of my most productive trips in the region. Sopc Hax waterhole and vicinity, *9284* to *9324g*. Between Sopc Hax and Hant Hax camp, *9325* to *9354*. Bajada between Hant Hax and Zozni Cmiipla, *9355* to *9360*.

San Pedro Nolasco, 26 November 1963, with Ike Russell and Alice Thomas.

We had been at Bahía San Pedro and early in the morning crossed the channel to Nolasco, anchoring at the southeast cove [Cala Güina]. I climbed the cliffs above the cove and collected plants along the way, but did not properly make note of my route and found it difficult to climb down the cliff to where Ike and Alice were waiting in the *Ofelia* [Ike's wooden boat]. They were only a few tens of feet away horizontally but hundreds of feet below. I yelled to Ike that I would meet him at the northeast canyon [Cañón el Farito] where the decent was not difficult. I then traversed the east side of the island, but unknowingly it necessitated numerous descents and ascents of intervening canyons. At one point a rock gave way and I started sliding toward a sea cliff. I grabbed a cholla (*Cylindropuntia fulgida* var. *fulgida*), spines and all, and held on to halt my fall. Due to the slow going, I did not arrive at the meeting place until the last light of day. We returned to Guaymas, and it was nearly midnight by the time we got to the once-elegant Hotel Casa Grande, but sleep was interrupted by bedbugs, so we passed the remaining hours of night in chairs in the interior garden patio. After sunrise I pressed the rest of the island plants, about 43 numbers, *9633* to *9675*.

Tiburón & Sonora, 30 April to 4 May 1964, with Ike and Jean Russell and John (dog).

30 April. Left Ryan Airfield 4 p.m. and landed at Nogales, Sonora. Couldn't get through, so back to Nogales, Arizona, and then straight to Hermosillo for more friendly customs and immigration.

Figure 2.4. Jesús Morales at Sopc Hax waterhole, Isla Tiburón, with bundles of carrizo (*Arundo donax*). RSF, 26 October 1963.

1 May. Leave Hermosillo 6 a.m. for Desemboque to visit the Mosers. Ike took Ed and Becky to his secret landing place near Sauzal on Tiburón so they could see the Sauzal waterhole. They brought back plant specimens in plastic bags, which I pressed and cataloged that evening as *Felger 9986* to *10008* (see entries for Mary Beck Moser and Edward Moser).

2 May. After breakfast in Desemboque, Ike, Jean, John dog, and I went to Palo Fierro. I collected from near the airfield westward to the upper edge of the bajada (*10011* to *10023*) and on the adjacent first rocky hills (*10025* to *10028*). Camped at Palo Fierro and the next day we went to Guaymas.

Tiburón, 15 and 18 July 1964, with Ike Russell and Nancy Thomas.

15 July. En route to Guaymas we landed at Ike's small, crummy airstrip near Arroyo Sauzal. It was hot and humid. Collected along Arroyo Sauzal and at the Xapij (Sauzal) waterhole, about an hour walk from the landing place, *10073* to *10119*.

18 July. Flying back to Tucson from Guaymas, we made a brief stop at Palo Fierro, *10134* to *10140*.

Tiburón & Patos, 10 and 11 August 1964, with Ike Russell, Joy Cooper, and Cathy Moser.

I went with Joy Cooper in Ike's plane from Tucson to Desemboque, where Cathy Moser joined us for a trip to Tiburón and Patos.

10 August. Tecomate, *10183* to *10231*.

11 August. We went to Ensenada Blanca, where summer annuals were just sprouting (*10239* to *10278b*) and headed back north, stopping at Palo Fierro (*10316* to *10350*). On the way back to Desemboque we landed on Patos: *10355* and *10356*. It was hot and humid, and we enjoyed a swim in the sea.

Tiburón, 20 and 21 October 1964, with C. H. Lowe, David Russell, and Ike Russell.

20 October. On the way back from the Sierra Madre Occidental in Ike's Cessna 180, we landed at Palo Fierro. I collected specimens in association with data recorded from 0.1 ha vegetation plots for my dissertation, *11053* to *11093*.

21 October. We also went to Tecomate, where I collected plants associated with additional plots, *11094* to *11151*.

San Pedro Nolasco, 12 November 1964, with Oda Kleine and Ike Russell.

Early in the morning we went in Ike's slow, wooden boat, the *Ofelia*, from Bahía San Pedro to Nolasco, at the northeast side, above landfall (the place later called Cañón el Farito—at that time there was not a lighting beacon). Oda and I explored the canyon while Ike kept the *Ofelia* nearby offshore, *11431* to *11453*.

San Pedro Nolasco, 18 January 1965, with Ike Russell and Robin Thomas.

From Bahía San Pedro, we went early in the morning in the *Ofelia* to the landfall at the northeast side of the island (Cañón el Farito). Robin and I spent a long day recording data from quadrats we set up between the shore and summit and one on the west side. A former U.S. Marine, Robin is Nancy Thomas's brother and son of Dr. Bob Thomas, who was a friend of Ike and Jean Russell and provided significant medical assistance to the Seris (see Bowen 2002). Robin was not a biologist but was amazingly tolerant, assisting me with quadrat studies for my dissertation. The vegetation was green and luxurious, *12066* to *12089E*.

Tiburón, 29 January to 2 February 1965, with Ike Russell and Robin Thomas.

28–30 January. On a field trip with Ike in the Cessna 180, we landed at Ensenada Blanca (Vaporeta), where Robin and I established five 0.1 ha plots, or quadrats, and collected associated plants, *12203* to *12252*.

31 January & 1 February. Arroyo Sauzal, five plots were established along with an extensive plant collection, *12254* to *12291c*.

1 and 2 February. Southwest and upper part of Central Valley, *12294* to *12436*. Abridged from my field notebook: "We flew around for about 40 min. and finally located some near pure stands of *Larrea* at the SW end of the upper Central Valley. The creosotebushes are rather short and Ike landed smack dab on top of them. Robin whomped out *Larrea* shrubs to make a place to take off. . . . Ike tried out his new airfield and it worked

71

OK. I spent the day collecting and made two 0.1 hectare vegetation plots."

Tiburón, 12–15 and 18 February 1965, with Ike Russell and Robin Thomas.

12 February. Went to Tecomate in Ike's Cessna 180, *12485* to *12532c*.

13–15 February. Palo Fierro, *12533* to *12562b*.

18 February. Southwest Central Valley in the vicinity of Ike's new landing place, *12621* to *12664*.

Alcatraz, San Esteban, & Tiburón, 25–27 February 1965, with Ike Russell and Jean Russell.

Went to several islands in the *Ofelia*, leaving from Kino.

Alcatraz: 25 February, *12700* to *12724*.

San Esteban: 26 & 27 February, Arroyo Limantour, *12728* to *12765b*.

Tiburón: 27 & 28 February, Arroyo de la Cruz, *12766* to *12805*.

Alcatraz, 20 March 1965, with Ike Russell.

I went with Ike from Bahía Kino in the *Ofelia* and collected at the southeast base of mountain, *12821* to *12827*.

Alcatraz, Cholludo, & Dátil, 2–4 December 1965, with Oscar Hommel Soule.

A fellow graduate student at the University of Arizona, Oscar subsequently taught environmental studies at Evergreen State College. We left Tucson on the morning of 1 December and drove to Hermosillo. On 2 December drove to Kino. Pilo and Catalina are expecting their fourth baby. Things seem better than last year, and Doña Gavina looks well. Arranged for a boat to Dátil the next day, at U.S. $20 per day. Gavina says her son Julio and his wife are on Tiburón with about a dozen men developing wells and water tanks—she said perhaps at Ensenada del Perro, but more likely they are at Tecomate where there is water development for a small military station.

Alcatraz: 2 December. Made arrangements in Kino to be dropped off on the island. They will pick us up tomorrow. Spent the afternoon on the island and camped for the night. Very windy. Plant collections, *13401* to *13410c*. Oscar set out live traps for rodents.

Alcatraz, Cholludo, & Dátil: 3 December. Oscar picked up the live traps in the early morning—no catch. Left Alcatraz 7:30 a.m. for Dátil and Cholludo.

Arrived at Cholludo 10:45 a.m. A bit rough on the way over, whitecaps. Stayed on Cholludo for several hours, *13411* to *13427*. Set out live traps.

Later in the afternoon we went to Dátil and camped on the narrow, cobble-rock beach at the usual place on the northeast side of the island. Collected plants on the northeast and northwest parts of the island, in the canyons and to peak elevation, and near the cobble beach, *13430* to *13476*. Oscar set out live traps.

Dátil: 4 December. Oscar caught some *Perignathus* mice in the traps. Spent the day on Dátil. Set up a quadrat study near the top of the island and hiked to the summit. Left Dátil at dusk and picked up the traps on Cholludo—nothing, and returned to Kino by full moon. A bit rough and we got soaked. Spent a nice dry night at a motel in Kino Nuevo.

Tiburón & Alcatraz, 8 October 1966, with Ike Russell and Kalman Muller.

On this trip we left from Tucson in Ike's plane for a trip to Sonora.

Alcatraz: After landing in Kino, we went in Ike's boat to the island, *14909* to *14928C*.

Tiburón: Back at Kino, we went by plane to Ensenada Blanca (Vaporeta), which was very green at this time, *14929* to *14979*.

Tiburón, San Esteban, & Dátil, 20–22 December 1966, with John Miller Cooper.

At Bahía Kino I hired a fisherman to take us to the islands, accompanied by John Miller Cooper from Boulder, where I was on the faculty of the University of Colorado. It was damp, cloudy, and rather cold most of the time.

Dátil: 20 December, northwest side of island, *15302* to *15354*.

Tiburón: 21 and 22 December. Coralitos, *15357* to *15394*. Tordillitos, spring with water about ¼ mi inland, *15466* to *15520*. El Monumento, *15522* to *15570b*.

San Esteban: 21 December, La Friedera and elsewhere on the north side of island, vegetation very green with ephemeral (annual plant) coverage 100 percent in places, *15396* to *15463*.

Tiburón, 28 April 1967, with Ike Russell and John Ahrens from Boulder.

Left Tucson mid-morning, in Ike's plane. After clearing customs and immigration at Nogales, Sonora, we made a few stops in Sonora, inspected Isla Patos from the air, and landed at Ensenada Blanca. Botanized and camped for the night, *15725* to *15764*. The next day we went to Guaymas.

San Esteban, 5 November 1967, from Bahía Kino in the *Ofelia* with Ike and Jean Russell and Nancy Thomas.

Arroyo Limantour, *16600* to *16615*. Canyon at southwest corner of island, above the Cascajal sand spit, *16616* to *16652*. San Pedro, rocky cove at the northwest side of the island, *16654* to *16675*.

Tiburón, 20 February 1968, with Ike Russell in his Cessna 185 and Joe Wallace Edmundson (a photographer friend from California).

Ensenada Blanca, *17240* to *17295f*. Southwest part of Central Valley, *17296* to *17359g*.

Dátil, San Esteban, Tiburón, & Cholludo, 8–12 April 1968, from Bahía Kino in the *Ofelia*, with Joe Edmundson, Ike and Jean Russell, and Nancy Thomas.

Dátil: 8 April, canyon at the southeast side of the island, *17490* to *17524*.
San Esteban: 9 April, Arroyo Limantour, vicinity of the beach, *17525* to *17529*. Climbed steep slopes to the south-central peak, *17530* to *17549b*.
10 April. Went around the island in the *Ofelia*, making several stops at places where we could go ashore: rocky canyon behind large cove at north side of island, *17550* to *17608*. Southwest end of island, rugged canyon, *17612* to *17646*.
11 April, vicinity of Arroyo Limantour, *17650* to *17684*.
Tiburón: 12 April, about 1 km north of Punta Willard, *17736* to *17757*. Ensenada del Perro, *17710* to *17735B*.
Cholludo: 12 April, *17700* to *17706*.

Tiburón, 4 December 1973, with Ike Russell in his plane.

Southwest part of Central Valley, *21285* to *21310*.

Tiburón & Patos, 19 December 1973, with Ike Russell in his plane.

Tiburón: 19 December, Punta Perla. Landed on the beach, walked to the estero (*21311* to *21317*) and then to a nearby low rock hill (*21318*).
Patos: 19 December, landed again on the island, *21319* to *21321*.

Tiburón, 4 April 1974, with Ike Russell in his plane, Richard Evans Schultes (Harvard University professor, 1915–2001), and Andrew Thomas Weil (Tucson physician).

We were visiting Ed and Becky Moser at Desemboque and made a short hop to Punta Perla, *74-10* to *74-15*. I photographed Professor Schultes sitting in a Seri vision circle at the top of a hill (Felger & Moser 1985:102, fig. 7.3).

Tiburón, 8 September 1974, with Ike Russell in his Cessna 185, Hank Gunn from Tucson, and Cayetano Montaño (1923 or 1924 to early 1990s) of Desemboque.

Landed at our improvised landing place at the southwest side of the Central Valley. We walked several hours to the southeast side of the Central Valley, near north side and base of basalt hills, vicinity of former Seri camp of Haap Caaizi Quih Yaii (tepary-bean-gathering camp), which I called Haap Hill on herbarium labels. The object was to find this bean, but none were found—it was too dry. Collected *Setaria liebmannii*, *T74-1*, and *Amoreuxia palmatifida*, *T74-2*.

Tiburón, 1 November 1974, with Ike Russell in his plane and Pedro Comito (ca. 1910–1990s).

We made another attempt to locate the beans at Haap Hill, landing at the southwest Central Valley and walked eastward to the bean campsite. Again, it was too dry, although I made a substantial collection of other plants, *74-T26* to *74-T65*.

Tiburón, 11 December 1976, with Ike Russell in his plane with Rosa Flores (ca. 1916–1993) and Cathy Moser (fig. 2.5).

Like the two earlier attempts to find the elusive bean at Haap Caaiji Quih Yaii, we landed at our remote landing

Figure 2.5. Cathy Moser and Rosa Flores, searching for the wild tepary (*Phaseolus acutifolias*) at Haap Caaizi Quih Yaii (Haap Hill) on Isla Tiburón, 11 December 1976. RSF.

place in the southwest and upper part of the Central Valley (see my notes for 1 and 2 February 1965). Rosa had not been to the bean camp for 36 years. This was her first ride in an airplane. She wore flip-flops and walked at a fast clip, without stopping, for several hours straight to the site. We found just a few intact pods of the bean, which was a tepary (*Phaseolus acutifolius*; Felger & Moser 1985). The photo of Rosa Flores on the back cover of the dust jacket and the end cover of the second printing in 1991 of *People of the Desert and Sea* was taken on the afternoon of this day (also see Felger 2000b). Collections *76-T1* to *76-T40*.

Tiburón, 22 April 1983, with Chris Bailey (later changed his name to Chris Baison; Senior Research Specialist at the Laboratory of Tree Ring Research, University of Arizona) and Pedro Comito (Felger & Moser 1985:412).

At Punta Chueca we hired a Seri panga to take us to Zozni Cmiipla and then walked to the Sopc Hax waterhole and canyons along the east side of the mountain. Photos of Pedro from this trip are in Felger and Moser (1985:256, 264, 322). Becky Moser and I had a near-completed text

for our ethnobotany book, and I made a trip to the Seri region with Chris mainly to obtain additional photos for the book. *83-127* to *83-137*.

San Pedro Nolasco, 11 February 2000. Cañón el Farito to the ridge crest, with Jeffrey A. Seminoff, Horacio Cabrera-Suarez, Juan Pablo Gallo-Reynoso, and Gabriella Suarez-Gracida.

We went to the island in a fiberglass panga, the *Mirounga*, the generic name of the elephant seal. We left the San Carlos marina at 7:30 a.m. and arrived at the island at 9 a.m. in moderately choppy sea. Our first landfall was the southeast cove, but seeing the cliffs and difficult assent, went instead to the northeast landfall at Cañón el Farito. Drought conditions prevailed, *2000-1* to *2000-4*.

Tiburón, 24 and 25 May 2006, with Humberto Romero and Ben Wilder.

This is my first trip to Tiburón since 1983. The collections are labeled as *Wilder 06-143* to *06-169*.

24 May, mid-morning. A short ride in a Seri fiberglass panga from Punta Chueca to Punta Tormenta, location

of the airstrip that Ike Russell and I called Palo Fierro. There is a small Mexican marine station here and near the beach the Seris have several four-wheel-drive vehicles showing extreme wear and tear. Marines ask what we are doing, write down our names, and inspect our collecting permits. Ben and I make a small plant collection along the shore. Battery and tires are switched between vehicles to get one that works—a red 1998 Dodge Durango that has been on the island for three years and is falling apart. Humberto drives on a rough, narrow dirt track that goes west from the long airfield to Caracol at the base of the mountains. The gauges do not work but the empty-gas red light is on. I am thinking about a long hot walk back—like my earlier fieldwork, before the roads, when it took most of the day to walk from Palo Fierro to these mountains, and that was not in summertime. Interesting plants as we rattle through a densely vegetated canyon.

11 a.m. Caracol. There are two houses, built as research stations, now used for bighorn hunters. Screens falling off, some trash, and smelling of packrats. Walking a few hundred meters west from the houses we find a substantial patch of buffelgrass. We then drive back toward the shore.

12:50 p.m. Back at Punta Tormenta. Humberto borrows 30 liters of gas from the marines, siphoned from a 50-liter plastic *bote*. It is getting hot as the day wears on. We drive southward to Zozni Cmiipla and continue on the dirt track near the shore. At one stop I start to collect a piece of sinita (*Lophocereus schottii*) but notice Humberto took four pieces of xoop (*Bursera microphylla*) and rubbed them between his hands. I quickly realize the cactus is culturally respected and do not collect it. He is appreciative and says, "You know." Photos are ok.

We drive west again into the foothills of Sierra Kunkaak, to Valle de Chalate, or Xpasni Quitalc, named for a large fig tree (*Ficus petiolaris*) on a canyon rim. Late lunch in the warm shade and I press plants. Humberto brought his copy of the Seri ethnobotany and looks up plants in the book (fig. 2.6). We backtrack onto the road to Pazj Hax. Some big signs admonish one to respect the water place and the wildlife, etc. It hardly cools down after sundown, and the place is full of kissing bugs (*Triatoma*), so we pack up and drive toward the coast, but not all the way because Humberto is concerned about sidewinders and mosquitoes along the beach dunes.

25 May. Last night I tried to sleep on the back seat of the car. *Ratoncitos* running around all night, chewing on something in the back of vehicle. Milky Way brilliant. Venus rises and a golden moon and the sky reddens. I hear shore birds a kilometer or so away. Flat tire and luckily

Figure 2.6. Humberto and Richard pressing plants under a *Ficus petiolaris*, vicinity of Pazj Hax, Isla Tiburón. BTW, 24 May 2006

the spare, tied on the roof, has air in it. Heat up water for my instant Italian espresso, but Humberto misses the sugar. No way to check water or engine because the hood-latch cable is broken.

Back at Zozni Cmiipla, Humberto finds tiny bleached whelks where the highest tides reached. We scoop the shells into small plastic bags for his wife for necklaces and for Cathy Marlett for her work on Seri knowledge of mollusks. We drive back northward. Honeybees get into the car at each stop. Bees are everywhere on this trip—they are persistent and are after water, even from our eyes.

9 a.m. Back at Punta Tortuga. The marines have a large, sturdy panga but no vehicles. They are out fishing for lunch. There are about eight of them, wearing heavy black combat uniforms, some with thick tight black simulated leather jackets, and it's already sweltering hot. Teenage marines carry their automatic weapons at all times. Humberto has good relations with them, and they take us across to Punta Chueca. They are happy for any diversion and to make calls since cell phones do not work at their island station.

Tiburón, 23–26 November 2006, with Edward "Ed" Erik Gilbert, Humberto, and Ben.

Our plant collections for this trip are numbered as *Wilder 06-345* to *06-513*.

23 November, Thanksgiving Day. Short ride across the Infiernillo in a Seri fiberglass panga; the cracked floorboard bends and groans as the boat slaps the water. At Punta Tormenta half a dozen deteriorating four-wheel-drive SUVs—no repair shop here: the best tires used to make the one working vehicle, the 1998 Dodge Durango. Humberto says the Dodge is *muy fuerte*, stronger than the dead Toyotas in the beachfront lineup. We find the first record of the invasive weed *Tribulus terrestris* on the island. We drive south near the coast to the vicinity of Zozni Quiipla at the base of Punta San Miguel and make additional collections.

In the late afternoon we turn back northward a short distance and head inland on the rough road westward into the east side of Sierra Kunkaak. The road narrows, and Humberto maneuvers the Dodge over rocks and through increasingly dense desert vegetation. This marginally passable dirt track has not been traversed for months and is nearly obscured by recent growth of shrubs. We drive through tunnels of *Hyptis albida* (*H. emoryi*), *Bursera microphylla, Colubrina viridis,* etc.

Leaves and spinescent branches whip through windows if not rolled up. A strange noise—from a wheel? Humberto is concerned. I am thinking of a possible long walk out. The vegetation is remarkably green due to Hurricane John and another rain event, portending favorable botanizing.

Our camp is at the end of the road at about 375 m and below the north side of Capxölim in the Sierra Kunkaak. Ben demonstrates he is the son of acclaimed chef Janos Wilder—Thanksgiving tacos of roast pork, cilantro, avocado, and habañero salsa. We press today's plant specimens.

24 November. We spend the day exploring densely vegetated canyons and rocky slopes. I remember years ago looking down from Ike's plane, wondering if I would ever be able to go to these mysterious mountain canyons. Many places have a buildup of leaf litter from small trees and shrubs, mostly on north-facing slopes and canyon bottoms—something generally not seen in deserts. Gravelly benches of old canyon-bottom floodplains support near 100 percent vegetation cover, mostly of *Ruellia californica,* a smelly glandular-glutinous small shrub that stains clothes. We return to camp near dusk with plastic bags full of plants. I press plants while Ben prepares another camp feast. After dinner we continue pressing the day's catch.

25 November. I wake early and finish processing the last of yesterday's specimens. Humberto leads us on an old trail about 3 km westward and deeper into the Sierra Kunkaak to Siimen Hax, a canyon-bottom bedrock tinaja with a large, dark, shaded pool. Little red-spotted toads (*Anaxyrus punctatus,* formerly *Bufo punctatus*) from this year's crop match the rock color. Nearby, carved into the hard, woody trunk of a saguaro is "José Torres 1949" (fig. 2.7; also see fig. 1.31B). After returning from the island I told Cathy Marlett about the name on the saguaro. She asked Angelita Torres, daughter of José, about the carved name, "and she remembered her father doing that" (Cathy Marlett, personal communication to Richard, 2006). (José Torres's grandson is José Ramón Torres, who went with us to the island 23–26 October 2007; see fig. 2.21.)

We return to camp in the late afternoon and begin pressing the day's collections. Humberto tells us about hehe pnaacol (*Sideroxylon leucophyllum*), pointing to the steep mountain slope opposite our camp. The late afternoon light casts a golden glow on the steep slopes of the isolated Capxölim mountain, highlighting unrecognized trees. Binoculars reveal trees with light-colored foliage

Figure 2.7. Vicinity of Siimen Hax, Tiburón, 24 May 2006. José Torres carved his name in this saguaro (*Carnegiea gigantea*) in 1949. BTW.

and thick trunks at the edge of the Capxölim rockslide, about 250 m above us and far across the canyon. We have to leave early tomorrow morning, and it seems too late in the day to climb up to the trees, but Ben and Ed make it to the top of the scree slope by dusk. Humberto is concerned they might not get back before dark—that he is responsible for our well-being. Returning in the last light, Ben and Ed bring specimens of spectacular new records. Near the top of the scree field of giant boulders they found a grove of about 20 hehe pnaacol trees with gnarled branches (fig. 2.8) and short, thick-trunked canyon-hackberry trees (*Celtis reticulata*).

26 November. We pack up early and head east to the shore at Punta Tortuga. At 9 a.m. the boat arrives from Punta Chueca just as agreed. Ben and Ed leave me in Hermosillo with Alberto Búrquez and Angelina Martínez, and the next day Alberto and I go to Guaymas and then to Nolasco on the 28th.

San Pedro Nolasco, 28 November 2006. Cañón el Farito to the ridge crest, with Alberto Búrquez-Montijo (Instituto de Ecología, UNAM, in Hermosillo), and Florencio Cota-Moreno and Jesús Ventura-Trejo (biologists with CONANP).

The island was green from recent rains, and I was exploring for new records and rewarded by finding *Cyperus squarrosus*. We saw honeybees visiting the *Bahiopsis*

Figure 2.8. *Sideroxylon leucophyllum* at Capxölim, Tiburón. BTW, 25 November 2006.

and other flowers, and Costa's hummingbirds visiting *Salvia similis*. The dense growth of *Vaseyanthus* made for slippery climbing over rocks. *06-73* to *06-111*.

San Pedro Mártir, Dátil, & Tiburón, 11 and 12 April 2007, with Miguel Durazo (boatman with CONANP), Exequiel Ezcurra, and Jesús Ventura-Trejo (biologist with CONANP), and Ben.

San Pedro Mártir: 11 April. We went from Kino in a CONANP fiberglass boat, the *Orca*, 25 feet long with a deep draft, and two outboard motors, one 50 hp and one 25 hp. Our first stop was a cove with guano-mining terraces, and then we went to the rough, rocky landing below the guano miners' village at east side of the island. From there we hiked to the plateau near the summit. Near the top of the island we photographed a large, black kingsnake (*Lampropeltis getula*), a species that had not been documented on the island for 40 years. We also encountered several large, very fat, light-colored rattlesnakes (*Crotalus atrox*). We catalogued 16 numbers, *07-09* to *07-24*, and returned to Kino at the end of the day.

Dátil & Tiburón: 12 April. Ensenada del Perro, collections numbered as *Wilder 07-143* to *07-165*. In the late afternoon we went to a canyon at the northeast side of Dátil. (See Ben's notes for description of the field trip and collection numbers.)

Tiburón & Patos, 29 and 30 September 2007, with Ben, Humberto, and Ernesto Molina in his panga. Ernesto is the nephew of Jesús Morales (see my notes for 26 and 27 October 1963).

Tiburón: 29 September. Leave Punta Chueca in Ernesto's fiberglass panga and go to Punta Tortuga to swap out a battery for the boat from the collection of batteries that the Seris have with their lineup of SUVs. Gun-toting black-uniformed marines take our names and want to know where we are going.

We go northward up the Infiernillo and across the north side of the island to the northwest side. Huge manta rays, porpoises, and some whales off the northwest coast. We travel near shore down the very arid west coast of the island to a place called Pooj Iime. This is the site where the elusive Tiburón palm (*Brahea armata*) is said to occur (see entry for José Juan Moreno). High sea cliffs and very steep slopes plunge into the sea, and there are few feasible landfalls. Pooj Iime is a large canyon complex in these steep, rugged mountains with very

sparse Sonoran desertscrub. Humberto remembers hearing about the palm from his late brother some years ago. We explore the canyon system and make a substantial plant collection but do not find a palm. The collections are cataloged as *Wilder 07-431* to *07-460*.

We return northward and camp at Tecomate at northwest side of the island. I have not been here since my travels with Ike Russell a few decades ago and am shocked to see unacceptable amounts of trash and garbage. The little marine station is in ruins, and the dock is gone except a few remnant pilings. The *Eucalyptus camaldulensis* planted here decades ago struggle on, but the *Tamarix aphylla* trees are thriving. The tamarisk trees produce dense shade and a blanketing ground cover of fallen twigs that prevents any other plants from growing. I collect a specimen of *Lycium brevipes*, *07-53*.

Patos: 30 September. Leaving Tecomate after coffee, we get to Patos at 9 a.m. and leave at noon. It is hot and humid, and I am drenched in sweat in spite of a mild breeze. Humberto and I collect over the entire island while Ben matches historical photographs. Among the nine plant collections, *07-54* to *07-62*, we add several species to the short list of plants known from the island.

The beach drift anchors this tiny uninhabited island to the world beyond its shores. Mollusk shells, entire and broken, are strewn along the south shore cobble beach just above high tide. Included are many pink murexes (*Chicoreus erythrostomus* [*Phyllonotus erythrostomus*], they are commercially harvested), black murexes (*Hexaplex nigritus*, also commercially harvested), reddish purple egg cases of a murex, turbo shells (*Turbo fluctuosus*), giant eggcockles (*Laevicardium elatum*), rock oysters, pearl oysters, etc. Other biodrift includes dry seaweeds such as *Sargassum* and other brown algae, gorgonian corals (sea fans), black coral, crab carapaces, clubspine sea urchin testa (*Eucidaris thouarsii*), dried trigger fishes (*cochi*, *Balistes polylepis*), horn shark egg cases (*Heterodontus* sp.), feathers, and bird, porpoise, and a few whale bones. There is abundant pumice in the drift but also a surprising array of trash in the beach drift and elsewhere along the shore.

Copious plastitrash includes plastic bottles of all shapes and sizes, plastic fishing nets, plastic rope, and pieces of thick plastic sheeting. Other trash includes aluminum cans, beer bottles, pieces of glass jars and bottles sea worn and freshly broken, cement blocks, hoop iron, flip-flops, men's underwear, shoes, pants and other pieces of clothing, pieces of rubber diving suits, cloth diving gloves, and crab or lobster traps. Other evidence of human activity on this tiny waterless isle includes a large and apparently old metate, numerous crude and recent-looking hearths of three or so rocks moved close together, a cement-walled pit near the shore about 1 m deep and 1.5 m in diameter next to a crude rock shelter, dirt pits including one 40 cm deep and 1 × 1.5 m wide, ruins of guano workers' shelters, low walls and dykes, some substantial rock walls, a pit dug on the southeast side of the island that is about 15 m × 5 m, elsewhere considerable amounts of earth and rock have been moved, two large stacks of large modern cement blocks, cement walls, and the ruined lighthouse at the summit, built in 1980.

We take a break in the shade of the cement lighthouse tower. Looking down onto the east shore are sea lions basking and barking and hundreds of pelicans and other seabirds on every wet algae rock, and cormorants swimming about. We see sideblotch lizards, *Uta stansburiana*, adults and hatchlings, across the entire island. The whole north side apron below the hill next to the sea is guano-white and devoid of vegetation. Most of the island is white with stinky bird guano, and *Atriplex barclayana* is the single ubiquitous plant, producing 100 percent cover across much the land surface.

Tiburón: 30 September. After spending the morning on Patos, we went back to the north shore of Tiburón, where we made a collection at the northwest side of the island at a place called Hee Inoohcö. We botanized on the bajada and high, steep slopes directly above the shore, finding several new records for the island. The collection is cataloged as *Felger 07-63* to *07-95*. (Our labels for these specimens have the place-name as Hee Imocö.) Arrived late in the day at Punta Chueca.

Tiburón, 23–26 October 2007, field trip to the Sierra Kunkaak, with Dennise Z. Avila-Jiménez, Brad Boyle, Exequiel Ezcurra, Pedro Ezcurra, Eduardo Gómez-Limón, Servando López-Monroy, Humberto, Jesús Sánchez-Escalante, José Ramón Torres, and Ben.

See Ben's notes for details of this trip; our collections are labeled as *Felger 07-96* to *07-126* and *Wilder 07-465* to *07-587*.

Alcatraz, 4 December 2007, with Ben and Miguel Durazo (boatman with CONANP).

Collected across the flats and lowland areas of the island, *07-158* to *07-192*.

San Pedro Mártir & Dátil, 5–7 December 2007, with Miguel Durazo and two crew members, Andrea Galindo (then a fellow at the Prescott College Kino Bay station), and Ben.

San Pedro Mártir: 5–7 December. I catalogued five numbers on December 5, at the highest elevations of the island, *07-194* to *07-198*. Collections from December 6 and 7 were cataloged as Ben's numbers. On this trip we obtained quantified vegetation data and detailed observations. We camped at the ruins of the old village site. Dawn and dusk was an especially noisy time with thousands of seabirds whirling overhead and jostling for a roosting place. On the second day we awoke to dense fog that did not lift until late in the day. Our attempted return to the mainland late in the day on December 7 was marred by engine failure and a non-functional radio in increasingly rough weather in open water south of Isla Dátil, resulting in a cold, wet, and dangerously challenging night at sea. (See Ben's account for this trip.)

Dátil: 8 December. We arrived at the northeast side of island in the early morning after a treacherous night at sea. We called this place Rescue Beach, and it was a great relief to be on land and get dry and reasonably warm. It was wonderful to see Tad Pfister and Cosme Becerra coming to pick us up a few hours later in a big white panga. We collected several specimens including *Ficus petiolaris* subsp. *palmeri*, *07-199*. (See Ben's notes for this place and date.)

Alcatraz, 28 January 2008, with Ben and Tom Fleishman's Prescott College class.

Collected across the lowland areas of the island, *08-01* to *08-08*.

Tiburón, 29 January 2008, with Ben, Cosme Becerra, and geologists Scott E.K. Bennett, Rebecca J. Dorsey, Michael E. Oskin, and Thomas C. Peryam.

Left from Kino for a day trip to Sauzal in Prescott College's large panga with Cosme piloting. We walked up Arroyo Sauzal to the waterhole, and returned to the terminus of Arroyo Sauzal at the shore. (See Ben's notes for description of the field trip and collection numbers.)

Tiburón, 30 January 2008, with Ben, Cosme Becerra, and Abram Fleishman.

Left from Kino in Prescott College's panga with Cosme piloting again, to the southeast part of the island including El Monumento and La Viga. See Ben's notes for description of the field trip and collection numbers. The next day Ben and I went to Guaymas for a trip to Isla San Pedro Nolasco.

San Pedro Nolasco, 2 and 3 February 2008, with Florencio Cota-Moreno, Jesús Ventura-Trejo, and Ben.

Both days we left from the marina at Bahía San Carlos about 7:30 a.m. and spent much of the day in the vicinity of Cañón el Farito. We set up permanent monitoring sites and obtained quantitative vegetation data and a collection of voucher specimens. The vegetation was moderately dry. See Ben's description for this field trip and collection numbers.

Tiburón, 26 September 2008, with Ben, Humberto, and members of the Comcaac "botany team" from Punta Chueca: Erik Salvador Barnett-F., Maximiliano "Max" López-Romero, and Servando López-Monroy.

Left from Punta Chueca by boat and went to Zozni Cmiipla. In that general vicinity we made a collection of the summer flora including *Tumamoca macdougalii, 08-111* to *08-134*. We pulled or dug up, bagged, and removed all known buffelgrass (*Cenchrus ciliaris*) from the area.

San Pedro Nolasco, 29 September 2008, with Juan Pablo Gallo-Reynoso and Ben.

We hired José Luis Ramirez-Zuñiga, from La Manga (northwest of San Carlos), to take us to the island in his fast, open boat. Travel time from La Manga to the southern tip of Nolasco was about 35 minutes. José Luis told us that about eight days earlier there was a *resaca* (described as a big tide) drifting onto and west of Nolasco, bringing "trash" from the tropical depression Lowell. This storm had made landfall at the Sonora-Sinaloa border. José Luis said that flotsam from the *resaca* included chairs, tables, part of a house, plastic items, and tree branches. Much of the "trash" drifted on past Nolasco, including tree branches containing a large green iguana (*Iguana iguana*) and another branch with a large black snake. Such events could be significant for biological colonization of the island.

We proceeded northward along the west side of the island, photographing and recording plants seen from the boat, arriving at Cañón de Mellink at 10 a.m. Ben, Juan Pablo, and I went ashore and climbed to 115 m in

the canyon, and Ben went on to the ridge crest. Juan Pablo and I went to the next canyon to the north and explored steep slopes high above the sea. The island was verdant due to recent rains and the day was very hot (36–39°C) and humid with a high, scattered cloud cover that gave some protection from the harsh September sun. We encountered adult as well as some hatchling spiny-tail iguanas, and sideblotch, spiny, and whiptail lizards. Adult iguanas were active, numerous, and easily approached. We also saw many desiccated carcasses of adult iguanas along the canyon bottom. We did not see honeybees on this day. We left Cañón de Mellink at 4 p.m. This trip allowed us to make the first plant collections from the west side of the island. Collections from the base and middle elevation of the canyon are catalogued as *Felger 08-135* to *08-148*. See Wilder, 29 September 2008, for the rest of the collections from this trip.

Juan Pablo Gallo-Reynoso.
San Pedro Nolasco, including 2000 & 2008.

Gallo-Reynoso has made numerous trips to the island including 2000 with Richard and 2008 with Richard and Ben. He is an ecologist studying the avifauna and mammals of northwestern Mexico, especially seabirds and aquatic mammals including river otters and sea lions. He is a research scientist at the Centro de Investigación en Alimentación y Desarrollo in Guaymas, Sonora. He has conducted extensive studies on Nolasco and collaborated with Felger and Wilder on the flora and vegetation of the island (Felger et al. 2011).

Howard Scott Gentry (1903–1993).
San Pedro Nolasco, 1951.

Recognized as the leading authority on the agaves, Gentry worked for the U.S. Department of Agriculture. After retiring in 1971 he was a research botanist at the Desert Botanical Garden in Phoenix, and in his later years he and his wife, Marie, moved to Tucson. He made extensive collections in Sonora and elsewhere in Mexico. His publication on Gulf of California plants (Gentry 1949), focusing largely on the Allan Hancock Foundation expeditions in the 1930s and 1940s, is one of the cornerstones of botanical research in the region. See also entries for E. Yale Dawson and Peter J. Rempel.

He visited San Pedro Nolasco on 16 December 1951 and collected eight numbers (*Gentry 11351* to *11358*), as evidenced by his collection notebook at ARIZ. He collected on the east side of the island in the vicinity of Cañón el Farito, as shown by a photo on his herbarium specimen of *Agave chrysoglossa*. His extensive herbarium collections are at ARIZ, DES, and elsewhere.

Charles Edward Glass (1934–1998) & Robert Foster. San Pedro Nolasco, 1975.

In November Glass and Foster collected the type specimen of *Mammillaria tayloriorum* and named the species, differentiating it from *M. evermanniana*. For a number of years they edited the *Cactus and Succulent Journal* (U.S.) and wrote extensively on cactus and succulents (Mitich 1993).

Dudley B. Gold (1897–1990). Tiburón, 1968.

Gold lived most of his life in Mexico City and was one of the founders of the Sociedad Mexicana de Cactología in 1951, publishers of *Cactáceas y Suculentas Mexicanas* (Mitich 2000). On 24 March he collected *Acacia willardiana*, *Bursera microphylla*, and *Caesalpinia intricata* at Tecomate. The majority of his collections are deposited at MEXU.

Charles F. "Harbie" Harbison (1904–1989). Tiburón, 1962.

Harbison was a noted entomologist and curator of entomology at the San Diego Natural History Museum. Harbie's extensive biological collections from California, Arizona, and northwestern Mexico included herbarium specimens. He collected *Cylindropuntia versicolor* at Punta Willard on 19 March 1962 and *Peniocereus striatus* at El Sauz on 20 March 1962.

James Rodney (Rod) Hastings (1923–1974). San Esteban & San Pedro Mártir, 1971 (fig. 2.9).

Hastings was born in Hayden, Arizona. He was in the army during WWII and later was mayor of Hayden while getting a PhD at the University of Arizona, where he later became professor of atmospheric science. His dissertation, *Historical Changes in the Vegetation of a Desert Region* (Hastings 1963), became the basis for *The Changing Mile*, co-authored with Ray Turner (Hastings & Turner 1965), his major professor. Their joint efforts included establishing cardón study plots in Sonora and Baja California.

Rod went on a substantial tour of the Gulf of California islands from 12 to 28 March 1971 on Antero Díaz's *San Agustín II*. His detailed notes, including those from this trip, are preserved in his field notebook, provided to us by Ray Turner.

Figure 2.9. Rod Hastings (left) and Ray Turner (right) at the University of Arizona, Tucson. Douglas K. Warren, ca. 1965, courtesy of Ray Turner.

San Esteban: Rod was on San Esteban 16 and 17 March. His collection numbers are *71-47* to *71-57*. "3/16/71: Proceeded to S. Esteban, making anchorage on the E side of the point that marks the SW corner of the island [Based on description and several photos he took, he was most likely at the southeast side at Arroyo Limantour. It is surprising that Rod was incorrect in his direction as he was a keen observer.] . . . Landfall was at a floristically rich region, the broad mouth of a large wash reaching deep into the interior and draining an obviously extensive area. Fairly steep basalt hills, and a stony mesa bordered the wash, so there was a variety of habitats. . . . I hiked up the wash about 2–2½ miles. . . . A curious feature of the island is the absence of an understory of semishrubs. No ENFA [*Encelia farinosa*] or FRDE [*Franseria deltoides*, synonym of *Ambrosia deltoides*] FRDU [*Franseria dumosa*, synonym of *Ambrosia dumosa*], or anything morphologically corresponding."

San Pedro Mártir: 17 March, *71-58* to *71-67*. "Proceeded then to San Pedro Martir, a giant rookery for the blue-footed booby. The island is extremely mountainous and rocky with only 2 fairly level spots. See that birdshit! A curious flora, evidently selected for its ability to tolerate high P and N levels. PAPR [*Pachycereus pringlei*] is the dominant, and superficially the island resembles the Islas Melisas [*sic*] in PAPR density. But virtually the entire stand is even-aged, about 3–4 m in ht. Only in north slopes was I able to find any individuals under 1 dm in ht; very few of them. A scattering between 1 dm and 3 m. The second dominant is *Sphaeralcea hainesii*!; a woody perennial. *Ficus palmeri* occurs as widely scattered individuals. Comps comprise most of the species list."

Ivan Murray Johnston (1898–1960).
San Pedro Nolasco, San Pedro Mártir,
San Esteban, Patos, & Tiburón, 1921.

Johnston had a long and illustrious botanical career. He was the botanist on the California Academy of Sciences expedition to the Gulf of California in 1921, an 87-day tour, 13 April to 13 July, in which "collections were made on all the 30 odd important islands in the gulf, at five localities in Sonora, and at 14 localities on the peninsula of Lower California" (Johnston 1924:951). His 1924 publication remains one of the finest treatises on the botany of the Gulf of California region. The majority of Johnston's collections from the expedition are at CAS in San Francisco, with many duplicates at UC in Berkeley

and some at ARIZ. He described a substantial number of new taxa from his 1921 collections, and numerous species have been named in honor of Johnston, largely based on his collections. He also collected marine algae, which were studied by William A. Setchell and Nathanial L. Gardner of the University of California, Berkeley. They published papers based on the Johnston specimens, naming a number of taxa in his honor (Readdie et al. 2006; also see entry for E. Yale Dawson).

San Pedro Nolasco: The party visited this island on 17 April, when Johnston collected 13 numbers, between *3112* and *3144* (the numbers for Nolasco are not all consecutive). He described two new taxa based on these collections: *Agave chrysoglossa* and *Hofmeisteria pluriseta* var. *pauciseta* (*Pleurocoronis laphamioides*).

San Pedro Mártir: 18 April. This should have been in the middle of the seabird nesting season, but as Joseph Slevin reports in the general account for the expedition, "It was formerly a great sea-bird rookery but appears to have been long deserted, probably due to the depredations of the guano hunters" (Slevin 1923:57). Johnston collected 20 numbers of 15 species with another species observed, *3145* to *3164*.

San Esteban: He was in the vicinity of Arroyo Limantour on 19 and 20 April. Gentry (1949) reports that Johnston collected 37 species on the island—we located *Johnston 3169* to *3208*.

Patos: On 23 April, en route to Freshwater Bay on Tiburón, the party made a brief stop at this small guano island where Johnston made seven collections, *3239* to *3245*.

Tiburón: The party stopped at four localities on the island: Freshwater Bay (Bahía Agua Dulce), Willard Point, Monument Point (the dot on Johnston's map indicates Ensenada del Perro, which is just north of El Monumento), and an unnamed locality north of Ensenada del Perro (Johnston 1924, insert map opposite p. 1218). Gentry (1949:99) reports that Johnston collected 60 taxa on Tiburón. We have only seen collections from 23 April at Freshwater Bay, *3250* to *3270*, and Willard's Point on 3 July 1921, *4245* to *4268*.

Bete Jones. Alcatraz, 2001.

Jones collected *Pectis papposa* on 26 October with Carmen Gabriela Suárez and Aaron Wickenheiser, as a student at the Prescott College field station at Bahía Kino.

John Kipping. San Pedro Mártir, 1975.

Kipping collected *Sphaeralcea hainesii* (CAS).

Paul Knight. Tiburón, 1979.

A New Mexico botanist, Knight made a trip to Tiburón in 1979 when he was the second biologist (see entry for Norman Scott) to make plant collections from the higher elevations in Sierra Kunkaak. Paul was invited by the U.S. Fish and Wildlife Service to participate on a scientific expedition to the island focused on the large population of desert tortoises. Paul and his Comcaac guide, Francisco "Chapo" Barnett, ascended the high peaks of the Sierra Kunkaak twice and went to the Sopc Hax waterhole on the final day. The first set of his collections from this trip is at UNM in Albuquerque, with a large set of duplicates at ARIZ. Associate collectors were Jean M. Duke and Sandra Limerick.

The party was on the island 23–28 October, and Paul's collection, catalogued as his numbers *915* to *1124*, include 59 species. One of these, *Chioccoa petrina*, is known from the island only by his collection. Paul kindly provided us with a copy of his detailed field notes, which are paraphrased here.

23 October. After transporting a ¾ ton U.S. Fish and Wildlife pickup to the island using a pontoon raft (later, after much mezcal, this truck gained the name "La Reina de Tiburón"), they made camp at the Caracol research station, which was then in use and occupied by a resident biologist. About an hour was spent exploring the area around the station.

24 October. The day was spent scouting a route to the top of the Sierra Kunkaak at the western base of the mountain. About 20 collections were made.

25 October. Paul climbed the high Sierra but not to the peak. "Exploration of northern slope [of] Sierra Kunkook [*sic*]. Area 30–40° slope, facing due north. . . . Extremely dry. Dominants are *Bursera, Lysiloma*, cardon [*Pachycereus pringlei*], *Cercidium, Fouquieria*. Also *Ferocactus wislizeni* [*F. tiburonensis*]. Area heavily grazed by mountain sheep [bighorn] and deer. Most specimens are from an extremely rocky area with an abundance of talus slopes. Area about 4–5 km south of Caracol."

26 October. On the following day, Paul and his companions ascended San Miguel Peak. "We ultimately approached the mountain from the west face and climbed up a gentle ridge maneuvering around the south slope. The thorn scrub was merciless." One of the primary collecting

localities Paul visited in the Sierra was an area he termed "the bowl." "From the top of the Sierra San Miguel we headed eastward along the ridgetop that formed the crest of the mountain. A large bowl and associated canyon occurred on the south side of the mountain, which formed a saddle between the two mountain peaks of the Sierra San Miguel chain. I had noticed that every morning clouds form over this bowl-like canyon from sea moisture."

Paul's "San Miguel Peak," based on his field notes, an associated photo, and a map he drew, is probably not the one visited by Ben in October of 2007, but a nearly equally high peak to the south. Precise locations of Paul's sites are inconclusive—at that time there were no available accurate maps of the island.

October 27. Chapo took Paul to the Sopc Hax waterhole, which Paul called "Carrizo Canyon," on the eastern side of the Sierra.

Irving William "Knobby" Knobloch (1907–1999). San Esteban, San Pedro Mártir, 1969.

In the 1930s Knobloch worked as a naturalist for the Civilian Conservation Corps. He collected extensively in Mexico, beginning in 1937 in Chihuahua and from 1938 to 1940 managed a copper mine in Chihuahua, amassing a large herbarium collection. He joined the faculty of Michigan State University in East Lansing in 1945 and specialized in the study of ferns. He co-authored *Ferns and Fern Allies of Chihuahua, Mexico* (Knobloch & Correll 1962) and published an extensive listing of plant collectors in Mexico (Knobloch 1983). He collected on a number of Gulf of California islands during June and July. The first set of his specimens is at the Michigan State University Herbarium (MSC). (See In Memoriam, Irving W. Knobloch 2000.)

San Esteban: "E bay" [Limantour], *Knobloch 2390* to *2397* (MSC, mostly with duplicates at SD).

San Pedro Mártir: *2398* to *2402* (MSC, duplicates at SD except *Ficus petiolaris* subsp. *palmeri, 2399*).

George Edmund Lindsay (1916–2002).
San Pedro Nolasco, 1947, 1952, 1961.
San Esteban, 1947, 1952, 1961, 1962.
Tiburón, 1952, 1962 (fig. 2.10).

Lindsay earned his undergraduate and doctoral degrees at Stanford University. He was founding director (1939 and 1940) of the Desert Botanical Garden in Phoenix.

Figure 2.10. George Lindsay (left) and Reid Moran (right) next to a giant barrel cactus (*Ferocactus diguetii*) on Isla Catalina, Gulf of California. 15 April 1952 or 9 April 1962, courtesy San Diego Natural History Museum.

From 1956 to 1963 he was director of the San Diego Natural History Museum, and from 1963 to 1982 he was director of the California Academy of Sciences. He published a number of works on cactus and succulents, especially of the Baja California Peninsula and the Gulf of California. He monographed *Ferocactus* (barrel cacti) for his PhD dissertation (Lindsay 1955b; Lindsay et al. 1996). The majority of his collections from the islands were of cacti, several of which are the types of endemic and near endemic species on the islands treated here. The first set of his specimens is at SD, and some are at CAS and also DES in Phoenix.

He organized a series of explorations to Gulf Islands in the 1950s and 1960s and helped lay the groundwork for major conservation efforts by the Mexican federal government. Lindsay also organized expeditions in 1964 (20 June to 4 July) and 1965 (between 9 and 20 August) but focused on the southern islands of the Gulf (Lindsay 1966).

A joint venture between the California Academy of Sciences, the San Diego Natural History Museum, and the Instituto de Biología of Mexico undertook a 10-day cruise (19–29 April) to 13 islands in the central Gulf of California in 1966 (Lindsay 1966). The *San Agustín II* of Antero Díaz from Bahía de Los Angeles was chartered with Reid Moran as the principal botanist of the trip (see Reid's entry for 1966). Additional participants were Richard P. Phillips, geologist and director of the San Diego Natural History Museum; Alejandro Villalobos F., Chief, Sección de Hidrobiología, Instituto de Biología and invertebrate zoologist; Robert T. Orr, mammalogist and ornithologist; Tom Tilton, trustee of the California Academy of Sciences; Richard C. Banks, mammologist and ornithologist; Allan J. Sloan, herpetologist; Dustin Chivers, invertebrate zoologist; Virgilio Arenas F., assistant to Dr. Villalobos; Raymond Bandar, assistant to Dr. Orr; Ken Lucas, collector; and Luis Baptista, ornithologist and good friend of Richard Felger.

A trip in 1973 via a chartered flying boat through the Gulf of California with famed aviator Charles Lindbergh ultimately led to greater conservation for the Gulf of California. On a trip several months later to Mexico City, Lindbergh with Lindsay spoke passionately to Mexican elected officials and the press about the natural wealth of the Gulf. Four years later all islands in the Gulf were federally protected (Ezcurra et al. 2002).

San Pedro Nolasco, 24 February 1947, with Herbert Bool.

After several attempts to reach the island, Lindsay and Bool finally made it and avidly collected and photographed for a brief period of time, but not making it to the crest of the island. To Lindsay's great exhilaration, "it seemed that every foot of the island capable of supporting a cactus had one!" (Lindsay 1947:75). They obtained the type collection of *Echinocereus websterianus, Lindsay & Bool 498.*

San Pedro Nolasco & San Esteban, 1947.

"In the spring of 1947, George E. Lindsay was invited on a month's cactus-hunting tour of the Gulf islands by Wilson and Lynne Long, who were spending a year roaming the seas on the sailing yacht *Adventurous*" (Bowen 2000:297).

San Pedro Nolasco: Late March: On this trip to the island (Lindsay 1955a) he collected at least *Mammillaria tayloriorum, 505.*

San Esteban: 3 April. "They stayed only a day . . . spent the night in the lee of El Cascajal" (Bowen

2000:397). Lindsay collected at least two specimens: *Chylismia cardiophylla* subsp. *cardiophylla* and *Physalis crassifolia* var. *infundibularis* (CAS).

Tiburón, San Pedro Nolasco, & San Esteban, 1952.

The Sefton Foundation-Stanford University Expedition to the Gulf of California in spring 1952 included an array of biologists collecting specimens (Lindsay 1955a). The expedition was out 61 days and visited 23 islands, and included Lindsay, Reid Moran, and Jon Lindbergh (son of Charles Lindbergh), then an undergraduate at Stanford University (photos from the expedition are at the San Diego Natural History Museum). Lindsay collected 152 numbers of cacti by the end of the expedition.

Tiburón: Between the end of April and beginning of May, Lindsay was at Ensenada del Perro and perhaps elsewhere on the island. On 30 April at Ensenada del Perro he made the type collection of *Ferocactus wislizeni* var. *tiburonensis*, *2229*.

San Pedro Nolasco: 2 and 3 May. Photos at the San Diego Natural History Museum show the island and many cacti, as well as Jon Lindbergh capturing *Ctenosaura* specimens. Lindsay collected cacti, *2225* to *2228*.

San Esteban: 6 May. Lindsay collected the two endemic cacti, *2232* to *2235*.

San Esteban: 13 January 1961.

Lindsay collected numbers *3002* to *3004*, including the type collection of *Mammillaria estebanensis*.

San Pedro Nolasco: 30 December 1961.

He made a trip to Nolasco with John and Priscilla Sloan and Chris Parrish (Lindsay 1962), collecting *Mammillaria tayloriorum*, *3228*.

San Esteban & Tiburón, 1962.

The Belvedere Scientific Fund provided the San Diego Natural History Museum with a grant for a six-week collecting trip to the islands of the Gulf of California (fig. 2.11). The museum charted the *San Agustín II* from Antero Díaz at Bahía de Los Angeles. The ship was an 85-foot diesel-powered vessel built for air-sea rescue.

San Esteban: 22 February, his final trip to the island, he collected *Echinocereus grandis*, *3252*.

Tiburón: 18 March, Agua Dulce Bay, *Mammillaria grahamii*, *3246*. 19 March, Willard Point, *Echinocereus scopulorum*, *3247* and [Willard Point?] *3249*.

Figure 2.11. Participants of the 1962 Belvedere Scientific Fund–San Diego Natural History Museum expedition. Back row, left to right: Richard Banks, Michael Soulé, unidentified person, Chris Parrish, Ira Wiggins, Chuck Shaw, and Reid Moran. Front row, left to right: Bill Emerson, Dennis Bostic, and Charles Harbison. GEL, March–April 1962, courtesy San Diego Natural History Museum.

Servando López-Monroy. San Esteban, 2008.

López-Monroy is a member of the Comcaac community at Punta Chueca. He is well versed in the local ecology and worked with Ben, Richard, and Humberto on Tiburón in 2007, 2008, and 2009. 3 September at Arroyo Limantour, *López-Monroy 08-01* to *08-04*.

Maximiliano "Max" Damián López-Romero. San Esteban, 2008.

López-Romero is a member of the Comcaac community at Punta Chueca. 3 September at Arroyo Limantour, *López-Romero 08-01* to *08-04*.

Emily Jane Lott & Thomas Harris Atkinson.
San Pedro Mártir, San Esteban, & Tiburón, 1985.

Lott is a botanist and her husband, Atkinson, an entomologist. At the time of their collections they were at the Instituto de Biología in Mexico City. On 22 July 2009, Emily wrote to Richard, "We were on an Instituto de Biología UNAM trip (on a Mexican Coast Guard ship) with a bunch of other biologists—herpetologists, ornithologists, and fish people. Several trips were made but Tom and I only went on one. The project was organized in conjunction with the establishment of seabird preserves. Enriqueta Velarde was the leader." It was early summer during drought conditions, and the specimens thus are not in excellent condition.

The collections are mainly at MEXU, and some duplicates are at UC in Berkeley and CAS in San Francisco. Collection information below is from Emily's field notes.

San Pedro Mártir: 4 May, *2426* to *2438*.

San Esteban: 5 May, "cañón en el SE de la isla," at least *Agave cerulata, 2464*.

Tiburón: 13 May, probably the south side of the island, *2513* to *2520*.

Charles Herbert Lowe Jr. (1920–2002).
Tiburón, San Esteban, & Dátil, 1958.
Alcatraz, 1969.

Herpetologist and desert ecologist at the University of Arizona, with a colorful reputation, Lowe published a few papers on the herpetofauna of the Midriff Islands (e.g.,

Lowe 1955) and made several trips to the region, some with Ken Norris from UCLA and Richard. However, he usually did not collect plants. On 10 November 1969, he made 13 plant collections on Alcatraz.

Thomas Dwight Mallery & William "Bill" V. Turnage. Tiburón, 1937.

At the time, Mallery and Turnage were both working with Forrest Shreve at the Desert Laboratory in Tucson (Bowers 1988). Their collections are primarily at ARIZ. Some duplicates and additional specimens are at CAS (DS), and may have been sent to Professor Wiggins at DS for identification or confirmation.

Tiburón: On 3 May they made a small collection of plants from "the N coast" as their labels read. They most likely went to Tecomate, where there was a vibrant Seri community. They also collected at Arroyo Desemboque and Sargento on the mainland on 2 and 4 May 1937. They left in late April and journeyed from Tucson to the island via Sasabe, Altar, Dátil, Libertad, and Sargento (Turnage & Hinckley 1938). We have located five of their Tiburón collections at ARIZ and UC in Berkeley, none of which have collection numbers.

Cathy Moser Marlett (fig. 2.12A,B).

As the daughter of Ed and Becky Moser, Marlett spent much of her childhood in El Desemboque, speaking Seri and Spanish in addition to her native English. Early in life Cathy learned from her Seri friends the invaluable arts of catching octopuses and making Seri baskets (three that she completed remain her treasures).

Cathy graduated from Wheaton College (Wheaton, IL) with a BA in art and biology, later doing graduate studies in linguistics at the University of North Dakota, where she met her husband, Steve. They were married in late December 1976 and have two sons. She contributed "A Desemboque Childhood" in a special issue of the Journal of the Southwest (Marlett 2000) and "A Good Day for Playing Hooky" (Marlett 2002) in Bowen's *Backcountry Pilot: Flying Adventures with Ike Russell*. Her illustrations have appeared in numerous publications, including the Seri ethnobotany (Felger & Moser 1985) and *Unknown Island* (Bowen 2000), and she contributed more than six hundred drawings to the trilingual Seri dictionary (Moser & Marlett 2005, 2010).

(A)

Figure 2.12. Cathy Moser Marlett. (A) A Seri child and Cathy Moser in the arms of her mother, Mary Beck "Becky" Moser, Desemboque. EWM, spring 1955, courtesy of Cathy Moser Marlett. (B) In front of her childhood home in Desemboque. BTW, 5 April 2008.

(B)

Cathy works with SIL International as a graphic artist and has illustrated scores of literacy books for Mexico, Pakistan, and elsewhere. She has completed an extensive study of Seri knowledge and use of mollusks, a work that combines her love of biology and scientific illustration with her interest in the Seri people. "I am never happier than when exploring isolated beaches in the Seri area, or visiting with Seri friends, enjoying their songs and oral histories" (personal communication to Richard, July 2009).

Stephen Alan Marlett.

Marlett is a linguistic consultant with SIL International and adjunct professor of linguistics at the University of North Dakota. Steve grew up in New York state and has lived for substantial periods of time in Argentina, Mexico, and Peru. It was while he was studying linguistics in North Dakota that he met Ed and Becky Moser, and their daughter Cathy. Ed first exposed Steve to the challenges of the Seri language during the summers when

Steve was completing an MA. Steve worked on the Seri language during his doctoral studies at the University of California, San Diego. He has published extensively on the language, helped complete the trilingual dictionary (Moser & Marlett 2005), and has worked on a Seri reference grammar as a National Endowment for the Humanities fellowship for documenting endangered languages. He has also helped to see that texts recorded and translated by Roberto Marcos Herrera and Ed Moser have been interlinearized, glossed, archived (along with recordings), and made accessible to the Seri people as well as the general public.

Jaime Armando Maya.
San Pedro Nolasco, 1963.

Maya collected plant specimens on 1 November above Cala Güina at the southeast side of the island during fieldwork for his PhD dissertation on fish-eating bats (Maya 1968).

William John "WJ" McGee (1853–1912).
Tiburón, 1895.

McGee published the first extensive ethnographic work on the Seris, *The Seri Indians* (1898), which has high-quality illustrations, historical photographs, and significant information if you read past McGee's "racist clap-trap and junk about unilineal cultural evolution, but his diary tended to include only what he observed" (Bernard L. Fontana, personal communication to Richard, 2000; Fontana & Fontana 2000).

The McGee expedition was on Tiburón between 19 and 28 December 1895 (fig. 2.13). He collected about 13 plant specimens on the island, which are the earliest known herbarium specimens from Tiburón. In his diary of 23 December, McGee wrote, "Some coffee was made at the Tinaja, and I collected some plants: a, The large white bark tree with large, fleshy palmate leaves, having a small fig-like fruit, the same seen as sending roots so far to water at Bacuachi [the fig, *Ficus petiolaris*]; b, paloblanco; c, carrizal [*Arundo donax*]; d, a large mimosa (manta), with light gray bark [*Lysiloma microphyllum*]" (Fontana & Fontana 2000:89). McGee named this waterhole "Tinaja Anita," which we now know is Pazj Hax (see part IV, "Gazetteer"). The plant names in brackets were added by Bernard Fontana. The plants are (a) *Ficus*

Figure 2.13. WJ McGee with six Seri boys, Encinas Ranch, Sonora, 1894. William Dinwiddie, printed with permission from the National Anthropological Archives, Smithsonian Institution (BAE GN 4276 C).

petiolaris subsp. *palmeri* (*McGee 11*, UC), (b) *Acacia willardiana* (*McGee 13*, UC), (c) *Phragmites australis*, (d) *Lysiloma divaricatum*—"manta" is probably a misinterpretation of *mauto*, the usual name for this tree in Sonora. On 26 December he also collected *Frankenia palmeri* and *Maytenus phyllanthoides* along the coast south of Punta San Miguel (Fontana & Fontana 2000). The first set of his collections is at US.

Raúl Eduardo Molina-Martínez.
Tiburón & Cholludo, 2009.

A member of the Comcaac community at Punta Chueca, Molina-Martínez accompanied Humberto and Ben on a trip to the islands.

Tiburón: 28 February. Raúl discovered and collected *Tiquilia canescens* at Ensenada de la Cruz, previously only known in the region from Isla Dátil (fig. 2.14). Ben

Figure 2.14. Raúl Eduardo Molina-Martínez at Ensenada de la Cruz with *Tiquilia canescens*, a new plant record for Tiburón. BTW, 28 February 2009.

and Richard had searched fairly hard for this plant on the south part of Tiburón and were prepared to treat it as a doubtful and excluded species on the island. Raúl found a large population on a hill behind the abandoned marine station building, a population overlooked by Richard, Reid Moran, and Ben. Cataloged as *Wilder 09-50*.

Cholludo: 1 March. Raúl collected *Chenopodium murale*, a new record for the island. Cataloged as *Wilder 09-69*.

Nestor Cristóbal Montaño-Herrera.
San Esteban, 2008.

Montaño-Herrera is a member of the Comcaac community at Punta Chueca. 3 September at Arroyo Limantour, *Montaño-Herrera 08-01* to *08-03*.

Reid Venable Moran (1916–2010).
Tiburón, 1952, 1962, 1966. San Esteban, 1952,
1962, 1975, 1980. Dátil, 1966. Patos, 1966.
San Pedro Mártir, 1952, 1962, 1975.
San Pedro Nolasco, 1952 (see figs. 2.10 and 2.15).

Moran was born and raised in northern California. He is the single most intrepid botanical collector ever to work on the Gulf Islands and the Baja California Peninsula. During the early 1950s he was at the L. H. Bailey Hortorium, Cornell University. His extensive collections

are primarily at the San Diego Natural History Museum, where for 25 years (1957–1982) he was curator of the herbarium. Numerous duplicates of his specimens are at other herbaria, including ARIZ. He published extensively on the flora of the region, describing numerous species new to science. On many expeditions to the islands, Reid would climb to the highest peaks to search for interesting specimens. He kept detailed notes of his field travels and botanical collections in clear, legible handwriting. His notebooks are at the San Diego Natural History Museum (Moran 1936–1993).

Moran specialized in the systematics of the Crassulaceae (Stonecrop family) and the flora of the Baja California Peninsula. He wrote *The Flora of Guadalupe Island, Mexico* (Moran 1996) and the treatment of the Crassulaceae for the *Flora of North America* (Moran 2009), and also co-authored *The Grasses of Baja California, Mexico* with Frank W. Gould (Gould & Moran 1981) and *The Vascular Flora of Isla Socorro, Mexico* with Geoffrey A. Levin, his successor as curator (Levin & Moran 1989). In addition to Reid's botanical achievements, his practical jokes and wry sense of humor are legendary. He once mailed a fresh egg complete with his return address, carefully lowering it on a string into a mailbox. The U.S. Postal Service was not amused and took him to court but was only able to convict him of improper packaging.

Reid made numerous trips to the Baja California Peninsula and the islands in the Gulf and on the Pacific side

Figure 2.15. Reid Moran pressing plants on Isla Ángel de la Guarda, 16 March 1962. Courtesy San Diego Natural History Museum.

of the Peninsula. In the Midriff Region he especially focused on Isla Ángel de la Guarda, where he made multiple trips, covering much of the island, including its high peaks. This work culminated in the flora of the island (Moran 1983a). He also made a number of trips to the Sonoran Midriff Islands, principally on scientific Gulf Island cruises. Reid made collections of the utmost quality. His keen sense of the flora is evident in his collection lists, which highlight many of the unique, disjunct, and otherwise fascinating components of the flora.

Reid was a member of five Gulf of California expeditions that included stops on the Sonoran Islands. A general description is presented here.

San Pedro Nolasco, San Pedro Mártir, Tiburón, & San Esteban, 2–6 May 1952.

Reid's first collection from the Sonoran Midriff Islands was as a member of the Sefton-Stanford Gulf of California Expedition aboard the *Orca* between 25 March and 19 May. The expedition left from San Diego, sailed to the tip of the Baja California Peninsula, and made its way up the Gulf, finishing at Cabo San Lucas. George Lindsay was in charge of scientific work, and the rest of

the scientists consisted of a professor and six students from Stanford: "Dr. Wm. C. Steere, bryologist; Jim Boelke and Dan Cohen, fish; Jay Savage and Frank Cliff, reptiles; John Figg-Hoblyn, invertebrates including insects; and Jon Lindberg, the only undergraduate of undecided specialty" (Moran field book 3, p. 52).

San Pedro Nolasco: 2 May. Reid collected near the middle of the east side of the island, which is Cañón el Farito, *4040* to *4052*. In his field notes he wrote, "The flora is rather poor but is in good shape at present, the cacti flowering" (Moran field notebook 3, p. 111). On 2 May he spent the afternoon on the island and all day on 3 May.

San Pedro Mártir: 4 May, *4053* to *4057*. "Sailed at 0500 [from Isla San Pedro Nolasco] for San Pedro Martir Island. We got ashore at noon and had to be back by 1400. This is a thousand-foot dome of rock with sheer cliffs on most sides, well whitewashed below. The top and the more gently sloping east side are forested with cardon. We landed on a broad sheltered ledge on the south end and found a trail thence to a deserted village of stone houses or half houses. It was very hot. I climbed to the top and collected a few plants, leaving half an hour to get back down. On the way down, I heard a rattlesnake buzz, and managed to capture it with a couple of cardon sticks. I got down almost in time and apologized for being late, saying that since there were no snakes on the island I knew nobody would believe me unless I brought this one. The herpetologists had come back early, saying that they had gotten all the reptiles of the island" (Moran field notebook 3, pp. 111–112).

Tiburón: 5 May, Perro Bay [Ensenada del Perro], *4058* to *4074*. "Spent the day in the immediate vicinity of the anchorage. Collected a few plants and took pictures. Although Tiburon is scarcely a garden spot, there were many things in flower, and it was probably just about at its best" (Moran field notebook 3, p. 113).

San Esteban: 6 May, [Arroyo Limantour], *4075* to *4080*. "I walked up a canyon and made a few collections" (Moran field notebook 3, p. 115).

Tiburón, San Pedro Mártir, & San Esteban, 18–23 March 1962.

The Belvedere Scientific Fund provided the San Diego Natural History Museum with a grant for a six-week collecting trip to the islands of the Gulf of California from 14 March to 28 April. The scientific personnel were George Lindsay, director; Richard Banks, birds and mammals; William K. Emerson, mollusks and other invertebrates, living and fossil; Charles F. Harbison, insects; Charles Shaw and Michael Soulé, reptiles; and Reid Moran and Ira Wiggins, plants. In addition, Ambrosio Gonzales C. of Recursos Naturales, interested in *Opuntia*, was with the boat for the first three weeks (see fig. 2.11; Reid Moran field book 7, p. 1).

Tiburón: 18 March, Bahía Agua Dulce, *8694* to *8716*. On 19 March, Moran wrote "The plant press was left covered by a tarp to protect it from dew. When the generator was started with the press still covered, it heated up and caught fire. There was a great commotion, and the plant press ended up in the Gulf. Only a few of yesterday's collections were rescued" (Moran field notebook 7, p. 16).

19 March, north of Punta Willard, *8717* to *8751*. "We sailed south to the canyon mouth just north of Willard Point, where we landed at 1130. I collected mostly near the shore, then started inland when a rainstorm came up at 1500, I returned to the ship" (Moran field notebook 7, p. 16).

20 March, El Sauz, *8752* to *8800*.

San Pedro Mártir: 21 March, north slope and summit, *8801* to *8820*. "With an early start we landed at 07:45 in a cove on the north side of San Pedro Martir Island. We climbed the trail remaining from guano harvesting days, up the steep lower slope, into the canyon above. There were many nesting pelicans and boobys. I found three rattlesnakes. . . . I went to the top of this rather dome-shaped island, where middle slopes cut off any view of the shore on most sides. We were to be back at noon, but I got in the wrong canyon and was well down before I was sure of my error. . . . I was down by 12:10 with rest of the stragglers." (Moran field notebook 7, pp. 24–25).

San Esteban: 21 March, [Arroyo Limantour], canyon mouth, southeast corner of island, *8821* to *8827*.

22 March, [Arroyo Limantour], south fork canyon, near southeast corner, *8828* to *8836*. High peak of southeast corner of the island, *8837* to *8844*. North slope of high peak, near southeast corner of island, *8845* to *8860*. [Arroyo Limantour], near southeast corner of island, by arroyo, 200 m elevation, *Janusia gracilis* [*Cotsia gracilis*], *8861*. Near southeast corner of island, in dry arroyo bed, 100 m elevation, *Lupinus arizonicus*, *8862*.

Tiburón, Patos, Dátil, San Pedro Mártir, & San Esteban, 23–26 April 1966.

"George Lindsay promoted a trip to the northern islands of the Gulf of California, sponsored jointly by the Museum [San Diego Natural History Museum] and the California Academy. . . . Our trip was too late for the best collecting at lower elevations, but it was better for plants of high north slopes. My best collecting was on a high peak near the north end of Angel and on a peak near the northeast corner of San Esteban. . . . I carried a list for each island to avoid collecting the common plants at the same localities as before." The trip was taken on the *San Agustin II*, 19–29 April, departing from Bahía de los Angeles and consisted of "Dick Phillips, Dick Banks, John Sloan, and I from the Museum; Bob Orr, Dustin Chivars, Ray Bandar, and Tom Titon (trustee) from the Academy; and Dr. Alejandro Villalobos and Virgilio Arenas from the Instituto de Biologia, UNAM." (Moran field book 9, p. 102).

Tiburón: 23 April, Agua Dulce Bay, *12985* to *13003*. Tecomate and the north slope of a small volcanic mountain west of Tecomate to 200 m elevation. "Although I collected here in 1962, most of the specimens were lost in the fire. This time the vegetation was dry and in poor condition on the flat, but I found a north slope where several plants were still in fair shape. Although I don't usually risk impalement on cacti by chasing lizards, I couldn't resist a blue-green collared lizard [*Crotaphytus dickersonae*] and caught it twice" (Moran field notebook 9, p. 113).

Patos: 23 April. *Atriplex barclayana*, *13004*.

Tiburón: 24 April, Ensenada de la Cruz, *13005* to *13018*.

Dátil (Turners): 25 April, *13019* to *13037*. "I was ashore on Turners Island from 0800 to 1100. Having no list for this island, I tried to collect everything that was in any condition for collecting and so far as possible to list everything else" (Moran field notebook 9, p. 116).

San Pedro Mártir: 25 April. The party stopped on Mártir but Reid collected no plants. "I climbed to the summit (305 m = 1000 ft). Finding no plants that I hadn't already collected, I took a half holiday, enjoying the view and photographing and making notes concerning the cardons so abundant there. Mature buds of the cardon, placed in a paper sack, opened well before those left on the plants, which waited for nightfall" (Moran field notebook 9, p. 116).

San Esteban: 26 April, [Arroyo Limantour], main arroyo, 50 m elevation, *13039* & *13040*. Northeast peak, 350–475 m elevation, *13040* to *13053*. [Arroyo Limantour] main arroyo, 100 m elevation, *Cercidium microphyllum* [*Parkinsonia microphylla*], *13054*. 27 April, Arroyo at southwest corner of island, 25 m elevation, *13056* & *13057*.

San Pedro Mártir & San Esteban, 16 and 17 April 1975.

Reid served as a leader for a Nature Expeditions International cruise in the Gulf in mid-April. The vegetation was extremely dry, so as Reid put it, "I had a vacation" (Moran field notebook 14, p. 77).

San Pedro Mártir: 16 April, north slope, *21745* to *21747*. On this trip Reid and others shipped out of Bahía de Los Angeles on April 13 aboard Antero Díaz's boat. Reid wrote, "We circumnavigated ISPM in a panga from 9:30 to 11:30. Our 1952 landing place turns out to be on the NW side. [Undoubtedly this is an error on Reid's part and their 1952 landing spot is indeed the SE side.] At 13:50 I went ashore at the usual landing cove on the north side and climbed up to the top." (Moran field notebook 14, p. 78).

San Esteban: 17 April.

[Arroyo Limantour] "Ashore at the east side, . . . went up the arroyo looking for agaves" (Moran field notebook 14, p. 78). *Agave cerulata* subsp. *dentiens*, *21748a & b*.

San Esteban, 29–31 March 1980.

"Having been hoping to visit IAG [Isla Ángel de la Guarda] this spring, I accepted an invitation from Charlie Sylber to go to San Esteban, San Lorenzo, and IAG. Because of weather and other problems we got no farther than San Esteban, and much time was wasted" (Moran field book 17, p. 108).

29 March, [Arroyo Limantour] near mouth of main arroyo, southeast side of the island, 10 m, *28164* to *28171*.

30 March, [Arroyo Limantour] near mouth of main arroyo, southeast side of the island, 25–150 m elevation, *28172* to *28173*.

31 March, southwest corner of the island, 10 m elevation, *28175* to *28177*.

José Juan Moreno (1925–2007). Tiburón, 1964.

Moreno was born on a ranch near Bahía Kino and was a member of the Comcaac community at El Desemboque. José Juan brought Mary Beck Moser a leaf from a palm tree from a canyon on the northwest coast of Tiburón, confirming its presence on the island: *Brahea armata*, *Moreno 20 May 1964*.

Edward "Ed" W. Moser (1924–1976). Tiburón, 1964 (figs. 2.3, 2.16).

Edward Moser was born in Joliet, Illinois. He finished his studies at Wheaton College in 1948 after serving in the U.S. Navy during World War II. He married Mary Beck in 1946 and studied descriptive linguistics with her at the Summer Institute of Linguistics program at the University of Oklahoma. In 1951 they went to live in the Seri region to learn the language. During the twenty-five years after that date, he completed his master's degree in linguistics at the University of Pennsylvania and used his linguistic abilities to learn the Seri language and collect cultural information. He spent many summers at the University of North Dakota giving linguistics courses. Ed published a number of papers of Seri culture, often co-authored with specialists. Together with his wife and various Seri co-workers (especially Roberto Herrera Marcos) and under the auspices of the SIL International (formerly Summer Institute of Linguistics) and the Secretaría de Educación Pública, he did the first linguistic analysis of the Seri language and developed an orthography to represent it adequately. His interest in the well being of the Seri people and in their rich culture and history was unflagging until his sudden and unexpected death in 1976. (Summer Institute of Linguistics in Mexico 2008)

1 May, Sauzal, collection by Ed Moser and Becky Moser, including *Phragmites australis*. Ike Russell took Ed and Becky to the Sauzal landing strip so they could see the famous Xapij waterhole, about an hour's walk up Arroyo Sauzal. They left El Desemboque in the morning and returned later the same day. The specimens were pressed and cataloged that evening in Desemboque by Richard as *Felger 9986* to *10008*.

Figure 2.16. Edward Moser collecting Seri language data from his consultant Pancho Contreras, Bahía Kino. MBM, 1952, courtesy of Cathy Moser Marlett.

Mary Beck "Becky" Moser.
(see figs. 2.3, 2.12A).

Moser and her husband, Ed Moser, moved to the Seri region in 1951, first at Kino and then making their home at El Desemboque. Becky earned a degree in social sciences and linguistics from the University of North Dakota in 1982. Her long and illustrious career includes a number of publications on Comcaac culture. Among her outstanding works are her collaborations on the Seri ethnobotany (Felger & Moser 1985) and the Seri dictionary (Moser & Marlett 2005). The Seris called her "Singing Woman" or "Blue-Skirted Woman," because she was often singing and wearing a blue denim skirt, or she was known as "Rebeca."

Richard wrote,

> It is important to keep in mind what it was like on the Seri coast only a few decades ago. There was no electricity, no running water, and the outside world was far away. (Later in the story there was running water of sorts in Desemboque. But when Ed Moser was not around, someone had to remember or bother to put diesel in the pump.) The road to Desemboque was rough and slow.
>
> From Tucson I communicated with Ed and Becky by sending letters in care of a friend of theirs in Hermosillo who received their mail, and every few weeks they made the eight-hour drive from Desemboque to Hermosillo. Try as they would to explain that they were in Desemboque to work on Seri linguistics—to make it a written language, teach reading and writing in Seri and Spanish, and translate the New Testament into Seri—there was the belief by some of the Seris that the real reason was that they were working on a gold mine or getting pearls. When they made the infrequent several-day trip to Hermosillo to purchase provisions, it was rumored that they were really checking on the their gold mine or collecting pearls. (Felger 2000b:524)

Becky prepared meals for some of the old people. Once, by the back door, old Antonio Burgos said he had not eaten in three days. He had just finished his breakfast of eggs, toast and coffee at the front door. But if confronted he would just laugh. Becky once told me that if she had enough money she would open up a seniors' home in Desemboque.

Old, infirm and blind Catalina lived in a little hovel in front of the Mosers' house. Becky fixed meals for her, cleaned her up, brought and emptied her plastic toilet bags. The poverty and troubles were sometimes heart wrenching, which stimulated tourists to bring piles of used clothes. Much of the time the Seris knew these were

just old clothes and tossed them out. There was a lot of dignity and laughing and fun. (Felger 2000b:531)

Becky collected a few plants on Tiburón and many more on the Sonoran mainland.

Gary Paul Nabhan.

Nabhan is a well-known writer, food and farming advocate, and conservationist whose work is rooted in the U.S./Mexico borderlands region he affectionately calls "the stinkin' hot desert." His connection and involvement in the Gulf of California, especially the Seri region, are long-standing, dating from his participation in the inaugural class of Prescott College in the Bahía Kino region in 1969 while he was an undergraduate at Prescott College. His interest in the knowledge of the natural world held by the Seris and how they have interacted with the desert landscape of the Gulf has led to multiple publications (e.g., Nabhan 2000, 2003) and close interaction with the Seri community. Gary initiated his work in desert ecology with Richard.

Bibiano Fernández Osorio-Tafall.
(1903–1990). Patos, 1946.

In the spring of 1944, Osorio-Tafall, a biological oceanographer of the Escuela Nacional de Ciencias Biológias in Mexico City, was sent by the Mexican government to investigate the prospects for guano mining on Isla Patos. The expedition visited many islands in the Gulf. He also was on Patos in 1946 and recorded the vegetation in several photographs prior to the clearing of the vegetation for the short-lived guano-mining venture (Bowen 2000; Osorio-Tafall 1944, 1946). Osorio-Tafall was a Spaniard who left Spain for Mexico during the Spanish Civil War. He had an illustrious career, producing numerous publications, and became active in international affairs with the United Nations (Giral 1994:162–164).

Edward Palmer (1831–1911).
San Pedro Mártir, 1887, 1890.

During summer and fall of 1887 and late winter of 1890, Palmer, an intrepid plant collector, focused on the poorly known flora of the Gulf of California region at Guaymas,

Mulegé, Bahía de Los Ángeles, and Isla San Pedro Mártir (Beaty 1964; McVaugh 1956; Watson 1889). Of the 415 native species collected, 89 were deemed new to science in Sereno Watson's (1889) report on the collection. Palmer collected multiple sets of specimens, which are deposited at various herbaria around the world. Many of his collections from the Gulf region can be found at the University of California Herbarium in Berkeley.

24 October to 5 November 1887. Palmer spent 8 days on San Pedro Mártir and collected 18 species. He called the island San Pedro Martin. (Watson 1889; Wilder & Felger 2010).

13 February 1890. Palmer spent one day on the island and collected three species of Asteraceae (Vasey & Rose 1890; Wilder & Felger 2010).

Sidney H. Parsons. Tiburón, 1932.

Parsons published articles on cactus and made a number of field trips to the Seri region, as early as 1910, discovering and naming several species including *Opuntia marenae* (*Grusonia marenae*). He collected a few *Nicotiana obtusifolia* specimens on Tiburón on 31 May and 6 June 1932, *Parsons 32-37 & 32-38*. The specimens are at UC in Berkeley and were probably collected for Dr. Thomas Harper Goodspeed, renowned expert on tobacco at the University of California. Parsons published articles for travel magazines, including one on the Seris (Parsons 1937).

Robert "Bob" H. Perrill. Tiburón, 1982.

29 May. Punta Willard, *5111* to *5116*. At that time he worked at the Arizona-Sonora Desert Museum and later as a technician at the Medical School of the University of Arizona.

H. Ronald Pulliam & Michael Rosenzweig. San Pedro Nolasco, 1974.

20–22 March. Pulliam and Rosenzweig led a University of Arizona advanced field class to the island to see how different an island might be from a nearby mainland (Michael Rosenzweig, personnel communication, 2008). Ron and Mike made a collection of 18 plant specimens of 17 species above Cala Güina to the crest and west side of the

island, including some of the rare species on the island. The primary set is at ARIZ. Ron was then on the faculty of the University of Arizona and later became Regents Professor at the Odum School of Ecology, University of Georgia. Mike is a quantitative theoretical ecologist and professor in the department of Ecology and Evolutionary Biology, University of Arizona. He is director of the University of Arizona Desert Laboratory on Tumamoc Hill, where he runs the Alliance for Reconciliation Ecology. Mike told us that Nolasco is one of the few places he has experienced "true wilderness."

Adrián Quijada-Mascareñas. Tiburón, 1990, 1991.

Quijada-Mascareñas is a herpetologist, originally from Hermosillo and later at the University of Arizona, with a strong interest in the biogeography and phylogenetic history of species in the Sonoran Desert region. He made two trips to the Sierra Kunkaak, both to the rugged peak Capxölim, labeled Cerro San Miguel by Adrian. The specimens are at ARIZ.

9 November 1990, *90T001* to *90T013*, and 8–10 March 1991, *91T001* to *91T024*. Adrian found several species not previously recorded for Tiburón, such as *Ambrosia carduacea*, *Cnidoscolus palmeri*, and *Sideroxylon leucophyllum* (Wilder et al. 2007b).

Peter J. Rempel. Tiburón, San Esteban, San Pedro Nolasco, & Patos, 1937.

Rempel was a member of the Allan Hancock Pacific Expedition to the Gulf of California in 1937 (Gentry 1949:99) and was at the University of Southern California associated with the Allan Hancock Foundation. The majority of his collections are cacti and are at Rancho Santa Ana Botanic Garden (RSA).

Tiburón: 10 March, "Agua Verde Bay," probably Bahía Agua Dulce on the north side of the island, at least number *118* (*Lycium brevipes*).

San Esteban: 27 March, "south end of island," maybe Arroyo Limantour area, *291* to *294*.

Tiburón: 27 March, Arroyo Sauzal, *296* to *298*.

San Pedro Nolasco: 29 March, he collected 10 numbers (Gentry 1949:99). We have seen *301* to *307*.

Patos: Gentry (1949:99) reports that Rempel made a single collection on the island.

Humberto Romero-Morales.
Tiburón, 2006-2009 (fig. 2.17).

Romero-Morales has lived his whole life in the Seri region: first in Desemboque, where he was born, and later he made his home in Punta Chueca. Humberto is a descendant of the Comcaac who lived on Isla Tiburón. From an early age he learned from his mother the cultural significance and knowledge held in the Comcaac community of the desert plants of the Sonoran Gulf Coast region. He is one of the most knowledgeable members of the community concerning Isla Tiburón and the distribution and uses of plants. He has worked with a number of scientists in the past several decades and has gained considerable knowledge from such opportunities. He is well versed in Comcaac plant knowledge and scientific information for the flora. Since about 1998 he has been the head guide for the bighorn sheep hunting operation on Tiburón and is active in community programs, including head of security for Punta Chueca, one of the two Comcaac villages.

His trips to Tiburón are numerous, and he has explored the majority of the island. Since the beginning of this project he has made more than 50 collections of plants, some of which represent the first records for the island. The collections, processed and labeled by Ben and Richard, are at ARIZ and USON.

2006: Collections from the Arroyo Sauzal area, *06-01* to *06-03*.

Figure 2.17. Humberto Romero-Morales, northwest coast, Isla Tiburón. RSF, 29 September 2007.

2007: Twenty-seven collections from across the island, primarily on the east side of the island around the Sierra Kunkaak, *07-01* to *07-27*.

2008: The southern part of the Central Valley in the vicinity of Satóocj, *08-01* to *08-04*. In and around the Sierra Kunkaak, *08-05* to *08-09*. Caracol, *08-14* and *08-15*.

2009: Collections in early spring from the southern part of the Central Valley and the northwestern foothills of the Sierra Kunkaak, *09-01* to *09-03*.

Joseph Nelson Rose (1862–1928).
Tiburón, Cholludo, & San Esteban, 1911.

Rose was the botanist on the cruise of the *Albatross* to Baja California and the Gulf, from San Diego, California, from 28 February to 23 April 1911 (Townsend 1916). Rose, with the assistance of crew members, collected some 1800 plant specimens, including more than 1000 living cacti. "The results of the studies of the cacti were incorporated in the *Cactaceae*, a four volume monograph of the family which was then in preparation" (Lindsay 1955a:43). These volumes were the first unifying work on the cactus family and the first modern classification of the family with descriptions for all known species. This massive work remains the foundation of cactus literature (Britton & Rose 1919–1923). Rose's collections and field books from the 1911 expedition are at the U.S. National Herbarium (US).

Tiburón: 11 April, southeastern part of the island, *16741* to *16810*.

Cholludo: 13 April, *16811* to *16818*, including the type collection of *Agave subsimplex*. Rose's field book and labels read "Seal Island." Mexican fishermen called the island "Lobos," their name for sea lions.

San Esteban: 13 & 14 April, *16819* to *16829*, including the type collection of *Echinocereus grandis*. They were in the Arroyo Limantour area.

Alexander "Ike" Russell
(1916–1980). (fig. 2.18).

Ike led a heroic life in spite of terrific health problems (Bowen 2002, 2008). He was a great friend and mentor to Richard—his first-hand knowledge of the Sonoran region was vast. On numerous occasions during the 1960s and

Figure 2.18. Alexander "Ike" Russell and the Cessna 185, at his landing place in the Central Valley, Tiburón. RSF, 8 September 1974.

1970s Richard went with Ike to the Sonoran Islands and many other places in northwestern Mexico. They had wonderful adventures, and Richard would not have accomplished his work without the generous assistance and friendship of Ike and Jean Russell (Felger 2000b, 2002). The trips from Tucson often included his wife, Jean, and one of their sons, Bob or David, as well as John (dog) and then Wennie or another Great Pyrenees. Ike was responsible for Richard meeting Ed and Becky Moser, which led to their work on the Seri ethnobotany and his work with sea turtles (e.g., Cliffton et al. 1982; Felger et al. 1976, 2005; Felger & Moser 1985).

Richard writes: My first plane trips with Ike were in his Aeronca Sedan. It had a 145 hp engine and a wide wingspan, which allowed for a slow landing, but it was not especially fast. In 1962 he purchased a Cessna 180 and in November 1967 he bought a faster plane: "That airplane was a Cessna 185, with a monster 300 horsepower Continental engine that delivered a frightful amount of power. It was one of the features Ike loved about it" (David Russell 2002:163). Ike would fly low enough so that I could identify plants, and there were innumerable places in Sonora where we would land. Ike would land at six places on Tiburón. The landing places at Tecomate and Palo Fierro (Punta Tormenta) were fine, long dirt airstrips. The one at Vaporeta (Ensenada Blanca), apparently only used by Ike, was not excellent. The one near the shore on a little mesa next to Arroyo Sauzal became riddled with rodent burrows, and I went there only a few times with Ike. We made the one at the upper (south) end of the Central Valley, near the middle of the island, by landing in a place where there was a relatively sparse cover of creosotebushes. Several times we landed on the beach at Punta Perla at the northeast corner of the island. Ike also flew to Isla Ángel de la Guarda on several occasions, the last resulting in a serious crash and several days stranded on the island (Bowen 2008).

We also went to the islands in Ike's pangas, or open skiffs. The first was a scarcely seaworthy Columbia River dory that Bob Russell and his brother Luke towed to Tucson from Oregon behind a Model A Ford in 1959. It was about 21 feet long and made of plywood and had a Chrysler six-cylinder flathead engine. This was the boat Bob took out to the islands with Armando Maya in the summer of 1962. The boat was wrecked in a *chubasco* (a violent summer storm) on the south shore of Tiburón. After that Ike had the *Ofelia*, named in honor of a friend in Guaymas and built for him by Leonardo Ruiz, a boat builder in Guaymas. It was a wooden inboard about 25

feet long based on a Mexican fishing panga. Ike made many trips in his plane from Tucson to Guaymas to work on the *Ofelia*. During some of these flights I would hitch a ride with Ike, and we often stopped to camp on Tiburón, or just about anywhere in the desert in Sonora. The *Ofelia* was slow but sure, and noisy, hot, and stinky with a 1927 Dodge engine—that engine was pretty primitive, with no valve lubrication, but it was economical on gas and just the sort of thing Ike liked to tinker with. "Before I drove the *Ofelia* around the Gulf with Nancy Thomas and friend in early '69, I installed the Buda diesel lifeboat engine, which gave it considerably better range and safe operation. I don't know if Ike ever sailed it after I put the new engine in—I think it was too much for him: he appreciated the lower-tech Dodge engine and the lack of a transmission" (Robert Russell, personal communication to Richard, October 2009). Ike passed away on 20 July 1980 after a long struggle with cancer.

Ike, Jean, and Robert (Bob) Russell each collected various plant specimens from the islands, most of which were prepared by Richard and cataloged with his numbers. For example, someone had landed a plane at an earlier date on a nearly barren area on Isla Patos, and after reconnoitering the site, Ike landed in the same place on 29 March 1963. He brought back two plants, which were processed and cataloged as *Felger 7012* (*Cylindropuntia fulgida*) and *7013* (*Atriplex barclayana*).

Norman J. Scott Jr. Tiburón, 1977, 1978.

A biologist at the University of New Mexico in the 1970s, Scott was the leader of a U.S. Fish and Wildlife team in conjunction with the Mexican Fauna Silvestre in the late 1970s, conducting ecological research focusing on vertebrates on Tiburón. Jean Duke was the primary botanist for the project, but she was not on the trip when Norman ascended the Sierra Kunkaak. In addition to collections, they collected data from vegetation phenology plots on the eastern bajada of the island. The primary set of Scott's collections from Tiburón are at UNM. Collections from the phenology plots are labeled "P" plus the collection number.

7 January 1977, ca. 10 km west of Punta Tormenta, 2 km E Caracol, *Scott P1 & P2.* 9 January, phenology plots ca. 3 km west of Punta Tormenta, *P5–P13*; 10 January, *P14–P24.*

11 and 12 April 1978. Norman Scott ascended the Sierra Kunkaak with Luis Roberto Muñoz and Tom Fritts.

They collected multiple specimens, *Scott 1–17* (these numbers are not reflected on the herbarium specimens but are present in the field notes, a copy of which was graciously provided to us by Norman Scott). "Our base camp, Caracól, was 12 km W of Punta Tormenta. We walked from Caracól about 6 km S to the base of Cerro Kunkaak and camped for the night. Starting at 0700, 11 April 1978, we reached the top of the northernmost peak at 1000. We came down the large, east-facing canyon that starts among the peaks. . . . I believe that nos. 1 & 2 were collected on the way up, 3–7 on top, and 8–17 on the way down" (Norman Scott, personal communication to Ben, 13 August 2009).

Wade Cutting Sherbrooke.
San Pedro Nolasco, 1964.

Sherbrooke was a graduate student at the University of Arizona when he visited the island. He was later the director of the Southwestern Research Station of the American Museum of Natural History at Portal, Arizona. His research centers on horned lizards, *Phrynosoma*.

Wade collected plant specimens above Cala Güina at the southeast side of the island. The specimens were given to Richard to process and catalog: 29 March 1964, *Felger 9852 & 9853*; 31 October 1964, *Felger 11232* to *11247*.

Richard A. "Rick" Smartt. Tiburón, 1976.

A wildlife biologist, Smartt received his PhD in zoology and botany from the University of New Mexico, Albuquerque. He was later a zoologist at the University of Texas at El Paso and is executive director of the Wildlife Experience near Denver, Colorado.

13 October, collections in the vicinity of Punta Tormenta (without collection numbers). The specimens are at UNM.

Marjorie L. Stinson. San Pedro Nolasco, 1974.

9 April. Stinson collected several valuable specimens on the island with Michael David Robinson. The plant specimens are at SD. Mike, a fellow graduate student at the University of Arizona with Richard, wrote his PhD dissertation on chuckwalla lizards (*Sauromalus*) of the Gulf Islands. "As I recall we arrived and left the island using a Mexican fisherman's boat. What I most remember from the island was watching a group of whales glide by snorting, breaching, and diving. First time I had seen them in the Gulf" (Michael Robinson, personal communication to Richard, May 2008).

Pedro Tenorio-Lezama.
San Pedro Mártir, 1985. Tiburón, 1985, 1986.

Tenorio-Lezama has been one of the most prolific and well-respected botanical collectors in Mexico, providing thousands of specimens for MEXU, with duplicates to many other herbaria, and works on invasive plants (weeds) of Mexico. His trips to the Gulf Islands were as a member of the UNAM-led expeditions.

San Pedro Mártir: 9 August 1985, with Fernando Chiang-Cabrera, Emily Lott, Alfonso Valiente, et al. They made a small collection on the island, and we have seen two of their collections (*Tenorio 9491* and *Valiente 604*).

Tiburón: August 1985, with Fernando Chiang-Cabrera, Emily Lott, Alfonso Valiente, et al. 10 August, Ensenada del Perro, *9495* to *9530*. 10 and 11 August, Bahía Agua Dulce, *9531* to *9541*.

Tiburón: 9 February 1986, Bahía Agua Dulce with T. P. Ramamoorthy; we have seen *Errazurizia megacarpa*, *Tenorio 10862*. On the same day he also collected at Ensenada del Perro with C. Romero de T.; we have seen *Acacia willardiana*, *Tenorio 10894*.

Jorge Torre-Cosío. Tiburón, 1999.

Torre-Cosío focused on *Zostera marina* for his dissertation (Torre-Cosío 2002) and provided us with information on the seagrasses *Halodule wrightii* and *Zostera marina*, including locality data for 17 June 1999, Bahía San Miguel. In 2009 he became executive director of Comunidad y Biodiversidad in Guaymas.

Raymond Marriner Turner.
San Pedro Nolasco, 1979. San Esteban, 1983
(see fig. 2.9).

Turner taught at the University of Arizona (1954–1962) before joining the U.S. Geological Survey and moving to the Desert Laboratory on Tumamoc Hill until his "retirement" in 1989. Ray's interest in desert vegetation dynamics has resulted in long-term permanent vegetation studies, and he is author or co-author of many peer-reviewed publications and books on vegetation changes and Sonoran Desert plants. His specimens are at ARIZ, with duplicates distributed to other regional herbaria, including SD. His work is careful, reasoned, and well documented, allowing for the next generation of desert ecologists to continue the long-term investigations he either began or continued from the generation before him. An indelible image of Ray is him in the field, behind his camera matching a historical photograph, lining up the shot, taking a Polaroid to make sure it is in line, assessing the changes in the scene, and whistling all the time, fully engrossed in the moment.

San Pedro Nolasco: 28–30 September 1979, vicinity of Cañón el Farito, *79-248* to *79-255*; collected and photographed, highlighting the rich grass flora; with Matt Gilligan, Richard Inouye, Rodrigo Medellín (then a graduate student at UNAM), Jim Munger, Bernardo Villa (Instituto de Biología, UNAM), and Oscar G. Ward.

Ray recorded the following observations in his field notebook: "San Pedro Nolasco seems remarkable in two respects: (1) the dense coverage of grass (mainly *Setaria palmeri* [*S. macrostachya*]) and annuals (*Vaseyanthus* and *Perityle*, are the main ones and mostly dead now) and (2) the absence of all woody legumes, save *Acacia willardiana*, although at the same latitude on the mainland they are abundant. These two features may be related to absence of many herbivores. Also, *Opuntia* (*bravoana*?) has weak spine development. The dominant shrub species is probably *Viguiera deltoidea* [*Bahiopsis triangularis*] with *Simmondsia* and *Jatropha cuneata* subdominants."

San Esteban. 20 April 1983, Arroyo Limantour. As part of an expedition to several Midriff Islands on the ship *Don José*, Ray collected numbers *83-25* to *83-29* and noted the presence of 41 species (Turner field notes) for the database he was constructing for the plant atlas project (Turner et al. 1995).

Alfonso Valiente-Banuet.
Tiburón, San Esteban, 1985.

Valiente-Banuet is a noted ecologist at the Instituto de Ecología, Universidad Nacional Autónoma de México. His publications include a co-edited book on columnar cacti (Fleming & Valiente-Banuet 2002). He collected on Tiburón and also on San Esteban in August.

Tiburón: 9 August, on an expedition of the Instituto de Biología with Fernando Chiang-Cabrera, Emily Lott, et al. He collected on the south side of the island, at least numbers *572* to *639*. The specimens are primarily at MEXU (see entry for Emily Lott and Pedro Tenorio). The expedition also visited Ángel de la Guarda, Salsipuedes, Partida, and San Lorenzo.

San Esteban: 15 August 1985. Collected at least the endemic *Agave cerulata* subsp. *dentiens* (*Valiente 672*, MEXU).

Thomas R. Van Devender.
San Esteban, 1990, 1992.

Van Devender works with the Sky Island Alliance in Tucson and was senior research scientist at the Arizona-Sonora Desert Museum from 1983 to 2009. His extensive list of publications includes books on desert grassland, the cacti of Sonora, the Sonoran desert tortoise, and packrat middens and paleoecology of the southwestern deserts. He is co-editor of the *Diversidad Biológica del Estado de Sonora* (Molina-Freaner & Van Devender 2009) and also with Richard and others on the vascular flora of the state of Sonora (Van Devender et al. 2010).

Tom has collected over 20,000 herbarium specimens, largely from Sonora, deposited in herbaria including the University of Arizona (ARIZ), Universidad de Sonora (USON), and UNAM (MEXU). He and his wife, Ana Lilia Reina-Guerrero, have a special interest in the flora of *La Frontera* in northern Sonora just south of the Arizona border. He also has a strong interest in the biogeography of the Sky Island Region in the Madrean Archipelago.

13–15 September 1990. Arroyo Limantour, *90-524* to *90-543*, with Stephen Hale and Howard E. Lawler and other researchers from the Arizona-Sonora Desert Museum. During this trip they were hit by Tropical Storm Norbert, with horizontal rain blowing through tents in gale force winds of 70 mph winds, which damaged their boat (Bowen 2002:279, 312–313).

29 April 1992. Arroyo Limantour, *92-477* to *92-491*, with Carlos Castillo-Sánchez (Centro Ecológico de Sonora in Hermosillo), J. Mario Cirett-Galán (SEMARNAT in Hermosillo), and Howard E. Lawler and others from the Arizona-Sonora Desert Museum.

From Tom's notes for this day: "*Ctenosaura* abundant. Often on top of shrubs. Saw a pair on top of a cardón arm. Carlos Castillo saw nine animals in a single *Olneya*. I saw them in two adjacent rock shelters at night—a large male alone in one and seven (at least) others crammed into the other hole."

Oscar G. Ward. San Pedro Nolasco, 1974, 1979.

In October 1974 Ward collected several plant specimens on Nolasco. He also went to Nolasco with Ray Turner from 28 to 30 September 1979. A faculty member of Department of Ecology and Evolutionary Biology at the University of Arizona (1966–1995), he worked on genetics and small Sonoran Desert mammals.

Alfred Frank Whiting (1912–1978). Tiburón, 1951.

Whiting was a noted ethnobotanist, anthropologist, professor, and a museum administrator (Severson 2007). Among his publications is the classic ethnobotany of the Hopi (Whiting 1939).

Whiting spent over a week at Tecomate, the largest Comcaac settlement on the island. His time on the island was within several years of when the last families left Tiburón to establish residence on the mainland. His substantial collection was made during the height of the early summer drought and the specimens reflect such conditions. On his labels he transcribed the Seri names of plants. Whiting traveled to Tiburón with William "Bill" Smith, and some records and associated photographs are archived in the Papers of William N. Smith (MS 316, University of Arizona Library Special Collections). It is important, however, to understand that Whiting's field notes include disgraceful, fictional accounts concerning Comcaac customs that could only have come from Bill Smith's warped perception.

14–22 June, Tecomate and vicinity, *Whiting 9000* to *9069*. Four sets of specimens, one at ARIZ.

Aaron Wickenheiser. Alcatraz, 2001.

Wickenheiser documented *Kallstroemia californica* on Alcatraz on 26 October with Bete Jones and Carmen Gabriela Suárez while at the Prescott College field station at Bahía Kino. *Wickenheiser 17.*

Ira Loren Wiggins (1899–1987). San Pedro Mártir & San Esteban, 1962.

Wiggins is the author of *Flora of the Sonoran Desert*, published in the *Vegetation and Flora of the Sonoran Desert* (Shreve & Wiggins 1964), which continues to be the foremost floristic work of the region. Wiggins published extensively on the flora of the region, during which time he was the director of the Dudley Herbarium (DS) at Stanford University, subsequently combined with CAS in San Francisco.

Wiggins's only collecting trip to the islands was in 1962 as a member of the Belvedere scientific expedition, 14 March to 26 April, of which he and Reid Moran were the botanists (see fig. 2.11 and entries for George Lindsay and Reid Moran). Wiggins made an extensive collection on this expedition; his field notebook (at CAS, the California Academy of Sciences Herbarium in San Francisco) includes collection numbers *16973* to *17860* for different Gulf Islands.

San Pedro Mártir: Wiggins was on the island on 21 March (see entry for Reid Moran) and collected 16 numbers, *17176* to *17191*, representing 12 species, as determined from his field notebook.

San Esteban: 22 March, "near south end of Isla San Esteban in large arroyo running inland from east side," which is Arroyo Limantour. He collected 51 numbers that day, *17192* to *17242*.

Benjamin Wilder. Tiburón, 2005–2009. San Esteban, 2007, 2008. Dátil, 2007, 2009. Patos, 2007. Alcatraz, 2007, 2008. Cholludo, 2008, 2009. San Pedro Mártir, 2007. San Pedro Nolasco, 2008, 2009.

Tiburón: 29 December 2005 and 1–2 January 2006, with Humberto and Edward Gilbert.

In Kino the night of the 28th at the Prescott College field station, I lay in my cot filled with tense nerves and anxious thoughts. Actually seeing the island, the massive Sierra Kunkaak rising from the water as we neared Kino Bay and the full extent of the island not visible, I wondered just how foolish were my aspirations to complete a flora of this island with almost no formal botanical skills. The next morning we arrived in Punta Chueca and were directed to Humberto's house. Cathy and Steve Marlett told me that Humberto was the person to work with and has a great knowledge and passion for working with plants. We arrived at his house, I introduced myself and Ed, told him of my goal to catalogue the flora of the island, and mentioned that Cathy and Steve recommended I work with him. Richard had given me a copy of the out-of-print *Ethnobotany of the Seri Indians* to give to Humberto. He lit up when I handed it to him and instantly started looking through it and relating information to us as he saw plants he knew. He then prepared for a trip to the island, which took him no time at all compared to my prolonged process of assembling the loads of gear that accompanied me. Two hours later we were at the base of Capxölim in the middle of the Sierra Kunkaak, having followed a road from Punta Tormenta.

Abridged from my journal:

29 December 2005. "The road came to an end and we started hiking, we got two huge bags and started collecting a sample of every plant in sight. Humberto set a blistering pace. We were near a waterhole but Humberto really wanted to take us up to see some of the higher elevation vegetation. Along the path Ed saw a smoothed rock and picked it up. Humberto said, 'es de personas antes'—it was a grinding stone. After we crossed the wash and were on the other side I saw a flat area above the wash. I asked Humberto what it was and he said it is a path used by the ancient ones as well as in the present. I was impressed at how flat it was compared to the surrounding area. We went a ways through some dense brush and up steep hillsides and emerged upon a beautiful view of the arroyo and Sierra. It was quite green compared to the rest of the island's vegetation. Humberto stopped and I could tell we had reached our destination. He said there is a song in Cmiique Iitom [Seri language] for the vegetation. As he sang I looked around at the land and had a new sense of understanding of the unique existence the Seri have had. He finished and said, 'OK, where do you want to go.'" Collections from this day are from a canyon at the base of Capxölim, *05-01* to *05-54*.

We arranged to return to the island with Humberto to spend one night to allow us to get to a more remote area. The 30th and 31st were spent processing the previous collections and exploring the desert around Kino.

1 January 2006. The first day of the new year was spent at the Pazj Hax waterhole from where we followed a road

into a larger arroyo that led to a valley at the base of the Sierra Kunkaak on the west side of Hast Cacöla, where we made camp.

2 January. We climbed a south-facing slope of the Sierra to about 500 m, *06-01* to *06-52*. We then returned to the mainland.

Tiburón, 16–19 March 2006, with Dave Bertelsen (Tucson-based botanist and expert of the Catalina Mountain flora).

16 March. We were dropped off at Sauzal by Larry Johnson, a local boatman and Gulf adventurer from Colorado based part of the year in Kino. He pointed us in the general direction of the waterhole, and I had assumed it was an easy and direct walk from the shore. Dave and I ended up spending a day and a half going up every wash in search of the waterhole, eventually narrowing down the areas it could be until at last we found our way. Time was thus limited at our destination, and later from looking at satellite images I realized we had only made it to the lower portion of the waterhole, and the main extent remained just beyond us around a low ridge. However, our wanderings allowed us to explore the majority of the Sauzal region.

17 and 18 March. We explored and collected at Coralitos on the 17th (*06-54* to *06-71*) and then at Arroyo Sauzal, including part of Xapij waterhole on the 17th and 18th (*06-73* to *06-129*).

When Larry came to pick us up at Sauzal the water was quite choppy and he was forced to beach his modified panga on the shore for us to embark. In doing this, waves started crashing over the stern and we furiously tried to push off from the beach before the boat become too inundated with seawater. Finally we were successful, and Larry said, "Two more waves and we would have been stuck." The bilge pump was working overtime the whole way back to Kino, and I had had my first experience of the trials associated with working on the Gulf Islands.

Tiburón, 23–25 May 2006, with Richard and Humberto.

This was Richard's first trip back to the Seri region in more than a decade (fig. 2.19). We went to the Caracol research station and the eastern bajada in the vicinity of Pazj Hax: *06-143* to *06-169*. See Richard's account for more information.

Tiburón: 2 September 2006, with Humberto, Mikhal Gold, Gloria Guadalupe (Lupita) Morales, and Jesús Sánchez-Escalante (curator of the USON herbarium in Hermosillo).

Went to Estero/Punta San Miguel for the day, *06-274* to *06-288*, where pitaya agria (*Stenocereus gummosus*) fruits were abundant and perfectly ripe. Humberto harvested several bright red fist-sized fruits and flicked off the spines from the outer walls, and we savored what is to me the most delectable fruit in the world. The taste of a mix between strawberry and watermelon is still with me. The trip was shortened to just that day due to the imminent arrival of Hurricane John, which caused extreme flooding and debris flows in the Sierra Seri on the mainland. The roads on either side of Desemboque were washed out and supplies had to be boated in for weeks.

Tiburón: 23–26 November 2006, Thanksgiving trip, with Richard, Humberto, and Edward Gilbert.

Due to Hurricane John and another rain event, the island's vegetation was luxuriously green and prime for collecting. We focused in the Sierra Kunkaak, with our camp at the base of Capxölim and explorations into the deep canyons of the Sierra Kunkaak beyond the Siimen Hax waterhole. Out of 169 collections, labeled *Wilder 06-345* to *06-513*, 22 were new additions to the flora of the island (Wilder et al. 2007b).

San Esteban & Dátil: 8–11 March 2007, with Seth Marriner Turner (Ray Turner's grandson and Ben's good friend since first grade) in the panga of Ernesto Inojosa from Kino Nuevo.

San Esteban: Spent two nights and two full days exploring San Esteban, camping at Arroyo Limantour.

8 March. The first day we went up the main arroyo about 6 km and then climbed the western slopes of the central peak, Icámajoj Zaaj, *07-38* to *07-69*.

9 March. We again went far into the center of the island along Arroyo Limantour, and then went to canyons on the far central-west side of the island, where among many collections, *07-70* to *07-93*, we found two new records: the fern *Notholaena californica* and the mesic grass *Bothriochloa barbinodis*, both surprising finds on such an arid island. At the end of a long and exhilarating

Figure 2.19. Richard Felger, Humberto Romero-Morales, and Benjamin Wilder, Punta Chueca, Sonora, BTW, 25 May 2006.

day of exploring the canyons on the west side of the island we found ourselves more than 6 km away from camp when darkness fell. While walking back in the arroyo we came upon numerous rattlesnakes (*Crotalus molossus estebanensis*) coming to the arroyo to warm themselves. The final 3 km back to camp went very slowly!

Dátil: 10 March. After two days on San Esteban we departed for Dátil. On Dátil we climbed and explored the northernmost high peak, and also collected in the canyon below, *07-94* to *07-131*.

11 March. We explored the lower hills on the north part of the island, *07-132* to *07-142*.

San Pedro Mártir, Tiburón, & Dátil: 11 and 12 April 2007, with Richard, Miguel Durazo (CONANP), Exequiel Ezcurra, and Jesús Ventura-Trejo (CONANP).

San Pedro Mártir: 11 April. Spent the day exploring the higher elevations of the island. Collections were catalogued under Richard's numbers (see Richard's account for this day).

Tiburón: 12 April. Ensenada del Perro, Tiburón. Conditions were quite dry, but we made collections of some of the unique aspects of the flora from this part of the island, including *Calliandra californica* and *Euphorbia magdalenae*, *07-143* to *07-165*.

Dátil: 2 April. After collecting at Ensenada del Perro on Tiburón, we made a brief visit in the late afternoon to a small canyon at the northeast side of Dátil, *07-166* to *07-183*. (This is the same place that Seth and Ben went to in March 2007.)

Tiburón, 2–5 May 2007, with Humberto and Seth Marriner Turner.

2 May. Punta San Miguel, *07-184* to *07-203*. Sopc Hax waterhole and vicinity, *07-204* to *07-228*.

3 May. Caracol, where we removed the infestation of buffelgrass by hand, *07-229* to *07-231*. Eastern bajada at Valle de Águila, *07-232* to *07-238*. Tecomate, *07-239* to *07-251*.

4 May. Northern part of the Central Valley at Hast Iif, *07-252* to *07-262*. Tecomate, *07-263* to *07-265*. Eastern bajada at Valle de Águila, *07-266* to *07-272*. Eastern bajada inland from Zozni Cmiipla, where we removed a patch of buffelgrass, *07-273*.

Tiburón & Alcatraz, 15 and 16 September 2007, with Humberto, Alexander Swanson (entomology student at the University of Arizona), and Francisco Barnett-Morales (member of the Comcaac community).

Tiburón: 15 September. On this sweltering late-summer day we went to Hast Coopol, the large black volcanic hill in the eastern foothills of the Sierra Kunkaak. Collected on the hill, *07-372* to *07-385*.

Figure 2.20. Whale shark, Gulf of California. BTW.

Alcatraz: 16 September. Collected over the entirety of the island on a hot and humid late-summer day, *07-388 to 07-430*. In the water right in front of Kino Bay on our way back to the boat ramp Ernesto Inojosa (our boatman) spotted a large object in the water—to the great surprise and excitement of all it was a juvenile whale shark (*Rhincodon typus*; fig. 2.20).

Tiburón and Patos: 29 and 30 September 2007, with Richard, Humberto, and Ernesto Molina.

Tiburón: 29 September. The goal of this trip was to find the reported palm on Tiburón (see entry for José Juan Moreno). We went in Ernesto's panga to Pooj Iime on the northwest coast where we made a substantial collection, *07-431 to 07-460*, but no palm tree. We camped that night at Tecomate.

Tiburón & Patos: 30 September. From Tecomate we went to Patos and then to the northeast corner of Tiburón, collections for the 30th were catalogued under Richard's numbers.

Tiburón, 23–26 October 2007, with Richard, Humberto, and Brad Boyle (botanist, ornithologist, and informatics specialist, University of Arizona), Exequiel Ezcurra (then at the San Diego Natural History Museum), Pedro Ezcurra (Exequiel's son), Eduardo Gómez-Limón (ornithologist and noted photographer from Hermosillo), Dennise Z. Avila-Jiménez (Universidad de Sonora, working at USON), Servando López-Monroy and José Ramón Torres (members of the Comcaac community), and Jesús Sánchez-Escalante (curator of the USON herbarium, Universidad de Sonora, Hermosillo). Also see Richard's account for these days.

Based on previous trips to the island, it was clear that the high elevations of the Sierra Kunkaak contain many plant species not previously recorded for the island. The intention of this trip was to make an extensive collection at the highest elevations. Brad, Servando, José Ramón, and I explored the highest elevations of the Sierra, and the rest of the group focused on the canyons and foothills of the Sierra (collections labeled *Felger 07-07* to *07-126*).

23 October. A base camp was made in the canyon near the north side of the base of Capöxlim, from where we mounted trips into the nearby canyons and peaks.

24 October. Brad, Servando, and I climbed up the steep north-facing slope of Capxölim to near peak elevation, *07-465 to 07-484*.

25 & 26 October. Brad, José Ramon, and I made a two-day trip into the deep canyons of the Sierra and then to the upper ridge of the island, where we went to the northernmost peak, Cójocam iti Yaii, and the island's summit, *07-485 to 07-587* (fig. 2.21). A base camp was made in the canyon below the upper ridge to enable us to have enough time to climb to the top and make adequate collections. Despite the vegetation being extremely dry, six new records were found, extending the known ranges of some of those species several hundred kilometers to the south. These disjunct occurrences seem best explained as Pleistocene relics (Wilder et al. 2008b). This was a highly successful trip; however, more collections are needed from the high elevations of the Sierra after times of good rain.

Alcatraz: 4 December 2007, with Richard and Miguel Durazo.

Spent part of the day on the island, departing from Kino. Collections catalogued under Richard's numbers.

San Pedro Mártir & Dátil: 5–8 December 2007, with Richard, Andrea Galindo (then a fellow at the Prescott College Kino Bay station), Miguel Durazo, and two crew members.

San Pedro Mártir: 5–7 December. Higher elevations on the island, *07-601 to 07-607*. It was on this trip that we had a challenging boat mishap on our way back to the mainland. The story from my viewpoint is related here.

5 December. Mártir is a remarkable place to call your office for several days. I was able to connect to wonders I miss while carrying out my days in the city. The radiance of the stars in a moonless and cloudless sky cast a faint light that painted thousands of seabirds and cardons with a pale glow. Using binoculars I scanned the sky, finding hundreds of stars in clusters where my naked eye found one or two. Shooting stars streamed above us at frequent intervals. The waters below the cliff glowed with occasional ghost-like whitish patches—sea lions outlined by bioluminescence.

6 December, the first full day of fieldwork. The entire upper part of the island was socked in by dense moisture-laden cold fog combined with a stiff wind. We established a quadrat and six transects at the top of the island.

7 December. After a windy night, we established a cardón plot near the top of the island. Andrea Galindo was the data taker and ran the show, ensuring that we

Figure 2.21. José Ramón Torres and Benjamin Wilder on the summit of Isla Tiburón. BB, 26 October 2007.

finished the task. Six hours and 209 cardons later, it was time to head back to the mainland. Looking out at the water, the wind, which had been strong earlier, had died down and conditions looked calm. I expected Miguel Durazo, our boatman, would want to return that night.

We departed in the *Orca* as rays of golden light broke through low clouds and the sun set behind the Baja Peninsula, expecting to soon be back in Kino enjoying a seafood dinner. After about an hour and a half Miguel slowed the boat and said that we still had a long way to Kino, we could keep going, but it would be better to head for a protected bay on the southeast coast of Isla Tiburón and get into Kino in the morning. We listened to our captain and he changed direction.

7:07 p.m. The engine stopped. An uncomforting sound followed by a discomforting statement from Miguel, "No hay más gasolina." There was a leak and all the gas dumped into the ocean. To make the situation worse the radio was not functioning and my Mexican cell phone received no coverage at sea level.

It was at this point I was most afraid. The story of acclaimed scientist Gary Polis and his fatal trip in the Bahía de Los Ángeles region of the Gulf haunted me. If we drifted to the south our fate might have been the same. I quickly tried to push this out of my consciousness with deep breaths.

Because Miguel changed course, we were in sight of Isla Dátil to the south of Tiburón. I got my GPS and marked our location after the gas went out. With the "tracks" setting I was able to gauge our progress or movement away from Dátil. To our great relief after fifteen minutes it showed progress. However, we had a long way to go.

More attempts with the radio, flashing SOS with lights, a flare casting eerie red glare against our faces—an image seared into my memory—all in vain. Andrea, Richard, and I held up Richard's large tarp and created a forward sail that caught a good deal of wind. Miguel got out two oars and began rowing to the island. Dátil was getting closer and closer; checking the GPS, I confirmed this. The deep breaths I used earlier to calm myself were replaced with a sense that we would gain the island.

Dátil: After five hours of rowing and bailing, we set anchor about 20 m from the high cliffs of Dátil and it was time for sleep, if you can call it that. We were cold and wet. We settled in, not easy with six people and three hundred pounds of jumbled gear in a small boat. We tried to divert our minds from focusing on reality and each found a quiet space that allowed us a bit of rest.

I realized how critical it is to have a clear and competent leader in situations like this. Miguel excelled in the role. He stayed up the entire night, made oar locks from plant press ropes, remained calm, knew the area and what needed to be done, asked for and allowed others to help, had confidence that we could draw from, and kept perspective. It was also extremely beneficial that all of us remained calm and optimistic. Lying in a disabled, water-soaked panga, desperately cold, next to a good friend and my mentor with the entire universe of stars spread above and a dancing sea of phosphorescent lights below, I had to pause and take in the beauty, breathing it in through the cold and fear.

8 December. As the world rotated over and the sun crossed the horizon we saw we were on the west side of Dátil and could see Islas Cholludo and Tiburón to the north. We spotted a beach with an accessible route to the island ridge. I needed to get up there and make a call with the cell phone. Miguel and I jumped into the surf and ran onto the beach and up the loose rock. Solid ground, even if a 60° angle, was a true relief. We made it to the top and I anxiously turned on the phone. The seconds passed . . . three bars! Miguel called his wife Maria Jesús and delivered the message of our location. She said she was headed over to tell Tadeo (Tad Pfister), the Prescott College station manager and good friend of us all.

We had gained safety and finally rowed through strong waves around the north end of the island into a gentle beach at the northeast side with minimal wave action. Fourteen hours since the engine quit. Time to wait and get dry. Then it was time for breakfast. The menu consisted of fish caught at least a day ago, fried in oil over a wood flame, with seawater-soaked tortillas from Anita Street Market in Tucson. Soon Tad and Cosme Becerra arrived in the Prescott College super panga and fit the role of heroes. We called this place Rescue Beach and collected *Lycium brevipes*, *Wilder 07-608*. We headed for Kino, with the *Orca*, refueled by Tad, alongside. We were all given intense respect for traveling in the Gulf and were on the good side of a great deal of luck. Richard may have summed up our feelings after this whole experience the best. "I feel like I have been in a great fight. I won, but I am wounded."

Alcatraz & Tiburon, 28–30 January 2008.

Alcatraz: 28 January, with Richard and Tom Fleishman's Prescott College coastal ecology class. Collections catalogued under Richard's numbers.

Tiburón: 29 January, with Richard, Cosme Becerra, and geologists Mike Oskin and his student Scott Bennett from the University of California at Davis, and Becky Dorsey and her student Tom Peryam from the University of Oregon.

We went to Arroyo Sauzal and the Xapij waterhole with geologists studying the Late Miocene faulting that led to continental rupture and opening of the Gulf of California. We collected in the Arroyo Sauzal area and at Xapij, *08-96* to *08-140*, and later in the day at the terminus of Arroyo Sauzal at the shore *08-141* and *08-142*.

30 January, with Richard, Cosme Becerra, and Abram Fleishman (ornithologist and biologist at Prescott College). First collections were from the rocky hills in the vicinity of El Monumento, at the southeast corner of Tiburón (*Felger 08-39* to *08-76*). We then went by boat a few kilometers to the west to a placid bay called La Viga, the portion of Tiburón closest to Dátil and Cholludo, and explored the low mountains and slopes surrounding the bay, *08-143* to *08-152*. The next day Richard and I went to Guaymas for a trip to Isla San Pedro Nolasco.

San Pedro Nolasco: 2 and 3 February 2008, with Richard, Florencio Cota-Moreno, and Jesús Ventura-Trejo.

We spent two days working on the east side of the island in the vicinity of Cañón el Farito. We established permanent vegetation monitoring sites and collected 32 numbers: 2 February, *Wilder 08-153* to *08-171*. 3 February, *Wilder 08-172* to *08-185*.

Tiburón: 6 and 7 April 2008, with Humberto and Brigitte Marazzi, a botanist from Switzerland who is an expert on the legume genus *Senna* who did her postdoctoral studies at the University of Arizona with Michael Sanderson.

6 April. We went with Ernesto Molina in his panga to Zozni Cacösxaj at the northern part of the Canal del Infiernillo where many Seri families were harvesting *callo de hacha* (probably *Atrina tuberculosa* and/or *Pinna rugosa*). The goal for this day was to find caal oohit, the

blue sand lily (*Triteleiopsis palmeri*), which I was told occurred in the sand near the coast on the northeast part of the island. We found *Triteleiopsis* and made additional collections, *08-196* to *08-207*. We then went to Hahjöaacöl, a locality just to the south, where we searched for *nopal* (*Opuntia* sp.). None were found but we made collections of other plants, *08-202* to *08-221*. That evening after arriving via panga at Punta Tormenta we drove to Caracol and the northern portion of the Sierra Kunkaak in order to depart early the next morning for exploration in the mountains. A few collections were made at Zoojapa, *08-222* to *08-225*.

7 April. The day was spent exploring Hast Cacöla, the front range of the Sierra Kunkaak. We made a large collection including *Mirabilis tenuiloba*, a new record for the island, *08-226* to *08-275*. I saw two golden eagles ("zep" in Cmiique Iitom) and a family of bighorn sheep high up in the mountain.

Tiburón, Cholludo, & San Esteban: 2–5 September 2008, with Humberto and nine Comcaac ecologists—the plant team—on a trip to Tiburón, Cholludo, and San Esteban to search for non-native plants (fig. 2.22).

The members of the team are Erik Salvador Barnett, Juan Alfredo Barnett-Díaz, Raymundo Barnett-Morales, Ivan Eliseo Barnett-Romero, Servando López-Monroy, Maximiliano Damián López-Romero, Nestor Cristóbal Montaño-Herrera, Raúl Eduardo Molina-Martínez, and Adán Humberto Romero-Morales.

Tiburón: 2 September. We stopped at Ensenada del Perro before heading to Cholludo, *08-299* to *08-311*.

Cholludo: 2 September. Covered the north-central portion of the island to peak, *08-312* to *08-320*.

San Esteban: 3 September. Ran transects in Arroyo Limantour, from the coast to about 2 km inland. Collec-

Figure 2.22. Humberto Romero-Morales, Benjamin Wilder, and the Seri Plant Team, at Caracol, Isla Tiburón. From left to right: Raymundo Barnett-Morales, Humberto Romero-Morales, Ivan Eliseo Barnett-Romero, Juan Alfredo Barnett-Díaz, Maximiliano Damián López-Romero, Benjamin Wilder, Servando López-Monroy, Nestor Cristóbal Montaño-Herrera, Raúl Eduardo Molina-Martínez, Erik Salvador Barnett, and Francisco Barnett-Morales. Francisco Barnett-Morales is not shown since he snapped the shutter of Ben's camera. 5 September 2008.

tions were made by several members of the plant team (see Raymundo Barnett-Morales, Juan Alfredo Barnett-Díaz, Maximiliano Damián López-Romero, Nestor Cristóbal Montaño-Herrera, and Servando López-Monroy).

Tiburón: 4 and 5 September. Two days were spent at the Caracol research station, where we removed buffelgrass plants that had re-established after the control effort made in May 2007 and searched for other populations, and I conducted a plant-pressing workshop for the plant team. I made a few collections from an outlying peak in the foothills of the Sierra Kunkaak, *08-321* to *08-323*, in the vicinity of Caracol, *08-324* to *08-327*, and in the estero at Punta Tormenta *08-328* & *08-329*.

Tiburón, 26 September 2008, trip to the vicinity of Zozni Cmiipla with Richard, Humberto, et al. (see Richard's entry for this trip).

San Pedro Nolasco: 29 September 2008, with Richard and Juan Pablo Gallo-Reynoso (see Richard's entry for a description of the trip).

In addition to exploring the lower half of Cañón Mellink with Richard and Juan Pablo, I ascended the canyon to the upper ridge at 260 m and was rewarded by finding a number of species that occur only at high elevations on the island. In addition to collections numbered by Richard, my collections are *08-344* to *08-361*.

Figure 2.23. En route to Isla Dátil. From left to right: Humberto Romero-Morales, Servando López-Monroy, Raúl Eduardo Molina-Martínez, and Seth Turner. BTW, 27 February 2009.

Tiburón: 8 and 9 November 2008, with Humberto, Servando López-Monroy, and Lorayne Meltzer and the Prescott College marine studies class.

After spending the 7th on the mainland coast at Campo Egipto establishing plots along a gradient from the coast and inland up the bajada, we went to Tiburón to repeat this process to gain insights into differences between Tiburón and the mainland.

8 November. We worked on the northeast part of Tiburón, at Iifa Hamoiij Quih Iti Ihiij along the Canal del Infiernillo, *08-369* to *08-371*.

9 November. We went by boat from Iifa Hamoiij Quih Iti Ihiij to Punta Perla and explored the estero and sand spit, *08-371* to *08-379*, and a nearby north-facing bajada, *08-380* to *08-382*.

Tiburón, Dátil, & Cholludo, 27 February to 1 March 2009, with Humberto, Raúl Eduardo López-Martínez, Servando López-Monroy, and Seth Marriner Turner (fig. 2.23).

We spent these three days in the vicinity of the southern part of Tiburón. Due to strong north winds we focused our exploration on the southeast corner of Tiburón and neighboring Dátil and Cholludo on 1 March.

Dátil: 27 February. Explored the east-central side of the island, ascending a canyon of the central mountain ridge of the island, *09-01* to *09-44*.

Tiburón: 28 February. We explored Arroyo de la Cruz near the abandoned Mexican Marine base and up to an unreliable but occasionally full tinaja, Hax Hasoj, *09-45* to *09-67*.

Cholludo: 1 March. Explored the vegetated part of the island except the northeast-facing slope, *09-68* to *09-81*.

San Pedro Nolasco: 11 November 2009, with Jesús Ventura-Trejo in the panga of Javier Cordoba of La Manga.

In early September the Tropical Storm Jimena settled over the San Carlos/Guaymas region and delivered about 700 mm of rainfall in 36 hours, a record for the state of Sonora. I made a trip to explore the island as well as the neighboring canyons of the Sierra el Aguaje to take advantage of the verdant vegetation. On Nolasco we landed at Cañón el Farito and from there ascended to the ridge crest and negotiated our way along the precipitous upper crest to the summit of the island (ca. 305 m). The canyon was lush with *Vaseyanthus insularis* covered in dodder (*Cuscuta corymbosa*) and occasional rock depressions still contained water. The slopes above the canyon had a dense cover of ephemerals (e.g., *Boerhavia triquetra, Muhlenbergia microsperma, Perityle californica*) that were already dry. One new record, the perennial grass *Heteropogon contortus*, was discovered on this trip. Collections from the canyon and ridge crest are numbered *09-122* to *09-142*.

We spent the 12th and 13th exploring Cañón Las Barajitas in the Sierra el Aguaje with Juan Pablo Gallo-Reyonso and Juan Ramón Castillo-Villaseñor. Floods from Tropical Storm Jimena had scoured the canyon's palm oases and lush vegetation, bringing in thousands of tons of alluvium and resetting the ecological clock of this diverse canyon (similar floods occurred in Cañón del Nacapule). Collections from Barajitas are numbered *09-143* to *09-245*.

Robert Curtis Wilkinson.
Tiburón & San Esteban, 1977.

Wilkinson was a student at the University of Arizona and worked with Richard.

Tiburón: 11 October. Coralitos, *Wilkinson 188* to *216.* La Pescadita, *Vaseyanthus insularis, Wilkinson 218* and *Setaria macrostachya,* not numbered.

San Esteban: 13 October. [Arroyo Limantour], *Wilkinson 150* to *182.* Also from San Esteban on this day, but not numbered are *Aristida adscensionis, Digitaria californica, Enteropogon chlorideus, Muhlenbergia microsperma, Setaria macrostachya,* and *Solanum hindsianum.*

PART III
The Flora

Species Accounts

The species accounts are organized as pteridophytes (ferns), then the one gymnosperm (*Ephedra*), and then the angiosperms (flowering plants; the dicots, including magnoliads and eudicots, and monocots are not separated). Within these major groups, all entries are listed alphabetically by family, genus, and species.

Plant family designations follow the APG III (Angiosperm Phylogeny Group; Stevens 2008) classifications, reflecting current knowledge of evolutionary relationships. Taxonomy is undergoing rapid and ample reclassifications with insights gained from fields such as genetic data analyses. Science marches on, and future changes in scientific names and even family alignments are expected in a time of a comprehensive re-examination of the evolutionary relationships of the diversity of life. The botanical nomenclature in this book represents our understanding and opinions as of 1 June 2011. (See appendix C for a listing of nomenclature changes that have occurred for the species covered in this work since their treatment by Felger and Moser in 1985.)

Descriptions and identification keys for the various taxa (e.g., families, genera, and species) are based on plants from the flora area and are not necessarily applicable to other regions—descriptions and measurements pertain only to plants and populations from islands in the flora area unless otherwise stated. For the sake of convenience, the qualifications *about* and *approximately* are generally omitted, with the obvious understanding that such quantitative values are, to varying degrees, seldom exact. The brief descriptions generally emphasize characters that seem important to understanding the variation and adaptations of plants in this insular, arid region. For this reason, there is sometimes more emphasis on vegetative characters and less on certain characters emphasized in other floras. More detailed description, discussion, and interpretation are offered for plants of special interest, uniqueness, or significance in the flora area, or those that we most like.

The accepted names are in boldface font. The few non-native plants are marked with an asterisk (*). Some selected pertinent synonyms are in square brackets following the accepted scientific name. Authors of scientific names follow the listings of Brummitt and Powell (1992), the International Plant Names Index (2011), and Tropicos (2011). Place and date of publication are provided for taxa based on type specimens from the flora area. Common names, when available, follow the scientific name, with the names(s) in Cmiique Iitom (the language of the Comcaac, or the Seri people) in Roman font, listed first, then the Spanish-language (Mexican) common name(s) italicized, followed by the English language (American) common name(s) in Roman font. Analyzable Seri names are translated into English. Sources of the Seri names are from Felger and Moser (1985), Moser and Marlett (2005, 2010), and information provided by Humberto Romero and various Seri consultants.

Unless otherwise specified, the color given for a flower or other structure is the most conspicuous or dominant color. In addition to the usual or commonly encountered range of variation, extreme or uncommon variation is often provided in parentheses, for example, (5) 10–15 (20). Overall plant sizes, especially shrubs and trees, are mostly rough estimates. Special attention is given to geographic ranges not covered in other, readily available regional floras. For example, the local distributions in Sonora, especially in western Sonora, are often poorly known but are significant to the understanding of relationships among taxa in the flora area. When a species is present on one of the islands treated here, the specimens documenting its occurrence are listed after the bolded name of the island where it occurs. For each species in the flora, a distribution map for the Sonoran Islands is presented (see base map, fig. 3.0). Each of the specimens cited corresponds to a dot on the map. Islands are listed in the following order: Tiburón, San Esteban, Dátil, Alcatraz, Patos, Cholludo, Mártir, and Nolasco. The occurrences of species on other Gulf of California islands are listed under the "Other Gulf Islands" heading and are based on the most recent checklists of plants for the larger Gulf of California islands (Rebman et al. 2002) and for some smaller islands, such as Islas Partida Norte, Salsipuedes, and Rasa (Moran & Rebman 2002). A few additional "Other Gulf Islands" records are presented that are not reported in the literature, and in such cases the island name is followed by specimen information, for example, Tortuga (*Johnston 4185*, CAS). The "Other Gulf Islands" are listed in order of decreasing latitude: Ángel de la Guarda, Rasa, Salsipuedes, Partida Norte, Las Ánimas (San Lorenzo "Norte"), San Lorenzo, Tortuga, San Marcos, San Ildefonso, Coronados, Carmen, Danzante, Monserrat, Santa Catalina, Santa Cruz, San Diego,

San José, San Francisco, Espíritu Santo, and Cerralvo. The Seri village of El Desemboque del Río San Ignacio is usually identified as El Desemboque.

Our flora is specimen based. Specific herbarium specimens are cited in order to provide information on distribution and habitat, as well as to provide verification. Felger and/or Wilder have inspected all specimens cited except those noted as "not seen." Geographic information on specimen labels is often shortened; place-names are provided in the gazetteer in this volume. We report only a single specimen per locality, unless a collection is of special interest (i.e., historical collections). Thus, there are more specimens from the islands than reported here.

In most cases, the first set of collections is deposited at ARIZ. Additional sets are mostly at SD and USON; many of our duplicate specimens are variously and mostly at ASU, BRIT, CAS, DES, MEXU, MO, RSA, SD, TEX, UC, and USON. All specimens cited are at ARIZ unless otherwise indicated by the abbreviations for herbaria (Index Herbariorum 2009). If a specimen is at ARIZ, or cited for another herbarium, we generally do not cite duplicates at other herbaria. In cases where more than

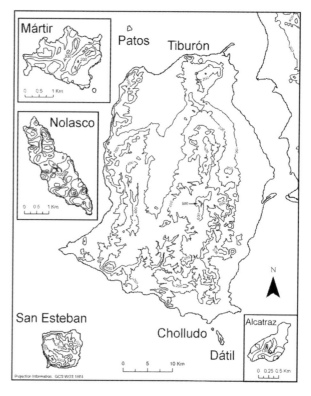

Figure 3.0 Base distribution map of the eight Sonoran Islands treated in the flora. Projection information GCS WGS 1984.

one collector is listed on the label, generally only the first collector's name is given. For example, Felger and Wilder made many collections together, but only the name associated with the collection number is cited (we often traded off on alternate days for name and collection numbers). If no collection number is given on the label, the specimen is identified by the date of collection, for example, *R. Russell 6 Aug 1962.* "Specimen not located" is used for cases in which a specimen is recorded in our field notes as having been collected and given a collection number, but despite searching for it, the specimen has not been found and is likely lost. When the date of collection is significant, such as collections of historical interest or type collections, both the collection number and date are given. In a few cases the herbarium accession number follows the herbarium abbreviation to avoid confusion, especially in the case of multiple specimens of type collections.

Key to the Major Plant Groups

1. Plants spore bearing, without flowers or seeds; leaves those of typical ferns, less than 40 cm long. _____ _____ **PTERIDOPHYTES – FERNS** (page 113)
1' Seed-bearing plants with or without flowers; leaves various or none.
    2. Cone-bearing plants with seeds but not "usual" flowers; shrubs with paired scale leaves at nodes; male and female cones on separate plants, the female cones producing a single seed. _____ _____ **GYMNOSPERMS – NON-FLOWERING SEED PLANTS** (page 117)
    2' Plants with flowers and seeds; leaves various or none. _____ _____**ANGIOSPERMS – FLOWERING PLANTS** (page 117)

PTERIDOPHYTES – FERNS

Pteridaceae – Brake Fern Family

The five species of "desert" or cheilanthoid ferns in the flora area have leaf segments that curl up tightly during dry periods and expand during moist conditions to reveal the green surfaces. All five are found in the Sierra Kunkaak on Tiburón. Only *Notholaena californica* occurs elsewhere on the island and on San Esteban, and *N. lemmonii* is also on San Pedro Nolasco. All grow sheltered among rocks, often on the shaded northern or eastern side of their rock shelters, or from crevices on cliffs.

Ferns are few in the Sonoran Desert and deserts in general. Tiburón is comparable to other large, mountain areas in Sonora bordering the Gulf of California, such as the Pinacate/Gran Desierto region, with 8 species of ferns and fern relatives, 6 of them in the Pteridaceae (Felger 2000a), and the Sierra el Aguaje, north of Guaymas, with 11 ferns and fern relatives, 7 of them in the Pteridaceae (Felger 1999 and unpublished notes).

Figure 3.1. *Cheilanthes wrightii.* Capxölim, Isla Tiburón. BTW, 25 November 2006.

Figure 3.2. *Notholaena californica* subsp. *californica*. Isla San Esteban. BTW, 9 March 2007.

1. Upper (adaxial) surfaces of leaf segments with often-deciduous stellate-pectinate scales (star shaped and comb-like). _____**Astrolepis**
1' Upper surfaces of leaf segments glabrous or sparsely gland dotted, without stellate or pectinate scales.
   2. Leaf blades about as wide as long, or less than twice as long as wide. _____ **Notholaena**
   2' Leaf blades at least twice as long as wide.
      3. Leaf blades 1–2 (3) times longer than wide; leaf segments green or brownish below, not farinose. **Cheilanthes**
      3' Leaf blades usually more than 3 times longer than wide; leaf segments white or yellow (farinose) below. ____
        _____**Notholaena lemmonii**

Astrolepis sinuata (Lagasca ex Swartz) D.M. Benham & Windham subsp. **sinuata** [*Notholaena sinuata* (Lagasca ex Swartz) Kaulfuss var. *sinuata*]

Star-scaled cloak fern

Description: Tufted ferns. Leaves sword shaped, to at least 37 cm long. Larger leaflets with 3 or 4 conspicuous lobes on each side; lower (abaxial) surfaces obscured by dense, overlapping scales 1.6–2.1 mm long, these scales narrowly lance-attenuate, brown with white ciliate-fringed broad membranous margins. This is the largest of the five ferns in the flora.

Local Distribution: Sierra Kunkaak at higher elevations in canyons and north-facing cliffs or shaded slopes. The nearest known populations are 130 km to the east in the vicinity of Hermosillo (*Van Devender 8 Feb 1978*) and in the Sierra el Aguaje north of Guaymas, 150 km

to the southeast (Felger 1999), although we predict its occurrence at higher elevations in the Sierra Seri.

Other Gulf Islands: None.

Geographic Range: Subsp. *sinuata* occurs from Arizona to Texas to South America, the Baja California Peninsula, Georgia, and the West Indies. It is an apogamous triploid. Another subspecies, a sexually reproducing diploid, ranges from western Texas and southeastern New Mexico to Central America.

Tiburón: San Miguel Peak, 2800 ft, *Knight 1034* (SD, UNM). Cerro San Miguel, common in canyons and humid areas, *Quijada-Mascareñas 91T012*. SW and up canyon from Siimen Hax, *Wilder 06-472*. Capxölim, *Wilder 07-467* (duplicate at MO verified by G. Yatskievych, 2008). Steep, protected canyon, 720 m, N interior of Sierra Kunkaak, rare, base of cliffs, *Wilder 07-516*.

Cheilanthes wrightii Hooker

Wright's lip fern

Description: Small rhizomatous ferns. Leaf blades dark green, often 1–2 (3) times longer than wide, 2 (3) times divided, glabrous, the ultimate leaf segments more or less ovate to linear and not bead-like. The Tiburón specimens are from drought-stunted plants with leaves 2–8 cm long (including the petiole or stipe); mainland specimens have leaves often 6–16 (21) cm long.

Local Distribution: Sheltered niches among rocks on north-facing slopes and canyon walls in the Sierra Kunkaak. The nearest known population occurs in the Sierra del Viejo south of Caborca.

Other Gulf Islands: None.

Geographic Range: Southern and central Arizona to western Texas, and Sonora to Coahuila and Durango, and Baja California Sur.

Tiburón: San Miguel Peak, *Knight 1029* (UNM). Capxölim, *Wilder 07-470* (duplicate at MO verified by G. Yatskievych, 2008). Interior of Sierra Kunkaak, E-facing slope above canyon and just below highest ridge on island, *Wilder 07-531*.

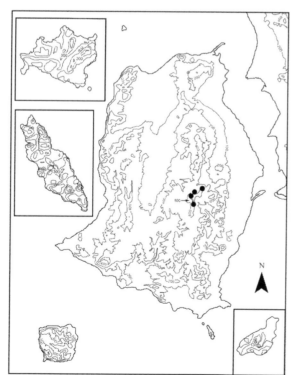

Astrolepis sinuata ssp. *sinuata*

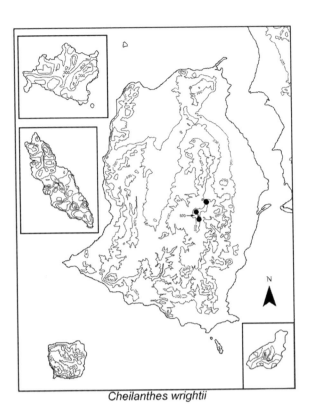

Cheilanthes wrightii

Notholaena – Cloak Fern

The three species in the flora have conspicuous farinose exudate coating the lower leaf surfaces, presumably a water-conserving adaptation (Wollenweber 1984).

1. Leaf blades at least twice as long as wide, with many pinnately arranged segments. _____ **N. lemmonii**
1' Leaf blades about as wide as long.
   2. Leaf blades divided into numerous minute, bead-like segments. _____ **N. californica**
   2' Leaf blades divided into (3) 5 major segments, not bead-like. _____ **N. standleyi**

Notholaena californica D.C. Eaton subsp. **californica** [*Cheilanthes deserti* Mickel]

Hehe quina "hairy plant"; California cloak fern

Description: Small tufted ferns or with very short rhizomes. Leaf blades triangular-ovate to pentagonal, nearly as wide at base as long, 2–5.5 cm long, 3-times divided into separate small, bead-like segments, the upper surfaces bright green to olive green, dotted with small yellowish glands, the lower surfaces obscured by golden-yellow farina.

Local Distribution: Tiburón at scattered, localized sites on larger hills and mountains, mostly north facing, across the island and to higher elevations in the Sierra Kunkaak, in seasonally moist niches among rocks. It is the only fern on San Esteban, where it occurs in sheltered north-facing canyons in the southern part of the island. This is the most widely distributed fern on Tiburón and other Gulf Islands (Johnston 1924:980), and one of the most xeric-inhabiting ferns in the Sonoran Desert (Felger 2000a).

Other Gulf Islands: Ángel de la Guarda, Carmen, Monserrat, Santa Catalina, Santa Cruz, San Diego, Espíritu Santo, Cerralvo.

Geographic Range: Southern California to the Cape Region of Baja California Sur, southern Arizona, and Sonora south to the vicinity of Bahía Kino. This species is apogamous. Subsp. *leucophylla* Windham, in California and northwestern Mexico, has white instead of yellow farina.

Tiburón: El Sauz, 200 m, one colony on N slope, *Moran 8789* (SD). Haap Hill, *Felger 74-T56*. Cerro San Miguel, común en lomeríos y cañones del cerro, *Quijada-Mascareñas 91T007*. San Miguel Peak, *Knight 991* (UNM). Capxölim, *Wilder 07-469* (duplicate at MO verified by G. Yatskievych, 2008). Sierra Caracol, *Knight 1061* (UNM). Bahía Agua Dulce, rocky hill W of Tecomate (Felger 1966:263). Pooj Iime, large canyon, NW shore, *Wilder 07-456*.

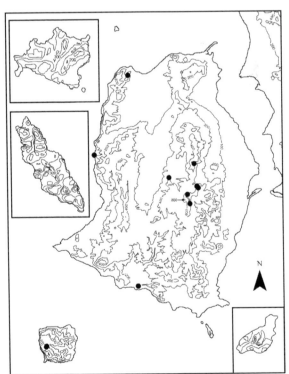

Notholaena californica ssp. *californica*
Hehe quina

San Esteban: SW corner of island: sheltered canyon branched off from main drainage, in sheltered nooks below rocks, *Wilder 07-86*; W-most N-trending canyon in area, sheltered nooks below rocks, occasional, *Wilder 07-93*.

Notholaena lemmonii D.C. Eaton var. **lemmonii**

Lemmon's cloak fern

Description: Small tufted ferns. Leaf blades usually about 3 times longer than wide, 2-times divided, olive green and glabrous above (adaxially), the lower surfaces densely farinose. The Tiburón specimens have white farina. The Nolasco specimen is the rare form with yellow farina. Farina, a powdery wax-like exudate on the abaxial ("lower") surface of the fronds of various cheilanthoid ferns, is formed by glandular trichomes and consists almost exclusively of flavonoid aglycones. In dry periods, when the fronds are desiccated and "rolled up," the abaxial surface is exposed and the farina reportedly reduces transpiration by its lipophilic nature, as well as by reflecting excess insolation. The biogeographic significance of the golden form on Nolasco is that "all the flavonoids exudate by a certain fern form a flavonoid pattern, which often is rather consistent, i.e., subjected to quantitative variation only, and can be characteristic for a species, for a variety or for a distinct chemical race" (Wollenweber 1984:4).

Local Distribution: Sheltered sites at higher elevations in the Sierra Kunkaak and Nolasco at higher elevation at the southeast side of the island. The nearest mainland population is in the Sierra el Aguaje, including the opposite shore at Bahía San Pedro, but those plants have white rather than yellow farina. It is expected but not recorded in the Sierra Seri opposite Tiburón.

Other Gulf Islands: Cerralvo.

Geographic Range: Mountains in Arizona and through much of Sonora and Chihuahua to Jalisco, and Baja California Sur. A second variety is disjunct in south-central Mexico and may represent a distinct species (Windham 1993).

Tiburón: Hast Cacöla, E flank of Sierra Kunkaak, top of E-facing slope at base of cliff, ca. 540 m, *Wilder 08-271*. Capxölim, *Wilder 07-468* (duplicate at MO verified by G. Yatskievych, 2008). Sierra Caracol, in rock, *Knight 1062* (UNM).

Nolasco: E side, talus area, 900–1000 ft, E exposure, *Pulliam & Rosenzweig 21 Mar 1974* (annotated as "rare form with golden farina," G. Yatskievych, 1994).

Sonora: San Pedro Bay, locally abundant about rocks on a N-facing hillside, 7 Jul 1921, *Johnston 4336* (CAS, UC, white farina).

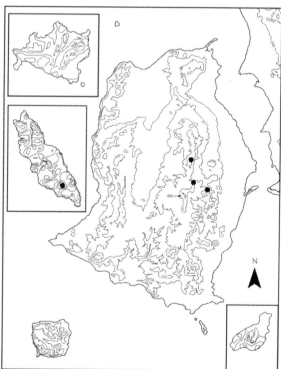

Notholaena lemmonii var. *lemmonii*

Notholaena standleyi Maxon

Hehe quina; star cloak fern

Description: Small tufted ferns. Leaf blades pentagonal in outline, more or less as wide as long, 3–7 cm wide, divided into (3) 5 major, deeply cleft pinnae or segments; upper surfaces olive green and glabrous, the lower surfaces obscured by golden yellow farina. Seigler and Wollenweber (1983) found 3 geographic chemical races. The Tiburón population belongs to the western "golden race."

Local Distribution: Higher elevations in the Sierra Kunkaak. Also in the Sierra Seri in adjacent Sonora. Extending into slightly more exposed niches than *N. lemmonii*, including south-facing slopes at high elevations.

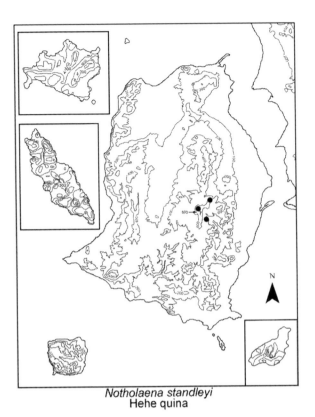
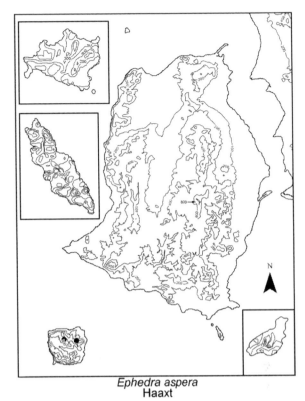

Notholaena standleyi
Hehe quina

Ephedra aspera
Haaxt

Other Gulf Islands: Tortuga (*Johnston 4185*, CAS).

Geographic Range: Southwestern United States and northern Mexico to Puebla, and both Baja California states. The Sonoran Desert plants are usually smaller than those in more mesic regions. Probably sexual diploids.

Tiburón: S slope of Sierra Kunkaak, sheltered rock crevice, ca. 350 m, rare, 1 Jan 2006, *Wilder* (photo). Cerro San Miguel, *Quijada-Mascareñas 91T013*. Cójocam Iti Yaii, main peak, common on S-face of peak in nooks in rocks, *Wilder 07-542* (duplicate at MO verified by G. Yatskievych, 2008).

GYMNOSPERMS – NON-FLOWERING SEED PLANTS

Ephedraceae – Joint-Fir Family

Ephedra aspera Engelmann ex S. Watson

Haaxt; *tepopote*; boundary ephedra

Haaxt is also the walkingstick insect (Phasmatidae).

Description: Shrubs about as broad as tall, the lower limbs woody. Twigs straight and green; each node bears 2 scale leaves, fused at their base, obtuse, and soon fraying and deciduous. Seed cones 2 to several per node, sessile, 1-seeded.

Local Distribution: Known from high elevations of the central and northeast peaks on San Esteban.

Other Gulf Islands: San Lorenzo, Carmen.

Geographic Range: Sonora from the Sonoyta/Pinacate region to the Sierra del Viejo southwest of Caborca. Also Baja California Norte, southeastern California to Utah and southwestern Texas, and southward to Zacatecas.

San Esteban: W slope of NE peak, 450 m, compact prostrate shrub, *Moran 13043* (SD). Sub-peak of central mountain, rocky upper N face of mountain, *Wilder 07-57*.

ANGIOSPERMS – FLOWERING PLANTS

A. Agaves, cattails, grasses, sedges, sedge-like plants, palms, submerged aquatic plants, succulent or large non-succulent rosette plants. _____**MONOCOTS**

A' All other plants, not as above. _____ **DICOTS**

A. MONOCOTS

1. Palms: a prominent, single woody trunk and large fan-shaped leaves (probably extirpated from the flora area).**Arecaceae**
1' Not palms, not developing a conspicuous trunk (in the flora area).
  2. Seagrasses, submerged in shallow seawater, leaves linear to thread-like.
    3. Leaves usually more than 15 cm long, 3.2–5 mm wide; fruits flask shaped, 3 mm long. _____ **Zosteraceae**
    3' Leaves less than 15 cm long, to ca. 1 mm wide; fruits rounded, ca. 2 mm long, or not found.
      4. Usually stemless or appearing stemless; leaves 0.5–1 mm wide, the leaf tip blunt, often with 2 or 3 minute teeth or points; fruits not present or hidden beneath the substrate. _____**Cymodaceae**
      4' Stems usually prominent; leaves ca. 1 mm wide, the tip acute (pointed) to obtuse (rather blunt) with microscopic teeth; inflorescences terminal, the peduncles (flowering stalk) at first enclosed in the leaf sheath and greatly elongating as the fruits develop, the fruits of 4 rounded nutlets. _____**Ruppiaceae**
  2' Plants terrestrial or emergent from shallow freshwater; leaves variable.
    5. Leaves in rosettes, thick, tough, and succulent with a stout terminal spine, or linear, relatively flat, and with many marginal spines. _____ **Asparagaceae** (*Agave, Dasylirion*)
    5' Leaves not in rosettes and not as above.
      6. Perennials from underground scaly bulbs; flowers blue and showy; desert habitats._____**Asparagaceae** (*Triteleiopsis*)
      6' Annuals or perennials but not from underground scaly bulbs; individual flowers often small, not blue, the perianth membranous or scale-like, or reduced or absent; plants of desert or wetland habitats.
        7. Cattails; leaves linear, erect, more than 1 m tall, thickened and pithy; wetland habitats. __ **Typhaceae**
        7' Grasses and sedges; leaves less than 1 m long, not thickened and pithy; various habitats.
          8. Sedges and spikerush; stems solid (pithy); leaf sheaths usually closed; each flower subtended by a single bract; mostly in wetland habitats or at least temporarily wet soils._____**Cyperaceae**
          8' Grasses; stems hollow or solid; leaf sheaths often open; each flower subtended by 2 bracts (the lemma and palea); wetland and desert habitats. _____**Poaceae**

A' DICOTS

1. Plants parasitic; without chlorophyll, or if they have chlorophyll (green) then the parasitic attachment is well above the ground (stem parasites). _____ **Key 1**
1' Plants with chlorophyll and not obviously parasitic.
  2. Composites; individual flowers (florets) generally small and borne in a head resembling a single large flower, the head surrounded by a series of somewhat sepal-like bracts (phyllaries) forming an involucre, the heads often sunflower-like or daisy-like, with a central group of tubular disk florets often surrounded by a ring of ray florets, or the heads with disk florets only, or all florets ray-like with a strap-shaped corolla; ovary of individual florets inferior, the fruit a cypsela (achene), topped by scales or bristles forming the pappus, or the pappus sometimes not present.
    _____ **Asteraceae**
  2' Inflorescence and flowers various but not grouped into heads subtended by a series of bracts forming an involucre, often with green sepals instead of a pappus.
    3. Plants leafless or essentially so, or with scale leaves or few soon-deciduous and reduced leaves (includes trees, shrubs, and some succulents such as cacti but not annuals and generally not plants that are seasonally leafy).
    _____ **Key 2**
    3' Plants leafy, at least seasonally, or the young growth leafy (when plants are dormant, check for leaf scars or dried leaves).
      4. Vines, the stems twining on other plants or sprawling across the ground. _____ **Key 3**
      4' Plants not vining.
        5. Plants conspicuously woody, trees and shrubs; perennials.
          6. Leaves compound. _____ **Key 4**
          6' Leaves simple.
            7. Leaves opposite (and sometimes whorled). _____ **Key 5**
            7' Leaves alternate. _____ **Key 6**
        5' Herbaceous and non-woody plants, or only moderately woody at base; annuals and perennials.
          8. Plants with conspicuously succulent leaves and/or stems (borderline cases key out in both couplets).
          _____ **Key 7**
          8' Plants not conspicuously succulent.

9. Leaves compound. _____ **Key 8**
9' Leaves simple.
    10. Larger leaves basal or in a conspicuous basal rosette (basal rosette leaves may be deciduous), the flowering stems leafless or stems with leaves substantially reduced in size and number (borderline cases key out in both couplets). _____ **Key 9**
    10' Midstem leaves well developed; if a basal rosette present, then the stem leaves conspicuous and usually at least ¼ to ⅓ as large as the basal leaves.
        11. Small herbs; leaves alternate or appearing whorled, or subopposite; stems slender, tough and wiry, flower in spikes on slender, scaly bracted stems; flowers bilaterally symmetrical, blue, ca. 6–8 mm long, the stamens 2, hidden in corollas. _____ **Acanthaceae** (*Elytraria*)
        11' Plants otherwise, the flowering stems not scaly; flowers various colors.
            12. Midstem leaves opposite (a few nodes may bear alternate leaves) or whorled. _____ **Key 10**
            12' Midstem leaves alternate. _____ **Key 11**

Key 1. Parasitic Plants

1. Dodder; stems vining, thread-like, and uniformly yellow or orange; flowers white. _____ **Convolvulaceae** (*Cuscuta*)
1' Plants not vining, the stems not thread-like, not yellow or orange; flowers not white.
    2. Broomrape; root parasites, attached to host roots underground. _____ **Orobanchaceae**
    2' Mistletoe; stem parasites, attached to shrub or tree branches. _____ **Santalaceae**

Key 2. Plants Leafless or Essentially So, or with Scale Leaves (note: excluded are most seasonally leafless plants and annuals that have shed their leaves)

1. Plants armed with spines, thorns, or spinescent twigs.
    2. Cacti; stems succulent, with spine-bearing areoles; leaves none or reduced, succulent, and soon deciduous. _____ **Cactaceae**
    2' Not cacti; trees or shrubs; stems not noticeably succulent, without areoles, the nodes spineless or with only 1 spine; leaves none or not succulent; trees and shrubs.
        3. Ocotillos; shrubs, with wand-like stems mostly unbranched from a common base; nodes regularly spaced and bearing a single stout spine or some stems without spines (leaves quickly drought deciduous); flowers red-orange and tubular. _____ **Fouquieriaceae**
        3' Stems not as above; nodes unarmed (or paired spines in *Parkinsonia florida*), the plants with spinescent-tipped twigs and/or irregularly spaced thorns, or both; flowers not red-orange (or reddish but not tubular in Simaroubaceae).
            4. Stems remaining green for multiple years.
                5. Stems laterally compressed (flattened in cross section); flowers reddish; fruits of small semi-fleshy drupes not persistent. _____ **Simaroubaceae**
                5' Stems terete (rounded in cross section); flowers white or yellow; fruits dry at maturity, persistent or not.
                    6. Fruits persistent, soon becoming woody capsules with 5 awn-tipped valves. **Celastraceae** (*Canotia*)
                    6' Fruits generally not persistent, not awn tipped.
                        7. Palo verde; flowers yellow or yellow and white, and bilaterally symmetrical; fruits of pods more than 2 cm long. _____ **Fabaceae** (*Parkinsonia* when leafless)
                        7' Flowers pale yellowish white and radially symmetrical; fruits rounded, 3–3.5 mm diameter. _____ **Koeberliniaceae**
            4' Stems not remaining green for multiple years.
                8. Shrubs usually less than 1 m tall; sharp points generally on long shoots; fruits bur-like and spiny. _____ **Krameriaceae**
                8' Shrubs generally more than 1 m tall; often with rigid thorn-tipped short shoots or short branches; fruits fleshy, not bur-like and not spiny. _____ **Rhamnaceae** (*Ziziphus*)
1' Plants unarmed.
    9. Stems succulent; sap milky or not.
        10. Sap not milky; halophytes generally near the shore. _____ **Amaranthaceae** (*Allenrolfea, Salicornia*)
        10' Sap milky; not halophytes, generally inland habitats.
            11. Stems thick (5–15 mm diameter); flower structures red; fruits reddish, angled, and spurred; Nolasco. _____ **Euphorbiaceae** (*Euphorbia lomelii*)

11' Stems not especially thick; flowers or flower structures whitish, pink, or yellowish; fruits greenish, smooth surfaced; not on Nolasco.

    12. Plants reed-like, the branches erect or conspicuously ascending; fruits 5 or more cm long. _____ _____ **Apocynaceae** (*Asclepias*)

    12' Plants not especially reed-like, the branches often spreading or not sharply ascending; fruits less than 0.5 cm long. _____**Euphorbiaceae** (*Euphorbia xanti*)

9' Stems not succulent; sap not milky.

    13. Nodes with semi-persistent scale leaves.

        14. Scale leaves opposite. _____ **GYMNOSPERMS: Ephedraceae**

        14' Scale leaves alternate. _____**Tamaricaceae**

    13' Leaves soon deciduous, not scale-like (plants seasonally leafless or nearly so; leafy phases also keyed elsewhere).

        15. Stems pale bluish green, not striate; herbage not scabrous; flowers red-orange; fruits club shaped; widespread. _____**Acanthaceae** (*Justicia californica*)

        15' Stems striate (with longitudinal grooves) and often yellowish; herbage scabrous; flowers bright yellow; fruits of 2 rounded lobes; higher elevations on Sierra Kunkaak. _____ **Oleaceae** (*Menodora*)

Key 3. Vines and Plants with Trailing Stems (stems climbing, twining, or trailing like vines; does not include plants that merely have prostrate stems)

1. Plants with tendrils.

    2. Tendrils at the base of flowering clusters; fruits 3-lobed and 3-celled, inflated like a miniature balloon or paper lantern. _____ **Sapindaceae** (*Cardiospermum*)

    2' Tendrils at nodes along stems; fruits rounded, not lobed.

        3. Herbage glabrous or scabrous; flowers unisexual, yellowish, less than 2 cm wide; fruits to ca. 1.5 cm wide. _____ **Cucurbitaceae**

        3' Herbage conspicuously pubescent but not scabrous; flowers more than 4 cm wide and complex; fruits 3 cm wide. _____**Passifloraceae**

1' Plants without tendrils.

    4. Plants have stinging hairs; flowers unisexual and inconspicuous. _____ **Euphorbiaceae** (*Tragia*)

    4' Plants do not have stinging hairs; flowers bisexual and conspicuous.

        5. Leaves compound.

            6. Leaves alternate, with 3 leaflets; flowers bilateral. _____**Fabaceae** (*Phaseolus, Rhynchosia*)

            6' Leaves opposite, pinnately compound with 3 or more leaflet pairs; flowers radial. _____ _____ **Zygophyllaceae** (*Kallstroemia, Tribulus*)

        5' Leaves simple.

            7. Leaves alternate; fruits dry (capsules).

                8. Leaves arrow shaped (hastate); ovary inferior; perianth 3.5–5 cm long, of a single, bilateral segment, tooth-like above and funnel-like below._____**Aristolochiaceae**

                8' Leaves arrow shaped or not; ovary superior; sepals and petals 5-merous or 5-parted.

                    9. Flowers bilateral, snapdragon-like; seeds numerous, minute. _____ **Plantaginaceae** (*Sairocarpus*)

                    9' Flowers radial; seeds not as above.

                        10. Corollas sympetalous, blue or white; stamens 5, the filaments not fused; fruits of capsules, not noticeably inflated._____**Convolvulaceae** (*Ipomoea, Jacquemontia*)

                        10' Corollas of 5 separate petals, pale yellow; stamens many, the filaments united below into a tube; fruits inflated like a miniature paper lantern. _____ **Malvaceae** (*Herissantia*)

        7' Leaves opposite; fruits dry or fleshy (not capsules).

            11. Sap often milky; flowers not yellow; fruits 3 cm or more long, fleshy, elongated, and not winged. _____ **Apocynaceae** (*Funastrum, Marsdenia, Matelea, Metastelma*)

            11' Sap not milky; flowers yellow or not; fruits not fleshy, winged or not.

                12. Plants with 2-armed hairs (hairs attached at middle with 2 branches extending in opposite directions); flowers bright yellow; fruits dry, with 2–4 papery wings. _____**Malpighiaceae** (*Callaeum, Cottsia*)

                12' Plants glabrous or with simple (not branched) hairs; flowers not yellow; fruits not papery winged. _____**Nyctaginaceae** (*Abronia, Allionia, Boerhavia*)

Key 4. Trees and Shrubs with Compound Leaves

1. Leaves opposite.
   2. Lenticels on stems conspicuous; flowers unisexual, without petals; fruits of single samaras with papery wings; interior of Sierra Kunkaak. _____ **Oleaceae** (*Fraxinus*)
   2' Lenticels not conspicuous; flowers bisexual, with showy petals; fruits of 5-lobed capsules, breaking into 1-seeded segments; widespread. _____ **Zygophyllaceae** (*Fagonia, Guaiacum, Larrea*)
1' Leaves alternate.
   3. Plants armed or unarmed; trunks and limbs not unusually thick or semi-succulent; stipules present and often well developed; herbage not strongly aromatic; leafstalks sometimes with a prominent gland; fruits of pods, multiple seeded (*Errazurizia* and some *Olneya* pods 1-seeded). _____**Fabaceae** (*Acacia, Calliandra, Coursetia, Desmanthus, Ebenopsis, Errazurizia, Hoffmannseggia, Lysiloma, Mimosa, Olneya, Parkinsonia, Pithecellobium, Prosopis, Psorothamnus, Tephrosia, Zapoteca*)
   3' Plants unarmed; trunks and limbs often thick, the wood pithy, soft, and semi-succulent or not; stipules none; herbage strongly aromatic; leafstalks without glands; fruits globose or nearly so, 1-seeded. _____ **Burseraceae**

Key 5. Trees and Shrubs with Simple and Opposite Leaves (sometimes with some alternate leaves, or whorled)

1. Leaf margins not entire: serrated, toothed, undulate (wavy), or the leaf tip conspicuously notched.
   2. Sap milky; leaves to 20 mm long, the surfaces dull, the margins serrated; fruits 3-lobed and 3-seeded; widespread. _____**Euphorbiaceae** (*Euphorbia tomentulosa*)
   2' Sap not milky; fruits not 3-lobed, seeds 4 or more per fruit.
      3. Creosotebush; herbage resinous; leaves deeply notched (actually 2 fused leaflets with entire margins that appear to form a simple leaf); flowers yellow. _____**Zygophyllaceae** (*Larrea*)
      3' Herbage not resinous; leaf tips acute to obtuse or rounded, the leaf margins wavy to toothed; flowers not yellow. _____**Lamiaceae**
1' Leaves entire.
   4. Shrubs or small trees, usually 2 or more m tall; mangroves in tidal lagoons.
      5. Plants forming prominent branched stilt roots descending from branches; "fruits" more than 10 cm long. _____**Rhizophoraceae**
      5' Plants not forming stilt roots but breathing roots emerging from the substrate; fruits less than 3 cm long.
         6. Forming numerous slender, erect pneumatophores; leaves dull, often encrusted with salt crystals, lacking petiole glands; fruits asymmetric and laterally compressed._____ **Acanthaceae** (*Avicennia*)
         6' Forming thick, knobby pneumatophores; leaves shiny green, without salt crystals, bearing a pair of conspicuous petiole glands near the leaf blade; fruits symmetric, terete (not laterally compressed). __ **Combretaceae**
   4' Shrubs usually less than 2 m tall; not mangroves and not in tidal lagoons.
      7. Flowers bright yellow, rose-pink, red, or purple; fruits dehiscent, with 4 to many seeds.
         8. Leaves mostly alternate, some opposite; flowers radially symmetrical, bright yellow; fruits very slender, more than 10 times longer than wide, the seeds with tufts of white hairs at each end. **Apocynaceae** (*Haplophyton*)
         8' Leaves opposite or some may be whorled; flowers bilaterally symmetrical, rose-pink, red, or purple; fruits not especially slender, less than 5 times longer than wide, the seeds without hairs.
            9. Leaves opposite; flowers tubular or not, orange or red-orange, rose-pink, or purple; capsules longer than wide, elastically dehiscent, seeds 4 (sometimes fewer by abortion). _ **Acanthaceae** (*Holographis, Justicia* in part, *Ruellia*)
            9' Leaves opposite or whorled; flowers tubular, bright red; capsules about as wide as long, the seeds minute and numerous. _____**Plantaginaceae** (*Gambelia*)
      7' Flowers or flowering structures green, white, or dull yellow; fruits indehiscent (not capsules), with 1–3 seeds.
         10. Flowers or flowering structures white or yellowish; fruits 3-seeded.
            11. Sap milky; leaves dull; flowers unisexual and in cyathia; fruits of 3-seeded capsules._____ _____**Euphorbiaceae** (*Euphorbia magdalenae*)
            11' Sap not milky; leaves shiny, the margins probably entire or serrated; flowers bisexual and separate (not in cyathia); fruits fleshy._____ **Rhamnaceae** (*Sageretia*)
         10' Flowers pale yellowish or yellow-green; fruits 1-seeded.

12. Stems slender and brittle; leaves semi-succulent; flowers bisexual, on long pedicels in umbels; fruits 0.8–1 cm long, with large sticky glands. _____**Nyctaginaceae** (*Commicarpus*)

12' Stems tough and not brittle; leaves not succulent; flowers unisexual, sessile or short pedicelled, not in umbels; fruits 1–5 cm long.

    13. Leaves thin, not leathery, to ca. 2 cm long, oblong-elliptic to narrowly oblanceolate; fruits to 1 cm long. _____**Oleaceae** (*Forestiera*)

    13' Leaves thick and leathery, 2–5 cm long, elliptic to ovate; fruits 1.5–2 cm long. __**Simmondsiaceae**

Key 6. Trees and Shrubs with Simple and Alternate Leaves

1. Shrubs with prickly-spiny herbage and inflorescences,

    2. Flowers ca. 10 cm wide, with separate white petals, the stamens numerous; fruits of prickly-spiny capsules; San Esteban. _____ **Papaveraceae** (*Argemone*)

    2' Flowers to 6 cm wide, white or lavender, the stamens 5 or 10; fruits smooth or prickly.

        3. Plants with stinging hairs and spines; leaf blades crinkled, about as wide as long; flowers unisexual, white, probably less than 1 cm wide, the stamens 10; fruits of spiny capsules. _____**Euphorbiaceae** (*Cnidoscolus*)

        3' Hairs or spines not stinging; leaf blades relatively flat, longer than wide; flowers bisexual, lavender (corollas), 3–6 cm wide, the petals fused (sympetalous), the stamens 5; fruits of smooth berries._____ _____ **Solanaceae** (*Solanum hindsianum*)

1' Herbage and inflorescences not prickly-spiny (except Kramericeae with bristly fruits and Loasaceae with bristly hairs).

    4. Plants with branched hairs (2-armed, dendritic [tree or candelabra shaped], or stellate [star shaped]), sac-like white hairs, or conspicuous scales.

        5. Plants scabrous, the herbage like sandpaper in texture and adhering like Velcro, the hairs white and firm, with one or more whorls of tiny grappling hooks. ____ **Loasaceae** (*Eucnide cordata* has simple hairs topped by a whorl of recurved barbs, *Petalonyx*)

        5' Plants not scabrous, the herbage not like sandpaper and not adhering; leaves petioled.

            6. Plants with sac-like white hairs, or conspicuous scales.

                7. Herbage bearing sac-like white hairs; leaves not bi-colored; flowers minute, unisexual; fruits 1-seeded, enclosed in bracts. _____**Amaranthaceae** (*Atriplex*)

                7' Leaves markedly bi-colored: dark green above; lower leaf surfaces, inflorescences, and calyces with silvery to brownish scales._____ **Capparaceae**

            6' Plants with branched hairs (2-armed, dendritic, or stellate).

                8. Plants with 2-armed (sometimes T-shaped) hairs.

                    9. Large shrubs and trees with thick, hardwood trunks and branches; fruits 1-seeded, fleshy. **Sapotaceae**

                    9' Small shrubs or subshrubs, with slender, scarcely woody stems; fruits of 3-seeded capsules. _____ _____ **Euphorbiaceae** (*Ditaxis lanceolata*)

            8' Plants with star-shaped (stellate) hairs.

                10. Flowers unisexual, inconspicuous, the corollas white or absent; fruits 2- or 3-seeded capsules. _____ **Euphorbiaceae** (*Bernardia, Croton*)

                10' Flowers bisexual, conspicuous, the corollas pink, yellow, orange, or lavender (rarely white); fruits multiple seeded (1-seeded in *Waltheria*).

                    11. Plants unarmed; corollas various colors but not lavender (rose-purple in *Melochia*), the petals separate or united only at base; stamens 5 or numerous, the filaments united at least below, the anthers small and without pores; fruits dry (capsules or schizocarps). ____ **Malvaceae** (*Abutilon, Hibiscus* in part, *Gossypium, Horsfordia, Melochia, Sphaeralcea* in part, *Waltheria*)

                    11' Plants sometimes unarmed but usually with spines or prickles on stems, leaves, and calyces; corollas lavender, 3–6 cm wide, sympetalous (petals fused); stamens 5, the filaments separate, the anthers large (7–10 mm long) with terminal pores; fruit a many-seeded berry, globose, and smooth. _____ _____ **Solanaceae** (*Solanum hindsianum*)

    4' Plants glabrous or with simple hairs or hairs with tiny hooks (hairs not branched, stellate, or sac-like, and the surfaces without scales).

        12. Leaf margins entire (margins sometimes wavy)

            13. Plants armed with spines or thorns (including thorn-tipped twigs).

      14. Ocotillos; spines regularly spaced and similar in size, or some stems without spines; flowers red-orange in terminal inflorescences. _____ **Fouquieriaceae**

14' Plants otherwise; flowers not red or orange.

    15. Twigs often with single or paired spines at nodes; leaf blades scabrous; fruits orange and fleshy. _____
        _____**Cannabaceae** (*Celtis pallida*)

    15' Leaves not scabrous; fruits not orange.

        16. Flowers bilateral; fruits of bristly burs. _____ **Krameriaceae** (*Krameria bicolor*)

        16' Flowers radial; fruits various but not bristly burs.

            17. Leaves with 3 conspicuous veins originating from the leaf base, especially on the lower surface.
            _____ **Rhamnaceae** (*Condalia, Ziziphus*)

            17' Leaves without conspicuous veins or with a conspicuous midvein.

                18. Flowers lavender or white; fruits fleshy, orange, with multiple seeds (more than 5). _____
                _____**Solanaceae** (*Lycium*)

                18' Flowers small, inconspicuous, petals yellowish or none; fruits 1- or 3-seeded, fleshy or dry, not orange.

                    19. Leaves narrowly spatulate, often semi-succulent, and glaucous, the leaf veins inconspicuous; flowers unisexual, inconspicuous, without petals; fruits fleshy, whitish or blackish, 1-seeded._____ **Achatocarpaceae**

                    19' Leaves obovate to orbicular, thin, and with a prominent midvein; flowers bisexual, yellowish; fruits of 3-seeded capsules. _____**Rhamnaceae** (*Colubrina*)

13' Plants unarmed, without spines and thorns.

    20. Leaves scabrous (rough like sandpaper when you rub your finger backward on the leaf).

        21. Large perennial herbs or slender-stemmed shrubs, without a hardwood trunk; inflorescences branches coiled at tips, fruits 4-seeded, fleshy, white-waxy. _____ **Boraginaceae** (*Tournefortia*)

        21' Hardwood trees or large shrubs often with substantial branches or a (short) trunk; flowers unisexual, solitary or in small clusters; fruits 1-seeded, hard and reddish or brownish. ____ **Cannabaceae** (*Celtis reticulata*)

    20' Leaves not scabrous.

        22. Trees and shrubs with milky sap; bark whitish or pale yellowish white, the roots growing over rocks; fruits of small figs. _____ **Moraceae**

        22' Sap not milky; bark and roots not as above; fruits various but not figs.

            23. Stems relatively thick and notably flexible, often with knobby short shoots; sap copious, watery or blood-like; flowers unisexual. _____ **Euphorbiaceae** (*Jatropha*)

            23' Stems not unusually thick and not especially flexible; sap not especially watery and not blood-like; flowers bisexual (often also unisexual in Sapindaceae).

            24. Leaves relatively thick and semi-succulent or firm, leathery, and tough.

                25. Shrubs with semi-succulent leaves.

                    26. Leaves green; flowers 3–4 mm wide, greenish yellow, solitary or small axillary clusters shorter than the leaves; fruits at first longer than wide, 9–12 mm long, the valves spreading open at maturity; coastal. _____ **Celastraceae** (*Maytenus*)

                    26' Leaves moderately glaucous; flowers 1 cm wide, whitish, in short axillary clusters or terminal racemes longer than the leaves; fruits 6–7 mm diameter, drying as ovoid, reddish capsules. _____ **Stegnospermataceae**

                25' Trees and shrubs with firm, coriaceous, tough leaves, not at all succulent.

                    27. Leaves falcate (moderately sickle shaped), usually more than 10 cm long; flowers white; rare, not reproducing, planted at abandoned military stations. _____**Myrtaceae**

                    27' Leaves symmetrical, to 6 cm long, native and well established.

                      28. Leaves mostly less than 1.5 cm long, glaucous, not spine tipped, essentially sessile or petioles less than 1 mm long; flowers white, with several (rarely 1) separate pistils.
                      _____ **Crossosomataceae**

                      28' Leaves more than 2 cm long, dark green and not glaucous, spine tipped, and petioled; flowers bright orange, and with a single pistil. _____**Theophrastaceae**

            24' Leaves relatively thin and flexible, not tough or semi-succulent.

            29. Herbage viscid-sticky; leaf blades narrowly oblong to lanceolate or oblanceolate, more than three times as long as wide; fruits with 3 papery wings. _____ **Sapindaceae** (*Dodonaea*)

            29' Herbage smooth or pubescent, not sticky; leaves mostly elliptic to ovate, less than three times as long as wide (except Plumbaginaceae); fruits not winged.

30. Flowers bilateral, bright yellow or purple.
   31. Leaves mostly less than 1.5 cm long; flowers purple, sepals petal-like, the actual petals very dissimilar; fruits of spiny burs. _____ **Krameriaceae**
   31' Leaves often more than 1.5 cm long; sepals green, the petals bright yellow with 4 similar petals and one lip petal; fruits of three small nutlets, not spiny. _____ **Malpighiaceae** (*Echinopterys*)
30' Flowers radial, not purple (except in Phytolaccaceae); fruits not spiny.
   32. Flowers about 3–5 mm wide; fruits fleshy, 1-seeded.
      33. Herbage green; leaves lanceolate; flowers in short inflorescences at nodes; flowers white, with a 5-merous calyx and corolla; fruits white, ovoid, 10–12 mm long. _____ **Apocynaceae** (*Vallesia*)
      33' Herbage reddish green; leaves ovate; flowers in slender, terminal racemes; flowers white or pink, the sepals 4, petals none; fruits red, 3–4 mm diameter._____ **Phytolaccaceae**
   32' Flower size various; fruits dry, or at least not fleshy; seeds 1 or more.
      34. Flowers white with blue anthers; fruits sticky by means of stalked calyx glands, the calyx enclosing the 1-seeded capsule opening circumscissle near the base. _____ **Plumbaginaceae**
      34' Flowers not pure white, the anthers not blue; fruits not sticky, with 3 or more seeds, not circumscissle.
         35. Herbage glabrous or sparsely pubescent, not velvety; leaves 0.5–3 cm long; flowers 0.5–0.6 cm wide, yellow; fruits rounded, 0.5–0.6 cm long, 3-seeded. _____ **Rhamnaceae** (*Colubrina*)
         35' Herbage velvety pubescent; leaves 2.5–4.5 cm long; flowers 2 cm wide, whitish or pale yellow; fruits ovoid, more than 1 cm long, multiple seeded. _____**Zygophyllaceae** (*Viscainoa*)
12' Leaves serrated, toothed, lobed, or divided.
   36. Leaf blades scabrous (rough like sandpaper).
      37. Leaf blades asymmetric at base; flowers unisexual, corollas none. _____ **Cannabaceae** (*Celtis reticulata*)
      37' Leaf blades symmetric; flowers bisexual, corollas prominent. _____ **Boraginaceae** (*Cordia, Tiquilia*)
   36' Leaf blades not scabrous, generally symmetric.
      38. Flowers unisexual, with male and female flowers very different.
         39. Not wetland plants; male and female flowers on the same plant; seeds not minute, without hairs. ___ _____ **Euphorbiaceae** (*Acalypha, Jatropha, Sebastiania*)
         39' Willows, wetland habitats with permanent water (Sauzal waterhole); male and female flowers on separate plants; seeds minute, each with a tuft of long, silky hairs. _____**Salicaceae**
      38' Flowers bisexual.
         40. Plants with simple hairs topped with a whorl of recurved barbs. _____ **Loasaceae** (*Eucnide cordata*)
         40' Hairs simple or stellate, not barbed.
            41. Plants with simple hairs; stamens 5, separate. _____ **Boraginaceae** (*Cordia, Tiquilia*)
            41' Plants with stellate hairs; stamens many, the filaments united below into a column. _____ _____**Malvaceae** (*Abutilon, Gossypium, Hibiscus denudatus, Horsfordia, Melochia, Sphaeralcea, Waltheria*)

Key 7. Non-woody, Leafy Succulents (leaves and/or stems succulent)

1. Perennials.
   2. Stems semi-succulent, slender, erect, reed- or rush-like, and with milky sap; leaves opposite and very slender (thread-like), few and quickly deciduous. _____ **Apocynaceae** (*Asclepias*)
   2' Stems not as above, the sap not milky; leaves well developed, not thread-like and not quickly deciduous (or present during at least one rainy season).
      3. Leaves all opposite; low, spreading or mat-forming; coastal habitats.
         4. Petioles often about as long as the blades, the blades less than twice as long as wide; flowers bright purple-magenta. _____ **Nyctaginaceae** (*Abronia*)
         4' Leaves sessile or short petioled, more than twice as long as wide; flowers greenish or pink.
            5. Flowers bisexual, pink, solitary in leaf axils. _____**Aizoaceae** (*Sesuvium*)
            5' Flowers greenish, inconspicuous, several or more in unisexual cone-like succulent structures. **Bataceae**
      3' Leaves alternate (or lower leaves sometimes opposite or subopposite); coastal or not.
         6. Shrubs, often 0.5 m or more in height.
            7. Leaves and stems both succulent or semi-succulent. ___ **Amaranthaceae** (*Atriplex barclayana, Suaeda nigra*)

7' Leaves or stems succulent or semi-succulent but not both.

    8. Stems woody, the leaves semi-succulent; flowers bisexual; fruits with a fleshy red aril. _____
_____ **Celastraceae** (*Maytenus*)

    8' Stems semi-succulent, the leaves not succulent; flowers unisexual; fruits green, drying as brownish capsules, without an aril. _____ **Euphorbiaceae** (*Jatropha*)

  6' Herbaceous perennials, mostly less than 0.5 m tall.

    9. Flowers in terminal spike-like inflorescences coiled (helicoid) at the tip; corollas white (center often becoming purplish); fruits breaking into four 1-seeded nutlets. _____ **Boraginaceae** (*Heliotropium*)

    9' Inflorescences not coiled; flowers not white; fruits not of 4 nutlets.

      10. Roots not tuberous; leaves mostly less than 3.5 cm long; flowers inconspicuous, greenish, in densely flowered clusters or inflorescences; fruits 1-seeded, indehiscent. _____**Amaranthaceae** (*Atriplex barclayana*, *Suaeda esteroa*)

      10' Plants appearing with summer rains from tuberous roots; leaves 2.5–11+ cm long; flowers pink, on openly branched panicles 30 or more cm long; fruits of multiple-seeded capsules. _____
_____ **Portulacaceae** (*Talinum*)

1' Annuals.

  11. Stems and petioles semi-succulent, with conspicuous large hairs topped by a whorl of recurved barbs, causing the plants to stick like Velcro, the herbage shinny; corollas tubular, yellow with green lobes. _____
_____ **Loasaceae** (*Eucnide rupestris*)

  11' Plants glabrous or glabrate; flower small, not tubular, corollas none or not as above.

    12. Leaves opposite. _____ **Aizoaceae** (*Mesembryanthemum*, *Trianthema*)

    12' Leaves alternate; fruits various.

      13. Herbage generally bluish green; inflorescences coiled (helicoid or scorpioid) at the tip; flowers white with a yellow or purplish center; mature fruits separating into four 1-seeded nutlets. _____
_____ **Boraginaceae** (*Heliotropium*)

      13' Herbage not bluish; inflorescences not coiled at tip; flowers not white.

        14. Halophytes at esteros; leaves thick and awl shaped; flowers greenish, inconspicuous; fruits 1-seeded.
_____ **Amaranthaceae** (*Suaeda esteroa*)

        14' Inland habitats; leaves succulent but flattened; flowers bright yellow or red-pink, fading with daytime heat; fruits of capsules opening around the middle (circumscissile), with several to many seeds.
_____ **Portulacaceae** (*Portulaca*)

Key 8. Herbaceous Plants with Compound Leaves (includes plants with highly divided leaves that may appear compound)

1. Leaves opposite (sometimes alternate on first or lower nodes); flowers radial._____**Zygophyllaceae** (*Fagonia*, *Kallstroemia*, *Tribulus*)

1' Leaves alternate or in a basal rosette, not obviously opposite; flowers radial or not.

  2. Leaves highly dissected with slender segments; flowers minute, with quickly deciduous white petals, in compact, densely flowered umbels; fruits 3 mm long, bearing spines with minute barbs at the tips._____ **Apiaceae**

  2' Flowers not in umbellate clusters; fruits otherwise.

    3. Leaflets 3–9, palmately (digitately) arranged.

      4. Leaflets 3–5 per leaf; flowers pale yellow, 3–5 mm long._____ **Cleomaceae**

      4' Leaflets (5) 7–9 per leaf; flowers lavender-pink with a yellow spot on the banner petal, 7–9 mm long. ___
_____**Fabaceae** (*Lupinus*).

    3' Leaflets pinnately arranged.

      5. Sepals and petals each 4; stamens 6._____ **Brassicaceae** (leaves deeply parted or divided, *Descurainia*, *Lyrocarpa*)

      5' Sepals or calyx lobes and petals or corolla lobes each 5; stamens 5 or 10.

        6. Stipules none; flowers radial, the corollas sympetalous; stamens 5; styles 2 or 2-branched; fruits of globose capsules. _____ **Boraginaceae** (leaves deeply parted or divided, *Eucrypta*, *Phacelia*)

        6' Stipules usually present; flowers bilateral, some or all petals separate; stamens 10; styles 1, the stigma unbranched; fruits of pods, not globose. _____ **Fabaceae** (*Acmispon*, *Dalea*, *Desmodium*, *Lupinus*, *Marina*, *Senna*, *Tephrosia*)

Key 9. Herbaceous Plants with Simple, Basal Leaves (larger leaves in basal rosettes, the stem leaves smaller, reduced or absent. Many of these plants can also be identified in other keys)

1. Perennials from a hardened, somewhat woody base, the stems often swollen; perianth parts 6, in 2 separate and similar whorls, surrounded by bracts; fruits 1-seeded. _____ **Polygonaceae** (*Eriogonum*)
1' Cool-season annuals (except *Mollugo*); stems not swollen.
   2. Plants glabrous, mat-like or prostrate; leaves opposite but crowded, the stipules and sepals white and papery.
     _____ **Caryophyllaceae** (*Achyronychia*)
   2' Plants glabrous or hairy; sepals and stipules (if present) not white and papery.
     3. Flowers with perianth parts each 4 (sepals or calyx lobes, and petals or corolla lobes).
       4. Perianth brown and papery when dry; fruits of globose capsules opening around the middle. _____
         _____ **Plantaginaceae** (*Plantago*)
       4' Perianth not brown and papery; fruits longer than wide, not opening around the middle.
         5. Leaves broadly obovate to 4 cm long, with forked and stellate hairs; petals white, less than 5 mm long and soon deciduous; stamens 6; ovary superior._____ **Brassicaceae** (*Draba*)
         5' Leaves various but not obovate, often more than 4 cm long, glabrate or with simple hairs; petals 5 or more mm long; stamens 8; ovary inferior._____ **Onagraceae**
     3' Flowers 5-merous (sepals, petals, and stamens 5) or the petals and stamens many.
       6. Plants with rough, branched hairs adhering like Velcro; flower parts mostly more than 5. _____ **Loasaceae** (*Mentzelia*)
       6' Plants glabrate, glabrous, or the hairs not sticking like Velcro; flowers 5-merous.
         7. Plants glabrous or sparsely pubescent.
           8. Cool-season annuals, glabrous or with few soft hairs at base; flowering stems zigzag; flowers with red-tipped white petals._____**Campanulaceae**
           8' Warm-weather annuals, glabrous; flowering stems not zigzag; petals none. _____**Molluginaceae**
         7' Plants conspicuously hairy.
           9. Plants with stiff, calcified or silicified hairs, soft hairs, or glandular hairs; flowers blue, lavender, or pink, or if white then less 10 mm long, open in daytime; fruits separating into 4 or fewer 1-seeded nutlets or capsules with few to many seeds. _____**Boraginaceae** (*Cryptantha, Eucrypta, Nama, Phacelia*)
           9' Plants sticky with glandular hairs; flowers white, opening at night and closing in daytime or remaining open in cool weather, corollas 12–20 mm long; fruits of capsules with numerous minute seeds.
           _____ **Solanaceae** (*Nicotiana clevelandii*)

Key 10. Herbaceous Plants, the Midstem Leaves Simple and Opposite or Whorled (some nodes may have alternate leaves)

1. Flowers unisexual, inconspicuous (non-floral appendages may be white or pink).
   2. Sap not milky; leaves scurfy gray or whitish due to inflated hairs that collapse upon drying; female flowers and fruits enclosed in a pair of sepal-like bracts; fruits 1-seeded, indehiscent, not 3-lobed. _____**Amaranthaceae** (*Atriplex* in part)
   2' Sap milky; leaves glabrous or hairy but not scurfy gray; female flowers and fruits not enclosed in bracts; fruits 3-lobed and 3-seeded. _____ **Euphorbiaceae** (*Euphorbia* subgenus *Chamaesyce*)
1' Flowers bisexual, inconspicuous or not.
   3. Leaves whorled or appearing whorled or densely clustered; flowers small, green and/or white.
     4. Leaves broadly elliptic, often semi-succulent, in clusters of 4–6 on slender stems; flowers small, with white petals._____ **Caryophyllaceae** (*Drymaria*)
     4' Leaves much longer than wide, linear or spatulate.
       5. Warm-weather annuals, usually 3–14 cm tall; leaves 3.5–11 mm long; stems thread-like; ovary and capsule not open at apex. _____**Molluginaceae**
       5' Non-seasonal annuals 5–40 cm tall; leaves 15–40+ mm long; stems slender but not thread-like; ovary and capsule open and gaping at apex. _____**Resedaceae**
   3' Leaves opposite (sometimes also with some alternate or basal-rosette leaves); flowers of various sizes, not green and/or all white (except *Mirabilis*, Nyctaginaceae).
     6. Herbage glandular pubescent, usually sticky.

7. Leaves opposite below, alternate above, often 6–15 cm long, the blades broadly ovate to orbicular or kidney shaped, and shallowly lobed; flowers bilateral, 3–4 cm long; fruits of woody capsules 4–6 cm long with 2 claws 9–14 cm long. _____**Martyniaceae**

7' Leaves mostly not more than 6 cm long; flowers and fruits not more than 2 cm long, fruits not woody and not clawed.

    8. Leaves petioled, the margins entire or wavy/crenate but not serrated; flowers radial and white, pink, lavender-pink, or bilateral, purple, and in clusters of 3 resembling a single flower; fruits indehiscent, 1-seeded. _____ **Nyctaginaceae** (*Allionia, Boerhavia, Mirabilis*)

    8' Leaves sessile, the margins serrated; flowers bilateral, dark blue; fruits of many seeded capsules. ____ _____ **Plantaginaceae** (*Stemodia*)

6' Herbage glabrous or with non-glandular hairs.

  9. Plants often with some alternate leaves; moderately to most often densely white woolly with tree-like (dendritic) hairs; flowers minute, with yellow sepals, petals none. _____ **Amaranthaceae** (*Tidestromia*)

  9' Plants glabrous or hairy but not densely woolly, the hairs not tree-like; flowers yellow or not, with or without petals.

    10. Opposite leaves of each pair unequal in size; flowers solitary, the calyx pink, petals none; capsules opening around the middle (circumscissile). _____**Aizoaceae** (*Trianthema*)

    10' Opposite leaves similar in size; flowers several or more per inflorescence, the calyx not pink, petals present; fruits not circumscissile.

      11. Flowers bright yellow, becoming orange or red with age; fruits 3-lobed, separating into 3 segments. _____**Malpighiaceae** (*Galphimia*)

      11' Flowers not as above; fruits of 2-valved capsules elastically splitting longitudinally. _____ _____ **Acanthaceae** (*Carlowrightia, Dicliptera, Justicia longii, Tetramerium*)

Key 11. Herbaceous Plants, the Midstem Leaves Simple and Alternate

1. Flowers small and unisexual (except Urticaceae with unisexual and bisexual flowers).

  2. Fruits 3-lobed, each lobe with a single seed (note: *Euphorbia eriantha* flowers are in a flower-like cyathium or cluster; this species differs from all other plants in Key 11 in having milky sap). _____ _____**Euphorbiaceae** (*Ditaxis, Croton, Euphorbia eriantha*)

2' Fruits 1-seeded and not lobed.

  3. Annuals or perennials; stems usually not weak and not semi-succulent; flowers all unisexual; sepal 5 or none; fruits of capsules opening around the middle (*Amaranthus*), or fruits tightly enclosed in bracts (*Atriplex*). __ _____ **Amaranthaceae**

  3' Cool-season, delicate annuals; stems weak and semi-succulent, the leaves thin and soft; flowers in small clusters with both unisexual and bisexual flowers; sepals 4; fruits of achenes, not enclosed by bracts. ____**Urticaceae**

1' Flowers bisexual (Urticaceae with bisexual and unisexual flowers), of various sizes.

  4. Plants glabrous; leaves linear-filiform; flowers sessile; petals minute and white; stamens 3; ovary and fruit gaping open at apex. _____**Resedaceae**

4' Plants glabrous or pubescent; leaves not linear-filiform (or if very slender, then not glabrous); flowers sessile or not; stamens various but not 3; ovary and fruit not gaping open at apex.

  5. Stems weak and trailing; leaves arrow shaped (hastate or sagittate); perianth bilateral with a single tooth-like segment 3.5–5 cm long, maroon and yellow speckled. _____ **Aristolochiaceae**

5' Stems mostly not trailing; leaves not hastate; perianth radial or bilateral with 4 or more segments.

  6. Plants with star-shaped (stellate) or tree-like (dendritic) hairs.

    7. Hairs harsh, dendritic, and minutely hooked, the leaves and capsules adhering like Velcro. _____ _____**Loasaceae** (*Eucnide rupestris; Mentzelia*)

    7' Hairs branched but not harsh, not hooked—not sticking like Velcro.

      8. Sepals and petals each 4 and separate; stamens 6; fruits laterally compressed, longer than wide. __ _____**Brassicaceae** (*Descurainia, Lyrocarpa*)

      8' Sepals or calyx lobes 5 and petals or corolla lobes 5 or more, or petals none; stamens 5 or more but not 6; fruits rounded or radially symmetrical.

        9. Corollas lavender, rotate (dish shaped and sympetalous), 3–5 cm wide; stamens 5, anthers large and with terminal pores; fruits fleshy, globose berries. _____ **Solanaceae** (*Solanum*)

        9' Corollas not rotate (petals separate, or united at base or at the top in *Ayenia*); stamens 5 or more, the anthers small and not opening at a pore; fruits of capsules.

10. Plants often white woolly (less so in shade and when well watered); hairs branched above the base (dendritic); leaves opposite and alternate on the same plant; flowers small and yellow, each enclosed in a cup-shaped involucre becoming hard as the fruit matures, the fruit indehiscent and inconspicuous, 1-seeded. _____ **Amaranthaceae** (*Tidestromia*)

10' Hairs branched from the base (stellate); leaves alternate; flowers of various colors as well as yellow, not enclosed in a cup-shaped involucre; filaments united below into a column; fruits of conspicuous capsules or schizocarps, multiple seeded._____ **Malvaceae** (in part)

6' Plants glabrous or with simple (unbranched) hairs.

11. Flowers small and inconspicuous, with a calyx but without a corolla; fruits indehiscent (not capsules) and 1-seeded.

12. Leaf blades about as long as wide; sepals usually 5._____ **Amaranthaceae** (*Chenopodium*)

12' Leaves at least twice as long as wide; sepals 4 _____ **Urticaceae**

11' Flowers with a calyx and corolla; fruits of capsules with 1 or more seeds.

13. Flowers 4-merous, white, yellow, orange, etc., the ovary inferior: calyx and corolla segments each 4 and attached to the top of the ovary, stamens 8; fruits of capsules with numerous minute seeds. _ **Onagraceae**

13' Flowers 4- or 5-merous, ovary superior.

14. Perennials from a single, thick, brown, tuberous root, responding to summer rains and dormant during the rest of the year; flowers 5-merous, large and showy, yellow or yellow and orange; fruits of capsules 3 or more cm long.

15. Herbage not mucilaginous; leaves deeply and palmately lobed, the lobes toothed; flowers bright orange, ca. 5 cm wide, the stamens with apical pores; capsules ovoid and herbaceous, 3–4 cm long._____ **Cochlospermaceae**

15' Herbage mucilaginous; leaves opposite below, alternate above, and shallowly lobed; flowers yellow, and orange and brown speckled, ca. 4 cm long, the stamens without pores; capsules becoming woody, the body more than 4 cm long, the apex curved into a long beak splitting into 2 long, hooked, claws. _____ **Martyniaceae**

14' Annuals or perennials but not root perennials from a single thick tuberous root (*Tiquilia palmeri* has several long, thickened black roots); flowers 4- or 5-merous, small to large, white, pink, pale yellow, or blue; fruits less than 3 cm long (except *Datura* with a globose, prickly capsule).

16. Fruits with 1–4 seeds.

17. Annuals; herbage glandular, stinky, and irritating; inflorescence branches coiled at tips (helicoid or scorpioid). _____**Boraginaceae** (*Phacelia*)

17' Annuals or perennials; herbage not glandular and stinky; flowers not on coiled flowering stems.

18. Tufted herbaceous perennials, stems erect and yellow-green; larger leaves lanceolate-elliptic and toothed, smaller leaves narrower with smaller teeth or sometimes entire; flowers axillary among uppermost leaves, petals white, to 3 mm long; capsules 3-seeded. __ _____ **Violaceae**

18' Plants not as above; fruits with 1–4 seeds, usually not 3.

19. Perennials, prostrate or mat-like or subshrubs to ca. 20 (40) cm tall; leaves petioled, the leaf veins conspicuously incised or the leaves firm, grayish, and finely woolly and with coarse hairs, the margins revolute (inrolled); flowers 4- or 5-merous, often in forks of branches, the corollas lavender; fruits separating into 4 nutlets. _____ _____ **Boraginaceae** (*Tiquilia*)

19' Annuals or perennials; leaves sessile or short petioled, the margins not revolute; flowers 5-merous, white or blue; capsules with 1–4 seeds._____ **Convolvulaceae** (*Jacquemontia agrestis, Cressa, Evolvulus*)

16' Fruits with more than 4 seeds.

20. Annuals; flowering branches coiled at tips (helicoid or scorpioid). _**Boraginaceae** (*Phacelia*)

20' Annuals or perennials; flowering branches not coiled.

21. Delicate annuals; leaves opposite below, alternate and reduced above, pinnatifid, 1.5–5 cm long; flowers radially symmetrical, whitish or pale lavender, ca. 3–4 mm wide. ____ _____ **Boraginaceae** (*Eucrypta*)

21' Annuals or perennials; leaves alternate, the margins entire or shallowly lobed or wavy.

22. Annuals; leaves linear to ovate; flowers blue or purplish, bilaterally symmetrical, stamens 4. _____ **Plantaginaceae** (*Nuttallanthus, Pseudorontium*)

22' Annuals or perennials; leaves lanceolate to ovate; flowers white or yellowish, radially symmetrical, stamens 5. ____ **Solanaceae** (*Datura, Nicotiana obtusifolia, Physalis*)

Acanthaceae – Acanthus Family
(includes Avicenniaceae)

Perennial herbs to shrubs and one tree. Leaves opposite (except *Elytraria*), simple, entire; stipules none. Flowers subtended by bracts. Fruit a capsule, usually elastically dehiscent (except *Avicennia*). The island and Sonoran acanths respond most vigorously during the warmer seasons if there is sufficient soil moisture. Some, however, cease flowering during the summer, such as *Holographis virgata*, *Justicia californica*, and *J. candicans*.

Acanth diversity in northwestern Mexico decreases sharply from tropical deciduous forest and thornscrub into the Sonoran Desert (Felger 2000a). A similar trend is evident on the Sonoran Islands, where acanth diversity is greatest in the Sierra Kunkaak and decreases sharply in more arid sites and islands. Only *Carlowrightia arizonica* and *Justicia californica* penetrate into the driest regions of the desert, such as the Pinacate region of northwestern Sonora, and are the only acanths on San Esteban. An enigma to this general trend is the absence of acanths from Isla San Pedro Nolasco.

The monographs of the acanths of Arizona, the Baja California Peninsula, and Sonora by Tom Daniel (1984, 1997, 2004) served as valuable resources for our report on this family.

Figure 3.4. *Carlowrightia arizonica*. FR (© James Henrickson).

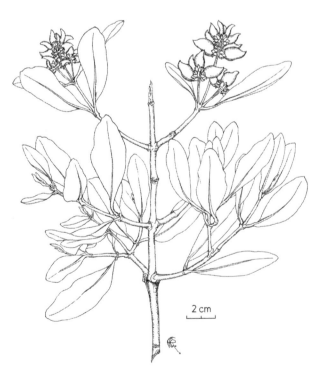

Figure 3.3. *Avicennia germinans*. FR.

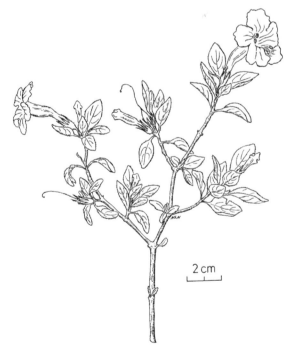

Figure 3.5. *Ruellia californica*. NLN.

Figure 3.6. *Avicennia germinans.* Punta San Miguel, Isla Tiburón. *Batis maritima* and juvenile *Rhizophora mangle* in foreground. BTW, 2 May 2007.

Figure 3.7. *Justicia candicans.* Interior canyons of Sierra Kunkaak, Isla Tiburón. BTW, 24 November 2006.

1. Small herbs; leaves alternate or appearing whorled; flower spikes on slender, scaly stems; flowers blue. ___ **Elytraria**
1' Leaves opposite; flowers not on slender scaly stems, not blue.
    2. Hardwood shrubs or small trees; mangroves with roots in tidal seawater. _____ **Avicennia**
    2' Herbaceous perennial or scarcely woody bushes; not mangroves.
        3. Shrubs or dense, mound-shaped bushes.
            4. Leaves conspicuously glandular-sticky; corollas lavender, large and showy, about as wide as long. __ **Ruellia**
            4' Leaves not glandular-sticky; corollas as long as or longer than wide, not lavender.
                5. Petioles very short, the leaves subsessile; leaf blades generally at least twice as long as wide, more or less oblong to linear; corollas rose-pink; fertile stamens 4. _____**Holographis**
                5' Petioles short to well developed; leaf blades generally about as long as wide, more or less ovate; corollas red-orange or red; fertile stamens 2. _____ **Justicia californica, J. candicans**
        3' Perennial herbs.
            6. Floral bracts large and conspicuous and leaf-like or heart shaped.
                7. Young stems 6-sided; flowers clustered but not in dense spikes, the flower bases enclosed by a pair of cordate to obovate bracts, these bracts to 7 mm long; corollas rose-purple. _____ **Dicliptera**
                7' Young stems moderately 4-sided or rounded in cross section; flowers in dense, 4-sided spikes, the floral bracts oblong-ovate (not cordate at base), 7–15 mm long; corollas cream color._____ **Tetramerium**
            6' Floral bracts not noticeably leafy or large and conspicuous.
                8. Plants not rhizomatous; flowers diurnal (corollas generally fall away with daytime heat); corollas whitish with a yellow eye and purplish markings on the upper lip, about as wide as long, ca. 1.5 cm long; capsules about as wide as long._____ **Carlowrightia**
                8' Multiple stems from short, rhizomatous rootstocks; flowers nocturnal; corollas pure white, longer than wide, with a tube 3.5 cm long; capsules longer than wide. _____**Justicia longii**

Avicennia germinans (Linnaeus) Linnaeus

Pnacojiscl "drab mangrove"; *mangle dulce, mangle negro*; black mangrove

Description: Hardwood large shrubs or small trees sometimes to more than 4 m tall. There are no resting terminal buds, although growth slows or ceases during colder weather. Rapid growth occurs during the long, hot summer.

The complex root system consists of cable roots, anchoring roots, and pneumatophores (breathing roots). The pneumatophores form miniature "forests" sticking up from the mud beneath the branches—at low tide small fields of pneumatophores extend across the shaded wet mud beneath the shrubs; they take up oxygen from the air into a root system growing in oxygen poor, often anaerobic soils. The pneumatophores are erect, often 20–30 cm tall, rather straight and pencil-like, usually unbranched, corky, and tapering to a blunt tip. The cable roots are slender and extend out horizontally beneath the mud often many meters from the base of the trunk. Both the pneumatophores and anchoring roots arise alternately at rather regular intervals from the cable roots.

Bark gray and rather smooth, with age becoming rough. Stems and leaves opposite and decussate (each leaf and branch pair at right angles to the one above and the one below). Leaves evergreen, often 6–12 cm long, encrusted with salt, the blades rather leathery, mostly narrowly to broadly elliptic or ovate with blunt tips, dull gray-green, the upper surface darker, and glabrous with age, the midrib prominent below; petioles 5–20 mm long, the petiole base has a deep groove, this groove blackish inside with a dense, brush-like margin of tawny yellowish white hairs.

The xylem sap has high salt concentrations and excess salt is exuded through microscopic salt glands in both leaf surfaces, but especially the upper surfaces and encrusting the leaves. The leaf buds as well as the axillary meristems are protected by the closely appressed petiole margins and the brush of dense hairs on the surrounding basal petiole groove of the ultimate leaf pair.

Inflorescences of crowded, terminal and axillary simple or compound spikes shorter than the leaves. Flowers white, 13–15 mm wide, bilateral, producing copious nectar, and remaining open several days. On calm warm nights the moist sea air in the esteros is filled with sweet scent of the flowers. Corollas thick, with a short tube and 4 lobes. Stamens 4, attached to the corolla base and bent to orient the anthers to one side above the larger corolla lobe. Stigmas with 2 spreading lobes, developing before the stamens, the lobes closing apparently after pollination and before the anthers open. Fruits of leathery capsules, 1–2 cm long, asymmetric, laterally compressed, pointed at the tip (a short lateral beak), 1-seeded.

Apparently flowering most heavily in late spring and early summer, with large numbers of seedlings produced in summer and early fall. Fruits ripen rapidly during the summer and at maturity consist of the thin fruit wall enclosing a single embryo. There is no true seed in the usual sense since the embryo germinates while the fruit is still on the tree. When the fruit falls in the water, often washing ashore, the fruit wall splits almost immediately, freeing the compact seedling, which consists of folded green cotyledons encasing a thick radicle densely covered by root hairs.

Local Distribution: *Avicennia* and the other two mangroves form impenetrable thickets in shallow tidal waters of the three mangrove esteros on the Infiernillo coast of Tiburón (Punta San Miguel, Palo Fierro or Punta Tormenta, and Punta Perla). *Avicennia* reaches maximum density on the landward side of the *Laguncularia* and *Rhizophora* zones. The roots and lower branches are flooded and exposed daily by the tide. *Avicennia* reaches the greatest height of the three mangrove species in the

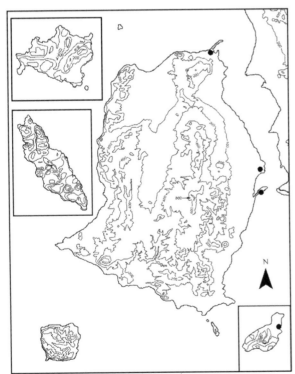

Avicennia germinans
Pnacojiscl

region, and occasional trees rise to about 2 m above the surrounding mangroves. In 2007 two juvenile plants were found in the *Allenrolfea* flat on Alcatraz.

As with the other mangroves, black mangroves are frost sensitive, and the ones at Estero Sargento (mainland opposite Punta Perla) are sometimes severely damaged by freezing weather.

Other Gulf Islands: Coronados, Carmen, Danzante, San José, San Francisco, Espíritu Santo.

Geographic Range: Both shores of the Gulf of California and the Pacific coast of southern Baja California Sur; Sonora northward to Estero Sargento and a small population of stunted plants farther north at Puerto Lobos. Coastal areas southward to Peru; Bahamas, Florida, the Caribbean, and Texas to Brazil, and West Africa. In the tidal brackish water of the wet tropics it develops into a tree sometimes 25 m in height with a massive bole. In dry regions it remains stunted.

Tiburón: Estero San Miguel, *Wilder 06-9.* Palo Fierro, estero, *Felger 12558.* Punta Perla, estero, *Felger 21314.*

Alcatraz: *Allenrolfea* flat, rare, in wet soil, ca. 1 m tall, *Wilder 07-408.*

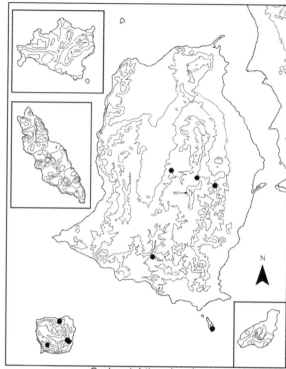

Carlowrightia arizonica

Carlowrightia arizonica A. Gray [*C. californica* var. *pallida* I.M. Johnston, Proc. Calif. Acad. Sci. Ser. 4, 12:1169, 1924]

Ramoneada flor blanca, lemilla

Description: Suffrutescent perennials or subshrubs with slender stems. Leaves gradually drought deciduous. The flowers snap open around sunrise, and the corollas and attached stamens fall as a unit with daytime heat. Corollas whitish with a yellow eye and purplish nectar guide lines on the upper lip (upper lip formed by two fused petals). Flowering primarily late spring and during summer rainy season.

Local Distribution: Common on San Esteban, mostly along gravely-rocky washes of Arroyo Limantour, in canyons, and on rocky slopes. Also on Dátil but not as common or widespread as on San Esteban. The Tiburón records are widely separated, and it is generally not common. In the Pinacate region of northwestern Sonora, Felger (2000a:64) observed that "in open places it is almost always grazed by rabbits, rodents, and especially chuckwallas, which reduce the plant to a mass of short, stubby stems."

Other Gulf Islands: San Marcos, Carmen, Monserrat, Santa Catalina, Santa Cruz, San Diego, San José, Espíritu Santo, Cerralvo.

Geographic Range: Southern Arizona to west Texas, widespread in Sonora, both states of Baja California, disjunct in Anza-Borrego in San Diego County, California, and through Mexico to Costa Rica. The geographic range essentially spans that of the entire genus. It is a "taxonomically complex species with numerous diverse morphological forms" (Daniel 1984:167).

Tiburón: Haap Hill, *Felger T74-9.* Xapij (Sauzal), *Wilder 08-123.* Arroyo bottom, NE base of Sierra Kunkaak, *Wilder 06-388.* Hast Cacöla, E flank of Sierra Kunkaak, 380 m, *Wilder 08-235.*

San Esteban: 20 Apr 1921, *Johnston 3195* (holotype of *C. californica* var. *pallida,* CAS). Arroyo Limantour, *Felger 7076.* Steep slopes of S-central peak, *Felger 17544.* N side of island, narrow rocky canyon, *Felger 17594.*

Dátil: SE side, steep narrow canyon, *Felger 17511.*

Dicliptera resupinata (Vahl) Jussieu

Alfalfilla, ramoneada flor morada

Description: Herbaceous perennials, often bushy, and also flowering in the first season or year. Flowers subtended by a pair of conspicuous, usually heart-shaped

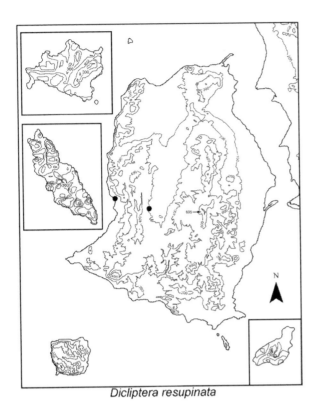

Dicliptera resupinata

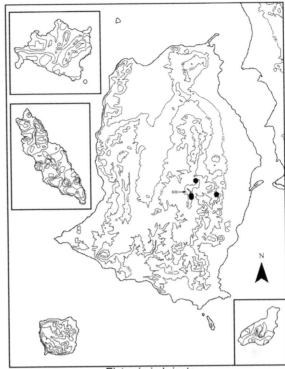

Elytraria imbricata

bracts. Corollas white with pale lavender lip markings. Flowering response non-seasonal.

Local Distribution: Known from Tiburón by two records at low elevations in the western margin and central parts of the island. We predict it is actually more widespread on the island.

Other Gulf Islands: San José, Cerralvo.

Geographic Range: Widespread in Sonora from deserts and thornscrub to oak woodland. Arizona and New Mexico, and western Mexico southward to Michoacán and Zacatecas, and in Baja California Sur.

Tiburón: Ensenada Blanca (Vaporeta), 20 Feb 1968, *Felger 17275*. SW Central Valley, scattered but locally common, *Felger 17309*.

Elytraria imbricata (Vahl) Persoon

Cordoncillo

Description: Small tufted perennials, the stems slender, tough and wiry. Leaves drought deciduous, alternate or appearing whorled, or subopposite. Flowers in spikes on slender, scaly-bracted stems, the corollas small and bright blue.

Local Distribution: Tiburón in the Sierra Kunkaak in canyons and mostly north-facing slopes. Growing and flowering with rains during the warmer seasons, especially with summer-fall rains and sporadically at other seasons.

Other Gulf Islands: None.

Geographic Range: Sonora from desert and thornscrub to pine forest. Arizona to Texas and southward to South America and the West Indies, and in Baja California Sur; introduced into the Old World.

Tiburón: On top of San Miguel Peak, 3000 ft, on cliff in the bowl region, *Knight 961* (UNM). On top of Sierra Kunkook [sic] in a large open bowl on S side of peak, *Knight 1016* (UNM). Sopc Hax, infrequent, *Felger 9315*. Canyon bottom, N base of Sierra Kunkaak, *Wilder 06-395*.

Holographis virgata (Harvey ex Bentham & Hooker f.) T.F. Daniel subsp. **virgata** [*Berginia virgata* Harvey ex Bentham & Hooker f. var. *virgata*]

Hooinalca "low hills"

Description: Densely branched, mound-shaped shrubs often 0.5–1 (1.5) m tall, or occasionally to 2 m growing through other shrubs. Leaves subsessile, narrowly oblong to linear-lanceolate, often moderately

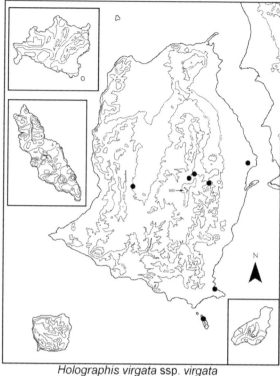

Holographis virgata ssp. *virgata*
Hooinalca

Justicia

1. Plants herbaceous; flowers nocturnal, the corollas pure white. _____ **J. longii**
1' Plants shrubby; flowers diurnal, the corollas red or orange.
    2. Stem surfaces densely covered with microscopic hairs of a single size; flowers stalked (pedicelled). _____ **J. californica**
    2' Stems with minute and larger hairs, not entirely covering the stem surfaces; flowers sessile or very short stalked. _____ **J. candicans**

Justicia californica (Bentham) D.N. Gibson [*Beloperone californica* Brandegee]

Nojoopis "what hummingbirds suck out"; *chuparrosa*; desert hummingbird bush

Description: Sprawling shrubs, often 1–1.5 (2+) m tall. Foliage usually sparse, the leaves highly variable in size, best developed with late summer–early fall rains, the larger leaves quickly drought deciduous, the plants often leafless or nearly so. Corollas tubular, red-orange. Flowering at any season except summer, often with

glaucous. Corollas rose-pink; flowering with moisture except summer time (July–Sep), peak flowering often March–May.

Local Distribution: Widespread and often abundant on Tiburón and the east side of Dátil; bajadas, arroyos, canyons, and rocky slopes.

Other Gulf Islands: None for var. *virgata*. Var. *glandulifera* is on San Marcos, Carmen, Danzante, San José, Espíritu Santo, Cerralvo.

Geographic Range: Western and central Sonora except the far northwestern and southwestern parts of the state, in Sonoran desertscrub and thornscrub. This subspecies also occurs in Baja California; another subspecies and two varieties occur on the Baja California Peninsula (Daniel 1997).

    Tiburón: Ensenada del Perro, *Felger 17730*. SW Central Valley, 1200–1400 ft, *Felger 12379*. Palo Fierro, 1 km inland, Felger 12555. Hast Cacöla, E flank of Sierra Kunkaak, very common in foothills, *Wilder 08-231*. Foothills, NE portion of Sierra Kunkaak, *Wilder 06-442*. 0.5 km E of Siimen Hax, *Wilder 06-501*.

    Dátil: NE side, steep rocky slopes, *Felger 15320*. E side, halfway up N-facing slope of peak, *Wilder 07-102*.

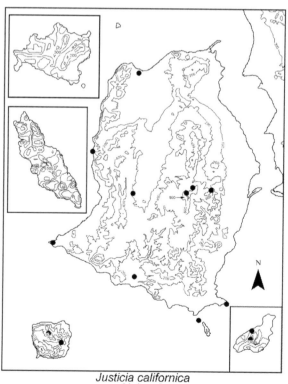

Justicia californica
Nojoopis

massive flowering in March and attracting hoards of hummingbirds.

Local Distribution: Often along arroyos and uplands including steep slopes. Across Tiburón including Sierra Kunkaak to higher elevations. Widespread on San Esteban. Common on Cholludo but no record from Dátil. Alcatraz on the east side of the mountain.

Other Gulf Islands: Ángel de la Guarda, Santa Cruz, San Diego, Espíritu Santo, Cerralvo.

Geographic Range: Western Sonora, Arizona, California, both Baja California states, and Sinaloa; mostly in Sonoran desertscrub and thornscrub.

Tiburón: 3 mi N of Willard's Point, *Johnston 4245* (CAS). Sauzal, *Felger 6436.* El Monumento, *Felger 08-48.* SW Central Valley, *Felger 12356.* E base of Sierra Kunkaak, *Wilder 05-29.* Hast Cacöla, E flank of Sierra Kunkaak, 380 m, *Wilder 08-239.* Mountain slope at N base of Sierra Kunkaak, *Wilder 06-426.* Steep, protected canyon, 720 m, N interior of Sierra Kunkaak, *Wilder 07-510.* Freshwater Bay, frequent in sandy draw near sea, *Johnston 3250* (CAS). Pooj lime, large canyon, NW shore, *Wilder 07-451.*

San Esteban: E-central side of island, canyon bottom, *Felger 12763.* S-facing canyon, central mountain, *Wilder 07-65.*

Alcatraz: E side of mountain, locally common on N- and NE-facing steep rocky slopes near top of island, *Felger 12723.* Near (above) foot of rock slope on N side, near cove on E side, *Lowe 10 Nov 1969.*

Cholludo: Scattered, *Felger 13414.*

Justicia candicans (Nees) L.D. Benson [*Jacobinia ovata* A. Gray]

Nojoopis caacöl "large *chuparrosa*"

Description: Understory shrubs to 1+ m tall. Stems slender, the leaves thin and gradually drought deciduous. Corollas bright red-orange, with a slender tube and white markings on the lower lip. Flowering profusely at various seasons except summer.

Local Distribution: Widespread in Sierra Kunkaak to higher elevations, especially in canyons and north-facing rocky slopes.

Other Gulf Islands: None.

Geographic Range: Widespread in Sonora but no records near the coast north of the Guaymas-Sierra el Aguaje region; generally extending only to the margin of the Sonoran Desert. Southern Arizona to Oaxaca and Baja California Sur.

Tiburón: Foothills of Sierra Kunkaak, *Felger 6972.* Hast Cacöla, E flank of Sierra Kunkaak, 380 m, *Wilder 08-237.*

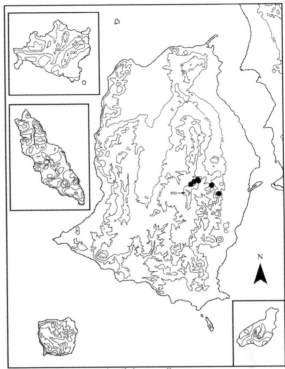

Justicia candicans
Nojoopis caacöl

Cerro San Miguel, 200–300 m, abundant in canyons, *Quijada-Mascareñas 90T002.* E base of Sierra Kunkaak, *Wilder 05-25.* Deep canyon, SW and up canyon from Siimen Hax, *Wilder 06-481.* Top of Sierra Kunkaak Segundo, E peak of Sierra Kunkaak, *Wilder 06-502.*

Justicia longii Hilsenbeck [*Siphonoglossa longiflora* (Torrey) A. Gray]

Description: Herbaceous perennials from rhizomatous rootstocks. Leaves and smaller stems drought deciduous. Corollas tubular, pure white, 3.5 cm long, opening in the early evening and falling shortly after dawn. Flowering mostly during the summer rainy season. Capsules 6.5–10 mm long.

Local Distribution: Sierra Kunkaak and adjacent sites on either side of the Sierra, and also recorded on the east slope of Sierra Menor. Mostly in canyons and protected niches on rock slopes.

Other Gulf Islands: None.

Geographic Range: Sonora south to about 28°30'N, desert and thornscrub, but not in the very arid northwestern part of the state; also in Arizona and Texas.

Tiburón: SW part of Central Valley, E-facing slopes of Sierra Menor, rare, several plants seen beneath *Olneya*, rocky

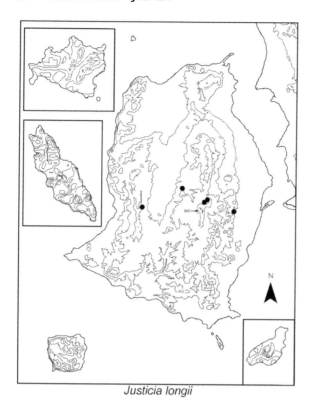

Justicia longii

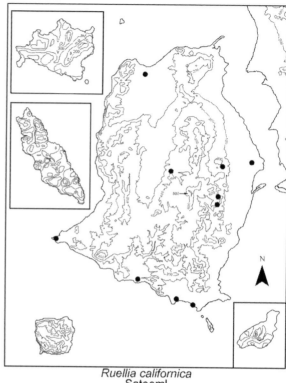

Ruellia californica
Satooml

soil, 4 Dec 1973, *Felger 21291* (specimen not located). Haap Hill, *Felger T74-19*. Hant Hax, *Felger 9349*. Base of Sierra Kunkaak, head of arroyo on N-facing exposure, *Wilder 05-33*. Deep canyon on N slope, SW and up canyon from Siimen Hax, *Wilder 06-482*.

Ruellia californica (Rose) I.M. Johnston

Satooml; *rama parda, tronadora*

Description: Dense, globose bushes often 0.8–1.5 m tall, the herbage sticky resinous and pungently aromatic, filling the air with a characteristic odor. Corollas lavender, large and showy. Flowering response non-seasonal, sometimes even in the most arid seasons, and with mass flowering following rains. When the dry, ripe capsules get wet they explosively and audibly snap open, flinging the seeds.

Local Distribution: Tiburón, widespread and abundant in xeric places, often forming 100 percent ground cover on benches along margins of larger arroyos and canyons. When walking through such dense patches your clothes become sticky with the resin from the leaves. Also on bajadas and dry slopes.

Other Gulf Islands: Coronados, Carmen, Danzante, Espíritu Santo.

Geographic Range: This subspecies occurs in central and southern Sonora, Tiburón, and both states of Baja California. Another subspecies occurs in Baja California Sur.

Tiburón: SW corner of island, *Johnston 4268* (UC). Arroyo Sauzal, 0.5 km inland, *E. Moser 1 May 1964*. Arroyo de la Cruz, *Moran 13005*. Coralitos, *Wilkinson 193*. Haap Hill, *Felger 74-T59*. Palo Fierro, 1.5 km inland, *Felger 8933*. Pazj Hax waterhole, *Wilder 06-44*. Sopc Hax, *Felger 9328*. Camino de Caracol, broad arroyo entering foothills, *Wilder 05-6*. Tecomate, *Felger 11126*.

Tetramerium fruticosum Brandegee

Description: Herbaceous perennials, the stems slender, the older ones with peeling bark. Leaves petiolate, the blades broadly ovate. Inflorescences of compact spikes with leafy bracts. Corollas cream colored with a maroon and purplish chevron on the lower lip as a nectar guide; opening at dawn and falling with daytime heat.

Local Distribution: Common in deep, protected canyons in the interior of Sierra Kunkaak to at least 720 m.

Other Gulf Islands: None.

Geographic Range: Widespread in Baja California Sur, known from mainland Sonora by a single collection

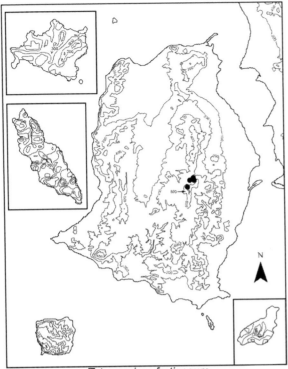

Tetramerium fruticosum

(Picu Mountains, southwest of Caborca; Daniel 2004). The Tiburón specimens provide disjunct, stepping-stone records.

Tiburón: Base of Sierra Kunkaak, head of arroyo, *Wilder 05-46*. Base of canyon, N slope of Sierra Kunkaak, *Wilder 06-386*. Siimen Hax, *Wilder 06-449*. Steep, protected canyon, 720 m, N interior of Sierra Kunkaak, canyon bottom, *Wilder 07-581*.

Sonora: Rocky canyon at Tinaja Picu in Picu Mountains, 2 mi N of concrete monument set on N side of Libertad Road, 24 Oct 1932, *Wiggins 6055* (CAS).

Achatocarpaceae

Phaulothamnus spinescens A. Gray

Aniicös "thorny inside"; *putilla*; snake-eyes

Description: Spinescent, lycium-like shrubs with hard wood, reaching 2–2.5 m tall. Leaves alternate, gradually drought deciduous, more or less narrowly spatulate, semi-fleshy and glaucous. Flowers small and inconspicuous, the fruits globose, 5 mm wide, fleshy, translucent white, and 1-seeded.

Local Distribution: Widespread on Tiburón from low to high elevations in the Sierra Kunkaak and prominent

along the Infiernillo bajada between the coastal *Frankenia* zone and mixed desertscrub, growing with other spinescent shrubs such as *Castela*, *Lycium*, and *Ziziphus* (Felger 1966:199). Also on the east side of Dátil and common on Cholludo, and rare on Alcatraz. Nolasco at scattered localities, mostly in canyons, and not common.

In 2007, José Ramón Torres told us that the stems are "medicina para los ojos. El tallo lo quema. Se usa la ceniza—lo muella y echa en agua, y ponen gotas en los ojos—pone poquito . . . en dentro de los ojos, una gota—cada vez que molesta [los ojos]. Todavia Don Armando Torres y otros se usan." (The stems are used as medicine for the eyes. Burn the stem. Use the the ashes—grind it and add water, and put drops in the eyes—use very little—one drop inside the eyes every time afflicted. Don Armando Torres and others still use it.)

Other Gulf Islands: Coronados, Monserrat, Santa Catalina, Santa Cruz, San José, San Francisco, Cerralvo.

Geographic Range: Widespread across much of Sonora nearly to the Arizona border (Rancho Arizonac, west of Nogales) but not in the extremely arid northwest and not at higher elevations; Sonoran desertscrub, thornscrub, and lowest margin of oak woodland. Southern Texas and northern Mexico including Baja California Sur.

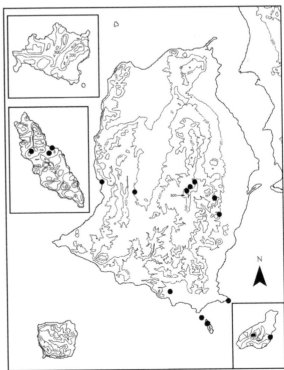

Phaulothamnus spinescens
Aniicös

Tiburón: Ensenada Blanca (Vaporeta), canyon, 29 Jan 1965, *Felger*, observation (Felger 1966:179). Arroyo de la Cruz, *Wilder 09-65*. El Monumento, *Felger 08-61*. SW Central Valley, arroyo, 200 m, 1 Feb 1965, *Felger*, observation (Felger 1966:239). Hast Coopol, *Wilder 07-376*. Carrizo Canyon, *Knight 1090,1106* (UNM). Base of Capxölim, N-facing rock slope, *Felger 07-105*. Deep sheltered canyon, N portion of Sierra Kunkaak, *Felger 07-120*. Steep, protected canyon, 720 m, N interior of Sierra Kunkaak, *Wilder 07-507*.

Dátil: NW side, *Felger 15345*. NE side, arroyo bottom, *Felger 9087*.

Alcatraz: Rocky gully at base of E slope of mountain, *Felger 12704*. S-central coastal side of island [base of mountain], few and scattered shrubs ca 1.7 m tall, *Felger 07-170*.

Cholludo: N side of island, scattered, a large shrub near shore, *Felger 13418*. Near beach and also at top of island, *Felger 9161*.

Nolasco: E side, *Felger 9641*. Cañón el Faro, one large shrub seen, canyon bottom at ca. 6 m elev., *Felger 06-78*. Cañón de Mellink, 110 m, canyon bottom, *Felger 08-145*.

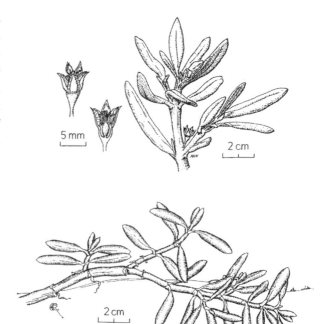

Figure 3.8. *Sesuvium portulacastrum*. FR below, NLN above.

Aizoaceae – Aizoon Family

Annuals, succulent or not, or succulent perennials. Leaves opposite.

1. Succulent perennials; leaves thick, more than twice as long as wide, the petioles short and indistinct. _____
    **Sesuvium**
1' Annuals; leaves about as wide as long, the petioles prominent.
   2. Cool-season annuals; plants extremely succulent, the surfaces covered with watery, bead-like vesicles. _
   _____ **Mesembryanthemum**
   2' Hot-weather annuals; plants semi-succulent or not succulent, the surfaces smooth. ____ **Trianthema**

Mesembryanthemum crystallinum Linnaeus [*Cryophytum crystallinum* (Linnaeus) N.E. Brown. *Gasoul crystallinum* (Linnaeus) Rothmaler]

Hax ixapz iti yacp "what has ice 'growing' on it"; *hielitos*; crystal iceplant

Description: Winter-spring annuals; highly variable in size and spectacularly fleshy, the surfaces covered with large, watery, crystal-like vesicles. Leaves often 2–10 cm long, the leaf blades about as wide as long. Flowers white, 1 cm wide with many slender "petals" (modified filaments, or staminodes) and many fertile stamens. Fruits fleshy at first, becoming dry capsules with valves at the flattened top opening when moist.

Local Distribution: Known from the Sonoran Islands by a single specimen from Dátil in 1998. It is seasonally and locally common on the beach at El Desemboque but not common southward to Bahía Kino on the Sonoran coast. It is likely to reach Tiburón, where it could become an invasive species that probably could not be controlled.

Geographic Range: Native of South Africa and extensively naturalized along the Pacific Coast of the Californias and South America. It has been in southern Arizona since the early twentieth century and coastal Sonora since at least the 1970s (Felger 2000a).

Dátil: Isla just S of Tiburon, seen only one on this island, *Canright 3 Nov 1998* (ASU; a photo with the specimen shows rocks that cannot be from Cholludo, so the island is Dátil). **Sonora:** El Desemboque, *Wilder 08-185*.

Sesuvium portulacastrum Linnaeus

Spitj caacöl "large spitj (coastal saltbush)"; spitj ctamcö "male coastal saltbush"; sea purslane

Description: Halophytic succulent perennials, often forming expansive, spreading mats. Leaves thick and succulent. Flowers pink.

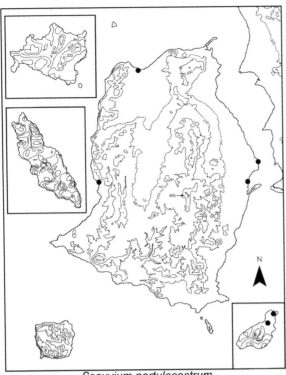

Sesuvium portulacastrum
Spitj caacöl, spitj ctamcö

bases and stipules form a membranous sheath around the stem. Flowers pink.

Local Distribution: Widespread in the lowlands across Tiburón and sandy flats on Alcatraz. Widespread on Patos, on the hillside and among *Atriplex barclayana* on the flats, where it was found in 2007 for the first time since Johnston's observation in 1921.

Other Gulf Islands: San Marcos, Coronados, Carmen, Danzante, Monserrat (observation), San Francisco, Espíritu Santo, Cerralvo.

Geographic Range: Across Sonora in lowlands, especially on semi-saline silty to sandy soils, also on gravelly and rocky soils, and often weedy. Worldwide.

Tiburón: Ensenada de la Cruz, *Felger 9230.* Coralitos, *Felger 15366.* Ensenada del Perro, *Wilder 08-307.* Haap Hill, *Felger T74-22.* Zozni Cmiipla, *Wilder 06-282.* Palo Fierro, *Felger 8929.* 3 km W of Punta Tormenta, *Smartt July 1993* (UNM).

Alcatraz: E side, sandy soil, *Felger 14925.* E side, flat, *Wilder 07-389.*

Patos: "Common on Patos Island growing on the guano flats with *Atriplex*" (Johnston 1924:1022). Mountain peak, *Felger 07-55.*

Local Distribution: Shores of Tiburón; common in wet, saline soils bordering the inland margins of mangrove and other estero margins, and upper beaches including low dunes. Often tidally inundated. Alcatraz on low, sandy flats and beaches.

Other Gulf Islands: Ángel de la Guarda, Rasa, San Lorenzo, Coronados, Carmen, Danzante, Monserrat, Santa Catalina, San José, San Francisco, Cerralvo.

Geographic Range: Shores of the Gulf of California and Pacific Coast of Baja California Sur; northward in Sonora to Puerto Lobos. Tropical and subtropical coasts in many regions of the world.

Tiburón: Ensenada Blanca (Vaporeta), upper beach just above high tide zone, *Felger 14978.* Beach 1 km N of Estero San Miguel, *Wilder 06-6.* Palo Fierro, beach at edge of mangroves, *Felger 6320a.* Tecomate, coastal, *Wilder 07-248.* **Alcatraz:** E side, sand beach, *Felger 14927.* E side, flat, *Wilder 07-388.*

Trianthema portulacastrum Linnaeus

Comaacöl; *verdolaga de cochi*; horse purslane

Description: Hot-weather annuals, often semi-succulent, the leaves highly variable in size. Expanded leaf

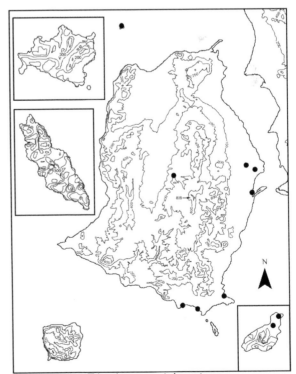

Trianthema portulacastrum
Comaacöl

Amaranthaceae – Amaranth Family
(includes Chenopodiaceae)

Annuals or perennial herbs or shrubs, often halophytic. Leaves simple, entire; stipules none, or plants appearing leafless. Flowers small, unisexual or bisexual, sepals 4 or 5, petals none. Fruits 1-seeded.

Figure 3.10. *Chenopodium murale.* FR.

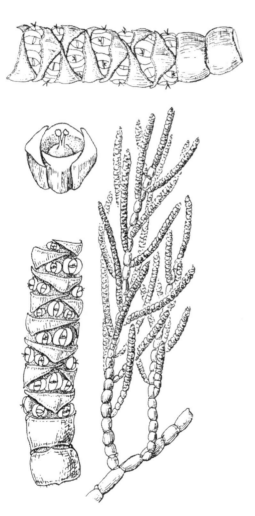

Figure 3.9. *Allenrolfea occidentalis.* BA (© James Henrickson 1999).

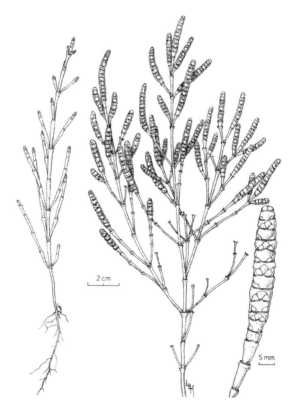

Figure 3.11. *Salicornia bigelovii.* LBH.

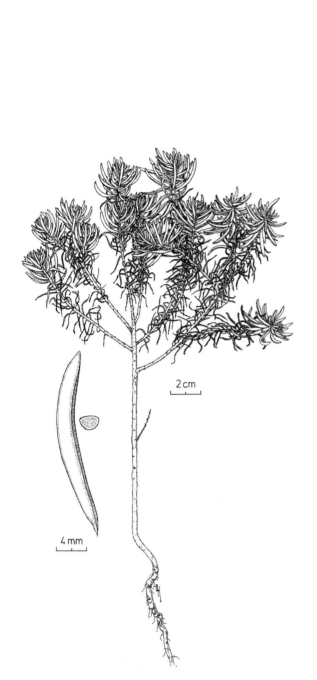

Figure 3.12. *Suaeda esteroa*. LBH.

Figure 3.13. *Suaeda nigra*. FR.

Figure 3.14. *Amaranthus fimbriatus*. Isla San Pedro Nolasco. BTW, 11 November 2009.

1. Plants appearing leafless; stems succulent and jointed.
    2. Branching alternate. _____ **Allenrolfea**
    2' Branching opposite. _____ **Salicornia**
1' Plants with obvious leaves; stems not jointed.
    3. Leaves opposite or alternate, usually densely white woolly with branched hairs obscuring the surfaces; flowers bisexual. _____ **Tidestromia**
    3' Leaves alternate or sometimes opposite below; glabrous or with short simple hairs or glands, or inflated hairs collapsing on leaf surfaces; flowers unisexual.
        4. Stems and leaves succulent, leaves linear to bead-like. _____ **Suaeda**
        4' Leaves with expanded, flattened blades, usually not especially fleshy and not linear.
            5. Female flowers and fruits enclosed in sepal-like bracts enlarging as the fruit develops. _____ **Atriplex**
            5' Female flowers and fruits not enclosed in sepal-like bracts.
                6. Sepals spinescent or with fringed margins, conspicuously longer than the seed. _____ **Amaranthus**
                6' Sepals pointed and not spinescent or fringed, not extending beyond the fruit. _____ **Chenopodium**

Allenrolfea occidentalis (S. Watson) Kuntze

Tacs; *chamizo verde, chamizo de agua*; iodine bush

Description: Succulent shrubs with jointed, alternately branched stems, the younger stems succulent, divided into green to red-orange bead-like segments and appearing leafless. Flowering observed in October, the plants yellow with copious pollen, and prodigious quantities of seeds produced after the winter solstice.

Allenrolfea superficially resembles perennial species of *Salicornia* but has alternate branching, often becomes more shrub-like, and grows on higher ground. Fleshy tissue on *Allenrolfea* and *Salicornia* stems is derived from reduced, modified leaf bases, one per node for *Allenrolfea* and two for *Salicornia*.

Local Distribution: Coastal margins on Tiburón, common on lowlying saline soils near the shore, often on ground slightly higher and drier than the mangrove and tidal mud zone where the salicornias thrive. It is the dominant plant in the low-lying interior flat on Alcatraz.

Other Gulf Islands: Ángel de la Guarda, Coronados (observation), Carmen, Monserrat, San José, San Francisco, Espíritu Santo, Cerralvo.

Geographic Range: Widespread in coastal Sonora. Coastal deserts and inland alkali sinks; Oregon and California to Texas and Baja California Sur, southward to northwestern Sinaloa, and the Chihuahuan Desert in Mexico.

Tiburón: Ensenada Blanca (Vaporeta), *Felger 7138*. Ensenada de la Cruz, *Felger 12770*. Ensenada del Perro, *Wilder 07-161*. 1 km N of Estero San Miguel, *Wilder 06-23*. Palo Fierro, *Felger 10328*. Valle de Águila, *Wilder 07-268*. Punta Perla, 19 Dec 1973, Felger, observation. Tecomate, abundant at mouth of drainages, *Whiting 9029*. Bahía Agua Dulce, dunes at beach, *Felger 6820*.

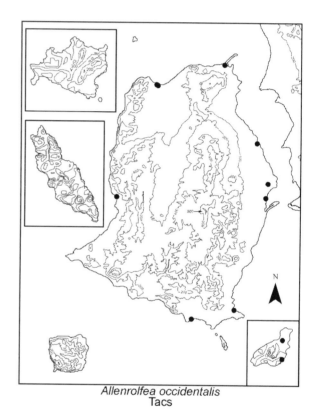

Allenrolfea occidentalis
Tacs

Alcatraz: Low areas (saline flats), E side of island, up to 95 percent local coverage, *Felger 12718*. Base of rocky mountain, SE side of island, *Felger 12825*.

Amaranthus

Fruits of small capsules splitting around the middle (circumscissle). Seeds lens shaped, smooth, and reddish brown to blackish.

1. Plants glabrous; male and female flowers on the same plant; inflorescences "soft," the bracts not spiny or stiff; sepals of female flowers conspicuously ragged or fringed; stamens 3. _____ **A. fimbriatus**
1' Plants glandular pubescent; male and female flowers on separate plants; female inflorescence bracts and sepals stiff and often sharp; female sepals entire; stamens 5._
_____ **A. watsonii**

Amaranthus fimbriatus (Torrey) Bentham

Ziim caaitic "soft ziim"; *bledo*; fringed amaranth

Description: Warm weather annuals, usually less than 1 m tall, glabrous, pale green or often reddish, especially late in the season. Leaves narrowly lanceolate. Inflorescences terminal and axillary; bracts herbaceous and not prickly. Male and female flowers on the same plant. Female flowers green and white, urn shaped, the sepals 5 and fringed (fimbriate, rarely nearly entire). Stamens 3. Seeds 0.85–1 mm wide, lens shaped, red-brown to blackish.

Local Distribution: Widespread on Tiburón, especially on sandy soils, including dunes near the shore around the island, also inland, especially along washes. Also on Dátil in close proximity to the beach. Alcatraz on the east-side flats, especially near the shore in sandy soils. Ivan Johnston found it on Patos in spring 1921, and Richard found it there in summer 1966. Widespread and common on Nolasco.

Other Gulf Islands: Ángel de la Guarda, Carmen, Monserrat, San Diego, San José, Espíritu Santo, Cerralvo.

Geographic Range: Sinaloa and Baja California Sur to southwestern United States.

Tiburón: Ensenada de la Cruz, *Felger 9226*. Xapij (Sauzal), *Wilder 08-131*. Ensenada del Perro, *Wilder 08-303*. Palo Fierro, *Felger 12538*. Tecomate, *Felger 8883*.

Dátil: NE side, arroyo bottom near beach, *Felger 9095*. E-central side, beach, *Wilder 09-02*.

Alcatraz: Edge of flat, E side of island, localized, *Felger 12715*.

Patos: Johnston (1924:1018) reported this species as common with *Atriplex barclayana* on the low guano flat in the spring of 1921. On 11 August 1964, Richard found a small localized population of diminutive plants near the ruins of a rock shelter built by guano workers. We have not, however, located specimens.

Nolasco: NE side, abundant on N-facing grassy slopes, *Felger 11449*. Base of Cañón el Farito, not common, *Wilder 08-168*.

Amaranthus watsonii Standley

Ziim quicös "prickly ziim"; *bledo, quelite de las aguas*; careless weed

Description: Annuals responding to both winter-spring and summer-fall rains. Highly variable in size depending upon soil moisture and temperature. Herbage glandular, the glands concentrated along leaf veins and often imparting a brownish color to the plant. Male and female flowers on separate plants. Inflorescences terminal and axillary; female plants with spinescent bracts and sepals, the sepals 5. Stamens 5. Seeds 1 mm wide, lens shaped, red-brown to blackish.

The Comcaac used the seeds for food.

Local Distribution: Tiburón, San Esteban, Dátil, and Alcatraz; often seasonally abundant, especially at lower elevations. Lowlands across Tiburón from beach dunes, bajadas, and valley plains to mountain slopes but probably not at higher elevations. Widespread on San Esteban and Dátil. Seasonally abundant on the vegetated areas of Alcatraz, imparting a green color to the island during rainy seasons and brown most of the remaining part of the year, and probably one of the largest contributors to biomass on the island.

Other Gulf Islands: Ángel de la Guarda, Partida Norte, Salsipuedes, San Lorenzo, San Ildefonso, Coronados, Carmen, Danzante, Monserrat, San Francisco, Espíritu Santo, Cerralvo.

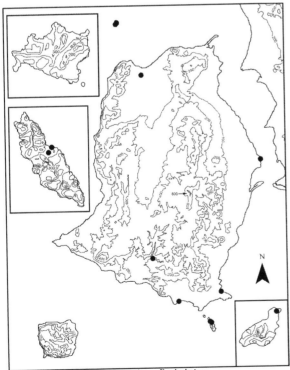

Amaranthus fimbriatus
Ziim caaitic

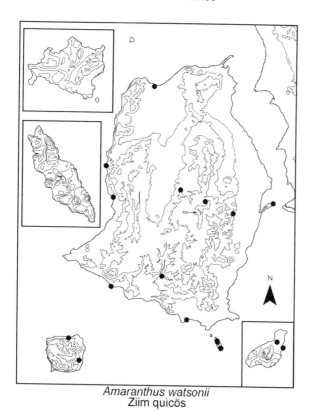

Amaranthus watsonii
Ziim quicös

bracts enlarge, thicken, and often grow ornamentations such as wings and crests.

1. Herbaceous perennials or subshrubs, sometimes flowering in first year, usually less than 0.5 m tall and as broad or broader than tall; bark at base becoming fissured and corky; wood usually not hard; stems herbaceous; fruiting bracts obovate, with unequal teeth. ___ **A. barclayana**
1' Woody shrubs, often to 1+ m tall, or if less then usually as tall or taller than wide; bark at base smooth; wood hard.
  2. Fruiting bracts 4-winged. _____ **A. linearis**
  2' Fruiting bracts with many teeth or wings. _____
  _____ **A. polycarpa**

Atriplex barclayana (Bentham) D. Dietrich

Spitj; *chamizo, saladillo, quelite blanco*; coastal saltbush

Description: Suffrutescent perennials, probably not long-lived and sometimes flowering the first year; commonly 0.5–1+ m across and usually broader than tall. Lower stems woody, the wood soft, the bark fissured. Herbage semi-succulent, often silvery gray-green.

Geographic Range: Endemic to the Gulf of California region; Sonora, both states of Baja California, and Gulf Islands.

  Tiburón: Ensenada Blanca (Vaporeta), *Felger 12222*. Tordillitos, *Felger 15472*. Ensenada de la Cruz, *Felger 9138*. Xapij (Sauzal) waterhole, *Wilder 06-120*. Punta San Miguel, *Wilder 07-188*. Hant Hax camp, *Felger 9354*. Haap Hill, *Felger 76-T2*. Canyon bottom, N base of Sierra Kunkaak, *Wilder 06-424*. Tecomate, *Felger 11114*. Pooj Iime, large canyon, NW shore, *Wilder 07-453*.

  San Esteban: N side, *Felger 15409A*. Limantour, *Felger 17664*.

  Dátil: SE side, *Felger 17494*. NE side, *Felger 9095B*. E side, ridge crest, *Wilder 07-115*.

  Alcatraz: E side of island, *Felger 14910b*. S-central side, *Felger 07-160*.

Atriplex – Saltbush

Herbaceous perennials and woody shrubs; herbage usually scurfy gray or whitish due to inflated hairs that collapse upon drying. Leaves mostly alternate (lower ones sometimes opposite). Female flowers and fruits enclosed in a pair of sepal-like bracts; as the fruits develop the

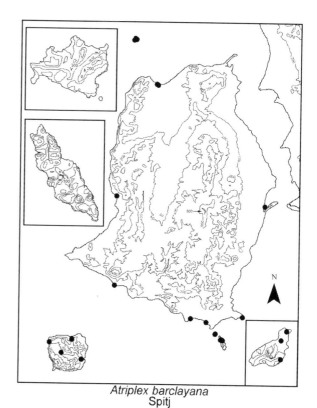

Atriplex barclayana
Spitj

Leaves highly variable, mostly ovate to obovate, entire or sometimes with a few broad shallow teeth or shallowly lobed. Usually dioecious but occasionally monoecious. Fruiting bracts highly variable in size, shape, and ornamentation; mealy, often thick and somewhat spongy; with 3 to about one dozen apical teeth. Flowering at various seasons.

Local Distribution: Common on Tiburón and neighboring islands. Abundant along the shores of Tiburón, including sandy beaches, strand and dune habitats, and upper margins of estero vegetation, sometimes in the tidal exchange zone, and often on rocky slopes. Coastal habitats and inland on San Esteban to the center of the island. Widespread and abundant on Alcatraz and Cholludo. Dátil at the shore and less common inland. It is the only abundant plant on Patos and often blankets the island.

Tender, young shoots of *quelite blanco* are edible, usually cooked with onions and tomatoes (Miguel Durazo, personal communication, 4 Dec 2007).

Other Gulf Islands: Ángel de la Guarda, Partida Norte, Rasa, Salsipuedes, San Lorenzo, Tortuga, San Marcos, San Ildefonso, Coronados (observation), Carmen, Danzante, Monserrat, Santa Catalina, Santa Cruz, San Diego, San José, San Francisco, Espíritu Santo, Cerralvo.

Geographic Range: Shores of islands and both coasts of the Gulf of California to Sinaloa, and the Pacific Coast of the Baja California Peninsula.

Tiburón: Ensenada Blanca (Vaporeta), *Felger 15747.* Tordillitos, *Felger 15480.* Ensenada de la Cruz, *Felger 9143B.* Coralitos, *Wilder 06-70.* El Monumento, *Felger 15531.* Zozni Cmiipla, *Wilder 06-38.* Tecomate, abundant, at shore, *Whiting 9017.* Bahía Agua Dulce, *Felger 6846.*

San Esteban: N side of island, narrow rocky canyon behind large cove, *Felger 17587.* San Pedro, *Felger 16671.* Arroyo Limantour: near shore, *Felger 7083;* 3 mi inland, 5 Dec 2008, *Wilder,* observation.

Dátil: NE side of island, near beach, hillslopes, and arroyo bottom, *Felger 9122.* NW side of island, *Felger 15332.* E-central side, beach, *Wilder 09-01.*

Alcatraz: Sandy flat, E side of island, *Felger 14924B.* SE side, base of mountain, *Felger 12824.* Interior flats with *Allenrolfea, Felger 07-165.*

Patos: In a protected draw, 23 Apr 1921, *Johnston 3242* (CAS, UC). Salty flat near S end of island, 23 Apr 1921, *Johnston 3244* (CAS). Collected by Luis Baptista, 23 Apr 1966, *Moran 13004* (CAS, UC, pistillate). Abundant on island, *Felger 7013, 21320* (CAS), *07-62.*

Cholludo: *Felger 13419, 17704.* Near shore, also found throughout the island, *Felger 9163.* N-central slope, common from base to top of island, *Wilder 08-316.*

Atriplex linearis S. Watson [*A. canescens* subsp. *linearis* (S. Watson) Hall & Clements]

Hatajixp, hatajísijc; *chamizo cenizo;* narrow-leaf saltbush

Description: Shrubs 0.5–1 m tall. Leaves narrow, several times longer than wide. Male and female flowers on separate plants. Fruiting bracts four winged.

Atriplex linearis is sometimes treated as a subspecies of *A. canescens.* The leaves and fruits of *A. linearis* are similar but smaller than those of *A. canescens.* They often occur in the same geographic region but occupy distinctive habitats or niches in many places in the Sonoran Desert. *Atriplex canescens* occurs on the mainland in the Kino region but has not been found on Tiburón.

Local Distribution: Known from scattered coastal localities around Tiburón.

Other Gulf Islands: San Lorenzo (observation).

Geographic Range: Common especially on saline silty flats and sand soils near the coast, and also inland in

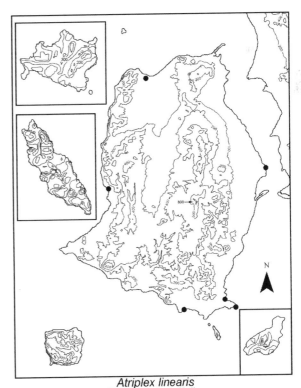

Atriplex linearis
Hatajixp, hatajísijc

deserts. Southeastern California, both Baja California states, southwestern Arizona, and coastal Sonora south to the Río Mayo delta and perhaps northwestern Sinaloa.

Tiburón: Ensenada Blanca (Vaporeta), near mouth of canyon near shore, *Felger 12250*. Ensenada de la Cruz, arroyo bed near shore, *Felger 12774*. El Monumento, *Felger 15524*. Ensenada del Perro, 200 m inland, *Wilder 07-144*. Punta Tormenta, *Wilder 06-347*. Bahía Agua Dulce, *Felger 8868*.

Atriplex polycarpa (Torrey) S. Watson

Hatajixp, hatajísijc; *chamizo cenizo*; desert saltbush

Description: Shrubs with woody branches and slender, brittle stems. Leaves small, gray-green, often about as wide as long. Male and female flowers on separate plants. Fruiting bracts usually 4–6 mm wide, somewhat orbicular to obdeltoid, with 7–17 finger-like blunt teeth, varying in size and shape and often densely covered by scurfy white hairs. Often with characteristic pink galls in the upper branches and flowering branches. Flowering in various seasons, especially following rains.

Local Distribution: Xeric habitats on Tiburón including arroyos, rocky slopes, and silty or sandy flats, and common across a wide variety of habitats on San Esteban and also on Dátil.

Other Gulf Islands: Ángel de la Guarda, San Lorenzo, Espíritu Santo.

Geographic Range: Northward in Sonora to western Arizona, southern Nevada, southern California, and the Baja California Peninsula.

Tiburón: La Sauzal, *Wiggins 17166* (CAS). Ensenada de la Cruz, *Felger 12772*. Ensenada del Perro, abundant, *Tenorio 9525*. 3 km W [of] Punta Tormenta, *Smartt 13 Oct 1976* (UNM). Tecomate, *Whiting 9049, Wilder 07-249*. N side, *Mallery & Turnage 3 May 1937* (CAS).

San Esteban: Canyon bottoms, hilltops, and some slopes, *Felger 7044*. SW part of island, W-facing side canyon with dense vegetation, *Wilder 07-90*. Limantour, *Felger 16604*. Along low bluff at edge of wash, *Johnston 3193* (CAS, UC). Colony, N slope of high peak near SE corner, 400 m, *Moran 8850* (CAS, UC).

Dátil: NE side of island, S-facing rocky slopes and edge of beach at base of hills, common, *Felger 9131*.

*Chenopodium murale Linnaeus

Ziim xat "ziim hail"; *chual*; net-leaf goosefoot

Description: Winter-spring annuals. Herbage often becoming reddish and semi-succulent. Leaves glabrous

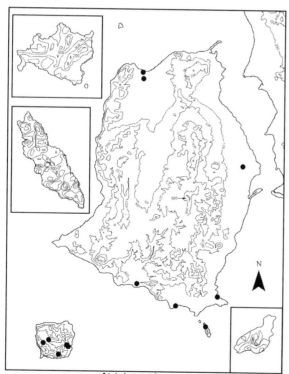

Atriplex polycarpa
Hatajixp, hatajísijc

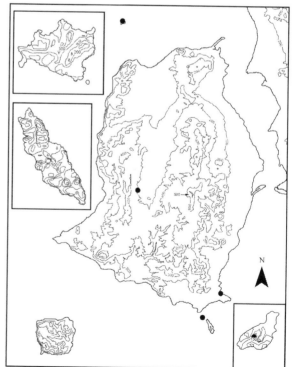

Chenopodium murale
Ziim xat

or grayish scurfy below. Seeds 1.2–1.4 mm wide, blackish, lens shaped, the margins acute with a thin rim. In April 2007 we saw unusually large, robust plants to 1 m tall behind the beach at Ensenada del Perro.

Local Distribution: Locally and seasonally common at scattered sites on Tiburón; although documented by only two collections, it is probably more widespread on the island. Also seasonally abundant on Alcatraz. Found on Patos for the first time in 2007 and on Cholludo in 2009.

Other Gulf Islands: San Jorge (southeast of Puerto Peñasco, where it is the only vascular plant recorded on the island; Felger 2000a), Ángel de la Guarda, Cerralvo.

Geographic Range: Reported in the literature as native to the Old World and adventive from Canada to Guatemala. This assumption warrants investigation. It has long been established and widespread in the Sonoran Desert, including remote places where few if any other non-native plants have established.

On the Sonoran mainland and in Arizona it is especially common as an urban and agricultural weed, in heavily grazed and other disturbed habitats and in fine-texture soils, including coastal plains, as well as many natural habitats, including canyons, floodplains, and bajadas.

Tiburón: Ensenada del Perro, behind beach, *Wilder 07-164*. Central Valley, valley plain, *Felger 17320*.

Alcatraz: Abundant through much of island, especially near top on E-facing slopes, *Felger 12714*.

Patos: Rock hill, steep slope, N-facing, in dense cover of *Atriplex barclayana*, *Felger 07-56*.

Cholludo: NW-facing slope, rare, at base of cardón forest in exposed guano soil, *Molina-Martínez*, 1 Mar 2009 (cataloged as *Wilder 09-69*).

Salicornia – Pickleweed, glasswort, samphire

Annual or perennial halophytes, herbaceous to shrubby. Stems succulent, jointed, green or turning red or orange-green, with opposite branching; fleshy part of stems actually reduced and modified leaves with 2 scale leaves per node. Flowers bisexual or unisexual, minute, in clusters of 3 (those in the flora area) on opposite sides of each node and sunken in fleshy tissue; flowering joints forming cylindrical terminal or subterminal spikes. Calyx fleshy, nearly surrounding the flower and fruit, the fruiting calyx dry and spongy but readily soaking up water.

1. Annuals, usually with a single main axis and branching above the lower half of the plant, the branches widely spaced and diffuse; middle flower conspicuously taller and inserted higher than the lateral ones. _ **S. bigelovii**
1' Perennials with semi-woody stems below, densely branched throughout, forming dense, low, and spreading shrubs; the 3 flowers of each cluster about equal height and inserted at about the same level. **S. subterminalis**

Salicornia bigelovii Torrey

Xnaa caaa "what calls the south wind"; pickleweed, glasswort

Description: Erect annuals often 20–50 cm tall, usually with a single prominent main axis, often taller than wide, sparsely to moderately branched. There can be more than one generation per year. Stems (actually modified leaf tissue) bright yellow-green, very succulent and glassy looking; almost every branch ending in a flower spike. Flowering at least in spring and late fall and winter. Seeds 1.8–2 mm wide, light tan, opaque due to lack of perisperm, nearly filled by the "folded" embryo, the seed surfaces with short, thickish, translucent, and often curled hairs; seeds usually ripening as the plants mature and die.

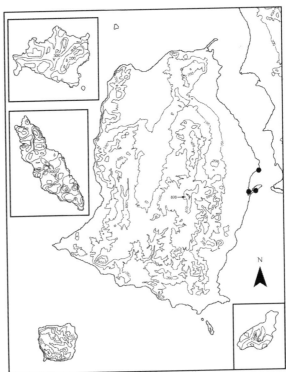

Salicornia bigelovii
Xnaa caaa

Commonly germinating with highest tides and rapidly growing to full size; 2 or more size classes may be present. Flowering and seed maturation often variable from low to high tidal zones; often about 1 month later in the higher tidal zones.

Local Distribution: Common on Tiburón in tidally wet sandy-muddy edges of mangroves on the Infiernillo shore; often partially to fully submerged at higher tides.

Other Gulf Islands: Ángel de la Guarda, San José.

Geographic Range: Shores of the Gulf of California southward to Sinaloa, and northward on the Pacific Coast to southern California. *Salicornia bigelovii* appears to be the only annual member of the genus in the Gulf of California.

Tiburón: Estero San Miguel, *Wilder 06-279*. Zozni Cmiipla, *Wilder 06-363*. Palo Fierro, *Felger 12560*.

Salicornia subterminalis Parish [*Arthrocnemum subterminalis* (Parish) Standley]

Description: Densely branched, mound-like, low sprawling shrubs, about 0.5 (1) m tall. Stems often glaucous and older succulent branches sometimes become pale reddish. Seeds glabrous, pale brown and with endosperm, the curved embryo visible through the seed surface.

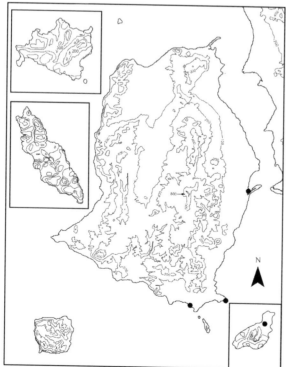

Salicornia subterminalis

Local Distribution: Scattered localities along southern and eastern shores of Tiburón, mostly at the inland margins of halophytic scrub. Common across the flats on Alcatraz.

Other Gulf Islands: Ángel de la Guarda, Rasa, Coronados, Carmen, Danzante, San Francisco, Espíritu Santo.

Geographic Range: Shores of the Gulf of California to Sinaloa, Pacific Coast of Baja California Sur to California, and inland areas of California.

Tiburón: Coralitos, shore just above high tide zone, *Felger 15379*. Shore at El Monumento, *Felger 08-76*. Zozni Cmiipla, *Wilder 06-364*.

Alcatraz: Interior salt flat dominated by *Allenrolfea occidentalis*, *Felger 07-164*.

Suaeda – Seablite, seepweed

Annuals or shrubs. Branches and leaves alternate (those in the flora area); leaves, young stems, and sepals succulent. Flowers usually bisexual, sometimes intermixed with unisexual female flowers.

1. Annuals, less than 0.5 m tall; leaves flattened above (awl shaped in cross section). _____ **S. esteroa**
1' Bushy or shrubby perennials, often 1 m or more tall; short-shoot leaves rounded in cross section.__ **S. nigra**

Suaeda esteroa Ferren & S.A. Whitmore

Sipjö yaneaax "what the cardinal washes his hands with"; estuary seablite

Description: Annual herbs (15) 20–30+ cm tall, glabrous, and often with a well-developed main axis. Leaves mostly 2–2.5 (4) cm long, sessile, green, often glaucous, becoming yellowish to red to red-purple, linear, semi-cylindrical with the upper surfaces flattened to slightly concave. Seeds 1.5–2 mm wide, of two kinds on the same plant, dull brown and sometimes also with some shiny blackish seeds.

Local Distribution: Mangrove esteros on the Infiernillo shore of Tiburón, including Punta San Miguel and Punta Perla, where it is often inundated by the highest tides.

Other Gulf Islands: None.

Geographic Range: Mangrove and salt scrub esteros in the Gulf of California and Pacific Coast of the Californias. The Gulf of California plants are annuals while those on the Pacific Coast generally are perennials.

Tiburón: Zozni Cmiipla, tidal zone with halophytes, *Wilder 06-362*. Punta Perla, estero, common at edge of mangroves on S side of sand spit, *Wilder 08-373*.

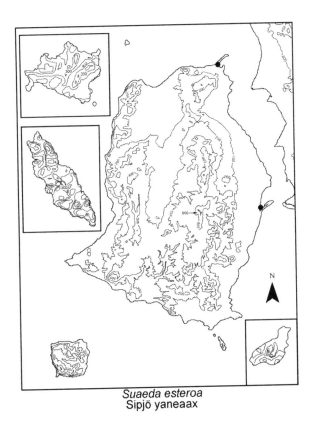

Suaeda esteroa
Sipjö yaneaax

There has been confusion in nomenclature in previous checklists. Moran (1983b) and Rebman et al. (2002) reported *S. nigra* from Tiburón and San Esteban as *S. californica* S. Watson and indicated three *Suaeda* species for Tiburón and San Esteban, but we are quite certain there are only two on Tiburón and one on San Esteban.

Other Gulf Islands: Ángel de la Guarda, Carmen, Danzante, San José, Espíritu Santo.

Geographic Range: Salt scrub and mangrove esteros throughout the Gulf of California and apparently southward to Altata, Sinaloa; also the Pacific Coast of the Baja California Peninsula and southern California. Alberta and the western United States to northern and central Mexico.

Tiburón: Ensenada Blanca (Vaporeta), *Felger 14934*. Ensenada de la Cruz, near beach and low places, *Felger 9210*. Punta San Miguel, edge of mangroves, *Wilder 07-202*. Zozni Cmiipla, *Wilder 06-365*. Palo Fierro, edge of mangroves, beach, and adjacent wet saline soil, *Felger 6421*. Bahía Agua Dulce, low saline flats near beach, *Felger 6847*. N side, *Mallery & Turnage 3 May 1937* (CAS).

San Esteban: Cascajal, SW side of island, near rocky beach, *Felger 9199*. La Freidera, *Felger 15462*.

Suaeda nigra (Rafinesque) J.F. Macbride [*S. moquinii* (Torrey) Greene. *S. ramosissima* (Standley) I.M. Johnston. *S. torreyana* S. Watson]

Hatajipol; *sosa*; bush seepweed

Description: Shrubs, often 1–2 m tall, semi-hemispherical, much branched throughout, the branches often interlacing. Stems slender and brittle. Herbage and calyces minutely and densely pubescent to sometimes essentially or completely glabrous, green to glaucous blue-green and often reddish purple. Long-shoot leaves moderately flattened with rounded margins, often 1–3 cm long, the short-shoot leaves usually crowded, often 3–8 mm long, terete, and bead-like. Flowering branches slender and paniculate. Flowers 1–ca. 10 per cluster, often functionally unisexual. Seeds 1–1.5 mm wide, erect, shiny, and blackish. Flowering mostly during warmer months.

Local Distribution: Common in coastal halophyte communities, saline flats, and arroyo bottoms near the shore around much of Tiburón and coastal areas on San Esteban, and also common on low-lying areas of Alcatraz. In 2007 we found it for the first time on Patos, where a few plants were near the shore on the south side of the island.

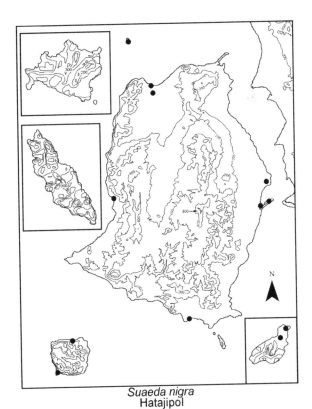

Suaeda nigra
Hatajipol

Alcatraz: Near shore and scattered on flat E part of island, *Felger 13410B*. Interior flats among *Allenrolfea occidentalis*, *Felger 07-163*.

Patos: Beach at S side, in dense stand of *Atriplex barclayana* among cobble rocks just above high tide line—three healthy young, reproductive plants plus two large dead ones, *Felger 07-58*.

Tidestromia lanuginosa (Nuttall) Standley subsp. **eliassoniana** Sánchez-del Pino & Flores Olvera [*T. eliassoniana* (Sánchez-del Pino & Flores Olvera) Sánchez-del Pino]

Halít an caascl "what causes dandruff"; *hierba lanuda*; honeysweet

Description: Summer-fall annuals, sometimes persisting until December; often low and spreading. Stems red, herbage usually densely scurfy white with branched (dendritic) hairs, although well-watered plants, especially when shaded, are greener and not as densely pubescent. Flowers minute and yellow.

Local Distribution: Widespread and often seasonally abundant across the lowlands of Tiburón. It seems strange that it is not known from any other Gulf of California island.

Other Gulf Islands: None.

Geographic Range: Seasonally common and widespread across most of the Sonoran Desert, as well as thornscrub, tropical deciduous forest, and sometimes in oak woodland: Baja California Sur and northern Sinaloa to southern California, Arizona, and southwestern New Mexico. Subsp. *eliassoniana* occurs west of the continental divide in southwestern North America and subsp. *lanuginosa* occurs east of the divide in the United States and Mexico.

The two subspecies are distinguished largely by presence or absence of decorations on the ends of the terminal arms of the branched hairs and pollen differences. Although the distinctions may seem minor, they are geographically segregated. Sánchez-del Pino and Motley (2010) recognize the two subspecies as distinct species, an opinion we do not share.

Tiburón: Ensenada Blanca (Vaporeta), *Felger 14964*. Tordillitos, 22 Dec 1966, *Felger 15476* (specimen not located). Sauzal, vicinity of landing field, forming ca. 80 percent ground cover with dead or nearly dead individuals, 1 Feb 1965, *Felger 12275* (specimen not located). El Monumento, 22 Dec 1966, *Felger 15529* (specimen not located). SW part of Central Valley, dry and dead, 2 Feb 1965, *Felger 12421* (specimen not located). Haap Hill, *Felger T74-23*. Hant Hax, 26 Oct 1963, *Felger 9352* (specimen not located). Lower slopes of San Miguel Peak, *Knight 944* (UNM). Caracol, *Romero-Morales 07-8*. Tecomate, *Felger 8875*.

Apiaceae (Umbelliferae) – Carrot Family

Daucus pusillus Michaux

Zanahoria silvestre; wild carrot

Description: Winter-spring annuals with stiff white (hispid) hairs. Stems slender, the leaves highly dissected. Flowers minute, white, in compact, densely flowered clusters (umbels) on long stalks; sepals none, the petals 0.6 mm long, pale yellow. Fruits bur-like, 3 mm long, intricately sculptured, bearing straight spines with minute barbs at the tip (seen with 10× magnification).

Local Distribution: Known from Tiburón by a single record.

Other Gulf Islands: None.

Geographic Range: British Columbia to Baja California Sur, northern Mexico, and South America.

Tiburón: SW Central Valley, 400 ft, *Felger 17357*.

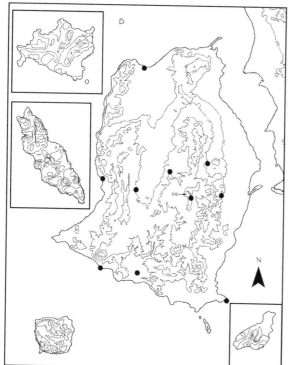

Tidestromia lanuginosa ssp. *eliassoniana*
Halít an caascl

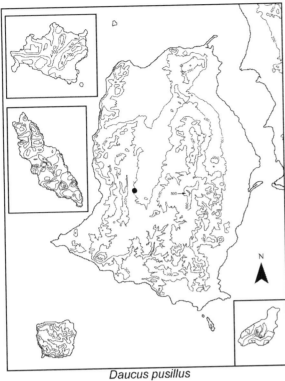

Daucus pusillus

Figure 3.15. *Asclepias albicans.* Hast Coopol, Isla Tiburón. BTW, 15 September 2007.

Apocynaceae – Dogbane Family
(includes Asclepiadaceae)

Perennial vines, herbs or shrubs; mostly with milky sap. The various milkweed vines, called *talayote*, have fruits that are edible, raw or cooked, when young and tender (Felger & Moser 1985).

1. Stems upright, not vining.
   2. Stems reed-like; leafless or leaves few, linear-filiform, less than 1 mm wide. _____ **Asclepias**
   2' Stems leafy and not reed-like, leaves more than 2 mm wide.
      3. Suffrutescent or herbaceous perennials usually less than 1 m tall; flowers bright yellow, 2 cm wide; fruits dry, multiple-seeded, slender capsules much longer than wide. _____ **Haplophyton**
      3' Woody shrubs usually more than 1.5 m tall; flowers white, less than 1 cm wide; fruits ovoid and fleshy, less than twice as long as wide, 1-seeded. _____ **Vallesia**
1' Stems generally vining, at least at the tips.
   4. Leaf blades linear to lanceolate or triangular-elongated, more than twice as long as wide.
      5. Stems essentially glabrous or pubescent, the hairs straight and not in a line; leaf margins not revolute; flowers more than 5 mm long or wide, the petal surfaces readily visible. _____ **Funastrum**
      5' Stems with a longitudinal line of small, usually curved hairs; leaf margins often inrolled (revolute); flowers 2–4 mm long, the inner petal surfaces obscured by hairs. _____ **Metastelma**
   4' Leaf blades broadly ovate to cordate, less than twice as long as wide.
      6. Herbage essentially glabrous; fruits at least 3 cm diameter, tough and leathery to hard shelled. _____ **Marsdenia**
      6' Herbage conspicuously pubescent; fruits less than 2 cm diameter, soft walled. _____ **Matelea**

Figure 3.16. *Matelea cordifolia.*
East of El Desemboque, Sonora.
RSF, March 1983.

Asclepias – Milkweed

The two species in the flora have reed-like, terete and semi-succulent stems with relatively few, narrowly linear, and quickly deciduous leaves. These two species are closely related (Fishbein et al. 2011).

1. Stems usually few, usually more than 1.5 m tall; flowers 8–9.5 mm long, the hoods 2–2.5 mm long and not exceeding the anther head. _____ **A. albicans**
1' Stems many, usually reaching 1 m; flowers 18–22 mm long, the hoods 6–10 mm long and longer than the anther head. _____ **A. subulata**

Asclepias albicans S. Watson

Najcaazjc; *yamate, candelilla*; white-stem milkweed, wax milkweed

Description: Stems reaching 2–2.5 m tall, mostly few and branching from near the base, erect and reed-like, whitish to glaucous bluish white and waxy. Flower clusters (umbels) 1 to several per stem, from the upper nodes. Flowers cream-white often suffused with pink, becoming pale yellow with age, 8–9.5 mm long. Fruits 7.5–15 cm long. Flowering various seasons.

Local Distribution: Scattered on arid, lowland areas on Tiburón, where it is documented from sparsely vegetated, hot, exposed rocky hills and mountains, often near the coast. Widespread and common on San Esteban, primarily on rocky outcrops, and widely scattered on Dátil.

Other Gulf Islands: Ángel de la Guarda, San Lorenzo, Tortuga, San Marcos, Coronados, Carmen, Danzante, Santa Catalina (observation), San Francisco, Espíritu Santo.

Geographic Range: Southeastern California, both Baja California states, western Arizona, and northwestern

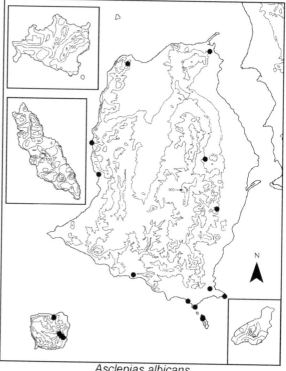

Asclepias albicans
Najcaazjc

Sonora south to Cerro Tepopa. It is characteristic of harsh desert sites, mostly on sparsely vegetated dry slopes.

Tiburón: Ensenada Blanca (Vaporeta), *Felger 14976*. Arroyo Sauzal, *Felger 9988*. Coralitos, small hill above beach, *Wilder 06-67*. La Viga, common, *Wilder 08-144*. El Monumento, *Felger 08-53*. Ensenada del Perro, base of hills, *Wilder 07-163*. Hast Coopol, rare, *Wilder 07-381*. E foothills of Sierra Kunkaak, ca. 280 m, occasional, *Wilder 08-322*. Punta Perla, rocky hillside, *Felger 74-15*. Bahía Agua Dulce, occasional on N slope, *Moran 12998* (SD). Pooj Iime, 29 Sep 2007, *Wilder*, observation.

San Esteban: E-central side, S-facing mountain side, 0.5 mi inland, *Felger 12760*. Main floodplain, 1 km inland, *Felger 466*. Narrow canyon at N side, *Felger 17555*. Large arroyo on SW corner, 50 m, *Hastings 71-52*.

Dátil: NW side, *Felger 15350*. E side, S-facing slope, 20 m below ridge crest, *Wilder 07-137*.

Asclepias subulata Decaisne

Najcaazjc; *jumete, mata candelilla*; reed-stem milkweed

Description: Plants forming a dense clump of many erect stems usually not more than 1 m tall. Flowers 18–22 mm long, waxy cream color, produced in profusion through much of the year and visited by large orange-winged spider wasps (*Hemipepsis* or *Pepsis*), the presumed pollinators.

Local Distribution: Known from a few widely scattered lowland sites on Tiburón and generally not common.

Other Gulf Islands: Cerralvo.

Geographic Range: Desert and thornscrub; Sinaloa and western Sonora to western Arizona and southern Nevada, and the Cape Region of Baja California Sur to southeastern California.

Tiburón: Arroyo Sauzal, 31 Jan 1965, *Felger*, observation (Felger 1966:221). Punta San Miguel, low dunes, *Wilder 07-187*. SW Central Valley, in arroyo, *Felger 12361*.

Funastrum hartwegii (Vail) Schlechter [*F. cynanchoides* (Decaisne) Schlechter var. *hartwegii* (Vail) Krings. *Sarcostemma cynanchoides* Decaisne subsp. *hartwegii* (Vail) R.W. Holm. *Funastrum heterophyllum* (Engelmann) Standley]

Hexe; *huirote, hierba lechosa*; climbing milkweed

Description: Robust vines with slender, twining or trailing stems. Facultatively drought deciduous, the stems dying back during severe drought. Leaves linear

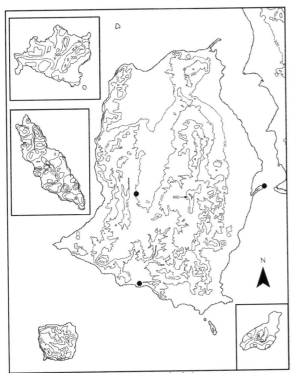

Asclepias subulata
Najcaazjc

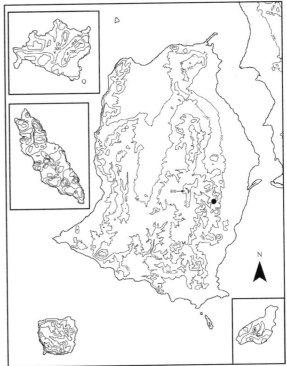

Funastrum hartwegii
Hexe

to linear-oblong or lanceolate and often hastate; usually with one or more conical yellow-brown glands (drying brown) 0.2–0.4 mm long at the base of the midrib on the upper surface of the blade. Corollas brownish purple with white, ciliate margins, the crown and vesicles white. Flowering at various times of the year including spring. Flowers edible with an onion-like flavor (Felger & Moser 1985).

Local Distribution: Documented on Tiburón only at the Pazj Hax waterhole at the base of Sierra Kunkaak. It is common on the nearby mainland, often in xeroriparian or semi-riparian localities such as major arroyos. Its scarcity on Tiburón seems unusual.

Other Gulf Islands: None.

Geographic Range: Southwest United States to Baja California Sur and central Mexico.

Tiburón: Pazj Hax, *Wilder 06-10.*

Haplophyton cimicidum A. de Candolle [*H. cimicidum* var. *crooksii* L.D. Benson]

Hierba de la cucaracha; cockroach plant

Description: Suffrutescent perennials with very slender, straight stems. Leaves opposite and/or alternate especially above. Flowers bright yellow and showy, 2 cm

wide; flowering following rainfall, mostly summer-fall. Fruits of paired very slender, multiple-seeded follicles 6–10 cm long, the seeds blackish with tufts of white hairs at each end.

Local Distribution: Tiburón in the interior of Sierra Kunkaak, sometimes locally common, and the adjacent southeast side of the Central Valley.

Other Gulf Islands: None.

Geographic Range: Southern Arizona to Guatemala and the Caribbean. The plant has been used as an insecticide, hence its common name. Lyman Benson segregated the northern populations as var. *crooksii*, differentiated mostly by smaller leaves and seeds, but the variety does not seem worthy of recognition.

Tiburón: Haap Hill, *Felger 74-T45.* Sopc Hax, abundant, *Felger 9287.* Hast Cacöla, E flank of Sierra Kunkaak, 450 m, *Wilder 08-262.* Canyon bottom at N base of Sierra Kunkaak, 395 m, *Wilder 06-434.* Base of Capxölim, N-facing rock slope, *Felger 07-107.*

Marsdenia edulis S. Watson

Xomee; *talayote*

Description: Robust, glabrous vines often growing on trees, with copious milky sap, the base often woody

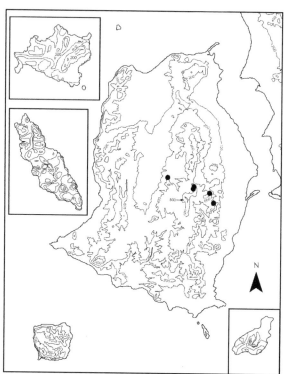

Haplophyton cimicidum

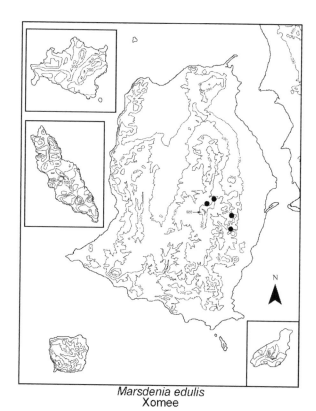

Marsdenia edulis
Xomee

with deeply fissured, corky bark. Leaves petioled, the blades ovate, to 15+ cm long. Flowers 7–8 mm long, white with a pale pink midstripe on the corolla lobes, and dense, stiff white hairs protruding into the corolla throat. Fruits green, smooth, and ovoid, 7.5–12 cm long, and hard shelled at maturity. Flowering mostly during the summer rainy season.

The unripe fruits are roasted and eaten (Felger & Moser 1985), and Palmer also reported that the green fruits are eaten (Watson 1889). Humberto reminds us that some Comcaac fear that the sap causes pain, which seems to be a reflection of a general aversion or fear of plants with milky sap (Felger & Moser 1985).

Local Distribution: Tiburón, in the Sierra Kunkaak, where it is often in arroyos and canyons and on brushy slopes from low to high elevations.

Other Gulf Islands: None.

Geographic Range: Northern Sonora, from near Caborca, Palm Canyon southeast of Magdalena, and the north end of the Sierra el Tigre to Sinaloa.

Tiburón: Arroyo midway between Pazj Hax and Hast Coopol, *Wilder 06-22.* Sopc Hax, canyon bottom, *Felger 9306.* SW and up canyon from Siimen Hax, *Wilder 06-475.* Top of Sierra Kunkaak Segundo, E peak of the Sierra Kunkaak, *Wilder 06-494.*

Matelea

1. Leaves to 10+ cm long; fruits smooth. _ **M. cordifolia**
1' Leaves to ca. 6 cm long; fruits with soft, blunt prickles.
_____ **M. pringlei**

Matelea cordifolia (A. Gray) Woodson

Comot

Description: Vines mostly more than 1.5 m long, with a semi-woody base, growing through desert shrubs and trees. Leaves drought deciduous, long petioled, the blades to 10+ cm long, thin and more or less heart shaped, and foul smelling when crushed. Corollas 15 mm long, lobed nearly to the base, cream colored and faintly lined, the corona green. Flowers, fruits, and foliage produced at various times of the year following rainy periods. Fruits 9–13 cm long, elongate-ellipsoid, smooth, mottled green and whitish.

Local Distribution: Tiburón, in arroyos, canyons, and rocky slopes of Sierra Kunkaak and at the south end of the Central Valley; apparently uncommon at lower elevations.

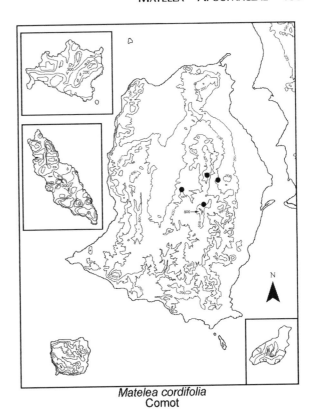

Matelea cordifolia
Comot

Other Gulf Islands: Carmen, Danzante, Cerralvo.

Geographic Range: Sinaloa northward through Sonora to the vicinity of Magdalena and the Sierra del Viejo near Caborca, Organ Pipe Cactus National Monument in southern Arizona, and Baja California Sur. Best developed in canyons, arroyos, and in protected, often-brushy places on mountain slopes, and less common on the desert flats.

Tiburón: Haap Hill, *Felger 74-T52.* Deep sheltered canyon, N portion of Sierra Kunkaak, *Felger 07-123.* Caracol, arroyo, *Wilder 07-230.* Top of Sierra Caracol, *Knight 1064* (UNM).

Matelea pringlei (A. Gray) Woodson

Nas, ziix is quicös "prickly-fruited thing"

Description: Small bushy perennials often to 50+ cm across or vining on shrubs. Leaves long petioled, the blades triangular-hastate, to 6 cm long but mostly smaller. Flowers 1 cm long, at first whitish, purplish maroon with age, and with meat-colored filaments wiggling in the wind and attracting flies. Fruits 5.5–9 cm long, mottled green and pale green, elongated-ellipsoid with blunt prickles to ca. 1 cm long. Reproductive at various seasons.

Local Distribution: Known from widely scattered localities on Tiburón. Mostly in arroyos and rocky, often north-facing slopes on Dátil.

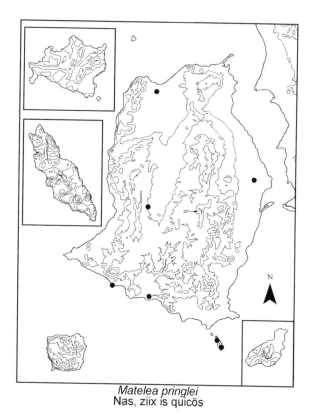

Matelea pringlei
Nas, ziix is quicös

Metastelma arizonicum A. Gray [*Cynanchum arizonicum* (A. Gray) Shinners]

Description: Flowers 3–4 mm long, green, and with very dense white villous hairs on the inner surfaces of the corolla lobes. These hairs point downward toward the center of the flower. A small insect attracted to the flower would be directed toward the center of the flower, and the downward-pointing hairs would impede its access elsewhere. Fruits 3.5–5.5 cm long.

Local Distribution: Sierra Kunkaak from interior canyons to higher elevations.

Other Gulf Islands: None.

Geographic Range: Durango and Sinaloa through most of Sonora to southwestern Arizona.

Tiburón: The "bowl" on top of the Sierra Kunkaak, *Knight 1020* (SD). Exposed upper ridge of island, between peak of Cójocam Iti Yaii and summit, *Wilder 07-550*. Base of Capxölim, N-facing rock slope, *Felger 07-104*.

Other Gulf Islands: Coronados, Monserrat, San Diego, San José, Cerralvo.

Geographic Range: Sonora from the vicinity of Caborca to the Guaymas region and much of the Baja California Peninsula.

Tiburón: Tordillitos, *Felger 15479*. El Sauz, in bush, one seen, *Moran 8775* (SD). SW Central Valley, not common, 450 ft, arroyo, *Felger 12366*. 3 km W [of] Punta Tormenta, *Scott P8* (UNM, 2 sheets). Arroyo Agua Dulce, *Felger 6843*.

Dátil: NW side of island, *Felger 15302*. SE side of island, steep narrow canyon, rare, N exposure, *Felger 17497*.

Metastelma

Small vines of many slender stems often twining together on shrubs. Glabrous except for a longitudinal line of usually curved hairs on the stem. Leaves linear and dark green, the margins often inrolled (revolute). Flowers very small. Fruits smooth and elongated.

1. Flowers 3–4 mm long, greenish, Tiburón. _____
_____**M. arizonicum**
1' Flowers 2.2–2.5 (3) mm long, white, Nolasco._____
_____ **M. californicum**

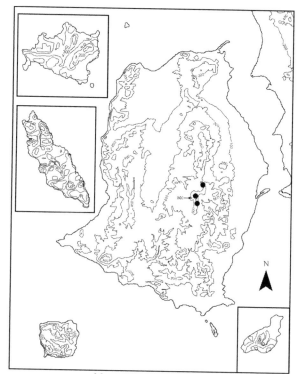

Metastelma arizonicum

Metastelma californicum Bentham

Description: Vines forming densely twining and tangled mats on shrubs such as *Bernardia*. Flowers white, 2.2–2.5 (3) mm long; corona scales comparatively long, the terminal appendage on anthers translucent and

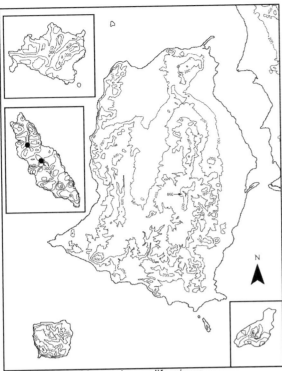

Metastelma californicum

Vallesia glabra (Cavanilles) Link

Tanoopa (variants: tinoopa, tonoopa, tenoopa); *citavaro, huevito*

Description: Multiple-stem shrubs often 1.5–2.5 m tall, glabrous or glabrate. Leaves alternate, tardily drought deciduous, lanceolate to elliptic, and shiny green. Inflorescences of small, several-flowered racemose or paniculate cymes, the flowers less than 1 cm wide, white, and not noticeably fragrant. Fruits 10–12 mm long, fleshy, opalescent white, and 1-seeded. Flowering at various seasons.

Local Distribution: Widespread on Tiburón but not known from the arid western margin, mostly along drainageways, canyon bottoms, and bajada slopes.

The fleshy pericarp is edible (Felger & Moser 1985), but there is evidence that eating too much of it is not advisable (Felger, field notes).

Other Gulf Islands: Carmen.

Geographic Range: Widespread across the southern two-thirds of Sonora, usually in low-lying, fine-textured heavy soils in riparian and semi-riparian habitats, Chihuahua, and southern Florida to South America, and southern Baja California Norte and Baja California Sur.

as broad or slightly broader than long, the appendage apex (tip) broadly obtuse, almost truncate. Inner surfaces of corolla lobes moderately hairy and nearly naked medially. Flowering recorded in September and November.

Local Distribution: Nolasco, common at higher elevations on the east side and along the ridge crest of island, especially in slightly shaded and sheltered sites among boulders. This species has not been found on the opposite mainland, where instead one finds *M. arizonicum* that is distinguished in part by its floral morphology including larger flowers. The report of *M. pringlei* A. Gray on Nolasco (Felger & Lowe 1976; Rebman et al. 2002) was based on incorrect identification of specimens from the island.

Other Gulf Islands: Santa Catalina.

Geographic Range: Coastal southwestern Sonora south of the desert, both states of Baja California, northwestern Sinaloa, and Islas Revillagigedos.

Nolasco: E-central side, N exposure below crest of island, 270 m, covering a large *Bernardia* shrub, seen only at higher elevations, *Felger 06-99.* Cañón de Mellink, ridgetop of island, *Wilder 08-360.*

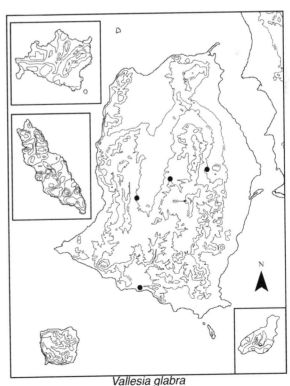

Vallesia glabra
Tanoopa

Tiburón: El Sauz, few, under trees, *Moran 8763* (SD). Haap Hill, *Felger 74-T65*. SW Central Valley, 400 ft, *Felger 17349*. Caracol, arroyo, *Wilder 07-229*.

Arecaceae (Palmae) – Palm Family

Brahea armata S. Watson

Zamij cmaam "female palm"; *palma azul*; blue hesper palm

Description: Fan-leaved palms with a single trunk. Leaves moderately costapalmate (the petiole extends into the leaf blade); petioles armed with rigid spines; blades relatively flat, tough, bluish glaucous with silvery-white lepidote scales. Fruits globose, 1.5 cm diameter, and rounded with a thin, fleshy, date-like pericarp.

Local Distribution: The report of a palm on Isla Tiburón had been enigmatic even though a number of Seris had information of a palm on the northwest side of the island (Felger & Moser 1985:351). In 1964, José Juan Moreno brought Mary Beck Moser a single leaf of a palm that he collected in a canyon along the northwest side of the island. Sr. Moreno had been on a fishing trip

and knew that Mary Beck and Richard were interested in the "palm from Tiburón" as they were beginning their studies of Seri knowledge and uses of plants. Soon thereafter Richard and Ike Russell made a number of searches by airplane along the west side of the island for the elusive palm. The search was futile even though Ike would fly low and close (Felger 2000b, 2002). Sometimes, however, shadows, other trees, and cliffs can make it difficult to spot plants from the air. In September 2007 we visited the "palm canyon," known as Pooj Iime, but no palm was found. Perhaps it has perished.

On 4 May 2007, Humberto confirmed that his late brother had seen a single palm at this site about 20 years earlier. The palm site is reported to be at or near a freshwater source in a canyon on the seaward side and base of high cliffs on the northwest side of the extremely arid Sierra Menor in the northwestern part of the island.

Other Gulf Islands: Ángel de la Guarda.

Geographic Range: The nearest locality for *B. armata* is at the southeast side of Isla Ángel de la Guarda, 65 km to the west of the Tiburón site (Johnston 1924; Moran 1983a; Turner et al. 1995). The vegetation and habitats of the Ángel palm canyon sites are quite similar to those of the northwest coast of Tiburón.

The nearest *Brahea* population from the Tiburón site is on the Sonora mainland 37.5 km to the northeast: "The most isolated and undoubtedly smallest palm population in Sonora occurs on a north-facing slope below Pico Johnson in the Sierra Seri. There are probably fewer than one dozen of these palms, hidden from view and far from any road" (Felger & Joyal 1999:10). The Pico Johnson population seems conspecific with the hesper palms in the Sierra el Aguaje, north of San Carlos and Guaymas, which have been treated as conspecific with *B. brandegeei* of Baja California Sur (Felger & Joyal 1999; Felger et al. 2001; Henderson et al. 1995). These western Sonora populations, however, appear to be closely related to *B. armata* of Baja California Norte and Isla Ángel de la Guarda, and the distinctions "have not been quantified or studied in depth" (Felger & Joyal 1999:11).

Tiburón: Canyon on NW side of the island, *José Juan Moreno 20 May 1964*.

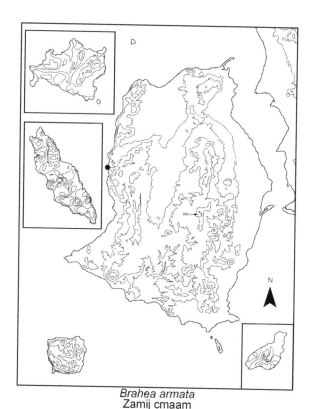

Brahea armata
Zamij cmaam

Aristolochiaceae – Pipevine Family

Aristolochia watsonii Wooton & Standley [*A. brevipes* Bentham var. *acuminata* S. Watson, not *A. acuminata* Lemaire. *A. porphyrophylla* H. Pfeifer]

Hataast an ihiih "what gets between the teeth"; *hierba del indio*; southwestern pipevine

Description: Perennials from a carrot-shaped root. Stems slender, herbaceous, sprawling, or weakly vining. Leaves narrowly to broadly triangular and often hastate lobed. Flowers bilateral, smelling of carrion, 3.5–5 cm long, solitary in leaf axils, the calyx tooth shaped, yellow-green with dark maroon spots, margins, and tip, this pattern and color apparently attracting flies; petals none. Fruits many-seeded capsules.

Local Distribution: Known from Tiburón by two records.

Other Gulf Islands: None.

Geographic Range: Widespread in Sonora, the growth and flowering non-seasonal but generally dormant during the cooler months; dying back to the root during drought. Southwestern New Mexico and Arizona to Nayarit, and Baja California Sur and southern Baja California Norte.

Tiburón: SW Central Valley, rare, gravelly arroyo bed, 600 ft, *Felger 17296*. Hast Coopol, occasional, *Wilder 07-380*.

Asclepiadaceae, see Apocynaceae

Asparagaceae – Asparagus Family
(includes Agavaceae, Nolinaceae, Ruscaceae, Themidaceae)

1. Leaves herbaceous, unarmed, from small, underground white bulbs; flowers blue and showy. ___ **Triteleiopsis**
1' Leaves in rosettes, thick, tough, and succulent with a stout terminal spine, or linear and relatively flat with marginal spines; flowers whitish, pink, or yellow, etc., but not blue.
2. Leaves thick and succulent with a stout terminal spine, the margins entire or with small to large prickles, these usually not curved. ___**Agave**
2' Leaves relatively flat, not succulent, with a row of curved spines on both margins. ___**Dasylirion**

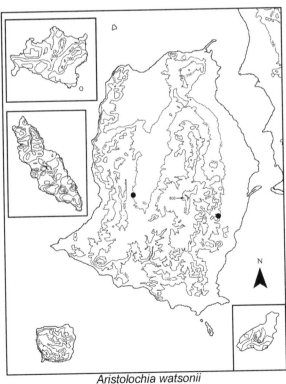

Aristolochia watsonii
Hatáast an ihíih

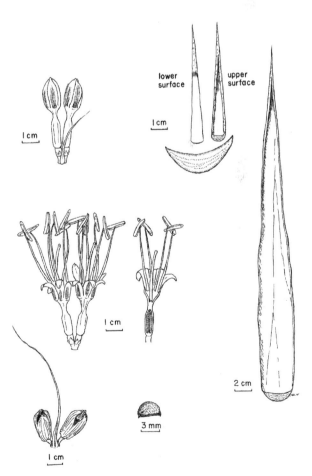

Figure 3.17. *Agave chrysoglossa*. NLN.

Figure 3.19. *Agave chrysoglossa.* Isla San Pedro Nolasco. BTW, 29 September 2008.

Figure 3.18. *Agave cerulata* subsp. *dentiens.* Isla San Esteban. BTW, 8 March 2007.

Figure 3.20. *Agave subsimplex,* Isla Dátil. BTW, 10 March 2007.

Figure 3.21. *Dasylirion gentryi.*
Near the summit of Sierra Kunkaak,
Isla Tiburón. BTW, 26 October 2007.

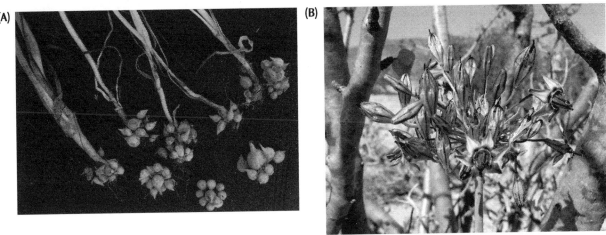

Figure 3.22. *Triteleiopsis palmeri.* (A) Bulblets at leaf bases, vicinity of
El Desemboque, Sonora, RSF, April 1983. (B) Zozni Cacösxaj, Isla Tiburón,
BM, 6 April 2008.

Agave – Maguey; century plant

The agaves are the most prominent leaf succulents in the flora.

1. Leaf margins entire with a terminal spine only; inflorescences spicate (without branches or the branches small and inconspicuous); Sierra Kunkaak, Tiburón, and Nolasco. _____**A. chrysoglossa**
1' Leaf margins usually spiny; inflorescences paniculate (branched).
    2. Flowers bright yellow; San Esteban. __ **A. cerulata**
    2' Flowers pale yellow to rose; Tiburón, Dátil, and Cholludo. _____ **A. subsimplex**

Agave cerulata Trelease subsp. **dentiens** (Trelease) Gentry, Occasional Papers of the California Academy of Sciences 130:43, 1978 [*A. dentiens* Trelease, Annual Report of the Missouri Botanical Garden 22:51, 1911]

Heme, xiica istj caaitic "things whose leaf is soft"; *maguey*; San Esteban century plant

Description: "This medium-sized century plant is considerably larger and more massive than *A. subsimplex*. . . . It forms expansive colonies on the coarse rocky slopes of San Esteban, and is one of the major landscape elements of the island" (Felger & Moser 1985:222). Leaves very thick, the marginal spines highly variable, relatively large on some clones and quite small or sometimes absent on others. Inflorescences branched, 3–4+ m tall; flowers yellow.

Across most of the island the plants have markedly gray- to bluish-glaucous leaves. Green-leaved clones, however, are common on the heights, where fog and clouds are more frequent than at lower elevations. Green- and grayish-leaved clones occur intermixed.

Local Distribution: Abundant across San Esteban from near sea level to the peak. "This highly variable subspecies is endemic to San Esteban Island, and is the only agave occurring on that island. . . . Unlike all other agaves in the Seri region, edible plants of this species could be found throughout the year. However, they are most flavorable during the latter part of January" (Felger & Moser 1985:222). Perhaps development of the flowering stalk (plants that are starting to become reproductive—the usual situation for agaves to be suitable for harvest because of accumulation of carbohydrates) becomes arrested due to drought and such plants might be suitable for harvesting and eating all year. This is an intriguing adaptation.

Other Gulf Islands: Subsp. *dentiens*: none. Subsp. *cerulata*: Ángel de la Guarda. Subsp. *subcerulata*: San Marcos.

Geographic Range: "It is possible that the agaves on Ángel de la Guarda Island should be aligned with *A. cerulata dentiens*" (Gentry 1982:369). Tom Bowen (personal communication, 2006) tells us that the Ángel de la Guarda agaves appear superficially indistinguishable from those on San Esteban. Agaves are not widespread on Ángel and at low elevation are seen only in a limited area on the west side of the island. Gentry recognized three other subspecies, all in the central part of the Baja California Peninsula. Investigation into the *A. deserti* complex to which *A. cerulata* subsp. *dentiens* belongs provides evidence for a lack of genetic differentiation from other forms of *A. cerulata* (Navarro-Quezada et al. 2003). However, based upon morphological features and the need for more genetic studies, the treatment of *A. cerulata* subsp. *dentiens* as a single island endemic remains valid (Francisco Molina-Freaner, personal communication, 2007).

San Esteban: "Lower California, San Esteban," 13 Apr 1911, *Rose 16819* (holotype US, images). Common in small colonies on the scoria-covered hillsides, stalks commonly 20 ft high, *Johnston 3194* (CAS). Arroyo Limantour, *Felger 77-11.* E side of island, large colonies common on hillside and in

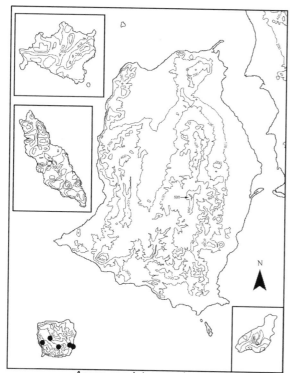

Agave cerulata ssp. *dentiens*
Heme, xiica istj caaitic

arroyo, *Moran 21748b*. Mountain slope, 120 m, *Felger 7048*. Cañones en el SE de la isla, infl. ca. 4 m, ca. 12 ramas laterals, botones verdes, *Lott & Atkinson 5 May 1985* (CAS).

Agave chrysoglossa I.M. Johnston, Proceedings of the California Academy of Sciences, IV, 12: 998, 1924

Hasot; *amole*

Description: Medium-sized agaves, suckering to form clusters of mostly several rosettes or sometimes solitary. Leaves 50–100 cm long, rather narrow and firm, thick, bright to dull green, the margins entire, the terminal spine stout. Flower stalks 2–4 m tall, inflorescens cylindrical and densely flowered. Flowers yellow; late spring. In the original description of this species, Ivan Johnston (1924:999) wrote, "When found it was enlivening the rocky mid-slopes of the island [San Pedro Nolasco] with spectacular, bright yellow tongues of color. The plants grew singly and produced dense elegant spicate floral clusters 1–2 m. long and 8–10 cm. broad which, due to their weight, almost invariably bent over with their tips nearly touching the ground."

Local Distribution: On Tiburón, common and widespread in the Sierra Kunkaak, from 350 m to the summit, on cliffs and rock slopes. Our observations from the 1990s to 2010 indicate the plants were heavily impacted by bighorn sheep grazing. On Nolasco, abundant on the east side of the island, canyons at the west side, and the ridge crest, often growing from rock crevices, most common toward higher elevations.

Other Gulf Islands: None.

Geographic Range: Mountains in west-central Sonora, mostly in coastal ranges from the Sierra Seri southward to Sierra el Aguaje (north of San Carlos), and Nolasco and Tiburón. Populations in eastern Sonora, near Bacanora and Sahuaripa, appear intermediate between *A. chrysoglossa* and *A. vilmoriniana* Berger (Gentry 1982; Turner et al. 1995).

Tiburón: S-facing slope of Sierra Kunkaak, 355 m, at lower limit of its occurrence on this slope, *Wilder 06-28*. Canyon at N base of Sierra Kunkaak, 385 m, *Wilder 06-504*. Cerro Kunkaak, 1200 m, *Scott 11 Apr 1978* (UNM, two sheets). E ridge of Capxölim, 400 m, *Ezcurra 07-2*. Sierra Kunkaak, arriba de Siimen Hax, 25 Oct 2007, *Romero-Morales 07-26*. Summit of island, not common on top but extremely common on slopes and cliffs just below summit with *Dasylirion gentryi*, *Wilder 07-563*.

Nolasco: Rocky slopes, 17 Apr 1921, *Johnston 3123* (holotype, CAS). Steep granitic slope, 16 Dec 1951, *Gentry 11349*. E side, common, *Felger 9664*. E side, very common, beautiful studding the extremely granitic slopes; the central flower stalk long, swooping in contorted curves generally toward the ocean, weighted down by literally hundreds of yellow flowers; thick central rosettes of leaves, *Stinson 29 Apr 1974* (SD). Ridge between Cañón de la Guacamaya and Cañón de Mellink, 3 May 2005, *Gallo-Reynoso* (photo).

Agave subsimplex Trelease, Annual Report of the Missouri Botanical Garden 22:60–61, plate 63 & 64, 1911

Haamjö; *maguey, mescal*

Description: Generally a small agave but quite variable in size, and plants in the mountains near Tecomate at the northwestern side of Tiburón can be relatively larger than elsewhere. Plants suckering, leaves bluish or gray-green, the marginal spines variable in size. Flowering stalks paniculate with short branches, sometimes unusually short on stunted plants. Flowering mostly March and April. Flowers pale yellow to pinkish, base of the tube pinkish white, tepals pale yellow to reddish or dusty rose, the filaments dusty rose, the anthers yellow.

Reid Moran recorded this detailed description based on a fresh specimen from Isla Dátil (25 Apr 1966, *Moran 13022*, SD): "Caudex short, ca. 5 cm, with slender sucker

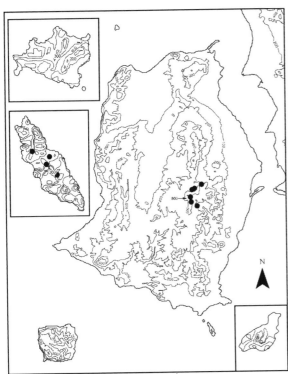

Agave chrysoglossa
Hasot

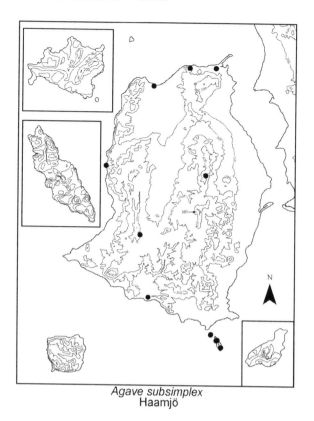

Agave subsimplex
Haamjö

Geographic Range: Western Sonora in coastal mountains from near Puerto Libertad in the Sierra Bacha southward to the vicinity of Bahía Kino and the three islands cited.

Tiburón: El Sauz, 150 m, rosette 6 dm, of ca. 60 leaves, suckers to ca. 0.5 m from base, leaves gray, floral stem 4 m tall, with ca. 5 umbels, *Moran 8786* (SD). Sur de Satoocj, 8 Jan 2008, *Romero-Morales 08-4*. Top of Sierra Caracol, *Knight 1050* (UNM). Punta Perla, rocky hill, *Felger 21318*. Hee Inoohcö, common, including lower slopes, 30 Sep 2007, *Wilder*, observation. Tecomate, abundant on rocky mountain slopes on W side of island, 14 Jun 1951, *Whiting 9047*. Pooj Iime, large canyon, NW shore, *Wilder 07-454*.

Dátil: Canyon, SE part of island, *Felger 2562*. NW part of island, *Felger 2781*. NE side, *Felger 13431*. Rocky slopes, *Moran 13022* (SD). E side of island, halfway up N-facing slope of peak, *Wilder 07-103*. E shore, canyon bottom, *Wilder 07-183*.

Cholludo: Lower California, Seal Island, 13 Apr 1911, *Rose 16811* (holotype US, isotype MO, images). Common and widespread, *Felger 2721*. Steep rocky slopes dominated by cacti, abundant on NE slope of isle but not on NW slope, *Felger 13425*. N-central slope, common and dense on lower ⅓ of island, not seen above ca. 40 m, often at base of cardons, *Wilder 08-319*.

Agave sp.

Haamjö caacöl "large *A. subsimplex*"

shoots underground; rosettes 3–4 dm with ca. 12–25 leaves plus the bud; leaves glaucous, oblong-oblanceolate or +/– rhombic, acuminate, 15–20 cm long, 4–6 cm wide, with ca. 8 teeth on each margin, the upper ones recurved, 10–13 mm long, ca. 2 mm wide at the base; floral stems 1½–4 m tall, 2–3 cm at the base; inflorescence 4–8 dm high, of ca. 7–10 branches, each 1–5 cm long and bearing an umbel of ca. 10–25 flowers; pedicels 2–3 mm long; ovary 2–3 cm long, 5–6 mm thick, narrowed at base and apex; tepals light yellow or reddish, nearly free, oblong, obtuse, 1½–2 cm long, 5–7 mm wide; filaments 2½–3½ cm long, pink to dark red; anthers linear, 15–18 mm long; styles elongating during anthesis, 1½–3½ cm long, colored like the filaments."

Local Distribution: Widely scattered on Tiburón, mostly on rocky slopes, including the northern and southern parts of the island, and in the Sierra Kunkaak. It is common and abundant on Dátil and Cholludo. On Dátil it occurs from near the shore and canyon bottoms to peak elevations but is generally not on south-facing slopes. The name Dátil was given to this island because of the prevalence of *A. subsimplex* (Felger 1966), although this name is usually applied to yuccas and more specifically to the date palm. On Cholludo it grows primarily on the northeast slope of the island.

Other Gulf Islands: None.

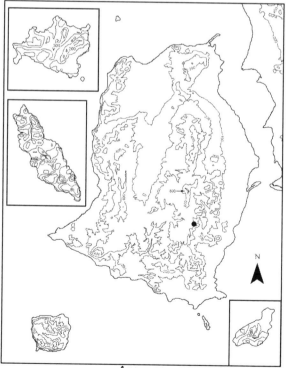

Agave sp.
Haamjö caacöl

Humberto describes another agave as having "hojas muy suculentas, mas cerca a *A. subsimplex*, pero mas grande." The leaf margins are spiny and the flowering stalks branched. It is found in the "parte sur de Sierra Kunkaak, en la cima y en el centro suroeste de Satoocj." The description indicates a large form of *A. subsimplex* or perhaps a different species such as *A. colorata*, known from coastal Sonora from near Tastiota to the Guaymas region and southward to Sinaloa.

Dasylirion gentryi Bogler

Istj ano caap "whose leaf stands inside"; *sotol*

Large rosette-forming plants with a short trunk. Leaves to about 1 m long, 12–17 mm wide not including the spines, green, relatively thin, flexible, and slightly channeled. Leaf margins with many sharp, recurved and mostly forward-projecting spines 2.3–4.5 mm long. Flower stalks to about 2 m tall. Male and female flowers on different plants. Fruits probably with 3 papery wings.

Local Distribution: Common in the Sierra Kunkaak from 350 m to the summit; mostly restricted to rocky slopes, canyon walls, and cliffs, commonest toward the summits of ridges and peaks.

Other Gulf Islands: None.

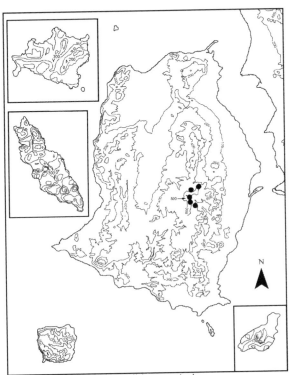

Dasylirion gentryi
Istj ano caap

Geographic Range: The Tiburón population shows close affinity to two isolated populations in mountains in western Sonora: the Sierra del Viejo southwest of Caborca and the Sierra el Aguaje north of San Carlos. Flowers and fruits from these three populations are unknown. Plants of these three populations resemble *D. wheeleri* S. Watson ex Rothrock but differ in part by having green and narrower leaves. David Bogler annotated Scott's UNM collection as *D. gentryi*, a species otherwise known from mountains in southwestern Chihuahua and southeastern Sonora, and we follow Bogler's opinion.

Tiburón: Cerro San Miguel, común en ladera de lomerios y cerro San Miguel, junto con *Agave chrysoglossa*, en un cerro pequeño, ca. 200 m [should be ca. 500 m; Adrian Quijada, personal communication, 2007], *Quijada-Mascareñas 91T018*. Cerro Kunkaak, 1200 m [*sic*], *Scott 11 Apr 1978* (UNM 53273 & 53278). S-facing slope of Sierra Kunkaak, 355 m, 2 Jan 2006, *Wilder* (photo). Capxölim, solitary peak NE of Sierra Kunkaak Mayor, *Wilder 07-483*. Sierra Kunkaak, arriba de Siimen Hax, 25 Oct 2007, *Romero-Morales 07-27*. Summit of island, common on cliffs, *Wilder 07-554*.

Triteleiopsis palmeri (S. Watson) Hoover [*Brodiaea palmeri* S. Watson]

Caal oohit "what the companion child eats"; blue sand lily

Description: Perennials from clusters of tiny cormlets (bulblets) produced on top of the previous year's bulbs and in axils at leaf bases; propagating (mostly or entirely?) by bulblets. Flowers produced in multiple-flowered umbels on leafless stalks to about 50 cm tall. Flowers about 1.5 cm wide, deep blue, and attractive on slender pedicels; the hypanthium and filaments form a scalloped collar with holes through which the floral tube appears white in the sun and probably functions as a nectar guide. Seeds blackish, flat. Growing in winter-spring and flowering in spring.

A chromosome count of a specimen from near Yuma, Arizona, indicates the plants are at least sometimes triploid ($2n = 33$) and therefore sterile (Lee Lenz, personal communication, 1987). The leaves are of 2 kinds. The first leaves of the season are green, stringy, and prostrate, resembling green spaghetti on the sand. Later in the season, as the days and sands become hot, the prostrate leaves wither and the newer leaves are held aloft, well above the substrate. The small "bulbs" are tasty, and Seri children at El Desemboque eagerly seek them (Felger & Moser 1985:343). Humberto remembers that when he was a child, Richard came back to

El Desemboque with fresh *caal oohit* and gave some of them to him and other children.

On 5 April 2008 at Punta Chueca, Humberto and Ben were brought a single flowering stem of *caal oohit* by Humberto's brother-in-law José Ramón Barnett, who found it at the historical Seri camp Zozni Cacösxaj, at the northeast part of Tiburón near the head of the Canal del Infiernillo. This was a new record for the island but a species we had expected because it is common on the adjacent mainland. We arrived at Zozni Cacösxaj the next day where several families were harvesting seeten (*callo de hacha*, pen shells, *Atrina tuberculosa* and *Pinna rugosa*; Basurto 2005, 2008). José Ramón, his wife, and their young children quickly took us to the plants, and another addition was made to the flora.

Local Distribution: In the flora area known only from the northeast coast of Tiburón, where it grows in sandy soil.

Other Gulf Islands: None.

Geographic Range: Sand flats and lower dunes and occasionally on gravelly granitic flats; southwestern Arizona, both Baja California states, and coastal dunes and sand flats in Sonora southward to Cerro Tepopa (29°24'N).

Tiburón: Zozni Cacösxaj, NE coast, 5 m elev., *Wilder 08-196.*

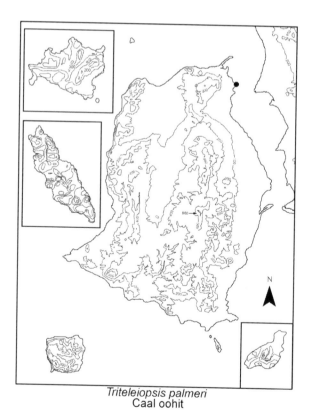

Triteleiopsis palmeri
Caal oohit

Asteraceae (Compositae) – Aster or Composite Family

This is the largest family of vascular plants, with more than 1530 genera and 22,750 species, or about 9 percent of the world flora. Mexico probably contains more than 10 percent of the world's species. The family is especially well developed and diverse in semi-arid regions. The composites and grasses (Poaceae) are the two largest families in the flora area. The composites are represented by 33 species in 26 genera, or 8.4 percent of the total flora of the Sonoran Islands. Their having evolved efficient chemical defenses against herbivores is believed to be a major factor in the worldwide success of the family (Cronquist 1981). Indeed, while studying Sonoran Desert composites one is impressed by the prevalence and diversity of glands on the stems, leaves, and especially on exposed surfaces of phyllaries and corollas (Felger 2000a). These glands often occur in combination with a great diversity of hairs.

The genera in the flora area are distributed in 6 tribes, the most diverse being the Heliantheae, with 14 species in 8 subtribes. *Ambrosia* has 6 species, and all other genera have 1 or 2 species.

ASTEREAE

Baccharis
Gymnosperma
Xanthisma
Xylothamia

CICHORIEAE

Stephanomeria

EUPATORIEAE

Brickellia
Chromolaena
Pleurocoronis
Hofmeisteria

HELENIEAE

Pelucha

HELIANTHEAE

Ambrosiinae
Ambrosia

Chaenctinidinae
Palafoxia
Peucephyllum

Coreopsidinae
Coreocarpus

Ecliptinae
Encelia
Heliopsis
Verbesina

Galinsoginae
Bebbia

Helianthinae
Bahiopsis
Helianthus

Pectidinae
Bajacalia
Pectis
Porophyllum

Peritylinae
Perityle
Thymophylla

MUTISEAE

Trixis

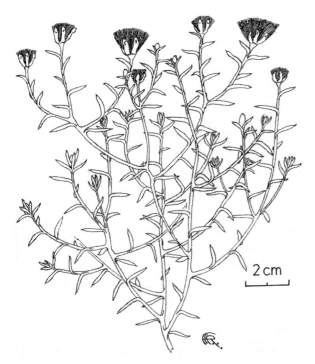

Figure 3.23. *Bajacalia crassifolia.* FR.

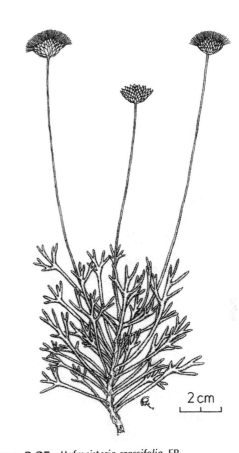

Figure 3.25. *Hofmeisteria crassifolia.* FR.

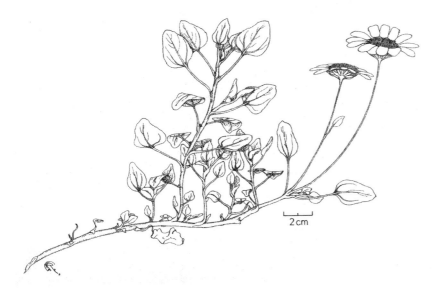

Figure 3.24. *Helianthus niveus.* FR.

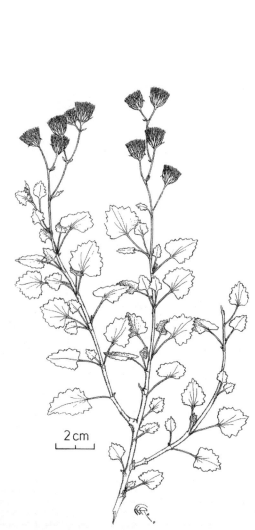

Figure 3.26. *Pleurocoronis laphamioides.* FR.

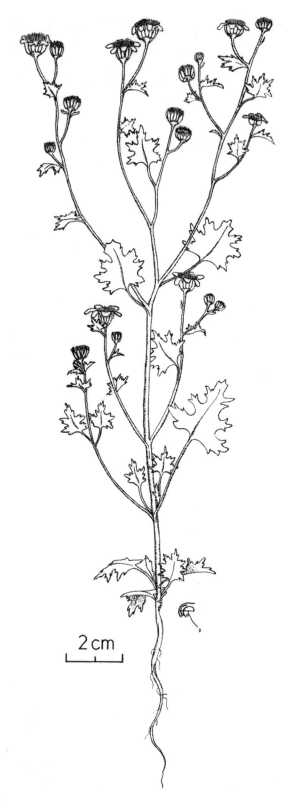

Figure 3.27. *Perityle emoryi.* FR.

Figure 3.28. *Ambrosia carduacea.* Interior canyon of Sierra Kunkaak, Isla Tiburón. BTW, 25 November 2006.

Figure 3.29. *Ambrosia ilicifolia.* Isla San Esteban. BTW, 8 March 2007.

Figure 3.30. *Bahiopsis triangularis.* (A and B). Isla San Pedro Nolasco. BTW, 2 February 2008.

Figure 3.31. *Bajacalia crassifolia.* Coralitos, Isla Tiburón. BTW, 2 September 2008.

Figure 3.32. *Coreocarpus sanpedroensis.* Isla San Pedro Nolasco. BTW, 2 February 2008.

Figure 3.33. *Hofmeisteria fasciculata.* Hast Cacöla, Isla Tiburón. BTW, 7 April 2008.

Figure 3.34. *Peucephyllum schottii*. Isla San Esteban, with Seri Plant Team members Adán Humberto Romero-Morales (right) and Ivan Eliseo Barnett-Romero (left) smelling the aromatic leaves of the shrub. BTW, 3 September 2008.

Figure 3.35. *Verbesina palmeri*. Sierra Kunkaak, Isla Tiburón. BTW, 24 October 2007.

Figure 3.36. *Xanthisma incisifolium.* Isla San Esteban. BTW, 9 March 2007.

1. Perennials; stems and especially the leaves succulent; flowers discoid.
    2. Leaves linear, more than 5 times longer than wide, 1–2.5 cm long, simple and sessile with entire margins; flowers yellow._____ **Bajacalia**
    2' Leaves about as wide as long, mostly more than 2.5 cm long, highly dissected, petioled; flowers lavender-pink. _ _____ **Hofmeisteria**
1' Annuals or perennials; not markedly succulent; flowers various (*Coreocarpus* sometimes semi-succulent, but with ray florets).
    3. Small shrubs or subshrubs; leaf bases persistent on stems as short, blunt projection; heads of bilabiate florets only; achenes expanded at apex into a disk bearing numerous pappus bristles._____ **Trixis**
    3' Herbs to shrubs; leaf bases not persistent as above; heads with ray and disk florets, or only ray or disk or disk-like florets, these not bilabiate, or corollas none or not evident; achenes various.
        4. Flowers heads unisexual, corollas none or not evident; female heads of burs. _____ **Ambrosia**
        4' Flowers heads not unisexual (except *Baccharis*), with at least some bisexual flowers or male and female flowers in the same flower head; corollas evident; not forming burs.
            5. Herbs or shrubs; heads with both ray and disk florets, or rays only, the rays usually obvious (taxa with small, inconspicuous, or early-deciduous rays will key out in either choice, or if in doubt go to 5').
                6. Pappus none (or microscopic in *Gymnosperma*)
                    7. Leaves opposite; rays yellow and showy, persistent and becoming green; mountains (Sierra Kunkaak, Tiburón). _____**Heliopsis**
                    7' Leaves alternate or sometimes opposite, or opposite below and alternate above; rays not persistent and not becoming green.
                        8. Annuals or short-lived herbaceous perennials; leaves dissected; rays white with dark longitudinal lines._____**Coreocarpus**
                        8' Annuals, subshrubs, or shrubs; leaves entire; rays yellow.
                            9. Herbage viscid-sticky; leaves linear; rays inconspicuous, less than 3 mm long. _____ _____ **Gymnosperma**
                            9' Herbage not viscid; leaves ovate; rays more than 10 mm long.
                                10. Rounded, perennial bushes; leaves mostly crowded (close together) at stem tips; achenes flattened, the margins outlined with white hairs. _____**Encelia**
                                10' Annuals to herbaceous perennials; leaves scattered along stems; achenes angular or only slightly compressed, the margins undifferentiated—not outlined (pappus deciduous, absent from older achenes). _____ **Helianthus**
                6' Pappus present, at least on disk achenes.
                    11. Plants glabrous, dotted with prominent oil glands and pungently aromatic.

12. Leaves deeply divided; rays white. _____ **Thymophylla**
12' Leaves entire but some with basal bristles; flowers yellow.
    13. Annuals; leaves with prominent bristles at base. _____**Pectis**
    13' Perennials; leaves without bristles (leaves filiform and soon deciduous on *P. gracile*). __
    _____ **Porophyllum**

11' Plants glabrous or hairy, not dotted with oil glands, aromatic or not.
    14. Annuals or perennials; leaves lobed and coarsely toothed.
        15. Cool season annuals; leaves palmately lobed and coarsely toothed; rays white or yellow. _
        _____ **Perityle**
        15' Non-seasonal annuals, perennials, or shrubs; leaves not palmately lobed.
            16. Sap milky; heads of 5 ray florets; pappus plumose. _____ **Stephanomeria**
            16' Sap not milky; heads with ray and disk florets and 4 to many florets; pappus not plumose.
                17. Shrubs; young stems white woolly; leaves opposite and often alternate above, scabrous, the petioles winged; heads with multiple disk and ray florets; pappus of 2 awns. _____ **Verbesina**
                17' Non-seasonal annuals to herbaceous perennials; stems not woolly; leaves alternate, not scabrous, the petioles not winged; heads with 4 to 7 florets including only 1 ray floret (or none); pappus of many slender bristles. _____ **Xanthisma**
    14' Annuals, perennial herbs, or shrubs, leaves entire or minutely toothed (serrulate); disk florets subtended by chaffy bracts, these enclosing the achenes and falling with them.
        18. Much-branched shrubs, the stems slender and straight; heads including rays 2.6–3.5 cm wide; pappus persistent. _____ **Bahiopsis**
        18' Sunflowers, annuals to short-lived perennials with relatively few, often rather thick and declining stems; heads including rays (3.5) 4–9 cm wide; pappus deciduous. **Helianthus**

5' Shrubs or subshrubs; heads of disk florets only, outer florets without an obvious ligule or ray, or if ray florets present then inconspicuous or reduced, or lacking a well-developed ligule (if in doubt about presence of rays then take this choice).
    19. Pappus none; shrubs or subshrubs with viscid-sticky herbage; leaves linear; all florets yellow._____
    _____ **Gymnosperma**

19' Pappus present.
    20. Plants highly aromatic; younger herbage white woolly; leaves gray-green or whitish, 3-toothed to deeply 3-lobed._____ **Pelucha**
    20' Plants aromatic or not; herbage not white woolly; leaves not 3-toothed or lobed.
        21. Plants resinous-glutinous and aromatic.
            22. Leaves not linear-filiform, 1–8 (12) cm long; male and female flowers on separate plants; pappus plumose, white, and uniform. _____ **Baccharis**
            22' Leaves linear-filiform, 1–2 cm long; flowers bisexual; pappus not plumose, brownish or not uniform.
                23. Flower heads 10–15 mm long, solitary at stem tips; involucres (phyllaries) 9–10 mm long; achenes 2.5–3 mm long; pappus of white bristles and scales, the scales 5–7 mm long._____ **Peucephyllum**
                23' Flower heads less than 8 mm long, in dense clusters; involucres less than 5 mm long; achenes 1.5–2 mm long; pappus of many brown bristles 4–5 mm long. **Xylothamia**
        21' Plants not resinous-glutinous.
            24. Stems and leaves scabrous; pappus bristles plumose. _____**Bebbia**
            24' Plants not scabrous; pappus bristles not plumose.
                25. Leaves sagittate, the margins otherwise entire, the petioles shorter than the leaf blades; flowers blue. _____ **Chromolaena**
                25' Leaves not sagittate, the margins scalloped to coarsely toothed, the petioles mostly as long or longer than the blades; flowers not blue.
                    26. Leaves coarsely toothed, the teeth with pointed tips; pappus of uniform, slender, capillary bristles. _____ **Brickellia**
                    26' Leaf margins crenate to toothed, the teeth rounded or blunt tipped; pappus of membranous scales and slender bristles. _____ **Pleurocoronis**

Ambrosia – *Chamizo*; ragweed, bursage

Shrubs or subshrubs, pubescent and glandular. Leaves alternate. Flower heads unisexual with disk florets only, male and female heads usually on the same plant; corollas reduced in male flowers and absent from female flowers.

Flowers wind-pollinated, causing hay fever. Achenes tightly enclosed in burs. Four of the six *Ambrosia* species in the flora have disjunct or limited occurrences in Sonora, being primarily distributed in the Baja California Peninsula: *A. camphorata, A. carduacea, A. divaricata,* and *A. ilicifolia*.

1. Leaves thread-like (linear-filiform) and simple, or especially the lower leaves often pinnatifid with thread-like divisions; burs with blunt, fan-shaped wings. _____ **A. salsola**
1' Leaves or their divisions not thread-like (not linear-filiform); burs with spines, hooked or not.
  2. Leaves 2- or 3-times pinnatifid and deeply dissected.
    3. Herbage exceptionally glandular-sticky; leaves often bi-colored, white woolly (tomentose) below, and green above with small clumps of bead-like woolly hairs (flocculose). _____ **A. camphorata**
    3' Herbages not exceptionally glandular-sticky; leaves generally whitish (tomentose) and of the same color on both surfaces. _____**A. dumosa**
  2' Leaves simple, toothed to lobed or cleft.
    4. Leaves sessile, firm, with spine-tipped teeth. _____ **A. ilicifolia**
    4' Leaves petioled, not firm, teeth or lobes not spine tipped.
      5. Robust shrubs (1) 1.5–3 m tall; leaves mostly 11–20+ cm long. _____ **A. carduacea**
      5' Shrubs or subshrubs less than 1.5 m tall; leaves mostly less than 10 cm long. _____ **A. divaricata**

Ambrosia camphorata (Greene) W.W. Payne var. leptophylla A. Gray

Estafiate

Description: Subshrubs probably to about 0.5 m tall. Leaves exceptionally glandular-sticky, 2- or 3-times dissected, bright green above with a bead-like texture due to clumps of hairs. Burs probably with mostly straight spines.

Local Distribution: Known from Tiburón by a single record in the southern part of the island in the low but rugged mountainous region that separates the Central Valley from the Sauzal watershed. The nearest Sonoran population is in the vicinity of Cerro Pelón about 8 km SE of El Desemboque, and another population occurs about 40 km N of Puerto Libertad on the road to Caborca.

Other Gulf Islands: Santa Catalina.

Geographic Range: This species is widespread through the Baja California Peninsula, with disjunct populations on Islas Cedros and Guadalupe and in west-central Sonora, and then scattered on the mainland as far south as San Luis Potosí (Payne 1964). The Sonoran and Tiburón populations and ones in northeastern to east-central Baja California are var. *leptophylla*.

Tiburón: Sur de Satoocj, 8 Jan 2008, *Romero-Morales 08-2*.

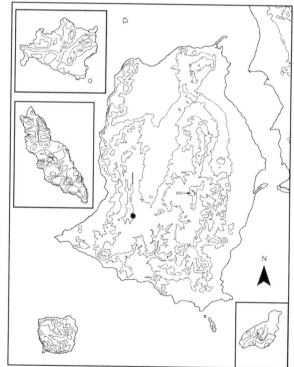

Ambrosia camphorata var. *leptophylla*

Ambrosia carduacea (Greene) W.W. Payne

Tincl caacöl "large *Ambrosia*"

Description: Robust shrubs (1) 1.5–2.5 (3) m tall. Herbage coarsely and roughly pubescent. Leaves ovate to lance-ovate, deeply cleft and coarsely and irregularly toothed, larger leaves often 11–20+ cm long. Inflorescence of panicles with staminate heads above the pistillate ones. Reproductive October–February. Burs 6–9 mm long, elliptic, with hooked spines.

Local Distribution: Sierra Kunkaak, fairly common and sometimes a locally abundant component of the canyon desertscrub/thornscrub community in foothill arroyos and on rocky, generally protected, slopes to higher elevations.

Other Gulf Islands: Cerralvo.

Geographic Range: Both states of Baja California, mostly in Sonoran desertscrub and extending into tropical deciduous forest in the Cape Region. Widely disjunct in east-central Sonora, where it is often arborescent to 4+ m tall and in the Sierra Seri, opposite Tiburón.

Tiburón: E base of Sierra Kunkaak, *Romero-Morales 07-10*. Top of San Miguel Peak in a sheltered canyon, *Knight 1019* (UNM). Cerro San Miguel near Sierra Kunkaak, upper bajada, 1 m tall, *Quijada-Mascareñas 90T007*. Sierra Kunkaak, *Romero-Morales 07-10*. Hast Cacöla, E flank of Sierra Kunkaak, 380 m, *Wilder 08-238*. Head of arroyo, base of Sierra Kunkaak, 1,280 ft, *Wilder 05-26*. Foothills of NE portion of the Sierra Kunkaak, *Wilder 06-376*. Capxölim, *Wilder 07-474*. Steep, protected canyon, 720 m, N interior of Sierra Kunkaak, abundant, *Wilder 07-500*.

Sonora Mainland: N side of Pico Johnson, Sierra Seri, *Wilder 11–223*.

Ambrosia divaricata (Brandegee) W.W. Payne [*Franseria divaricata* Brandegee]

An icoqueetc "what is dropped downward"; *chicurilla, huizapo*

Description: Openly branched small shrubs, the stems slender, brittle, and woody. Leaves mostly about as wide as long with few large, coarse teeth, the larger leaves often 5–7 cm long, the dry, dead leaves semi-persistent. Burs with numerous straight and some hooked spines.

Local Distribution: On Tiburón and Dátil it is often on rocky slopes and in arroyos, where it is quite common and occasionally the dominant shrub. Rare on San Esteban, known only from the central peak, where it occurs as a dwarf shrub 20 cm tall, tucked into the north-facing slope of the ridge.

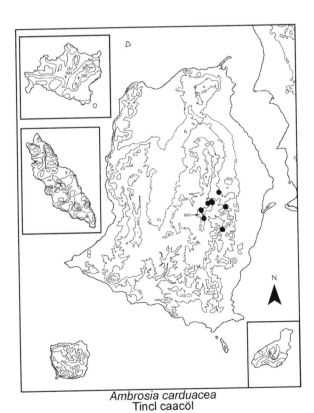

Ambrosia carduacea
Tincl caacöl

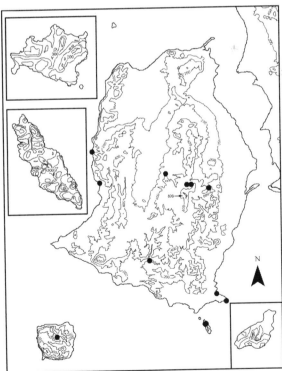

Ambrosia divaricata
An icoqueetc

Other Gulf Islands: None.

Geographic Range: Primarily in the central part of the Baja California Peninsula and scattered localities in coastal Sonora, and a few Gulf Islands. There seems to be northern and southern disjunct populations in Sonora. The northern one is between the vicinity of Puerto Libertad and Cerro Tepopa (south of El Desemboque San Ignacio) and on Isla Tiburón. The southern one was found on granitic slopes at Cerro Tastiota and arroyos and sandy-gravelly soil around Morro Colorado at El Colorado.

Notes: Payne (1964:442) reported that *A. divaricata* is "a species of great similarity to *A. magdalenae*, being distinguished primarily by the differences in leaf morphology." *Ambrosia divaricata* tends to have greener and simple but lobed leaf blades, while *A. magdalenae* has bipinnatifid and more densely pubescent leaf blades. The overall geographic ranges of the two taxa are rather similar, although *A. magdalenae* occurs on Ángel de la Guarda and *A. divaricata* does not. During a relatively wet spring the rapid, new growth of *A. magdalenae* in Sonora very much resembles that of *A. divaricata* in having greener, thinner leaf blades with reduced pubescence. Perhaps they should be regarded a single species, in which case *Franseria magdalenae* has priority since it was originally described on page 170 and *F. divaricata* on page 171 of the same publication (*Proceedings of the California Academy of Sciences*, II, 1889).

Tiburón: Ensenada Blanca (Vaporeta), arroyo bottom, *Felger 7115*. Xapij (Sauzal), *Wilder 08-136*. El Monumento, rocky hills, *Felger 15442*. Ensenada del Perro, rocky hillslopes, *Felger 17729*. Haap Hill, *Felger 76-T4*. Hast Cacöla, E flank of Sierra Kunkaak, 380 m, *Wilder 08-234*. Head of arroyo, base of Sierra Kunkaak, between Sierra Kunkaak Mayor and Sierra Kunkaak Segundo, *Wilder 05-45*. Siimen Hax, *Wilder 06-460*. Pooj Iime, large canyon, NW shore, *Wilder 07-439*.

San Esteban: Subpeak of central mountain on steep slopes of ridgetop, dwarf shrubs, *Wilder 07-59*.

Dátil: NW side of island, very common, arroyos, *Felger 15329*. E side of island, halfway up N-facing slope of peak, in dense vegetation, *Wilder 07-129*. E shore, canyon bottom, *Wilder 07-173*.

Ambrosia dumosa (A. Gray ex Torrey) W.W. Payne

Xcoctz "old"; *chamizo*; white bursage

Description: Dwarf shrubs; summer dormant. Twigs and leaves densely pubescent with white hairs, the twigs sometimes spinescent tipped. Leaves mostly 1–3 cm

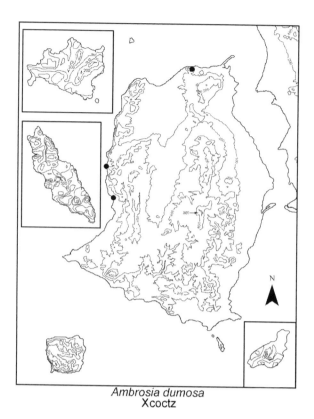

Ambrosia dumosa
Xcoctz

long; petioles prominent, often winged; leaf blades 1–3 times pinnately dissected into small segments variable in shape depending upon moisture conditions, grayish green, or whitish with drier conditions. Male and female heads often intermixed (unique within the genus); each female flower head occurs below a male head. Burs 7–9.5 mm wide and about as long, glandular, sometimes with sparse, slender white hairs, and armed with numerous spines, these straight, sharp, and flattened, the lower ones larger and often channeled.

Local Distribution: Arid western and northern margins of Tiburón, mostly on exposed rock slopes and in canyons.

Other Gulf Islands: Ángel de la Guarda.

Geographic Range: Sonoran and Mojave Deserts in southern California, southern Nevada and extreme southwestern Utah southward to Baja California Sur, and Sonora southward to the vicinity of Bahía Kino.

Tiburón: Ensenada Blanca (Vaporeta), *Felger 15752*. Hee Inoohcö, mountains at NE coast, *Felger 07-80*. Pooj Iime, large canyon, NW shore, *Wilder 07-460*.

Ambrosia ilicifolia (A. Gray) W.W. Payne

Tincl, xcoctz; holly-leaf bursage

Description: Broadly spreading shrubs or subshrubs, often 0.5–1 m tall and usually wider than tall; summer dormant. Stems thick but scarcely woody. Leaves tardily drought killed to partially evergreen, 3.5–10 cm long, sessile, dry and rough to the touch, relatively firm, and glandular hairy; margins with coarse, spine-tipped teeth; dead leaves persistent and turning white. Burs (10) 15–18 mm long, densely glandular with stalked glandular hairs, the spines many, curved and hooked with grooves above; burs resembling a cocklebur (*Xanthium*).

Local Distribution: Common on San Esteban, especially along arroyos. Also very common along the northwestern margin of Tiburón, from Pooj Iime to the northwestern tip of the island on sea cliffs, steep slopes, and canyon washes.

Other Gulf Islands: Ángel de la Guarda, San Lorenzo.

Geographic Range: Arid regions surrounding the northern part of the Gulf of California and the lower Colorado River Valley; northwestern Sonora, southwestern Arizona, southeastern California, northeastern Baja California, and on Gulf Islands.

Tiburón: NW corner of island, 29 Sep 2007, *Felger & Wilder,* observation. Pooj Iime, large canyon, NW shore, *Wilder 07-437.*

San Esteban: Stems erect or spreading to 8 dm, only upper leaves green, 6 May 1952, *Moran 4076* (SD). Arroyo Limantour, *Van Devender 92-486.* San Pedro, steep S-facing rocky slope, *Felger 16662.* SW corner of island, canyon, 11 Apr 1968, *Felger 17628* (specimen not located).

Ambrosia salsola (Torrey & A. Gray ex A. Gray) Strother & B.G. Baldwin var. **pentalepis** (Rydberg) Strother & B.G. Baldwin [*Hymenoclea salsola* Torrey & A. Gray ex A. Gray var. *pentalepis* (Rydberg) L.D. Benson]

Caasol cacat "bitter caasol," caasol coozlil "slippery caasol," caasol ziix iic cöihiipe "medicinal caasol"; *jécota*; burro bush, white burro bush

Description: Globose, aromatic, resinous shrubs, smelling like a dead animal when wet. Foliage usually sparse; leaves quickly drought deciduous, 1–7 cm long, thread-like (linear-filiform) and simple or, especially the lower leaves, often pinnatifid with thread-like segments. Burs 3–7 mm wide with one to several whorls of papery, blunt, fan-shaped wings at the middle spreading like tiny airplane propellers. Flowers and fruits in March and April.

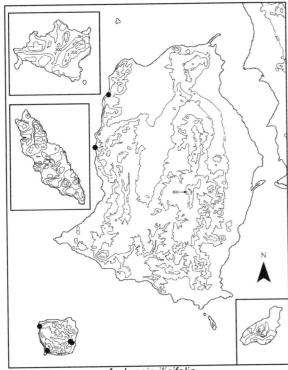

Ambrosia ilicifolia
Tincl, xcoctz

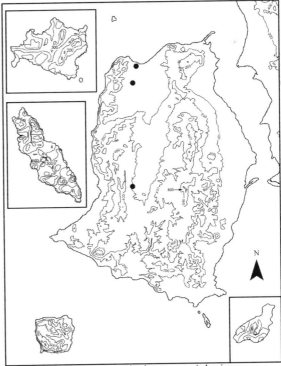

Ambrosia salsola var. *pentalepis*
Caasol cacat

Local Distribution: Tiburón, along the Agua Dulce/Central Valley, from the interior center of the island to the shore at Tecomate; washes, floodplains, and bajada-plains.

Other Gulf Islands: Ángel de la Guarda.

Geographic Range: Two varieties, both states of Baja California, Sinaloa to Utah and Nevada. Var. *pentalepis* is the southern and only variety in Sinaloa and across most of Sonora, and with minor exceptions the only variety known from the Baja California Peninsula.

Tiburón: SW Central Valley, 400 ft, *Felger 17346*. Bahía Agua Dulce, floodplain, *Felger 10208*. Valle de Agua Dulce, broad arroyo, *Wilder 07-252*.

Baccharis

Woody shrubs. Male and female flowers on separate plants, of disk florets only, the flowers white. Pappus bristles white and feathery.

1. Leafy shrubs, the leaves more than 4 cm long. _____
 _____ **B. salicifolia**
1' Broom-like shrubs with sparse foliage, often essentially leafless, leaves 1–3 cm long. _____ **B. sarothroides**

Baccharis salicifolia (Ruiz & Pavón) Persoon [*B. glutinosa* Persoon]

Caajö; *batamote*; seep willow

Description: Shrubs often 2–3 m tall, leafy and mostly evergreen, the leaves and young stems glutinous, the leaves lanceolate, 5–8 (12) cm long, somewhat willow-like, usually toothed.

Local Distribution: Localized on Tiburón at the Sauzal waterhole, where it grows at the edge of the water among *Phragmites australis*, *Salix exigua*, *Tamarix chinensis*, and other wetland plants. The nearest population is along drainageways and ditches in the vicinity of Bahía Kino.

Other Gulf Islands: None.

Geographic Range: South America to southwestern United States.

Tiburón: Sauzal waterhole, not common, *Felger 10090*. Xapij (Sauzal), *Wilder 08-134*.

Baccharis sarothroides A. Gray

Caasol caacöl "large caasol"; *escoba amargo, romerillo*; desert broom

Description: Broom-like shrubs. Foliage sparse; leaves small and quickly drought deciduous. Achenes

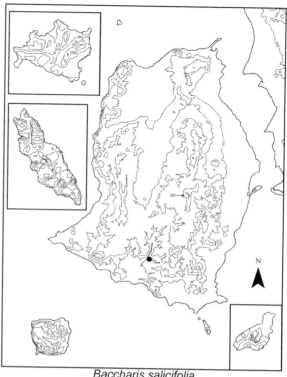

Baccharis salicifolia
Caajö

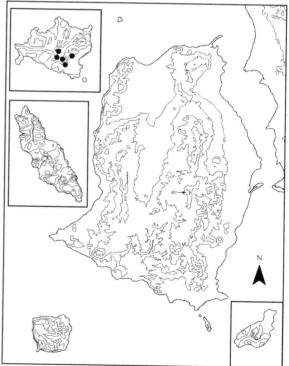

Baccharis sarothroides

probably 2.5 mm long, the pappus bristles white and feathery.

Local Distribution: Mártir, widely scattered but rather rare and mostly at higher elevations. Palmer reported it as "'Yerba del pasmo'; the twigs are used as a remedy for toothache. Rarely over 2 feet high" (Watson 1889:55). Johnston reported that it is "occasional in draws where it forms a bright-green, compact shrub 12–15 dm. high. It is reported (Proc. Am. Acad. 24:55, 1889) as growing only 6 dm high on the island, but all plants seen there were considerably taller" (Johnston 1924:1192–1193). It is strangely not known from Tiburón, although common on the adjacent mainland.

Various resinous and bushy composites are called *hierba de pasmo* in northwestern Mexico and southwestern United States and are highly esteemed for medicinal purposes (e.g., Felger 2007; Felger & Moser 1985). The branches would also serve well as roofing for the tiny rockwall huts on Mártir. Did Yoeme (Yaqui) guano workers bring it to the island for medicinal purposes and/or roofing material? It is the only dioecious species on the island and Johnston suggests an increase in plant size between 1887 and 1921. In 2007 we found it to be rather rare and widely scattered on the island and mostly at higher elevations.

Other Gulf Islands: Ángel de la Guarda (observation), San Lorenzo, San Ildefonso (observation).

Geographic Range: Sinaloa to southwestern United States and both states of Baja California, often in disturbed habitats.

Mártir: 1887, *Palmer 415* (UC; 3 specimens, 2 pistillate specimens with achenes 0.9–1.7 mm long, pappus 15–36 mm long, and 1 staminate). Occasional in draws, compact shrub 4 or 5 ft high, 18 Apr 1921, *Johnston 3159* (CAS). Shrub 1.5 m high, only 2 seen, on top of island, 4 May 1952, *Moran 4054* (CAS). Top of island, occasional, 21 Mar 1962, *Moran 8818* (SD). Near top of island, rare, two large shrubs seen, ca. 4–5 ft high, 25 Jan 1963, *Felger 6356*. Elevation 125 m, 17 Mar 1971, *Hastings 71-59* (pistillate). Small canyon at base of SE-facing slope that leads to island summit, only one plant seen [here], about 1.2 m tall, sheltered at the NE base of a large cardón, 11 Apr 2007, *Felger 07-24*.

Bahiopsis

Drought-deciduous shrubs, ultimately leafless or nearly so in extreme drought. Leaves, inflorescences, and number of flower heads highly variable in size depending on soil moisture. Flower heads daisy-like with yellow ray and disk florets. The two species in the flora area have achenes about 3–4 mm long and a pappus of 2 awns and 2 scales. *Bahiopsis*, formerly treated as a subgenus of *Viguiera*, includes 12 species in southwestern United States and northwestern Mexico, with greatest diversity in the Baja California Peninsula.

1. Leaves mostly silvery gray, often less than 3.5 cm long; Tiburón and Dátil. _____**B. chenopodina**
1' Leaves usually green, usually more than 3.5 cm long (except in drought); Nolasco and San Esteban._____ _____**B. triangularis**

Bahiopsis chenopodina (Greene) E.E. Schilling & Panero [*Viguiera chenopodina* Greene. *V. deltoidea* A. Gray var. *chenopodina* (Greene) S.F. Blake]

Hehe imoz coopol "black-hearted plant"; *ariosa*; goldeneye

Description: Shrubs reaching 1+ m tall but often smaller, with slender, brittle stems. Leaves triangular to ovate, 1–3.5 (5) cm long, canescent and often silvery gray, the leaf margins entire, serrated, or toothed. Flower heads about 2.5 cm wide.

Under dry conditions the plants have the small, grayish leaves typical of *B. chenopodina*, but with ample rains

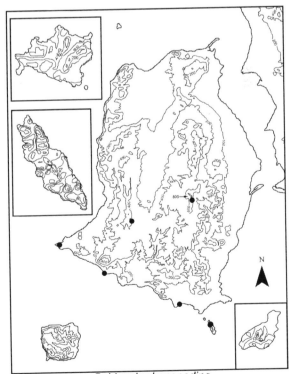

Bahiopsis chenopodina
Hehe imoz coopol

the plants tend to approach or can be identifiable as *B. triangularis*.

Local Distribution: Tiburón, known from a few widely scattered locations, sandy soils and rocky slopes, and to higher elevation in the Sierra Kunkaak. These shrubs are widespread and common on Dátil, where the specimens sometimes approach *B. triangularis* in appearance.

Other Gulf Islands: Tortuga, San Marcos, Carmen, Danzante, Monserrat, Santa Catalina, Santa Cruz, San Diego, San José, San Francisco, Espíritu Santo, Cerralvo.

Geographic Range: *B. chenopodina* is widespread through Baja California Sur and adjacent islands, and the Sonoran coast from the vicinity of Guaymas north at least to Tastiota. To the north it is replaced by the closely related *B. parishii* (Greene) E.E. Schilling & Panero.

Tiburón: 1 mi N of Punta Willard, *Felger 17740*. Tordillitos, *Felger 15497*. Ensenada de la Cruz, rocky hill near beach, *Felger 2574*. Sur de Satoocj, 8 Jan 2008, *Romero-Morales 08-3*. Cerro Kunkaak, 1100 m [11 Apr 1978; specimen without date or number], *Scott* (*UNM 53244*).

Dátil: NE side, *Felger 9090*. NW side, *Felger 15330*. NE shore, *Wilder 07-178*. E side, halfway up N-facing slope of peak, in dense vegetation, *Wilder 07-118*.

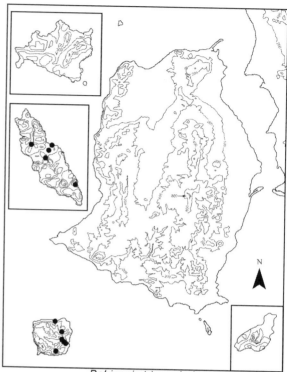

Bahiopsis triangularis

Bahiopsis triangularis (M.E. Jones) E.E. Schilling & Panero [*Viguiera triangularis* M.E. Jones]

Description: Open, more or less straggly shrubs often 1–2 m tall with slender, strigullose to hispid stems and peduncles. Leaves often 3.5–9 (13) cm long. Flower heads 2.5–3 cm wide.

Local Distribution: Widespread and common on San Esteban in a variety of habitats; from low elevation arroyos to high ridge crests. In some sites on the island, mainly the benches of arroyos and on north-facing slopes, it appears to fill the same niche as *Ruellia californica* on Tiburón as a dominant shrub. Abundant on Nolasco at all elevations on both sides of the island. The Nolasco plants are often exceptionally large and robust. "At the southern end of the range, the distinction between *V. triangularis* and the other polyploids, *V. chenopodina* and *V. deltoides*, is problematical. . . . Specimens from Isla San Pedro Nolasco are atypical in the small head size but are provisionally placed in *V. triangularis*; as with other samples from islands in the Sea of Cortéz [sic], assignment to a hexaploid species is somewhat arbitrary" (Schilling 1990).

Other Gulf Islands: Ángel de la Guarda.

Geographic Range: Central Baja California Norte to the Cape Region of Baja California Sur, and three Gulf Islands.

San Esteban: In a narrow canyon, open shrub 4–6 ft high, *Johnston 4379* (CAS). Arroyo Limantour, *Van Devender 90-535*. N side, *Felger 15440*. N slope of high peak near SE corner of island, steep arroyo, 350 m, *Moran 8854*. Summit on NE peak, 475 m, *Moran 13046* (SD).

Nolasco: SE side, 31 Oct 1964, *Sherbrooke* [*Felger 11234*]. Very common on all parts of the island, loose erect shrub, 4–5 ft high, 17 Apr 1921, *Johnston 3127* (CAS, UC, determined by E. E. Schilling, 1985, as *V. triangularis*, "not typical"). E side, near summit between the two landfalls, abundant, *Felger 9646*. N-facing slope above Cañón el Farito, *Wilder 08-165*. Cañón de Mellink, *Wilder 08-348*.

Bajacalia crassifolia (S. Watson) Loockerman, B.L. Turner & R.K. Jansen [*Porophyllum crassifolium* S. Watson. *P. tridentatum* var. *crassifolium* (S. Watson) I.M. Johnston]

Description: Low growing, aromatic subshrubs with a citrus-like scent, compact and mound shaped; herbage and phyllaries with oil glands. Stems short and much branched, brittle, glabrous and slightly glaucous, and often purplish. Leaves succulent, linear to narrowly elliptic, 0.5–2.5 cm long, sessile, tipped with a bright-red gland and a few rounded glands below. Flower heads

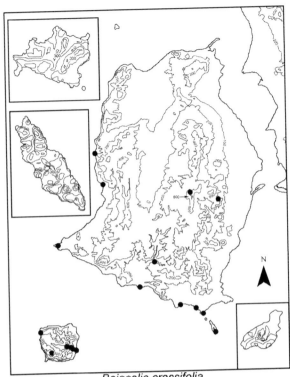

Bajacalia crassifolia

discoid, 10–13 mm long and about as broad, solitary at ends of short branches on stout peduncles 1 cm or less; flowers yellow.

Local Distribution: Common especially near the shore on rocky slopes, upper beaches and arroyos along the southern, western, and northern margins of Tiburón, and locally at higher elevations in the Sierra Kunkaak. Rocky shores and slopes to the peaks, arroyos and canyons on Dátil and San Esteban.

Other Gulf Islands: Ángel de la Guarda, San Lorenzo, San Marcos, Coronados, Carmen, Danzante, Monserrat.

Geographic Range: Gulf Coast of Baja California Sur and adjacent islands and Sonoran Midriff Islands. Not on the Sonora mainland.

Tiburón: Ensenada Blanca (Vaporeta), *Felger 15735*. 1 km N of Punta Willard, *Felger 17736*. Sauzal, near beach, among sea-worn rocks and boulders, *Felger 7019*. Sauzal, rocky soil near waterhole and infrequent along arroyo bed, *Felger 10088*. Ensenada de la Cruz, on rock, *Felger 2571*. Coralitos, hill above beach, *Wilder 06-68*. La Viga, *Wilder 08-149*. Carrizo Canyon [Sopc Hax], 1000 ft, *Knight 1100 & 1105* (UNM). Steep, protected canyon, 720 m, N interior of the Sierra Kunkaak, *Wilder 07-512*. Pooj Iime, large canyon, NW shore, *Wilder 07-431*.

San Esteban: Arroyo Limantour, *Van Devender 92-488*. Main arroyo at El Monumento, floodplain near beach, also

inland but less frequent, also seen on hilltops, *Felger 7061*. San Pedro, steep rocky slopes, *Felger 16669*. Steep slopes of S-central peak, *Felger 17543*. Near SE corner of island, rocky canyon sides 250 m and up, S fork canyon, *Moran 8836*. **Dátil:** SE side, steep, narrow canyon, *Felger 17515*.

Bebbia juncea (Bentham) Greene var. **aspera** Greene

Sapatx; *hierba ceniza*; sweetbush

Description: Rounded bushes to about 1 m tall with rough, scabrous herbage. Stems slender and brittle, the branches and leaves opposite, sometimes alternate above. Foliage sparse; leaves quickly drought deciduous. Heads with disk florets only, yellow and fragrant. Pappus bristles plumose. Flowering at almost any time of the year except extreme drought.

Local Distribution: On Tiburón, San Esteban, Dátil, and Nolasco; widespread in arid habitats including valley plains and bajadas, coastal dunes, and rocky slopes, especially south and west facing.

Other Gulf Islands: Var. *aspera*: Ángel de la Guarda, San Lorenzo (observation), San Marcos, Santa Catalina. Var. *juncea*: San Marcos, Coronados, Danzante, Monserrat, Santa Catalina, Santa Cruz, San Francisco.

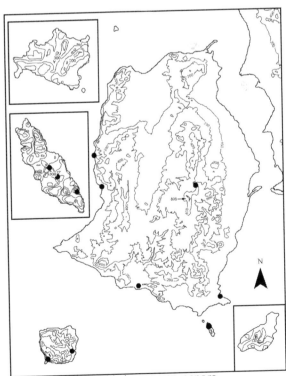

Bebbia juncea var. *aspera*
Sapatx

Geographic Range: Var. *aspera* from Baja California Norte and northwestern Sinaloa to southwestern United States. Var. *aspera*, the more widespread of the two varieties, is replaced by the often leafier, less xeromorphic var. *juncea* in southern Baja California Norte and Baja California Sur.

Tiburón: Ensenada Blanca (Vaporeta), *Felger 7107*. Sauzal, *Wilder 08-108*. Ensenada del Perro, 12 Apr 2007, *Felger*, observation. Canyon, base of N side of Sierra Kunkaak, 24 Nov 2006, *Felger*, observation. Pooj Iime, 29 Sep 2007, *Felger*, observation.

San Esteban: Arroyo Limantour, *Felger 75-119*. Canyon at SW corner, *Felger 16627*.

Dátil: NE side, *Felger 13454*. E side, *Wilder 07-139*. NW side, *Felger 15307*.

Nolasco: Above SE cove, *Cooper [Felger 10399]*. E side, mid-elevation, *Felger 9638*. Ridgetop above canyon just N of Cañón el Farito, *Wilder 09-130*.

Brickellia coulteri A. Gray var. **coulteri**

Comima; Coulter brickell-bush

Description: Shrubs or subshrubs, the stems slender and brittle. Leaves drought deciduous; long-shoot leaves 6–9 cm long, the short-shoot leaves 1.5–1.8 cm long; leaf blades relatively thin, triangular, and coarsely toothed. Flower heads of disk florets, not showy, the flowers yellow-green and purple. Phyllaries strongly graduated, the inner ones 8–11 mm long, long acuminate, the outer ones broader. Achenes (3) 3.5–4 mm long. Flowering at various seasons.

Local Distribution: Widespread on Tiburón; arroyos, hillsides, and mountain canyons and slopes; especially common in the Sierra Kunkaak.

Geographic Range: Var. *coulteri* mostly in the Sonoran Desert including most of Sonora. Two other varieties in southern Sonora and elsewhere in Mexico and in Texas.

Other Gulf Islands: Carmen (*Wiggins 17517*, CAS).

Tiburón: El Sauz, wash, *Moran 8764* (SD). SW Central Valley, arroyo, *Felger 17328*. Sopc Hax, *Felger 9307*. Hast Cacöla, E flank of Sierra Kunkaak, 380 m, *Wilder 08-242*. Arroyo at base of Capxölim, *Wilder 06-372*. Head of arroyo at base of Sierra Kunkaak, *Wilder 05-36*.

Chromolaena sagittata (A. Gray) R.M. King & H. Robinson [*Eupatorium sagittatum* A. Gray]

Comima

Description: Woody-based bushy perennials, often forming dense mounds 1.5–2 m tall. Stems slender and

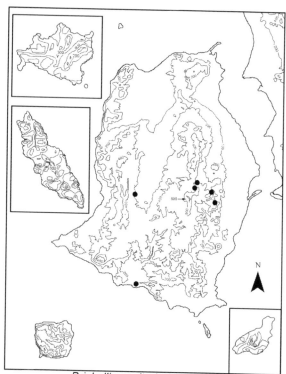

Brickellia coulteri var. *coulteri*
Comima

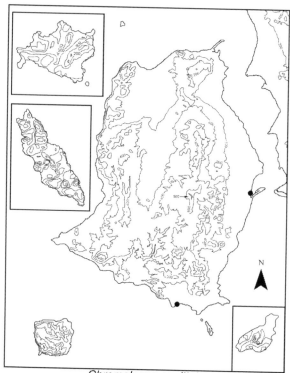

Chromolaena sagittata
Comima

brittle. Herbage glandular-pubescent. Leaves opposite, soon wilting and shriveling as drought conditions set in; leaf blades ovate to lanceolate, sagittate and thin. Flowers blue, of disk florets only; this is the only blue-flowered composite in the flora area.

Local Distribution: Coastal margins of Tiburón but not known on the west side; along washes and low-lying semi-saline soils, especially inland margins of mangroves.

Other Gulf Islands: None.

Geographic Range: Especially common along the coastal plains of Sonora but also inland as far as the Río Sonora at Hermosillo. Coastal Sonora from the vicinity of El Desemboque San Ignacio to the coastal lowlands of Sinaloa and Nayarit and both coasts of Baja California Sur.

> **Tiburón:** Ensenada de la Cruz, *Felger 9134.* Zozni Cmiipla, 100–300 m inland, at base of larger shrubs that create pockets of high diversity desert scrub species within the *Frankenia palmeri* zone, *Wilder 06-369.*

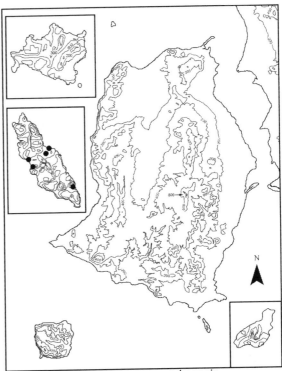

Coreocarpus sanpedroensis

Coreocarpus

Non-seasonal annuals to bushy herbaceous perennials, essentially glabrous. Leaves opposite, pinnately dissected. Flower heads widely spaced on slender-stemmed cymose panicles; rays 4 or 5 per head, pistillate, the ligules (rays) white with 4 purplish lines or veins on the ventral side; disk florets yellow. Achenes usually awnless, with a pale line near each margin and straw-colored corky wings dissected into several separate minute teeth or the teeth virtually absent. Responding to rainfall and soil moisture at any season.

1. Leaf segments linear; Nolasco. ____**C. sanpedroensis**
1' Leaf segments ovate to deltate; Tiburón and other islands but not Nolasco. _____ **C. sonoranus**

Coreocarpus sanpedroensis E.B. Smith, American Journal of Botany 72:626, 1985 [*C. arizonicus* (A. Gray) S.F. Blake var. *sanpedroensis* (E.B. Smith) B.L. Turner, Phytologia 80:136, 1996]

Description: Non-seasonal annuals to bushy herbaceous perennials to ca. 50 cm tall. Herbage sometimes semi-succulent. Lower and larger leaves to ca. 15 cm long; petioles shorter than the blades; blades pinnately dissected into 3–5 linear segments. Flower heads 8–10 mm wide including the rays, the ligules (rays) 3–5 mm

long. Achenes 2.3–3 mm long, obovate to spatulate, compressed, usually awnless, with corky wings dissected into several separate minute teeth or the teeth virtually absent. Responding to rainfall and soil moisture at any season.

"[In flower-]head characters it resembles *C. sonoranus* Sherff . . . but has a more reduced achene wing and more dissected leaves with much narrower leaf segments" (Smith 1985:627–628). The smaller achene wing is in line with reduced dispersal characteristic of island plants (see Carlquist 1965).

Local Distribution: Endemic to Nolasco. Seasonally abundant on both sides of the island at all elevations and exposures, especially common on the east side of the island in deeper soil pockets and sometimes growing from rock crevices.

Other Gulf Islands: None.

> **Nolasco:** 6 Feb 1940, *Dawson 1034* (holotype, UC 945977). Frequent on rocky slopes near sea, *Johnston 3144* (CAS). [Cañón el Farito], scattered trees and giant cactus, shade slope, *Gentry 11353.* Above SE cove, ca. 100 ft on cliff, common, *Cooper* [*Felger 10401*]. N-facing slope above Cañón el Farito, common, *Wilder 08-154.* Cañón de la Guacamaya, 14/15 Apr 2003, *Gallo-Reynoso* (photo). Cañón de Mellink, *Wilder 08-349.*

Coreocarpus sonoranus Sherff var. **sonoranus**
[*C. parthenioides* Bentham var. *parthenioides*]

Description: Facultative annuals to short-lived glabrous perennials. Leaf segments ovate to deltate; plants along the beaches and rocky shores often have succulent leaves. Achene margins winged, the wings incurved, corky, dissected into separate teeth like a series of little pillars.

Local Distribution: Tiburón, on rocky shores, including cobble and sandy beaches above the usual high tide zone, arroyos, and rocky slopes, and from the Central Valley to the Sierra Kunkaak. Widely scattered on Dátil. Reported for San Esteban (e.g., Felger & Lowe 1976; Rebman et al. 2002), but we have not located specimens.

Other Gulf Islands: None.

Geographic Range: Tiburón, the Sierra Seri and vicinity of Hermosillo southward to the Guaymas Region. Another variety occurs northeast of Guaymas.

Tiburón: Ensenada Blanca (Vaporeta), *Felger 15728.* Just N of Punta Willard, *Wiggins 17141* (CAS). Arroyo Sauzal, *Wilder 06-77.* Xapij, *Wilder 06-119.* El Monumento, *Felger 08-40.* SW Central Valley, 400 ft, arroyo, under trees and shrubs, *Felger 17301.* E foothills of Sierra Kunkaak, canyon bottom, *Felger 6971.* Top of Sierra Kunkaak Segundo, E peak of Sierra Kunkaak, 490 m, *Wilder 06-489.* San Miguel Peak,

Knight 1027 (UNM). "Bowl" on top of Sierra Kuncock [*sic*], *Knight 1021* (UNM). Hee Inoohcö, mountains at NE coast, *Felger 07-65.* Tecomate, floodplain, *Felger 11136.*

Dátil: SE side, abundant, *Felger 17493.* E side, *Wilder 07-100.* NE side, *Felger 9104.* NW side, *Felger 15308* (SD). Few seen, mostly dry, 25 Apr 1966, *Moran 13021* (SD).

Encelia farinosa A. Gray var. **farinosa** [*E. farinosa* var. *phenicodonta* (S.F. Blake) I.M. Johnston]

Cotx "what smells acrid"; *incienso, hierba del bazo, hierba ceniza, rama blanca*; brittlebush

Description: Bushy perennials, often 0.5–1 m tall, probably short-lived. Foliage seasonally dense, the leaves drying with drought and often semi-persistent. Leaves petioled, the blades mostly broadly ovate, whitish to gray-green, and highly variable depending on soil moisture. Flowering branches of slender, usually few-branched panicles generally raised well above the foliage. Rays yellow, the disk florets on different plants yellow or brownish purple. Sometimes producing mass displays of showy yellow, daisy-like flower heads in spring; also flowering with summer-fall rains. Achenes 3.5–5 mm long, flat, gray to blackish, obovate, rimmed with long white hairs. Achenes surrounded by glandular

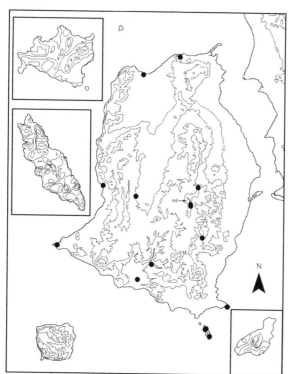

Coreocarpus sonoranus var. *sonoranus*

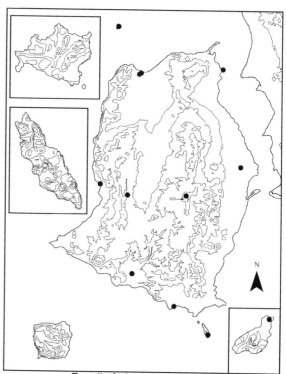

Encelia farinosa var. *farinosa*
Cotx

chaffy bracts slightly longer than the achene and falling with it; pappus none. Flower heads turn down to shed achenes as they ripen.

Local Distribution: Widespread and often abundant across Tiburón, especially at lower elevations such as the Arroyo Sauzal region and on the eastern bajada of the island, and locally to higher elevations in the Sierra Kunkaak. Also steep slopes on Dátil and on Alcatraz, where it is apparently rare. Ivan Johnston collected it on Patos in 1921, but it has not been recorded there since then.

Other Gulf Islands: Var. *farinosa*: Ángel de la Guarda, San Lorenzo, Tortuga, San Marcos, Monserrat, Santa Catalina, San José. Var. *radians*: Carmen, San José.

Geographic Range: Desert and semi-arid regions in the western half of Sonora, northwestern Sinaloa, both states of Baja California, and southwestern United States. Var. *radians* occurs in Baja California Sur.

> **Tiburón:** Ensenada Blanca (Vaporeta), *Felger 7135*. Just N of Punta Willard, *Wiggins 17136* (MEXU). Arroyo Sauzal, *Felger 12276*. Ensenada de la Cruz, *Felger 2584*. SW Central Valley, mountain bordering valley, 1200–1400 ft, *Felger 12422*. Cójocam Iti Yaii, saddle, occasional to common on SE-facing slope of this peak, *Wilder 07-537*. 3 km W [of] Punta Tormenta, *Scott P18* (UNM). Zozni Cacösxj, NE coast, *Wilder 08-207*. Tecomate, *Felger 6844*. Freshwater Bay, *Johnston 3254* (UC).
>
> **Dátil:** SE part, slopes and drainageways, *Felger 2561*.
>
> **Alcatraz:** NE end of island, near lighthouse, low flat, sand soil with cobble rock, rare, *Felger 07-184*.
>
> **Patos:** A single plant on the island about 5 ft high, 23 Apr 1921, *Johnston 3239* (UC, disk is dark in color, determined by Annetta Carter, 1966, as *E. farinosa* var. *phenicodonta*).

Gymnosperma glutinosum (Sprengel) Lessing

Gumhead

Description: Small shrubs or subshrubs, glabrate (minutely scabrous) and glistening with viscid, glandular exudate especially prominent in dry seasons. (This exudate is apparently water soluble because the surfaces are green and virtually devoid of visible exudate after rainy periods.) Leaves alternate, sessile, tardily drought deciduous, dark olive green, densely gland dotted, linear-lanceolate to linear-oblanceolate, 2–8 cm long, the midrib prominently keeled on the lower surface. Flower heads mostly 1.5 mm wide, in dense terminal clusters; phyllaries with a resin pocket near tip. Flowers bright yellow, with ray and disk florets, the rays small and

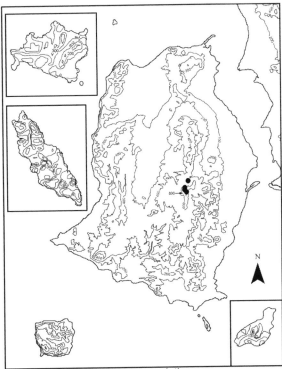

Gymnosperma glutinosum

inconspicuous. Achenes 1.3–2 mm long, the pappus an almost microscopic ring. Flowering much of the year except in extreme drought.

Local Distribution: Widely scattered in the Sierra Kunkaak, from 360 m to the summit, canyon bottoms and mostly on rocky slopes. The nearest populations are on the Sonoran mainland, where it occurs in scattered localities from the Guaymas region northward (Felger 1999, 2000a).

Other Gulf Islands: None.

Geographic Range: Southern Arizona to southern Texas and Guatemala.

> **Tiburón:** 0.5 km E of Siimen Hax, 360 m, broad canyon bottom, *Wilder 06-467*. SE-facing slope of deep canyon, N interior of Sierra Kunkaak, *Wilder 07-490*. E-facing slope above canyon and just below highest ridge on island, *Wilder 07-533*. Summit of island, not common, *Wilder 07-558*.

Helianthus niveus (Bentham) Brandegee var. **niveus**

Mirasol; sand sunflower

Description: Decumbent annuals to short-lived perennial herbs with erect flowering stalks. Leaves opposite below, sometimes alternate above. Showy sunflower

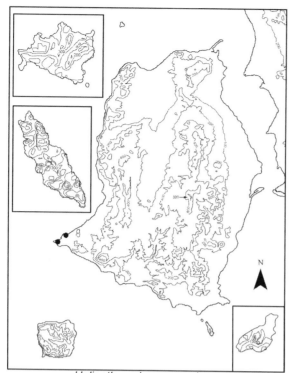

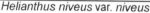

Helianthus niveus var. *niveus*

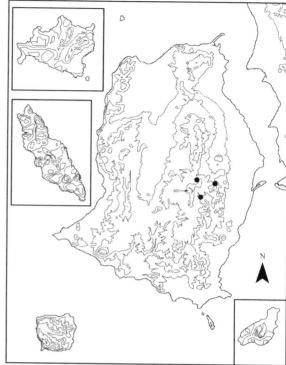

Heliopsis anomala

heads, the rays bright yellow, the disk dark brownish purple. Achenes 4–5 mm long, thick and mottled black and pale tan, with long, silky, forward-pointing hairs; pappus deciduous, of several shorter scales and 2 larger, awn-tipped scales. Flowering more or less throughout the year except during drought.

Local Distribution: Documented on dunes at the southwestern coast of Tiburón.

Other Gulf Islands: None.

Geographic Range: Baja California Peninsula and the Sonoran coast south to the Guaymas region. Common on coastal strands and dunes on Gulf of California shores, including Bahía Kino.

Tiburón: 3 mi N of Punta Willard, very common on sand dunes, *Johnston 4247* (CAS). Punta Willard, wind blown sand piled against rock hills, *Felger 17744.*

Heliopsis anomala (M.E. Jones) B.L. Turner [*H. parvifolia* A. Gray var. *rubra* (T.R. Fisher) Wiggins. *H. rubra* T.R. Fisher]

Description: Mostly perennials, probably short-lived, also flowering in the first year or season. Heads showy, solitary on long peduncles, the rays often 2–3 cm long, the ray and disk florets yellow. Rays, bracts, and disk with

age changing to green or green and yellow, and persisting for several weeks. Achenes thick, without pappus. Flowering October–April.

Local Distribution: Tiburón in deep canyons of Sierra Kunkaak, where it is sometimes locally common. Usually in shaded places, often north-facing slopes among rocks, and occasionally along open washes of large canyon drainages. The nearest population occurs in similar habitats in the Sierra Seri.

Other Gulf Islands: None.

Geographic Range: Western Sonora from the Sierra Bacha (south of Puerto Libertad) to the Guaymas region. Also both Baja California states.

Tiburón: Hast Cacöla, E flank of Sierra Kunkaak, 380 m, *Wilder 08-251.* Head of arroyo, E base of Sierra Kunkaak, *Wilder 06-32.* Canyon bottom, N base of Sierra Kunkaak, *Wilder 06-411.*

Hofmeisteria

Mound-shaped perennials with succulent stems and leaves. Nearly evergreen, the number and size of the leaves greatly reduced during drought. Flower heads solitary on long stalks; flowers fragrant and numerous, the corollas very slender, pale lavender and white.

1. Leaves or leaf segments terete; Nolasco. _____
   _____ **H. crassifolia**
1' Leaf segments flattened, wider than thick; Midriff, or
   northern islands. _____**H. fasciculata**

Hofmeisteria crassifolia S. Watson

Description: Leaves crowded at stem tips; leaf segments
thick, terete, glaucous, and very succulent. Flowering recorded February–April and in October; flowers fragrant,
the ray florets lavender-pink, the disk florets white.

Local Distribution: Growing from crevices on exposed rock surfaces on both sides of Nolasco and especially conspicuous on sea cliffs near the shore; scarce at
higher elevations.

Other Gulf Islands: None.

Geographic Range: Endemic to the Guaymas region
and Nolasco.

Nolasco: A brittle succulent perennial growing in rock crevices near ocean, 17 Apr 1921, *Johnston 3142* (CAS, UC). SE
side, sea cliffs and crevices in rocks, somewhat less common
on S-facing slopes than others, *Felger 9648*. Cañón el Faro,
exposed rock surfaces, mostly N- and E-facing, near shore,
Felger 06-80. Cañón de la Guacamaya, near shore, Jul 2003,
Gallo-Reynoso (photo).

Hofmeisteria fasciculata (Bentham) Walpers var. fasciculata

Taca imas "trigger fish's pubic hair"

Description: Several to many branches, forming dense
and often hemispherical plants 20–50 cm tall. Herbage
essentially glabrous. Leaves opposite below, alternate
above, 2.5–8 cm long, 3 or more times lobed or dissected
into small segments.

Local Distribution: Widespread on Tiburón, San Esteban, Dátil, and Cholludo. Common especially near the
shore, mostly on sea cliffs, canyon walls, and barren rock
slopes, often growing from rock crevices. Sporadically
extending to higher elevations on Tiburón.

Other Gulf Islands: Ángel de la Guarda, San Lorenzo,
Tortuga.

Geographic Range: Coastal Sonora from the Sierra
Bacha to mountains north of Bahía Kino, the Midriff
Islands, and the Gulf Coast of Baja California Norte and
Sur. Two other varieties occur in Baja California Sur.

Tiburón: Ensenada Blanca (Vaporeta), *Felger 15738*. N of
Punta Willard, *Perrill 5111*. Arroyo Sauzal, 4 km inland, *Fel-
ger 10100*. Sauzal waterhole, *Wilder 06-126*. El Monumento,
Felger 15562. Hast Cacöla, E flank of Sierra Kunkaak, 450 m,

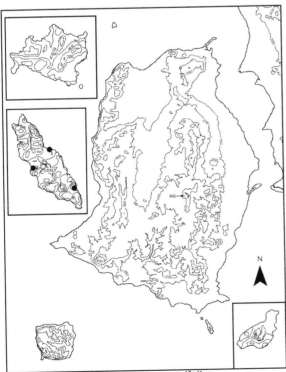

Hofmeisteria crassifolia

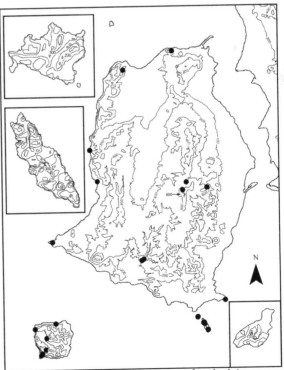

Hofmeisteria fasciculata var. *fasciculata*
Taca imas

Wilder 08-264. 0.5 km E of Siimen Hax, *Wilder 06-468*. Steep, protected canyon, 720 m, N interior of Sierra Kunkaak, common on cliff face, *Wilder 07-508*. Hee Inoohcö, mountains at NE coast, *Felger 07-91*. W of Tecomate, above 650 ft to top of mountain, *Felger 6270*. Pooj Iime, *Wilder 07-457*.

San Esteban: Canyon, SW corner of island, *Felger 17632*. Arroyo Limantour, 3.25 mi inland, *Wilder 07-56*. Steep slopes of S-central peak, *Felger 17547A*. San Pedro, *Felger 16667*. NE corner of island, *Felger 9198*. Canyon, N side, *Felger 17570*.

Dátil: NE side, sea cliffs and steep rocky slopes, *Felger 9132*. NW side, *Felger 15351*. Ridge crest, *Felger 13460*. SE side, canyon, *Felger 17506*.

Cholludo: Steep rocky slopes, *Felger 17700*. Common on sea cliffs and exposed rock outcrops, *Wilder 08-313*.

Palafoxia arida B.L. Turner & M.I. Morris var. **arida**

Moosni iiha "sea turtle's possessions," mosnoohit "what sea turtles eat"; Spanish needles

Description: Winter-spring annuals and sometimes also growing with summer-fall rains, often 30–90 cm tall; rarely surviving as short-lived perennials; erect, often with a single main axis and several or more branches from above middle; taproot well developed. Leaves with coarse, forward-pointing hairs, the flowering stems and phyllaries with glandular hairs. Leaves 3–9 cm long, linear to linear-lanceolate, sometimes semi-succulent, the apex blunt. Heads with disk florets only; corollas white to pink, the anthers purple.

In central Sonora, at about Hermosillo and Bahía Kino, *P. arida* grades into *P. linearis* (Cavanilles) Lagasca, which extends southward into Sinaloa, and a similar transition occurs on the Baja California Peninsula. Turner and Morris (1975:79) differentiate *P. linearis* as "sprawling shrublets having linear leaves with round or obtuse apices" and restrict it to "coastal sand dunes of southern Baja California." Plants from that region appear indistinguishable from those of coastal Sinaloa and southwestern Sonora. The Tiburón and Alcatraz plants appear intermediate between the two taxa.

Local Distribution: Documented near the coast on Tiburón and also on Alcatraz, especially on sandy soils, including beach dunes.

Other Gulf Islands: None.

Geographic Range: Sandy soils in the Mojave and Sonoran Deserts; southwestern United States, Baja California Norte, northern Baja California Sur, and western Sonora.

Tiburón: Ensenada Blanca (Vaporeta), *Felger 14955*. Just N of Punta Willard, *Wiggins 17127* (CAS). Punta San Miguel, low dunes 5–50 m inland, *Wilder 07-192*. Palo Fierro, beach dunes, *Felger 7010*. Bahía Agua Dulce, dunes at beach, *Felger 6818B*.

Alcatraz: Sand flat, NE end of island, *Felger 07-187*.

Pectis

The two species in the flora area are common summer-fall annuals and pungently aromatic, the leaves and phyllaries dotted with conspicuous oil glands. Leaves opposite, mostly linear-oblong, with conspicuous soft, marginal bristles at the leaf bases. Flower heads bright yellow with ray and disk florets.

1. Flower heads on peduncles mostly 1–3 cm long and usually shorter than the nearby leaves; achenes with unique, curled, bulbous-tipped hairs; pappus of ray achenes a crown of short scales. _____**P. papposa**
1' Flower heads on peduncles 1–8 cm long and usually longer than the nearby leaves; achene hairs not bulbous tipped; pappus of ray achenes with 2 fairly stout awns. _____ **P. rusbyi**

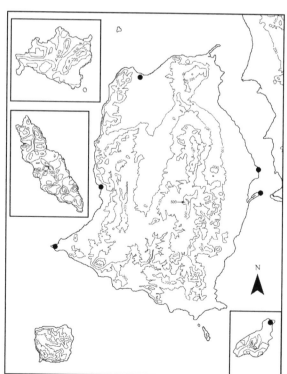

Palafoxia arida var. *arida*
Moosni iiha, mosnoohit

Pectis papposa Harvey & A. Gray var. **papposa**

Caasol heecto "small caasol," caasol ihasii quiipe "pleasant-smelling caasol," cacaatajc "what causes vomiting"; *manzanilla del coyote*; desert chinchweed

Description: Flower heads on peduncles usually shorter than the nearby leaves. Achenes with unique, curled, bulbous-tipped hairs; pappus of ray achenes a crown of short scales; pappus of disk achenes of sub-plumose or minutely barbed bristles.

Local Distribution: Widespread in the lowland desert habitats of Tiburón although not recorded on the west side of the island. Also on Alcatraz.

Other Gulf Islands: None.

Geographic Range: Abundant and widespread across most of the Sonoran Desert. Var. *papposa* primarily in the Great Basin, Mojave, and Sonoran Deserts, and also in thornscrub and tropical deciduous forest, from southeastern California and southwestern Utah to southern Baja California Sur, New Mexico, and central Sinaloa.

Tiburón: Ensenada de la Cruz, *Felger 9148*. Haap Hill, *Felger 76-T30*. Between Sopc Hax and Hant Hax camp, lower foothills of Sierra Kunkaak, *Felger 9330*. Zozni Cmiipla, 100–300 m inland, *Wilder 06-286*. Palo Fierro, *Felger 11071*. 3 km W [of] Punta Tormenta, without date or number, *Smartt*

(UNM 84139). Zoojapa, NE foothills of Sierra Kunkaak, 180 m, *Wilder 08-222*. Bahía Agua Dulce, *Felger 8863*.

Alcatraz: Rocky-sandy substrate, 28°49'22"N, 111°56'27"W, *Jones 16* (Prescott College Field Station Herbarium, Bahía Kino).

Pectis rusbyi Greene ex A. Gray [*P. palmeri* S. Watson]

Caasol heecto caacöl "large chinchweed"; *manzanilla del campo*; chinchweed

Description: Summer-fall annuals. Flower heads on slender peduncles usually longer than the nearby leaves. Phyllaries striate and broader than those of *P. papposa*. Pappus of ray achenes with 2 fairly stout awns; pappus of disk achenes with barbed bristles; achene hairs not bulbous tipped.

It is somewhat of an ecological counterpart of *P. papposa*, often replacing it in less arid habitats and regions of the Sonoran Desert in Sonora and southern Arizona. *P. rusbyi* is generally larger, more upright, with longer peduncles, larger flower heads, and different pappus bristles than *P. papposa*.

Local Distribution: Widespread and common on Tiburón, from low to mid-elevations. Also on Cholludo and predicted for Dátil.

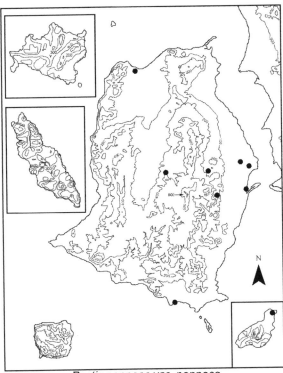

Pectis papposa var. *papposa*
Caasol heecto

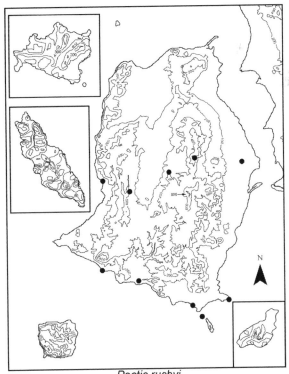

Pectis rusbyi
Caasol heecto caacöl

Other Gulf Islands: San José.

Geographic Range: Southern Arizona to northern Sinaloa, and Baja California Sur. One of the most abundant hot-weather wildflowers in the region, and the most common and widespread *Pectis* in the southern and eastern parts of the Sonoran Desert in Sonora.

> **Tiburón:** Ensenada Blanca (Vaporeta), *Felger 14956*. Tordillitos, *Felger 15515*. Arroyo Sauzal, *Felger 2751*. Coralitos, *Felger 15362*. El Monumento, *Felger 15523*. SW Central Valley, ca. 1200–1400 ft, upper rocky slope of Sierra Menor, *Felger 12412*. Haap Hill, *Felger T74-25*. 3 km W [of] Punta Tormenta, without date or number, *Smartt* (UNM 84138). Top of Sierra Caracol, 1500 ft, *Knight 1048* (UNM).
>
> **Cholludo:** Steep rocky slopes, *Felger 9153*.

Pelucha trifida S. Watson, Proceedings of the American Academy of Arts and Sciences 24:55, 1889

Description: Highly aromatic, scarcely woody shrubs or bushy perennials, 0.3–1 m tall. Herbage thinly gray- to white-woolly. Leaves alternate, often semi-succulent and 3-toothed or 3-lobed. Flower heads in terminal inflorescences; flowers discoid and golden yellow. Achenes 1.9–3.1 mm long, silky haired; pappus of many bristles, one set longer and the other shorter and somewhat fused

basally, the larger ones 5.5 mm long and minutely barbed. Edward Palmer recorded that "it has a very strong aroma as of cloves and cinnamon, and so powerful is this at times that it causes persons to sneeze and cough" (Vasey & Rose 1890:78) and Reid Moran noted that it has a turpentine odor.

Local Distribution: Tiburón on rocky slopes along the northwest shore and ridges at high elevations in the Sierra Kunkaak. On Mártir it is scattered and moderately common at higher elevations, often on rock outcrops, with one record at ca. 50 m elevation. Also reported for San Esteban but we have not found a specimen.

Other Gulf Islands: Ángel de la Guarda.

Geographic Range: Gulf coast areas of southern Baja California Norte and northern Baja California Sur and three Gulf Islands. Johnston (1924) pointed out that it seems to require open exposure to sea breezes.

"The shrubby habit of *Pelucha* is unusual in the otherwise mostly herbaceous Helenieae s.s. and is conceivably the result of evolution on islands in the Gulf of California" (Baldwin & Wessa 2000:522). Their DNA studies demonstrate that *Pelucha* is in a clade with *Psathyrotes* and *Trichoptilium*, small annuals of the drier, winter-rainfall regions in the northwestern parts of the Sonoran Desert (e.g., Felger 2000a).

> **Tiburón:** Rock outcrop above steep, protected canyon, 720 m, N interior of Sierra Kunkaak, base of cliff, rare, 30 cm tall, *Wilder 07-519*. Cójocam Iti Yaii, saddle, *Wilder 07-535*. Exposed upper ridge of island, between peak of Cójocam Iti Yaii and summit, not common, *Wilder 07-576*. Bahía Agua Dulce, occasional on N slope, ca. 200 m, *Moran 12994*. Pooj Iime, *Wilder 07-432*.
>
> **Mártir:** 1887, *Palmer 407* (isotype, US 47511, image). 1890, *Palmer 150* (UC 84228). Common along the rocky crest of the island, open irregularly branched shrub 2–3½ ft high, *Johnston 3151* (UC). Steep N slope, common above landing, 50 m elev., shrub ½ m high, with turpentine odor, heads yellow, $2n = 19$ pairs fide A. M. Powell, *Moran 21745*. Ridge just below top elevation of island, 285 m, uncommon shrub less than 1 m tall, herbage highly aromatic, seen only at upper elevations, *Felger 07-194*.

Perityle

The three species in the flora are cool-season annuals, highly variable in size, with ray and disk florets, the rays yellow or white, the disk yellow. Achenes flattened and black, the pappus of small scales and with or without a single, delicate awn.

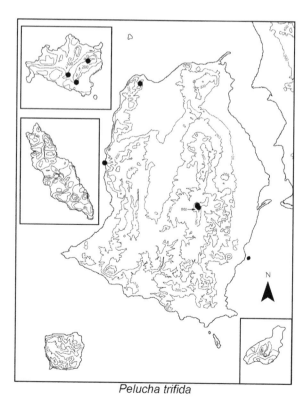

Pelucha trifida

1. Rays white, the disk flowers yellow; achenes awnless or the awn with only forward-pointing barbs. _**P. emoryi**
1' Ray and disk flowers yellow.
  2. Achenes awnless or the awn with forward-pointing barbs along the shaft plus backward-pointing and spreading barbs at the awn tip. _____ **P. aurea**
  2' Achenes awnless or the awn with only forward-pointing barbs. _____ **P. californica**

Perityle aurea Rose

Description: Plants delicate, with erect stems, or sometimes robust when well watered. Leaves alternate or the first leaves sometimes opposite. Ray and disk florets yellow. Achenes margins thin and densely white-ciliate (white hairs); the awn, when present, 1.5–2.5 mm long with forward pointing (antrorse) barbs (barbells) on the awn shaft and a few retrorse or spreading barbs at the awn tip.

Local Distribution: Widespread and common on San Esteban and the west and north sides of Tiburón, where it is often abundant.

Other Gulf Islands: Ángel de la Guarda, San Lorenzo, San Marcos, Danzante, Santa Catalina, Espíritu Santo, Cerralvo. Reported from Carmen and San José but probably not on those islands.

Geographic Range: Gulf Coast of Baja California Sur and Gulf Islands, and a few coastal localities on the Sonora mainland. Powell (1974:263) reports, "The island specimens . . . differ slightly from the mainland forms, most notably in achene morphology and in being awnless, although awned forms also occur on San Esteban Island."

Tiburón: Ensenada Blanca (Vaporeta), *Felger 15727* (ARIZ, UC, duplicates at SRSC determined by A. M. Powell). Tecomate, *Felger 12488*.

San Esteban: Summit and N slope of high peak near SE corner, *Moran 8845* (CAS). Summit of NE peak, *Moran 13044* (CAS). Canyon, N side of island, *Felger 17553A* (duplicate at SRSC determined by A. M. Powell). Steep slopes, S-central peak, *Felger 17546B*.

Sonora: 3 mi NE of Puerto Lobos, low hills of decomposed granite, *Felger 20817b*.

Perityle californica Bentham

Description: Plants highly variable in size, often with delicate stems. Leaves often opposite below and alternate above. Ray and disk flowers bright yellow. Achene margins usually prominently calloused; pappus awn, when present, 1.5–3.5 mm long with only forward-pointing barbs.

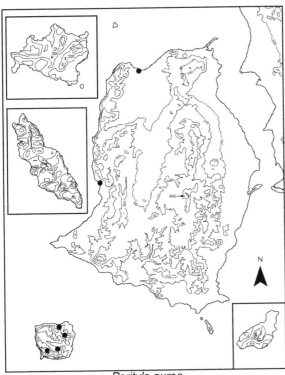

Perityle aurea

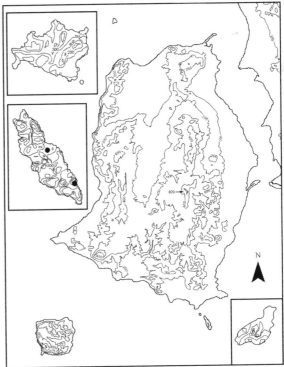

Perityle californica

Local Distribution: Seasonally common on Nolasco and often abundant at all elevations, especially on the east side of the island.

Other Gulf Islands: Coronados, Carmen (*Carter 3714*, UC), Danzante (*Wiggins 17554*, CAS), Santa Catalina, Santa Cruz, San José (*Moran 3785*, UC), San Francisco, Cerralvo.

Geographic Range: Central Sonora to Sinaloa, both Baja California states, and Gulf Islands adjacent to Baja California Sur. It is similar in appearance to *P. aurea*, from which it is generally distinguished by technical features of the achene awn.

Nolasco: E side, abundant, *Felger 9653* (duplicate determined by A. M. Powell). NE side, very common on N-facing slopes, less common on other slopes, rare on SW slopes, *Felger 12071* (determined by A.M. Powell). SE side of island, *Sherbrooke* 31 Oct 1964.

Perityle emoryi Torrey

Hee imcát "what the jackrabbit doesn't bite off," hehe cotopl "plant that clings," zaah coocta "what looks at the sun"; desert rock daisy

Description: Leaves mostly alternate. Rays white, the disk yellow. Achenes 2.1–3.0 mm long, the awn, when present, has only forward-pointing barbs.

Local Distribution: Widespread and common on all islands in the flora area except Nolasco. In many habitats including desert flats, drainageways, and rocky slopes.

Other Gulf Islands: Ángel de la Guarda, Partida Norte, Salsipuedes, San Lorenzo, Tortuga, San Marcos, San Ildefonso, Coronados, Carmen, Monserrat, Santa Cruz, San Francisco, Espíritu Santo, Cerralvo.

Geographic Range: Sonora south to the Guaymas region, Baja California Sur to southern California, southern Nevada, southwestern Utah, and western Arizona. Also Peru and Chile. This is one of the few species of *Perityle* found in South America and the only amphitropical species in the genus.

Tiburón: Ensenada Blanca (Vaporeta), *Felger 17242* (duplicate determined by A. M. Powell). Just N of Punta Willard, *Wiggins 19 Mar 1962* (CAS, determined by A. M. Powell). Tordillitos, *Felger 15483*. Arroyo Sauzal, 5 km inland, *Felger 12255*. El Monumento, *Felger 08-42*. Ensenada del Perro, *Felger 17735B* (duplicate determined by A. M. Powell). Sur de Satoocj, common, 12 Feb 2009, *Romero-Morales 09-3*. Rocky slope of mountain bordering SW Central Valley, 1200 ft, *Felger 12399* (duplicate determined by A. M. Powell). Palo

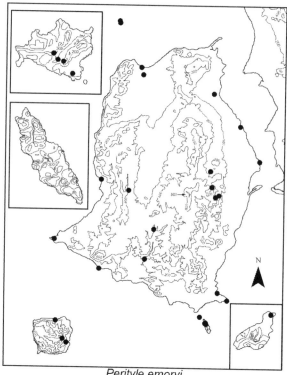

Perityle emoryi
Hee imcát, hehe cotopl

Fierro, Punta Tortuga, *Felger 12550* (duplicate determined by A. M. Powell). Hast Cacöla, E flank of Sierra Kunkaak, 380 m, *Wilder 08-258*. Zoojapa, NE foothills of Sierra Kunkaak, 180 m, *Wilder 08-223*. Sopc Hax, *Wilder 07-218*. Canyon, foothills of Sierra Kunkaak, *Felger 7007*. Valle de Águila, *Wilder 07-271*. Hahjöaacöl, NE coast, *Wilder 08-212*. Bahía Agua Dulce, *Felger 8863*. Freshwater Bay, frequent on a loamy bank near the sea, *Johnston 3268* (CAS).

San Esteban: E-central side of island, 0.5 km inland, rocky N-facing steep hillside and arroyo, *Felger 12758* (duplicate determined by A. M. Powell). N side of island, arroyo, *Felger 15396* (duplicate determined by A.M. Powell). Near S end, large arroyo running inland from E, *Wiggins 17218* (CAS, determined by A. M. Powell).

Dátil: NE side of island, rocky slopes, *Felger 9093* (duplicate determined by A. M. Powell). NW side, *Felger 15354*. E side, slightly N of center, narrow ridge crest, *Wilder 07-116*.

Alcatraz: Localized, rock and sand soil near beach (also rare on bajada at base of E side of mountain), *Felger 12700*.

Patos: Frequent in draw and on sea cliffs, 23 Apr 1921, *Johnston 3240* (CAS, determined by A. M. Powell). Rocky slope, mountain peak, *Felger 07-54*.

Cholludo: Most common annual seen here, 21 Oct 1963, *Felger 9152*. Abundant over island, 12 Apr 1968, *Felger 17705*.

Mártir: S end of island, 1890, *Palmer 149* (not seen, cited by Vasey & Rose 1890:79). Common along crest, growing between loose rocks, 18 Apr 1921, *Johnston 3148* (CAS, UC, determined by A. M. Powell). Small canyon at base of SE-facing slope that leads to island summit, most of the plants nearly smothered by *Vaseyanthus, Felger 07-20.* High elevation plateau, NW of highest point of island, 275 m, *Wilder 07-603.*

Peucephyllum schottii A. Gray

Romero; desert fir, pygmy cedar

Description: Much-branched shrubs 1–2 m tall, resembling a dwarf conifer with a well-developed stout, woody trunk, the bark twisted and shredding. Herbage conspicuously resinous with sessile, dot-like glistening glands on stems and leaves, and stalked glands on peduncles and sometimes on stem tips, phyllaries, and corollas. Leaves mostly evergreen, often 1–2 cm long, alternate, bright green, crowded at ends of twigs, sessile, linear-filiform. Flower heads 10–15 mm long, solitary at stem tips, of disk florets; flowers fragrant, at first pale yellow-green, the corolla lobes and upper tube becoming red-purple. Involucres (phyllaries) 9–10 mm long. Achenes 2.5–3 mm long; pappus of many bristles and scales, with slender white

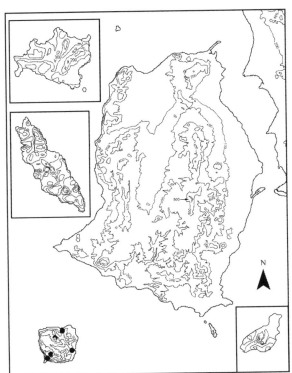

Peucephyllum schottii

bristles to the outside, and longer flat bristles or narrow scales 5–7 mm long to the inside.

Local Distribution: Widespread on San Esteban from arroyos and canyons to steep slopes and high ridges. The single specimen from Tiburón has not been verified by other records (see "Doubtful and Excluded Reports").

Other Gulf Islands: Ángel de la Guarda, San Lorenzo, San Marcos.

Geographic Range: Arid regions of the Sonoran and Mojave Deserts of Utah, southern Nevada, southeastern California, western Arizona, the gulf side of Baja California Norte and northern Baja California Sur, and northwestern Sonora.

San Esteban: Arroyo Limantour, *Felger 75-126.* Narrow, canyon behind cove, N side of island, *Felger 17551.* Canyon, SW corner of island, *Felger 16649.* Steep slopes, S-central peak, *Felger 17546A.* Subpeak of central mountain, *Wilder 07-64.*

Pleurocoronis laphamioides (Rose) R.M. King & H. Robinson [*Hofmeisteria laphamioides* Rose. *H. laphamioides* var. *pauciseta* (I.M. Johnston) S.F. Blake. *H. pluriseta* A. Gray var. *pauciseta* I.M. Johnston, Proceedings of the California Academy of Sciences, IV, 12:1187–1188, 1924]

Hamt inoosj "land's claw," hast yapxöt "rock flower"

Description: Globose small shrubs or bushy perennials often 0.5–1 m across. Leaves minutely glandular-pubescent, gradually and tardily drought deciduous; leaf blades green, semi-succulent, broadly ovate to orbicular with toothed or crenulate margins. Flowers white to pale yellow with purple stigmas; reproductive at various seasons. Achenes 2.3–3.4 mm long, blackish and slender, the pappus with hyaline scales alternating with slender awns, or awns absent.

Local Distribution: Tiburón, San Esteban, Dátil, Alcatraz, Mártir, and Nolasco; mostly on steep rocky slopes, headlands, and sea cliffs, from near the shore to higher elevations.

Other Gulf Islands: Tortuga, San Marcos, San Ildefonso, Danzante.

Geographic Range: Western Sonora, from east of the Pinacate region and adjacent Arizona southward to the Guaymas region, Gulf Islands, and Gulf Coast of both Baja California states.

This species shows clinal gradation into *H. pluriseta* northward near the United States border. As might be expected in an environmental gradient of increasing aridity

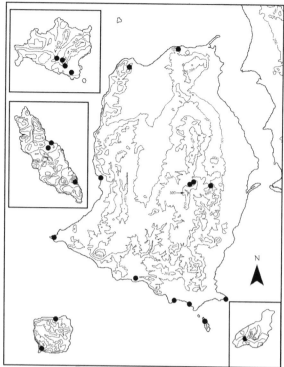

Pleurocoronis laphamioides
Hamt inoosj, hast yapxöt

from south to north, the leaf blades decrease in size and the ratio of petiole to blade increases (Felger 2000a).

Tiburón: Ensenada Blanca (Vaporeta), *Felger 15742*. 1 km N of Punta Willard, *Felger 17743*. Arroyo Sauzal, 400 m S of waterhole, *Felger 10101*. Ensenada de la Cruz, *Felger 2582*. Coralitos, hill above beach, *Wilder 06-54*. El Monumento, cliffs, *Felger 15564*. Hast Cacöla, E flank of Sierra Kunkaak, 380 m, *Wilder 08-245*. Canyon bottom, N base of Sierra Kunkaak, *Wilder 06-399*. Deep canyon, SW and up canyon from Siimen Hax, *Wilder 06-477*. Hee Inoohcö, mountains at NE coast, *Felger 07-64*. W of Tecomate, 650 ft to mountaintop, *Felger 6275*.
San Esteban: N side of island, *Felger 15430*. Canyon, SW corner of island, *Felger 17626*.
Dátil: NE side of island, steep rocky slopes, crevices on sheer rock sea cliffs, mostly N and E exposures, *Felger 13441*. E shore, N of island center, canyon bottom, *Wilder 07-182*.
Alcatraz: Mountain near top of island, NE exposure, localized on steep rocky slopes and cliffs, *Felger 12703*.
Mártir. Oct 1887, *Palmer 406* (GH, image). Abundant along crest of island, 18 Apr 1921, *Johnston 3157* (CAS). On talus at foot of sea cliff about 20 ft above high tide, only 2 shrubs seen here, these about 2½ ft high, 18 Apr 1921, *Johnston 3162* (CAS). Small canyon at base of SE-facing slope that leads to island summit, *Felger 07-23*.
Nolasco: Frequent on ledges, particularly near the sea, 17 Apr 1921, *Johnston 3134* (holotype of *H. pluriseta* var.

pauciseta, CAS 8723). SE side, *Sherbrooke* [*Felger 11237*]. Near middle, E side, *Moran 4043* (SD).

Porophyllum

Pungently aromatic, short-lived perennial herbs or subshrubs, with conspicuous oil glands. Flower heads more or less cylindrical, of disk florets.

1. Leaves filiform, mostly 0.5–1 mm wide; flower heads mostly solitary at stem tips; flowers white to purplish; not on Nolasco. _____ **P. gracile**
1' Leaves linear to narrowly oblanceolate, 2 mm or more wide; flower heads in dense clusters (cymes) at stem tips; flowers yellow; Nolasco. _____ **P. pausodynum**

Porophyllum gracile Bentham

Xtisil (variants: xtesel, xtisel); *hierba del venado*

Description: Short-lived perennials, often to 50 cm tall, glabrous and glaucous, sometimes purple-tinged, with translucent oil glands and a strong, pungent odor somewhat like that of marigolds (*Tagetes*), to which *Porophyllum* is related. We have found that people believe this plant smells like cilantro and find it pleasing (Richard), or find the smell repulsive (Ben). Stems straight, slender, and very brittle. Leaves opposite below, alternate above, 1.5–6 cm long, linear to thread-like, and quickly drought deciduous. Flower heads 12–20 mm long, the flowers white to pale purple with dark purple longitudinal glandular lines. Receptacles often persistent after achenes mature and fall, leaving a "stump" of brain-like convoluted, polished thickenings. Flowering at various seasons.

Local Distribution: Widely scattered on Tiburón, San Esteban, and Dátil, and generally not common. It is an important traditional Seri medicinal plant (Felger & Moser 1985).

Other Gulf Islands: Ángel de la Guarda, San Marcos, Coronados, Carmen, Danzante, Monserrat, Santa Catalina, Santa Cruz, San Diego, San José, Cerralvo.

Geographic Range: Southeastern California, southern Nevada, and southwestern Utah to western Texas, Baja California Sur, and much of Sonora to northwestern Sinaloa. This is one of the most variable and wide-ranging species in the genus and occurs in drier habitats than any other *Porophyllum*.

Tiburón: Arroyo Sauzal, *Felger 08-30*. La Viga, SW part of island just N of Islas Dátil and Cholludo, abundant, *Wilder 08-146*. SW Central Valley, *Felger 12327*. Sopc Hax, *Felger*

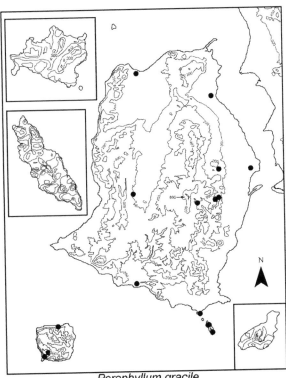

Porophyllum gracile
Xtisil (var. xtesel, xtisel)

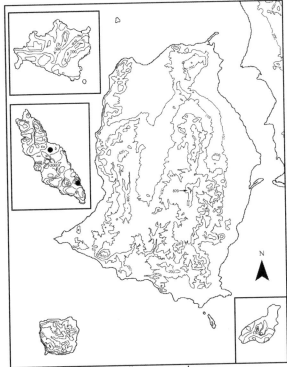

Porophyllum pausodynum

9292. Head of arroyo, E base of Sierra Kunkaak, *Wilder 06-33*. Foothills of Sierra Kunkaak, rocky arroyo walls, *Felger 6986*. Palo Fierro, *Felger 8924*. Camino de Caracol, broad arroyo entering foothills of Sierra Kunkaak, *Wilder 05-7*. Hahjöaacöl, NE coast, *Wilder 08-219*. Bahía Agua Dulce, *Felger 8903*.

San Esteban: N side of island, N-facing sea cliffs, *Felger 15422*. Canyon, SW corner of island, *Felger 16645*. Steep slopes, S-central peak, *Felger 17539*.

Dátil: Occasional on rocky slopes, *Moran 13027*. NW side of island, *Felger 15305*. SE side, steep narrow canyon, *Felger 17510*. E side of island, slightly N of center, *Wilder 07-106*.

Porophyllum pausodynum B.L. Robinson & Greenman

Description: Short-lived perennial herbs or sub-shrubs, probably also reproductive in the first season. Stems slender, often erect, and sparsely branched. Leaves linear and bright green. Flower heads in dense clusters (cymes) at stem tips; flowers pale yellow.

Local Distribution: Nolasco, known from the southeast and east-central side of the island, especially at higher elevations. The plants might not be seen during dry years or seasons, either dying back to their bases or surviving as seeds.

Other Gulf Islands: None.

Geographic Range: Endemic to the Guaymas region, Sierra Libre, and Nolasco.

Nolasco: E-central side of island, *Felger 9655*. Above SE cove, Cooper [*Felger 10402*].

Stephanomeria pauciflora (Torrey) A. Nelson

Hehe imixaa "rootless plant", posapatx camoz "what think it's a sweetbush (*Bebbia*)"; desert straw, desert wirelettuce

Description: Dense, globose or mound-shaped perennial bushes 0.5+ m tall, glabrous, and with milky sap. Leaves drought deciduous, variable in size, often about 5 cm long, pinnatifid with several narrow, spreading segments, the leaves reduced in size upward; first leaves (of first-season plants) in a basal rosette. Involucres 6.8–10 mm long, narrow and cylindrical, the phyllaries mostly 5, plus smaller, accessory bracts. Heads of ray flowers only, with 5 flowers, the rays 10–12 mm long and pale pink. Styles and stigmas prominent and purple. Achenes columnar; pappus of many slender plumose bristles, 6–8 mm long. Flowering at various seasons, mostly spring and fall with sufficient rainfall.

Stephanomeria is the only composite in the flora with milky sap and the only member of the tribe Cichorieae in the flora area. The Cichorieae, which includes lettuce,

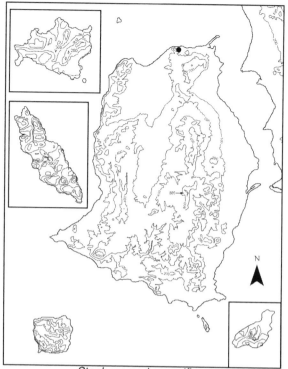

Stephanomeria pauciflora
Hehe imixáa, posapátx camoz

Thymophylla concinna
Cahaahazxot

is mostly in temperate latitudes and far more diverse in desert and non-desert regions to the north of the Midriff Island region. For example, there are 8 genera and 13 species of Cichorieae in the flora of the Gran Desierto of northwestern Sonora (Felger 2000a). *Sonchus oleraceus*, an introduced Eurasian species, is the only Cichorieae in the flora of the Guaymas region.

Local Distribution: Documented in mountains at the northeast coast of Tiburón.

Other Gulf Islands: None.

Geographic Range: California to Utah and Kansas, and southward to Baja California Norte, Chihuahua, and Coahuila, and southward in western Sonora to the vicinity of Bahía Kino.

> **Tiburón:** Hee Inoohcö, mountains at NE coast, adjacent to sea, 29.21843°N, 112.34568°W, NE-facing slope, base of drainage, 130 m, *Felger 07-88.*

Thymophylla concinna (A. Gray) Strother [*Dyssodia concinna* (A. Gray) B.L. Robinson]

Cahaahazxot "what causes sneezing"; *manzanilla del coyote*; dogweed

Description: Winter-spring annuals 3.5–10 cm tall, glabrous or sparsely hairy, pungently aromatic, dotted with small oil glands, the stems becoming semi-prostrate. Leaves opposite below, alternate above, 7–16 mm long, pinnate with slender segments. Flower heads often clustered at ends of leafy stems, the rays white, 4–5 mm long, the disk yellow.

Local Distribution: Tiburón, seasonally common in the Central Valley and the broad bajada along the Infiernillo side of the island, mostly on sandy or gravelly soils.

Other Gulf Islands: None.

Geographic Range: Sonoran Desert in southern Arizona and southward to central and western Sonora to the Guaymas region.

> **Tiburón:** SW Central Valley, *Felger 12301.* Palo Fierro, *Felger 7006.*

Trixis californica Kellogg var. **californica**

Cocaznootizx "rattlesnake's foreskin"; *plumilla, arnica*; trixis

Description: Small shrubs or subshrubs, essentially glabrous. Leaves gradually drought deciduous, mostly lanceolate, 3–8 cm long, thin and semi-persistent after drying, the leaf bases persistent. Foliage and flowers at various seasons, especially spring. (This variety is unique among at least the North American members of *Trixis*

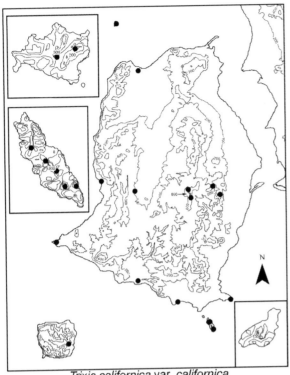

Trixis californica var. californica
Cocaznootizx

in having stomates on both surfaces of the leaves. This feature seems to be part of the character set involving ascending rather than spreading leaves, which is probably an adaptation to an arid or semi-arid environment. Var. *peninsularis* (S.F. Blake) C. Anderson, endemic to the Cape Region of Baja California Sur, is pubescent and has stomates restricted to the lower leaf surface.) Flower heads with bright yellow, bilabiate corollas; reproductive at various seasons, especially spring. Achenes 8–10 mm long, with a pappus of numerous soft, barbellate bristles about as long as the achenes. In the flora area this is the only composite with bilabiate flowers and the only member of the tribe Mutiseae, a polyphyletic group centered in the New World tropics.

Local Distribution: Widespread on Tiburón, San Esteban, Dátil, Mártir, and Nolasco; open desert, slopes, arroyos and canyons, often beneath trees and larger shrubs, and among rocks. Extending to high elevations in the Sierra Kunkaak. Widely scattered on Mártir, and Palmer reported it "common upon the summit" (Watson 1889:59), which was still the case in 2008. Also documented on Patos in 1921 and not found there since.

Other Gulf Islands: Ángel de la Guarda, San Lorenzo, Tortuga, San Marcos, Coronados, Carmen, Danzante, Monserrat, San Francisco.

Geographic Range: Southeastern California to western Texas and southward to Baja California Sur, Sinaloa, and San Luis Potosí, Zacatecas and Nuevo León; most of Sonora below the oak zones.

Tiburón: Ensenada Blanca (Vaporeta), arroyo, *Felger 7147*. Just N of Punta Willard, *Wiggins 17143* (CAS). Arroyo Sauzal, 15 m from shore, small wash, *Wilder 06-82*. Ensenada de la Cruz, rocky hill, *Felger 2581*. El Monumento, *Felger 08-46*. SW Central Valley, *Felger 17299*. Foothills of Sierra Kunkaak, arroyo, *Felger 6999*. Hast Cacöla, E flank of Sierra Kunkaak, 380 m, *Wilder 08-236*. Cerro Kunkaak, 1200 m, *Scott 11 Apr 1978* (UNM). Canyon, 720 m, N interior of Sierra Kunkaak, *Wilder 07-514*. Freshwater Bay, frequent shrub on sandy bottoms, *Johnston 3270* (CAS).

San Esteban: Arroyo Limantour, 1 km inland, *Felger 12740*.

Dátil: NE side of island, rocky hills, *Felger 2545*. NW side of island, *Felger 15347*. SE side, steep and narrow canyon, *Felger 17513*. E shore, N of island center, canyon bottom, *Wilder 07-183B*.

Patos: A few plants in a protected draw, shrub 2–2½ ft high, 23 Apr 1921, *Johnston 3246* (CAS).

Mártir: Summit: 1887, *Palmer 408* (GH, UC, images); *Felger 07-197*. Canyon on N slope, 100 m, *Moran 8801*.

Nolasco: SE side, *Sherbrooke* [*Felger 11232*]. W side, W exposure, 700 ft, *Pulliam & Rosenzweig 21 Mar 1974*. E-central side, N exposure, 270 m, below crest of island, *Felger 06-106*. Cañón de Mellink, upper reaches of canyon, *Wilder 08-354*. Infrequent shrub 1½–2 ft high on a N facing cliff near island crest, *Johnston 3135* (CAS).

Verbesina palmeri S. Watson [*V. oligocephala* I.M. Johnston subsp. *palmeri* (S. Watson) Felger & Lowe]

Cocaznootizx caacöl "large *Trixis*"

Description: Shrubs 0.5–1.5+ m tall, often forming rather compact clumps, with rough hairs (hispid-scabrous). Leaves opposite or uppermost sometimes alternate. Petioles comparatively short (5–15 mm long), broadly winged and clasping at base. Leaf blades deltoid-ovate or rhombic-ovate, 3–8 cm long, coarsely toothed to nearly entire. Flowers yellow, the heads with rays 8–10 m long and disk florets 4 mm long. Flowering at almost any season following favorable rainfall. Achenes with thin, ciliate wings and 2 prominent awns.

Local Distribution: Tiburón, in mountains at the northern and northwestern coasts and to the summit in the Sierra Kunkaak; canyons and rocky slopes.

Other Gulf Islands: Ángel de la Guarda, San Lorenzo, Santa Cruz.

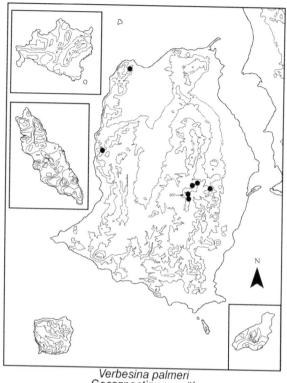

Verbesina palmeri
Cocaznootizx caacöl

Geographic Range: Gulf Coast of Baja California Norte, Gulf Islands, and the Sonora mainland in the Sierra Seri. It is replaced southward in Baja California Sur by the closely related *V. oligocephala*.

Tiburón: Hast Cacöla, E flank of Sierra Kunkaak, 450 m, *Wilder 08-266*. SE-facing mountain slope above canyon bottom, N base of Sierra Kunkaak, *Wilder 06-432*. Near the top of Sierra Kuncock [*sic*], *Knight 1007, 1041, 1044* (UNM). Capxölim, *Wilder 07-476*. Summit of island, rare, *Wilder 07-568*. Rocky igneous hill W of Tecomate, above 650 ft, *Felger 6271*. Pooj Iime, *Wilder 07-455*.

Xanthisma

Perennials and often flowering in the first season. Leaves alternate, dissected or lobed and bristle toothed. Flower heads with yellow ray and disk florets. Achenes with a pappus of many slender bristles. This genus is segregated from *Machaeranthera*.

1. Leaves with relatively broad teeth; flower stalks (peduncles) leafy, up to 2.5 cm above foliage; San Esteban. _____**X. incisifolium**
1' Leaves with slender teeth; flower stalks mostly more than 5 cm long, leafless (or leaves reduced) and elongated well above the foliage; Tiburón. _____**X. spinulosum**

Xanthisma incisifolium (I.M. Johnston) G.L. Nesom [*Haplopappus arenarius* Bentham var. *incisifolius* I.M. Johnston. *Machaeranthera pinnatifida* var. *incisifolia* (I.M. Johnston) B.L. Turner & R.L. Hartman. *Xanthisma spinulosum* var. *incisifolia* (I.M. Johnston) D.R. Morgan & R.L. Hartman]

Description: Compact perennial herbs, the herbage with glandular hairs. Stems leafy, the leaves often 2–3+ cm long, once or twice pinnately lobed with relatively broad, bristle-tipped lobes. Flower stalks (peduncles) short, leafy or bracted almost to the flower head; flower heads often 2–4.3 cm wide.

Local Distribution: San Esteban, often locally common on the north side of the island and at higher elevations. Moran (1983b) and Rebman et al. (2002) list it for Tiburón, but we have not seen a specimen from the island.

Other Gulf Islands: San Lorenzo (type locality), Coronados, Santa Catalina, Santa Cruz, San Diego, Cerralvo (*Carter 4272*, UC).

Geographic Range: A Gulf of California endemic: San Esteban, San Lorenzo, and islands along Baja California Sur and the nearby coast.

San Esteban: N side of island, N-facing cliff near shore, *Felger 15424*. S ridge and W slope, NE peak, *Moran 13041* (UC, CAS). SW part of island, W-facing side canyon with dense vegetation, in rocks on side of canyon, *Wilder 07-88*.

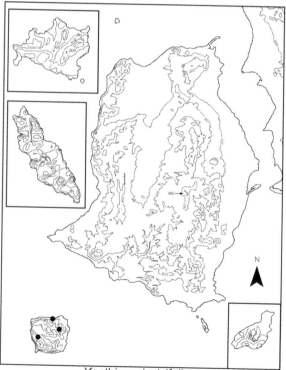

Xanthisma incisifolium

Xanthisma spinulosum (Pursh) D.R. Morgan & R.L. Hartman var. **scabrella** (Greene) D.R. Morgan & R.L. Hartman [*Haplopappus spinulosus* (Pursh) De Candolle subsp. *scabrellus* (Greene) H.M. Hall. *Machaeranthera pinnatifida* (Hooker) Shinners var. *scabrella* (Greene) B.L. Turner & R.L. Hartman]

Description: Perennial herbs, probably short-lived, ca. 30 cm tall, semi-woody at base. Herbage with glandular and non-glandular hairs. Lowermost or first leaves pinnatifid, often 1–2 cm long, with bristle-tipped teeth, upper leaves often reduced. Flowering stems (peduncles) nearly leafless. Flower heads probably to 3 cm wide.

Local Distribution: Known from a single locality in the interior of Tiburón.

Other Gulf Islands: Ángel de la Guarda (*Wilder 10-340*), San Marcos, Coronados (*Moran 3915*, UC), Carmen, Monserrat, San Francisco.

Geographic Range: Arizona to Colorado, Texas, Baja California Norte, Sonora, and north-central Mexico.

Tiburón: SW Central Valley, 450 ft, *Felger 12299*.

Xylothamia diffusa (Bentham) G.L Nesom [*Haplopappus sonoriensis* (A. Gray) S.F. Blake]

Caasol cacat "bitter caasol," caasol ziix iic cöihiipe "medicinal caasol"; *hierba de pasmo*

Description: Slender-stem shrubs 0.5–1 m tall, gland dotted and resinous. Leaves 5–25 mm long, slender and simple. Flowers bright yellow, the flower heads small and crowded at stem tips; ray florets 0–3, the disk florets 4 or 5. Achenes 1.5–2 mm long, pappus of many brownish bristles. Flowering at various seasons, especially in fall, the flowers visited by hoards of flies.

Local Distribution: Coastal areas on Tiburón; dunes, arroyos, arid canyon bottoms, and low hills. Also on sand flats on Alcatraz.

Other Gulf Islands: Ángel de la Guarda, San Lorenzo, San Marcos, San Ildefonso, Coronados, Carmen, San José, San Francisco, Espíritu Santo, Cerralvo.

Geographic Range: Coastal Sonora on sandy to silty and often alkaline soils of coastal plains, low coastal dunes and strands, and sometimes on rough, rocky slopes near the shore, from the vicinity of Bahía Kino southward to northwestern Sinaloa. Also Baja California Norte and Sur, and Gulf Islands.

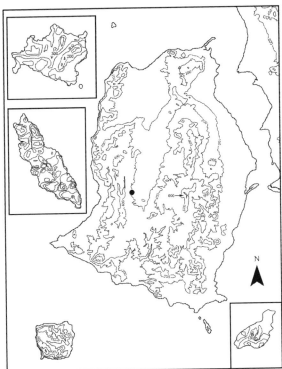

Xanthisma spinulosum var. *scabrella*

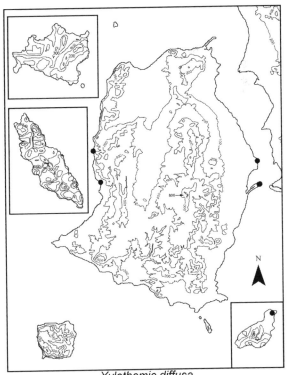

Xylothamia diffusa
Caasol cacat

Tiburón: Ensenada Blanca (Vaporeta), *Felger 17268*. Punta San Miguel, low dunes, 5–50 m inland, *Wilder 07-191*. Palo Fierro, *Felger 10323*. Pooj Iime, *Wilder 07-444*.

Alcatraz: NE end of island, near lighthouse, low flat, sand soil with cobble rock, *Felger 07-189*.

Avicenniaceae, *see* Acanthaceae

Bataceae – Saltwort Family

Batis maritima Linnaeus

Pajoocsim (variants: xpajoocsim, xpacoocsim); *dedito*; saltwort

Description: Succulent, glabrous perennials, mostly forming dense mats of trailing and scrambling stems often rooting at the nodes. Leaves opposite, often 1.3–4 cm long, linear-oblong to narrowly spatulate, thickly terete, yellow-green and extremely succulent. Male and female flowers on different plants. Inflorescences in leaf axils, of fleshy cone-like spikes or catkins often 7.5–15 mm long, the individual flowers greatly reduced and inconspicuous. The fruits float and often are seen in tidal drift in the Canal del Infiernillo in summer months.

The leaves are edible (Uphof 1968) but quite salty. The Comcaac used the roots to sweeten coffee before they had ready access to sugar (Felger & Moser 1985). "The roots are very sweet, used fresh; you just peel them and use as a sweetener" (Humberto).

Local Distribution: Tiburón, along shores in upper reaches of tidal wetlands including mangrove margins to saline wet but seldom-inundated sandy or muddy soils. Also common among halophytic vegetation on Alcatraz.

Other Gulf Islands: Ángel de la Guarda, Rasa, Coronados, Carmen, Danzante, Santa Catalina, San José, San Francisco, Espíritu Santo, Cerralvo.

Geographic Range: Gulf of California in saline wet mud and sand of bays, esteros, and tidal flats, often inundated by high tides and locally abundant; not on beaches facing the open sea. Coastal southern California to the Cape Region of Baja California Sur, the Gulf of California, Pacific and Atlantic coasts of tropical and subtropical America, Galapagos Islands, and Hawaii.

Tiburón: Ensenada de la Cruz, arroyo bed near shore, *Felger 12771*. Beach 1 km N of Estero San Miguel, *Wilder 06-7*. Punta Perla, estero, *Wilder 08-374*. Bahía Agua Dulce, mud flat near beach, *Felger 6819*.

Alcatraz: Interior of the low lying flat with *Allenrolfea occidentalis* dominant, *Trianthema portulacastrum*, *Atriplex barclayana*, *Abronia maritima*, *Felger 07-162*.

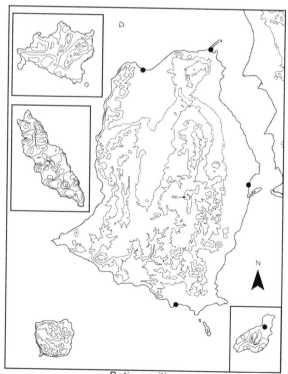

Batis maritima
Pajoocsim

Boraginaceae – Borage Family
(includes Hydrophyllaceae)

Plants of diverse growth forms: cool-season ephemerals, herbaceous perennials, and shrubs. Leaves simple, alternate except sometimes opposite in *Cryptantha*. Flowers radial, the sepals and petals each united at least basally. *Cordia*, *Heliotropium*, and *Tournefortia* mostly show "southern" tropical-subtropical affinity, while *Cryptantha*, *Nama*, *Phacelia*, and *Tiquilia* have a "northern" affinity.

Figure 3.37. *Cordia parvifolia.*
Vicinity of Zozni Cmiipla, Isla Tiburón.
BTW, 24 November 2006.

Figure 3.38. *Heliotropium curassavicum.* Xapij waterhole,
Arroyo Sauzal. BTW, 29 January 2008.

Figure 3.39. *Tiquilia canescens.* Isla Dátil. BTW, 27 February 2009.

1. Shrubs or subshrubs more than 50 cm tall (mostly 1+ m).
   2. Inflorescences not curled; flowers 2.5–3 cm wide; fruits dry and inconspicuous. _____**Cordia**
   2' Inflorescences curled like a scorpion tail ("scorpioid"); flowers 1–1.5 cm wide; fruits fleshy. ____ **Tournefortia**
1' Annual or perennial herbs, or subshrubs mostly less than 30 cm tall (sometimes to ca. 80 cm in *Cryptantha fastigiata*).
   3. Suffrutescent perennials, semi-woody at base; corollas lavender. _____**Tiquilia**
   3' Winter-spring annuals, corollas lavender or not, or if herbaceous perennials, then the corollas white.
      4. Plants glabrous, succulent or semi-succulent. _____**Heliotropium**
      4' Plants hairy, not succulent.
         5. Plants with rough, glassy hairs, flowers white. _____**Cryptantha**
         5' Plants with glandular or non-glandular hairs but not rough and glassy; flowers lavender or white.
            6. Leaves sessile or the blade tapering into the petiole, the margins entire (sometimes inrolled). _**Nama**
            6' At least the lower leaves petioled, the petiole and blade clearly differentiated, the blades pinnately lobed, pinnatifid or dissected, or toothed to wavy.
               7. Leaves opposite below, alternate above; ovary 1-chambered; calyx lobes cleft to about the middle or ¾ distance to base, the tube evident; stamens not exerted from corolla tube. _____**Eucrypta**
               7' Leaves alternate; ovary 2-chambered; calyx lobes divided nearly to base, the tube very short or none; stamens exerted or not. _____**Phacelia**

Cordia parvifolia A. de Candolle

Naz (variant: nooz), hehet inaail coopl "black-barked plant"; *vara prieta*; littleleaf cordia

Description: Shrubs with slender, flexible stems, the wood relatively hard, the bark dark colored. Leaves, mostly 1–3 cm long, alternate on long shoots and mostly clustered on short shoots and produced at various seasons following rains, and with rough, irritating hairs. Corollas often 2.5–3 cm wide, showy, pure white, produced in profusion following rains at almost any time of year except winter; flowers opening an hour or so after dawn, the corollas falling with midday or afternoon heat.

Local Distribution: Tiburón, locally abundant in creosotebush and mixed desertscrub communities in a limited area of the expansive bajada on the east side of the island. It was not found north of Punta Tormenta, but is one of the principal species on the bajada in the foothills of the Sierra Kunkaak in the vicinity of Sopc Hax waterhole. *Cordia parvifolia* shares dominance with *Larrea divaricata* "in a narrow zone parallel to the Infiernillo shore and situated between coastal desertscrub and the more inland and complex mixed desertscrub communities" (Felger & Lowe 1976:18). *Cordia* is likewise common on the coastal bajada on the opposite mainland.

Other Gulf Islands: None.

Geographic Range: Common and widespread across the southern half of the Sonoran Desert to Sinaloa, Baja California Sur, and the Chihuahuan Desert in Chihuahua, Coahuila, Durango, and Zacatecas. This drought-resistant shrub is widely cultivated in southern Arizona.

Tiburón: Palo Fierro, *Felger 10333*. Vicinity of Hant Hax, 2 May 2007, *Wilder*, observation.

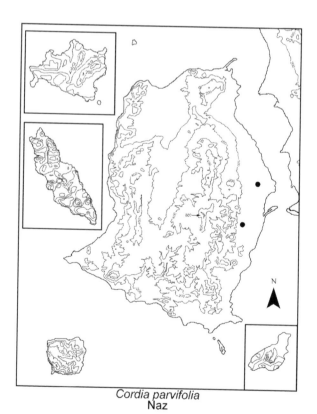

Cordia parvifolia
Naz

Cryptantha

Annuals and one facultatively short-lived perennial, growing and flowering only during the cooler seasons. Herbages and calyces rough, with firm and glassy hairs. Leaves linear, the first ones in a basal rosette, the lower leaves opposite, upper ones alternate. Flowers small and white. Fruits with 1 or 4 nutlets enclosed in a bristly (hispid) calyx.

The center of diversity for this large and diverse genus is in the dry regions of southwestern North America. The diversity drops off sharply from north to south across the Sonoran Desert: 16 species occur within the desert (Felger 2000a), 9 in the Pinacate/Gran Desierto Region, 4 are on Tiburón, and only 1 species, *C. grayi* (Vasey & Rose) J.F. Macbride, occurs in Sonora south of the Guaymas region. These distributions correlate with winter-spring rainfall patterns.

1. Nutlets 1 per fruit, the surfaces smooth. _____
_____ **C. maritima**
1. Nutlets 4 per fruit (heteromorphous: 3 smaller and similar, and one larger than the other 3); surfaces ornamented with tubercles, not smooth.
　　2. Nutlets margins rounded or angled but not sharp edged; largest nutlet 1–1.4 mm long. _____
_____ **C. angustifolia**
　　2' Nutlet margins sharp edged.
　　　　3. Largest nutlet ca. 0.7 mm long.___ **C. angelica**
　　　　3' Largest nutlet 1.5–1.9 mm long. 　**C. fastigiata**

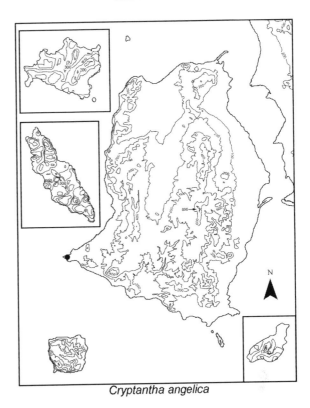

Cryptantha angelica

Cryptantha angelica I.M. Johnston

Description: Herbage gray green. Flowering branch tips curled (like a scorpion tail). Corollas minute and inconspicuous. Nutlets 4, of two sizes (heteromorphous), similar to those of *C. angustifolia* but smaller, sharp edged, and dark colored.

The two specimens from the vicinity of Punta Willard appear intermediate between *C. angelica* and *C. angustifolia*. They approach typical *C. angelica* in seemingly having sharper-edged nutlets than *C. angustifolia* but the distinction is not clear cut.

Local Distribution: Known from sandy areas at the southwestern corner of Tiburón.

Other Gulf Islands: Ángel de la Guarda (type locality), Partida Norte (*Wiggins 17300*, CAS), San José.

Geographic Range: Baja California Norte and several Gulf Islands.

Tiburón: Just N of Willard's Point in canyon, *Moran 8727* (SD). 1 km N of Punta Willard, *Felger 17753*.

Cryptantha angustifolia (Torrey) Greene

Hehe cotopl "plant that clings," hehe czatx "stickery plant"; *peludita*; popcorn flower, narrow leaf cryptantha, desert cryptantha

Description: Herbage gray green. Flowering branch tips curled (like a scorpion tail). Corollas minute and inconspicuous. Nutlets 4, of two sizes (heteromorphous).

Local Distribution: Seasonally widespread and often abundant across the lowlands of Tiburón. Also widespread on San Esteban. In many desert habitats, from the shore above the high tide zone on sandy soils to rocky slopes, and especially along major arroyos.

Other Gulf Islands: Ángel de la Guarda, San Marcos, Coronados, Carmen, San Francisco.

Geographic Range: This is the most common and widespread *Cryptantha* in the Sonoran Desert. It is one of those irritating, itchy plants that gets in your socks and sleeping bag, and people are likely agents of dissemination. Deserts and dry regions of southwestern United States, Baja California Norte, and Sonora south to the vicinity of Guaymas.

Tiburón: Ensenada Blanca (Vaporeta), *Felger 17251*. Arroyo Sauzal, *Felger 12280*. Ensenada de la Cruz, dunes near beach, *Felger 12796*. El Monumento, *Felger 08-43*. Ensenada del Perro, *Felger 17720*. Punta San Miguel, low dunes, *Wilder*

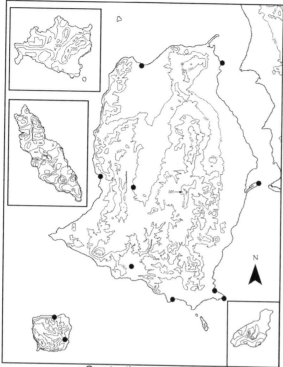

Cryptantha angustifolia
Hehe cotopl, hehe czatx

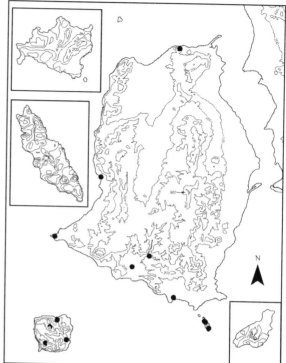

Cryptantha fastigiata

07-189. SW Central Valley, *Felger 12297.* Zozni Cacösxaj, NE coast, *Wilder 08-201.* Tecomate, *Wilder 07-239.*

San Esteban: Arroyo Limantour, near beach in rocks, *Van Devender 92-489.* N side of island, *Felger 17566.*

Cryptantha fastigiata I.M. Johnston

Description: Annuals to short-lived perennials, extremely variable in size, sometimes reproductive as single-stem ephemerals from about 10 cm tall to bushy plants to 80 cm tall, clothed with stiff, spreading and variously appressed hairs to 2 mm long. Nutlets 4, ovate-triangular, strongly convex on back, dark brown with pale tubercles and pale, sharply angled margins, of two sizes, the 3 smaller nutlets 1–1.5 mm long, the larger nutlet 1.5–1.9 mm long.

Local Distribution: Common and widespread on San Esteban, Dátil, and the southern, western, and northern parts of Tiburón; sandy washes, arroyos, plains, and rocky slopes.

Other Gulf Islands: Ángel de la Guarda, San Lorenzo, San Marcos, Carmen, Danzante.

Geographic Range: Gulf of California islands, the Baja California Peninsula, and Sonora in the Seri region.

Tiburón: Ensenada Blanca (Vaporeta), *Felger 15731.* 1 km N of Punta Willard, *Felger 17745.* Arroyo Sauzal, *Felger 12256.* Xapij (Sauzal) waterhole, *Wilder 06-111.* Arroyo de la Cruz, *Moran 13014.* Hee Inoohcö, mountains at NE coast, *Felger 07-92.*

San Esteban: N side of island, *Felger 17557.* Large arroyo, SW corner, *Hastings 71-54.* Arroyo Limantour, along wash, *Van Devender 92-477.* S-central peak, steep slopes, *Felger 17537.* Sub-peak of central mountain, *Wilder 07-63.*

Dátil: NW side of island, *Felger 15312.* SE side, *Felger 17499.* E side, ridge crest of peak, *Wilder 07-113.*

Cryptantha maritima (Greene) Greene [*C. maritima* var. *pilosa* I.M. Johnston]

Hehe cotopl "plant that clings," hehe czatx "stickery plant"; *peludita*; popcorn flower

Description: Plants generally erect with a main axis and branching mostly above middle, the stems dark to reddish brown, the dead leaves persistent and dark brown. Flowering branches scruffy looking and short-racemose (not curled). Fruiting sepals slender, bristly-hairy, and longer than the enclosed nutlet; corollas scarcely visible, the flowers presumably selfing. Nutlets 1 per fruit, 1.2–1.5 mm long, smooth, shiny, dark brown, and slender.

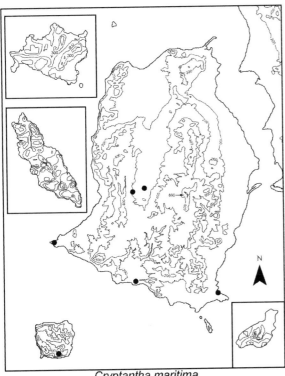

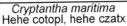

Cryptantha maritima
Hehe cotopl, hehe czatx

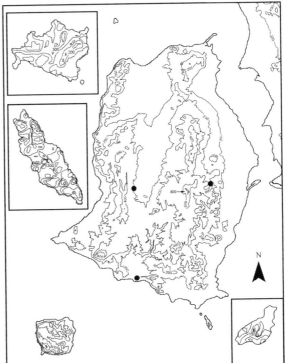

Eucrypta micrantha

Local Distribution: Widespread on Tiburón and also on San Esteban.

Other Gulf Islands: Ángel de la Guarda, Partida Norte, San Lorenzo, Tortuga, San Marcos (*Wiggins 17322*, CAS).

Geographic Range: Southern Nevada, southern California including Channel Islands, both states of Baja California, western and southern Arizona, and Sonora south to the Guaymas region. Next to *C. angustifolia*, this is the most widespread and common cryptantha in western Sonora. All Sonoran Island and mainland specimens examined have only 1 seed, while in non-desert regions there are often 2 seeds per fruit.

Tiburón: Just N of Willard's Point, *Moran 8727* (SD). Arroyo Sauzal, *Felger 08-17*. Ensenada del Perro, *Felger 17716*. Central Valley, base of mountain, *Felger 12659*. SW Central Valley, *Felger 12307*.

San Esteban: N slope of high peak, near SE corner of island, 420 m, *Moran 8847* (SD).

Eucrypta micrantha (Torrey) A. Heller

Description: Delicate, aromatic and glandular-viscid plants, with stalked glandular hairs intermixed with non-glandular hairs. Leaves pinnatifid, 1.5–5 cm long, opposite below, alternate and reduced upward with petiole bases clasping the stem. Flowers small, corolla lobes white to pale lavender, the throat with yellow nectaries and often nectar filled in the morning. Fruiting calyx 3–4.5 mm long, not spreading open at maturity (revealing only the tip of the capsule). Capsules splitting but the 2 carpels (halves) not falling free; seeds up to 16 per capsule, all alike, brownish black, 0.8–1 mm long, incurved and cylindrical, with sharply sculptured transverse ridges.

Local Distribution: Documented on Tiburón from three widely separated sites, along washes and beneath trees and shrubs, and among rocks; it probably occurs elsewhere on the island.

Other Gulf Islands: None.

Geographic Range: Southeastern California to Nevada, Utah, and Texas, and southward to Baja California Norte and Sonora south at least to the vicinity of Hermosillo.

Tiburón: Arroyo Sauzal, *Wilder 08-115*. SW Central Valley, along arroyo under trees and shrubs, *Felger 17341*. Hast Cacöla, E flank of Sierra Kunkaak, 380 m, *Wilder 08-248*.

Heliotropium curassavicum Linnaeus

Hant otopl "what the land sticks to," potacs camoz "what thinks it's an iodine bush"; *hierba del sapo*; alkali heliotrope

Description: Perennial herbs and sometimes facultative annuals; glabrous. Stems and leaves semi-succulent, bluish glaucous. Leaves mostly 2.5–7.5 cm long. Flowers in several spike-like, terminal, curved (scorpioid) branches. Corollas white with a yellowish center becoming purplish with age. Nutlets 4, ovoid. Growing and flowering with warm to hot weather any time of year.

Local Distribution: Tiburón, widely scattered along the southern and eastern coasts in low-lying saline to alkaline soils including upper beaches, beach dunes, arroyos, and brackish water of estero margins, and inland at the Sauzal waterhole. Also on Alcatraz in the low-lying *Allenrolfea* flat.

Other Gulf Islands: Ángel de la Guarda (*Wilder 10-382*), San Lorenzo, San Marcos, Carmen, Danzante, Monserrat, San Diego, San José, San Francisco, Espíritu Santo.

Geographic Range: Ranging through much of the warm portion of the Western Hemisphere.

Tiburón: Ensenada de la Cruz, low wet saline soil near beach, *Felger 9071*. Sauzal waterhole (Xapij): Damp saline soil near waterhole, *Felger 2766*; *Wilder 08-101*. Punta San Miguel, low dunes, *Wilder 07-203*. Palo Fierro, 100 m inland, *Wilder 06-149*.

Alcatraz: High moisture area of flat, *Wilder 07-420*.

Nama hispidum A. Gray

Hohroohit "what donkeys eat"; sand bells, bristly nama

Description: First leaves form a basal rosette, larger plants form leafy stems. Leaves 1.5–4 cm long, narrowly spatulate, the lower or first leaves gradually narrowed to a winged petiole, the upper leaves smaller, sessile. Corollas about 1 cm wide, lavender with a yellow throat, and readily deciduous.

Local Distribution: Widely scattered on Tiburón; on the north side of the island, the Central Valley, and along Arroyo Sauzal; sandy and fine-textured soils of arroyos and valley plains.

Other Gulf Islands: None.

Geographic Range: Southeastern California to Baja California Sur, Nevada and Utah to western Texas and northern Mexico including Sonora and Sinaloa.

Tiburón: Sauzal, sandy-gravelly wash, *Wilder 08-104*. SW Central Valley, locally common, *Felger 12294*. Hahjöaacöl, NE coast, *Wilder 08-211*. Arroyo Agua Dulce, localized, *Felger 6866*.

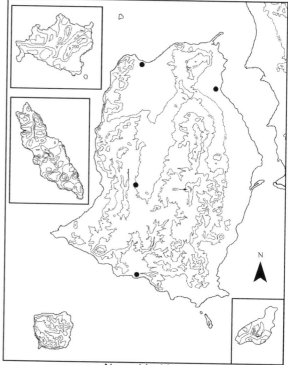

Heliotropium curassavicum
Hant otopl, potacs camoz

Nama hispidum
Hohroohit

Phacelia

Herbage conspicuously hairy and glandular, stinky and irritating, often causing dermatitis. Inflorescence branches moderately to strongly helicoid (curled at the tip like a scorpion tail—in the shape of a flattened spiral, straightening as it grows, the flowers on one side).

1. Stamens not protruding; seeds many per capsule, solid, more or less terete in cross section, similar on both sides. _____**P. affinis**
1' Stamens usually protruding, readily visible; seeds 4 per capsule, boat shaped, concave and sculptured on the ventral side (excavated on either side of central ridge), convex on other side.
   2. Fruiting pedicels mostly 1–1.5 mm long, shorter than the capsules; fruiting sepals 3–5 mm long, about as long as to ¼ longer than capsules._____**P. crenulata**
   2' Fruiting pedicels 4–7 mm long, longer than the capsules; fruiting sepals 5–6.5 mm long, about twice as long as the capsules. _____ **P. pedicellata**

Phacelia affinis A. Gray

Description: Small annuals, mostly several-branched from near the base. Herbage, inflorescences, and calyx densely pubescent with stiff white hairs and sessile glands. Leaves mostly basal and on lower part of stems (1.5) 3–6 cm long, pinnately lobed to pinnatifid, mostly narrowly oblong; upper leaves reduced. Flowering stalks (cymes) moderately helicoid, especially on larger plants. Calyx lobes oblanceolate. Corollas white (occasionally pale lavender) with a pale yellow-green throat. Filaments whitish, the anthers included (not protruding) and cream colored. Seeds many, nearly 1 mm long, solid, brown, reticulate, and transversely corrugated.

Local Distribution: Documented on Tiburón from a single record. The nearest known population is 150 km to the north in the Sierra del Viejo near Caborca.

Other Gulf Islands: None.

Geographic Range: Northwestern Sonora, Arizona, southwestern New Mexico, southwestern Utah, southern Nevada, and southeastern California to Baja California Sur.

Tiburón: SW Central Valley, *Felger 17337*.

Phacelia crenulata Torrey [*P. crenulata* var. *ambigua* (M.E. Jones) J.F. Macbride. *P. ambigua* M.E. Jones. *P. crenulata* var. *minutiflora* (J. Voss) Jepson. *P. ambigua* var. *minutiflora* (J. Voss) Atwood. *P. minutiflora* J. Voss]

Cahaahazxot ctam "what causes sneezing (*Baileya*) male," najmís; desert heliotrope

Description: Herbage stinky and irritating, sticky-viscid with spreading white hairs, and minutely glandular

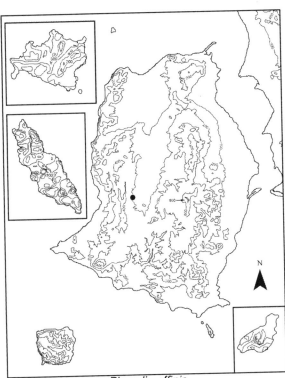

Phacelia affinis

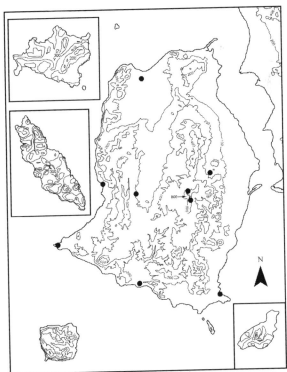

Phacelia crenulata
Cahaahazxot ctam, najmís

above. Stems often 10–30 (45+) cm long, mostly erect, simple to several-branched. First leaves in a prominent basal rosette, pinnatifid or the lower segments cleft to the midrib or into separate leaflets. Inflorescences helicoid, simple to several-branched. Flowering and fruiting pedicels often 1–1.5 mm long. Flowers moderately fragrant; corollas lavender; stamens protruding from corolla. Fruiting sepals 3–5 mm long, about as long as to ¼ longer than the capsules. Seeds 4 per capsule, 2.2–3 mm long, boat shaped, and red-brown.

Local Distribution: Widespread in the lowlands across Tiburón, especially in arroyos, also canyons, bajadas, and rocky slopes to higher elevations, and reported for San Esteban (see "Doubtful and Excluded Reports"). Some Tiburón and nearby mainland specimens have been identified as var. *ambigua* or var. *minutiflora*. We believe that only a single taxon occurs on the island.

Other Gulf Islands: None.

Geographic Range: Northwestern Mexico and southwestern United States.

Tiburón: Ensenada Blanca (Vaporeta), *Felger 7151*. Willard's Point, just N, *Moran 8729* (SD). Arroyo Sauzal, near beach to 2 mi inland, *Felger 7030*. Ensenada del Perro, *Felger 17733*. SW Central Valley, arroyo, *Felger 12335*. Deep canyon, N interior of the Sierra Kunkaak, 680 m, *Wilder 07-491*. Cerro Kunkaak, 800 m, *Scott 11 Apr 1978* (UNM). Zoojapa, NE foothills of Sierra Kunkaak, 180 m, *Wilder 08-225*. Tecomate, *Felger 12513*.

Phacelia pedicellata A. Gray

Description: Plants often robust, the stems 10–45+ cm long, relatively thick and semi-succulent, and usually leafy. Herbage viscid glandular hairy and stinky. Leaves often 5–16+ cm long; pinnatifid to pinnately compound, and semi-succulent. Inflorescences helicoid. Flowers on slender pedicels with dense, spreading hairs, the fruiting pedicels 4–7 mm long. Corollas pale lavender-blue. Stamens and styles protruding. Fruiting sepals 5–6.5 mm long, about twice as long as capsules. Seeds 4 per capsule, boat shaped.

Specimens from Tiburón and the opposite mainland appear to be intermediate between *P. pedicellata* of northwestern Sonora (the Gran Desierto region) and *P. scariosa* Brandegee, which occurs from the Guaymas region to northwestern Sinaloa and in Baja California Sur. The calyx lobes are somewhat elongated in the Gran Desierto and Tiburón specimens, approaching those of the closely related *P. scariosa* in length but narrower than in that species (Felger 2000a).

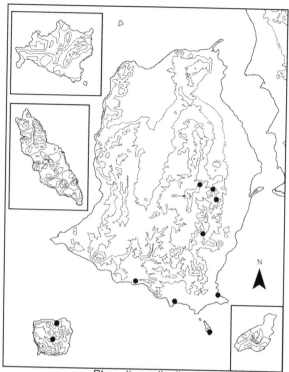

Phacelia pedicellata

Local Distribution: Widely scattered on Tiburón and San Esteban, from near the shore to higher elevations; often along arroyos and also canyons and rocky slopes to at least 400 m, and on Dátil.

Other Gulf Islands: Ángel de la Guarda.

Geographic Range: Southeastern California, eastern Baja California Norte, Arizona, and northwestern Sonora.

Tiburón: Arroyo Sauzal, arroyo and slopes, *Felger 9995*. Arroyo de la Cruz, on cobble berm, *Moran 13017* (SD). Ensenada del Perro, *Felger 17732*. E foothills of Sierra Kunkaak, *Felger 6974*. Vicinity of Sopc Hax, *Wilder 07-212*. Hast Cacöla, E flank of Sierra Kunkaak, 380 m, *Wilder 08-257*. Cerro San Miguel, 300–400 m, *Quijada-Mascareñas 91T008*.

San Esteban: Narrow rocky canyon behind cove, N side of island, *Felger 17584*. Arroyo Limantour, 2¼ mi inland, *Wilder 07-78*.

Dátil: SE side, steep and narrow canyon, *Felger 17501*.

Tiquilia

Perennials with firm hairs and bristles; leaf margins revolute (inrolled); flowers lavender, axillary, often in forks of upper stems; fruits rounded, with 4 nutlets.

1. Leaf veins inconspicuous or not evident and covered by a dense layer of hairs. _____ **T. canescens**
1' Leaves with 2 or 3 (4) pairs of conspicuous impressed veins. _____ **T. palmeri**

Tiquilia canescens (de Candolle) A. Richardson [*Coldenia canescens* de Candolle]

Description: Perennials forming low, spreading mats, the lower stems somewhat woody and tough. Foliage scruffy gray. Corollas pale lavender.

Local Distribution: Common on Dátil on north-facing slopes. On Tiburón known only from near the south shore about 7 km northward from Dátil—we did not find it elsewhere on the island even in areas closer to, and immediately north of Dátil.

Other Gulf Islands: Coronados, Carmen.

Geographic Range: The nearest population in Sonora occurs in the foothills of Sierra del Viejo southwest of Caborca and the nearest Baja California Norte population is at Bahía de Los Ángeles. Southeastern California to southwestern Utah, New Mexico, and Texas, northern Sonora and the Chihuahuan Desert in north-central Mexico.

Tiburón: Ensenada de la Cruz, locally common on N- and S-facing hillslope behind abandoned Mexican marine station building, collected by Molina-Martínez 28 Feb 2009 (cataloged as *Wilder 09-50*).

Dátil: NE side, steep rocky slopes, *Felger 13437*. Occasional locally, *Moran 13032* (SD). E side, slightly N of center, halfway up N-facing slope of peak, in dense vegetation, *Wilder 07-125*.

Tiquilia palmeri (A. Gray) A. Richardson [*Coldenia palmeri* A. Gray]

Hee ijoját "jackrabbit's *saya* (*Amoreuxia*)"

Description: Semi-prostrate perennial herbs from very deep, thickened, blackish roots. Herbage densely pubescent with white hairs and scattered bristles. Leaves broadly elliptic to ovate, rhombic, or nearly orbicular, with conspicuous but shallowly impressed veins on the upper surface; petioles prominent, often as long as or longer than the blades. Corollas readily falling away, lavender with a pale yellow throat. Flowering with spring and summer-fall rains.

Local Distribution: Common on the dune fields near Punta Willard at the southwestern corner of the Tiburón.

Other Gulf Islands: Ángel de la Guarda.

Geographic Range: Southeastern California, southern Nevada, western Arizona, Baja California Norte, and Sonora southward on coastal dunes to Bahía Kino.

Tiburón: Willard's Point, 3 km N, sand dunes, *Johnston 4248* (CAS). 1 km N of Punta Willard along coast, wind-blown sand piled against rock hills, *Felger 17739*.

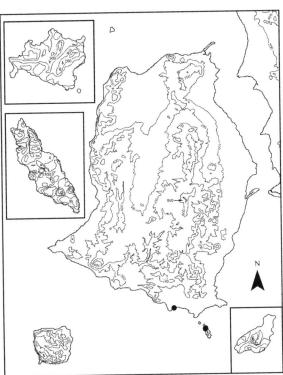

Tiquilia canescens

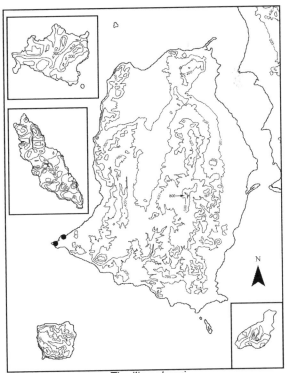

Tiquilia palmeri
Hee ijoját

Tournefortia hartwegiana Steudel

Description: Densely branched shrubs 1–1.5+ m tall with slender stems. Leaves and stems with rough, sandpaper-like hairs (scabrous). Leaves entire or nearly so, often 5–12 cm long, elliptic, lanceolate, or oblanceolate. Corollas white, 1–1.5 cm wide. Fruits fleshy, waxy-white, and rounded, 5–7 mm wide.

Local Distribution: Known from Tiburón from the middle of the Central Valley.

Other Gulf Islands: None.

Geographic Range: Tiburón and southern Sonora to northeastern and southern Mexico, and Baja California Sur. This species has subtropical affinities and is not found along the mainland Gulf Coast north of the Guaymas region but extends farther north inland along the eastern edge of the desert. It is the only borage on Tiburón with fleshy fruits, an attribute of tropical-inclined borages.

Tiburón: SW Central Valley, not common, in gravelly arroyo bank, *Felger 12314*. Central Valley, main arroyo, *Romero-Morales 06-1*. Satoocj, 10 Jan 2008, *Romero-Morales 08-1*.

Brassicaceae (Cruciferae) – Mustard Family

The Sonoran Desert crucifers are winter-spring annuals except *Lyrocarpa*. As with various other cool-season ephemerals, the diversity of the mustards quickly drops off south of northwestern Sonora—there are 10 native species in the Gran Desierto of northwestern Sonora, and only 4 make it to the Midriff Islands.

1. Perennials (sometimes flowering in first season); petals twisted, 12–20+ mm long; fruits usually widest well above the middle, more than 7 mm wide. _____
_____ **Lyrocarpa**
1. Winter-spring annuals; petals less than 2 mm long and not twisted; fruits widest at about the middle, less than 3 mm wide.
    2. Herbage with dendritic, candelabra-shaped hairs (stalked and branched above); plants with leafy stems; leaves finely divided into many small segments; fruits nearly terete._____ **Descurainia**
    2' Herbage with forked and stellate hairs; leaves in a basal rosette, stems leafless, leaves toothed but not divided into segments; fruits laterally flattened. _
_____**Draba**

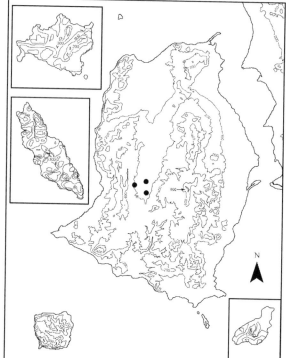

Tournefortia hartwegiana

Figure 3.40. *Lyrocarpa coulteri.* AE.

Descurainia pinnata (Walter) Britton

Cocool; *pamita, pamitón*; tansy mustard

Description: Plants 8–60 cm tall. Leaves and lower stems with branched (dendritic), soft white hairs (canescent). Leaves 1–3 times pinnately divided into small segments, the lower leaves petioled, the upper ones nearly sessile. Flowers pale yellow, 1.5 mm wide. Pedicels slender, the fruits narrowly club shaped, 4–7 mm long. Seeds mucilaginous when wet.

Local Distribution: Widely distributed on Tiburón in many habitats and common on Dátil.

Other Gulf Islands: None.

Geographic Range: North America except the Atlantic coast. A number of infraspecific taxa are described.

Tiburón: Ensenada Blanca (Vaporeta), slopes and arroyos, *Felger 17259.* El Sauz, *Moran 8762* (SD). Ensenada del Perro, 12 Apr 1968, *Felger 17728* (specimen not located). SW Central Valley, arroyo, *Felger 12337.* Cerro San Miguel, *Quijada-Mascareñas 91T024.* Foothills, E side of Sierra Kunkaak, *Felger 6975.* Bahía Agua Dulce, *Moran 8713* (SD).

Dátil: SE side, steep, narrow canyon, *Felger 17512.*

Draba cuneifolia Nuttall ex Torrey & A. Gray [*D. cuneifolia* var. *sonorae* (Greene) Parish]

Cocool cmaam "female cocool (*Descurainia*)"; wedge-leaf draba

Description: Small plants with forked and stellate hairs. Leaves in a basal rosette, broadly obovate, sessile, thin, to about 4 cm long with (1) 2 or 3 pairs of coarse teeth. Flowering stems leafless, mostly less than 10 (15) cm long. Flowers inconspicuous, white, the sepals and petals very quickly deciduous. Fruits laterally flattened. Seeds mucilaginous when wet. This species is distinguished from other Sonoran Desert crucifers by its broad and shallowly toothed basal-rosette leaves, small and leafless flowering stems, and relatively short, broad, and laterally flattened fruits about 1 cm long.

Local Distribution: Documented from the north coast of San Esteban and the interior of Sierra Kunkaak on Tiburón.

Other Gulf Islands: Ángel de la Guarda.

Geographic Range: Sonora from the vicinity of Alamos northward and both Baja California states; southwestern United States and Mexico southward to Zacatecas.

Tiburón: Hast Cacöla, E flank of Sierra Kunkaak, 380 m, *Wilder 08-254.*

San Esteban: N side of island, narrow rocky canyon behind large cove, *Felger 17569.*

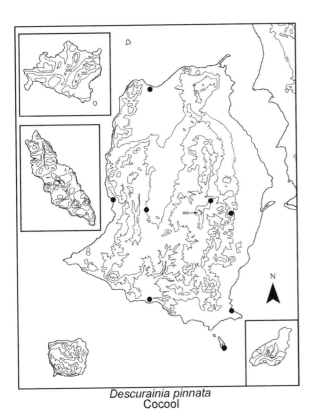

Descurainia pinnata
Cocool

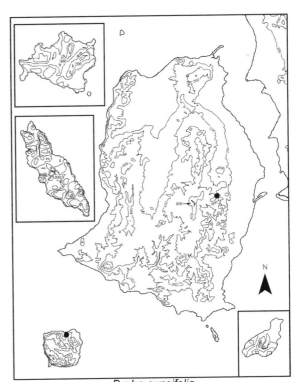

Draba cuneifolia
Cocool cmaam

Lyrocarpa – Lyre-pod

The generic name derives from the lyre-shaped fruits. Flowers on elongated racemes and highly fragrant at night. Petals elongated and usually twisted. Fruits relatively large and laterally flattened. This is the only summer-growing and perennial member of the mustard family in the Sonoran Desert, although a few other perennial mustards extend to the margins of the Sonoran Desert.

1. Leaves pinnate, the terminal lobe 4–15 mm wide; fruits 10–28 mm long; not on San Esteban. ____ **L. coulteri**
1' Leaves simple and 1–3 mm wide or pinnate with the terminal lobe 1–2 mm wide; fruits 7–10 mm long; San Esteban. _____ **L. linearifolia**

Lyrocarpa coulteri Hooker & Harvey ex Harvey

Ponás camoz "what thinks it's a *Matelea pringlei*"; lyre-pod

Description: Perennial herbs, sometimes flowering in the first season; densely covered with white, stellate or candelabra-shaped hairs. Stems slender and brittle. Leaves variable, the blades thinner during wet periods. Petals slender, generally twisted, 2–3 cm long, yellow- to purple-brown, often chartreuse at first or purple-brown at tips or along margins, becoming darker with age. Fruits lyre shaped, green, laterally compressed but relatively thick, 10–28 mm long, 8–14 mm wide. Seeds 2.4–2.8 mm long, brown, irregularly flattened, not mucilaginous. Growing and flowering at various seasons with sufficient moisture.

Local Distribution: Widely scattered across Tiburón but seldom very common. Also on Dátil, where it is seasonally common. Often among shrubs along arroyos and canyons and also on rocky slopes. The seemingly greater abundance on Dátil may be due to absence of larger herbivores on the island.

Other Gulf Islands: San Marcos, Monserrat, San José, Espíritu Santo.

Geographic Range: Western Sonora, southern Arizona, southeastern California, and both Baja California states.

Tiburón: Ensenada Blanca (Vaporeta), 7 Apr 1963, *Felger 7130* (specimen not located). Xapij (Sauzal), *Wilder 08-125*. El Monumento, *Felger 08-72*. Ensenada del Perro, *Felger 17717*. SW Central Valley, arroyo, *Felger 12355*. Hast Cacöla, E flank of Sierra Kunkaak, 380 m, *Wilder 08-247*. Tecomate, ¾ mi inland, *Felger 12516*.

Dátil: NE side, common, canyon bottom and N-facing slopes, 21 Oct 1963, *Felger 9097* (specimen not located). E-central side, canyon bottom, *Wilder 09-27*.

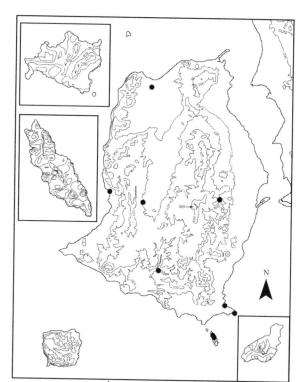

Lyrocarpa coulteri
Ponás camoz

Lyrocarpa linearifolia Rollins

Description: Small subshrubs with slender stems, the bark exfoliating in longitudinal strips. Leaves sparingly lobed with 1–3 pairs of lobes, the terminal lobe 1–2 mm wide or leaves entire and linear, 1–3 mm wide. Flowers 4–10 mm long, 5–12 mm wide, in short racemes or clustered at the top of a nearly leafless stem and approaching a sub-umbellate condition. Fruits 7–10 mm long, 9–13 mm wide. Seeds 1.5 mm long.

Lyrocarpa linearifolia is unmistakably related to *L. coulteri* but differs strikingly in being woodier and having narrowly linear leaves and leaf lobes, shorter inflorescences, and smaller flowers, fruits, and seeds.

Local Distribution: Documented on San Esteban by a few specimens.

Other Gulf Islands: Ángel de la Guarda.

Geographic Range: Islands and the adjacent Gulf Coast of Baja California Norte.

San Esteban: Colony on N slope of high peak near SE corner of island, 400 m, *Moran 8848* (SD). Near S end of island, large arroyo running inland from E side, *Wiggins 17235*.

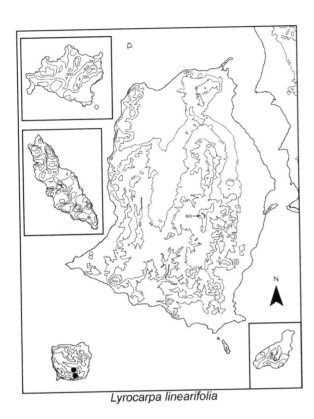

Lyrocarpa linearifolia

Burseraceae – Torchwood Family

Bursera

Trees and shrubs with resin ducts containing aromatic triterpenes and ethereal oils. Leaves drought decidu-ous, alternate, without stipules, often clustered on short shoots, and pinnate to tripinnate. Flowers small, radial, often unisexual, and present only for a short time, usually in early summer. Fruits small, 1-seeded, drupe-like but the exocarp at maturity separates into 2 or 3 segments or valves. Seeds partially to fully enveloped in a thin, pleasant-tasting, aril-like mesocarp (pseudoaril).

The closely related *B. fagaroides* and *B. microphylla* are in section *Bullockia*, the trivalvate-fruited species, and *B. hindsiana* and *B. laxiflora* are in section *Bursera*, the bivalvate-fruited species.

The family is tropical in affinity and all members are frost sensitive. Eighty species occur in southwestern Mexico, 9 in the tropical deciduous forest in south-eastern Sonora, 4 on Tiburón, and only *B. microphylla* makes it to the Pinacate region and southern Arizona and southeastern California. There is also a general trend for reduction in leaf and leaflet size from south to north.

2 cm

Figure 3.41 *Bursera hindsiana.* FR.

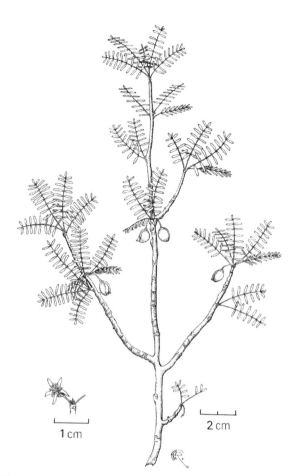

Figure 3.42. *Bursera microphylla.* FR.

Figure 3.43. *Bursera fagaroides* var. *elongata.* Capxölim, Isla Tiburón. BTW, 24 October 2007.

Figure 3.44. *Bursera microphylla.* Plant in the foothills of Sierra Kunkaak, Isla Tiburón, 24 November 2006, and flower at Bahía de Kino, 25 July 2007. BTW.

1. Outer bark exfoliating in papery flakes or sheets especially during the dry seasons; leaves once pinnate; fruits with a thin, papery aril covering the light-colored seed and with 3 fruit valves.
 2. Leaflets lanceolate to elliptic, 15–60 × 3–10 mm, the margins irregularly toothed or sometimes entire; higher elevations on Tiburón. _____ **B. fagaroides**
 2' Leaflets mostly linear, 5–25 mm × 1–2.5 mm, the margins entire or occasionally with a few small lobes; widespread at various elevations on multiple islands. _____ **B. microphylla**
1' Bark not exfoliating or peeling in sheets; leaves 1–3 times pinnate, occasionally reduced to a single leaflet; fruits with a thin fleshy orange aril covering one-half to two-thirds of the blackish seed and with 2 fruit valves.
 3. Leaves once pinnate with 3 (5) leaflets or reduced to a single leaflet. _____ **B. hindsiana**
 3' Leaves usually bipinnate (occasionally pinnate or tripinnate) with numerous small leaflets. _____ **B. laxiflora**

Bursera fagaroides (Kunth) Engler var. **elongata** McVaugh & Rzedowski [*B. confusa* (Rose) Engler, in part]

Xoop isoj "true elephant tree"; *torote de venado*

Description: Small trees 3–4 (5) m tall, with a well-developed trunk; bark papery and exfoliating during the dry seasons. Leaves once pinnate, with (5) 7–13 lanceolate to elliptic leaflets. The overall aspect of the tree is similar to that of *B. microphylla* but the trees, leaves, and leaflets are larger and it is in leaf only during the summer rainy period. The foliage briefly turns yellow to orange or reddish in September and is quickly shed, and the trees remain leafless until the next summer rains. Seeds 6 mm long, covered by a thin, papery, red-orange aril.

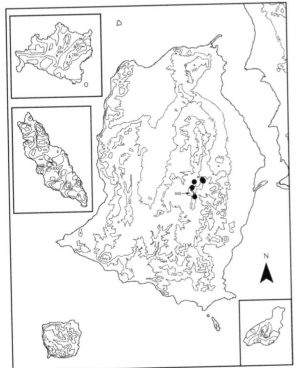

Bursera fagaroides var. *elongata*
Xoop isoj

Sepal tips reddish, the corollas yellow or greenish white. Flowering and leafing out June through July.

Several varieties of *B. fagaroides* are generally recognized for western Mexico. *B. fagaroides* and *B. microphylla* appear closely related, although Becerra and Venable (1999) indicate they are in separate but related clades. They are most easily distinguished by leaf differences. Leafless *B. fagaroides* often can be distinguished from a leafless *B. microphylla* by the larger sheets of exfoliating papery bark, better-developed trunk, and broader crown. *Bursera fagaroides* generally reaches higher densities in habitats slightly less xeric than does *B. microphylla*.

Bursera fagaroides and *B. microphylla* sometimes grow intermixed, but *B. fagaroides* has a far more restricted (less xeric) range on Tiburón than *B. microphylla*. Their flowers are remarkably similar. Apparent hybrids, or plants with intermediate characteristics, are occasionally encountered in places where both putative parents are present, for example, in the Guaymas region. As with *B. microphylla*, the putative hybrids leaf out at any time during hot weather if given ample water (Felger et al. 2001).

Local Distribution: Tiburón in the Sierra Kunkaak, mostly above mid-elevations, ca. 400 m; rocky slopes, canyons, and arroyos.

Other Gulf Islands: None.

Geographic Range: Nearly statewide in Sonora except north of about 30°30'N and the easternmost margin of the state. Nayarit to southwestern Chihuahua and Sonora, and a small, disjunct population in southern Arizona.

Tiburón: Siimen Hax, *Romero-Morales 07-4*. Cerro Kunkaak, 800 m, *Scott 11 Apr 1978* (UNM). Cerro San Miguel, *Quijada-Mascareñas 91T016*. Top of Sierra Kunkaak Segundo, E peak of Sierra Kunkaak, 490 m, 2 m tall, *Wilder 06-492*. Capxölim, occasional to common on slope, to ca. 3 m tall, the predominant *Bursera* in this area, replacing *B. microphylla* around 400 m and above, *Wilder 07-473*. SE-facing slope of deep canyon, N interior of Sierra Kunkaak, *Wilder 07-496*.

Bursera hindsiana (Bentham) Engler

Xopinl "elephant tree's fingers"; *torote prieto, copalquín, copal*; red elephant tree

Description: Mostly shrubs (small trees on the nearby mainland). Bark smooth, not peeling, dark reddish brown on small branches, gray on older branches and trunk; young branches pubescent, soon becoming glabrous. Leaves highly variable and produced at various seasons depending on soil moisture, to 5 cm long, with (1) 3 (5) leaflets, the leafstalk often winged; leaflets 8–35 × 6–25 mm, ovate, obovate or broadly lanceolate, with soft, velvety hairs on both surfaces, the margins irregularly crenate. Flowers on slender stalks 1–5 (8) cm long. Fruits globose, red and green, resembling a miniature apple. Seeds 7–8 mm long, blackish, two-thirds covered with a fleshy orange aril. Flowering August–October.

In cultivation the seedlings quickly develop a swollen stem above the soil line. Seed viability appears high, regardless of season of collection. Fully formed seeds are produced 3–4 weeks after flowering, and germination occurs in 5–12 days in hot weather.

This bursera is distinctive from other members of the genus in Sonora and seems most closely related to *B. cerasifolia* Brandegee and *B. epinnata* (Rose) Engler of Baja California Sur. Becerra and Venable (1999), however, indicate that *B. hindsiana* is most closely related to *B. stenophylla* and in the same clade as *B. penicillata*; these three species are, however, morphologically quite distinctive.

The wood has been used by the Comcaac for carving various objects because of its relatively soft but firm character, and it does not split on drying.

Local Distribution: Scattered on Tiburón, San Esteban, Dátil, and Cholludo but not the principal *Bursera* species on any island.

Other Gulf Islands: Ángel de la Guarda, Tortuga, San Marcos, San Ildefonso, Coronados, Carmen, Danzante, Monserrat (observation), Santa Catalina, Santa Cruz, San Diego, San José, San Francisco (observation), Espíritu Santo, Cerralvo.

Geographic Range: Through much of the Baja California Peninsula and the coast of Sonora from Puerto Lobos southward to the vicinity of San Carlos.

Tiburón: Ensenada Blanca (Vaporeta), *Felger 14970*. Arroyo Sauzal, ½ mi inland, arroyo floodplain, many have been heavily damaged by deer, *Wilder 06-92*. El Monumento, *Felger 15522*. Foothills of Sierra Kunkaak, ca. 1.25 mi SE from Sopc Hax waterhole, *Wilder 07-228*. Palo Fierro, *Felger 10320*. 3 km W [of] Punta Tormenta, *Scott P19* (UNM 53312). Hast

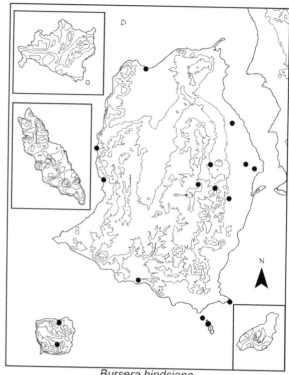

Bursera hindsiana
Xopinl

Cacöla, E flank of Sierra Kunkaak, 380 m, *Wilder 08-252* (USON). Cliff face of canyon, N base of Sierra Kunkaak, 415 m, *Wilder 06-402*. E foothills of Sierra Kunkaak, ca. 280 m, not common, *Wilder 08-323*. Bahía Agua Dulce, *Tenorio 9537* (MEXU). Valle de Águila, middle bajada, *Wilder 07-234*. Pooj Iime, large canyon, NW shore, *Wilder 07-448*.
San Esteban: N side, narrow, rocky canyon behind large cove, *Felger 17595*. Canyon, SE corner, *Felger 17624*.
Dátil: NE side, scattered and not common, canyons and steep rocky slopes, *Felger 13461*. NW side, *Felger 15325*.
Cholludo: Near center of island near N shore, rare, *Felger 9160*.

Bursera laxiflora S. Watson

Xoop caacöl "large elephant tree"; *torote prieto*; fern-leaf bursera

Description: Large shrubs or small trees; all parts of the plants aromatic when crushed. Bark reddish brown, not peeling. Twigs slender. Leaves fern-like, produced at almost any season and drought deciduous, highly variable, pinnate or bipinnate, primary leaflets 5–15, secondary leaflets (when present) 3–7 per pinnae; rachis and rachillas winged from excurrent leaflets. Inflorescences few-flowered, emerging with the new growth, the branches and peduncles slender even in fruit, sometimes

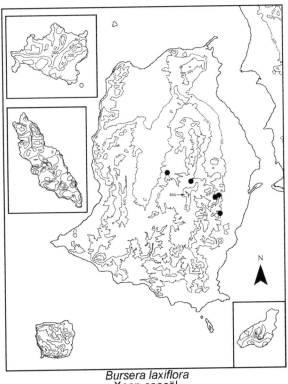

Bursera laxiflora
Xoop caacöl

with small leafy bracts at the base of the pedicels. Flowers white to cream yellow, the petals 4 mm long. Fruits hard, globose, 8 mm wide, shiny green and becoming reddish when fully ripe. Seeds blackish, the upper one-half to two-thirds covered by a fleshy orange aril. Flowering during the summer rainy season.

Local Distribution: Foothills and interior of Sierra Kunkaak, from the upper bajadas to high elevations.

Other Gulf Islands: None.

Geographic Range: Most of the southern two-thirds of Sonora, Sinaloa, southwestern Chihuahua, and Baja California Sur.

Tiburón: Haap Hill, locally common, *Felger T74-8*. Hast Coopol, common, *Wilder 07-373*. Foothills, E side of Sierra Kunkaak, "bench" just above canyon bottom and lower slopes, *Felger 6967*. 3 km E of Siimen Hax, *Wilder 06-499*. Sierra Kunkaak Mayor, carrizo (Sopc Hax), *Romero-Morales 06-3*.

Bursera microphylla A. Gray

Xoop; *torote blanco, torote colorado, torote prieto, copal*; elephant tree

Description: Large shrubs or small trees, often 2–3 (4+) m tall (generally larger on Tiburón than on San Esteban). Limbs and trunk fat and semi-succulent, the wood soft

and pithy. Bark becoming papery in late spring dry season, peeling off in puzzle-like pieces or papery sheets. Sap especially rich in terpines and the leaves highly aromatic when crushed. When a leaf is broken off a turgid stem, the stem often squirts a tiny jet of clear, aromatic sap. When cut deeply, the branches or roots ooze blood-red sap. Leaves bright green, once pinnate, 1.5–8 cm long, produced at almost any time of year, the leafstalk often narrowly winged; leaflets 7–31 in number, linear to narrowly oblong, or rarely lanceolate. Fruits dull purplish brown, 7–9 mm long. Seeds roughly triangular in cross section, enveloped in a thin red aril. Xoop is a major medicinal and ritual plant in the Seri culture (Felger & Moser 1985).

Local Distribution: Widespread and abundant on Tiburón, San Esteban, and Dátil, being the most prolific *Bursera* of these islands, ranging from near sea level to the peaks. Two individuals were found on Alcatraz in 1965 and 1966, and none were found in 2006. On ascending the north slopes of Sierra Kunkaak, *B. microphylla* is common to about 400 m, above which it is replaced by *B. fagaroides*, and *B. microphylla* is again common above 750 m and the only *Bursera* to the summit. It is common on Nolasco at higher elevations on the east side, and scattered elsewhere.

Other Gulf Islands: Ángel de la Guarda, San Lorenzo (observation), Tortuga (observation), San Marcos (observation), Coronados, Carmen, Danzante, Monserrat,

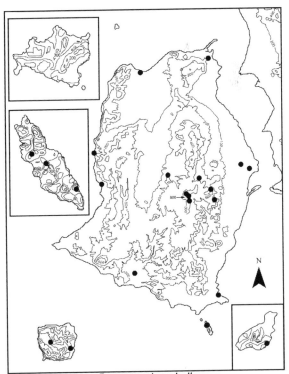

Bursera microphylla
Xoop

Santa Catalina, Santa Cruz, San Diego (observation), San José, San Francisco, Espíritu Santo, Cerralvo.

Geographic Range: Southern limits of the Guaymas region northward through western Sonora to southwestern Arizona, Gulf Islands, and most of the Baja California Peninsula to southeastern California. There seem to be several undescribed geographic subspecies or ecotypes, distinguished in part by growth form including trunk height and terpene content (e.g., Mooney & Emboden 1968).

Tiburón: Ensenada Blanca (Vaporeta), 29 Jan 1965, *Felger*, observation (Felger 1966:179). Sauzal, 31 Jan 1965, *Felger*, observation (Felger 1966:215). Ensenada de los Perros, lado E de la Isla Tiburón, *Tenorio 9511* (MEXU). Haap Hill, *Felger T74-32*. Palo Fierro, *Felger 11085*. 3 km W [of] Punta Tormenta, *Scott P7* (UNM 53355). Arroyo downstream of Sopc Hax, *Wilder 07-227*. Hast Cacöla, E flank of Sierra Kunkaak, 380 m, *Wilder 08-253* (USON). Foothills of Sierra Kunkaak, *Wilder 05-18*. San Miguel Peak, 1000 ft, N slope, *Knight 921* (UNM). Cójocam Iti Yaii, *Wilder 07-544*. Summit of island, not common, *Wilder 07-573*. Punta Perla, 19 Dec 1973, *Felger*, observation. Tecomate: *Gold 371* (MEXU); *Whiting 9035*. Pooj Iime, 29 Sep 2007, *Felger*, observation.

San Esteban: Limantour, *Felger 16608*. Near center of island, *Felger 469*.

Dátil: NE side, *Felger 9088*. NW end, *Felger 2772*.

Alcatraz: One seen, above rocky beach, near SE base of the mountain, a dome-shaped shrub ca. 2×2½ m across, ca. 1.8 m

tall, 2 Dec 1965, *Felger 13402*. Only one individual found on island, near beach at S side of flat terrain of E side of the island, 8 Oct 1966, *Felger 14920*.

Nolasco: SE side, *Sherbrooke* [*Felger 11240*]. NE side, common near the ridge crest, *Felger 06-110*. Cañón de Mellink, 110 m, canyon bottom, *Felger 08-144*.

Cactaceae – Cactus Family

There are 20 species of cacti in 9 genera in the flora area, ranging from plants about 10 centimeters tall to giant, arborescent columnar cacti. The cactus flora of the state of Sonora includes about 100 species (Paredes et al. 2000; Felger 2000a), while the Sonoran Desert as a whole includes about 140 species, or about 6 percent of the total native flora. Cacti represent about 5 percent of the flora of the Sonoran Islands.

The columnar cacti in the region produce edible fruits and were significant food and wine resources for the Comcaac in traditional times and continue to be harvested for their delicious fruits. In addition, certain cacti have been exploited for medicinal purposes (e.g., Felger & Moser 1985; Paredes et al. 2000; Felger 2000a).

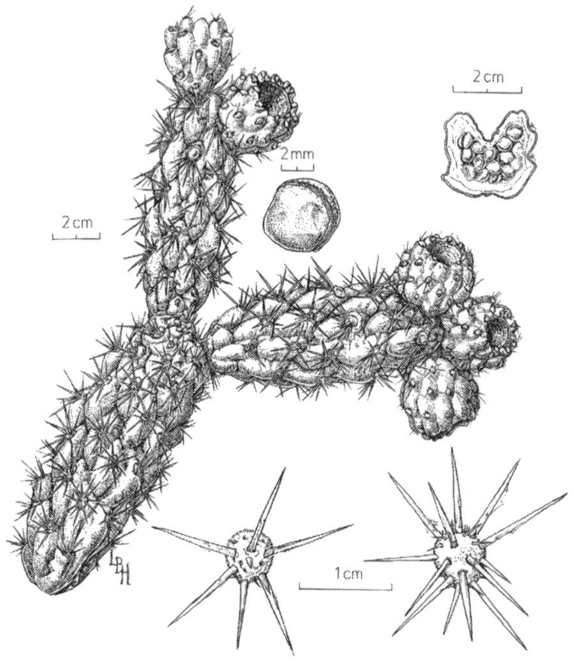

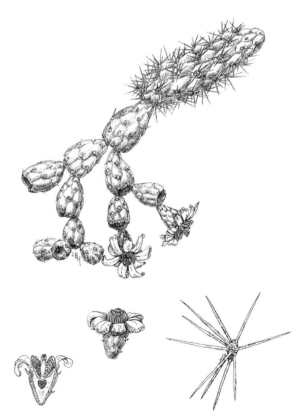

Figure 3.45. *Cylindropuntia alcahes* var. *alcahes.* LBH.

Figure 3.46. *Cylindropuntia fulgida* var. *fulgida.* LBH.

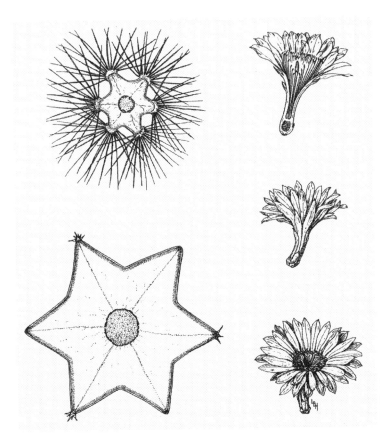

Figure 3.47. *Lophocereus schottii* var. *schottii*. Cross section of young stem, lower left, cross section of reproductive stem, upper left, and flowers, right. LBH.

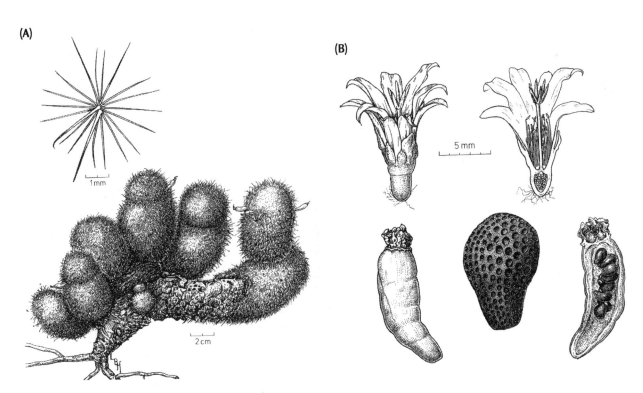

Figure 3.48. *Mammillaria estebanensis.* (A and B). LBH.

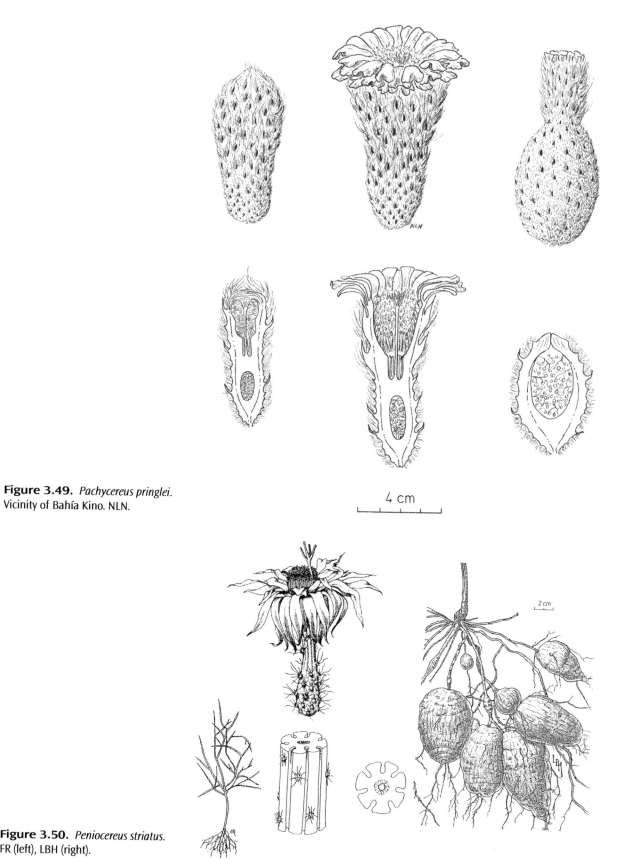

Figure 3.49. *Pachycereus pringlei.*
Vicinity of Bahía Kino. NLN.

4 cm

Figure 3.50. *Peniocereus striatus.*
FR (left), LBH (right).

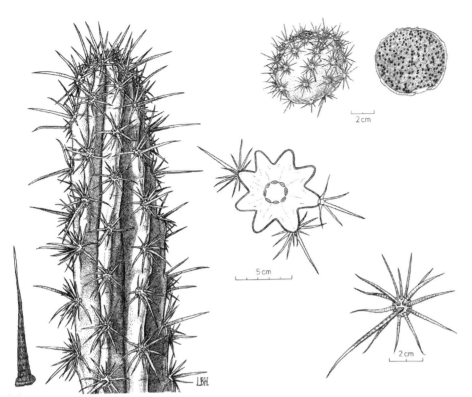

Figure 3.51. *Stenocereus gummosus.* LBH

Figure 3.52. *Echinocereus grandis.* Isla San Esteban. BTW, 8 March 2007.

Figure 3.53. *Echinocereus scopulorum.* Sierra Kunkaak, Isla Tiburón. BTW, 25 November 2006.

Figure 3.54. *Echinocereus websterianus.* Vicinity of Cañón el Farito, Isla San Pedro Nolasco. May 1952, GEL (courtesy San Diego Natural History Museum).

Figure 3.55. *Ferocactus emoryi.* Isla Alcatraz. BTW, 4 December 2007.

Figure 3.56. *Ferocactus tiburonensis.* Isla Tiburón, foothills of Sierra Kunkaak, half-grown plant. HRM, January 2009.

Figure 3.57. *Lophocereus schottii*
var. *schottii*. Eastern bajada, Isla Tiburón.
A mature plant with Humberto
Romero-Morales. BTW, 24 May 2006.

Figure 3.58. *Mammillaria
estebanensis*. Arroyo Limantour, Isla
San Esteban. BTW, 3 September 2008.

Figure 3.59. *Mammillaria multidigitata.* Cañón El Farito, Isla San Pedro Nolasco. BTW, 2 February 2008.

Figure 3.60. *Mammillaria tayloriorum.* Isla San Pedro Nolasco. BTW, 3 February 2008.

Figure 3.61. *Mammillaria* sp. Isla Cholludo. BTW, 2 September 2008.

Figure 3.62. *Opuntia engelmannii* var. *engelmannii*. Isla Alcatraz. BTW, 16 September 2007.

Figure 3.63. *Opuntia bravoana* with *Vaseyanthus insularis*. Isla San Pedro Nolasco. BTW, 2 February 2008.

Figure 3.64. *Pachycereus pringlei.* A venerable cardón with new and decaying osprey nests, Valle de Águila, Isla Tiburón. BTW, 8 November 2008.

(A)

(B)

Figure 3.65. *Stenocereus gummosus,*
Isla Tiburón. (A) Eastern bajada, 25 July
2007. (B) Fruit from the vicinity of Zozni
Cmiipla, 2 September 2006. BTW.

(A)

(B)

Figure 3.66. *Stenocereus thurberi.* (A) With Exequiel Ezcurra, Ensenada del Perro, Isla Tiburón, 12 April 2007. (B) Near ripe fruit, vicinity of Punta Chueca, Sonora, 24 July 2007. BTW.

Figure 3.67. Imám, the Seri term for ripe fruits of columnar cacti. White and pink forms of *Pachycereus pringlei* on left, *Carnegiea gigantea* on upper right, and *Stenocereus thurberi* on bottom right and center. Seri basket made from *Jatropha cuneata* with *Krameria bicolor* root-dye. JRT, July 2008.

1. Small leaves present on new growth, the spine clusters bearing glochids (small spines deciduous at a touch) in addition to the larger persistent spines.
    2. Chollas; stem segments ("joints" or cladodes) more or less rounded in cross section (cylindroid), often tuberculate; spines mostly with papery sheaths. _____ **Cylindropuntia**
    2' Prickly pears; stem segments flattened or compressed ("pads" or cladodes); surfaces relatively flat, not tuberculate; spines not sheathed. _____ **Opuntia**
1' Leaves and glochids not present.
    3. Columnar cacti, the stems more than 5 cm diameter, usually more than 1.5 m tall, the branches many or if unbranched then reaching more than 2 m tall.
        4. Stems with 6–8 ribs; spines of the adult (upper or fertile) portion of the stems twisted and much longer than the spines of the juvenile (lower or sterile) portion of the stems. _____ **Lophocereus**
        4' Stems with more than 10 ribs; spines of the fertile (upper) portion of the stems absent or not twisted and shorter than or equal to the spines of the sterile (lower) portion of the stems.
            5. Stems and spines not noticeably dimorphic, the spines similar on juvenile (lower or sterile portion) and adult (upper or fertile portion) stems; areoles of adult portion not different from those of the juvenile portion. _____ **Stenocereus**
            5' Stems dimorphic, the juvenile and adult stems markedly different (e.g., in stem diameter, distance between areoles, spine lengths and morphology, and rib numbers); areoles of adult growth coalesced or close together and spineless or with smaller, bristly spines.
                6. Lower branches mostly arising at or above mid-height of the plant; stem surfaces green, not glaucous; areoles of adult stems close but not coalesced, their spines bristly; flowers and fruits glabrous, spineless or with very few spines. _____ **Carnegiea**
                6' Lower branches mostly arising much below mid-height of the plants; stem surfaces pale glaucous blue-green; areoles of adult stems coalesced, lacking spines or with short, stout spines; flower bases and fruits spiny and/or covered with dense felt-like hair. _____ **Pachycereus**
    3' Not columnar cacti, stems mostly less than 1.5 m tall.
        7. Barrel cacti, the stem usually solitary and more than 15 cm thick, the spines mostly rigid. _____**Ferocactus**
        7' Not barrel cacti, the stems solitary to branched or clustered, less than 15 cm thick, the spines variable.
            8. Stems less than 1 cm diameter and more than 20 times longer than wide, the roots tuberous; flowers open at night. _____ **Peniocereus**
            8' Stems more than 2 cm diameter and usually less than 10 times as long as wide, the roots not tuberous; flowers open in daytime.
                9. Areoles arranged on ribs; spines all straight, the flower bases and fruits spiny. _____**Echinocereus**
                9' Areoles on separate tubercles; spines straight or hooked, the flowers and fruits spineless. **Mammillaria**

Carnegiea gigantea (Engelmann) Britton & Rose

Mojepe; *saguaro, sahuaro*; saguaro

Description: Giant columnar cacti, unbranched or with several or sometimes more, mostly erect branches. Saguaros tend to initiate branches higher off the ground than cardons (*Pachycereus pringlei*) and the mature saguaro stems are green rather than glaucous. Young plants are difficult to distinguish.

Flowers large and white, nocturnal, and often remaining open until about midday, depending on temperature. Flowering mostly late April and May, the first buds and flowers mostly emerging on east side of stems. Fruit rind (pericarpel) bright red inside, inedible, splitting open at maturity into 3 or 4 thick, spreading to recurved lobes. Fruit pulp juicy, bright red, sweet, and edible. Seeds edible (digestible when the seed coat is broken, e.g., ground on a grinding stone or *metate*). Fruits mostly ripe June and early July.

Local Distribution: Widespread on Tiburón, where the densest concentrations occur at lower elevations of the Sierra Kunkaak and the east side of the Central Valley on the bajada coming off the Sierra Kunkaak, and the Tecomate region. A few saguaros about 1 m tall were seen at the margin of the cactus forest on Cholludo in the 1960s, but none were seen by Ben or Humberto on their trips to the island in 2008 and 2009 despite searching for them.

For Alcatraz, Felger (1966:139) reports that "of the two columnar species of arborescent columnar cacti on the island, *Carnegiea* is poorly established. . . . Fruiting individuals of *Carnegiea* are present but most of the

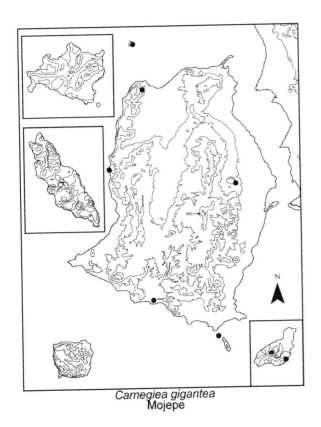

Carnegiea gigantea
Mojepe

unbranched plant about 3 to 4 m high and grew below the low rock ledge at the southeast shore. . . . The species no longer exists on the island."

Other Gulf Islands: None.

Geographic Range: Arizona to southwestern Sonora, and barely entering California near the Colorado River.

Tiburón: Arroyo Sauzal, 31 Jan 1965, *Felger*, observation (Felger 1966:221). 4 mi W Palo Fierro, granitic rocky hill, *Felger 10025*. Pooj Iime, 29 Sep 2007, *Felger & Wilder*, observation. Hill W of Tecomate, 12 Feb 1965, *Felger*, observation.

Alcatraz: E-central base of mountain, rare, 2 seen, ca. 7 m tall within a stand of five *P. pringlei*, *Wilder 07-423* (photo). SE side of mountain, 2 Dec 1965, *Felger 13409* (specimen not located).

Patos: SE shore, *Osario Tafal 1946* (photo, Felger 1966:353).

Cholludo: Several juvenile saguaros at the top of the island, along the margin of the cardón forest, "2 saguaros ca. 2 m (19 ribs) and ca. 0.8 m (17 ribs)" (Felger, 3 Dec 1965, field notes). "Edge of island, N-NW side, saguaro ca. 1 m (16 ribs), healthy" (Felger 1966:309).

population consists of immature individuals." There is no record of saguaro on Dátil.

One or more mature saguaros on Patos were cut down in 1945 when the vegetation was cleared by a Mexican guano mining company (Bowen 2000; Felger 1966; Gentry 1949). Ivan Johnston (1924:1108) discovered saguaro on Patos in 1921: "The single plant on Patos Island is over 12 m. high and has a single large branch." Felger (1966:358) reported that "another saguaro is evident in one of Osario Tafal's 1946 photographs. It was an

Cylindropuntia – *Choya*; cholla

Stems more or less cylindrical with determinate, rhythmic (seasonal or annual), constricted growth increments resulting in individual stem segments or joints (cladodes). Areoles with glochids (small, readily detachable spines with tiny retrorse barbs). Leaves present on young growing stem segments, 1 per areole, succulent, conical and soon deciduous. Stamens sensitive, the filaments close inward when touched, covering a visiting bee. "Seeds" relatively large, with a light tan, bony, aril-like structure covering the actual seed.

1. Flowers pink; fruits green and fleshy, present all year (perennial), sometimes solitary but mostly proliferating in chains of 3 or more.
    2. Fruits usually held upright, single or proliferating in chains of 2–5. _____ **C. cholla**
    2' Fruits usually proliferating in hanging chains, often with more than 5 in a chain. _____ **C. fulgida**
1' Flowers various colors but not pink; fruits single, not proliferating in chains (occasionally 2 or 3 fruits together), sometimes not present all year and sometimes not green.
    3. Stems very slender, mostly 4.5–6.5 mm diameter; ripe fruits bright red-orange. _____**C. leptocaulis**
    3' Stems more than 1 cm diameter (excluding spines); ripe fruits yellow or green, not red-orange.
        4. Stem segments often purplish in winter and dry seasons, the stem surfaces not obscured by spines; fruits green, fleshy, and persistent. _____ **C. versicolor**
        4' Stems not purplish; stem surfaces generally obscured by spines; fruits at first fleshy but usually not persistent, green or yellow.
            5. Trunk (main stem) seldom straight, without dead, persistent joints. _____ **C. alcahes**
            5' Trunk usually straight and erect with persistent, dead joints with dark-colored (blackish) spines. **C. bigelovii**

Cylindropuntia alcahes (F.A.C. Weber) F.M. Knuth var. **alcahes** [*Opuntia alcahes* F.A.C. Weber var. *alcahes*]

Heem icös cmasl "yellow-spined pencil cholla (*C. arbuscula*)"; *choya*; Baja California cholla

Description: Staghorn-like chollas, often 0.7–1.5 m tall. The terminal cladodes (joints) sometimes fall away and propagate clonally, although they are not nearly as fragile as those of *C. bigelovii* or *C. cholla*. Flowers yellowish or greenish and with a full complement of tepals. Fruits fleshy and yellow, persisting up to several months. The seeds are presumably fertile.

Local Distribution: On San Esteban, abundant at lower elevations, especially arroyo benches, and extending to ridge crests of island peaks. The large San Esteban chuckwallas (*Sauromalus varius*) are often seen along Arroyo Limantour under this cholla eating the flower buds, flowers, fruits, and joints, with cholla mucilage and spines around their mouths and faces. On Mártir, infrequent to locally common at higher elevations, where it grows intermixed with *C. cholla*, larger ones 0.7–1.1 m tall, often draped with *Vaseyanthus* vines.

Other Gulf Islands: Var. *alcahes*: Ángel de la Guarda, Partida Norte, Rasa, Salsipuedes, San Lorenzo, Tortuga, San Ildefonso, Carmen, Danzante, Monserrat, Santa Catalina, Santa Cruz, San Diego, San José, San Francisco, Espíritu Santo, Cerralvo (observation).

Geographic Range: This species occurs through most of the Baja California Peninsula, adjacent Gulf Islands, and on San Esteban and Mártir. There are four intergrading varieties, of which var. *alcahes* is the most widespread, extending across most of the range of the species.

San Esteban: *Lindsay 3004* (SD). Arroyo Limantour: *Bostic 21 Jun 1965* (SD); *Felger 2614, 12732*; *Van Devender Apr 1992*. **Mártir:** 18 Apr 1921, *Johnston*, observation (Johnston 1924:1115). Erect to 1 or rarely 2 m tall, stems to 6 cm diameter, petals green, erect, *Moran 4056* (CAS). Top of island, 300 m, occasional, 8 dm tall, stem to 5 cm diameter at base, *Moran 8815* (SD). Small canyon at base of SE-facing slope that leads to island summit, scattered, not common, nearly all draped with *Vaseyanthus* vines, *Felger 07-18* (MEXU, USON). *Wiggins 21 Mar 1962* (ASU).

Cylindropuntia bigelovii (Engelmann) F.M. Knuth [*Opuntia bigelovii* Engelmann]

Coote, sea; *choya, choya güera*; teddybear cholla

Description: Trunks stout and straight, beset with persistent, dead joints bearing blackened spines. Trunk and cladodes (joints) very densely spiny, the spines dull

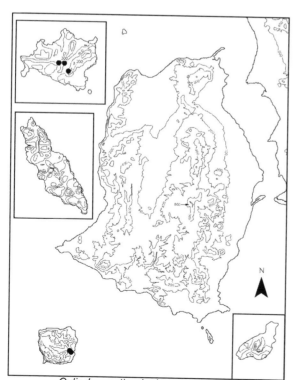

Cylindropuntia alcahes var. alcahes
Heem icös cmasl

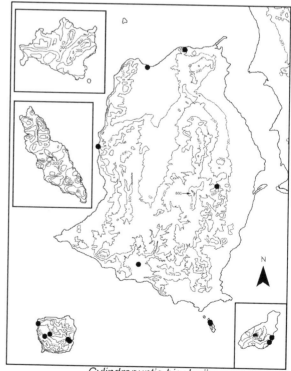

Cylindropuntia bigelovii
Coote, sea

yellow. Inner tepals greenish yellow. Fruits yellow, somewhat fleshy, usually soon drying and falling.

This clonal species is highly successful in terms of abundance and geographic distribution. The upper or younger joints fall at a touch, and the spines are difficult and painful to pull out of flesh due to the heavily barbed spines. The plants propagate prolifically from fallen joints. This species is generally triploid, consequently there is very little chance of the seeds being viable. Pollen production is highly variable—at least the triploid plants produce relatively few pollen grains. In the Tucson region pollen production may be higher in seasons of high rainfall and lower during drier periods (Mary Kay O'Rourke, personal communication). This species and *C. fulgida* have fewer inner tepals than most other chollas, a feature correlated with reliance on asexual reproduction.

Local Distribution: Widespread and common across Tiburón, especially at lower elevations, and scattered on San Esteban, Dátil, and Alcatraz. This cholla is notably dense at the northeast part of Tiburón at a place called Coote An.

Other Gulf Islands: Ángel de la Guarda, Salsipuedes, San Lorenzo (observation), Tortuga, Santa Cruz.

Geographic Range: Common and widespread in western Sonora, from south of Tastiota and Bahía Kino and northward to southern and western Arizona, southeastern California, southern Nevada, eastern Baja California Norte, and northeastern Baja California Sur.

Tiburón: Sauzal, 3 km inland, 31 Jan 1965, *Felger,* observation (Felger 1966:216). Hast Cacöla, E flank of Sierra Kunkaak, foothills at 200 m, *Wilder 08-232.* Hee Inoohcö, mountains at NE coast, 30 Sep 2007, *Felger,* observation. Tecomate, *Whiting 9064.* Pooj Iime, 29 Sep 2007, *Felger & Wilder,* observation.

San Esteban: [Arroyo Limantour], canyon bottom floodplain, locally common, *Felger 7047.* S end of island [Arroyo Limantour], *Rempel 292* (RSA). SW part of island, canyon at the bottom of talus slope, *Wilder 07-83B* (photo).

Dátil: *Dawson Jan 1940* (UC). NE side, canyon bottom, 11 Apr 2007, *Felger,* observation.

Alcatraz: SE base of mountain, *Felger 12827.* E-central base of mountain, ca. 20 m, among thickest stand of *Pachycereus pringlei* on island, with *Carnegiea gigantea, Cylindropuntia fulgida* var. *fulgida,* 4 individuals seen, ca. 0.5 m tall, *Wilder 07-422* (USON). S-central coastal side of island, short "bajada," above SE part of *Allenrolfea* flat, with *C. fulgida* var. *fulgida* and var. *mamillata,* 4 Dec 2007, *Felger & Wilder,* observation.

Cylindropuntia cholla (F.A.C. Weber) F.M. Knuth
[*Opuntia cholla* F.A.C. Weber]

Choya

Description: Chollas to about 1.5 m tall. Cladodes (joints) and fruits green, highly succulent, and readily detaching and potentially forming new plants and clonal colonies; spination variable, the plants rather sparsely to densely spiny, with reduced spine sheaths to broad, yellowish papery sheaths. Flowers pink, with relatively few tepals. Fruits globose, persistent, and solitary or proliferating into short chains of 2–5 fruits.

Local Distribution: This cholla is documented from Mártir since Palmer's collection in 1887. It is infrequent to locally common at higher elevations on the island, where it often grows intermixed with *C. alcahes.* Palmer reported it as "choyer" and common on the island (Watson 1889:52).

Other Gulf Islands: Ángel de la Guarda, Partida Norte, San Lorenzo (observation), Tortuga, San Marcos (observation), San Ildefonso (observation), Coronados, Carmen, Danzante, Monserrat (observation), Santa Catalina, Santa Cruz, San Diego, San José, San Francisco, Espíritu Santo, Cerralvo.

Geographic Range: Widespread through the Baja California Peninsula and a number of Pacific and Gulf Islands. It is closely related to *C. fulgida* of the mainland

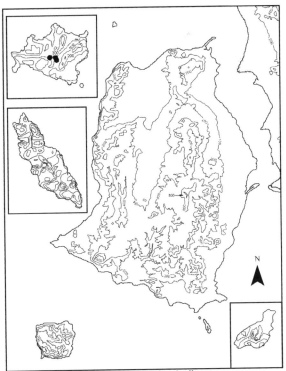

Cylindropuntia cholla

and other Sonoran Islands, the distinctions being subtle (see discussion below and Rebman 1995).

Mártir: *Palmer 419* in 1887 (specimen not seen, cited by Watson 1889). Top of island, occasional, broad yellowish papery sheaths, very spiny, *Moran 8814* (CAS, SD). Lower (first) "plateau," 500 ft, not common, 4–5 ft high, *Felger 6357*. Summit, 300 m, localized population (photo shows fruits in chains of 3–5), *Wilder 07-607* (USON).

Cylindropuntia fulgida (Engelmann) F.M. Kunth [*Opuntia fulgida* Engelmann]

Sea cotopl "clinging teddybear cholla," sea icös co-oxp "white-spined teddybear cholla"; *choya*; chain-fruit cholla

Description: Chollas 1.5+ m tall, forming clonal colonies from fallen joints and fruits that readily form roots. Trunks seldom straight, often with several major branches from about mid-height. Stems and fruits green all year. Fruits persisting for several years and forming pendulous chains with up to about a dozen fruits, but often fewer, especially near the shore. Flowers 4–4.5 cm wide, the inner tepals, filaments, and style deep pink-purple; stigma and anthers white. Ants, feeding on the nectaries, rush to attack an intruder such as a finger or pruning shears.

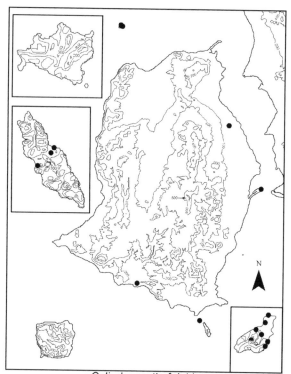

Cylindropuntia fulgida
Sea cotopl, sea icös cooxp

As with most chollas, the spines are armed with tiny reverse barbs, which inflict a painful wound when you try to pull them out. When an unwary animal, including a person, touches the plant it comes away with a joint stuck to it—an obvious adaptation for dispersal. This species is triploid and the seeds are likely not viable (Felger & Zimmerman 2000; Rebman 1995).

Local Distribution: Widespread across the lowlands on the south and east side of Tiburón, widely scattered through the lowland flats of Alcatraz, and locally common on Nolasco. The chollas on Patos seem to represent a single, variable population: "Colonies of mature individuals laden with pendulant chains of fruits occur at the rocky ledges near the southeast shore and also on the flats near the base of the hill. Before clearing of the island in 1946 it occurred at least in the former site and was common" (Felger 1966:359). In 2007 we found aggregations of this cholla in the same places, especially near the southern margin of the island. We also saw a rather large stand of dead chollas near the base of the hill at its south side. It is also common on Cholludo, hence the name of the island.

Other Gulf Islands: Rasa, Salsipuedes, San Lorenzo (observation).

Geographic Range: Southern Arizona through much of Sonora, especially the western part of the state, to northwestern Sinaloa.

Notes: Two varieties are often recognized. Var. *fulgida* has densely spiny cladodes, and the spines are enclosed by conspicuous and rather loose-fitting yellowish, papery sheaths. Var. *mamillata* (Schott ex Engelmann) Backeberg [*Opuntia fulgida* var. *mamillata* (Schott ex Engelmann) J.M. Coulter] is distinguished from var. *fulgida* by fewer and shorter spines (the stem surface is not obscured by the spines) and tight-fitting sheaths so the spine sheaths are not papery. The varieties are generally segregated on the mainland, but on the islands there is often the full range of variation locally and sometimes even on the same plant. The plants on Tiburón are generally spiny like var. *fulgida*, whereas the full variation may be seen on Nolasco, Alcatraz, Patos, and Cholludo.

Cylindropuntia fulgida is closely related to *C. cholla* and the differences are subtle. Their similarity has led to specimens from the Sonoran Islands (except Mártir) confusingly being identified as either or both species. Some plants or populations from the Sonoran Islands, however, might be intermediate between the two (Rebman 1995).

Tiburón: Arroyo Sauzal, interhill bench adjacent to waterhole, *Felger 10118*. S end of island, *Rempel 296* (RSA). Punta

San Miguel, *Wilder 2 May 2007* (photo). Valle de Águila, *Wilder 3 May 2007* (photo).

Alcatraz: Plants resembling var. *fulgida*: SE base of mountain, *Felger 12826*. E-base of mountain, *Wilder 07-428* (USON). S-central coastal side of island, short "bajada" above SW part of *Allenrolfea* flat, *C. bigelovii*, *C. fulgida* var. *fulgida*, and var. *mamillata* growing together, 28.81054°N, 111.96561°W, 4 Dec 2007, *Felger & Wilder*, observation. E-central side of island, base of slope, 10 m, just above *Allenrolfea* flat, much bird guano, vegetation sparse, many *C. fulgida* are dead, 28.81210°N, 111.96706°W, 4 Dec 2007, *Felger & Wilder*, observation. N-central base of mountain, upper flat, plants mostly dead, 28.81319°N, 111.96836°W, 4 Dec 2007, *Felger & Wilder*, observation.

Plants resembling var. *mamillata*: N-central portion of flat, ca. 2 m elev., rocky bench just inland from beach, on inland portion of bench, rocky with sandy soil, ca. 1 m tall, restricted to edge of flat, 28.814790°N, 111.966210°W, *Wilder 07-412*. N-most end of island, sand soil, 28.81647°N, 111.96549°W, 4 Dec 2007, *Felger & Wilder*, observation.

Patos: Near SE shore, *Osario Tafal 1946* (photo, Felger 1966:353). S shore, 4 m elevation, locally common on low sea ledges, ca. 2 m tall, *Felger 21319*. S side of island, near the shore, flat dominated by *Atriplex barclayana*, and isolated patches of chollas, locally common, 1.5 m tall (photo shows fruits in chains of 5 and 8), 29.26930°N, 112.45805°W, *Felger 07-61B* (USON).

Cholludo: N face of island, abundant, *Felger 13422*. Seal Island, near Tiburón, *Rose 16813* (NY, US, cited as var. *mamillata* by Rebman 1995:163).

Nolasco: Cañón de la Guacamaya, vicinity of cave near shore, pink flowers, 3 May 2005, *Gallo-Reynoso* (photo, resembling var. *fulgida*). Cañón el Farito, 18 Jan 1965, *Felger 12074* (appearing intermediate between var. *fulgida* and var. *mamillata*). Cañón el Farito, fruits up to 7 per chain, 2 & 3 Feb 2008, *Wilder*, photos. E side, common, *Felger 9665* (resembling var. *mamillata*).

Cylindropuntia leptocaulis (de Candolle) F.M. Knuth [*Opuntia leptocaulis* de Candolle]

Iipxö; *siviri*; desert Christmas cholla

Description: Slender-stem chollas to 1+m tall. Stems 4.5–6.5 mm diameter, green all year, and with large, yellowish spines or nearly spineless. Young plants produce a moderately thickened tuberous root. Flowers pale yellowish to cream, 1.5–2 cm wide, opening in late afternoon. Fruits fleshy, bright red or red-orange.

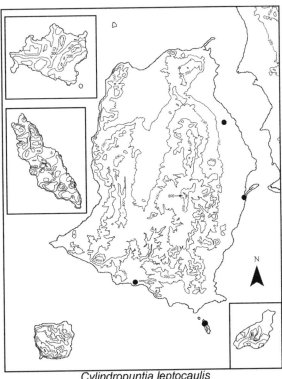

Cylindropuntia leptocaulis
Iipxö

Many plants on Dátil had proliferating chains of 2 or 3 fruits and the fruits appear to be unusually large for this species, features of *O. leptocaulis* var. *brittonii* J. González Ortega, which is common in coastal thornscrub in southern Sonora (Paredes et al. 2000).

Local Distribution: On Tiburón, documented by only a few records, near the shore, at the southern and eastern margins of the island, where it generally occurs with *Peniocereus striatus*. It seems strange that it has not been found elsewhere on the island. On Dátil, common along canyons at the northeast side of the island and rare on the northwest side.

Other Gulf Islands: None.

Geographic Range: Oklahoma and Texas to Arizona and southward to Puebla, and lowlands of Sonora and Sinaloa. This is the most wide-ranging cholla species.

Tiburón: *Dawson 25 Jan 1940* (RSA, UC; probably at the S shore of the island). 2 km S of Zozni Cmiipla and 75 m inland, rare, gowing with *Peniocereus striatus*, both have been eaten by rabbits, *Wilder 06-162* (USON). Valle de Águila, 4 May 2007, *Wilder*, observation.

Dátil: NW part, rare, steep rocky slope, *Felger 2777*. NE side: arroyo bottom, not common, *Felger 9118*; canyon, common, to 80 cm tall, fruits fleshy, red, many in proliferating chains of 2 or 3 fruits, 12 Apr 2007, *Felger*, observation. NE side, canyon bottom, *Wilder 07-142*.

Cylindropuntia versicolor (Engelmann ex J.M. Coulter) F.M. Knuth [*Opuntia versicolor* Engelmann ex J.M. Coulter. *O. thurberi* subsp. *versicolor* (Engelmann ex J.M. Coulter) Felger & Lowe]

Heem icös cmáxlilca "stiff-spined pencil cholla," hepem ihéem "white-tailed deer's pencil cholla"; *siviri*; staghorn cholla

Description: Chollas 1–1.5 (3) m tall with an upright trunk and main stems, and spreading branches, the stems purplish brown during drier, cooler months. Inner tepals greenish yellow with red-brown tips. Flowering March and early April. Fruits fleshy, green or yellow-green, usually persistent until the following year, often becoming enlarged and swollen, usually solitary or rarely in chains of 2 or 3 fruits.

Local Distribution: Widely scattered across Tiburón including higher elevations nearly to the summit. Also recorded from Dátil.

Other Gulf Islands: None.

Geographic Range: South-central Arizona to the Guaymas region in Sonora.

Tiburón: Punta Willard, *Harbison 19 Mar 1962* (2 sheets, SD *51614 & 84157*). Arroyo Sauzal, 28.83136°N, 112.42168°W, ca. 35 m, few scattered shrubs, this one ca. 3 m tall, 29 Jan 2008, *Felger & Wilder*, observation. E-facing slope of Sierra Kunkaak

above canyon and just below highest ridge on island, common, 1–1.3 m tall, *Wilder 07-530* (USON). Hee Inoohcö, mountains at NE coast, 30 Sep 2007, *Felger*, observation. Agua Dulce Bay, rather few, on volcanic N slope, shrub 2 m high × 5 m, jointed throughout; trunk 2 dm diameter with gray-brown spineless bark; the main branches sprawling; perianth greenish yellow, filaments green, fruits solitary, turgid, 23 Apr 1966, *Moran 12993* (SD). Tecomate, *Whiting 9016*. Pooj Iime, rare, 29 Sep 2007, *Felger & Wilder*, observation.

Dátil: NE side, NE-facing slope near crest, rare, steep and rocky slope, *Felger 13471*.

Echinocereus – Hedgehog cactus

Solitary or branched small to medium-sized cacti with straight spines. Stigmas green. Ovary spiny, the spines falling away as the fruit matures. Fruits fleshy, edible, and not persistent, falling soon after maturing. The three species on the Sonoran Islands are distinctive and part of the *E. rigidissimus* group (Taylor 1985), which consists of six species in the Sonoran Desert region and southeastern Arizona and southwestern New Mexico. They are densely covered with relatively short and stout but not especially "harmful" spines from a human standpoint.

The two distinctive island endemics, *E. grandis* and *E. websterianus*, share several gross morphological features. Both have unusually large, thick, and multiple stems arising from the base and lower third of the plant, and flowers reduced in size and brightness of color as compared with those of *E. scopulorum*, the presumed closest relative. The flowers of the two island endemics are also relatively smaller and not as showy compared to most other members of the genus in the Sonoran Desert. One might think that these island endemics have less showy flowers because of the relative lack of competition for pollinators among the few other local species of cacti (also see *Mammillaria estebanensis*). The flowers of *E. grandis* on Islas San Lorenzo and Las Ánimas ("San Lorenzo Norte"), however, are bright pink.

1. Stems mostly solitary or occasionally with 1–few branches; spines 8–17 mm long; flowers large and showy, 7.5–8 cm long, the perianth deep rose-pink; Tiburón. _____ **E. scopulorum**
1' Stems often branched; spines 3–15 mm long; flowers ca. 5–7 cm long, the perianth usually cream color or dull pink.
  2. Spines often whitish; flowers usually cream color; San Esteban. _____ **E. grandis**
  2' Spines often yellowish; flowers usually pale pink; Nolasco. _____ **E. websterianus**

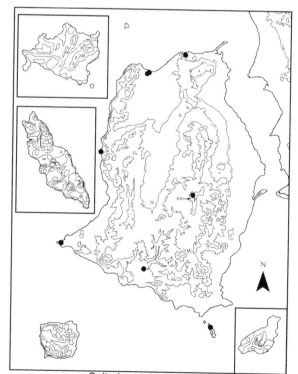

Cylindropuntia versicolor
Heem icös cmáxlilca

Echinocereus grandis Britton & Rose, Cactaceae 3:18, fig. 18, pl. 3.3, fig. 3, 1922

Hant ipzx íteja caacöl "large arroyo's bladder (*Mammillaria*)"; giant rainbow cactus

Description: Plants reaching 50 cm tall, the stems very thick, sometimes solitary, but mostly with a few branches. Spines dull whitish, the central spines 3–6 mm long, the radials 5–12 mm long. Flowers whitish (rarely pale pink), 5–7 cm long and about as wide, the outer tepals pink-tinged. Style and stigma branches green.

Local Distribution: Common and widespread through most of San Esteban at all elevations.

Other Gulf Islands: Las Ánimas, San Lorenzo.

Geographic Range: This species is endemic to Islas Las Ánimas ("San Lorenzo Norte"), San Lorenzo, and San Esteban. The flower color of the Las Ánimas/San Lorenzo plants is bright pink as opposed to the whitish flowers of the San Esteban plants, which indicates at least some distinction between the island populations (Wilder et al. 2008a).

San Esteban: San Esteban, 13 Apr 1911, *Rose 16823* (holotype, NY, image). *Dawson 1037* (ARIZ, CAS). SE end of island, main canyon, flowers white, but one plant found with pinkish flowers, *Felger 2610*. 6 May 1952, *Lindsay 2232* (CAS, SD). SE corner, rocky ridge of high peak, seen from shore to summit, *Moran 8838* (CAS, SD). S end of island [Arroyo Limantour], *Rempel 294*. Arroyo Limantour, *Van Devender [29] Apr 1992*. SW part of island, W-facing side canyon with dense vegetation, 175 m, *Wilder 07-87* (USON).

Echinocereus scopulorum Britton & Rose [*E. pectinatus* (Scheidweiler) Engelmann var. *scopulorum* (Britton & Rose) L.D. Benson, pro parte.]

Hant ipzx íteja caacöl "large arroyo's bladder (*Mammillaria*)"; Sonoran rainbow cactus

Description: Stems usually solitary, occasionally with 1–few branches, 20–44 cm tall, 5–8.5 cm diameter, with 13–15 slightly spiraled ribs; areoles closely set, 5–9 mm long, 3–5 mm wide. Plants in the southwestern part of Tiburón are often exceptionally large (e.g., to 44 cm). Spines about 25 per areole, 8–17 mm long, acicular, at first white with dark purplish red tips, often becoming yellowish and then dark with age, the radial spines spreading close to the stem. Flowers large and showy, 7.5–8 cm long, 6–8 cm wide, the perianth deep rose pink. Ovary 15–20 mm long, 12–16 mm wide. Larger perianth segments 28–32 in number, oblanceolate to spatulate, mucronate, 4–5 cm long, 8–17 mm wide above, 1.5–3 mm wide at base. Filaments yellowish, 9–12 mm long. Style 20–23 mm long, cream color, the stigma lobes bright green, 10–11 in number, 8–11 mm long. Flowering April, and July and August.

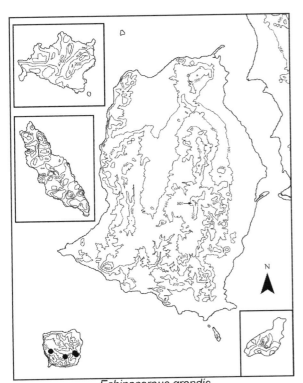

Echinocereus grandis
Hant ipzx íteja caacöl

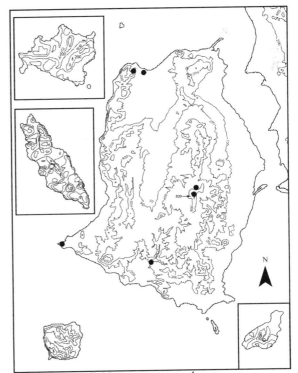

Echinocereus scopulorum
Hant ipzx íteja caacöl

Local Distribution: Widely but thinly scattered in rocky places on Tiburón from low elevations to the summit, where they almost always occur in protected sites out of the reach of hungry bighorn sheep. Not seen on the lowland bajadas and valley plains.

Other Gulf Islands: None.

Geographic Range: Western Sonora, from the Guaymas region northward to the Sierra Seri and Sierra del Viejo near Caborca.

Tiburón: Willard's Point, *Lindsay 3247* (SD). Arroyo Sauzal, rocky ledge adjacent to arroyo, ½ mi S of waterhole, 44 cm tall, *Felger 10096*. SW and up canyon from Siimen Hax, *Wilder 06-471* (USON). N base of summit of island, N-facing sheltered cliff, only one seen, *Wilder 07-588* (USON). Agua Dulce Bay, N slope, 200 m, *Moran 13000* (SD).

Echinocereus websterianus G.E. Lindsay, Cactus and Succulent Journal (US) 19:153–154, figs. 102 & 103, 1947

Description: Plants often 50–60 cm tall, producing several to sometimes 40 or more stems, the stems arising from the base to about one-third above the base of the parent stem. Spines pale golden yellow, 9–14.7 mm long. Flowers 4.5–6 cm long, the major tepals pink to rose. Flowering ovary ca. 25 mm long, the floral tube ca. 18 mm long. Filaments green, the anthers bright yellow. Stigma lobes

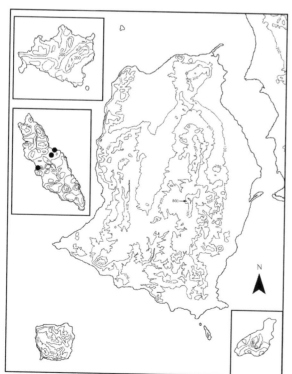

Echinocereus websterianus

green. Fruits succulent and spiny, the spine clusters falling away from ripe fruits. Flowering late April and early May.

Local Distribution: Endemic to Nolasco. Abundant on steep, rocky slopes, especially at higher elevations on east-facing slopes on the east side of the island, and canyons on the west side of the island. Often growing with *Agave chrysoglossa, Mammillaria multidigitata*, and *Opuntia bravoana. Echinocereus scopulorum* is the presumed closest relative.

Other Gulf Islands: None.

Nolasco: 27°50'N, 111°24'W, ca. 50 m, 24 Feb 1947, *Lindsay & Bool 498* (holotype, CAS/DS 314191; isotype, SD 45130). 29 Mar 1937, *Rempel 301, 303*. 6 Feb 1940, *Dawson 1040*. Ridge above Cañón de la Guacamaya, in flower, 3 May 2005, *Gallo-Reynoso* (photo).

Ferocactus – *Biznaga*; barrel cactus

Stems very thick, usually solitary (those in the flora area). Spines stout. Flowers yellow, orange, or red. Fruits yellow and fleshy.

1. All spines stout and firm, not bristly; flowers red. ___
 _____**F. emoryi**
1' Central spines stouter and firmer than the radials; flowers yellow or sometimes red or orange. ___ **F. tiburonensis**

Ferocactus emoryi (Engelmann) Orcutt [*F. covillei* Britton & Rose]

Siml caacöl "large barrel cactus," siml yapxöt cheel "red-flowered barrel cactus," siml cöquicöt "killer barrel cactus"; *biznaga*; Coville barrel cactus

Description: Robust barrel cactus (on the nearby mainland attaining heights of 1+ m). All spines thick and rigid—juvenile plants have markedly more rigid spines than adult plants (unlike other barrel cacti in the region, the radials are thick and not at all bristle-like). Flowers red, the fruits yellow and fleshy (nearby mainland population).

Local Distribution: In 2007 we found three robust juvenile plants near the shore on Alcatraz, which is the only record for this species on a Gulf of California island. We also found this barrel cactus on the nearby mainland on the sandy flats a few kilometers inland from the shore at Bahía Kino. It also occurs in the Sierra Seri.

Other Gulf Islands: None.

Geographic Range: Southwestern Arizona to the Guaymas/Yaqui region. The northern populations have red or red-orange flowers. The southern population,

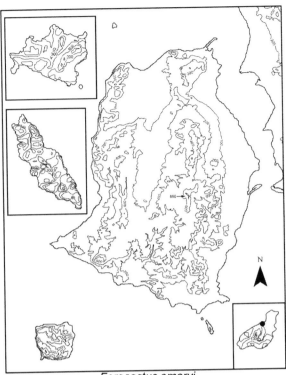

Ferocactus emoryi
Siml caacöl, siml yapxöt cheel

differentiated into radial and central series; the 4 most central spines terete, cruciform in arrangement, straight or somewhat twisted, the lower sometimes flattened and to 9 cm long; radial spines subulate, annulate, strongly resembling the centrals but not as heavy, though never setaceous" (Lindsay 1955b:166). On younger plants the central spine can be hooked and much stouter than the other spines. Flowers yellow or sometimes red or orange, mostly in spring. Fruits yellow and fleshy.

The distinctive firm outer radial spines, first pointed out by Johnston (1924), sets *F. tiburonensis* apart from *F. wislizeni*. Taylor (1984) pointed out the distinctive seed coat morphology of *F. tiburonensis* (he treated it at the varietal rank). Backeberg treated it as a distinct species, an opinion substantiated by Felger and Zimmerman (2000).

Local Distribution: Fairly widespread across Tiburón including the eastern bajada and ranging to higher elevations on Sierra Kunkaak. At higher elevations it is being heavily impacted by the introduced bighorn sheep.

Other Gulf Islands: None.

Geographic Range: On Tiburón and the opposite coastal mainland from south of Bahía Kino to the vicinity of Tastiota. Much of the probable mainland habitat has been converted to agriculture and the few remaining

from the vicinity of Guaymas southward to northern margin of the Yaqui lands, has bright yellow flowers. The population in the Sierra Libre, between Guaymas and Hermosillo, has the full span of yellow, orange, orange-red, or red flowers, and all three colors sometimes occur on a single plant (Paredes et al. 2000).

The southern, yellow-flowered population has been named *F. emoryi* subsp. *covillei* (Britton & Rose) D.R. Hunt & Dimmitt. The type locality for *F. covillei* Britton & Rose, however, is in northwestern Sonora, in midst of the range of the red-flowered, northern population ("Sonora, collected on hills and mesas near Altar," *C. G. Pringle, 11 August 1884*).

Alcatraz: NW side of island, cobble/shell rock berm along shore, 28.81373°N, 111.96785°W, three juvenile, healthy robust plants, 4 Dec 2007, *Felger and Wilder*, photo (see fig. 3.55).

Ferocactus tiburonensis (G.E. Lindsay) Backeberg [*F. wislizeni* var. *tiburonensis* G.E. Lindsay, Cactus and Succulent Journal [US] 27:166, fig. 155, 1955]

Siml, siml áa "true barrel cactus"; *biznaga*; Tiburón barrel cactus

Description: Stems to 1+ m tall, often with 19–21 ribs. "Spines usually heavily annulated ["ringed"], not clearly

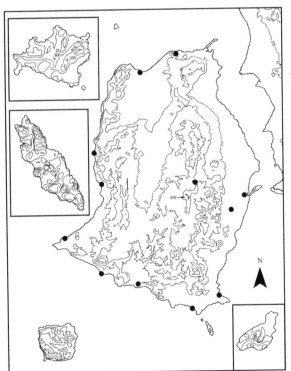

Ferocactus tiburonensis
Siml, siml áa

populations are rapidly being lost to roads, shrimp farming, and other developments. *F. tiburonensis* and *F. wislizeni* are clearly allopatric, although the distance separating them remains unknown.

Tiburón: [SE corner, Ensenada del Perro] Tiburón Island, 5 May 1952, *Lindsay 2229* (holotype, CAS; the protologue gives 30 Apr 1952 as the collection date). Ensenada Blanca (Vaporeta), *Felger 28 Apr 1967*, observation. Willard's Point, 3 mi N, frequent on rocky hillsides, *Johnston 4251* (CAS). Tordillitos, *Felger 15473*. Arroyo Sauzal, infrequent and scattered on rocky slopes, pediments, and benches adjacent or near arroyo, *Felger 10098*. Coralitos, *Felger 15373*. 0.5 mi NW of Zozni Cmiipla, scattered, *Wilder 06-160* (USON). 2 km inland from shore, road to Pazj Hax, scattered plants, *Wilder 06-164* (photo). Base of Capxölim, *Felger 07-100* (USON). Hee Inoohcö, mountains at NE coast, *Felger 07-86* (USON). Tecomate, W of village, *Whiting 9065*. Pooj Iime, 29 Sep 2007, *Felger & Wilder*, observation.

Lophocereus schottii (Engelmann) Britton & Rose var. schottii [*Pachycereus schottii* (Engelmann) D.R. Hunt]

Hasahcapjö, hehe is quizil "small-fruited plant"; *sinita, músaro*; senita, old man cactus

Description: Columnar cacti with many erect to arching branches. Stems strongly dimorphic: sterile (lower) stem portions relatively thick, with 5 and 6 stem ribs, widely spaced areoles, short and stout spines, and not producing flowers; fertile (upper) portion of stems relatively slender, with 6–8 stem ribs, a shaggy mane of long, slender and twisted spines, the areoles close together or confluent, and producing flowers. Fertile areoles producing more than one flower. (*Myrtillocactus* is one of the few other cacti to produce more than one flower per areole.) Flowers nocturnal, often 3–4 cm long and about as wide, whitish to pale pink. Flowering and fruiting especially during hot, humid weather. Fruits often 2.5–4 cm diameter, red, and spineless, ovoid to globose. The succulent pulp is red and sweet, but it is often difficult to find the fruits before birds and ants have hollowed them out and consumed the contents.

Local Distribution: Generally on finer textured soils of level or nearly level terrain, occasionally on rocky slopes. Plants on ridges and peaks often seem poorly developed and almost unhealthy. These seemingly out-of-place plants probably result from bird-dispersed seeds.

Sinita is widely but thinly scattered across the lowlands of Tiburón. It is much more common on the adjacent Sonora mainland. A few scattered sinitas are found on Dátil.

A few straggly sinitas were found near the top of Alcatraz in the 1960s and 1970s. In 2007 four healthy individuals were seen at the top of the mountain, in addition another plant was found at lower elevation just inland from the cobblestone beach in sand soil near the shore at the northern tip of the island. Several individuals are present near the top of Cholludo.

Two sinitas are evident on Patos in Osorio Tafal's 1946 photos near the southeast shore growing with *Atriplex barclayana, Carnegiea gigantea, Cylindropuntia fulgida,* and *Pachycereus pringlei*. One of the sinitas was still present in 1966 and was the only one then found on the island (Felger 1966:353, 359) but was not there in 2007. In 2007 we found four healthy juvenile sinitas near the base of the lone rocky hill in interior of the island.

This cactus is reported for San Esteban (Rebman et al. 2002), but we are unable to document it for the island. It is, however, on San Lorenzo, where there are a half dozen small colonies on the east side of the island near the northern tip (observations by Tom Bowen, 2006, and Wilder, 2010).

Hasahcapjö "was one of the first plants formed. . . . The spirit of vegetation, called *Icor*, caused the *senita* to have a very powerful spirit. . . . Anyone might seek the aid of the spirit of the *senita* against an enemy. . . . However, one knew that he got involved with the *senita* at great

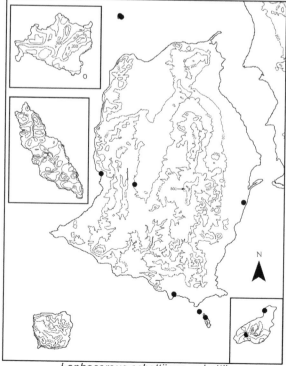

Lophocereus schottii var. *schottii*
Hasahcapjö, hehe is quizil

personal risk. If too frequent use was made of the cactus in placing a curse, that curse might backfire and affect him instead. . . . Good luck was solicited from the spirit of the senita by wedging clamshells, or sometimes twigs or other objects in the stem of the cactus. . . . It was believed that the senita was able to hear a conversation and people therefore often avoided it" (Felger & Moser 1985:248–249). Richard wrote his master's thesis on the evolution of the geometry of cactus using the sinita a few years before his introduction to the Comcaac and Becky and Ed Moser in El Desemboque (see Felger & Lowe 1967).

Other Gulf Islands: San Lorenzo (Tom Bowen, personal communication, 2006), Partida Norte, Rasa, Tortuga (observation), San Marcos, San Ildefonso (observation), Coronados, Carmen (observation), Danzante (observation), Monserrat (*José Juan Flores 20 Dec 2007*, MEXU), Santa Catalina, Santa Cruz (observation), San Diego, San José, San Francisco, Cerralvo.

Geographic Range: Northwestern Sinaloa to the borderlands in southwestern Arizona and through most of the Baja California Peninsula and many islands in the Gulf of California. Not known from Islas San Pedro Mártir and San Pedro Nolasco.

There are two geographic varieties. Var. *schottii* is replaced to the south by var. *australis* (K. Brandegee) Borg, 1937 [*L. schottii* var. *tenuis* G.E. Lindsay] in the Guaymas region, which ranges southward into northwestern Sinaloa. A similar scenario occurs in southern Baja California Sur, where the southern populations can be assigned to var. *australis*. Var. *australis*, primarily in thornscrub, usually forms a short trunk, a relatively open branching system, and the stems are more slender, with 6–10 ribs. The shift in architecture and morphology, from stems with a higher surface-to-volume ratio in the south to stems with a lower surface-to-volume ratio in the north, is interpreted as an adaptation to increased aridity northward. The northern distribution in Arizona and northern Sonora is probably limited by freezing weather. The Midriff Island populations are var. *schottii*, the northern variety.

Lophocereus schottii var. *australis* is based on *Cereus schottii* var. *australis* K. Brandegee published in 1900. Based on morphology, Felger believes that var. *tenuis* should be a synonym of var. *australis*. Molecular work, however, indicates significant differences between the populations on the Baja California Peninsula and those on the mainland (Nason et al. 2002).

The stems are highly esteemed in Mexico for medicinal purposes, especially stems with five ribs. Five-ribbed stems are the sterile or juvenile growth, and are thicker, "softer" or more flaccid and much more mucilaginous, and not as tough as fertile stems with a higher number of ribs.

Tiburón: Ensenada Blanca (Vaporeta), 7 Apr 1963, *Felger 7103* (specimen not located). Arroyo de la Cruz, 50 m elev., rather scarce, *Moran 13015* (SD). SW Central Valley, 2 Feb 1965, *Felger*, observation (Felger 1966:236). About $^1/_3$ mi inland from shore, road to Pazj Hax, *Wilder 06-163* (photo).

Dátil: NE side of island, widely scattered, *Felger 13469*.

Alcatraz: "Occurs here only in limited numbers near the top of the island, with birds seeming to be the agents of dispersal" (Felger 1966:137). Four individuals at the top of the island, 16 Sep 2007, *Wilder*, observation. One robust plant 2 m tall, near the shore at the N-most end of island, sand soil, with few adult *Pachycereus pringlei*, *Cylindropuntia fulgida* var. *mamillata*, 4 Dec 2007, *Felger*, observation.

Patos: SE shore, *Osario Tafal 1946* (photo, Felger 1966:353). About 25 m S of the rocky hill, flat area covered with *Atriplex barclayana*, 4 juvenile senitas, stem ribs 5 and some with 6 ribs, 30 Sep 2007, *Felger*, observation.

Cholludo: Few seen at top of isle, sickly looking, 4–6 feet tall, 30 May 1954, *Felger*, observation, in field notes.

Mammillaria

Small globose or short-cylindrical cacti. Spine clusters on separate tubercles usually arranged in a spiral pattern. Central spine(s) hooked or straight. Ovaries and fruits fleshy and spineless.

1. Plants solitary or forming dense, cespitose clusters of several mostly globose stems; sap milky; tubercle axils at top of stem densely woolly; tubercle axils without bristles; Nolasco. _____ **M. tayloriorum**
1' Plants solitary or branching, the stems globose as juveniles, becoming longer than wide at maturity; sap watery; tubercle axils not woolly, with or without some bristles.
  2. Plants forming clusters of numerous stems branching variously along the stem, the stems mostly 5 cm or less in width and generally longer than wide; Nolasco. _____ **M. multidigitata**
  2' Plants solitary or with 1 to usually few branches, the stems mostly more than 5 cm wide; not on Nolasco.
    3. Inner (larger) tepals cream-white; San Esteban. _____ **M. estebanensis**
    3' Inner tepals whitish with a pinkish midstripe; Tiburón and Alcatraz. _____ **M. grahamii**

Mammillaria estebanensis G.E. Lindsay, Cactus and Succulent Journal [US] 39:31, 1967 [*M. angelensis* R.T. Craig var. *estebanensis* (G.E. Lindsay) Reppenhaggen, Die Gattung *Mammillaria* nach dem heutigen Stand meines Wissens, 1988:34, 1981. *M. dioica* K. Brandegee subsp. *estebanensis* (G.E. Lindsay) D.R. Hunt, Mammillaria Postscripts 7:3, 1998]

Hant ipzx íteja caacöl "large arroyo's bladder (*Mammillaria*)"

Description: Plants often irregularly several branched. Tubercle axils bear multiple bristles sometimes about as long as the tubercle. Spines light colored and a few to many areoles on a plant bear a moderately hooked central spine.

Male and female flowers on separate plants (dioecious). Flowers in a ring near the stem tip, 22–26 mm wide, cream white; outer tepals each with a broad greenish red midstripe, inner (larger) tepals 8, whitish with greenish tinge and slightly darker along the midstripe. Filaments whitish, the anthers yellow. Flowering ovary 3 mm long, 3.7 mm wide; style and stigmas well exerted, the style white to whitish pink, 11 mm long; stigma lobes 5 or 6 in number, 3 mm long, linear, green. Fruits bright red or orange, fleshy, and delicious, often in a ring around the upper part of the stem. On 3 May 2008, Ben and Humberto observed cactus bees and honeybees visiting the flowers.

Local Distribution: Widespread on San Esteban, often growing with *Echinocereus grandis*.

Other Gulf Islands: San Lorenzo.

Geographic Range: Endemic to Islas San Esteban and San Lorenzo, and not seen on Las Ánimas ("San Lorenzo Norte"; Tom Bowen, personal communication, 2006). The San Esteban *Mammillaria* is part of the *M. dioica* complex centered in the Baja California Peninsula, as shown by Hunt (1998).

San Esteban: 13 Jan 1961, *Lindsay 3002* (holotype, SD). 20 Apr 1921, *Johnston 3198* (CAS). *Lindsay, 22356* (SD). *Moran 4078* (SD). Arroyo Limantour, common in rocks on gentle bajada and broad wash, *Van Devender 90-525*.

Mammillaria grahamii Engelmann subsp. **sheldonii** (Britton & Rose) D.R. Hunt [*M. swinglei* (Britton & Rose) Boedeker]

Hant ipzx íteja "arroyo's bladder"; *viejito, cabeza de viejo*; fishhook cactus

Description: Globose to cylindrical small cacti. Tubercle axils with or without bristles. Spine color variable, light to dark, perhaps correlating with the substrate;

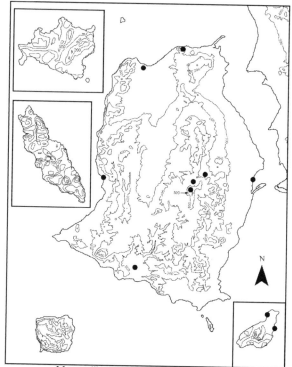

Mammillaria estebanensis
Hant iipzx iteja caacöl

Mammillaria grahamii ssp. *sheldonii*
Hant ipzx íteja

areoles bearing a single strongly hooked central spine. Flowers bisexual; inner (larger) tepals mostly white with a pinkish midstripe. Fruits globose to mostly club shaped, usually red or sometimes orange.

Local Distribution: Widely scattered across Tiburón but seldom common, from near the shore to high elevations in the Sierra Kunkaak. Also on Alcatraz where two small groups of robust plants with dark spines grow on the cobblestone margin of the east flat near the shore. *Mammillaria* was not seen on Alcatraz in the 1960s and presumably has colonized the island since that time.

Other Gulf Islands: None.

Geographic Range: As broadly interpreted, *M. grahamii* ranges from southeastern California to western Texas, northern and western Chihuahua and Sonora nearly to the Sinaloa border. Subsp. *sheldonii* occurs in coastal Sonora from about Puerto Libertad to the southwestern part of the state. Subsp. *sheldonii* is reported to have smaller flowers and fewer radial spines (12–15) than subsp. *grahamii* (20–30+), but other than flower color, consistent distinguishing features appear elusive.

Tiburón: Ensenada Blanca (Vaporeta), 160 m, *Felger 17625* (SD). Arroyo Sauzal, ca. 2 km inland, 29 Jan 2008, *Wilder*, observation. E bajada, vicinity of Zozni Cmiipla, *Wilder 07-205* (USON). E side of island, foothills of Sierra Kunkaak, *Wilder 05-16* (USON). 50 m E of Siimen Hax, *Wilder 06-446* (USON). Cójocam Iti Yaii, saddle, fairly common on ridge, almost entirely under nooks beneath rock, *Wilder 07-541* (USON). Hee Inoohcö, mountains at NE coast, *Felger 07-66A* (photo). Bahía Agua Dulce, 18 Mar 1962, *Lindsay 3246* (SD).

Alcatraz: N-central portion of flat, ca. 2 m elev., 28.81545°N, 111.96645°W (NAD 83), rocky bench just in from beach, on the inland portion of bench, rocky with sandy soil with *Amaranthus watsonii, Atriplex barclayana, Lycium brevipes, Sesuvium portulacastrum*, growing under *L. brevipes*, 2 plants seen on this side of island, just past flowering, spines moderately hooked, plant has a large cluster of 30–40 heads, ca. 0.3 m × 0.3 m × 15 cm tall, 16 Sep 2007, *Wilder 07-410* (photo). E-central portion of flat on rocky cobble bench just inland, population of ca. 15 individuals, some quite large (ca. 30 cm tall), 16 Sep 2007, *Wilder*, observation.

Mammillaria multidigitata W.T. Marshall ex G.E. Lindsay, Cactus and Succuluent Journal (US) 19:152, figs. 99–100, 1947

Description: Many-stemmed, low, cespitose-spreading plants. Stems elongated, about 4–5 cm diameter, often somewhat flaccid, the tubercles and spines short, the spines straight, or rarely a few areoles with the central spine curved or moderately hooked at the tip. Male and female flowers on different plants (dioecious). Tepals white to cream; outer tepals ciliate fringed and with a faint, pale pink midstripe; inner tepals with entire or essentially entire margins, the tips broadly obtuse except the 3 or so innermost tepals have acute tips. Stigma lobes 5 or 6, green; pistillate flowers with relatively thick lobes, the staminate flowers with slender stigma lobes. Anthers of pistillate plants small and indehiscent, those of the staminate plants larger, dehiscent and golden yellow. Fruits clavate, bright red to orange, and not very sweet. Flowering in early summer, beginning middle to late May.

Local Distribution: Endemic to Nolasco. Abundant on rocky ridges and slopes on both sides of the island, especially on east-facing slopes and cliffs on the east side of the island, and major canyons on the west side of the island. The density is sometimes extreme: "This plant was often so thick and in such large clumps as to make it difficult to avoid stepping on them" (Lindsay 1947:75). This species often is found growing with *Agave chrysoglossa, Echinocereus websterianus*, and *Opuntia bravoana*. The axillary bristles, dioecious habit, and flower color and structure point to a relationship with the Baja California

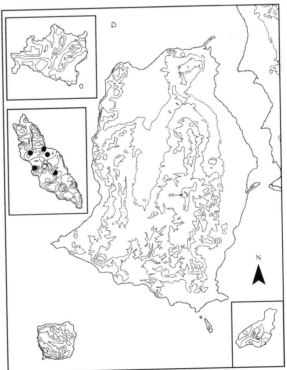

Mammillaria multidigitata

and Gulf Island complex of *M. dioica* K. Brandegee including *M. estebanensis*.

Other Gulf Islands: None.

Nolasco: 27°50'N, 111°24'W, ca. 50 m, 24 Feb 1947, *Lindsay 499* (holotype, CAS). Very common all over the island in dense mats of 30–40 heads, *Johnston 3112* (CAS). Cañón el Farito, 60 m, abundant on S-facing rock slopes, and some on other exposure, in places free of *Vaseyanthus* or mostly so, *Felger 06-82* (photos). Cañón de la Guacamaya, 100+ m, 14/15 Apr 2003, *Gallo-Reynoso* (photo). Canyon de Mellink, common, 30 Sep 2008, *Wilder* (photo).

Mammillaria tayloriorum C. Glass & R. Foster, Cactus and Succulent Journal (US) 47:175, 1975

Description: Plants single-stemmed or cespitose with several heads, branching from base, stems nearly globose to cylindroid, 15–25 cm tall, with milky sap. Areoles woolly when young, the tubercle axils densely woolly in the flowering zone at the stem apex. Spines to about 1 cm long. Observed in flower in mid-February and March when no other cactus on the island was seen in flower. Flowers bisexual and in a ring near the stem tip, fresh flowers (3 Feb 2008) 18–19 mm long, ca. 15 mm wide; outer, smaller tepals with margins short-fringed (fimbriate-ciliate); inner tepals bright pink-purple at base, with a broad, fuchsia-colored (dark pink) midstripe and translucent colorless margins, entire to mostly entire with a few very small shallow teeth, and a pronounced apiculate tip. Filaments and style white below, fuchsia colored above (dark pink same as tepal midstripe). Anthers yellowish white. Stigma lobes 5–7, pale amber (perhaps greenish when fresher). Fruits 1–1.5 cm long, elongated (clavate), succulent, and red.

Local Distribution: Endemic to Nolasco, where it generally is found at higher elevations. Abundant on canyon slopes on the northwest side of the island and less common along the east-facing side of the island near the ridge. Growing among rocks, often from rock crevices and in semi-shade beneath shrubs.

Other Gulf Islands: None.

Notes: Glass and Foster (1975) postulate that this species is most closely related to the *M. sonorensis* complex of montane east-central and southeastern Sonora and adjacent Chihuahua. They also indicate potential relationship with *M. johnstonii* (Britton & Rose) Orcutt of the Guaymas–San Carlos region and *M. bocensis* Craig, which replaces the latter to the south. In earlier publications, *M. tayloriorum* was known as *M. evermanniana* (Britton & Rose) Orcutt, which is found on Isla Cerralvo and adjacent Baja California Sur and differs in part from the Nolasco cactus by having larger and yellowish green flowers.

The species is named for "Bob and Suzanne Taylor of El Cajon, California, in recognition of their knowledge of Mexican cactus habitats" (Glass & Foster 1975:175).

Nolasco: "San Pedro Nolasco Island . . . near the highest points of the island along with *Mammillaria multidigitata* and *Echinocereus websterianus*. Note: in the original description the color plates were inadvertently transposed, and fig. 3, supposedly of *M. tayloriorum*, is *M. evermanniana*, and figure 4, listed as *M. evermanniana*, is *M. tayloriorum*," Nov 1975, *Glass & Foster 2686* (holotype, POM 325135; the specimen consists of only a few rather poorly pressed flowers). Seen only near summit crest, rock crevice of a ledge, lactiferous, 17 Apr 1921, *Johnston 3121* (CAS). 6 Feb 1940, *Dawson 1039* (RSA, UC). Found . . . only along rocky crest, 27 Mar 1947, *Lindsay 502* (CAS). 3 May 1952, *Lindsay 2228* (SD). Above Cañón el Farito, 11 Feb 2000, *Felger & Seminoff* (photos). Above Cañón el Farito, top of W slope, crest above drop-off, group of 6 heads, flowers pink with yellow/amber/greenish 5–7 parted stigma, multiple white/yellow stamens,

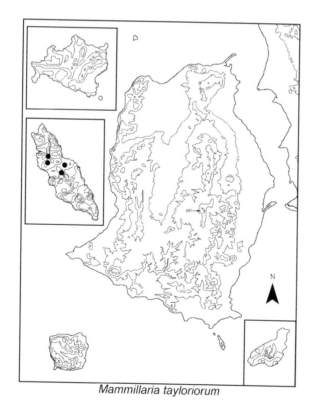

Mammillaria tayloriorum

3 Feb 2008, *Wilder 08-185* (photo). Cañón de Mellink, upper part of canyon, above 100 m, 29 Sep 2008, *Wilder & Felger* (photos).

Mammillaria sp.

Description: Stems rather slender, often branched at base (clustered), spines not hooked, reddish purple; sap watery. Tubercle axils bearing bristles; areoles bearing a slender, hooked central spine. Flowers bisexual, the inner (larger) tepals white with a prominent maroon midstripe. Fruits at first orange, becoming red at maturity.

Local Distribution: Widespread and abundant on Dátil and on the vegetated slope of Cholludo. Although these plants appear distinctive, they may represent only a minor variation of *M. grahamii* var. *sheldonii* on nearby Tiburón.

Other Gulf Islands: None.

Geographic Range: This *Mammillaria* appears to be endemic to these two small islands.

Dátil: NE side of island, *Felger 13430* (ARIZ, SD). E side, slightly N of center, halfway up N-facing slope of peak, *Wilder 07-121* (USON).

Cholludo: *Felger*, observation (Felger 1966:418). 2 Sep 2008, *Wilder* (photos).

Opuntia – *Nopal*; prickly pear

Stems with determinate, rhythmic (seasonal or annual), constricted growth increments resulting in individual stem segments or pads (cladodes). Glochids in at least the most recently formed areoles. Leaves on young growing stem segments, 1 per areole, succulent, conical, and soon deciduous. Stamens sensitive—the filaments close inward when touched. Fruits fleshy. "Seeds" relatively large, with a light tan, bony, aril-like structure covering the actual seed.

1. Cladodes (pads) bright green, often shiny, with relatively few spines; Nolasco. _____ **O. bravoana**
1' Cladodes dull green, conspicuously spiny; not on Nolasco._____ **O. engelmannii**

Opuntia bravoana E.M. Baxter

Description: Low, spreading to shrub-sized prickly pears, often reaching 1–1.5 m tall, few to nearly 2 m in favorable niches along canyon bottoms. Cladodes (pads) bright green, relatively flat-sided and very succulent, glabrous and often shiny green, and sometimes purplish at the areoles and margins of the pads. Areoles in 4–6 series (diagonal rows), with a partial to full circle of short white wool surrounding the glochids. Many or most areoles

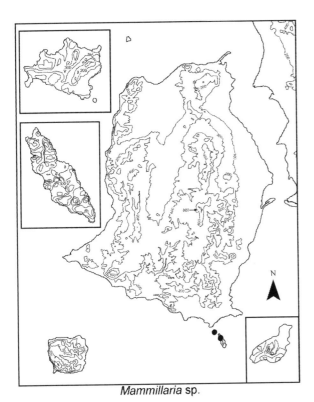

Mammillaria sp.

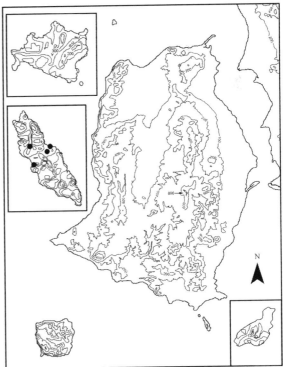

Opuntia bravoana

on each cladode spineless, and some areoles with one to few, usually deflexed spines. Cladodes becoming rather desiccated and yellowish green during extended drought. Flowers bright yellow and relatively large; flowering in spring. Fruits red-purple, bearing glochids (some rather long), the epidermis glabrous.

Local Distribution: Common on the east side of Nolasco and on the west side mostly in canyons and high ridges. On the east side it is abundant and best developed on relatively deep soil, as well as on rock faces. The population on this island seems to be dynamic. George Lindsay recorded the following information on the label of a herbarium specimen (*Lindsay 2225*): "A platyopuntia which in 1947 covered large areas of the island in impenetrable thickets, but at the present time [1952] is rare, . . . the large areas of dead opuntias are now covered with *Vaseyanthus insularis*, a few of the opuntias are left, one of which was in flower, these yellow." We found it abundant and healthy during our visits to the island, although some at low elevations on exposed rock were rather scraggly.

Other Gulf Islands: Cerralvo.

Geographic Range: This prickly pear occurs on islands and island-like habitats at Guaymas and Bahía San Carlos including Isla Santa Catalina at the southeast end of Punta Doble bordering the entrance to Bahía San Carlos, some rock slopes at Bahía San Carlos, islands in Guaymas Harbor, Isla San Pedro Nolasco, and the northern edge of the coastal plain near the vicinity of Las Guásimas southward at least to Culiacán, Sinaloa, and inland in southern Sonora and northern Sinaloa, and in the Cape Region of Baja California Sur and adjacent Isla Cerralvo.

Nolasco: 29 Mar 1937, *Rempel 306, 307* (RSA). 3 May 1952, *Lindsay 2225* (CAS). E side, common in granitic steep ravines, commonly observed *Ctenosaura* lizard feeding on flowers, *Stinson & Robinson 29 Apr 1974* (SD). E side, spreading prickly pear to 1 m tall, common on E-, SE-, and NE-facing rocky slopes and steep arroyos, absent from S-facing slopes, *Felger 9663*. Cañón de la Guacamaya, 14/15 Apr 2003, *Gallo-Reynoso* (photo). Cañón de Mellink, upper reaches of canyon, *Wilder 08-356*.

Opuntia engelmannii Salm-Dyck ex Engelmann var. **engelmannii** [*O. phaeacantha* Engelmann var. *discata* (Griffiths) L.D. Benson & Walkington]

Heel hayeen ipaii "prickly-pear used for face painting"; *nopal*; desert prickly pear

Description: Low, usually spreading or sprawling prickly pears. Cladodes (pads) obovate to elliptic, often

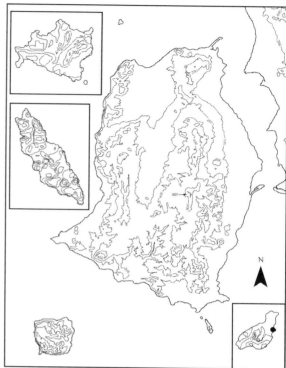

Opuntia engelmannii var. *engelmannii*
Heel hayeen ipaii

20+ cm long, with stout, spreading and deflexed whitish spines. Flowers relatively large, uniformly bright yellow, changing to yellow-orange (apricot color) as the flowers age (second day); flowering in spring. Fruits purple-red including pulp, ripening July and August, the pulp juicy and sweet (fruit description based on mainland populations).

Local Distribution: A small population on Alcatraz is the only prickly pear known from any of the Midriff Islands. It formerly occurred on the "east-facing slope of the mountain near peak elevation" (Felger 1966:138), but in 2007 it was found only near the shore at the south-central part of the island, where it grows with *Mammillaria grahamii*.

Another small population of this prickly pear occurs on the small rocky hill at the far end of Punta Sargento opposite the northeast corner of Tiburón (Felger & Moser 1985; also see appendix B). Although there are unconfirmed reports of prickly pear on Tiburón, none has been documented for the island.

Other Gulf Islands: None.

Geographic Range: Northern Mexico and southwestern United States. Parfitt and Pinkava (1988) recognized 6 varieties, with var. *engelmannii* being the most widespread one.

Alcatraz: S-central part of island near the shore, *Wilder 07-429*.

Pachycereus pringlei (S. Watson) Britton & Rose

Xaasj, xcocni; *cardón, sagüeso, sahueso*

Description: One of the most massive cacti anywhere and the largest cactus in the Sonoran Desert. Often 8–10+ m tall on the islands and reaching greater sizes in the Bahía Kino region but "dwarfed" on Mártir. Trunk short and enormous, with massive branches mostly arising from less than 1 or 2 m above the ground. Stems bluish glaucous, the mature stems with (12) 13–15 ribs. Juvenile plants are densely spiny with distinct areoles; the upper or reproductive stem portions of adult plants are spineless with coalesced areoles. The stems show constrictions and color differences for each growth increment, probably representing a single year or season except during extreme drought events.

As with saguaros, the flowers open in the evening and remain open part of the following day depending on temperature. Reid Moran (1936–1993:116) made the following observation from Isla San Pedro Mártir on 25 April 1966: "Mature buds of the cardon, placed in a paper sack, opened well before those left on the plants, which waited for nightfall." Flowers 7–10 cm long, the inner tepals white; March–May. The outside of the flower and the fruits are covered with dense, golden brown hairs. The

fruits, which ripen in early summer, are about as big as a medium-sized peach. At maturity the fruits split open to reveal crimson-purple to pinkish white or occasionally white pulp filled with small black seeds. The pulp is sweet and delicious. The fruits are densely covered with felt-like hairs and vary among individual plants from spineless to spiny.

There is considerable variation in the fruit in color of the pulp, taste, and spine cover, although an individual cactus bears fruit of only one color and general spine length. The Comcaac classify and have names for four different kinds of xaasj according to the color of the pulp and additionally classify and have names for three kinds of xaasj, distinguished by characteristics of the fruit involving combinations of spination, taste, and color (Felger & Moser 1985). The extensive research of Ted Fleming and colleagues has shown that cardons have complex breeding systems involving individual plants producing unisexual male or female, or bisexual flowers resulting in trioecy (e.g., Fleming 2002; Fleming et al. 2002).

Although normally slow growing and long-lived (Turner et al. 2003), under cultivation in western Sonora cardons may grow considerably faster than saguaros. Cardons are readily grown from seed and small plants can be successfully transplanted, but like saguaros, larger plants seldom survive transplantation, and stem cuttings rarely, if at all, form roots. Cardons are highly frost sensitive, unlike saguaros.

Local Distribution: Widespread on all the Sonoran Islands:

Tiburón: *Pachycereus* is distributed across the lowlands from near the shore to the lower foothills on arid slopes. Notable stands occur on the eastern bajada facing the Infiernillo such as in the Valle de Águila region at the northeast side of the island, as well as in the Sauzal area at the south side of the island.

San Esteban: Widespread and common throughout the island, extending to higher elevations.

Dátil: Densest on the northwest part of island but ranges throughout the island.

Alcatraz: Felger (1966:139) reports that, "of the two columnar species of arborescent columnar cacti on the island, *Carnegiea* is poorly established while *Pachycereus* is one of the more common and widespread species." *Pachycereus* is locally common at the northeast base of the mountain and extends to peak elevation. It is also scattered across the sandy flats surrounding the interior *Allenrolfea* flat.

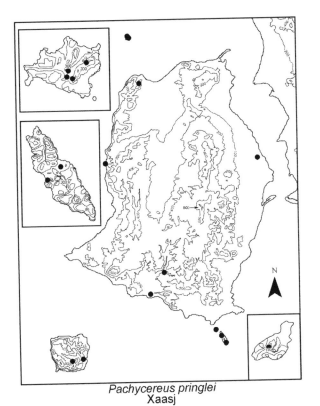

Pachycereus pringlei
Xaasj

Patos: "The 1946 photographs show a very thinly distributed stand across the flats where it is today absent except for one medium-sized plant. Two colonies are also visible in the older photographs, one at about halfway up the south side of the hillslope and another at the rocky ledges near the southeast shore. The hillside colony consisted of larger individuals and seems to have been slightly bigger than the one near the shore. Today *Pachycereus* occurs in the same places but the individual plants are markedly smaller than those shown in the 1946 photographs. The population is recovering but the plants have not yet reached maximum size" (Felger 1966:359–360). In 2007, the cardons appeared to be increasing, with juvenile plants and a few adult or subadult plants widely scattered, mostly on sea ledges at the south shore and also on the rocky peak. On the west side of the single hill, about 16–18 m below the summit, we counted 65 juvenile cardons, mostly less than 1 m tall, few to 1.2 m, plus 14 adult plants, and one plant about 70 cm tall at the south side of the hill.

Cholludo: This tiny isle has a dense "forest" of relatively short cardons but these are not the "dwarf" form found on Mártir. On Cholludo the branching is not basal—the majority of the branches arise well above ground level like those of the nearby Sonoran mainland and Isla Tiburón. The nearly impenetrable cactus forest on Cholludo includes cardón, organ pipe, and chainfruit cholla. All age classes of cardón are abundant, including seedlings or small plants and juveniles. Factors that likely contribute to their amazing density include a north-facing slope surrounded by warm water, absence of major herbivores, and rich soil nutrients (nitrogen) from bird guano (see Polis et al. 1997, 2002; Wait et al. 2005; Wilder et al. 2008a). Comparable dense cardón forests occur on Isla San Pedro Mártir and Islas Mellizas in Guaymas harbor (Turner et al. 2003).

Mártir: This cactus is the dominant landscape feature across the vegetated part of the island. The largest ones are at higher elevations, where the majority of the island surface is a veritable cardón forest. In 2008 we found juvenile to adult plants abundant throughout the island. Seedlings and small plants were found in the open, growing among large and small rocks that make up the island surface, and were not associated with nurse plants. Upon reaching maturity with many, massive branches the plants seems prone to toppling over, probably because of the loose rock and the effects of strong winds when the plants are well hydrated. In 2007 and 2008, we saw quite a few still "living" and green large cardons that had fallen over, probably as a result of the last hurricane.

The Mártir population consists of the strange "dwarf" form, reaching a maximum of about 5 m in height and are trunkless or with very short trunks, with the branches arising at or near ground level. The "dwarfed" and "trunkless" growth form is documented for certain from Mártir and Isla Partida Norte in the Midriff Region (a small island to the south of Ángel de la Guarda) (e.g., Lindsay 1948; Medel-Narvaez et al. 2006; Wilder & Felger 2010; Tom Bowen, personal communication, 2008). For more discussion of the Mártir cardón population, see Wilder and Felger (2010).

Palmer reported, "'Cardon'; forming a forest on the summit of S. Pedro Martin Island. The dead wood is much used for fuel and other purposes, and the seedy fruit is an article of food" (Watson 1889:52). In 1971 Rod Hastings saw few juvenile plants. He observed virtually the entire stand to be even-aged, about 3–4 m tall. Only on the north slopes did he find any individuals less than 10 cm tall and then very few, and only a scattering of plants between 10 cm and 3 m tall (Hastings, field notes).

On 11 April 2007, we noted, especially at the lower elevations and at the bases of slopes, that the short trunks were commonly buried in rock rubble, indicating that the land surface is rather unstable. Smaller, young plants were fairly common, especially at higher elevations, and most of them were at least partially covered by *Vaseyanthus* vines.

Nolasco: Common at all elevations on the west side of the island and less numerous on the east side, mostly on south- and east-facing rocky slopes. In November 2006 we observed some cardons near the top of the island above Cañon el Farito that were 1–1.2 m tall and had blackened, necrotic stem tips, but the meristems seemed intact.

Other Gulf Islands: Ángel de la Guarda, Rasa, Salsipuedes, Partida Norte, Las Ánimas (San Lorenzo "Norte"), San Lorenzo (observation), Tortuga, San Marcos (observation), San Ildefonso (observation), Coronados (observation), Carmen, Danzante (observation), Monserrat (observation), Santa Catalina, Santa Cruz (observation), San Diego (observation), San José, San Francisco (observation), Espíritu Santo, Cerralvo.

Geographic Range: Endemic to the Sonoran Desert and Cape Region of Baja California Sur: most of the Baja

California Peninsula and Sonora from near Empalme (south of Guaymas) northward along the coast to Puerto Lobos and inland to Pitiquito, and on most of the islands in the Gulf. The Sonoran distribution encompasses the original area of occupation of the various groups of Comcaac.

There are numerous published photos of cardons on the islands. Here we cite only a few archived photos and a miscellany of voucher specimens.

Tiburón: Arroyo Sauzal, rocky bench adjacent to waterhole, *Felger 10119*. S end of Tiburón, *Rempel 298* (cited by Gentry 1949, not located). Middle bajada near Valle de Águila, *Wilder 07-238*. W of Tecomate, rocky hill, *Felger 6244*. Pooj Iime, 29 Sep 2007, *Felger & Wilder*, observation.

San Esteban: SE corner of island, *Bostic 21 Jun 1965* (SD). Near S end, in large arroyo running inland from E side [Arroyo Limantour], *Wiggins 17227* (CAS).

Dátil: Occasional, *Moran 13036* (SD). NW part of island, steep rocky W-facing slope, location of densest stand, seen across island, *Felger 2553*.

Alcatraz: NE-facing slope of island, locally abundant, *Felger 13410*. E base of mountain, *Felger 08-08*.

Patos: *Osario Tafal 1946* (photos, Felger 1966:352, 353). Aug 1964 (photo, Felger 1966:356). Mountain peak, *Felger 07-57* (USON).

Cholludo: Abundant on the NE side of the island (Felger 1966:306–308).

Mártir: Summit, 1887, *Palmer 418* (GH, image). Forming a forest over upper part of island, branching from base and getting 25 ft or more high, 18 Apr 1921, *Johnston 3160* (CAS, image). "Forming forest on summit . . . Branching mostly near the ground, the branches 2–5 dm diameter, with 13–15 ribs. Buds sometimes within 1–2 m of the ground, 7½–8 × 3½–4 cm. Some mature buds placed in a paper sack in my pack began to open well before any on the plant; in fact, when I reached the shore at 1850 (just before dark) none were yet seen open. On one small plant even small buds had stigmas exerted; in most plants stigmas not exerted before anthesis," 25 Apr 1966, *Moran 13038* (SD, UC, images). Small canyon at base of SE-facing slope that leads to island summit, abundant and widespread, the specimen is taken from a fallen branch, the cut surface (cortex and pith) quickly turned orange-brown, *Felger 07-19* (USON).

Nolasco: Documented for the island by our quadrat studies (Felger 1966; Felger et al. 2011) and multiple observations (e.g., Lindsay 1962). Cañón de la Guacamaya, July 2003, Gallo-Reynoso (photo).

Peniocereus striatus (Brandegee) Buxbaum [*Wilcoxia striata* (Brandegee) Britton & Rose. *W. diguetii* (F.A.C. Weber) Diguet & Guillaumin. *Neoevansia striata* (Brandegee) Sánchez-Mejorada. *N. diguetii* (F.A.C. Weber) W.T. Marshall. *Peniocereus diguetii* (F.A.C. Weber) Backeberg]

Xtooxt; *sarramatraca, sacamatraca*; Sonoran night-blooming cereus

Description: Root system with about a dozen or more potato-like tuberous roots strung on clusters of slender connecting roots. Plants often 1–2.5 m tall and much branched above. Stems 4.5–8 mm diameter (1–2 years old), at first greenish, soon becoming brownish or grayish and glabrous, with 6–9 flat ribs separated by deep furrows or grooves, the stomata restricted to the grooves. Spines short, bristle-like, and ultimately deciduous. Flowers white, nocturnal, about 8 cm long and wide. Fruits 5 cm long, ovoid, the rind red when ripe, the ripe pulp red, juicy, sweet and edible, with short bristly spines on small, sparsely distributed deciduous areoles. Seeds 2–2.3 mm long, dark red-brown to blackish. Flowering in summer, the fruits ripening late summer or early fall.

The pencil-thin stems often scramble through spiny shrubs such as *Lycium*. The cactus stems seem to mimic

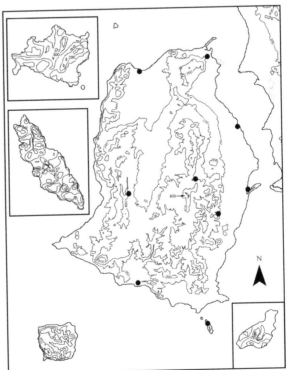

Peniocereus striatus
Xtooxt

those of their nurse plants. This plant is unusual compared to other cereoid cacti in the Sonoran Desert. The slender stems seem out of place due to their comparatively high surface-to-volume ratio. However, the stomata are in the furrows, or grooves, between the stem ribs, and as the stem shrinks during drought the ribs close off the furrows. The exposed surfaces of the ridges, or ribs, are devoid of stomata. Thus, during drought the stomata are closed off from the desert air. In addition, scanning electron microscope studies show that during drought the stomata are sealed by a coating of an amorphous-looking substance, perhaps a polysaccharide, which is water soluble (Felger & Henrickson 1997).

Local Distribution: On Tiburón, this rather cryptic cactus is often common in scattered localities including bajadas, valley plains, and rocky slopes. It has not been seen at the arid western margin of the island. It occurs to at least 400 m in the eastern slopes of the Sierra Menor (west side of the Central Valley) and similarly to about 400 m in the interior of Sierra Kunkaak but has not been found at higher elevations. It is also widely scattered on Dátil.

Other Gulf Islands: Tortuga, San Marcos, Coronados, Carmen, Danzante (observation), Monserrat, San José, San Francisco, Espíritu Santo.

Geographic Range: Near the Sonora border in Arizona in Organ Pipe Cactus National Monument and the O'odham lands, western Sonora, northwestern Sinaloa, both Baja California states, and a number of Gulf of California islands. It is especially common along the mainland Infiernillo coast opposite Isla Tiburón.

Tiburón: El Sauz, *Harbison 20 Mar 1962* (*Moran 8800½,* SD). SW Central Valley, upper rocky slopes of mountains bordering valley, 1200–1400 ft, *Felger 12434.* Hast Coopol, *Wilder 07-382.* Vicinity of Zozni Cmiipla, 26 Sep 2008, *Felger,* observation. Trail to Siimen Hax, *Felger 07-110.* Punta Perla, common on rocky hillside, *Felger 74-14.* Valle de Águila, within 50 m of shore, *Wilder 07-269.* Tecomate, *Felger 8877.* **Dátil:** NE side of island, steep rocky slopes, arroyo bottom near beach, *Felger 9129.*

Stenocereus

1. Stems often arching, inclined, or spreading, mostly not erect, with 8 or 9 ribs; spines stout and rigid, the larger ones with sharp angles. _____ **S. gummosus**
1' Stems mostly erect and nearly straight, with 13–18 ribs; spines very slender and not angled, rather bristly and flexible. _____ **S. thurberi**

Stenocereus gummosus (Engelmann) Gibson & Horak [*Macherocereus gummosus* (Engelmann) Britton & Rose]

Ziix is ccapxl "sour-fruited thing"; *pitaya agria*

Description: Large sprawling, multi-stemmed cactus often 1.5–4 m tall. The leaning, arching, or scrambling stems can create dense thickets. Stems that bend to the ground produce roots and can form clonal colonies. Stems 5–8 cm diameter with 8 to 9 somewhat indistinct ribs; spines stout, brittle, and extremely sharp, and with sharp angles—this cactus has the fiercest armature of any plant in the region. Flowers nocturnal and spectacular, white with pink or rose, 10–15 cm long and 6–8 cm wide. Flowering July and August, and occasional flowers found at other times of year. Fruits globose, 6–8 cm diameter, covered with sharp spines that fall away as the fruit ripens. The rind and pulp are red, and the pulp succulent and delicious. These sweet but tart fruits (hence the name *pitaya agria*) are among the most delectable fruits in the world. The fruits may be aborted in years of low rainfall.

Local Distribution: Common and widespread on bajadas, valley plains, and low hills nearly throughout Tiburón. Abundant on San Esteban, especially along

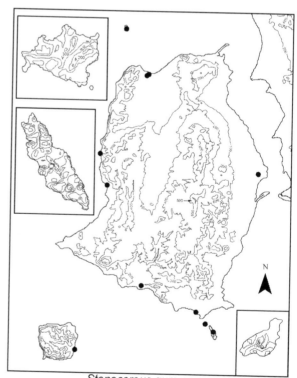

Stenocereus gummosus
Ziix is ccapxl

arroyos and adjacent lower hillsides. Also on Dátil and Cholludo.

A single mature plant or colony occurred on Patos in 1946 at the south edge of the island. It was no longer there in 1966 (Felger 1966:354, 359). In 2007 we found one small plant, also at the south edge of the island, but it was not at all vigorous.

Other Gulf Islands: Ángel de la Guarda, Rasa, Salsipuedes, San Lorenzo (observation), Tortuga, San Marcos (observation), San Ildefonso (observation), Coronados (observation), Carmen, Danzante (observation), Monserrat, Santa Catalina, Santa Cruz (observation), San Diego, San José, San Francisco, Espíritu Santo, Cerralvo.

Geographic Range: Both states of Baja California, many Gulf Islands, and the mainland coast of Sonora adjacent to Isla Tiburón. It is frost sensitive, and freezing weather is essentially absent throughout its range. The occurrence of this species through the length of the Baja California Peninsula is remarkable in that it transcends the seasonal precipitation gradient of the Peninsula. It ranges from the winter-rainfall–dominated north, near Ensenada, to the predominantly summer rainfall in the south in the Cape Region. Its presence on the Sonoran mainland may be due to migration via the Midriff Islands (Clark-Tapia & Molina-Freaner 2003; Cody et al. 1983). Although common along about 60 km of the seaward-facing mainland bajada, it does not range farther inland in Sonora. We cannot, however, rule out transportation by people, of this or various other species.

Tiburón: Ensenada Blanca (Vaporeta), 29 Jan 1965, *Felger*, observation (Felger 1966:173). Arroyo Sauzal, *Felger 2752.* Parte Sur de la Isla Tiburón, *Valiente-Banuet 614* (MEXU). Coralitos, *Wilder 06-63* (photo). Palo Fierro, lower bajada, edge of *Frankenia* zone, *Felger 10343.* Tecomate, SE of village, *Whiting 9042.* Pooj Iime, 29 Sep 2007, *Felger & Wilder*, observation.

San Esteban: Main arroyo at El Monumento, rocky hillsides and floodplain, *Felger 9187.*

Dátil: NE side of island, one pitaya agria seen, 21 Oct 1963, *Felger*, observation.

Patos: Near S edge of island, *Osario Tafal 1946* (photo, Felger 1966:354). S shore of island, base of sea ledge near shore, one juvenile about 30 cm tall with 3 stems, among *Atriplex barclayana*, 30 Sep 2007, *Felger*, observation.

Cholludo: Few seen, 21 Oct 1963, *Felger*, observation. A few individuals at the top of the island, 1 Mar 2009, *Wilder*, observation. 13 Apr 1911, *Rose 16815* (US, not seen, listed in Rose's field book).

Stenocereus thurberi (Engelmann) Buxbaum [*Lemaireocereus thurberi* (Engelmann) Britton & Rose]

Ool; *pitaya dulce*; organ pipe cactus

Description: Columnar cacti often 2–4 (4.5+) m tall, nearly trunkless or more often with several, short to ascending trunks, branching mostly from near the base, the stems with 13–18 ribs of low relief. Spines slender, grayish to nearly black, spreading in all directions, longer spines 1–4 (5) cm long, straight and needle-like (occasional spines are twisted). Areoles with dark red-brown glandular hairs, these turning black within one or two years, the exudate often producing dark red to blackish encrustations on the spines.

Flowers showy, 7–9 × 6–7 cm, nocturnal, closing at dawn, the inner (larger) tepals white or pale pinkish with white margins and bases; scales on tube purplish red; nectar chamber relatively large and producing copious nectar. Stamens, style, and stigma creamy white. Fruits globose, 4–6.5 cm diameter, very succulent, indehiscent or splitting irregularly; pericarpel (skin and cortex) thin, red when ripe. Unripe fruits green and spiny, the spines increasing in size before ripening; spines (areoles) readily deciduous from mature fruits. Seeds 2.2–2.4 × 1.5–1.8 mm, blackish. Flowering and fruiting mostly in summer; fruits

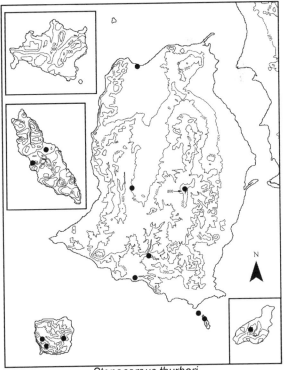

Stenocereus thurberi
Ool

mostly ripening July and early August, and sometimes a minor fruiting peak again in September.

Unlike other columnar cacti in the region, the fruit rind is relatively thin and edible. The fruit pulp is sweet, juicy, and super delicious. The small seeds are consumed along with the fruit pulp. Many Comcaac say it is the most delicious of all fruits. In traditional times the fruit was extensively utilized to make wine.

Local Distribution: Pitaya dulce is on all islands in the study area except Patos and Mártir. On Tiburón, widespread and common across most of the island, from near the shore to the peaks. On San Esteban, scattered and not common except in one canyon at the northwest part of the island (Bowen 2003). On Dátil, more common on the relatively arid west side of the island than on the east side. On Cholludo, a major component of the cactus forest dominating the isle. It is rare on Alcatraz. On Nolasco, abundant and widespread on the island including the west side but generally not on north-facing slopes on the east side of the island.

Other Gulf Islands: Tortuga (observation), San Marcos (observation), San Ildefonso (observation), Coronados (observation), Carmen (observation), Danzante (observation), Monserrat (observation), Santa Catalina, Santa Cruz (observation), San Diego (observation), San José, San Francisco, Espíritu Santo, Cerralvo.

Geographic Range: Organ pipe is the most widespread columnar cactus in Sonora and ranges from southwestern Arizona southward through the lowlands, foothills, and mountains below the oak-pine zones to Sinaloa and southwestern Chihuahua, and is on many Gulf Islands, and southern Baja California Norte and Baja California Sur.

Tiburón: S end of island, *Rempel 297*. Sauzal waterhole, locally common, rocky bench and pediment, also scattered along lower hills and intermountain valley, *Felger 10097*. SW Central Valley, 1 Feb 1965, *Felger*, observation (Felger 1966:239). Highest ridge of island, between Cójocam Iti Yaii and summit, 26 Oct 2007, *Wilder* (photo). Tecomate, *Whiting 9038*.

San Esteban: At about center of island, rocky basalt hills, 100 ft, *Felger 471*. Steep slopes of the S-central peak, *Felger 17549B*. SW part of island, W-facing side canyon with dense vegetation, in rocks on side of canyon, 2 m tall, rare, *Wilder 07-89b* (USON).

Dátil: Canyon at NE side of island (Felger 1966:299).

Alcatraz: E-central base of mountain, 28.81120°N, 111.96931°W (NAD 83), ca. 20 m, in thickest stand of *Pachycereus pringlei* on island, with *Carnegiea gigantea, Cylindropuntia bigelovii, C. fulgida* var. *fulgida*; only one individual seen (plus one skeleton), on cliff, ca. 4 m tall, 16 Sep 2007, *Wilder 07-424* (photo).

Cholludo: Common in the cactus forest, NE side of island (Felger 1966:306–308).

Nolasco: Documented for the island by our quadrat studies (Felger 1966; Felger et al. 2011) and multiple observations (e.g., Lindsay 1962) and photos. Cañón de la Guacamaya, July 2003, *Gallo-Reynoso* (photo).

Campanulaceae – Bellflower Family

Nemacladus orientalis (McVaugh) Morin [*N. glanduliferus* Jepson var. *orientalis* McVaugh]

Xtamaaija oohit "what mud turtles eat"; redtip thread plant

Description: Small winter-spring annuals mostly 4–15 cm tall. Stems wiry and thread-like, often much branched, the herbage usually dark olive green to purple-brown. Leaves 3–10 mm long, in a basal rosette and withering as the plant matures. Flowers less than 3 mm wide, bilateral, 2-lipped, white with pointed lobes and dark maroon-purple tips as if dipped in red wine. Fruits of capsules with many minute seeds.

Local Distribution: Widely scattered on San Esteban during favorable seasons and reported for Tiburón and Dátil but no specimens have been located (see "Doubtful and Excluded Reports").

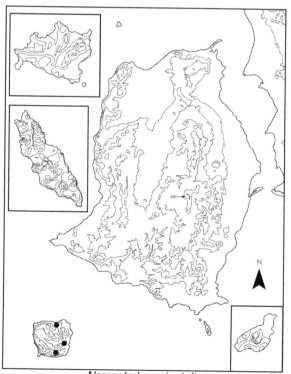

Nemacladus orientalis
Xtamaaija oohit

Other Gulf Islands: Ángel de la Guarda, San Lorenzo.

Geographic Range: Sonora from the coast in the vicinity of Bahía San Agustín [south of Tastiota] northward to Arizona, California, Nevada, Wyoming, Utah, and Baja California Norte.

San Esteban: Bed of main arroyo, *Moran 13039* (SD). E ridge of high peak near SE corner of island, 380 m, *Moran 8844* (SD). Narrow canyon, NW corner of island, *Felger 17605* (SD).

Cannabaceae – Hemp Family

Celtis – Hackberry

Hardwood trees and shrubs. Leaves scabrous (rough when you rub the leaf backwards), often asymmetric at the base, with 3 main veins and few lateral veins. Flowering with new growth, the flowers relatively inconspicuous, bisexual or unisexual, both often on the same branch. When the pollen ripens the filaments suddenly spring open and fling the dry, powdery pollen from the anthers. Fruit a drupe, the seeds primarily bird dispersed.

1. Stems and twigs often zigzag and mostly armed with thorns; leaf margins usually toothed; fruits with fleshy, orange pericarp. _____ **C. pallida**
1' Stems and twigs straight, unarmed; leaf margins mostly entire, some with a few apical teeth; fruits reddish, the pericarp scarcely fleshy. _____ **C. reticulata**

Celtis pallida Torrey subsp. **pallida** [*C. tala* Gillies var. *pallida* (Torrey) Planchon]

Ptaacal; *garambullo*; desert hackberry

Description: Briar-like thorny shrubs often 2–2.5 m tall. Leaves gradually drought deciduous, the margins usually toothed. Fruits 8–10 mm wide, round and orange with a thin, fleshy, edible and moderately sweet pericarp; often abundant in early fall and sparingly fruiting at other seasons.

Local Distribution: Tiburón, along washes in the Central Valley, along the eastern base of the Sierra Kunkaak, and in the Sierra mostly in canyon bottoms to at least 720 m.

Other Gulf Islands: None.

Geographic Range: Arizona to Texas and southward to Oaxaca, and southern Baja California Norte and Baja California Sur, and disjunct in Paraguay and Argentina. An additional subspecies occurs in South America.

Tiburón: SW Central Valley, 1 Feb 1965, *Felger*, observation (Felger 1966:239). Hast Cacöla, E flank of Sierra Kunkaak, 380 m, *Wilder 08-273*. Large wash in canyon in

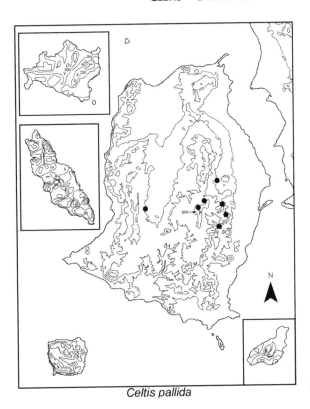

Celtis pallida

interior of the Sierra Kunkaak, past Pazj Hax waterhole, ca. 28°55'57.65"N, 112°17'27.25"W, *Romero-Morales 07-11*. Carrizo Canyon [Sopc Hax], near an active stream, *Knight 1119*. Caracol research station, *Knight 1099* (UNM). Canyon bottom, N base of Sierra Kunkaak, *Wilder 06-435*. Steep, protected canyon, 720 m, N interior of the Sierra Kunkaak, common in canyon bottom and throughout the canyon system, *Wilder 07-515*.

Celtis reticulata Torrey

Cumero; canyon hackberry, western hackberry

Description: Shrubs or occasionally small trees to 4+ m tall on Tiburón with smooth, gray bark. Leaves broadly lanceolate to mostly ovate and nearly as wide as long, often asymmetric, the margins mostly entire. Fruits reddish brown, hard-walled, globose, 7–8 mm diameter, the slightly fleshy mesocarp soon becoming dry.

Local Distribution: Tiburón in the interior of Sierra Kunkaak; canyon bottoms and on steep rocky slopes such as the talus slope of Capxölim. The nearest Sonoran populations are in deep riparian canyons in the Sierra el Aguaje north of Guaymas, such as Cañón Nacapule (Felger 1999) and the closest peninsular population is in the Sierra San Francisco. The Tiburón trees are relatively stunted compared to most other populations.

Other Gulf Islands: None.

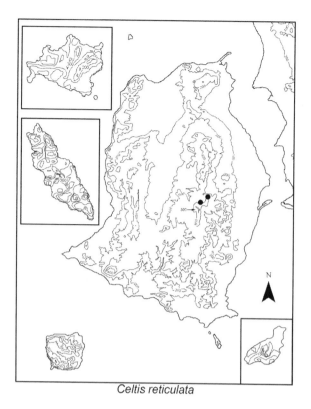

Celtis reticulata

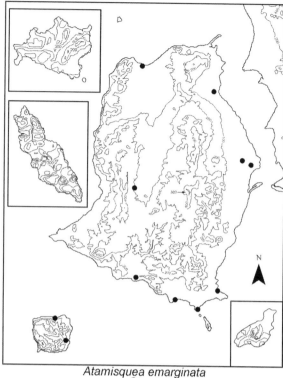

Atamisquea emarginata
Cöset

Geographic Range: Widespread trees in western North America including Baja California Sur. The type locality is considered to be somewhere in Colorado and trees from that region might not match those on Tiburón and in southern Sonora. The Arizona, and especially the Sonoran populations, have relatively thick leaves. The leaves of the southern Sonora trees are nearly as wide as long and usually with entire margins, indicating that they might be assigned to *C. laevigata* var. *brevipes* (S. Watson) Sargent. The Tiburón and southern Sonora populations are essentially evergreen rather than winter-deciduous and have relatively smooth bark rather than the irregular, warty-corky bark like those farther north.

Tiburón: Deep sheltered canyon in main complex of the Sierra Kunkaak, *Wilder 07-485*. Capxölim, 2 m tall, trunk not well developed, *Wilder 07-472*.

Capparaceae – Caper Family

Atamisquea emarginata Miers ex Hooker & Arnott [*Capparis atamisquea* Kuntze]

Cöset; *palo hediondo*

Description: Large shrubs, often 2–3 m tall and wider than tall, with rigid, woody branches; twigs straight, stout, and brittle, often branching at right angles and bluntly thorn tipped. Crushed leaves, flowers, and fruits foul smelling with mustard-oil glucosides, producing a strong, spicy odor like creosotebush and sharp horseradish. Young stems, lower leaf surfaces, outer surfaces of sepals, and pistils and fruits densely covered with translucent-winged silvery to yellowish peltate scales. Leaves evergreen or eventually leafless in extreme drought, alternate, tough and leathery, dark green above, dull silvery-gray below, 1–5 cm long, linear-oblong. Flowers white, moderately bilateral, the petals ca. 6.5 mm long. Fruits 1 cm long, oval, the exocarp splitting to reveal a bright red, fleshy aril containing usually 1 seed.

The flowers attract honeybees, native bees, large orange-winged tarantula hawk wasps (*Pepsis* and *Hemipepsis*), and many other insects. The red arils with the enclosed seeds, on the mainland are eaten by the house finch and verdin, the probable seed dispersers.

Local Distribution: Widespread in the lowlands of Tiburón at the southern, eastern, and northern parts of the island. San Esteban along Arroyo Limantour and on the north side of the island. A few large shrubs occur on Cholludo. Reported for Nolasco but not seen on the island and no specimens have been located.

Other Gulf Islands: Ángel de la Guarda (observation), Tortuga, Coronados (observation), Carmen, Danzante (observation), Monserrat (observation), San Diego, San José, San Francisco, Espíritu Santo.

Geographic Range: Arizona along the Sonora border in Organ Pipe Cactus National Monument and southward through western Sonora to northwestern Sinaloa, southern Baja California Norte and Baja California Sur, Gulf Islands, and in Bolivia, Chile, and Argentina.

Tiburón: Arroyo Sauzal, *Felger 08-27*. Ensenada de la Cruz, low arroyo, *Felger 2587*. La Viga, *Wilder 08-151*. Ensenada del Perro, *Tenorio 9528*. SW Central Valley, 2 Feb 1965, *Felger*, observation (Felger 1966:236). Palo Fierro, *Felger 10139*. 3 km W [of] Punta Tormenta, *Scott 9 Apr 1978* (UNM). Hahjöaacöl, NE coast, *Wilder 08-214*. Tecomate, *Whiting 9002*.

San Esteban: Arroyo Limantour, floodplain 100 m inland, *Felger 7036*. N side of island, *Felger 15443*.

Cholludo: N-central slope, one large shrub seen, ca. 2.3 m tall, near top of island, *Wilder 08-312*.

Caryophyllaceae – Pink Family

1. Stipules and flowers conspicuously papery white; internodes shorter than the leaves; leaves opposite, longer than wide. _____**Achyronychia**
1' Stipules and flowers not papery white; lower internodes much longer than the leaves; leaves fascicled, nearly as wide as long. _____ **Drymaria**

Achyronychia cooperi Torrey & A. Gray

Hant yapxöt "land's flower," tomitom hant cocpeetij *maybe* "tomitom circular on the ground," tomitom hant cocpeetij caacöl "large tomitom"; frost mat, onyx flower

Description: Small winter-spring annuals; glabrous, becoming prostrate and mat-like. Leaves opposite, bright

Figure 3.68. *Achyronychia cooperi.* Hahjöaacöl, Isla Tiburón. 6 April 2008. BTW.

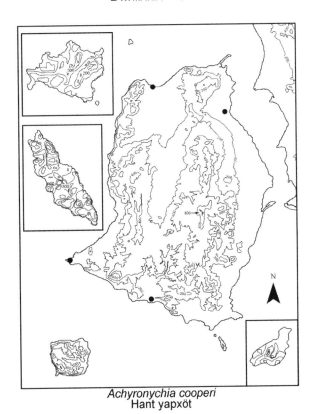

Achyronychia cooperi
Hant yapxöt

green, spatulate (widest at tip), those of each pair unequal in size; stipules thin, papery white. Flowers small, white, papery, and crowded in axils along the stem. Fruits 1-seeded, indehiscent.

Local Distribution: Recorded at widely scattered lowland sites on Tiburón, mostly in sandy-gravelly soils near the coast.

Other Gulf Islands: Ángel de la Guarda, San Lorenzo, San Marcos.

Geographic Range: Western Sonora from the Guaymas region northward, both Baja California states, southern California, western Arizona, and southern Nevada.

Tiburón: Ca. 1 km N of Punta Willard, 12 Apr 1968, *Felger 17741* (specimen not located). Arroyo Sauzal, broad sandy wash, *Felger 08-15*. Hahjöaacöl, NE coast, *Wilder 08-218*. Bahía Agua Dulce, dunes near beach, *Felger 6818*.

Drymaria holosteoides Bentham var. **holosteoides**

Description: Winter-spring annuals. Leaves 4–10 mm long, broadly elliptic and often semi-succulent, clustered 4–6 on slender stems. Flowers in short dense racemes in leaf axils. Flowers small, with white petals.

Local Distribution: Common along the western and northern parts of Tiburón and locally in the Sierra Kunkaak; arroyos, valley plains, and sometimes on rocky slopes.

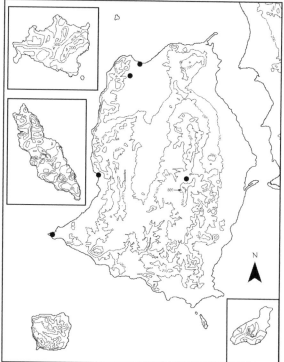

Drymaria holosteoides var. *holosteoides*

Other Gulf Islands: Ángel de la Guarda, San Marcos, Coronados, Carmen, San José, San Francisco, Espíritu Santo.

Geographic Range: Baja California Sur and southern part of Baja California Norte, and Gulf Islands.

Tiburón: Ensenada Blanca (Vaporeta), W-facing rocky hillside, *Felger 12204* (UC). Near Willard's Point, common on sandy wash just back from ocean, *Johnston 4263* (CAS, UC). Arriba de Siimen Hax, en una mesa, Dec 2002, *Romero-Morales* (photo). Valle de Agua Dulce, *Wilder 07-260*. Tecomate, flat terrain, *Felger 6836* (SD).

Celastraceae – Staff-Tree Family

1. Branches essentially leafless and spinescent._ **Canotia**
1' Branches conspicuously leafy and unarmed. **Maytenus**

Canotia holacantha Torrey

Xooml icös caacöl "*Koeberlinia* with large spines"; canotia, crucifixion thorn

Description: Hardwood and essentially leafless shrubs 0.5–1.3 m tall on exposed ridges to sometimes small trees to ca. 3 m tall in more favorable niches. Leaves scale-like, minute, very quickly deciduous, present only on fresh, new growth. Smaller branches and twigs pale green to bluish green, rigid, mostly thorn tipped; dark

areas of densely crowded, dark red-brown stalked glands are conspicuous on stems just above the nodes. Flowers white, in small clusters, 5-merous, 8–9 mm wide, the sepals and petals with slightly ragged margins. Fruits persistent, 1.5–2 cm long, at first fleshy, soon becoming woody capsules with 5 carpels splitting near the tip (apically) into awned valves.

Local Distribution: *Canotia* is localized at the top of Tiburón on exposed ridges that link the highest elevations of the Sierra Kunkaak. Two subpopulations were encountered. The subpopulation on the ridge near Cójocam Iti Yaii discovered in 2007, has two aggregations, each with several shrubs (*Wilder 07-528 & 07-549*), which are notably stunted. Browsing by the large population of bighorn sheep might be responsible for the relatively dwarfed stature of these shrubs, but more likely they are dwarfed by the harsh conditions on windswept rock ridges with minimal soil. The second subpopulation, discovered in 2008 by Humberto, is on the southern portion of the highest elevations of the Sierra and has some individuals that reach about 3 m in height.

Other Gulf Islands: None.

Geographic Range: The nearest known *Canotia* plants occur 230 km to the northeast in northern Sonora at 610–1525 m in the Altar-Tubutama area and in the foothills southeast of Magdalena de Kino to the nearby Sierra Baviso and Sierra Madera near Imuris (Felger et al. 2001;

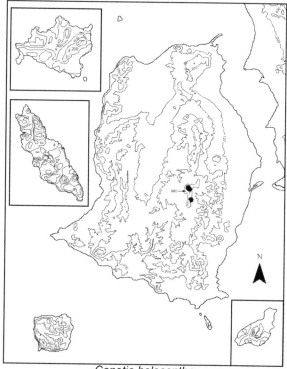

Canotia holacantha
Xooml icös caacöl

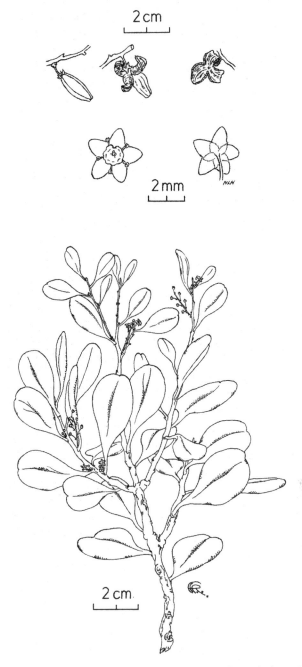

Figure 3.69. *Canotia holacantha.* LBH.

Figure 3.70. *Maytenus phyllanthoides.* Branch, FR; fruits and flowers, NLN.

Figure 3.71. *Canotia holacantha.* José Ramón Torres, left, and Brad Boyle, right. Upper ridge of Sierra Kunkaak, Isla Tiburón. BTW, 26 October 2007.

Wilder et al. 2008b). Turner et al. (1995:145) report that this species is a widespread and occasional dominant in central Arizona, where it is often a thick-trunked "shrub or small tree 2–6 (occasionally 10) m tall." *Canotia* is characteristic of elevations above and at the uppermost limits of the desert. Although sometimes reported for southern Utah, there are no records for it in the state (Welsh et al. 2008; Wilder et al. 2008b).

Canotia is a genus of two species and one of three genera of "crucifixion-thorns" in the Sonoran Desert (see *Castela* and *Koeberlinia*). It somewhat resembles a leafless foothill palo verde (*Parkinsonia microphylla*), from which it is readily distinguished by the very different, persistent capsules, shredding bark, and somewhat different growth form.

Turner et al. (1995:145) reported *Canotia* on the east-central coast of Tiburón based on collections made by Felger. Via conversations and data checking with Ray Turner (personal communication, 2007) it became clear that collections of *Castela polyandra* (*Felger 9359, 10135*) were incorrectly ascribed to *C. holacantha* and were represented as such in *Sonoran Desert Plants: An Ecological*

Atlas (Turner et al. 1995). The presence of *C. holacantha* on an island in the Gulf of California is best explained as a Pleistocene relict (Wilder et al. 2008b).

Tiburón: Exposed upper ridges of the Sierra Kunkaak, 825 m, 1.3 m tall, and about as wide, the tallest individual seen, 28°57'49.08"N, 112°19'40.37"W, *Wilder 07-549*. E-facing slope just below high ridge, 800 m, ca. ½ m tall, population of ca. 10 individuals, 28°57'57.27"N, 112°19'47.75"W, *Wilder 07-528*. Sierra Kunkaak, parte sur, c. 3 m de altura, trono c. 10 cm de diámetro, 22 Jan 2008, ca. 28°56'37.35"N, 112°19'29.56"W, *Romero-Morales 08-7*.

Maytenus phyllanthoides Bentham [*Tricerma phyllanthoides* (Bentham) Lundell. *Maytenus texana* Lundell]

Cos; *mangle dulce*

Description: Mound-shaped, unarmed shrubs often 1–3 m tall, densely and sinuously branched, with hard wood and brittle twigs; often forming impenetrable thickets along the coast. Roots reddish with elastic fibers (a common trait in stems and roots in this family).

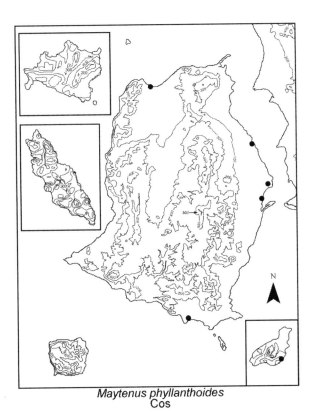

Maytenus phyllanthoides
Cos

Glabrous except minute hairs on youngest herbage, inflorescences, and flowers. Leaves alternate, 2–6 cm long, semi-succulent, and obovate. Flowers 3–4 mm wide, greenish yellow, among the leaves in small axillary clusters or solitary, (4) 5-merous; unisexual or sometimes appearing bisexual, the plants perhaps often dioecious. Fruits 9–12 mm long, splitting to reveal a sticky, bright red, fleshy, and moderately sweet aril. Flowering and fruiting during the warmer months.

Local Distribution: Northern, eastern, and southern coastal margins of Tiburón in saline soils; best developed where the roots are flooded at highest tides. Also in arroyo beds and low-lying areas ranging into the lower margins of mixed desert scrub. Especially common along the Infiernillo shore bordering mangroves and a major component of the shoreline halophytic scrub. One large group of plants seen on Alcatraz in 2007, the only record for the island. Roots used in remedies for sore throat.

Other Gulf Islands: Tortuga, San Marcos (observation), Coronados, Carmen (observation), Danzante, Santa Catalina, San José, San Francisco, Espíritu Santo.

Geographic Range: Coastal habitats in Sonora from Estero Sargento and Tiburón to Sinaloa and Baja California Sur. The northernmost *Maytenus* is reported from Puerto Libertad, but it apparently has been extirpated there (Turner et al. 1995). Also Texas, Florida, and Mexico on the Gulf of Mexico and the Caribbean.

Tiburón: Ensenada de la Cruz, 0.5 km inland, *Felger 12794.* Beach 1 km N of Estero San Miguel, *Wilder 06-8.* Palo Fierro, *Russell 15 Mar 1961.* Valle de Águila, within 50 m of shore, *Wilder 07-267.* Bahía Agua Dulce, arroyo bottom, *Felger 10210.*

Alcatraz: SE side of island, coast at base of mountain, *Felger 07-168.*

Chenopodiaceae, *see* Amaranthaceae

Cleomaceae – Cleome Family

Cleome tenuis S. Watson subsp. **tenuis**

Cocool; *cleome*

Description: Delicate hot-weather annuals, glabrous, with slender, upright stems. Leaves palmately compound, with 3–5 leaflets. Flowers inconspicuous, 3–5 mm long, the corollas pale yellow; pedicels with small leafy bracts. Fruit an elongated, many-seeded capsule.

Local Distribution: Tiburón in the Central Valley and in the Sierra Kunkaak.

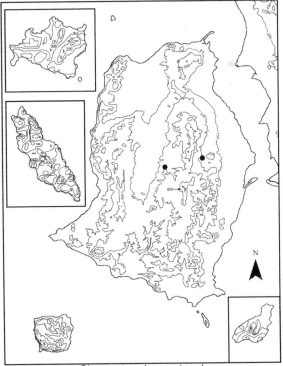

Cleome tenuis ssp. *tenuis*
Cocool

Other Gulf Islands: Carmen, Danzante, San José.

Geographic Range: Sinaloa northward at least to Hermosillo, and Baja California Sur. Another subspecies occurs southward in Mexico to Guatemala.

Tiburón: Haap Hill, *Felger T74-11*. Near Caracol research station, *Knight 955* (UNM).

Cochlospermaceae – Cochlospermum Family

Amoreuxia palmatifida Sessé & Moçiño ex de Candolle

Joját (variant: xoját); *saya*

Description: Herbaceous perennials from a single, thick and succulent, tuberous root, responding to summer rains and dormant during the rest of the year. Stems seasonal, reaching 20–30 cm tall but often shorter, herbaceous and leafy; leaves palmately lobed. Flowers bilateral, ca. 5 cm wide, the stamens and petals in loose opposite bunches, forming a cradle-like arrangement, the stamens with apical pores. Petals bright orange, the four lower petals each with a large maroon blotch. The flowers are buzz pollinated: large bees alight in the open center of the flower and vibrate, or buzz vigorously, causing the pollen to shoot out of the anther pores. Upper stamens with maroon anthers and orange filaments, the lower stamens with orange anthers of the same color as the petals. Fruits 3–4 cm long, ovoid, green, fleshy until the seeds ripen, at which time the fruit dries and splits. Seeds 4–5 mm long, kidney shaped, and blackish.

Flowering with the new shoots during hot, humid weather, beginning with the first rains, sometimes as late as August or early September. Flowers open in the early morning and fade as the heat mounts later in the morning. The flowers seem to mimic those of *Kallstroemia grandiflora*, which is much more common and offers insect visitors a nectar reward—*Amoreuxia* flowers provide no nectar.

The roots were an important food resource for the Comcaac and other Sonoran people, and have been seasonally offered for sale in local Sonoran marketplaces. Apparently all parts of the plant are edible. According to Howard Scott Gentry (personal communication, 1990), the best part is the green, unripe fruit.

Local Distribution: Tiburón in the Sierra Kunkaak and hills and valley plains at the south end of the Central Valley. It also occurs in mountains on the adjacent Sonora mainland.

Other Gulf Islands: None.

Geographic Range: South-central Arizona to Colombia.

Tiburón: Haap Hill, *Felger T74-2*.

Combretaceae – Combretum Family

Laguncularia racemosa (Linnaeus) Gaertner

Pnaacoj hacaaiz "mangrove spear"; *mangle blanco*; white mangrove

Description: Evergreen shrubs or small trees 2–4 m tall, the wood relatively hard. Thick, knobby, branched roots (pneumatophores) develop from the shallow, horizontal root system. Leaves evergreen, opposite and decussate, the blades 3.2–8.7 cm long, shiny green, semi-succulent and brittle; stipules none.

The leaves have several kinds of secretory structures: (1) A conspicuous pair of circular petiole glands near the leaf blade functioning as extrafloral nectaries, secreting a sweet solution. (2) Dot-like or pore-like submarginal glands on the lower surface at the junctions of minor

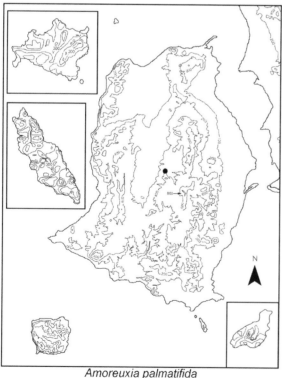

Amoreuxia palmatifida
Joját (var. xoját)

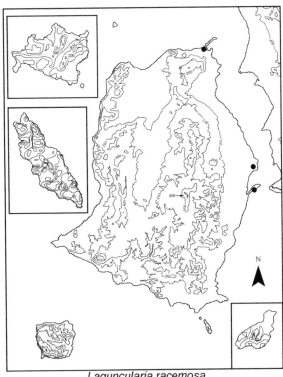

Laguncularia racemosa
Pnaacoj hacaaiz

Figure 3.72. *Laguncularia racemosa*. **LBH.**

veins; in young leaves these seem to function as hydathodes, secreting water or mucilage. (3) Microscopic, translucent dot-like glands in small pits excreting salt. (4) Inconspicuous glandular hairs on the midrib and young leaves.

Flowers bisexual, or unisexual and male and female flowers on different plants. Flowers small and greenish white and visited by bees and other insects. Fruits of 1-seeded nutlets 1.4–2 cm long, with a spongy outer portion facilitating water dispersal. Seedling development begins while the fruit is still on the tree, the seedling emerges rapidly after the fruit falls and becomes stranded in shallow tidal water. Flowering and fruiting during warmer months; fruits probably ripening 2 or 3 months after flowering.

Local Distribution: Infiernillo coast of Tiburón in the three mangrove esteros. The roots tidally inundated daily. Growing in close association with the other two mangroves. White mangrove is densest and most abundant in the zone between the red (*Rhizophora*) and black (*Avicennia*) mangroves.

Other Gulf Islands: Carmen, Danzante, San José, San Francisco, Espíritu Santo.

Geographic Range: Coastal Sonora from Estero Sargento and Tiburón southward, and both shores of Baja California Sur. The northernmost Sonoran white mangroves are sometimes severely frost damaged. Mexico to Panama, Gulf of Mexico and Caribbean to Florida, northern South America, and West Africa. In the tropics it develops into a tree sometimes 20 m in height.

Tiburón: Estero San Miguel, *Wilder 06-275*. Punta Tormenta, estero, *Wilder 08-328*. Punta Perla, estero, Cyazim It, *Wilder 08-375*.

Convolvulaceae – Morning Glory Family

Annual or perennial herbs, mostly vines, or obligate parasites with little or no chlorophyll and vegetative parts much reduced in *Cuscuta*. Leaves alternate, simple, entire to deeply parted, lacking stipules. Flowers radial and often showy. Fruits (of species in the flora area) of dry capsules with 1–4 (6) seeds.

Figure 3.73. *Jacquemontia abutiloides.* Foothills of Sierra Kunkaak, Isla Tiburón. BTW, 25 November 2006.

1. Parasitic plants without chlorophyll; stems uniformly yellow or orange; leaves reduced to scales. _____**Cuscuta**
1' Not parasitic, with chlorophyll, stems green or brown; leaves well developed.
   2. Stems conspicuously vining.
      3. Herbage sparsely pubescent with simple hairs; stigma lobes globose. _____ **Ipomoea**
      3' Herbage moderately to usually densely pubescent with stellate hairs; stigma lobes oblong. ___**Jacquemontia**
   2' Stems not vining.
      4. Flowers solitary in leaf axils, the pedicels appearing continuous with the calyx, shorter than flowers or capsules; corollas white; coastal halophyte. _____ **Cressa**
      4' Flowers 1 or more on distinct, slender peduncles or pedicels much longer than the flowers; corollas blue; not halophytes.
         5. Perennial herbs; peduncles 1- or 2-flowered, with a slender bract 2 mm or less in length; styles 2, each bifid, the 4 stigmas long and thread-like. _____ **Evolvulus**
         5' Annuals; peduncles 1- to many-flowered, with leafy bracts more than 2 mm long; style 1, the stigma with 2 ovate to oblong lobes. _____ **Jacquemontia agrestis**

Cressa truxillensis Kunth

Ziix casa insii "what the one who is stingy does not smell"; alkali weed

Description: Herbaceous perennials, dying back during adverse times to thickened, underground stems and rootstocks including rhizomes. Herbage silvery gray-green with appressed, silky hairs. Stems erect to spreading, 8–20 cm long, not twining. Leaves less than 1 cm long, sessile or nearly so, mostly ovate to elliptic, often reduced above. Flowers white, 6–8 mm long including the short pedicel, solitary in upper stem axils. Calyx leafy, persistent, the corolla also persistent but shriveling, the styles 2, separate, the stigmas knob-like. Capsules 5 mm long, ovoid, usually 1-seeded, shiny brown.

Local Distribution: Scattered coastal localities on Tiburón; saline soils near the shore including mangrove margins and coastal halophyte communities, and sometimes on upper beaches and lower beach dunes. Also common and widespread near the shore on Alcatraz.

Other Gulf Islands: Ángel de la Guarda, Rasa, Salsipuedes, Monserrat, San José, San Francisco, Cerralvo.

Geographic Range: Western North America, Mexico, and South America.

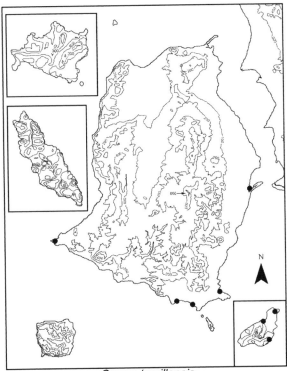

Cressa truxillensis
Ziix casa insii

Tiburón: Coralitos, *Wilkinson 215*. 1 km N of Punta Willard, wind-blown sand piled against rock hills, *Felger 17746*. Ensenada de la Cruz, low wet saline soil, *Felger 9068*. Ensenada del Perro, *Wilder 07-158*. Zozni Cmiipla, *Wilder 06-366*.
Alcatraz: NE shore, sandy-rocky soil and *Allenrolfea* flats, *Felger 12712*. Rocky beach, among rounded stones, *Felger 14928b*. S-central coastal side of island, with *Atriplex barclayana*, *Cylindropuntia fulgida* var. *fulgida*, shell rubble at upper beach, *Felger 07-167*.

Cuscuta – Hamt itoozj (variant: hant itoozj) "land's intestines"; *fideo*; dodder

Obligate parasites. Annual vines, glabrous, the stems orange or yellow-orange, slender and thread-like, attached to host plants by numerous small haustoria. Leaves reduced to minute scales. Flowers 4- or 5-merous, small, fleshy, and white. Stamens alternating with the corolla lobes; infrastaminal scales commonly present and with toothed or fringed margins. Capsules globose, membranous when mature, separating near the base (circumscissle, those in the flora area). Seeds 1–4 per capsule.

Identification of most *Cuscuta* species is not easy and examination under a microscope is recommended. Measurements of floral parts were done on rehydrated

herbarium material. Length of flowers was measured from the base of calyx to the tip of straightened corolla lobes. Images with details of dissected flowers are available from a digital atlas of *Cuscuta* (Costea 2009). Stem diameters on living plants are described as "slender" with the diameter of 0.35–0.4 mm and "medium" with the diameter of 0.4–0.6 mm (Yuncker 1921). Species with "coarse" stem diameters, greater than 0.6 mm, are not known from the Sonoran Islands or the adjacent mainland.

1. Stems of medium diameter; calyx lobes wider than long, with obtuse or rounded tips (apices); seeds 1 or 4 per capsule.
    2. Flowers 2.5–4.2 mm long; capsules globose-ovoid to ovoid, seeds 1 per capsule. _____ **C. americana**
    2' Flowers 4.5–6.5 (7) mm long; capsules globose to globose-depressed, seeds 4 per capsule._____
    _____**C. corymbosa**
1' Stems of slender diameter; calyx lobes longer than wide, with acute tips (apices); seeds 2–4 per capsule.
    3. Flowers 5-merous, 2–3 mm long; calyx equaling or somewhat longer than corolla tube. **C. desmouliniana**
    3' Flowers 4-merous, 3.5–4.5(5) mm long; calyx ¼ to ¾ as long as the corolla tube. _____ **C. leptantha**

Cuscuta americana Linnaeus

Hamt itoozj "land's intestines"

Description: On perennial vines and shrubs including *Colubrina viridis*, *Cottsia gracilis*, *Lantana velutina*, *Melochia tomentosa*, and *Solanum hindsianum*. One of the two "larger" *Cuscuta* on the Midriff Islands, the stems medium diameter, yellow-orange. Flowers 5-merous, 2.5–4.2 mm long, white when fresh, brownish upon drying. Calyx 2.4–3.3 mm long, ¾ to as long as the corolla tube. Corollas 2–3.3 mm long, the tube 1.7–2.5 mm long, the lobes 0.5–0.8 mm long. Stamens included (not protruding), shorter than the corolla lobes. Capsules 1.8–3 × 0.8–2 mm, globose-ovoid to elliptic. Seeds 1 per capsule.

Local Distribution: Common in canyons and slopes in the Sierra Kunkaak.

Other Gulf Islands: None.

Geographic Range: The nearest known localities in Sonora are just south of Punta Chueca and in the Guaymas region; it is widespread in Sonora in non-desert regions. Florida and Mexico to South America and the West Indies.

Tiburón: E side of island, foothills of NE portion of the Sierra Kunkaak, *Wilder 06-373* (determined by Mihai Costea,

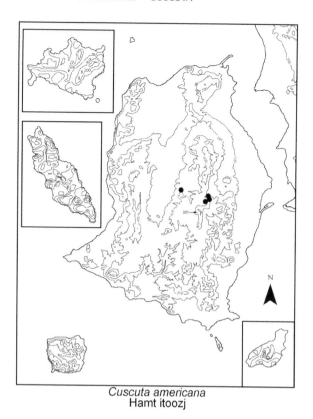

Cuscuta americana
Hamt itoozj

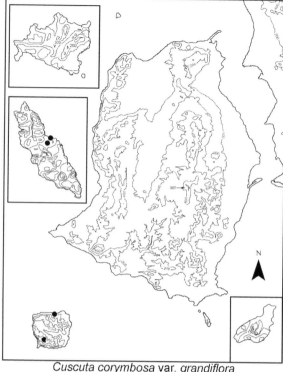

Cuscuta corymbosa var. grandiflora
Hamt itoozj

2008). Canyon at NE base of Sierra Kunkaak, *Wilder 06-428*. N base of Sierra Kunkaak, *Felger 07-98* (determined by Mihai Costea, 2008). Top of Sierra Kunkaak Segundo, E peak of Sierra Kunkaak, *Wilder 06-495* (determined by Mihai Costea, 2008). SE side of Agua Dulce Valley, 12 mi S from Tecomate, 280 m, *Felger 76-T14*.

Cuscuta corymbosa Ruiz & Pavón var. **grandiflora** Engelmann

Hamt itoozj "land's intestines"

Description: Stems medium diameter, orange. Flowers 5-merous, 4.5–6.5 (7) mm long, white when fresh, brownish when dried. Calyx 2–2.5 mm long, ½ to ¾ as long as the corolla tube. Corollas 4–6 mm long, the tube 3–5 mm long. Stamens included (not protruding), shorter than the corolla lobes. Capsules 2–2.9 × 2.2–2.6 mm, globose to slightly depressed. Seeds 4 per capsule.

Local Distribution: On San Esteban, apparently widespread during times of favorable rains. On Nolasco, seasonally common on the east side of island, parasitic on *Vaseyanthus insularis* as well as many shrubs, grasses, and herbaceous annuals and perennials.

Other Gulf Islands: Ángel de la Guarda, Partida Norte, Carmen, Danzante, Espíritu Santo, Cerralvo (*Johnston 4070*, CAS, determined to var. *grandiflora* by Yunker, 1948; listed in Rebman et al. 2002 as the species but not as the variety).

Geographic Range: Also on the opposite Sonora mainland. Mexico to South America. Four varieties are recognized; var. *grandiflora* extends across most of the range of the species.

San Esteban: N side of island, arroyo, *Felger 15405*. Steep slopes of S-central peak, *Felger 17531*. Steep N slope of NE peak, 28°42'N, 112°35'W, 450 m, 26 Apr 1966, on *Mirabilis laevis* var. *crassifolia* (*Moran 13052*).

Nolasco: Steep granitic slope, *Gentry 11354* (determined by M. Costea, 2005). NE side of island, N-facing slope, parasitic vine on shrubs and herbs, many on *Perityle californica*, also on *Vaseyanthus insularis*, *Felger 12082* (determined by M. Costea, 2005, "intermediate between var. *stylosa* (Choisy) Engelmann and var. *grandiflora*, but closer to *grandiflora* because calyx not reaching beyond middle of corolla tube and corolla bulging, but these are dried"). NE side, halfway to summit, 150 m, abundant at all elevations, mostly on *Vaseyanthus insularis*, also on grasses and many shrubs, *Felger 06-91*.

Cuscuta desmouliniana Yuncker

Hamt itoozj "land's intestines"

Description: Growing on various annuals including *Amaranthus watsonii, Boerhavia, Pectis*, and herbaceous *Euphorbia* subgenus *Chamaesyce*. Stems slender, yellow-orange. Flowers 5-merous, 2–3 mm long, white when fresh, creamy white when dried. Calyx 1–1.4 mm long, brownish yellow, equaling or somewhat longer than the corolla tube, the calyx lobes acute at apex. Corollas 1.8–2.8 mm long, the tube cylindric, 1.2–1.5 mm long, the lobes 1.2–1.5 mm long, erect to spreading or reflexed, as long as the tube; stamens short-exerted, shorter than corolla lobes. Capsules 1.5–2 × 0.9–1.7 mm, globose, capped by the withered corolla. Seeds 2–4 per capsule.

Local Distribution: Recorded from Tiburón and Dátil by a single record from each island but probably more widespread. Common on the adjacent Sonora mainland.

Other Gulf Islands: None.

Geographic Range: Western and central Sonora, Sinaloa, and the Baja California Peninsula.

Tiburón: Zozni Cmiipla, at base and N side of Punta San Miguel, *Felger 08-120*.

Dátil: NW side of island, *Felger 15313A*.

Cuscuta leptantha Engelmann

Hamt itoozj "land's intestines"

Description: Generally growing on herbaceous species of *Euphorbia*, subgenus *Chamaesyce*, especially *E. polycarpa*. Stems slender, yellow-orange. Flowers 4-merous, 3.5–4.5 (5) mm long, white when fresh, creamy white when dried. Calyx ⅓ to ½ as long as the corolla tube, calyx lobes acute at apex. Corolla 3–4 mm long; corolla tube 1.5–2.5 mm long; corolla lobes 1.5–2 mm long, about equal to the tube, spreading, triangular oblong and acute. Stamens short-exerted, shorter than corolla lobes. Capsules 1.5–2 × 1.6–1.9 mm, globose, capped by the withered corolla. Seeds 2–4 per capsule.

Local Distribution: Documented for Tiburón in the Central Valley, the Sierra Kunkaak, and along the Infiernillo coast, and predicted to be more widespread.

Other Gulf Islands: None.

Geographic Range: Western Sonora from the vicinity of El Desemboque southward, Sinaloa, the Baja California Peninsula, New Mexico, and Texas.

Tiburón: SW Central Valley, *Felger 17342*. 1 km inland at Zozni Cmiipla, at base and N side of Punta San Miguel, *Wilder 06-368*. Canyon at base of Capxölim, *Wilder 06-381*.

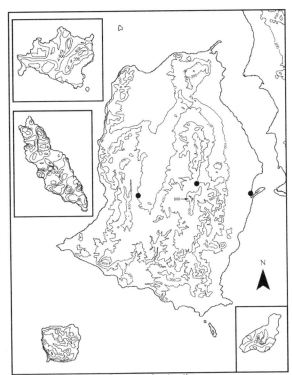

Cuscuta desmouliniana
Hamt itoozj

Cuscuta leptantha
Hamt itoozj

Evolvulus alsinoides Linnaeus [*E. alsinoides* var. *acapulcensis* (Willdenow) van Ooststroom. *E. alsinoides* var. *angustifolia* Torrey]

Oreja de ratón

Description: Small perennials, also flowering in the first season, with slender stems and pedicels, and narrow leaves. Flowers open in the morning and withering with daytime heat, the corollas pale blue. Styles 2, bifid, the 4 stigmas long and thread-like. Flowering at various seasons.

Local Distribution: Widespread in the Sierra Kunkaak, from canyons and hills at the base of the mountains to the summit.

Other Gulf Islands: Cerralvo.

Geographic Range: Texas to southern Arizona and Baja California Sur to South America. This species, native to the New World, is now a pantropical weed. *E. alsinoides* is highly polymorphic, and at least a few populations in Asia and Australia appear to have differentiated in post-Columbian times into new infraspecific taxa, in a manner akin to the English Sparrow in North America (Austin 2008).

Tiburón: Sopc Hax, large rocky canyon, *Felger 9324g.* Mountain above valley, E base of Sierra Kunkaak, 310 m, *Wilder 06-4.* Canyon bottom, N base of Sierra Kunkaak, *Wilder 06-408.* Top of Sierra Kunkaak, just within the "bowl," *Knight 1005.*

Ipomoea hederacea Jacquin

Hehe quiijam "entwining plant," hataaij "what is spun (like a top)"; *trompillo azul*; morning glory

Description: Short-lived annual vines growing with summer rains, often robust and growing over shrubs and small trees. Leaves mostly broadly ovate, entire or usually broadly 3 (5)-lobed. Sepal bases with large, coarse hairs, the sepal lobes attenuate tipped; corollas opening wide, the color unknown for the Tiburón plants but this species has bluish or pinkish to lavender corollas, generally with a white base. The flowers open in the early morning and wither with mid-morning heat. Capsules in this species can have up to 6 seeds.

Local Distribution: Documented in widely separated areas on Tiburón and likely to be seasonally more widespread. Not known from the hyper-arid western margin of the island.

Other Gulf Islands: None.

Geographic Range: Widespread in the Sonoran Desert including western Sonora and Baja California Sur in the Sierra San Francisco. North and South America, and adventive in the Old World.

Tiburón: Coralitos, *Wilkinson 188.* Ensenada del Perro, *Romero 06-2A.* SW part of Central Valley, dead vine in shrub, 1 Feb 1965, *Felger 12312* (specimen not located). Haap Hill, *Felger 76-T23.* Zozni Cmiipla, *Wilder 06-349.*

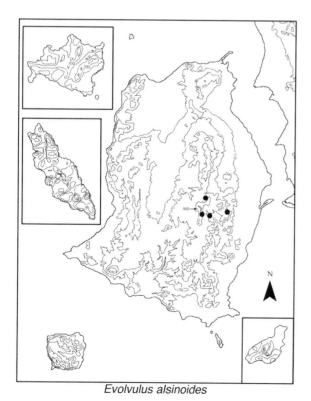

Evolvulus alsinoides

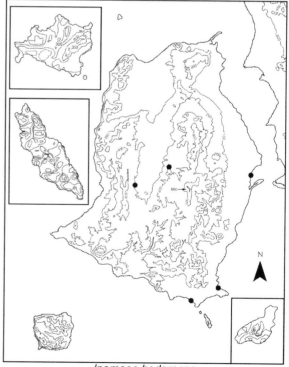

Ipomoea hederacea
Hehe quiijam, hataaij

Jacquemontia

1. Perennials, usually with a woody or semi-woody base; calyx 6–9 mm long; corollas (18) 20–35 mm wide. ___
_____ **J. abutiloides**
1' Annuals, growing during summer rainy season; calyx 5–7 mm long; corollas probably 10–15 mm wide. ___
_____ **J. agrestis**

Jacquemontia abutiloides Bentham

Description: Scrambling and twining perennial vines with slender branches growing into and overtopping shrubs; younger stems and herbage densely pubescent with crowded, several-rayed, stellate hairs nearly white when young and golden brown with age. Leaves broadly ovate. Inflorescences mostly 3- to 7-flowered, the upper and outer stems bear numerous crowded inflorescence clusters. Corollas blue (18) 20–35 mm wide.

Local Distribution: Widespread in the Sierra Kunkaak and its eastern bajada to the south shore of the island; especially along arroyos and canyons, and also on desert plains and rocky slopes.

Other Gulf Islands: Tortuga, San Marcos, San Ildefonso (type locality), Carmen, Danzante, Santa Catalina, Santa Cruz, San José, San Francisco, Espíritu Santo, Cerralvo.

Geographic Range: Central Baja California Norte to the Cape Region in Baja California Sur and on Tiburón. Apparently not on the Sonora mainland, depending on taxonomic interpretations.

Notes: *Jacquemontia abutiloides* is apparently closely related to *J. eastwoodiana* of the Baja California Peninsula and they are questionably distinct. According to Robertson (1971), *J. pentantha* (Jacquin) G. Don is at the center of a group of species that includes *J. abutiloides*, *J. albida*, *J. eastwoodiana*, *J. polyantha*, *J. pringlei*, and five others. This group of taxa ranges from Arizona to Central America. The relationships of these taxa in the Sonoran Desert region remain confused (Dan Austin, personal communication, 2008).

Tiburón: Arroyo Sauzal, 2 mi inland, *Felger 12266*. Ensenada de la Cruz, *Felger 2569*. Coralitos, *Wilder 06-66*. Upper bajada E of Pazj Hax, *Romero-Morales 07-12*. Zozni Cmiipla, 1 km inland, *Wilder 06-370*. Between shore and Sierra Kunkaak, bajada flats, *Wilder 06-21*. Between Sopc Hax and Hant Hax camp, lower foothills, *Felger 9335*. Camino de Caracol, broad arroyo entering foothills of Sierra Kunkaak, *Wilder 05-1*. Cerro San Miguel, común sobre arbustos en el pie de monte, *Quijada-Mascareñas 90T013*. San Miguel Peak, 2000 ft, rough brushy SW-facing slope, *Knight 958* (UNM). Top of Sierra Kunkaak Segundo, 490 m, *Wilder 06-487*.

Jacquemontia agrestis (Martius ex Choisy) Meisner [*J. palmeri* S. Watson]

Description: Short-lived summer annuals (ephemerals) with 3-armed stellate hairs. Stems not vining or perhaps forming a small vine. Corollas blue. Calyx 3.5–6.5 mm long; corollas 7-8 mm wide; capsules rounded, 4–5 mm wide (measurements based on specimens from elsewhere in the Sonoran Desert region).

Local Distribution: Documented on Tiburón from a single collection of dry, dead specimens in the Central Valley but probably more widespread.

Other Gulf Islands: None.

Geographic Range: South-central Arizona through Mexico to Honduras and South America; annuals or perennials and often a vining weed in cultivated fields. Widespread in Baja California Sur and Sonora mostly east and south of the Sonoran Desert.

Tiburón: Haap Hill, 11 Dec 1976, *Felger 76-T24*.

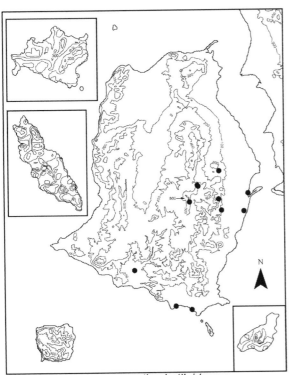

Jacquemontia abutiloides

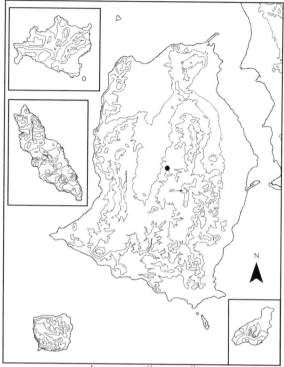

Jacquemontia agrestis

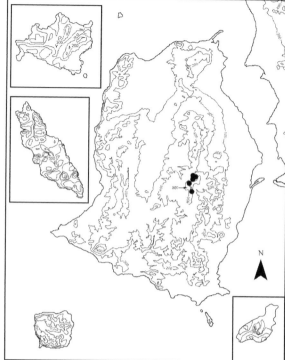

Crossosoma bigelovii

Crossosomataceae – Crossosoma Family

Crossosoma bigelovii S. Watson

Ragged rock-flower

Description: Small shrubs often 1–1.5 m tall. Leaves simple, mostly 7–15 mm long, subsessile, alternate on long shoots, clustered on short shoots, thickish, grayish to glaucous green, elliptic to oblong or obovate and entire. Flowers fragrant with 5 white, propeller-like petals each 12.5–13.5 mm long; stamens numerous; pistils 2–5 and separate from each other. Flowering November–February.

Local Distribution: Tiburón in the Sierra Kunkaak; canyon bottoms and often growing from rock crevices and mostly on north-facing slopes to higher elevations.

Other Gulf Islands: None.

Geographic Range: Deserts and desert-woodland ecotone across much of northern Sonora southward to the Sierra el Aguaje, northwestern Chihuahua, western and southern Arizona, southern Nevada, inland southern California, and Baja California Norte.

Tiburón: San Miguel Peak, *Knight 987, 1032* (UNM). Canyon bottom at N base of Sierra Kunkaak, *Wilder 06-414.* 0.5 km E of Siimen Hax, *Wilder 06-455.* Deep sheltered canyon, N

portion of Sierra Kunkaak, *Felger 07-112.* Steep, protected canyon, 720 m, N interior of Sierra Kunkaak, *Wilder 07-582.*

Cucurbitaceae – Gourd Family

Annuals or perennials with seasonal vines, the stems with tendrils and climbing or trailing. Leaves alternate, petioled, and palmately veined or lobed. Flowers unisexual.

1. Perennials from tuberous roots, summer growing; tendrils simple; fruits globose to ovoid, smooth, fleshy and bright red. _____ **Tumamoca**
1' Annuals, generally not present during summer; tendrils forked; fruits globose, prickly or smooth, dry with large air spaces when ripe, green when fresh, brown when dry. _____ **Vaseyanthus**

Tumamoca macdougalii Rose

Hatoj caaihjö "what causes red eye"; Tumamoc globe-berry

Description: Perennials; stems vining from a thickened underground tuberous root (or perhaps a cluster of several roots, as is common on the mainland). Vegetative growth and flowers appear with summer rains

Figure 3.74. *Vaseyanthus insularis.* Isla San Esteban. BTW, 8 March 2007.

and quickly perish with drying conditions. Leaves often about 4 × 4 cm including the petiole. Male and female flowers on the same plant. Fruits globose, 1 cm wide, glabrous and not firm, at first green with white mottling, becoming bright red-orange when fully ripe, usually in late summer and early fall. The flowers are apparently self-compatible and moth pollinated (Reichenbacher 1990).

Local Distribution: We found it on the lower bajada on the east side of Tiburón, where it is locally common in the *Frankenia* zone, especially at the margins of *Frankenia* and mixed desertscrub. The plants are generally found beneath small shrubs and grow into their spiny nurse shrubs. *Tumamoca* is rather seldom seen because the aerial portions disappear soon after summer rains cease.

Other Gulf Islands: None.

Geographic Range: Common in many parts of Sonora, such as the sandy or fine-textured silty-sandy soils in the vicinity of Bahía Kino and the coastal plain southeast of Guaymas nearly to the Sinaloa border, and northward through central Sonora to south-central Arizona.

This monotypic genus is named for Tumamoc Hill in Tucson, and the specific name honors Dr. Daniel Trembly MacDougal, founder of the Carnegie Desert Laboratory on the hill and collector of the type specimen on Tumamoc Hill in 1908.

Tiburón: Zozni Cmiipla, in *Frankenia* zone, many young seedlings with small tubers, locally common, 26 Sep 2008,

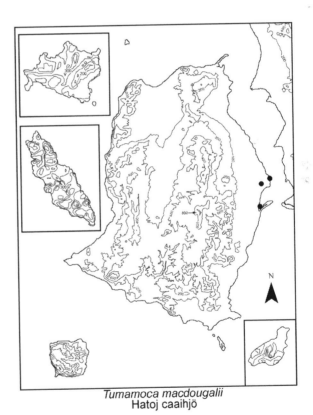

Tumamoca macdougalii
Hatoj caaihjö

Felger 08-111. Growing near the beach in the vicinity of Punta Tormenta under *Frankenia palmeri*, single seed per fruit, 24 Oct 1979, *Knight 909* (UNM). Punta Tormenta, 14 Sep 2007, *Romero-Morales 07-19.*

Vaseyanthus insularis (S. Watson) Rose [*Echinopepon insularis* S. Watson, Proceedings of the American Academy of Arts and Sciences 24:51, 1889. *E. palmeri* S. Watson. *Brandegea palmeri* (S. Watson) Rose. *Vaseyanthus insularis* var. *palmeri* (S. Watson) Gentry. *V. insularis* var. *inermis* I.M. Johnston]

Hant caitoj "land creeper"

Description: Annual vines, herbaceous and often robust, growing luxuriantly with late summer or more often fall to spring rains; sometimes carpeting otherwise barren rocky slopes, forming dense intertwining, sprawling mats and often festooning trees and shrubs in green curtains; with a thick, carrot-shaped, fleshy, white taproot. Stems slender, the tendrils usually forked. Leaves pale green, the blades relatively thin and highly variable depending on shading, position on the vine, and moisture; shallowly to deeply palmately lobed and parted. Male and female flowers occur on the same plant; flowers small, the male flowers white. Fruits with a globose spiny or smooth body, and a slender, smooth, seedless beak longer than the body or base; newly ripe or near ripe fruits bright green and fleshy, the body 12.7–14.3 × 11–12.4 mm, the beak green and succulent. As the fruits mature the beak falls away, leaving a dry and brown, globose, and corky structure with one or two light seeds and large air pockets.

Gentry (1950) recognized three varieties: Var. *inermis* is based on smooth-fruited specimens, and var. *insularis* has echinate (spiny-prickly) fruits. Both smooth and prickly fruits may occur on same plant. Var. *palmeri*, based on thin, sparsely pubescent to glabrate leaves with mostly lanceolate, acuminate, and aristate lobes, seems to be based on plants responding to high soil moisture and perhaps shaded conditions. These varieties do not seem worthy of taxonomic recognition. A second but rather weakly distinguished species occurs in the southern part of Baja California Sur.

It seems strange that *Vaseyanthus* is restricted to the Gulf of California and the Pacific shore of the southern part of Baja California Sur. The ocean-dispersed fruits should allow wider dispersal. The distribution may relate to its winter-spring (cool-season) habitat. Cool-season rains become reduced south of the Guaymas region, but suitable conditions seem available farther north on the Pacific side of the Baja California Peninsula, where it likewise does not occur. There are only a few summer records for *Vaseyanthus*; it apparently perishes during the pre-summer drought. There is one August collection (9 Aug 1985, *Tenorio 9491*) and in late September 2008 we found luxuriantly growing mature plants on Nolasco. Does the thick root sometimes survive the summer drought?

The Seris prepared liquid shampoo by boiling the green leaves. It was said to cause abundant hair growth. They also used the plant in vision quests. The roots were broken up, placed in water, and allowed to stand for several days, resulting in a bitter drink. During a vision quest four sips were drunk many times over three or four days of fasting. It was said to make one "like a drunk person." This liquid was kept in a pottery vessel (Felger & Moser 1985).

Vaseyanthus and *Brandegea* seem to be sister genera. *Brandegea* is a small genus of southwestern United States and northwestern Mexico. It generally replaces *Vaseyanthus* to the north in the Sonoran Desert surrounding the northern part of the Gulf of California. Both thrive with cool season rains and succumb or cease growing during the hottest time of the year, and are similar in general size and habit of the root, stems, and leaves. They have similar-sized flowers, are monoecious (male and female flowers occur on the same plant), and have small, beaked fruits. *Brandegea* has thin-walled, dehiscent fruits, while *Vaseyanthus* fruits are indehiscent and substantially different.

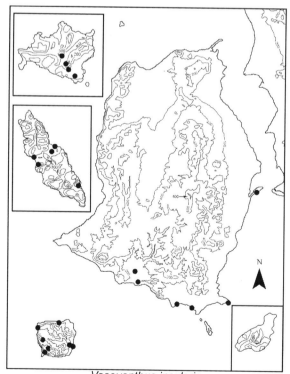

Vaseyanthus insularis
Hant caitoj

Local Distribution: Above high tide zones on rocky beaches on Tiburón and San Esteban. On Tiburón, found on the south side of the island, extending inland several kilometers in Arroyo Sauzal, and at Punta San Miguel but does not reach the density on Tiburón that is seen on the other islands. The reduced vigor of *Vaseyanthus* on Tiburón may reflect the decreased role of maritime climate over the island. Throughout San Esteban including rocky slopes and cliffs to the upper elevations, with dense growth after heavy rains.

This is one of the most abundant and widespread plants on Mártir. During favorable seasons it carpets otherwise barren rocky slopes, forming dense intertwining, sprawling mats on the ground and over cacti and shrubs, preventing other plants from growing. At such times the island appears green when viewed from the sea, and after the *Vaseyanthus* dries the island will be brown. We observed the vines to generally be absent or at least only sparse and infrequent in guano-covered places such as the lower elevations, sea cliffs, and precipitous edges of the island, and some high, exposed places. We observed that seabirds do not nest in areas of dense *Vaseyanthus* cover.

It is often seasonally abundant on Nolasco, occurring throughout the island at all elevations, but it is generally less dense on south-facing slopes. *Vaseyanthus* often blankets the ground and rocks, sometime making it slippery and dangerous to climb. It is apparent that this rank-growing vine can smother other plants and substantially influence their local distributions. *Vaseyanthus* may have a strong influence on nesting seabirds, as it does on Mártir.

Other Gulf Islands: Ángel de la Guarda, Partida Norte, San Lorenzo, Tortuga, San Marcos, Coronados, Monserrat, Santa Catalina, Santa Cruz, San Diego, San José, Espíritu Santo, Cerralvo.

Geographic Range: Coastal Sonora from the vicinity of El Desemboque to the vicinity of Guaymas, most of the islands in the Gulf of California, and Baja California Sur. The fruits, which readily float, are well adapted to sea dispersal, and also become windblown on land.

Tiburón: Arroyo Sauzal: Near beach, 5 Apr 1963, *Felger 7016*; 1¼ mi inland, 18 Mar 2006, *Wilder 06-80*. Ensenada de la Cruz, at beach, 23 Oct 1963, *Felger 9212*. Coralitos, among sea-worn cobblestones at back of beach, 21 Dec 1966, *Felger 15357*. El Monumento, 30 Jan 2008, *Felger 08-71*. Vicinity of Estero San Miguel [beach cobble, new growth and some longer old stems], 2 Sep 2006, *Wilder 06-277*.

San Esteban: Arroyo Limantour, 13–15 Sep 1990, *Van Devender 90-526*. Canyon, SW corner of island, 5 Nov 1967,

Felger 16630. El Monumento, beach just above high tide zone, among sea-worn rocks, 29 May 1954, *Felger 470*. San Pedro, upper beach and steep slopes of arroyo, 5 Nov 1967, *Felger 16673*. Steep slopes, S of S-central peak, 9 Apr 1968, *Felger 17530*. S-facing canyon on central mountain, 8 Mar 2007, *Wilder 07-66*. N side of island, 21–22 Dec 1966, *Felger 15405A*.

Mártir: Common, 1887, *Palmer 409* (type collection of *Echinopepon insularis*, US 228679, image). Common, running up cereus trunks or matting rocks, 18 Apr 1921, *Johnston 3146* (CAS, image). Abundante, 9 Aug 1985, *Tenorio 9491*. Small canyon at base of SE-facing slope that leads to island summit, abundant, 10 Apr 2007, *Felger 07-21*.

Nolasco: In a gulch near sea, covering rocks and shrubs with a very dense, thick mat of stems, growing interlaced with #3131, a smooth fruited plant, 17 Apr 1921, *Johnston 3132* (UC, image). Steep granitic slope, green ground vine and low climber, smooth and bristly fruits intertwined, apparently annual, 16 Dec 1951, *Gentry 11351*. Above SE cove, cliff at 100 ft, 11 Aug 1964, *Cooper* [*Felger 10404*]. Base of Cañón el Faro, the most abundant and extensive ground cover on the island, 100 percent cover in arroyo bottom and on some slopes, in many places covering a dense, decaying mat of perennial grasses, fruits smooth or echinate, 28 Nov 2006, *Felger 06-73*. Cañón de la Guacamaya, dry, dead plant, 3 May 2005, *Gallo-Reynoso* (photo). Cañón de Mellink, 10 m elev., 29 Sep 2008, *Felger 08-141*.

Cymodaceae – Manatee Grass Family

Halodule wrightii Ascherson [*H. beaudettei* (Hartog) Hartog]

Zimjötaa (also used for *Ruppia maritima*); shoal grass

Description: Small, delicate, submerged seagrass, perennials with short-creeping rhizomes and short stems or appearing stemless. Leaves flattened, 8–12 cm long, 0.5–1 mm wide; leaf base a flared sheath usually persisting after the rest of the leaf perishes; leaf tips blunt and variable, often with 2 or 3 minute teeth or points.

This species is known to be highly clonal. Flowers and fruits are unknown from Gulf of California plants—the flowers and fruits might be readily lost when the plants are collected. The flowers are minute and cryptic and lack petals and sepals. The female flowers are produced below the substrate surface, and the style elongates, projecting the stigma above the substrate. The male flowers produce elongated pollen grains that join to form even

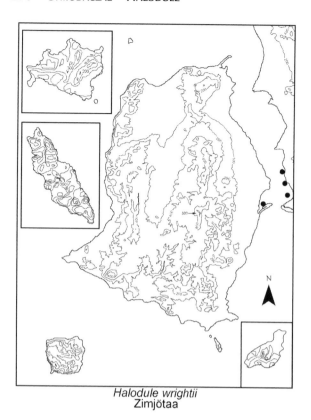

Halodule wrightii
Zimjötaa

longer "search vehicles" that can be as much as 5 mm long and contact the stigma. The fruits, probably 2 mm long, develop below the substrate surface.

Local Distribution: Common in the Canal del Infiernillo in warm, shallow seawater, 15–150 cm below low tide level during the summer/early fall months, when it replaces *Zostera marina* (see Felger 2004; McMillan & Phillips 1979; Meling-López & Ibarra-Obando 1999; Torre-Cosío 2002). *Halodule* is seasonally abundant in Bahía San Miguel, between Punta San Miguel and Punta Tormenta. This seagrass is a favorite food plant of the local sea turtles (*Chelonia*)—Richard observed *Halodule* in the stomachs of sea turtles butchered by Seris along the Infiernillo coast of Tiburón and at Punta Chueca in summer and fall in the 1970s and 1980s. (Also see *Ruppia marina*).

Other Gulf Islands: None.

Geographic Range: Shallow, protected seawater; intermittently from the Canal del Infiernillo, Sonora to Central America, and Atlantic waters from southeastern United States to South America and the West Indies.

Tiburón: Bahía San Miguel, water 1 m deep, 30°C, "abundant and growing in areas where [there] were remnants of *Zostera*" almost entirely seasonal dead and disintegrated,

17 Jun 1999, *Jorge Torre*, personal communication, and label information for *Zostera marina* (see *Z. marina*, collection of same date).

Canal del Infiernillo: S of Punta Chueca, recovered from stomach content of a *Chelonia mydas* harpooned and butchered at Punta Chueca, stomach full, ca. 99 percent comprised of this seagrass, 16 Oct 1973, *Felger 21205*. Punta Chueca: Canal del Infiernillo, 29°01′N, 112°11′W, rooted in sandy substrate at 1.3 m depth, in patches among marine algae, seawater 26°C, 16 Oct 1973, *Felger 21261*; Water depth of 6 inches to 1.5 m, 16°C, *Phillips & McMillan 5–7 Jun 1979*. 2 km N of Punta Chueca, 25 Jul 2007, *Wilder 07-370*.

Cyperaceae – Sedge Family

The few sedges in the region are known as hasoj an hehe, which refers to plants that have a relationship with water.

1. Leaf blades present; inflorescence of dense clusters of spikelets. _____**Cyperus**
1' Leaves reduced to basal sheaths, blades lacking; inflorescence a single, terminal spikelet. _____**Eleocharis**

Cyperus

1. Plants sticky viscid; annuals or perennials, mostly (15) 30+ cm tall; spikelet scales straight, the tips awnless and not recurved. _____**C. elegans**
1' Plants not sticky viscid; annuals (ephemerals), mostly 10 cm or less in height; spikelet scales with recurved awn tips. _____**C. squarrosus**

Cyperus elegans Linnaeus

Sticky sedge

Description: Tufted annuals (on Tiburón) or possibly perennials on Nolasco (generally perennials elsewhere), rather pale in color and notably sticky-viscid, even the spikelets; birds are the likely seed dispersers. Inflorescence bracts leafy. Spikelets in dense, sometimes globose clusters, the scales eventually deciduous. Stamens 3. Style branches 3, the achenes 3-angled, black with a whitish cellular covering.

Local Distribution: On Tiburón, sandy-muddy wet soils at several waterholes. On Nolasco, east side of the island. On February 2008, we were able to clearly view this sedge using binoculars from a boat just below the Agua Amarga water seep and obtain photos. This is the only record for this species on the island since Dawson

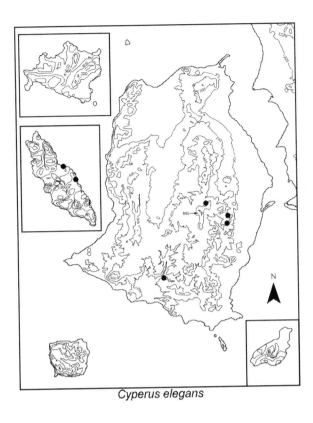

Cyperus elegans

known locality on island, several plants ca. 30 cm tall, clustered at seep, 3 Feb 2008, *Wilder 08-184* (photos).

Cyperus squarrosus Linnaeus [*C. aristatus* Rottbøell. *Mariscus squarrosus* (Linnaeus) C.B. Clarke]

Dwarf sedge

Description: Non-seasonal annuals (1.5) 3–10 cm tall and tufted (this is the smallest *Cyperus* in the Sonoran Desert). Leaves few, soft, basal or nearly so, usually less than 1 mm wide. Each spikelet scale has a prominent recurved awn-like tip giving a "fringed" appearance to the spikelets; scales often reddish bronze or yellowish with green margins. Stamen 1, sometimes with an additional 1 or 2 stamens or staminodes. Style branches 3, the achene 3-sided.

Local Distribution: Known from Mártir only by Palmer's collection in 1887 and from Nolasco at two localized sites. This species may exist on these islands within a metapopulation dynamic.

Other Gulf Islands: Espíritu Santo (Tucker 1994:90).

Geographic Range: Widespread in the Sonoran Desert, including the Guaymas region, in permanently to temporarily wet soils. Worldwide in temperate and tropical regions.

collected it in 1940. It is highly unlikely that Dawson got it from Agua Amarga. Cañón el Farito, where we presume he collected, is about 1 km northward from Agua Amarga. Since birds are presumed seed/disseminule dispersers, it is plausible that one or more sedge plants grew in a water catchment, probably temporary and no longer present, along the canyon bottom like the one reported by Ray Turner in 1979 (see *Eragrostis pectinacea, Turner 79-248*). The nearest known localities are on the nearby mainland in the vicinities of Bahía San Carlos and Tastiota.

Other Gulf Islands: None.

Geographic Range: Southern United States (New Mexico to Florida) to South America. This species occurs in scattered wetland sites, natural and disturbed, in western Sonora from near Ures (on the Río Sonora) and near Tastiota southward. Also in Baja California Sur.

Tiburón: Sauzal waterhole, common, *Felger 10084*. Pazj Hax, *Wilder 06-50*. Sopc Hax, *Felger 9324B*. N base of Sierra Kunkaak, upper reach of canyon next to small ephemeral tinaja, *Wilder 06-437*.

Nolasco: 6 Feb 1940, *Dawson 1036* (CAS; also NY, verified by Gordon Tucker, personal communication, 2007; *also see* Tucker 1994 and Wilder et al. 2007a). Agua Amarga, only

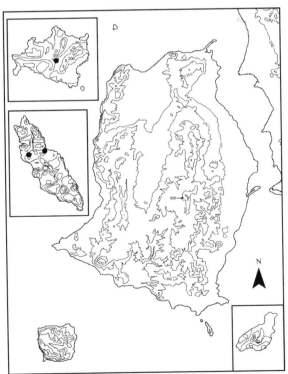

Cyperus squarrosus

Mártir: On the summit of the island, small grassy annual, 1887, *Palmer 417* (UC, US, both cited by Tucker 1994:201). **Nolasco:** Cañón el Faro, 60 m, open area near canyon bottom, E-facing exposure in local area mostly free of *Vaseyanthus*, with *Boerhavia, Coreocarpus, Eragrostis, Setaria liebmannii*; localized population with several hundred plants (not seen elsewhere), 28 Nov 2006, *Felger 06-87*. Cañón de Mellink, locally on a steep grassy slope, 29 Sep 2008, *Felger 08-147*.

Eleocharis geniculata (Linnaeus) Roemer & Schultes [*E. caribaea* (Rottbøell) S.F. Blake]

Trujillo; spikerush

Description: Small, delicate grassy annuals, the stems densely tufted, very slender and bright green. Style branches 2, the achenes lens shaped and black at maturity.

Local Distribution: Locally at three waterholes on Tiburón; emergent from shallow water and wet soil at edge of water.

Other Gulf Islands: San Marcos.

Geographic Range: Warmer regions of the world. Widespread in Sonoran Desert wetlands.

Tiburón: Sauzal waterhole (Xapij): Locally common, *Felger 10085*; Small pocket of ground water, ¼ mi S of main waterhole, *Wilder 06-127*. Sopc Hax, *Wilder 07-210*. Pazj Hax, *Wilder 06-45*.

Euphorbiaceae – Spurge Family

Diverse plants, herbs to shrubs, many with milky sap. Flowers unisexual. Different flower parts often reduced, sometimes greatly so; perianth often inconspicuous or none. Styles usually 3, simple or branched. Fruits of capsules with 3 lobes (or 1 or 2 by abortion), with 1 or sometimes 2 seeds per chamber or lobe. Seeds often with a knob-like appendage (caruncle) at base.

There are 28 species in 12 genera of euphorbs in the flora area, representing 7 percent of the Sonoran Island flora and making it the fourth most diverse family in the flora, as well as in the Sonoran Desert and the state of Sonora (Felger 2000a; Steinmann & Felger 1997).

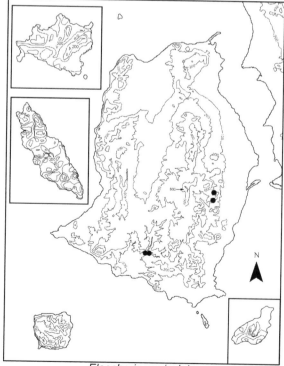

Eleocharis geniculata

Figure 3.75. *Ditaxis neomexicana*. FR.

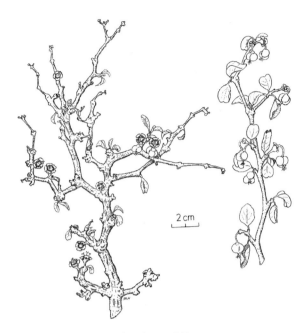

Figure 3.76. *Euphorbia misera*. NLN.

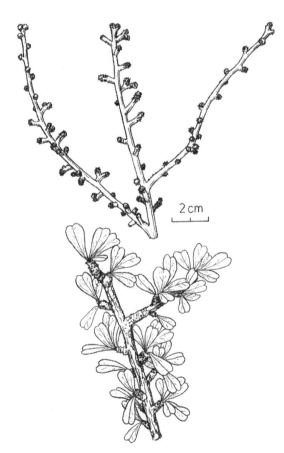

Figure 3.77. *Jatropha cuneata*. Punta Santa Rosa, Sonora. Leafless dry season aspect and leaves on short shoots after a rainy season. LBH.

(A)

(B)

Figure 3.78. *Cnidoscolus palmeri*. (A) Foothills of Sierra Kunkaak, Isla Tiburón, BTW, 4 September 2008. (B) Rosa Flores with recently dug roots on Cerro Dos Hermanos, ca. 5 km east of Estero Sargento, Sonora, RSF, March 1983.

Figure 3.79. *Croton magdalenae.*
Interior canyon of Sierra Kunkaak. Isla
Tiburón. BTW, 25 November 2006.

Figure 3.80. *Euphorbia lomelii.* Isla San Pedro Nolasco. Large colony in Cañón de
Mellink, 29 September 2008, and developing fruits and withered staminate flowers,
2 February 2008. BTW.

Figure 3.81. *Euphorbia tomentulosa.* Sierra Kunkaak, Isla Tiburón. BTW, 25 November 2006.

Figure 3.82. *Euphorbia xanti.* Near Punta Chueca, Sonora. BTW, 7 March 2007.

Figure 3.83. *Jatropha cinerea.* Staminate flowers, vicinity of Punta Tormenta, Isla Tiburón, BTW, 5 September 2008.

1. Stems white-waxy, thick, succulent, and terete, without spur-branches, the leaves few and quickly drought deciduous. _____**Euphorbia lomelii**

1' Stems not as above, not succulent or if so then with spur branches; stems leafy.
    2. Plants usually scandent or vining, mostly herbaceous, with stinging hairs. _____ **Tragia**
    2' Plants not scandent or vining.
        3. Sap milky; leaves alternate, opposite, or whorled; flowers enclosed in a cup-like involucre (cyathium) of gland-bearing, united bracts, the whole structure simulating a bisexual flower; male (staminate) flowers of a single, naked, pedicelled stamen; female (pistillate) flower of a single naked pedicelled ovary. _____**Euphorbia**
        3' Sap milky or not; leaves alternate; flowers not enclosed in a cup-like, gland-bearing involucre (cyathium); at least the male (staminate) flowers with a perianth.

4. Annuals and herbaceous perennials, usually less than 1 m tall (borderline cases key out both ways).

 5. Plants densely glandular-pubescent, the hairs tack shaped and unbranched. _____

                Andrachne (go to Phyllanthaceae)

 5' Plants not densely glandular, the hairs 2-armed (2-branches). _____ **Ditaxis**

4' Shrubs.

 6. Plants with harsh, stinging hairs; sap of milky latex; sepals petal-like; petals none. _____ **Cnidoscolus**

 6' Plants without stinging hairs; sap milky or not; petals present or absent.

  7. Pubescence of stellate (star-shaped) hairs or lepidote scales.

   8. Leaf margins toothed; Nolasco. _____ **Bernardia**

   8' Leaf margins entire, wavy, or shallowly lobed, not toothed; Tiburón. _____ **Croton**

  7' Plants glabrous or with simple or 2-armed (2-branched) hairs.

   9. Female flowers enclosed or subtended by a toothed or lobed, leaf-like and accrescent (enlarging in fruit) bract; anthers elongate and narrow. _____ **Acalypha**

   9' Female flowers not enclosed by such a bract; anthers globose or oblong.

    10. Plants with milky sap; leaves glabrous; petioles bearing conspicuous glands below the base of the blade. _____ **Sebastiania**

    10' Sap not milky; leaves pubescent or glabrous; petioles without glands.

     11. Herbs or subshrubs mostly less than 1 m tall; pubescence, at least in part, of 2-armed hairs; petals separate (distinct); seeds sculptured, less than 5 mm wide. _____ **Ditaxis**

     11' Shrubs often to 1 m or more tall; glabrous or with simple hairs; petals connate (joined); seeds smooth, 9 mm or more in width. _____ **Jatropha**

Acalypha californica Bentham [*A. pringlei* S. Watson]

Queeejam iti hacniiix "what is piled up out-of-season"; *hierba del cancer*; copper leaf

Description: Small shrubs with slender stems; with glandular and non-glandular hairs, or glandular hairs few or absent except on the margins of the pistillate bracts. Leaves highly variable with moisture conditions, often ovate to triangular-ovate, gradually drought deciduous, the dry-season herbage usually viscid-sticky and tawny brownish, the wet season leaves green, larger, thinner, and often not viscid. Female flowers with a fringed, feather-like, red or white stigma. Flowering various seasons but most luxuriantly with summer-fall rains.

Local Distribution: Widespread across Tiburón to the summit on Sierra Kunkaak. Canyons on San Esteban, and common on the east side of Dátil from canyon bottoms to the ridge crest.

Other Gulf Islands: Santa Cruz, San José, Espíritu Santo, Cerralvo.

Geographic Range: Northwest Sinaloa to southwest Arizona, Gulf Islands, and Baja California Sur to southern California.

Tiburón: 1 km N of Punta Willard, *Felger 17754*. Sauzal waterhole, *Felger 2743*. El Monumento, *Felger 08-63*. Haap Hill, *Felger T74-26* (discarded). Hast Cacöla, E flank of Sierra Kunkaak, 380 m, *Wilder 08-249*. Cerro San Miguel, 200–300 m, *Quijada-Mascareñas 90T005*. Canyon bottom at N base

of Sierra Kunkaak, *Wilder 06-427*. Steep, protected canyon, 720 m, N interior of Sierra Kunkaak, *Wilder 07-509*. N base of summit of island, N-facing sheltered cliff, *Wilder 07-552*. Caracol, arroyo, *Wilder 08-326*. E bajada, Iifa Hamoiij Quih

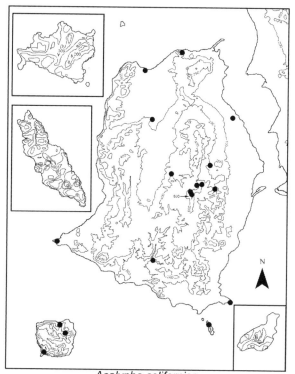

Acalypha californica
Queeejam iti hacniiix

Iti Ihiij, *Wilder 08-370.* Hee Inoohcö, mountains at NE coast, *Felger 07-89.* Tecomate, *Felger 11128.* Hast Iif (Ojo de Puma), Central Valley, *Wilder 07-257.*

San Esteban: Narrow rocky canyon behind large cove, N side of island, *Felger 15576.* Canyon, SW corner, *Felger 17619.* N slope of NE peak, in steep arroyo, *Moran 13048.*

Dátil: NE side of island, *Felger 15340.* E shore, N of island center, canyon bottom, *Wilder 07-170.*

Andrachne microphylla, see **Phyllanthaceae**

Bernardia viridis Millspaugh

Description: Shrubs 1.6–2.2 m tall, with rigid, woody branches. Leaves and young stems densely covered with stellate hairs. Leaves drought deciduous, short petioled, the blades about as wide as long and with toothed margins. Flowers small and inconspicuous; male and female flowers on separate plants. Seeds obovoid, 6.7–7.5 mm long, 5.5–6.7 mm wide, without a caruncle.

Local Distribution: Nolasco on steep, mostly north- and east-facing slopes on both sides of the island, mostly near the crest and scattered in canyon bottoms.

This species and jojoba (*Simmondsia chinensis*) are the only dioecious species on the island, and thus colonization would require establishment of at least two individuals. The closest known populations of *B. viridis* are in thornscrub and tropical deciduous forest in southern Sonora (Steinmann & Felger 1997).

Other Gulf Islands: Monserrat, Santa Cruz, Espíritu Santo, Cerralvo.

Geographic Range: Also Sinaloa and widespread along the Gulf side of Baja California Sur and adjacent islands.

This species was previously treated as a synonym of *B. mexicana* (Hooker & Arnott) Müller Argoviensis (e.g., Wiggins 1964:790). However, they seem to be distinct species and *B. mexicana* does not range farther north than central Sinaloa (Steinmann & Felger 1997).

Nolasco: NE side, canyon near crestline, *Felger 2000-3.* E-central side, just below upper crest of island, 215 m, scattered, *Wilder 08-177.* Cañón de Mellink, upper reaches of canyon, *Wilder 08-355.*

Cnidoscolus palmeri (S. Watson) Rose

Coaap; mala mujer, ortiguilla

Description: Shrubs often 1–1.5 m tall, beset with stout, stinging hairs or spines 1.6–7.5 mm long inflicting a painful but short-lasting throbbing pain. The hypodermic-like stinging hairs deliver a combination of

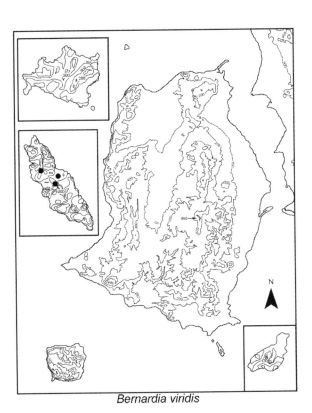

Bernardia viridis

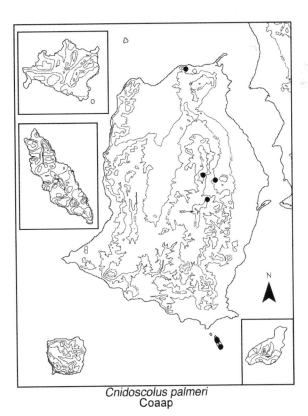

Cnidoscolus palmeri
Coaap

three chemical compounds, acetylcholine, serotonin, and histamine, that produce the intense reaction. This trio of chemicals has been found to make up the toxin present in multiple and distantly related plant families with stinging hairs, or trichomes (Euphorbiaceae, Loasaceae, Hydrophyllaceae now placed within Boraginaceae, and Urticaceae), a remarkable example of convergent evolution (Emmelin & Feldberg 1949; Thurston & Lersten 1969). Each plant may produce up to several dozen fleshy, potato-like tuberous roots, each 5–20 cm long, often compressed as they grow wedged between rocks. Leaves gradually drought deciduous, the blades 2–5.5 cm long and about as wide, relatively thick, crisped ("crinkled") with prominent milky-white veins, the margins usually coarsely toothed, each tooth spine tipped, the petioles 5–18 mm long. Young leaves velvety pubescent between the large stinging hairs and becoming sparsely pubescent or glabrate (between the spines) with age. Flowers white, the sepals petal-like; petals none; stamens 10. Seeds 10–11.5 mm long. Flowering at least May–September and seeds ripening during the same summer or early fall.

The tuberous roots, edible raw or cooked, were an important staple for the Comcaac (Felger & Moser 1985). *Cnidoscolus palmeri* seems to have its closest relative in *C. shrevei* I.M. Johnston from Durango (Johnston 1940:261). These are the smallest leaved members of the genus.

Local Distribution: Mountains on Tiburón at higher elevations in the Sierra Kunkaak and the northeast coast. Scattered along the east side of Dátil but not in the most arid sites.

Other Gulf Islands: Carmen (?), Danzante, Santa Cruz, Espíritu Santo, Cerralvo.

Geographic Range: Coastal Sonora, in the Sierra Seri and the Sierra el Aguaje and mountains around Guaymas, and Baja California Sur.

Tiburón: Cerro San Miguel, hasta cerca de la cima hasta 400 m, abundante en las laderas rocosas, *Quijada-Mascareñas 90T009*. E foothills of Sierra Kunkaak, ca. 280 m, occasional, *Wilder 08-321*. Top of Sierra Caracol, *Knight 1058* (UNM). Hee Inoohcö, mountains at NE coast, *Felger 07-78*.

Dátil: NE side of island, canyon bottoms, and N- and E-facing steep rocky slopes, rarely on SE-facing slopes, *Felger 13457*. E side of island, N of center, ¾ way up N-facing peak, sharp ridge, *Wilder 07-110*. SE side of island, steep and narrow canyon, *Felger 17503*.

Croton

Plants often strong scented and with stellate hairs.

1. Herbaceous perennials, often bushy but not woody; male and female flowers on separate plants; flowers without petals; style branches 12 or more. _____
   _____ **C. californicus**
1' Woody shrubs; male and female flowers on the same plant; flowers with well-developed petals (although sometimes deciduous); style branches 3, each bifid.
   2. Leaves more than 2.5 cm wide, about as wide as long, the tip mostly blunt. _____ **C. magdalenae**
   2' Leaves less than 2.5 cm wide, longer than wide, the tip pointed. _____ **C. sonorae**

Croton californicus Müller Argoviensis

Hacaain cooscl "drab windbreak," moosni iti hateepx "what sea turtle meat rests on"; sand croton

Description: Openly branched perennials, often to 75 cm tall, densely silvery gray pubescent. Flowering at various seasons. Male and female flowers on separate plants.

Populations in Tiburón and in western Sonora key out to var. *californicus*. However, the varieties as well as the

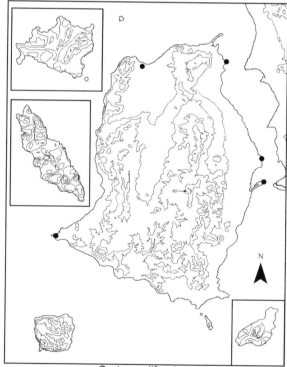

Croton californicus
Hacaain cooscl, moosni iti hateepx

relationships with the several closely related species are poorly defined.

Local Distribution: Localized on beach dunes and sandy soils along the coast of Tiburón.

Other Gulf Islands: Coronados, Monserrat.

Geographic Range: Sand soils of coastal dunes and strand habitats of western Sonora to Sinaloa. Also Baja California Sur to southern California, Arizona, southwestern Utah, southern Nevada.

Tiburón: Canyon just N of Willard's Point, dunes near beach, *Moran 8718* (SD). Punta San Miguel, low dunes 5–50 m inland, *Wilder 07-194.* Palo Fierro, beach dunes, *Felger 12533.* Zozni Cacösxj, NE coast, *Wilder 08-203.* Bahía Agua Dulce, abundant, *Tenorio 9541.*

Croton magdalenae Millspaugh

Hehe ziix capete "plant that causes swelling"; Baja California croton

Description: Shrubs often 1.2–1.7 m tall, "fuzzy" with a dense pubescence of stellate hairs. Leaves relatively large, greenish white, often turning orange before falling. Reproductive after the summer rains as well as in spring.

Local Distribution: Sheltered canyons in the Sierra Kunkaak to the summit.

Other Gulf Islands: San Marcos, Carmen, Danzante, Monserrat (observation), Santa Catalina, Santa Cruz, San Diego (observation), San José, San Francisco (observation), Espíritu Santo, Cerralvo.

Geographic Range: Sonora mainland in riparian canyons in the Sierra el Aguaje and widespread on the Baja California Peninsula.

Tiburón: Foothills of Sierra Kunkaak, 3–5 mi W of Punta Narragansett, SE part of island, *Felger 6965.* Near Sopc Hax, *Felger 9286.* Hast Cacöla, E flank of Sierra Kunkaak, 380 m, *Wilder 08-240* (USON). Canyon bottom, N base of Sierra Kunkaak, *Wilder 06-429.* Summit of island, not common, *Wilder 07-566.* Steep, protected canyon, 720 m, N interior of Sierra Kunkaak, common to dominant in canyon bottom, *Wilder 07-580.*

Croton sonorae Torrey

Hooinalca "low hills"; Sonoran croton

Description: Small woody shrubs to about 1.2 m tall, not glandular, the leaves become orange as they

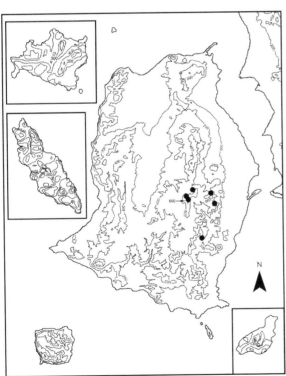

Croton magdalenae
Hehe ziix capete

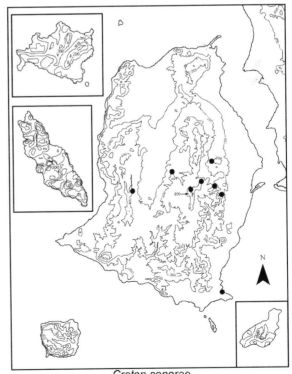

Croton sonorae
Hooinalca

age and fall. Flowering mostly during the summer rainy season.

Local Distribution: Widespread on Tiburón including the Sierra Kunkaak to higher elevation; rocky slopes, bajadas, canyons and arroyos, and desert plains.

Other Gulf Islands: None.

Geographic Range: Northwestern Sinaloa to southern Arizona, and Baja California Sur.

Tiburón: Ensenada del Perro, 12 Apr 2007, *Felger*, observation. SW part of Central Valley, E-facing slopes of Sierra Menor, rocky N-facing slope, 4 Dec 1973, *Felger 21307* (specimen not located). Haap Hill, *Felger T74-36*. Between Sopc Hax and Hant Hax camp, lower foothills, *Felger 9329*. Hast Cacöla, E flank of Sierra Kunkaak, foothills, *Wilder 08-228*. Capxölim, *Wilder 07-479*. Steep, protected canyon, 720 m, N interior of Sierra Kunkaak, *Wilder 07-511*. SE of Caracol, *Wilder 08-324*.

Ditaxis – Wild mercury

Annual or perennial herbs to subshrubs, the sap not milky, usually densely pubescent with two-armed hairs. Both the perennials and annuals are non-seasonal, capable of responding quickly to soil moisture during warm weather at any time of the year. Flowers green and white, small and rather inconspicuous.

1. Subshrubs; stems mostly erect and straight; leaf margins entire; petals united to the staminal column at base, appearing to arise above the glands; style branches sometimes dilated and flattened at the apex. _____
   _____ **D. lanceolata**
1' Plants herbaceous; stems mostly ascending to decumbent or spreading or sometimes the main axis at first erect; leaf margins often with at least some teeth; petals free from the staminal column, appearing to arise between and alternating with the glands; styles branches terete at the apex.
   2. Leaves mostly ovate-elliptic (widest at middle), the tips mostly pointed, not truncate; seeds usually with a reticulate pattern of shallow craters with fine radiating lines, hairs (if present) not sac-like and papilla-based. _____**D. neomexicana**
   2' Leaves mostly obovate to spatulate (widest above the middle), the tips mostly more or less truncate; seeds smooth to sometimes faintly patterned, the narrower end often with sac-like or papilla-based white hairs. _____ **D. serrata**

Ditaxis lanceolata (Bentham) Pax & K. Hoffmann [*Argythamnia lanceolata* (Bentham) Müller Argoviensis]

Hehe czatx caacöl "large stickery plant"

Description: Perennials, sparsely to densely branched bushes, mostly less than 1 m, or occasionally taller and scrambling through shrubs. Stems and leaves silvery pubescent, although during hot, wet weather the herbage becomes much greener and the leaves larger and more luxuriant. Leaves linear-lanceolate to broadly lanceolate.

Local Distribution: Widespread on Tiburón, San Esteban, and Dátil, including arroyos, canyons, bajadas, and steep slopes. Often growing as an undershrub within the protection of a spiny shrub.

Other Gulf Islands: Ángel de la Guarda, Tortuga, San Marcos, Coronados, Carmen, Danzante, Monserrat, Santa Catalina, Santa Cruz, San José, San Francisco, Espíritu Santo, Cerralvo.

Geographic Range: Nearly throughout the Sonoran Desert; southeastern California and western Arizona through the Baja California Peninsula and Sonora south to the Guaymas region.

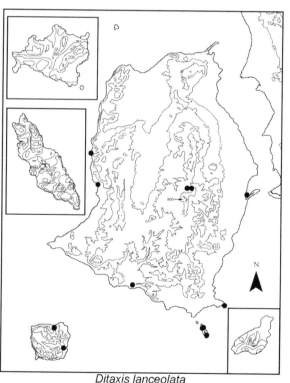

Ditaxis lanceolata
Hehe czatx caacöl

Tiburón: Ensenada Blanca (Vaporeta), *Felger 14940*. Sauzal, *Wilder 08-122*. El Monumento, *Felger 08-45*. Zozni Cmiipla, *Felger 08-116*. Head of arroyo at base of Sierra Kunkaak, between Sierra Kunkaak Mayor and Sierra Kunkaak Segundo, *Wilder 05-32*. Siimen Hax, *Wilder 06-461*. Pooj Iime, *Wilder 07-447*.

San Esteban: Arroyo Limantour, *Van Devender 92-479*. Narrow rocky canyon behind large cove, N side of island, *Felger 17598*.

Dátil: NW side of island, *Felger 15346*. SE side of island, steep narrow canyon, *Felger 17516*. E side of island, slightly N of center, canyon bottom, *Wilder 07-99*.

Ditaxis neomexicana (Müller Argoviensis) A. Heller [*Argythamnia neomexicana* Müller Argoviensis]

Hehe czatx caacöl "large stickery plant"

Description: Non-seasonal annuals to short-lived herbaceous perennials; densely to sometimes sparsely hairy. Leaves mostly 13–34 mm long, elliptic to oblanceolate, the apex acute to obtuse (pointed, not truncate), the margins entire or with a few small teeth.

Local Distribution: Widespread across Tiburón and Dátil, from near the shore to high elevations.

Other Gulf Islands: Ángel de la Guarda.

Geographic Range: Lowlands of southwestern North America.

Tiburón: Coralitos, small hill just above beach, *Wilder 06-60*. El Monumento, *Felger 08-44*. Ensenada del Perro, *Wilder 08-304*. Zozni Cmiipla, *Felger 08-115*. Haap Hill, *Felger T74-39*. Vicinity of Sopc Hax, *Romero-Morales 07-16*. Cerro Kunkaak, 1200 m, *Scott 11 Apr 1978* (UNM). Bahía Agua Dulce, *Tenorio 9531*. Pooj Iime, *Wilder 07-459*.

Dátil: NE side of island, steep rocky slopes, near summit, ridge crest, *Felger 13462*. NW side, *Felger 15341*. E-central side, beach, *Wilder 09-04*.

Ditaxis serrata (Torrey) A. Heller [*Argythamnia serrata* (Torrey) Müller Argoviensis]

Description: Robust non-seasonal ephemerals to short-lived herbaceous perennials with a well-developed taproot. Herbage densely pubescent with silvery hairs (strigose). Stems with age spreading to decumbent. Leaves 7–50 mm long, often broadly cuneate-spatulate, sometimes obovate to ovate, the leaf tip commonly

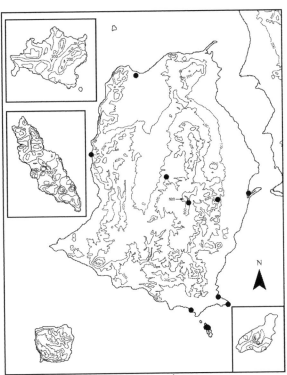

Ditaxis neomexicana
Hehe czatx caacöl

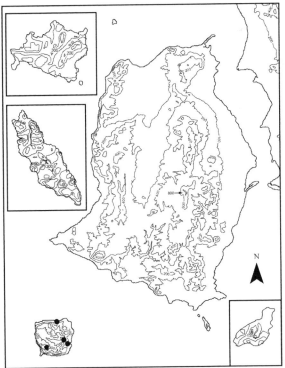

Ditaxis serrata

somewhat truncate and toothed, the margins with few small teeth.

Ditaxis serrata is distinguishable from *D. neomexicana* by its more robust habit of growth, stouter and deeper taproot, lighter-colored foliage, usually broader and blunter (often truncate) leaves. In addition minor differences in the flowers and seed surfaces are reported. The two may not be distinct species; *D. serrata* occurs in more arid habitats than the closely related *D. neomexicana* (Felger 2000a; Steinmann & Felger 1997).

Local Distribution: San Esteban along large washes and their floodplains and less common on rocky slopes.

Other Gulf Islands: San Marcos.

Geographic Range: Northwestern Sonora on sandy soils southward to the vicinity of El Desemboque del Río de la Concepción, southwestern Arizona, southeastern California, Baja California Norte, and perhaps Baja California Sur and some Gulf Islands.

San Esteban: E side of island, rocky hills, *Felger 7066.* Main arroyo [Limantour] at El Monumento, floodplain,

Felger 9178. N side of island, steep N-facing slope, near shore, *Felger 15432.* Steep slopes of S-central peak, *Felger 17534.*

Euphorbia – Spurge

Ephemerals to shrubs, with milky sap. Flowers greatly reduced, in cyathia that have petal-like appendages simulating petals (except *E. lomelii*), or the appendages greatly reduced or absent. The shape, size, and surface features of the seeds can be diagnostic. Seeds of the *Chamaesyce* are often mucilaginous when wet, and adhere tenaciously when dry, an effective adaptation for dispersal.

This is the most diverse genus in the flora area, with 15 species, 14 of which occur on Tiburón. Eleven are in the subgenus *Chamaesyce*, which has its greatest diversity in arid and semi-arid regions of Mexico. Most of the *Chamaesyce* are collectively known as tomitom hant cocpeetij, while the shrubby *E. tomentulosa* is distinguished as caacöl (large).

1. Stems white-waxy, thick, succulent, and terete, without spur-branches, the leaves few, alternate, and quickly drought deciduous; flowers and flowering structure (cyathium) more than 2 cm long and red-orange; Nolasco. __ **E. lomelii**
1' Stems not as above, usually leafy, not succulent or if so then not on Nolasco; flowers and cyathia not red-orange, less than 1.5 cm wide.
  2. Leaves symmetrical, alternate (may be crowded on short shoots), or alternate below and whorled above. Subgenus **Agaloma**:
    3. Annuals; leaves alternate below, whorled above; seeds appendaged (with a caruncle). _____ **E. eriantha**
    3' Shrubs; leaves alternate (crowded on short shoots) or whorled; seeds without a caruncle.
      4. Leaves alternate or crowded in short shoots; stems gnarly with many thick, short shoots. _____ **E. misera**
      4' Leaves whorled (3 per node); stems straight, without short shoots. _____ **E. xanti**
  2' Leaves opposite, asymmetric toward the base; seeds lacking a caruncle. Subgenus **Chamaesyce**:
    5. Shrubs, often reaching 0.7–1+ m tall.
      6. Leaf margins serrated. _____ **E. magdalenae**
      6' Leaf margins entire. _____ **E. tomentulosa**
    5' Herbaceous annuals or perennials, not shrubs.
      7. Petaloid appendages absent.
        8. Capsules glabrous; involucral glands usually oval, 0.3–0.6 mm wide; seeds smooth. ____ **E. polycarpa**
        8' Capsules glabrous or hairy; involucral glands rounded to elliptic, less than 0.3 mm wide; seeds with transverse ridges.
          9. Capsules glabrous. _____ **E. abramsiana**
          9' Capsules hairy, at least partially.
            10. Capsules densely hairy throughout. _____ **E. petrina**
            10' Capsules hairy only or mostly on ridges, especially toward capsule base. _____ **E. prostrata**
      7' Petaloid appendages present (sometimes not developed in young cyathia or drought-stressed plants; these plants key out in both choices).
        11. Cyathia narrowed or constricted at apex, urn shaped or narrowly turbinate (shaped like a top).
          12. Ephemerals to perennials; petaloid appendages entire to slightly lobed (rounded); glandular hairs moderately enlarged at tip (club shaped). _____ **E. arizonica**
          12' Ephemerals; petaloid appendages with 7 triangular, pointed segments, giving the cyathia a star-shaped appearance; glandular hairs not enlarged at tip. _____ **E. setiloba**

11' Cyathia not narrowed or constricted at apex.
    13. Perennials with conspicuously thickened roots; leaf margins serrated; capsules uniformly pubescent. _____ **E. leucophylla**
    13' Annuals or perennials, the roots not especially thick; leaves entire; capsules glabrous or hairy.
        14. Seeds smooth. _____ **E. polycarpa**
        14' Seeds with transverse ridges and grooves.
            15. Leaves very narrowly linear, often inrolled at margins; capsules glabrous; seeds 1.7–2 mm long, with 1–few transverse ridges. _____ **E. florida**
            15' Leaves ovate to obovate or oblong; capsules glabrous or hairy; seeds 1–1.2 mm long, with several transverse ridges.
                16. Involucral appendages minute; capsules glabrous. _____ **E. abramsiana**
                16' Involucral appendages conspicuous, showy; capsules hairy._____ **E. pediculifera**

Euphorbia abramsiana L.C. Wheeler [*Chamaesyce abramsiana* (L.C. Wheeler) Koutnik. *Euphorbia pediculifera* var. *abramsiana* Ewan]

Description: Small warm-weather annuals, mostly growing with summer-fall rains, often forming prostrate mats. Herbage often red-brown; herbage and involucres pubescent or glabrous. Cyathia minute and inconspicuous, 0.4–0.5 mm wide, the involucral glands dot-like, rounded or nearly so, 0.1 (0.15) mm wide, the appendages absent or to 0.2 mm wide, white to pink. Capsules glabrous, bright green with red margins (angles) and furrows, the angles rather sharp. Seeds resembling a mealy bug (Pseudococcidae), 1.0–1.2 mm long, ashy grayish white to tan, with a sharply angled crest and transversely ridged; mucilaginous when moistened.

The seeds are conspicuously cross-ridged, similar to those of *E. pediculifera*, but the angles are sharper than on *E. pediculifera* and they do not seem to be closely related.

Local Distribution: Known from Tiburón at two localities on the bajada near the east shore of the island.

Other Gulf Islands: San José.

Geographic Range: Recorded from widely separated localities in western Sonora. Southeastern California to Baja California Sur and southern Arizona to Sinaloa.

Tiburón: Near airstrip [landing field at Punta Tormenta], 25 Oct 1979, *Knight 906* (UNM). Zozni Cmiipla, *Felger 08-119*.

Euphorbia arizonica Engelmann [*Chamaesyce arizonica* (Engelmann) Arthur]

Arizona spurge

Description: Herbaceous perennials also flowering in the first season; reproductive at any season. Herbage glandular pubescent, often reddish, the petaloid appendages pink. Seeds 0.9–1.1 mm long, transversely ridged, not conspicuously mucilaginous when wet but adhering tenaciously after drying.

Local Distribution: Widely scattered on Tiburón; arroyos, canyons, and rocky slopes, especially in mountains.

Other Gulf Islands: Monserrat, San José.

Geographic Range: On the Sonora mainland reaching its southernmost limit in the Guaymas region. Western Texas to southern California, Baja California Sur and Norte, and Sonora, Chihuahua, Durango, and Coahuila.

Tiburón: Arroyo Sauzal, *Felger 08-14*. SW Central Valley, mountain [Sierra Menor] bordering valley, 1200–1400 ft, S-facing slope, *Felger 12423*. Haap Hill, *Felger 74-T40*. Between Sopc Hax and Hant Hax camp, *Felger 9339*. Large arroyo heading to a valley at E base of Sierra Kunkaak, *Wilder 06-52*.

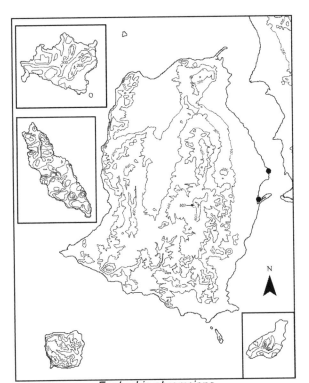

Euphorbia abramsiana

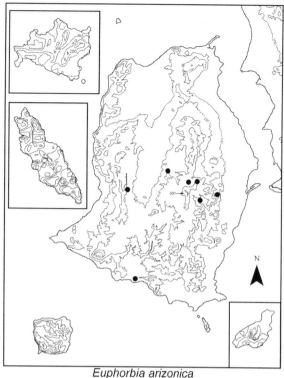

Euphorbia arizonica

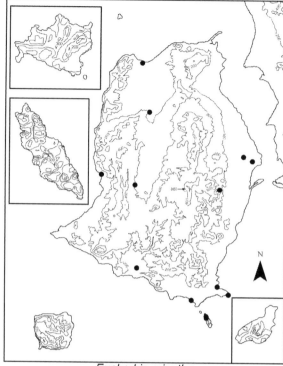

Euphorbia eriantha
Pteept

Siimen Hax, *Wilder 06-448*. Cerro San Miguel, *Quijada-Mascareñas 91T005*.

Euphorbia eriantha Bentham

Pteept (variants: taapt, teept); beetle spurge

Description: Non-seasonal annuals, usually erect-growing and taller than broad, mostly branched from the upper part of the plant. Stems terete and semi-succulent. Leaves relatively few and linear, the lower ones alternate, the upper ones forming a whorl beneath the inflorescences.

Local Distribution: Tiburón, widely scattered at lower elevations across most of the island including arroyos, bajadas, valley plains, and low hills. Common on Dátil, especially on north-facing slopes.

Other Gulf Islands: Ángel de la Guarda, San Marcos, Espíritu Santo.

Geographic Range: Widespread in lowland desert habitats. Sonora throughout the Sonoran Desert to northwestern Sinaloa, and Baja California Sur to southeastern California and eastward to southwestern Texas and Coahuila.

Tiburón: Ensenada Blanca (Vaporeta), *Felger 14936*. Sauzal landing field, *Felger 6471*. Coralitos, *Wilkinson 197*. El

Monumento, *Felger 08-68*. Ensenada del Perro, *Wilder 07-145*. SW Central Valley, *Felger 17326*. Vicinity of Sopc Hax, *Romero-Morales 07-13*. Palo Fierro, *Felger 11084*. 3 km W [of] Punta Tormenta, *Smartt 13 Oct 1976* (UNM). Tecomate, *Felger 8878*. Hast Iif (Ojo de Puma), N part of Central Valley, *Wilder 07-259*.

Dátil: S-facing slope of ridge just N of center of island, less common than on N-facing slopes, *Wilder 07-140*. E side of island, slightly N of center, halfway up N-facing slope of peak, *Wilder 07-124*.

Euphorbia florida Engelmann [*Chamaesyce florida* (Engelmann) Millspaugh]

Description: Warm weather annuals, glabrous, erect to ascending, the stems very slender. Leaves extremely narrow, becoming revolute in age. Appendages white and showy.

Local Distribution: Documented on Tiburón from sandy soils in the Central Valley.

Other Gulf Islands: None.

Geographic Range: Sinaloa to southern Arizona.

Tiburón: Haap Hill, near N side in vicinity of Haap Caaizi Quih Yaii (former tepary bean-gathering camp), *Felger T74-14*.

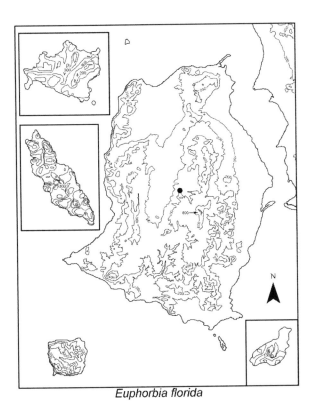

Euphorbia florida

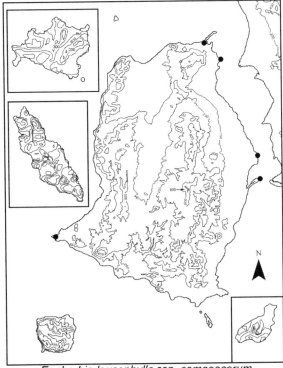

Euphorbia leucophylla ssp. *comcaacorum*

Euphorbia leucophylla Bentham subsp. **comcaacorum** V.W. Steinmann & Felger [*Chamaesyce leucophylla* Bentham]

Description: Herbaceous perennials with deep, thick, and often gnarled roots. Stems decumbent to prostrate, often partially buried in drifting sand. Leaves somewhat glaucous, nearly orbicular, the margins serrated. Appendages white and conspicuous. Reproductive at almost any season.

Local Distribution: Beach dunes on the eastern and southwestern coasts of Tiburón.

Other Gulf Islands: None.

Geographic Range: Subsp. *comcaacorum* is endemic to the east side of the Gulf of California: on beach dunes on Tiburón and the opposite Infiernillo coast southward to Tastiota. Subsp. *leucophylla* occurs on the Baja California Peninsula and on Isla Cerralvo. Plants of subsp. *comcaacorum* consistently differ from those of the Baja California Peninsula by possessing larger seeds (1.5–1.7 mm long vs. 0.9–1.3 mm). The leaves of Tiburón–Sonoran plants have well-developed petioles and blades with oblique bases. These leaf characteristics are only rarely encountered on Baja California plants, on which the leaves lack conspicuous petioles and usually are

amplexicaulous with cordate bases. Both subspecies are restricted to coastal dunes.

Tiburón: 1 km N of Punta Willard, *Perrill 5114*. Punta San Miguel, *Wilder 07-190*. Palo Fierro, *Felger 6424*. Zozni Cacösxaj, NE coast, *Wilder 08-200*. Cyazim It (Punta Perla), *Wilder 08-379*.

Euphorbia lomelii V.W. Steinmann [*Pedilanthus macrocarpus* Bentham]

Candelilla

Description: Rhizomatous perennials; stems often 0.5–1+ m tall, several to many, thick (5–15 mm diameter) and succulent, mostly straight, upright, and unbranched or few-branched, with a white-waxy coating and copious white latex. Leaves few, ca. 1 cm long, and very quickly drought deciduous. Cyathia about 2–2.5 cm long, bilateral, the cyathia and flowers bright red-orange, visited by Costa's hummingbirds. The gland chamber of the involucre overflows with nectar that often drips down the flowers. Female flower appearing before the male flowers; when the female flower emerges it sticks straight out but bends downward and out of the way as the many male flowers emerge in the

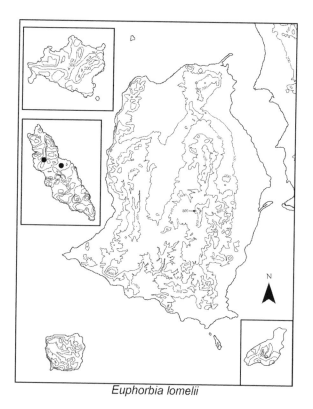

Euphorbia lomelii

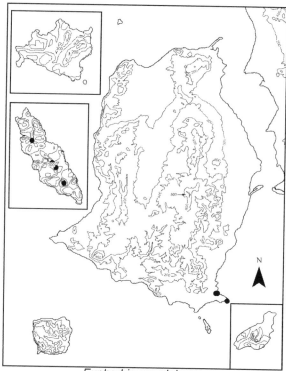

Euphorbia magdalenae

former position of the female flower, thus reducing the chance for self-fertilization. Fruits angled, spurred, and essentially indehiscent. Seeds often 8–8.5 mm diameter, nearly round.

Local Distribution: Common on Nolasco on steep, rocky slopes at all elevations, especially on slopes with *Echinocereus websterianus* and *Mammillaria multidigitata*.

Other Gulf Islands: Carmen, Espíritu Santo, Cerralvo.

Geographic Range: Western Sonora, in the vicinity of the coastal towns of El Cholludo and Tastiota, and the coastal plain southeast from Guaymas to northwestern Sinaloa. Also Baja California Norte and Baja California Sur.

Nolasco: Cañón el Faro, 60 m, forming colonies on S- and E-facing rock slopes, seen at all elevations, especially on S-facing rock slopes, *Felger 06-85*. Cañón de Mellink, 29 Sep 2008, *Wilder* (photo).

Euphorbia magdalenae Bentham [*Chamaesyce magdalenae* (Bentham) Millspaugh]

Description: Shrubs mostly 0.3–1.3+ m tall, glabrous or essentially so, the branches often erect and not intricately branched. Leaves about twice as long as wide, oblong-obovate, the margins entire or sometimes faintly serrulate. Petaloid appendages white or perhaps reduced in drought. This species and *E. tomentulosa* are the only shrubs in the subgenus *Chamaesyce* in northwestern Mexico.

Local Distribution: Southeastern Tiburón, corner of the island on rocky hillsides and arroyo margins near the coast. On Nolasco, known from several sites near the ridge crest of the island.

Other Gulf Islands: San Marcos, Coronados, Carmen, Danzante, Monserrat (observation), Santa Catalina, Santa Cruz, San Diego, San José, San Francisco, Espíritu Santo, Cerralvo.

Geographic Range: Widespread on the Baja California Peninsula. Johnston (1924:1069) reported it at Bahía San Pedro on the Sonora mainland opposite Nolasco, but we have not been able to locate it there or find a herbarium specimen.

Tiburón: Vicinity of El Monumento, *Felger 15530*. Ensenada del Perro: *Romero-Morales 06-2*; arroyo and S-facing slope, *Wilder 07-157*.

Nolasco: E-central side of island, rocky ridge ca. 20 m below summit, small shrub, *Felger 9657*. Summit of island, especially on N side of peak, with several hundred plants ca. 30 cm tall, *Wilder 09-138*. [Above SE side] E exposure, in talus at 900–1000 ft, rare, 0.6–1 m tall, *Pulliam & Rosenzweig 21 Mar 1974*. Cañón de Mellink, E-facing slope just below highest point on N side of island, group of ca. 50 shrubs ca. 30–40 cm tall, *Wilder 08-358*.

Euphorbia misera Bentham

Hamacj "fires"; *jumetón*; cliff spurge

Description: Shrubs with copious milky sap, often 0.7–1.4 m tall, the stems thick, semi-succulent, and flexible, appearing gnarled due to the knobby short shoots; multiple stemmed from the base, resembling *Jatropha cuneata* in architecture. Herbage with short hairs. Leaves alternate but crowded on short shoots, drought deciduous, the blades ovate-orbicular, about as wide as long. Cyathia predominantly yellow-green, 2–3.8 mm wide, the appendages conspicuous, white to yellow-white (green flies observed feeding from the cyathia disks). Seeds 3–3.5 mm long, ovoid globose. Growing and flowering after rains at various seasons.

Local Distribution: Rocky hills and mountains along the western and northern sides of Tiburón. Common on Dátil on north- and east-facing slopes on the east side of the island, and north-facing slopes and steep canyons near the summit on San Esteban.

Other Gulf Islands: Ángel de la Guarda, San Marcos, Espíritu Santo.

Geographic Range: Southern California to mid-Baja California Norte and Sonora southward along the coast to Cerro Tepopa. Farther south in Baja California and in the Hermosillo-Guaymas region in Sonora it is replaced by the closely related *E. californica* Bentham.

Tiburón: Ensenada Blanca (Vaporeta), *Felger 17272.* Hee Inoohcö, mountains, *Felger 07-91.* Pooj Iime, *Wilder 07-441.*

San Esteban: N slope of high peak near SE corner of island, 400 m, *Moran 8852* (SD). N slope of NE peak, steep arroyo, 450 m, *Moran 13049* (SD).

Dátil: NE side of island, steep rocky slopes, absent from S-facing hills, *Felger 9105.*

Euphorbia pediculifera Engelmann var. **pediculifera**
[*Chamaesyce pediculifera* (Engelmann) Rose & Standley]

Description: Non-seasonal annuals to short-lived perennials. Young herbage, capsules, and cyathia densely pubescent, the hairs thickish, short, white, and appressed to spreading. Stems usually erect when young, then spreading to prostrate. Herbage often red-brown to gray-brown. Leaves at least twice as long as wide, ovate to obovate or oblong, the margins entire or sometimes with a few small, irregular teeth. Cyathia 1.2–1.5 mm wide, the glands maroon, oval; appendages rather showy, white, becoming pink with age. Seeds 1–1.2 mm long, chunky, conspicuously cross-ridged, mucilaginous when wet.

Local Distribution: Common and widely distributed on Tiburón, San Esteban, and Dátil; upper beaches to higher elevations including dunes, bajadas, washes and canyons, and rocky slopes.

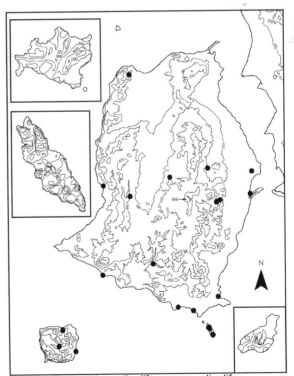

Euphorbia misera
Hamacj

Euphorbia pediculifera var. *pediculifera*

Other Gulf Islands: Ángel de la Guarda, San Lorenzo, Tortuga, San Marcos, Coronados, Carmen, Danzante, Monserrat, San José, Espíritu Santo.

Geographic Range: Sonoran Desert in southeastern California and Arizona, to Sinaloa and the Cape Region of Baja California Sur. Var. *linearifolia* S. Watson, from the Guaymas region, differs in having narrower and often longer leaves.

> **Tiburón:** Ensenada Blanca (Vaporeta), *Felger 714*. Tordillitos, *Felger 15470*. Sauzal waterhole, *Felger 10081*. Ensenada de la Cruz, *Felger 9228*. Coralitos, *Felger 15368*. Ensenada del Perro, *Wilder 07-153*. SW Central Valley, mountain bordering valley, *Felger 12404*. Haap Hill, *Felger 74-T41*. Zozni Cmiipla, *Felger 08-132*. SE base of Sierra Kunkaak, *Felger 9324*. Between Sopc Hax and Hant Hax camp, *Felger 9346*. Palo Fierro, *Felger 8934*. Caracol, *Knight 952* (UNM). Tecomate, *Felger 12491*.

> **San Esteban:** E-central side of island, near shore among sea-worn boulders, just N of El Monumento, *Felger 12735*. Arroyo Limantour, 2.25 mi inland, *Wilder 07-51*. Narrow canyon behind cove at N side of island, *Felger 17600*.

> **Dátil:** NE side of island, just above cobblestone beach, *Felger 9094*. NW side of island, *Felger 15304*. E side of island, N of center, ridge crest on peak, *Wilder 07-114*. SE side of island, steep and narrow canyon, *Felger 17524*.

Euphorbia petrina S. Watson, Proccedings of the American Academy of Arts and Sciences 24: 75, 1889 [*Chamaesyce petrina* (S. Watson) Millspaugh]

Tomitom hant cocpeetij *maybe* "tomitom that is circular"; *golondrina*; desert spurge

Description: Non-seasonal annuals, the plants often prostrate and reddish brown, moderately pubescent with spreading hairs. Involucral glands without appendages. Seeds 1 mm long, mucilaginous when wet.

Superficially resembling *E. polycarpa* but distinguished even in the field by the brownish color of the plants and the often larger leaves. The sure way to always tell them apart is by the seeds. *E. petrina* and *E. polycarpa* have seeds about the same size and shape, but *E. petrina* has low, transverse ridges or bumps across its surface. They do not seem to be closely related.

Local Distribution: Tiburón, sandy and rocky beaches and dunes. Alcatraz, sandy flats. San Esteban, large, gravelly washes and floodplains. San Pedro Mártir, seasonally common at higher elevations.

Other Gulf Islands: Ángel de la Guarda, Partida Norte, Salsipuedes, San Lorenzo (Johnston 1924:1072), San Francisco, Espíritu Santo.

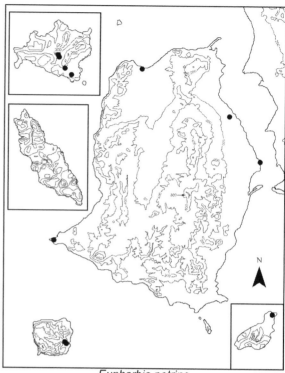

Euphorbia petrina
Tomitom hant cocpeetij

Geographic Range: El Golfo at the mouth of the Río Colorado and Puerto Peñasco southward in western Sonora to Sinaloa, Gulf Islands, and the Baja California Peninsula.

> **Tiburón:** Punta Willard, 1 km N, *Felger 17737*. Palo Fierro, dunes at beach, *Felger 12535*. Iifa Hamoiij Quih Iti Ihiij, *Wilder 08-371*. Tecomate, dunes at beach, *Felger 8908*.

> **San Esteban:** *Johnston 3169* (Johnston 1924:1073, specimen not located). Near S end of Isla, large arroyo running inland from E side [Arroyo Limantour], *Wiggins 17241* (CAS). [Arroyo Limantour], *Wilkinson 171*.

> **Alcatraz:** NE side of island, locally common on sandy flat, *Felger 07-186*.

> **Mártir:** Oct 1887, *Palmer 412* (GH, holotype; see Steinmann & Felger 1997. Apparent isotypes from the type collection: Two sheets, UC 17728 & 110399. There are also four specimens of *Palmer 412 in 1887* at US (images): three are labeled San Pedro Martin Island (US 47790, 798887, 7988880), and one (US 47789) bears two labels, one with "Flora of Lower California" and the other with "Los Angeles Bay"; presumably all specimens of *Palmer 412 in 1887* are from Isla San Pedro Mártir and are probably isotypes). Very common in the cordonada [*sic*] crowning the island, where it grows between the loose rocks, *Johnston 3155* (CAS). Top of island, 300 m, rather common, *Moran 8809* (SD, UC). 21 Mar 1962, *Wiggins 17189* (CAS). Among rocks, very common, *Felger 6348*.

Euphorbia polycarpa Bentham [*E. polycarpa* var. *hirtella* Boissier. *E. intermixta* S. Watson. *Chamaesyce polycarpa* (Bentham) Millspaugh]

Tomitom hant cocpeetij *maybe* "tomitom that is circular"; *golondrina*; desert spurge

Description: Non-seasonal annuals to small perennial herbs, the taproot well developed (first season roots may be more than 30 cm deep). Herbage and capsules glabrous or hairy. Stems prostrate to ascending, the larger plants much branched. Involucral glands 0.3–0.6 mm wide, dark maroon to almost black (or yellow in summertime), oval (wider than long), with conspicuous white petaloid appendages, or the appendages sometimes reduced or absent on drought-stressed plants. Seeds 0.8–1 mm long, smooth-surfaced, mucilaginous when wet. Flowering at any time of year.

Local Distribution: This is the most common and widespread *Euphorbia* on Tiburón, San Esteban, and Dátil, ranging from upper beaches to arroyos, canyons, bajadas, valley plains, hills and steep slopes including higher elevations. Common on the Sonoran coast opposite Tiburón.

Other Gulf Islands: Ángel de la Guarda, Tortuga, San Marcos, Coronados, Carmen, Monserrat, San Diego, San José, San Francisco, Espíritu Santo, Cerralvo.

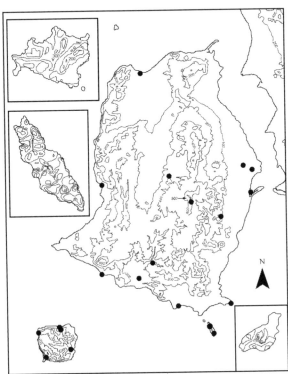

Euphorbia polycarpa
Tomitom hant cocpeetij

Geographic Range: Widespread and common throughout the Sonoran Desert and also in adjacent regions of subtropical scrub and the Mojave Desert. Southern California to Baja California Sur, and southern Nevada and Arizona southward in Sonora at least to the Río Mayo region.

Tiburón: Ensenada Blanca (Vaporeta), *Felger 12228.* Tordillitos, *Felger 15507.* Sauzal airfield, *Felger 6463.* Xapij (Sauzal), *Wilder 08-135.* Ensenada de la Cruz, *Felger 9146.* El Monumento, *Felger 15536.* Haap Hill, *Felger 76-T20.* Hast Coopol, *Wilder 07-384.* Zozni Cmiipla, *Felger 08-118.* Palo Fierro, *Felger 8922.* 3 km W [of] Punta Tormenta, *Smartt 13 Oct 1976* (UNM). San Miguel Peak, *Knight 942* (UNM). Tecomate, *Felger 8870.*

San Esteban: Narrow rocky canyon behind large cove at N side of island, *Felger 17558.* N side of island, on N-facing sea cliffs, *Felger 15423.* Canyon at SW corner of island, *Felger 17640.* San Pedro, upper beach and steep slopes of gully-like arroyo, *Felger 16675.* Arroyo Limantour, *Van Devender 90-524.*

Dátil: E side of island, halfway up N-facing slope of peak, *Wilder 07-130.* NE side of island, *Felger 21 Oct 1963.* NW side of island, *Felger 15324.* SE side of island, *Felger 17496.*

Euphorbia prostrata Aiton [*Chamaesyce prostrata* (Aiton) Small]

Description: Warm-weather annuals (often short-lived perennials in non-desert regions); upright to mostly prostrate. Stems, capsules, and young leaves moderately hairy with short, crinkled, white hairs. Cyathia minute, 0.4 mm wide, the glands pink, 0.2 mm wide, without appendages (elsewhere with short appendages). Capsules green with white hairs on the angles, especially near the capsule base. Seeds 0.7–0.8 mm long, ashy gray-white to tan, with a sharply angled crest and relatively short transverse ridges; not conspicuously mucilaginous when wet but adhering tenaciously after drying.

Local Distribution: Known from a single collection along the eastern shore of Tiburón, which is the only record for this species on a Gulf Island.

Other Gulf Islands: None.

Geographic Range: Widespread in the Americas and naturalized in warm regions of the Old World. The plants might be confused with *E. abramsiana*, from which they can be distinguished by the usually larger, broader leaves, pubescent capsule margins, and smaller seeds with sharper ridges. *E. prostrata* has been incorrectly called *E. chamaesyce* Linnaeus, which is an Old World plant.

Tiburón: Zozni Cmiipla, sandy soil, *Wilder 06-513.*

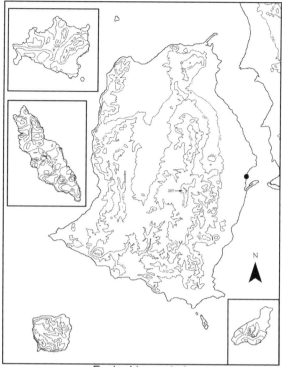

Euphorbia prostrata

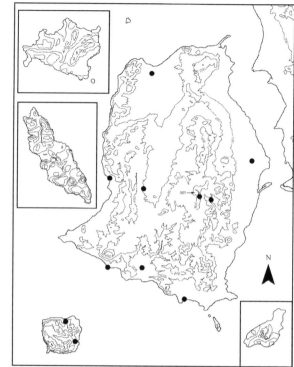

Euphorbia setiloba
Tomitom hant cocpeetij

Euphorbia setiloba Engelmann ex Torrey [*Chamaesyce setiloba* (Engelmann ex Torrey) Norton]

Tomitom hant cocpeetij *maybe* "tomitom that is circular"; *golondrina*; desert spurge

Description: Non-seasonal annuals, conspicuously pubescent, the herbage reddish most of the year, yellow-green in summer, the stems slender and often blackish. Cyathia white to pink, very small, the involucral appendages white, becoming pink, deeply divided into 7 pointed segments resulting in the cyathium looking like a tiny star-shaped "flower." The star-shaped cyathia are unique among the euphorbias of the Sonoran Desert region and northwestern Mexico. Seeds 1.0–1.2 mm long, chunky, conspicuously cross-ridged, the dorsal ridge rounded and somewhat flattened; mucilaginous when wet.

Local Distribution: On Tiburón, common and widespread from near the shore to higher elevations on Sierra Kunkaak. On San Esteban, seasonally common in washes or gravelly soil.

Other Gulf Islands: Ángel de la Guarda, Carmen, Monserrat, Santa Catalina, Cerralvo.

Geographic Range: Southeastern California, southern Nevada, and southwestern Utah to western Texas and southward into Baja California Sur and Sinaloa.

Tiburón: Ensenada Blanca (Vaporeta), *Felger 17280*. Tordillitos, *Felger 15509*. Arroyo Sauzal, *Felger 12287*. Arroyo de la Cruz, *Wilder 09-51*. SW Central Valley, arroyos, *Felger 12296*. Palo Fierro, bajada 2 mi inland, *Felger 12545*. E base of Sierra Kunkaak, *Wilder 06-20*. San Miguel Peak, rocky ledge abutting one of the major N drainages, *Knight 958* (UNM). Tecomate, arroyo, *Felger 1117*.
San Esteban: Limantour, *Felger 17672*. Narrow canyon at NW corner of island, *Felger 17601*.

Euphorbia tomentulosa S. Watson [*Chamaesyce tomentulosa* (S. Watson) Millspaugh]

Tomitom hant cocpeetij caacöl "large tomitom that is circular"; *jumetón*

Description: Shrubs reaching 1–1.5 m tall, the branches forked, the herbage pubescent. Leaves nearly orbicular with serrated margins. Cyathia in dense clusters at ends of branchlets, the appendages white.

Local Distribution: Especially common in the Sierra Kunkaak on Tiburón and the southern part of the island but not known on the western margin and north side of the island; washes, bajadas, and canyons to mountain slopes at the highest elevations.

Other Gulf Islands: Coronados, Carmen, Espíritu Santo.

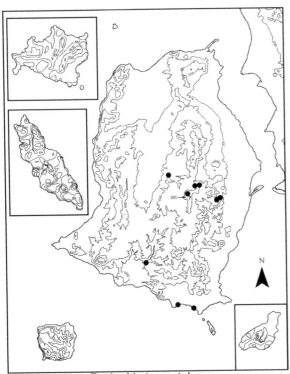

Euphorbia tomentulosa
Tomitom hant cocpeetij caacöl

Geographic Range: Western Sonora from near Altar and Caborca southward to northwestern Sinaloa, islands in the Gulf, and much of the Baja California Peninsula. One of the few shrubby species of subgenus *Chamaesyce* in Mexico and the only one in mainland Sonora.

Tiburón: Arroyo Sauzal, 3 mi inland, *Wilder 06-106*. Ensenada de la Cruz, *Felger 2586*. Coralitos, *Wilder 06-69*. Haap Hill, *Felger 74-T42*. Sopc Hax, *Felger 9305*. E foothills of Sierra Kunkaak, *Felger 6956*. Cerro San Miguel, *Quijada-Mascareñas 91T001*. San Miguel Peak, upper W slope, 1500 ft, *Knight 973* (UNM). Cójocam Iti Yaii, 855 m, *Wilder 07-546*. Head of arroyo, base of Sierra Kunkaak, *Wilder 05-39*.

Euphorbia xanti Engelmann ex Boissier

Hehe ix cooxp "plant whose sap is white"; *indita, jumetón*

Description: Shrubs 1.2–2.5 m tall, often propagating by rhizomes—sometimes from long, lateral roots that produce sprouts. Stems terete, pale green to bluish green with a thin waxy coating. Leaves 3 per node, linear to broadly elliptic; sparsely leafy following rains, the leaves quickly drought deciduous. Involucral glands maroon, the petaloid appendages showy, white, becoming pink. Filaments white, becoming pink, the anthers yellow. Flowering profusely in early to middle March and probably other times of the year.

Local Distribution: Common on the bajada of the northeastern side of Tiburón and canyons in the interior of Sierra Kunkaak.

Cody et al. (1983) hypothesized this species migrated from the Baja California Peninsula to the Sonoran mainland via the Midriff Islands. Many other species show a similar distributional pattern; for example, Steinmann and Felger (1997) list six euphorbs with this pattern. However, in a thesis on the group (section *Alectoroctonum*) to which *E. xanti* belongs, *E. peganoides* Boissier is given as the most similar relative of *E. xanti* (Ramírez 1996). Victor Steinmann (personal communication, 2007) treats both *E. peganoides* and *E. colletioides* Bentham as synonyms of *E. cymosa* Poiret as eluded to by Steinmann and Felger (1997). *E. xanti* is the only member of section *Alectoroctonum* occurring on the Baja California Peninsula, so the relatives of *E. xanti* are mainland taxa.

Other Gulf Islands: Tortuga, Monserrat.

Geographic Range: Predominately in Baja California Sur and disjunct pockets along 215 km of the Sonoran coast from near El Desemboque southward to Bahía Kino and farther south near Guaymas.

Gentry's (1949) report of *E. xanti* in Sinaloa was based on plants he collected and initially identified as that species, which is now known as *E. gentryi* V.W. Steinmann & T.F. Daniel.

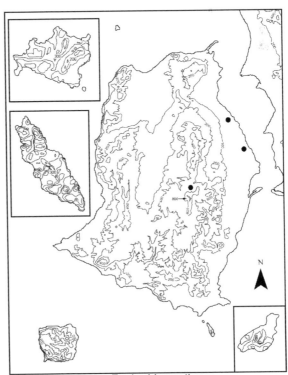

Euphorbia xanti
Hehe ix cooxp

Tiburón: Canyon, vicinity of Siimen Hax waterhole, cluster of 5 plants on N-facing canyon wall, *Wilder 06-453.* Lower E bajada, 4 km N of Punta Tormenta, *Wilder 07-233.* Iifa Hamoiij Quih Iti Ihiij, ca. 1 km inland, *Wilder 08-369.*

Jatropha

Shrubs with flexible stems and clear or colored sap (not milky latex).

1. Petioles about as long as the blades; leaf blades about as broad or broader than long, mostly more than 2 cm long. _____**J. cinerea**
1' Leaves sessile to subsessile, or the long-shoot leaves with petioles less than half as long as blades; leaf blades about twice as long as wide, mostly less than 2 cm long. ___ _____ **J. cuneata**

Jatropha cinerea (Ortega) Müller Argoviensis [*J. canescens* (Bentham) Müller Argoviensis]

Hamisj, oot iquejöc "coyote's firewood"; *sangrengrado*; ashy jatropha

Description: Multiple-stem shrubs often 1–2.3 (3.4) m tall. Leaves appearing with rains at various seasons but most vigorously with summer rains, and quickly drought deciduous. Corollas of female flowers dark pink inside; corollas of male flowers pinkish white to rose, with age becoming white. Flowering late summer and early fall. Fruits usually 2-lobed and 2-seeded, the seeds rounded, 1 cm diameter.

Local Distribution: Common and widespread in lowland habitats on Tiburón, especially on bajadas and valley plains.

Other Gulf Islands: Danzante, Monserrat, Santa Cruz, San José, Espíritu Santo, Cerralvo.

Geographic Range: Southwestern Arizona (Organ Pipe Cactus National Monument) to Sinaloa, both states of Baja California and Gulf Islands.

Tiburón: Ensenada del Perro, 12 Apr 2007, *Felger*, observation. Palo Fierro, *Felger 10138.* Tecomate, *Felger 10195.*

Jatropha cuneata Wiggins & Rollins

Haat; *torote, matacora, sangrengrado*; desert limberbush

Description: Trunkless, many-stemmed shrubs often 1–2 (2.5) m tall, the roots thick and somewhat tuberous. Roots and stems oozing blood-like sap when cut. The

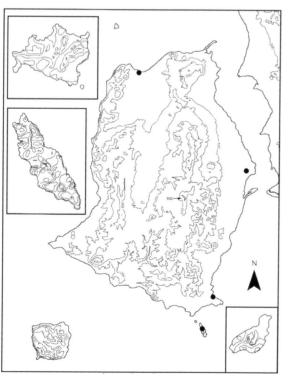

Jatropha cinerea
Hamisj, oot iquejöc

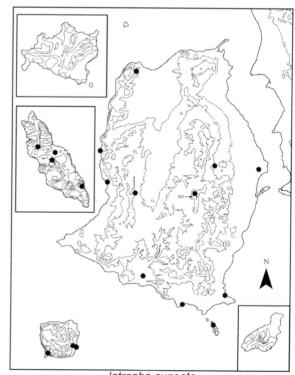

Jatropha cuneata
Haat

sap from even slight bruising of the stems permanently stains clothing brownish red resembling blood stains (the dye becomes fixed after washing). Leaves quickly drought deciduous. Short-shoot leaves often 5–20 mm long, entire, and sessile, appearing after rains at almost any time of the year; long shoots, or primary growth including seedlings, growing with sufficient soil moisture during hot, humid weather, and mostly producing larger leaves, mostly 20–50 mm long, lobed, and short petioled. Female flowers solitary or in pairs, with green sepals and white petals; male flowers white, several in subsessile to short inflorescences. Flowering and fruiting during hot, humid summer rainy season. Fruits rounded, 1-seeded, the seeds 9–10 mm wide.

Seri baskets are made from carefully selected and prepared stems of this shrub, the youngest, straight stems are the ones collected (e.g., Felger & Moser 1985; Moser 1973).

Local Distribution: Widespread on Tiburón, San Esteban, Dátil, and Nolasco; especially common on rocky slopes and south- and west-facing exposures, arid bajadas, desert flats, and washes. On Tiburón it extends to higher elevations on exposed sites in the Sierra Kunkaak and to the ridge crest of Sierra Menor. This is one of the most abundant shrubs in the region.

Other Gulf Islands: Ángel de la Guarda, San Lorenzo (observation), Tortuga, San Marcos (observation), Coronados (observation), Carmen, Danzante, Monserrat (observation), Santa Catalina, Santa Cruz, San Diego, San José, San Francisco, Espíritu Santo, Cerralvo.

Geographic Range: Northwestern Sinaloa, western Sonora, and southwestern Arizona, most of the Baja California Peninsula, and Gulf Islands.

Tiburón: Ensenada Blanca (Vaporeta), *Felger 10249*. Ensenada de la Cruz, *Felger 2590*. Ensenada del Perro, 12 Apr 2007, *Felger*, observation. Sierra Menor, 450 m, 2 Feb 1965, *Felger*, observation (Felger 1966:247). Palo Fierro, *Felger 8927*. Cójocam Iti Yaii, occasional at saddle, *Wilder 07-538*. Caracol, *Wilder 08-327*. Hill 0.5 km W of Tecomate, 12 Feb 65, *Felger*, observation (Felger 1966:265). Pooj Iime, 29 Sep 2007, *Felger & Wilder*, observation.

San Esteban: Canyon at SW corner of island, *Felger 16616*. Limantour, *Felger 16611*. Main arroyo at El Monumento, *Felger 9173*.

Dátil: NE end of island, *Felger 9123*. E-central side, E-W trending canyon, *Wilder 09-38*.

Nolasco: Above SE cove, *Maya 1 Nov 1963*. Center of island, W side, 800 ft, *Felger 12089*. Cañón el Faro, 60 m, *Felger 06-83*. Cañón de Mellink, upper reaches of canyon, *Wilder 08-352*.

Sebastiania bilocularis S. Watson [*Sapium biloculare* (S. Watson) Pax]

Hehe coaanj "poisonous plant"; *hierba de la flecha*; Mexican jumping bean

Description: Many-stemmed shrubs often 2–2.5 m tall, with copious milky sap. Leaves shiny green, often becoming reddish; nearly evergreen in riparian canyons, elsewhere gradually drought deciduous. Staminate flowers yellow.

This shrub is widely regarded as highly poisonous and is generally avoided by the Comcaac. The sap was used as arrow poison.

Local Distribution: Widespread across Tiburón except the arid western margin; mostly in washes and canyons, and also on rocky slopes including the east side of the Sierra Menor and the Sierra Kunkaak to the summit.

Other Gulf Islands: None.

Geographic Range: Southwestern Arizona to the Guaymas region and eastward to the Río Bavispe region and east of Hermosillo, and both states of Baja California.

Tiburón: Arroyo Sauzal, *Felger 10091*. Arroyo de la Cruz, *Moran 13009*. SW part of Agua Dulce Valley (Central Valley): Washes, *Felger 21286*; 1200–1400 ft, mountain bordering valley [Sierra Menor], *Felger 12395*. 3 km W of Palo Fierro,

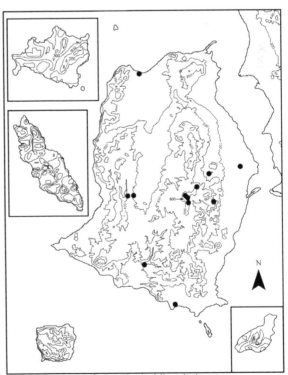

Sebastiania bilocularis
Hehe coaanj

bajada, *Felger 6978*. Sopc Hax, canyon, *Felger 9291*. Cerro San Miguel, *Quijada-Mascareñas 90T012*. Peak of Mount San Miguel, 2800 ft, *Knight 1013* (UNM). Foothills of Sierra Kunkaak, *Wilder 05-13*. Cójocam Iti Yaii, main peak, *Wilder 07-545*. Summit of island, occasional, *Wilder 07-562*. Tecomate, arroyo, *Felger 10190*.

Tragia jonesii Radcliffe-Smith & R. Govaerts

Hehe cocozxlim "plant that causes rash"; *ortiguilla*; noseburn

Description: Delicate twining or semi-vining herbaceous perennials; drought deciduous with moderately to strongly stinging, simple hairs. Leaves 1.5–5.5 cm long, the blades ovate and thin, the margins toothed. Inflorescence rachis with conspicuous tack-shaped, stalked glands. Flowers inconspicuous. Reproductive at various seasons.

Local Distribution: On Tiburón it is most common in canyons of the Sierra Kunkaak and also occurs in the Central Valley. On Dátil it is known from a single record.

Other Gulf Islands: None.

Geographic Range: Widespread in many desert habitats in northwestern Sonora from the vicinity of Altar and coastal areas near El Desemboque San Ignacio to the southwest portion of the state. Also Baja California Sur, and to southern Mexico.

Tiburón: SW Central Valley, 400 ft, arroyo, *Felger 17330*. Haap Hill, *Felger 74-T64*. S-facing slope of Sierra Kunkaak, 488 m, *Wilder 06-2*. Base of canyon on N slope of Sierra Kunkaak, 300 m, *Wilder 06-389*. Deep sheltered canyon, N portion of Sierra Kunkaak, *Felger 07-117*.

Dátil: NE side of island, *Felger 13438*.

Fabaceae (Leguminosae) – Legume or Bean Family

There are 24 genera with 35 species of legumes in the flora area, representing 8.7 percent of the flora. The legume genera have only one or two species in the flora area, except *Marina* with three species, and are native to the area with the exception of a single immature *guamúchil*, *Pithecellobium dulce*. None of the 34 native species appear to be rare in the flora area and, with three exceptions, are documented with multiple collections. The tepary, *Phaseolus acutifolius*, has been collected only once and has a limited distribution on Tiburón. *Marina evanescens* is known from a single collection on the north side of San Esteban, but that area is not well explored botanically. The precatory bean, *Rhynchosia precatoria*, has been found in one area in a large interior canyon in the Sierra Kunkaak, but it is likely to be more widespread in the Sierra.

The legumes in the flora area are distributed among the three major subfamilies:

Mimosoideae	**Papilionoideae**
Acacia, 2 species	*Acmispon*, 2
Calliandra, 2	*Coursetia*, 1
Desmanthus, 2	*Dalea*, 2
Ebenopsis, 1	*Desmodium*, 1
Lysiloma, 1	*Errazurizia*, 1
Mimosa, 1	*Lupinus*, 1
Pithecellobium, 1	*Marina*, 3
Prosopis, 1	*Olneya*, 1
Zapoteca, 1	*Phaseolus*, 2
	Psorothamnus, 1
Caesalpinioideae	*Rhynchosia*, 1
Hoffmannseggia, 1	*Tephrosia*, 2
Parkinsonia, 2	
Senna, 2	

Legumes are prominent in the vegetation and flora of Tiburón and Dátil, and less so on San Esteban. Four of the five genera of trees, *Acacia*, *Lysiloma*, *Olneya*, and

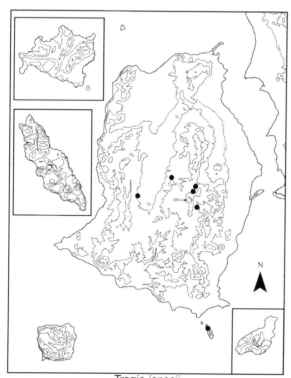

Tragia jonesii
Hehe cocozxlim

Parkinsonia, are prominent in the vegetation of different regions of the flora area. Mesquite trees, *Prosopis glandulosa*, are surprisingly uncommon, although stunted, shrubby forms are locally common at some shoreline sites. The remaining legumes are more or less evenly represented among annuals (winter-spring, summer, or non-seasonal species), herbaceous perennials, and shrubs.

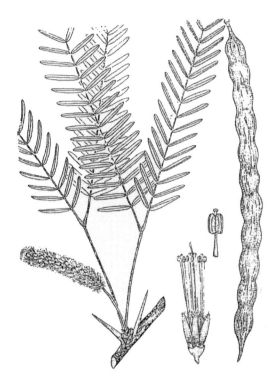

Figure 3.85. *Prosopis glandulosa* var. *torreyana*. FR.

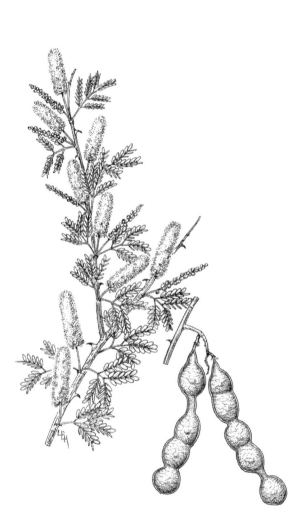

Figure 3.84. *Acacia greggii*. LBH.

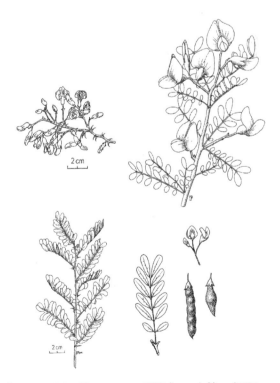

Figure 3.86. *Olneya tesota*. CMM (lower left) and MBJ.

Figure 3.87. *Parkinsonia florida*. MBJ.

Figure 3.88. *Parkinsonia microphylla*. MBJ.

Figure 3.89. *Acacia greggii*.
Vicinity of Caracol, Isla Tiburón.
BTW, 24 May 2006.

Figure 3.90. *Acacia willardiana*. Foothills of Sierra Kunkaak, Isla Tiburón.
PRG, October 2001.

Figure 3.91. *Calliandra californica*. Isla Dátil.
BTW, 10 March 2007.

Figure 3.92. *Calliandra eriophylla*
var. *eriophylla*. Foothills of Sierra
Kunkaak, Isla Tiburón.
JRT, January 2008.

Figure 3.93. *Dalea mollis.* Vicinity of Punta Chueca, Sonora. BM, 5 April 2008.

Figure 3.94. *Olneya tesota.* Isla Cholludo. BTW, 1 March 2009.

Figure 3.95. *Phaseolus acutifolius.* Pods and seeds from cultivated plants of the wild tepary collected on Haap Hill, Isla Tiburón in 1976. RSF.

Figure 3.96. *Psorothamnus emoryi* var. *emoryi*. Zozni Cacösxaj, Isla Tiburón. BTW, 6 April 2008.

1. Trees more than 2 m tall with whitish, papery, peeling bark. _____ **Acacia willardiana**
1' Annuals to shrubs, or trees but the bark not as above.
  2. All leaves with 3 leaflets.
    3. Delicate annuals, not vining; pods segmented into odd-shaped single-seeded segments separating at maturity, the valves not dehiscent. _____ **Desmodium**
    3' Vines, annuals or perennials; pods not segmented, the valves dehiscent.
      4. Annuals; leaflets not gland dotted; flowers pink; seeds mottled dull gray and brown._____ **Phaseolus**
      4' Perennials; lower leaflet surfaces conspicuously gland dotted; flowers yellow; seeds shiny red and black._
        _____ **Rhynchosia**
  2' All or most leaves with more than 3 leaflets.
    5. Trees or woody shrubs usually more than (1) 1.5 m tall.
      6. Trees and shrubs armed with spines, prickles, or thorns on at least some branches or stems, or twigs spinescent tipped.
        7. Spines (prickles) on twigs between the nodes.
          8. Leaflets 3.5–6 (8) mm long; flowers cream color; pods usually 5–15 cm long, 1 cm or more in width. _____ **Acacia greggii**
          8' Leaflets 6–12 mm long; flowers lavender-pink; pods 4.5–5.5 cm long, 0.5–0.6 cm wide.__**Mimosa**
        7' Spines or thorns at nodes (not internodal) or twigs ending in a sharp point.
          9. Paired spines persisting to form unique horizontal ring-like ridges on the bark; leaves with one pair of pinnae, each with 2 broad leaflets commonly 2–5 cm long, 1+ cm wide; flowers mimosoid. ___
            _____ **Pithecellobium**
          9' Spines or thorns not forming ridges as in *Pithecellobium*; pinnae or once-pinnate leaves with more than 2 leaflets, the leaflets usually less than 2 cm long, not more than 8 mm wide.
            10. Leaves once pinnate, or appearing so.
              11. Bark remaining green for multiple years; nodes without spines, many twigs ending in a spinescent point; leaves actually of two pinnae but these sessile and appearing as two leaves; leaflets 1–3 mm long; flowers caesalpinioid. _____**Parkinsonia microphylla**
              11' Bark not remaining green; nodes often bearing spines, twigs not ending in a sharp point; leaflets 10–20 mm long; flowers papilionoid._____**Olneya**
            10' Leaves twice pinnate.
              12. Pods thick and woody, 2 cm or more in width and semi-persistent._____ **Ebenopsis**
              12' Pods not woody, 1.5 cm or less in width and mostly not persistent.
                13. Bark remaining green multiple years; leaflets 5–10 mm long; pods dry, less than 10 cm long; flowers caesalpinioid._____**Parkinsonia florida**
                13' Bark gray or brown; leaflets 4.5–31 mm long; pods with mesocarp (pulp), (7) 10–20 cm long; flowers mimosoid. _____ **Prosopis**
      6' Trees and shrubs unarmed (without spines, prickles, or thorns).
        14. Bark whitish or pale tan, peeling in papery sheets; leafstalks 10–20 cm long; pinnae 1 or 2 pairs, at or near the tip of the leafstalk and soon drought deciduous. _____ **Acacia willardiana**
        14' Bark not white, papery and peeling; leafstalks less than 9 cm long; pinnae or leaflets not only at the tip of the leafstalk.
          15. Leaves once pinnate; flowers papilionoid.
            16. Multiple-stem woody shrubs often 2–3 m tall; leaflet tips with a short, terminal bristle or mucro; flowers pink and yellow; pods multiple seeded, gradually dehiscent._____ **Coursetia**
            16' Scarcely woody small shrubs or subshrubs often less than 1 (1.5) m tall; leaflet tips entire; flowers bright yellow or pink and white; pods 1(2) seeded, indehiscent.
              17. Plants often taller than wide, the stems slender; flowers rose-pink and white; pods less than 3 mm long, hidden within the calyx; Sierra Kunkaak. _____ **Dalea bicolor**
              17' Plants often as wide or wider than tall, the stems often rather thick; flowers yellow; pods 8–11 mm long; widespread at low elevations on Tiburón. _____ **Errazurizia**
          15' Leaves 2 or more times pinnate; flowers mimosoid.
            18. Multiple-trunk large shrubs or small trees often 3–4+ m tall; pods 8–15 cm long, 10–20 mm wide, with a narrow cord-like rim separating from the rest of the pod at maturity.___
              _____ **Lysiloma**

18' Shrubs, sometimes to 3 m tall; pods 3–10 cm long, 3–10 mm wide, the rim or margin not separating from the rest of the pod.

    19. Leafstalks with prominent crater-form glands; valves of pods remaining straight. _____ _____ **Desmanthus fruticosus**

    19' Leafstalks without glands; valves of pods curling back when separating at maturity.

        20. Stipules 2–4 mm long, slender and stiff (setaceous to subulate); leaflets 5–15 pairs per pinna, 3–6 mm long; flowers pink or red; pods 3–7 cm long, 3–7 mm wide. __ _____ **Calliandra**

        20' Stipules 4–5 mm, ovate, leaflet-like; leaflets 5–7 pairs per pinna, 5–15 (20) mm long; flowers probably reddish; pods 5–10 cm long, 5–10 mm wide. _____ **Zapoteca**

5' Plants herbaceous or if woody then shrubs mostly less than 1 m tall and not heavy wooded.

  21. Cool season annuals; leaflets digitately arranged, the leaflets widest above the middle. _____ **Lupinus**

  21' Annuals and perennials; leaflets pinnately arranged, the widest part variable.

    22. Leaves 2 or more times pinnate.

      23. Pods slender and without a cord-like rim, the valves parting after opening but straight and not curling or twisting. _____ **Desmanthus**

      23' Pods with or without a rim, the valves curling or twisting after opening.

        24. Flowers caesalpinioid; pods gland dotted, without a cord-like rim, the valves tightly twisting after opening. _____ **Hoffmannseggia**

        24' Flowers mimosoid; pods not gland dotted, with a cord-like rim, the valves curling back after opening but not twisting.

          25. Stipules 2–4 mm long, slender and stiff (setaceous to subulate); leaflets 5–15 pairs per pinna, 3–6 mm long; flowers pink or red. _____ **Calliandra**

          25' Stipules 4–5 mm, ovate, leaflet-like; leaflets 5–7 pairs per pinna, 5–15 (20) mm long; flowers probably reddish. _____ **Zapoteca**

    22' Leaves once pinnate.

      26. Leaves even pinnate with (4) 6 (10) leaflets; flowers caesalpinioid; anthers large, with terminal pores. _____ **Senna**

      26' Leaves odd pinnate with 3–23 leaflets; flowers papilionoid; anthers small, opening longitudinally, without pores.

        27. Herbage not gland dotted; leaves with (3) 5–7 leaflets in number; pods elastically dehiscent, the valves twisting after opening, pods with more than 2 seeds.

          28. Leaves 0.7–2.3 cm long, leaflets 3–7 in number, 4.5–13 mm long, often thick and sometimes semi-succulent; stipules represented by small but conspicuous dark glands; flowers yellow or orange, 5.5–9 mm long; pods 1.5–2.7 cm long. _____ **Acmispon**

          28' Leaves 2–15 cm long, leaflets 7–21 in number, 15–55 mm long, not thickened; stipules slender and pointed, not glandular; flowers lavender to purple, 8–15+ mm long; pods 3–6 cm long. _____ **Tephrosia**

        27' Herbage gland dotted; leaves with (3) 5–35 or more leaflets; pods indehiscent, 1(2)-seeded.

          29. Herbage not woolly; flowers blue or violet with white; pods less than 3 mm long.

            30. Leaflets 7–13 in number, 4–7 mm long, widest near the tip (truncate to notched); flowers subtended by teardrop-shaped, dark reddish brown, persistent glands; corolla about as long as to scarcely longer than the calyx; midrib of calyx lobes extending in awn-like plumose bristles. _____ **Dalea**

            30' Leaflets (4) 6–29 in number, 1–6 mm long, orbicular to obovate; flowers not subtended by dark glands; corolla longer than the calyx; calyx lobes not bristle tipped. ___ **Marina**

          29' Herbage woolly; flowers yellow or purple; pods 2.5–10 mm long.

            31. Herbage white-woolly; flowers yellow; pods 8–10 mm long. _____ **Errazurizia**

            31' Herbage gray-woolly; flowers purple; pods probably 2.5–4 mm long. __ **Psorothamnus**

Acacia

Hardwood trees and shrubs; flowers mimosoid.

1. Trees and shrubs armed with sharp, recurved prickles; bark brown, not peeling. _____ **A. greggii**
1' Trees unarmed; bark whitish and often peeling in papery sheets. _____ **A. willardiana**

Acacia greggii A. Gray [*Senegalia greggii* (A. Gray) Britton & Rose]

Tis; *uña de gato*; catclaw acacia

Description: Shrubs or small trees, the trunk often irregular and crooked; wood reddish brown. Stems armed with sharp, recurved, usually laterally compressed internodal prickles, or some stems may be unarmed. Leaves twice pinnate with many small leaflets, gradually winter deciduous. Flowers cream color, fragrant, in short, cylindrical spike-like inflorescences often 1.5–3.5 cm long including the peduncle; flowering in April. Pods flat, usually curved, without mesocarp, indehiscent to tardily semi-dehiscent (the whole pod usually falls and is chewed or eaten by rodents).

Local Distribution: Widespread on Tiburón except the western margin and especially common at the base of the Sierra Kunkaak; arroyos and canyon bottoms.

Other Gulf Islands: Ángel de la Guarda.

Geographic Range: Durango and Tamaulipas north to Nevada, Utah, and New Mexico, and the northern two-thirds of Sonora and both Baja California states but rare in Baja California Sur.

Tiburón: Arroyo Sauzal, *Felger 10111*. Foothills of Sierra Kunkaak, road to Caracol station, *Wilder 06-156*. Tecomate, along drainage W of village, *Whiting 9024*. Hast Iif (Ojo de Puma), Central Valley, broad arroyo, *Wilder 07-261*.

Acacia willardiana Rose [*Mariosousa willardiana* (Rose) Seigler & Ebinger]

Cap; *palo blanco*

Description: Handsome and unique, unarmed, wispy trees 3–7+ m tall, the trunk and major limbs slender to robust, the branches flexible, with drooping, leafy twigs. Bark cream white and peeling away in papery sheets during dry seasons. Leaves 10–27 cm long, with 1 (2) pair(s) of small pinnae, the leaflets and pinnae quickly

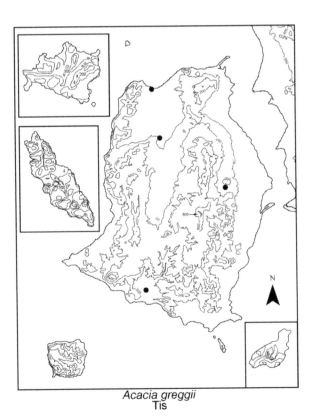

Acacia greggii
Tis

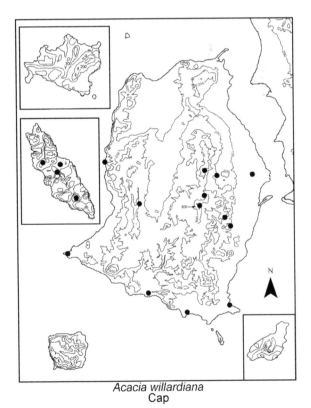

Acacia willardiana
Cap

drought deciduous, leaving the long, slender, and flattened leafstalk to function as a phyllode. Flowers pale yellow in dense cylindrical spikes. Pods flat, straw colored, without mesocarp, tardily dehiscent. Flowering at least February–May and October.

The leafstalks or phyllodes, which resembles those of some Australian acacias, make this species unique in the New World. Although it has no obvious close relatives it seems to be closer to *A. russelliana* (Britton & Rose) Lundell of central and eastern Sonora and Sinaloa than any other acacia.

Local Distribution: Tiburón, widespread and common across most of the island, including mountains to their summits; arid rocky slopes, ridges and cliffs, as well as upper bajadas, arroyos, and canyons. Some stands of palo blanco on the island are notably tall, such as in the main arroyo just south of Pazj Hax at east base of Sierra Kunkaak. Nolasco, widespread through the island, especially at higher elevations. It is the only legume on Nolasco.

Other Gulf Islands: None.

Geographic Range: Western Sonora, from the Cerros Anacoretas southward, to northwestern Sinaloa (Díaz et al. 2008).

Tiburón: Pooj Iime, 29 Sep 2007, *Felger & Wilder*, observation. Near Willard's Point, a few trees at edge of draw, *Johnston 4252* (UC). Arroyo Sauzal, *Felger 9989*. Arroyo de la Cruz, *Moran 13006*. Ensenada del Perro, *Tenorio 10894*. SW part of Central Valley, E-facing slopes of Sierra Menor, major washes and N-facing slopes, trees to 25 ft, trunk 10 inches diameter, 4 Dec 1973, *Felger 21287* (specimen not located). Hast Coopol, *Wilder 07-374*. W of Palo Fierro, arroyo in upper bajada, *Felger 10023*. Tiburon Island [Tinaja Anita, 23 Dec 1895, see "Gazetteer" and "Botanical Explorations"], *McGee 13* (UC 81155). Head of arroyo, base of Sierra Kunkaak, *Wilder 05-42*. Summit of island, rare, *Wilder 07-557*. Caracol, *Gold 397* (MEXU). Top of Sierra Caracol, 1500 ft, *Knight 1060* (UNM).

Nolasco: Common over higher parts of the island, *Johnston 3125* (CAS, UC). [S]E side, eastern exposure, 700–800 ft, *Pulliam & Rosenzweig 21 Mar 1974*. Cañón el Faro, 60 m, mostly along canyon and on S-facing slopes, and ridge crests, *Felger 06-86*. Cañón de Mellink, 110 m, canyon bottom, *Felger 08-143*.

Acmispon

Low-growing winter-spring annuals (those in the flora area); leaves odd pinnate, stipules (Sonoran Desert species) represented by small but conspicuous, dark glands.

Flowers papilionoid. Pods elastically dehiscent, the valves coiling.

1. Leaflets broadly elliptic or nearly so, widest at or slightly above the middle, the tip rounded to pointed (acute). _____ **A. maritimus**
1' Leaflets obovate, broadest at or near apex and gradually tapered to base, the tip blunt (truncate) and sometimes notched. _____ **A. strigosus**

Acmispon maritimus (Nuttall) D.D. Sokoloff var. **brevivexillus** (Ottley) Brouillet [*Lotus salsuginosus* Greene var. *brevivexillus* Ottley]

Hee inoosj "jackrabbit's claw"; desert lotus

Description: Plants glabrous or sparsely hairy except for short, white hairs on youngest stem tips. Leafstalks moderately flattened, the leaflets 3–7 in number, rather thick and often semi-succulent, broadly elliptic, 6–12 mm long, the apex broadly rounded to acute plus a small, blunt point. Peduncles slender, bearing 2–4 flowers, and a leaflet-like bract on the joint (node) just below the flowers. Flowers 5.5–7 mm long, the corollas bright yellow, or the banner becoming pale orange with age. Pods 10.6–16.7 mm long.

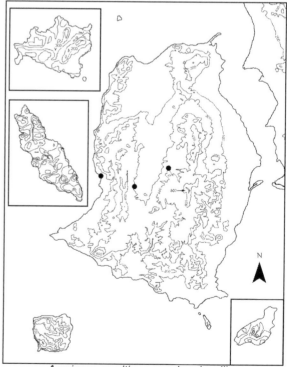

Acmispon maritimus var. *brevivexillus*
Hee inoosj

Local Distribution: Documented from the western and central parts of Tiburón and expected elsewhere on the island; valley plains, arroyos, and low hills.

Other Gulf Islands: Ángel de la Guarda, San Lorenzo.

Geographic Range: Northern Baja California Sur to southern California, Arizona, and northwestern Sonora including the Sierra Seri.

Tiburón: Ensenada Blanca (Vaporeta), *Felger 17241*. SW Central Valley, *Felger 17317*. Haap Hill, *Felger 76-T11*.

Acmispon strigosus (Nuttall) Brouillet [*Hosackia strigosa* Nuttall. *Lotus strigosus* (Nuttall) Greene var. *tomentellus* (Greene) Isely. *L. tomentellus* Greene. *Ottleya strigosa* (Nuttall) D.D. Sokoloff]

Hee inoosj "jackrabbit's claw"; hairy lotus

Description: Roots conspicuously nodulated. Herbage sparsely to densely hairy (cinereous to strigose). Leaf-stalks flattened, the leaflets (3) 5–7, relatively thick, obovate, 4.5–13 mm long, the apex truncate (blunt) and sometimes notched. Flowers solitary and sessile in axils, or 1 or 2 on slender peduncles with a leaflet-like, or sometimes 3-foliolate, bract on the joint (node) just below the flower(s). Flowers 7.5–9 mm long, the corollas bright

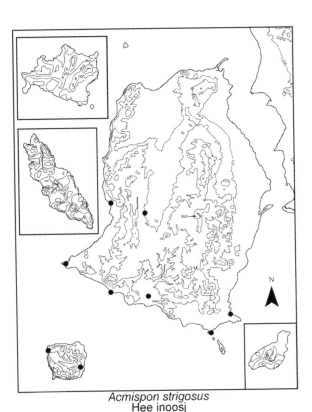

Acmispon strigosus
Hee inoosj

yellow with a red nectar guideline on the banner. Pods 12–27 mm long.

Local Distribution: Widespread in the lowlands of Tiburón; not documented for the Sierra Kunkaak. Also on San Esteban, including Arroyo Limantour and rocky slopes at least along the north side of the island.

Other Gulf Islands: Ángel de la Guarda, San Lorenzo.

Geographic Range: Sonora south to Bahía Kino, and Baja California Norte to southeastern California and Arizona.

Tiburón: Ensenada Blanca (Vaporeta), *Felger 17276*. Canyon just N of Willard's Point, arroyo, *Moran 8731* (CAS). Tordillitos, *Felger 15470*. Arroyo Sauzal, *Felger 08-10*. La Viga, occasional, *Wilder 08-143*. Ensenada del Perro, *Felger 17724*. SW Central Valley, *Felger 17353*.

San Esteban: Limantour, *Felger 17663*. San Pedro, rocky slopes, *Felger 16660*.

Calliandra

Unarmed shrubs. Leaves gradually drought deciduous, twice pinnate, the leaflets small and numerous. Flowers mimosoid. Stamens numerous, large and showy. Pods flattened with thick, cord-like margins (rim), and dehiscent with the valves separating elastically and curling back after splitting apart.

1. Flowers bright red. _____ **C. californica**
1' Flowers pink to pale rose._____ **C. eriophylla**

Calliandra californica Bentham

Hast iti heepol cheel "red *Krameria bicolor* on the rocks"; *tabardillo, zapotillo*

Description: Many-branched shrubs often 0.6–1 m tall, resembling *C. eriophylla*. The easiest way to distinguish it from *C. eriophylla* is by its bright red flowers. Although *C. californica* and *C. eriophylla* may be difficult to distinguish vegetatively, there are recognizable differences that set them apart: *C. californica* is usually a taller and larger shrub, and tends to have larger leaves and leaflets. In cultivation in southern Arizona it is much faster growing than *C. eriophylla*.

Local Distribution: Tiburón, near the southeastern corner of the island, almost always in association with larger plants such as ironwood (*Olneya*) or cacti such as organ pipe (*Stenocereus thurberi*), and in the Sierra Kunkaak. Common on Dátil, especially on protected N- and E-facing slopes and canyon bottoms on the E side of the island.

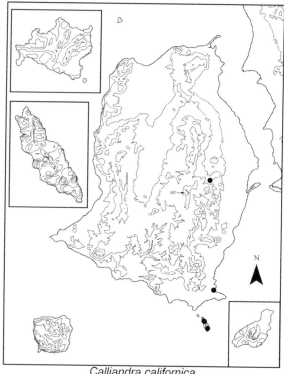

Calliandra californica
Hast iti hepool cheel

twigs, peduncles, and leafstalks densely to moderately pubescent with short white hairs, the leaflets sparsely pubescent. Flower clusters with showy, whitish to pinkish stamens, opening at night, drooping with daytime heat. The 8-grained pollen packets adhere to butterfly wings (Turner et al. 1995). Pods often 4–5.6 cm long. Flowering after rains, mostly in spring.

Local Distribution: Tiburón in the Sierra Kunkaak and south end of the Agua Dulce/Central Valley. Often along small arroyos and canyons.

Other Gulf Islands: None.

Geographic Range: Through much of the Sonoran Desert, mostly in less xeric habitats and especially at elevations just above the desert. Southeastern California to northern Baja California Norte and southern Arizona to southwestern New Mexico and southward to Chiapas. A second variety occurs in southern Texas.

Tiburón: Haap Hill, *Felger 76-T15*. N base of Sierra Kunkaak, between Sierra Kunkaak Mayor and Sierra Kunkaak Segundo, *Wilder 06-391*. Base of Capxölim, *Felger 07-101*. Hast Cacöla, base of mountain, 20 Jan 2008, *Romero-Morales 08-6*.

Other Gulf Islands: San Marcos, Coronados, Carmen, San José (observation), San Francisco, Espíritu Santo.

Geographic Range: Centered on the Baja California Peninsula, where it is the most common *Calliandra*. Sonora in the vicinity of San Carlos and the Sierra el Aguaje, and near Puerto Libertad.

In Sonora *C. californica* occurs in pockets within the general distribution of *C. eriophylla* but only in coastal habitats with maritime influence. They do not occur intermixed. The presence of the disjunct populations of *C. californica* supports the hypothesis that in more mesic times certain peninsular species may have spread to the mainland via the Midriff Islands (Cody et al. 1983; Turner et al. 1995).

Tiburón: Ensenada del Perro, *Wilder 07-156*. Hast Cacöla, E flank of Sierra Kunkaak, foothills at 200 m, *Wilder 08-230*.
Dátil: NE side, *Felger 13447*. SE side, *Felger 17504*. E side, N-facing slope of peak, *Wilder 07-108*.

Calliandra eriophylla Bentham var. eriophylla

Haxz iztim "dog's hip bone"; *huajillo*; fairy duster

Description: Much-branched dwarf woody shrubs, with firm but flexible stems and grayish bark. Young

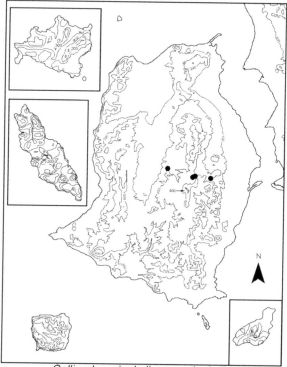

Calliandra eriophylla var. *eriophylla*
Haxz iztim

Coursetia glandulosa A. Gray

Cmapxöquij "what bursts open," hehe ctoozi "resilient plant"; *sámota*

Description: Hardwood, multi-stem shrubs often 1.5–3 m tall, the stems and branches slender and flexible. Foliage unfolding after flowering if there is sufficient soil moisture; leaves odd pinnate, luxuriant with larger, thin leaflets during the summerfall rainy season and much smaller leaflets in drier seasons; leaflets with a short bristle-tip; leaves gradually shed during fall and winter. Flowers papilionoid; petals pale yellow and white with faint red tinges; flowering in spring, usually when the plants are nearly leafless. Pods densely glandular-pubescent, with a septum and constriction between each seed, and gradually dehiscent.

The stems are sometimes encrusted with ant-tended, orange-colored lac produced by the scale insect *Tachardiella*. This lac was used as an all-purpose adhesive by the Comcaac (Felger & Moser 1985).

Local Distribution: Tiburón in the Sierra Kunkaak to the south end of the Agua Dulce Valley and adjacent east side of the Sierra Menor. Generally in arroyos, canyons, and rocky slopes.

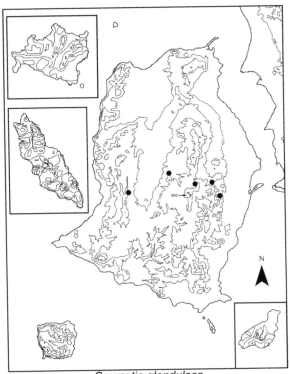

Coursetia glandulosa
Cmapxöquij, hehe ctoozi

Other Gulf Islands: None.

Geographic Range: Southern Arizona to Oaxaca and the Cape Region of Baja California Sur.

Tiburón: SW part of Central Valley, E-facing slopes of Sierra Menor, canyons and N-facing slopes, 4 Dec 1973, *Felger 21309* (specimen not located). Haap Hill, *Felger T74-12*. Head of arroyo, base of Sierra Kunkaak, between Sierra Kunkaak Mayor and Sierra Kunkaak Segundo, *Wilder 05-52*. Foothills of Sierra Kunkaak, *Felger 6951*. Hast Cacöla, E flank of Sierra Kunkaak, foothills to mid-elevation, *Wilder 08-233*.

Dalea

Annuals or small shrubs, with spirally twisted and usually reddish brown hairs and dotted with blister-like glands (secretory vesicles); herbage aromatic when bruised. Leaves odd-pinnate. Flowers papilionoid. Pods 1-seeded, indehiscent.

1. Perennials—subshrub or shrubs often to 1 m or more tall; calyx teeth triangular, not extending into bristles. _____ **D. bicolor**
1' Annuals, less than 30 cm tall; calyx teeth ending into feathery (plumose) bristles. _____ **D. mollis**

Dalea bicolor Humboldt & Bonpland ex Willdenow var. **orcuttiana** Barneby [*D. megalostachys* (Rose ex Rydberg) Wiggins]

Description: Subshrubs or shrubs 0.6–1.5 m tall, with upright branches and slender stems; the crushed foliage smells like fruit punch. Leaves highly variable in size and pubescence depending on soil moisture, often less than 1 cm long in dry seasons to perhaps 4 cm long with favorable conditions; leaflets at least 5 or 7 in number, 2.5–6+ mm long, obovate. Inflorescences 1–7.5 cm long, short and cone-like in dry seasons and longer and spicate in wetter seasons. Calyx teeth more or less triangular (deltoid). Petals lavender-pink except the banner petal at first white, becoming rose-pink. Pods less than 3 mm long.

Dalea bicolor is distinguished by its relatively short calyx lobes (shorter than to about equal to the calyx tube); tuberculate stems; almost always less than 10 pairs of leaflets, usually 4–6 pairs; and a banner petal that is largely or entirely a different color than the other petals, at least when young.

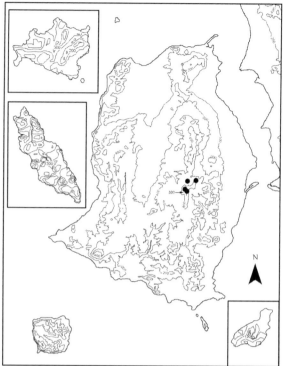

Dalea bicolor var. *orcuttiana*

Local Distribution: Canyons and steep slopes in the Sierra Kunkaak to the summit; sometimes abundant at higher elevations. The nearest population occurs on the opposite mainland near the western base of the Sierra Seri range.

Other Gulf Islands: None (the species and its varieties).

Geographic Range: This species ranges from New Mexico and Texas to southern Mexico. Var. *orcuttiana* is largely limited to the Baja California Peninsula, where it is widespread in both states, and also occurs on the Sonoran mainland southeast of El Desemboque (Wilder et al. 2007b). This species has several varieties and is extremely variable, and taxonomic realignments are expected.

> **Tiburón:** E of Siimen Hax, *Romero-Morales 07-3*. Cerro San Miguel, común en la cima, *Quijada-Mascareñas 91T019*. Cójocam Iti Yaii, saddle, abundant, 60 cm tall, *Wilder 07-536*. Summit of island, occasional, *Wilder 07-559*.
>
> **Sonora Mainland:** Vicinity of Cerro Pelón, 29°34'N, 112°09'W, 5 mi SE of El Desemboque, *Felger 17936*.

Dalea mollis Bentham

Hanaj itaamt "raven's sandals"; silky dalea

Description: Non-seasonal annuals, mostly found during spring, highly variable in size, with a relatively deep, orange-colored taproot. Leaves 0.9–4.6 cm long; leaflets (7) 11–13 in number, 2.5–8.2 mm long, broadly obovate, the tips blunt (truncate or notched), the lower surfaces gland dotted, the upper surfaces without glands. Inflorescences and pedicels with conspicuous teardrop-shaped dark maroon-brown glands remaining even after the flowers or fruits have fallen. Flowering at a very early age and apparently continuing as long as the soil moisture holds out. Calyx ribbed with rows of iridescent orange glands, the calyx lobes extending into feathery (plumose) bristles. Petals violet, drying red-purple. Pods 2.5 mm long.

Local Distribution: Widespread through the lowland regions of Tiburón, especially along arroyos, bajadas, and valley plains. Also on Dátil and the sandy lowland flat on Alcatraz.

Other Gulf Islands: Ángel de la Guarda, Tortuga, San Marcos.

Geographic Range: Widespread across the Sonoran Desert. Sonora from the vicinity of Guaymas northward, southwestern Arizona, southeastern California, and both Baja California states.

> **Tiburón:** El Sauz, gravelly flat of broad arroyo, *Moran 8757* (SD). La Viga, *Wilder 08-148*. Punta San Miguel, low dunes, *Wilder 07-186*. Vicinity of Sopc Hax, canyon, *Romero-Morales 07-17*. Hahjöaacöl, NE coast, *Wilder*

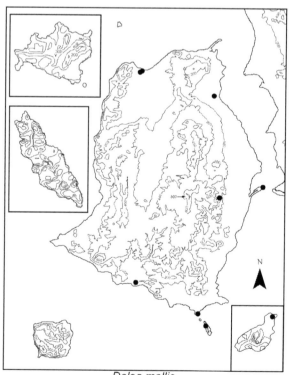

Dalea mollis
Hanaj itaamt

08-215. Tecomate, *Wilder 07-243*. Bahía Agua Dulce, on flat, *Moran 12989* (SD).

Dátil: E side, halfway up N-facing slope of peak, *Wilder 07-120*.

Alcatraz: E side, locally common on sandy low shell mound, *Felger 12724*.

Desmanthus

Unarmed shrubs. Leaves twice pinnate, the leaflets small and numerous, and folding together at night. Flowers mimosoid, white, 3–4 mm long. Pods slender, dry and dehiscent, in digitate clusters.

1. Shrubs to 1 m tall, much branched below and above; pods to 4 mm wide. _____ **D. covillei**
1' Shrubs often 2–3 m tall, sparsely branched, mostly above, and with a slender trunk; pods 5–6 mm wide. _____ **D. fruticosus**

Desmanthus covillei (Britton & Rose) Wiggins

Pohaas camoz "what thinks it's a mesquite"

Description: Slender shrubs mostly to 1 m tall, lacking a well-defined trunk, and with feathery, drought deciduous foliage. Pods 6.3–11.5 cm long, 3–4 mm wide.

Local Distribution: Upper bajadas of the Sierra Kunkaak and benches above arroyos, where it is sometimes the dominant shrub.

Other Gulf Islands: None.

Geographic Range: South-central Arizona to Sinaloa and Gulf Coast of Baja California Sur.

Tiburón: 3 km W of Punta Tormenta, *Smartt 13 Oct 1976* (UNM). Canyon, ¼ mi below Sopc Hax waterhole, *Felger 9324*. Bench above arroyo bottom at NE base of the Sierra Kunkaak, *Wilder 06-392*.

Desmanthus fruticosus Rose

Hehe casa "putrid plant"

Description: Slender shrubs often 2–3 m tall with a slender, flexible trunk and sparsely branched above. Leaves with 2–9 pairs of pinnae, each with 10–20 pairs of leaflets. Flowers small, in axillary heads or short spikes, which may contain bisexual, male, and sterile flowers. Pods 5–9 cm long, 5–6 mm wide.

This species is much taller and more robust than *D. covillei*, with larger and more robust leaves and pods.

Local Distribution: Widespread across Tiburón, including arroyos, canyons, bajadas, valley plains, and

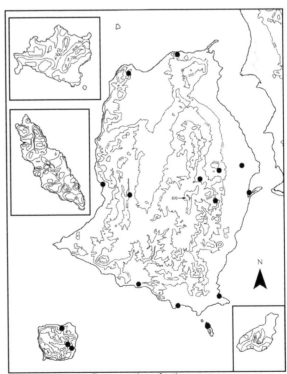

Desmanthus covillei
Pohaas camoz

Desmanthus fruticosus
Hehe casa

hillsides and foothills of the Sierra Kunkaak, but not seen in the higher mountains. Common on San Esteban, mostly along arroyos and canyons, and Dátil on steep, rocky slopes, and ridge crests.

Other Gulf Islands: Ángel de la Guarda, Tortuga, San Marcos, Coronados, Carmen, Danzante, Monserrat, Santa Catalina, Santa Cruz, San José, San Francisco, Espíritu Santo, Cerralvo.

Geographic Range: Baja California Peninsula from 28.5°N southward to Cabo San Lucas, coastal Sonora from north of Bahía Kino to north of Guaymas, and Gulf Islands.

Tiburón: Ensenada Blanca (Vaporeta), *Felger 14939*. Arroyo Sauzal, *Felger 10103*. Ensenada de la Cruz, *Felger 9222*. Ensenada del Perro, *Tenorio 9506*. SW Central Valley, upper rocky slope of mountain bordering valley, 1200–1400 ft, S-facing slope, *Felger 12425*. Zozni Cmiipla, *Felger 08-134*. Palo Fierro, bajada 2 mi inland, *Felger 12542*. Sopc Hax, *Felger 9313*. Camino de Caracol, arroyo entering foothills of Sierra Kunkaak, *Wilder 05-5*. Foothills of Sierra Kunkaak, *Wilder 05-19*. Hee Inoohcö, mountains at NE coast, *Felger 07-76*. Rocky hill W of Tecomate, *Felger 6273*.

San Esteban: Limantour, *Felger 17653*. E-central side of island, hillside, 400 m inland, *Felger 12749*. N side of island, narrow rocky canyon behind large cove, *Felger 17590*.

Dátil: NE side of island, steep rocky slopes and arroyo bottom, scattered, *Felger 9081*. E side, slightly N of center, S-facing slope, *Wilder 07-133*. NW end of island, steep rocky slopes, *Felger 2779*.

Desmodium procumbens (Miller) Hitchcock

Tick clover

Description: Delicate, short-lived summer annuals with very slender stems. Leaves with 3 leaflets, rough and dry to the touch with minute hooked hairs. Flowers papilionoid, yellow to greenish, 3–4 mm long. Pods slender, resembling a string of cutouts, the segments 1-seeded, triangular to somewhat square, often twisted or curled, and breaking apart between the segments.

Local Distribution: Tiburón on Sierra Kunkaak and southern part of the Central Valley; usually in washes and canyon bottoms, often among leaf litter from the shrub and tree over-story. The nearest known populations are in the vicinity of Hermosillo and the Guaymas region but it is expected in the Sierra Seri; the closest peninsular population is in the Sierra San Francisco.

Other Gulf Islands: Cerralvo.

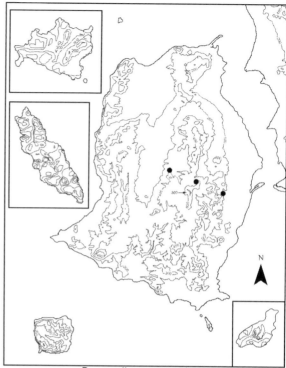

Desmodium procumbens

Geographic Range: Baja California Sur and southern Arizona to northern South America and West Indies; also in the Old World, where it is probably introduced.

Tiburón: Haap Hill, *Felger T74-38*. Canyon bottom, N base of Sierra Kunkaak, between Sierra Kunkaak Mayor and Sierra Kunkaak Segundo, 395 m, *Wilder 06-409*. Hant Hax camp, on way to Sopc Hax from Zozni Cmiipla, arroyo bottom, *Felger 9348*.

Ebenopsis confinis (Standley) Britton & Rose [*Pithecellobium confine* Standley]

Heejac

Description: Intricately branched shrubs to ca. 2+ m tall with very rigid, zigzag twigs, and very sharp, stipular spines. Leaves twice pinnate with 1 (2) pairs of pinnae, each with 2–5 pairs of small, circular leaflets. Flowers mimosoid, pale purple to rose. Pods dry and eventually splitting apart, with very thick, semi-persistent woody valves, 3.7–10 cm long, 1.5–3.5 cm wide (based on island and adjacent mainland specimens), dark brown or blackish, crusty and scaly.

Local Distribution: Tiburón, along arroyos, canyons, upper bajadas, and rocky slopes; Sierra Kunkaak and

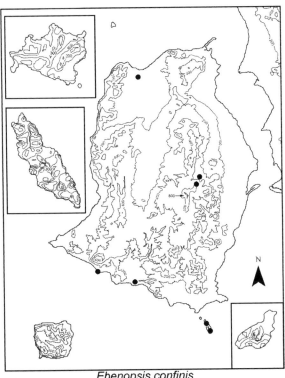

Ebenopsis confinis
Heejac

cm long, white tomentose, pinnate with 9–17 leaflets. Flowers papilionoid, in spikes 3–13 cm long, elongating to 15 cm in fruit; corollas 6–7 mm long, bright yellow, and semi-fleshy. Pods 8–10 mm long, 1- or 2-seeded and indehiscent.

Local Distribution: Widespread at lower elevations across Tiburón, including coastal dunes, bajadas, valley plains, desert pavement ridges, and rocky hills.

Other Gulf Islands: Ángel de la Guarda, San Marcos.

Geographic Range: Gulf Coast of Baja California Norte and northern Baja California Sur, northern Gulf Islands, and Gulf Coast of Sonora from Puerto Libertad to Bahía Kino.

Tiburón: Ensenada Blanca (Vaporeta), *Felger 14938.* 1 km N of Punta Willard, *Felger 17757.* Sauzal, *Wilder 08-109.* Ensenada de la Cruz, *Felger 12769.* Coralitos, small hill above beach, *Wilder 06-65.* El Monumento, *Felger 15528.* Punta San Miguel, low dunes 5–50 m inland, *Wilder 07-185.* Palo Fierro, 14 Feb 1965, *Felger,* observation (Felger 1966:190). E bajada near Valle de Águila, middle bajada, *Wilder 07-236.* Hahjöaacöl, NE coast, *Wilder 08-217.* Tecomate, *Whiting 9005.*

southern part of the island. Dátil, in canyons and on steep slopes.

Other Gulf Islands: San Marcos (observation), Monserrat (observation), Santa Catalina, Santa Cruz, San Diego (observation), San José (observation), Espíritu Santo, Cerralvo.

Geographic Range: Widespread in Baja California Sur and the southern part of Baja California Norte, Gulf Islands, and Gulf Coast of Sonora from El Desemboque to Bahía Kino and mountains north of San Carlos (Turner et al.1995).

Tiburón: Tordillitos, *Felger 15486.* Sauzal, granite hills and canyons, *Wiggins 17173* (CAS). Arroyo [Ensenada] de la Cruz, 28 Feb 2009, *Wilder,* observation. Foothills of Sierra Kunkaak, *Wilder 05-17.* Capxölim, 24 Oct 2007, *Wilder,* observation. **Dátil:** NE side of island, steep rocky slopes, SE exposure, localized colony, *Felger 13449.* SE side of island, steep and narrow canyon, *Felger 17523.*

Errazurizia megacarpa (S. Watson) I.M. Johnston

Cooxi iheet "dead man's gambling sticks"

Description: Small, mound-shaped shrubs, white tomentose with glandular-pustulate hairs. Leaves 3–7

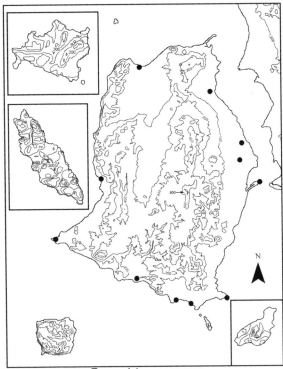

Errazurizia megacarpa
Cooxi iheet

Hoffmannseggia intricata Brandegee [*Caesalpinia intricata* (Brandegee) E.M. Fisher]

Haxz iztim "dog's hip bone"

Description: Intricately branched dwarf bushes to 60 cm tall, the twigs with weakly spinescent tips. Foliage sparse, the leaves twice pinnate with three more or less equal pinnae each with 4–10 pairs small leaflets. Flowers caesalpinioid, gland dotted, the petals bright yellow, and red veined. Pods 1–2 cm long, gland dotted, and dehiscent.

Local Distribution: Widespread at lower elevations on Tiburón; bajadas, arroyos, and rocky hills and mesas. Scattered on Dátil. Common on San Esteban along Arroyo Limantour to high elevation of the central peak.

Other Gulf Islands: San Marcos.

Geographic Range: Central Gulf Coast of Sonora from El Desemboque to Bahía Kino, the Baja California Peninsula, and Gulf Islands.

Tiburón: Ensenada Blanca (Vaporeta), arroyo, *Felger 7111.* Arroyo Sauzal, 2¾ mi inland, *Wilder 06-89.* Ensenada de la Cruz, rocky hill, *Felger 9223.* La Viga, *Wilder 08-150.* Ensenada del Perro, *Tenorio 9501.* Bajada between Hant Hax camp and Zozni Cmiipla, *Felger 9355.* Palo Fierro, *Felger 10325.* E bajada near Valle de Águila, *Wilder 07-235.* Hahjöaacöl, NE coast,

Wilder 08-220. N coast, *Mallery & Turnage 3 May 1937* (CAS). Bahía Agua Dulce, floodplain and arroyo, *Felger 10200.* Tecomate, 24 Mar 1968, *Gold 373* (CAS).

San Esteban: Limantour, canyon bottom (floodplain), near beach and less frequent farther inland, also on hillslopes including S slopes, *Felger 17676.* N face of sub-peak of central mountain, *Wilder 07-58.*

Dátil: NW side, *Felger 15306.* E side, N of center, *Wilder 07-180.* SE side, *Felger 17520.*

Lotus, see *Acmispon*

Lupinus arizonicus (S. Watson) S. Watson

Zaah coocta "what looks at the sun"; *trébola;* Arizona lupine

Description: Winter-spring annuals. Herbage sometimes semi-succulent, with relatively coarse hairs, the leaflets moderately hairy or glabrate. Leaflets digitate (5) 7–9 per leaf, the tips mostly blunt to broadly rounded and sometimes notched to apiculate (with a soft, blunt projection at the tip). Flowers papilionoid. Corollas 7–9 mm long, pale lavender-pink with a yellow spot on the banner petal. Pods dehiscent.

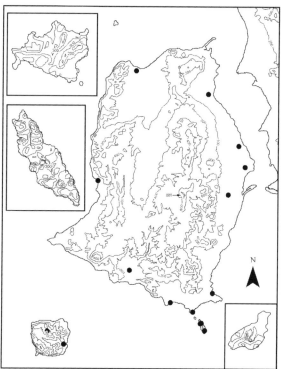

Hoffmannseggia intricata
Haxz iztim

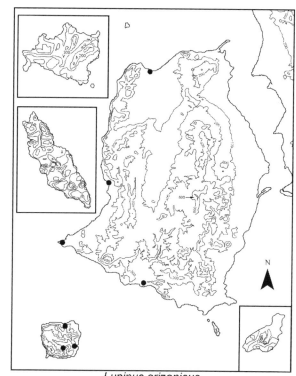

Lupinus arizonicus
Zaah coocta

Local Distribution: Documented at widely scattered lowland localities on Tiburón but not on the eastern side of the island. San Esteban in arroyos and canyons.

Other Gulf Islands: San Ildefonso, Santa Catalina, Santa Cruz, Cerralvo.

Geographic Range: Arizona and southeastern California to the Cape Region of Baja California Sur and Sonora southward to the vicinity of Tastiota.

Tiburón: Ensenada Blanca (Vaporeta), *Felger 17240*. Willard's Point, canyon just N, sandy arroyo bottom, *Moran 8730* (SD). El Sauz, *Moran 8768* (SD). Tecomate, few scattered individuals, 12 Feb 1965, *Felger 12495* (specimen not located). **San Esteban:** Arroyo Limantour, *Felger 17665*. NW corner of island, narrow, rocky canyon behind cove, *Felger 17604*. SE corner of island, dry arroyo bed, *Moran 8862* (SD).

Lysiloma divaricatum (Jacquin) J.F. Macbride [*L. microphyllum* Bentham]

Hehe ctoozi "resilient plant"; *mauto*

Ctoozi is the term for resilient or flexible (Felger & Moser 1985; Moser & Marlett 2005).

Description: Unarmed large shrubs or trees 4–6 m tall, multiple-stemmed with flexible hardwood branches. Leaves twice pinnate with numerous small leaflets, the leafstalks usually with a conspicuous nectary gland; stipules leafy, larger than the leaflets and soon deciduous. New growth mostly produced in early summer or with summer rains. Most of the foliage usually shed in early fall. Flowers mimosoid; white, in globose heads, April and May. Pods flat, dry, and dehiscent, the rim separating from the rest of the pod; ripening late summer to October and conspicuously semi-persistent.

Local Distribution: Sierra Kunkaak in arroyos, canyons, and slopes to peak elevation. It is one of the more common trees in canyons and at higher elevations on Tiburón.

Other Gulf Islands: None.

Geographic Range: Costa Rica to northern Mexico in Baja California Sur, Sonora, Chihuahua, Tamaulipas, and San Luis Potosí.

Tiburón: Sopc Hax, canyon bottom, *Felger 9300*. Near Caracol station, *Knight 1094*. Cerro San Miguel, 300 m, *Quijada-Mascareñas 90T010*. Deep canyon, SW and up canyon from Siimen Hax, *Wilder 06-476*. Steep, protected canyon, 720 m, N interior of Sierra Kunkaak, large population at base of cliffs, trees 4–5 m tall, *Wilder 07-517*. Summit of island, rare, *Wilder 07-571*. Exposed upper ridge of island, between peak of Cójocam Iti Yaii and summit, occasional shrub, 2–4 m tall, *Wilder 07-578*.

Marina

Gland-dotted herbs with relatively firm, short, straight or curly hairs that remain unchanged by drying (not spirally twisted and reddish brown as in *Dalea*). Leaves odd-pinnate, the terminal leaflet larger than the others; leaflets marked on upper surface with tiny wavy lines representing veins (sometimes difficult to see on small leaflets of drought-stressed plants). Flowers papilionoid, small and brightly colored. Calyx 5-toothed and 10-ribbed, with prominent glands between the ribs. Pod small, 1-seeded.

1. Leaflets (4) 7–13 (15), the flowers minute and bright magenta; stamens 5 or 6. _____ **M. evanescens**
1' Leaflets 13–29, the flowers dark blue or purplish; stamens 10.
    2. Leaflets 1.4–7 mm long; calyx lobes shorter than the tube. _____**M. parryi**
    2' Leaflets 1–2 mm long; calyx lobes longer than the tube. _____**M. vetula**

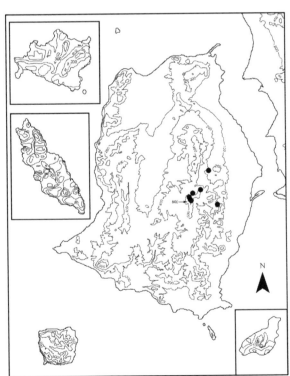

Lysiloma divaricatum
Hehe ctoozi

Marina evanescens (Brandegee) Barneby [*Dalea evanescens* Brandegee]

Description: Non-seasonal annuals, with a yellowish taproot. Stems very slender, spreading, decumbent, or semi-prostrate. Leaflets (4) 7–13 (15) per leaf, often less than 2.5 mm long (considerably larger with ample soil moisture), flat and oval. Flowers minute, the corollas bright purple magenta. (Description based on specimens from the Guaymas region.)

Local Distribution: Known from the flora area by a single record from the north side of San Esteban.

Other Gulf Islands: None.

Geographic Range: Sonora in the Guaymas region and the Baja California Peninsula.

San Esteban: La Freidera, arroyo bottom, dense growth of ephemerals, with *Amaranthus watsonii, Perityle emoryi, Atriplex barclayana, Vaseyanthus insularis, Muhlenbergia microsperma, Phaseolus filiformis*, 21 Dec 1966, *Felger 15445*.

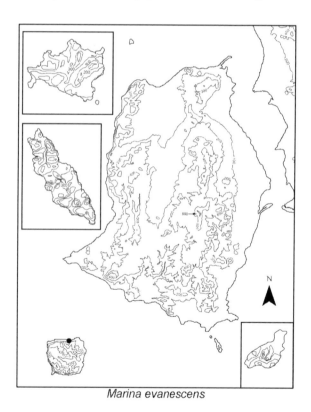

Marina evanescens

Marina parryi (Torrey & A. Gray ex A. Gray) Barneby [*Dalea parryi* (A. Gray) Torrey & A. Gray ex A. Gray]

Hanaj iit ixac "nits of the raven's lice"

Description: Non-seasonal annuals and sometimes short-lived perennials, gland dotted, mostly growing with

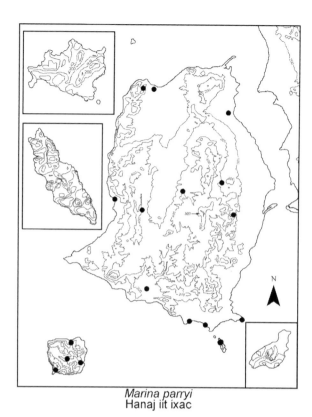

Marina parryi
Hanaj iit ixac

winter-spring rains, with a yellow taproot. Leaflets 13–29 per leaf, 1.4–7 mm long, orbicular to obovate. Corollas dark blue or purplish blue and white.

Local Distribution: Widespread and common on Tiburón, San Esteban, and Dátil from arroyo bottoms to steep slopes. On San Esteban it extends from near the shore to the summit.

Other Gulf Islands: Ángel de la Guarda, Tortuga, San Marcos, Coronados, Carmen, Danzante, Monserrat, Santa Cruz, San José, San Francisco, Espíritu Santo.

Geographic Range: Nearly throughout the Sonoran Desert; Sonora southward to the vicinity of Ciudad Obregón.

Tiburón: Ensenada Blanca (Vaporeta), arroyo, *Felger 7102*. Arroyo Sauzal, 2 mi inland, rocky arroyo, *Felger 12254*. Ensenada de la Cruz, rocky hill, *Felger 9225*. Coralitos, *Felger 15360*. El Monumento, *Felger 15560*. SW Central Valley, upper rocky slope of mountain [Sierra Menor] bordering valley, *Felger 12403*. Haap Hill, *Felger 74-T51*. Vicinity of Sopc Hax, *Romero-Morales 07-15*. Near Caracol station, *Knight 956*. Hahjöaacöl, NE coast, *Wilder 08-213*. Bahía Agua Dulce, *Tenorio 9538*. Tecomate, rocky hill, E-facing slope, *Felger 6265*.

San Esteban: Arroyo Limantour: Canyon bottom, *Felger 12731*; 2.25 mi inland, *Wilder 07-50*. Canyon at SW corner of island, *Felger 17617.* N side of island, narrow, rocky canyon behind cove, *Felger 17564.*

Dátil: NE side of island, *Felger 9109.* NW side of island, *Felger 15331.* E side, slightly N of center: Halfway up N-facing slope of peak, *Wilder 07-123*; Canyon bottom, *Wilder 07-171.*

Marina vetula (Brandegee) Barneby [*Dalea vetula* Brandegee]

Description: Annuals, found during cool seasons but probably non-seasonal. Roots orange. Leaflets 19–25 per leaf, 1–2 mm long, orbicular. Flowers pale purplish or pink, with exerted yellow stamens. Plants similar in appearance to the much more common *M. parryi* but with smaller leaflets, conspicuously larger and "fuzzier" inflorescences, larger bracts, and the flowers more densely crowded, the calyx larger, and the corollas nearly hidden within the calyx.

Local Distribution: Scattered on Tiburón, most prevalent in the Sierra Kunkaak.

Other Gulf Islands: Tortuga, San José.

Geographic Range: Central part of the Baja California Peninsula from Calmallí to Comondú and San Gregorio islands, and coastal Sonora north of Bahía San Carlos.

Tiburón: SW side, near El Sauz, *Moran 8794.* Arroyo bottom, NE base of Sierra Kunkaak, *Wilder 06-380.* Siimen Hax, *Wilder 06-447.* Foothills of Sierra Kunkaak, *Wilder 06-506.*

Mimosa distachya Cavanilles var. **laxiflora** (Bentham) Barneby [*M. laxiflora* Bentham]

Hehe cotázita "plant that pinches"; *uña de gato, garabatillo*

Description: Woody shrubs, 2–3 m tall; bark light tan; twigs with sharp, clothes-ripping recurved spines. Leaves twice pinnate (3.5) 6–8 cm long; drought- and winter-deciduous; leaflets 6–12 mm long and numerous. Flowers mimosoid, lavender-pink, in cylindrical spike-like racemes 2–6 cm including the peduncle. Flowering August and September with summer rains. Pods 4.5–5.5 cm long, 5–6 mm wide, multiple seeded, breaking into 1-seeded, quadrangular segments.

Local Distribution: Eastern and central part of Tiburón including Sierra Kunkaak to higher elevations; arroyos, canyons, foothills, and rock slopes.

Other Gulf Islands: San José, Espíritu Santo.

Geographic Range: Var. *laxiflora* from Sinaloa and western Chihuahua to southern Arizona. This species

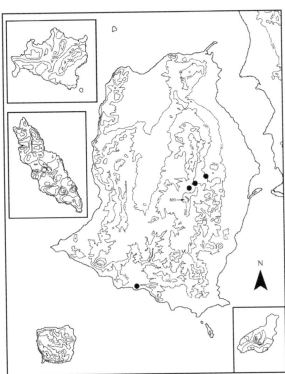

Marina vetula

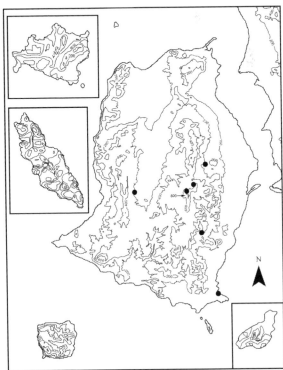

Mimosa distachya var. *laxiflora*
Hehe cotázita

with four varieties ranges from Arizona to southern Mexico.

Tiburón: Ensenada del Perro, *Tenorio 9520*. SW Central Valley, arroyo, *Felger 12352*. E foothills of Sierra Kunkaak, arroyo, *Felger 6976*. Canyon bottom, N base of Sierra Kunkaak, between Sierra Kunkaak Mayor and Sierra Kunkaak Segundo, 395 m, *Wilder 06-412*. Steep, protected canyon, 720 m, N interior of the Sierra Kunkaak, occasional, together with *Celtis pallida* it gives the vegetation a very thorny character, *Wilder 07-584*. Caracol, *Romero-Morales 08-14*.

Olneya tesota A. Gray

Comitin, hesen; *palo fierro*; desert ironwood

Description: Shrub or trees, often with one or several massive trunks, shredding gray bark, dense gray-green foliage, and short, sharp, paired, stipular spines or some branches spineless. These trees are long-lived and generally slow growing. The wood is extremely hard, does not float, and burns with a hot flame. Leaves pinnate. Flowers papilionoid, 1.5 cm long; calyx purple-brown, the petals mostly pink to lavender or purple, varying from tree to tree. Banner petal with a white, succulent callus at the base of the blade, a chartreuse nectar guide above the callus, a whitish band highlighting the nectar guide, and the flat blade of the banner folding back from the nectar

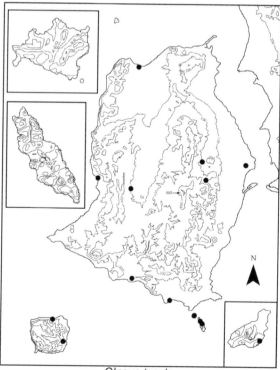

Olneya tesota
Comítin, hesen

guide usually toward the end of the day. Pods generally 2–4 cm long, thick, dark red-brown with glandular and non-glandular hairs, and tardily dehiscent (often opening after falling). Seeds 7–8 mm long, 1–2 (3–4) per pod. Mass flowering, attracting myriads of bees and other insects, usually late April and May; pods ripening June and July.

Local Distribution: Tiburón, especially at lower elevations, and all neighboring islands except Patos. This is the most common and widespread tree on Tiburón and San Esteban; best developed along arroyos, and sometimes also on rocky slopes usually as shrubs. Through time the meandering and anastomosing shallow washes writhing across the expansive Central Valley of Tiburón support irregular, low galleries of ironwood trees. As the shallow watercourses shift, the ironwoods along their margins perish: "Long lines of dead and eroded ironwood stumps are often present in the creosotebush flats. That these stumps, seemingly very old, were once growing along an arroyo margin of slightly lower elevation is evidenced by the fact that crown of the stump is usually buried 2 to 4 decimeters. Every stage in this dramatic erosion-directed cycle is present" (Felger 1966:243). After one or more favorable rainy seasons ironwood seedlings are seen along the active washes, as well as across the valley floor, but after only a single unfavorably dry season nearly all of them perish except an occasional one along an active drainageway.

Other Gulf Islands: Ángel de la Guarda, Tortuga (observation), San Marcos (observation), Coronados (observation), Carmen, Danzante (observation), Monserrat (observation), Santa Catalina, Santa Cruz (observation), San Diego (observation), San José (observation), San Francisco (observation), Espíritu Santo, Cerralvo.

Geographic Range: Southeast California, both Baja California states, southern Arizona, and Sonora. Essentially endemic to the Sonoran Desert, although it extends southward beyond the desert in thornscrub in southwestern Sonora and the lowlands of the Cape Region of Baja California Sur.

Tiburón: Ensenada Blanca (Vaporeta), *Felger 10265*. Arroyo Sauzal, 31 Jan 1965, *Felger*, observation (Felger 1966:221). Ensenada de la Cruz, *Felger 2598*. SW Central Valley, 2 Feb 1965, *Felger*, observation (Felger 1966:236). Palo Fierro, *Felger 10329*. Hast Cacöla, foothills, 7 Apr 2008, *Wilder*, observation. Caracol station, *Wilder 06-152*. Tecomate, *Felger 11118*.

San Esteban: Arroyo Limantour, common throughout island, canyon bottom and hillslopes, *Felger 7090*. N side of island, *Felger 15441*.

Dátil: E side, on top of ridge facing sea, tree flattened and spreading, 3 m across and 1 m tall, *R. Russell 6 Aug 1962*.

Canyon at Rescue Beach, NE side of island, 12 Apr 2007, *Felger*, observation.

Alcatraz: SE side of mountain, steep rock slope facing sea, much guano, localized, *Felger 13405*.

Cholludo: NW corner of island, localized at summit, rare, *Felger 13416*. N-central slope, several small trees at top of island, *Wilder 08-318*.

Parkinsonia – Palo verde

Trees and large, heavy-branched shrubs with green bark and relatively soft wood. Leaflets small, the leaflets and leaves quickly drought deciduous, the leaflets often falling independently; leaves twice pinnate or appearing once pinnate. Flowers caesalpinioid, yellow or yellow and white, produced in prodigious quantities in spring. Pods ripening in early summer, indehiscent to tardily partially dehiscent, 1- to several-seeded.

1. Twigs not spinescent at tip; solitary or paired axillary spines often present; leaves petioled, the leaflets mostly 3–7.8 mm long. _____ **P. florida**
1' Twigs spinescent at tip; axillary spines none; petiole absent, leaves with two sessile pinnae, the leaflets mostly 1–3.3 mm long. _____ **P. microphylla**

Parkinsonia florida (Bentham ex A. Gray) S. Watson
[*Cercidium floridum* Bentham ex A. Gray subsp. *floridum*]

Ziij, iiz; *palo verde*; blue palo verde

Description: Common desert trees often reaching 7–8+ m tall with a well-developed trunk. Leaves short petioled, usually with 1 pair of pinnae, each usually with 3 or 4 pairs of leaflets, the leaflets 3–7.8 mm long. Flowers bright yellow. Blazing masses of flowers in spring color the eastern bajada of Tiburón yellow, the peak flowering usually in late March and early April (about two weeks in advance of peak flowering for *P. microphylla*). Pods flattened, usually ripening in May and June.

Local Distribution: Widespread at lower elevations across most of Tiburón; mostly in arroyos and canyon bottoms.

Other Gulf Islands: None.

Geographic Range: Northwestern Sinaloa to Arizona and southeastern California. *Cercidium floridum* subsp. *peninsulare* (Rose) Carter occurs in Baja California Sur and the trees on the southern coast of Tiburón approach subsp. *peninsulare* in having characteristic large leaflets.

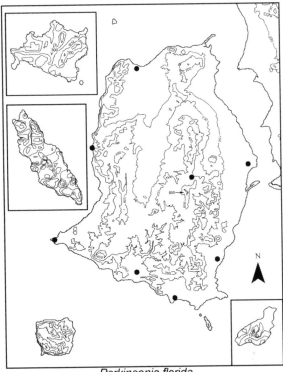

Parkinsonia florida
Ziij, iiz

Tiburón: Canyon just N of Willard's Point, *Moran 8723* (SD). Arroyo Sauzal, *Wilder 06-75*. Ensenada de la Cruz, *Felger 2606*. E bajada below S portion of Sierra Kunkaak, mass flowering of trees, 12 Apr 2007, *Felger & Wilder*, observation. Bajada, 1 mi W of Palo Fierro, *Felger 10015*. NE base of Sierra Kunkaak, *Wilder 06-385*. Tecomate, Arroyo Agua Dulce, 12 Feb 1965, *Felger*, observation (Felger 1966:261). Pooj Iime, *Wilder 07-446*.

Parkinsonia microphylla Torrey [*Cercidium microphyllum* (Torrey) Rose & I.M. Johnston]

Ziipxöl; *palo verde*; foothill palo verde

Description: Large shrubs or small trees, usually branching near the base from a short, thick trunk. Leaves each with two pinnae and lacking a petiole, the leaflets minute; individual leaflets and ultimately the rachis drought deciduous. (Because the leaves lack petioles, each sessile pinna might be confused for a single, once-pinnate leaf.) Banner or upper petal whitish, becoming orange-flecked after pollination, the other 4 petals pale yellow. Pods indehiscent to tardily partially dehiscent, constricted between the seeds, the apex often extending into a slender sterile snout. Masses of pale yellow flowers in spring, the pods ripening in late May and June.

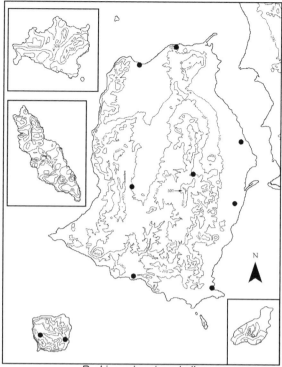

Parkinsonia microphylla
Ziipxöl

Leaflets 3. Flowers papilionoid, small and pink, the keel petal bent across the flower and twisted above the bend. Pods multiple-seeded, flattened, and elastically dehiscent.

1. Leaves "rough" to the touch; pods 5+ cm long, about the same width throughout. _____ **P. acutifolius**
1' Leaves smooth; pods 1.5–3 cm long, narrowest at base, broader toward the tip._____**P. filiformis**

Phaseolus acutifolius A. Gray var. **latifolius** G.F. Freeman

Haap; *tépari del monte*; tepary

Description: This wild tepary bean grows with summer and early fall rains. It is an ephemeral vine climbing through desert shrubs and small trees. The pods are commonly 5–5.5 cm long, and each contains several seeds, which are relatively large for a wild tepary. Seeds mottled gray and blackish or yellowish, resembling gravel, 5.8–6.4 mm long × 4.1–4.8 mm wide. The leaflets are broad, aligning the Tiburón population with var. *latifolius*.

The wild tepary, unlike the domesticated, cultivated tepary, has pods that burst open explosively when ripe, and the seeds are smaller.

Haap was an important wild food crop for the Comcaac who lived in the interior of Tiburón. The pods,

Local Distribution: Widespread on Tiburón; bajadas, valley floors, arroyos, canyons, and rocky slopes. San Esteban along major washes and canyons.

Other Gulf Islands: Ángel de la Guarda, Tortuga, San Marcos, Carmen, Danzante, Monserrat (observation), Santa Catalina, San José.

Geographic Range: Endemic to the Sonoran Desert; widespread and common in Sonora south to the Guaymas region, southwestern Arizona, both states of Baja California, and barely entering southeastern California.

Tiburón: Arroyo Sauzal, 31 Jan 1965, *Felger*, observation (Felger 1966:221). Ensenada del Perro, *Wilder 07-148*. SW Central Valley, 2 Feb 1965, *Felger*, observation (Felger 1966:236). 1.5 km inland on road to Pazj Hax, 24 May 2006, *Felger*, observation. 2 km E of Siimen Hax, 25 Nov 2006, *Felger*, observation. E bajada near Valle de Águila, *Wilder 07-237*. Hee Inoohcö, NE coast, near shore, *Felger 07-81*. Tecomate, *Whiting 9014*. **San Esteban:** Limantour, rare, floodplain, *Felger 16603*. Narrow canyon, SW corner of island, *Felger 17603*.

Phaseolus – *Frijoles*; beans

Ephemerals (the two species in the flora area), the stems usually vining. Pubescence of both small straight hairs and small, hooked hairs (visible with 10–20× magnification).

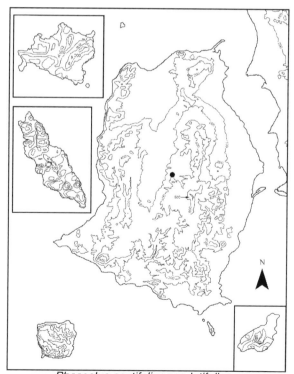

Phaseolus acutifolius var. *latifolius*
Haap

which ripen at the end of the short summer monsoon, were gathered in the early morning while it was still relatively damp and not too hot. If harvested at midday the pods would dry quickly and burst open, scattering the seeds. The seeds were boiled alone or with meat, if it was available—especially deer meat or bones. These beans were said to taste like the Mexican pinto bean but cook faster.

Local Distribution: In the flora area it is known from *Haap Caaizi Quih Yaii*. For convenience, we refer to this place as Haap Hill. It is in the southeastern part of the Central Valley. This tepary population, well known in Seri oral history, is on the north-facing slope of a large hill of volcanic scoriae. One tepary plant was found about 100 m above the arroyo bed.

Richard described finding the *haap* in a special issue of the *Journal of the Southwest* (Felger 2000b:533–534):

We had long heard about a certain bean, called *haap*, from the middle of Tiburón Island. The beans were one of the major harvests, but the region seemed too arid for any ordinary bean. There was a campsite, called *Haap Caaizi Quih Yaii* or "Haap-bean Users' Place," and I made several futile trips to the island to find it. Once I even went to the right place, but the Seri man who took me there could not locate the plant—it was too dry. On the final trip to locate the elusive bean, on December 11, 1976, Rosa Flores was our guide. Rosa, Cathy Moser, and I climbed into Ike Russell's airplane in Desemboque and went to our rough little landing place near the middle of the island. It was hot at midday and Rosa wore the jacket I had given Jesús Morales the previous winter and it had somehow ended up in her possession. She wore a long, full skirt and flip-flops and headed straight east to the *haap* bean site. The rest of us had a hard time keeping up with her as she marched between the creosote bushes, mesquites, ironwoods, and palo verdes—she did not stop until reaching the site several hours later. She had not been to this place for more than 34 years, not since the early 1940s when she was about 25 years old.

We found only part of one vine, one dried stem with a few shriveled leaves and one dry, intact pod. It was a wild tepary bean, no doubt about it, and the descriptions we had been given were indeed accurate. It was just that no tepary had ever been documented from such an arid habitat, so I was uncertain about what to expect. We were about two weeks too late; all of the other tepary vines had dried up. Before I picked

that one intact pod, I reached over and pinched it together and carefully placed it in a small brown paper seed envelope. When I let go it went "pop."

The place where this tepary grows is a huge, nearly barren, north-facing hillslope of large, rough, black lava rocks—really just a big rock pile. The beans grow up through the rocks, which seem to act like giant mulch. I knew that if I even touched the pod without firmly holding it together that it would explosively separate— the valves would elastically and violently twist apart, flinging out the seeds with no hope of recovering them. There were five small, gray and blackish mottled seeds. I brought them back to Tucson and grew them, eventually getting a good seed stock. The next year I brought Rosa a full bag of *haap* beans, and also gave a seed stock to Native Seed/SEARCH to grow and distribute.

Other Gulf Islands: Espíritu Santo (as var. *acutifolius*, Rebman et al. 2002).

Geographic Range: The nearest known wild tepary is documented from near Bahía San Carlos (*Gentry 4728*; Nabhan & Felger 1978). Wild teparies are found along eastern Sonora at elevations above the desert. Northwestern Mexico to Jalisco and southwestern United States.

Tiburón: Near N side and base of basalt hill, vicinity of former Seri tepary-bean-gathering camp: Haap Caaizi Quih Yaii (donde estaban los que trabajan con el frijol tepari), N-facing slope of volcanic scoriae of rounded volcanic rocks, about 100 m above the arroyo bed; rare, late in the season; with *Carnegiea*, *Bursera microphylla*, *Acacia willardiana*, *Ruellia californica*, *Fouquieria splendens*, *Larrea*, *Jatropha cuneata*, etc.; leaves mostly withered and pods dehisced; vine to ca. 1.5 m in a *Lycium* shrub (Rosa says this is a small plant; these seeds are black, but Rosa says some are gray and some are yellow), 11 Dec 1976, *Felger 76-T32 with Rosa Flores, Cathy Moser, & Alexander Russell*.

Phaseolus filiformis Bentham

Haamoja ihaap "pronghorn's tepary"; *frijol del desierto*; desert bean

Description: Delicate vines that appear at various seasons following sufficient rainfall. It has twining stems and pink, pea-like flowers, and pods 1.5–3 cm long, slightly curved and laterally flattened.

Local Distribution: Widespread on Tiburón, San Esteban, and Dátil. On Dátil it is seasonally the dominant vine and covers substantial areas of vegetation after sufficient

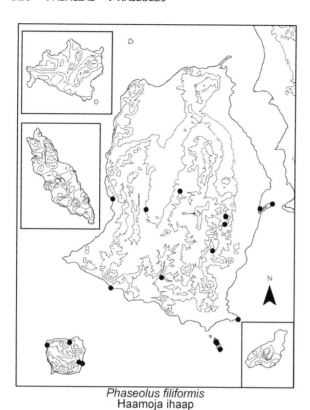

Phaseolus filiformis
Haamoja ihaap

*Pithecellobium dulce (Roxburgh) Bentham

Camootzila; *guamúchil*; Manila tamarind

Description: These large trees have paired spines persisting to form unique horizontal ring-like ridges on the smooth gray bark. The leaves have a pair of pinnae, each with 2 broad leaflets often 2–5 cm long. Flowers mimosoid, cream white, the stamens 1 cm long; flowers crowded in heads on racemose inflorescences. Pods curved or coiled, partially splitting open to reveal black seeds embedded in either pinkish or white, soft spongy, edible and sweet aril, or pulp. Flowering in late winter and early spring; pods ripe May and June. (Description based on mainland trees.)

Local Distribution: We found one young tree in 2007 that had been planted at Sopc Hax in the Sierra Kunkaak. Humberto learned that a man from Guaymas, engaged in illegal deer hunting on Tiburón, planted guamúchil seeds at the waterhole.

Other Gulf Islands: Cerralvo.

Geographic Range: Native to the hot lowlands of Mexico and northern South America, *guamúchil* trees are extensively planted for their edible fruits throughout lowland Sonora including at Punta Chueca and Bahía

rains. Its role on this island is analogous to that of *Vaseyanthus insularis* on San Esteban.

Other Gulf Islands: Ángel de la Guarda, Partida Norte, San Lorenzo, Tortuga, San Marcos, San Ildefonso, Coronados, Carmen, Danzante, Monserrat, Santa Catalina (observation), San José, Espíritu Santo, Cerralvo.

Geographic Range: Sonora from the Guaymas region northward, both Baja California states, and Arizona to Texas.

> **Tiburón:** Ensenada Blanca (Vaporeta), *Felger 14957.* Tordillitos, *Felger 15505.* Xapij (Sauzal) waterhole, *Wilder 06-116.* El Monumento, *Felger 15523B.* SW Central Valley, arroyo, *Felger 12333.* Haap Hill, *Felger 76-T33.* E foothills of Sierra Kunkaak, *Felger 6990.* Punta San Miguel, *Wilder 07-197.* Zozni Cmiipla, *Felger 08-112.* Pazj Hax, *Wilder 06-11.* Sopc Hax, *Wilder 07-211.*
>
> **San Esteban:** Arroyo Limantour, *Turner 83-26.* Just N of El Monumento, just above shoreline, *Felger 12743.* N side of island, narrow rocky canyon behind cove, *Felger 17563.* San Pedro, *Felger 16670.*
>
> **Dátil:** E side, slightly N of center, narrow ridge crest on peak, *Wilder 07-119.* NE side, *Felger 9098.* NW side of island, *Felger 15342.* SE side, *Felger 17498.*

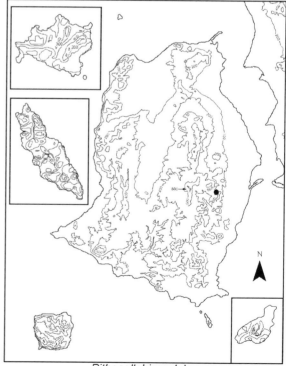

Pithecellobium dulce
Camootzila

Kino. These trees are not known to persist untended in the Sonoran Desert, with a few notable exceptions. Reid Moran documented a tree on Isla Cerralvo (16 Apr 1962, tree 5 m tall × 10 m, *Moran 9511*, SD). We also found reproductive, feral trees in a weedy arroyo at Bahía Kino Nuevo (*Wilder 07-461*). The manila tamarind is planted and naturalized in many tropical parts of the world.

Tiburón: Sopc Hax, tree 3 m tall, 2 May 2007, *Wilder 07-215*.

Prosopis glandulosa Torrey var. **torreyana** (L.D. Benson) M.C. Johnston [*P. juliflora* (Swartz) de Candolle var. *torreyana* L.D. Benson. *P. odorata* Torrey & Frémont, see Palacios 2006.]

Haas; *mezquite*; western honey mesquite

Description: Shrubs or trees. Trunks usually crooked or irregular, the roots of soft wood, fibrous, and pliable, while the wood of the trunks and branches is hard, heavy, and strong although somewhat brittle, and very resistant to decay. Long-shoot nodes mostly armed with straight, stout thorns. Leaves generally glabrous or sparsely pubescent, twice pinnate with one pair of pinnae, the leaflets usually more than twice as long as wide; gradually winter deciduous. Brilliant green new foliage appearing with warm weather in March and April. The new growth can be exceedingly rapid, and there is considerable variation in leaf and leaflet size. Flowers mimosoid; small and pale yellow in dense, cylindrical spike-like racemes. Pods mostly (7) 10–20 cm long, multiple seeded, indehiscent with sweet pulp.

Local Distribution: Widely scattered across Tiburón; lowland habitats from near the shore as shrubs and inland as shrubs or small trees in arroyos and canyons, and occasionally on slopes. Mesquite shrubs and trees are decidedly uncommon in the interior of the island, mostly occurring as solitary plants in widely scattered sites along canyons and extending to higher elevations in the Sierra Kunkaak. On San Esteban it is especially common near the shore in Arroyo Limantour, where it forms low, briar-like shrubs that often have very large, stout thorns; also as larger shrubs farther inland and in canyon bottoms, and occasionally on rocky slopes. Also common on Dátil. It was near the shore on Alcatraz in the 1960s, but none were seen in 2007.

Other Gulf Islands: San Marcos, Carmen, San José, Espíritu Santo.

Geographic Range: Var. *torreyana* occurs in many habitats in Sonora but is most numerous and largest along drainageways, riverine floodplains, and low-lying, coastal plains with fine-textured soils. Vast bosques of *mezquital* once covered the coastal plains from near Bahía Kino to Tastiota and southeast of Guaymas. It ranges from Sinaloa, Chihuahua, and Baja California Sur to Arizona, New Mexico, southwest Utah, and southern California. In Sonora it occurs in the lowland western part of the Sonoran Desert, the Chihuahuan Desert of northeastern Sonora, and is common in the thornscrub and tropical deciduous forest regions of southern and eastern Sonora. Var. *glandulosa* occurs east of the continental divide primarily from New Mexico and Texas to Kansas and northeastern Mexico primarily east of the Chihuahuan Desert region.

Tiburón: Arroyo Sauzal, 2 km inland, *Felger 10078* (SD). Vicinity of Sopc Hax, *Wilder 07-217*. Rock outcrop above steep, protected canyon, 720 m, N interior of Sierra Kunkaak, only one seen in entire canyon, 3 m tall, *Wilder 07-520*. San Miguel Peak, 3000 ft, *Knight 1010* (UNM). Zozni Cacösxj, NE coast, *Wilder 08-205*. Tecomate, *Whiting 9061*. Pooj lime, large canyon, NW shore, *Wilder 07-435*.

San Esteban: Large arroyo, SW corner of island, *Hastings 71-57* (SD). Canyon at SW corner of island, *Felger 16620*. N side of island, N-facing slope near shore, *Felger 15434* (SD).

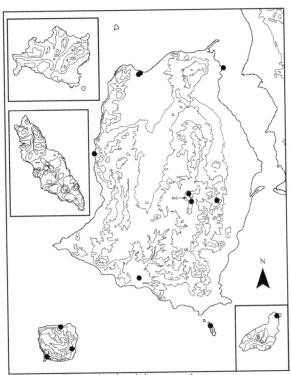

Prosopis glandulosa var. *torreyana*
Haas

Dátil: NE side of island, locally common, rocky soil behind rock beach, base of steep slope, *Felger 13470*. E shore, N of island center, canyon bottom, *Wilder 07-168*.

Alcatraz: Only near the shore, *Felger*, observation (Felger 1966:137).

Psorothamnus emoryi (A. Gray) Rydberg var. emoryi
[*Dalea emoryi* A. Gray]

Xomeete; *ojo de venado*; Emory indigo bush

Description: Small shrubs to 1+ m tall. Herbage tomentose, grayish white with very dense tangled hairs obscuring the surfaces, and sparsely dotted with minute red-orange glands; crushed herbage and flowers sweet smelling and imparting a fugitive yellow-orange stain. Leaves odd pinnate; leaflets (3) 5–11, the terminal leaflet largest. Flowers papilionoid, dark purple, in dense capitate clusters; corollas dark purple, longer than the calyx. Pods 1- or 2-seeded. Flowering after rains, spring and summer.

Local Distribution: Coastal dunes of Tiburón and especially common along Bahía Agua Dulce at the north side of the island, and sometimes spreading to adjacent washes.

Other Gulf Islands: Ángel de la Guarda, San Lorenzo, Santa Catalina.

Geographic Range: Southeastern California to northern Baja California Sur, southwestern Arizona, and coastal Sonora. Southward in Baja California Sur var. *emoryi* intergrades with var. *arenarius* (Brandegee) Barneby. This species is characteristically on coastal dunes in the Gulf of California region.

Tiburón: N of Punta Willard, *Perrill 5116*. Punta San Miguel, low dunes, *Wilder 07-193*. Zozni Cacösxaj, NE coast, *Wilder 08-206*. Tecomate: floodplain of arroyo, *Felger 12529*; Edge of sand dune, common, *Whiting 9044*.

Rhynchosia precatoria (Humboldt & Bonpland ex Willdenow) de Candolle

Ojo de pajarito; rosary pea, precatory bean

Description: Perennial vines growing over shrubs. Leaflets 3, velvety pubescent, large, and gland dotted on the lower surfaces. Flowers papilionoid, 8–9 mm long, yellow striped with brown. Pods short pubescent even in age, constricted between the seeds, becoming contorted and twisted in age. Seeds shiny, half black and half red, and tending to remain attached to the pod by a conspicuous funiculus.

Local Distribution: Deep canyons in the interior of the Sierra Kunkaak. The nearest known localities are 150 km to the southeast in the Sierra el Aguaje north of

Psorothamnus emoryi var. emoryi
Xomeete

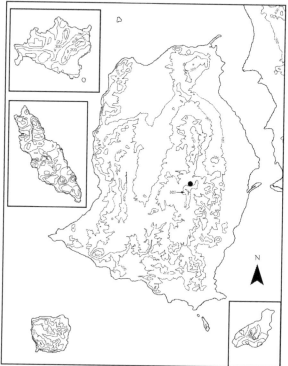

Rhynchosia precatoria

Guaymas (Felger 1999) and 160 km to the southwest in the Sierra San Francisco of Baja California Sur.

Other Gulf Islands: None.

Geographic Range: Extreme southern Arizona to northern South America.

Tiburón: Deep canyon, SW and up canyon from Siimen Hax, *Wilder 06-480*.

Senna

Herbaceous or suffrutescent perennials, probably short-lived (those in the flora area). Leaves once pinnate. Flowers caesalpinioid and yellow; anthers opening by terminal pores. Pods dry, partially dehiscent by opening at the apex (those in the flora area).

1. Pods 6–7 mm wide, with larger spreading hairs (1–1.5 mm long) among much shorter hairs pressed close to the surface. _____ **S. confinis**
1'. Pods 4–6 mm wide, with only short hairs pressed close to the surface. _____ **S. covesii**

Senna confinis (Greene) H.S. Irwin & Barneby [*Cassia confinis* Greene]

Description: Plants similar to *S. covesii* but generally somewhat larger in most of its parts and with larger as

well as smaller hairs. Foliage canescent-tomentulose. Leaves with (1) 2 (3) pairs of leaflets.

Local Distribution: On San Esteban at Arroyo Limantour and the western margin of Tiburón at Ensenada Blanca (Vaporeta), where it was found growing with *Frankenia palmeri*.

Other Gulf Islands: Ángel de la Guarda, San Lorenzo (*Lott & Atkinson 2466*, MEXU), San Marcos, Coronados, Carmen, Danzante, Santa Catalina, Santa Cruz (observation), San Diego (observation), San José, San Francisco, Espíritu Santo, Cerralvo.

Geographic Range: Baja California Sur and islands mostly along the gulf side of the Peninsula.

Tiburón: Ensenada Blanca (Vaporeta), *Felger 14969*.

San Esteban: SE side, few in bottom of main arroyo near mouth, *Moran 28168*.

Senna covesii (A. Gray) H.S. Irwin & Barneby [*Cassia covesii* A. Gray]

Hehe quiinla "plant that rings"; *hojasén, daisillo*; desert senna

Description: Perennials to 0.5 m tall, probably not long-lived; dormant during colder and drier months, dying back in drought; herbage densely velvety hairy. Leaves drought

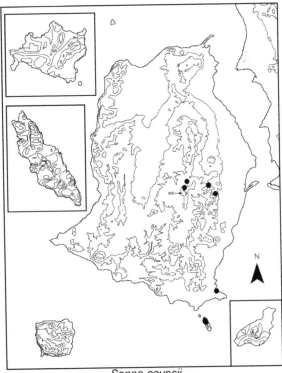

Senna confinis

Senna covesii
Hehe quiinla

deciduous and gray-green, the leafstalk with a long-stalked orange nectary gland between the leaflets of each pair; leaflets (2) 3 (4) pairs. Pods 2.4–3.5 cm long. Flowering response non-seasonal during warmer weather.

Local Distribution: This senna generally occurs in localized pockets of sparsely vegetated sites on Tiburón, where it is known from the southern part of the island and in the Sierra Kunkaak. It does not occur on the arid western margin of the island. It is fairly common on Dátil. This species is reported for San Esteban, but we have not seen specimens (see "Doubtful and Excluded Reports").

Other Gulf Islands: Tortuga, Monserrat, Santa Catalina, Santa Cruz, San José, San Francisco, Espíritu Santo, Cerralvo.

Geographic Range: Sinaloa to Arizona, Nevada, southern California, and Baja California Norte.

Tiburón: Ensenada del Perro, *Tenorio 9495*. Vicinity of Sopc Hax, *Romero-Morales 07-14*. Hast Cacöla, E flank of Sierra Kunkaak, 380 m, *Wilder 08-260*. Vicinity of Siimen Hax, *Wilder 06-451*. SE-facing slope of deep canyon, N interior of Sierra Kunkaak, *Wilder 07-492*.

Dátil: NE side of island, S-facing rocky slopes, *Felger 9127*. NW end, steep rocky slope, *Felger 15326*. Steep rocky slopes near summit, *Felger 13436*. E side, slightly N of center, S-facing slope, *Wilder 07-132*.

Tephrosia

Annuals or perennial herbs. Leaves odd pinnate; stipules slender and pointed. Flowers papilionoid. Pods elastically dehiscent, the valves coiling.

1. Herbage densely silvery pubescent; flowers 1.5+ cm long; pods 4–6 cm long, 4–5 mm wide. ___ **T. palmeri**
1' Herbage pale green, sparsely pubescent or glabrous; flowers less than 1 cm long; pods 3–4.5 cm long, 3–4 mm wide. _____ **T. vicioides**

Tephrosia palmeri S. Watson

Cozi hax ihapooin "*Condalia* that is used to cover [water in an *olla*]"

Description: Bushy perennials, sometimes flowering in the first season; generally densely silvery pubescent. Leaves 8–15 cm long; leaflets silvery silky, 7–15 in number, 1.5–5 cm long. Flowers pale lavender, 14–15+ mm long. Style barbinate (hairs distributed all along the shaft of the style; this constant character also can be seen on the young pods since the style is persistent). Flowering at various seasons with sufficient soil moisture.

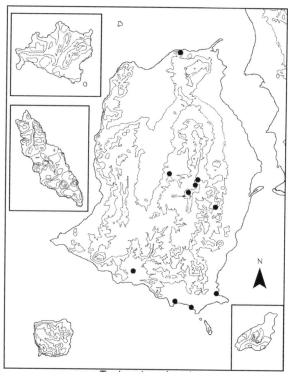

Tephrosia palmeri
Cozi hax ihapooin

Local Distribution: Widespread on Tiburón including the Central Valley, southern and eastern parts of the island, and lower elevations in the Sierra Kunkaak and one record at high elevation.

Other Gulf Islands: San Marcos, Carmen, San José, Cerralvo.

Geographic Range: The type collection is the only report for this species from mainland Sonora: "In the mountains near Guaymas," 1887, *Palmer 246* (GH, not seen). This species is known for certain on Tiburón and the Baja California Peninsula, where it extends from mid-peninsula to its southern tip. Palmer's collection probably is actually from the Baja California Peninsula.

Tiburón: Arroyo Sauzal, *Felger 10087*. Arroyo de la Cruz, *Moran 13008* (SD). Coralitos, *Wilkinson 202*. Ensenada del Perro, *Wilder 07-150*. Haap Hill, *Felger 76-T37*. Pazj Hax, *Wilder 06-47*. Base of Sierra Kunkaak, 3 km S of Caracol station, *Knight 947*. Base of Capxölim, *Wilder 5-20*. San Miguel Peak, *Knight 917* (UNM). Hee Inoohcö, mountains at NE coast, *Felger 07-74*.

Tephrosia vicioides Schlechtendal

Description: Non-seasonal annuals on Tiburón. Herbage sparsely pubescent or essentially glabrous. Leaves 2–10 cm long; leaflets pale green, 7–21 in number,

1.5–3.5 cm long. Flowers deep purple, drying wine colored, less than 8 mm long. Hairs penicillate on the style. (The shaft is smooth and the hairs clustered at the style tip. This useful character can often be seen on young pods. However, the cluster of hairs at the end of the style is sometimes difficult to see or not present on both fresh specimens and herbarium material, and in such cases the style appears glabrous, but it will not be barbinate as in *T. palmeri*.) The plants are generally smaller than *T. palmeri* and have weaker stems, smaller parts, and greener herbage.

Local Distribution: Widely scattered on Tiburón from low to high elevations, often occurring with *T. palmeri*; often along sandy-gravelly washes at lower elevations and rocky slopes at higher elevations. Also on the opposite mainland in the Sierra Seri.

Other Gulf Islands: None.

Geographic Range: A somewhat variable species; Baja California Sur and southern Arizona to western Texas, through most of Mexico, the Caribbean and South America.

Tiburón: Arroyo Sauzal, 2¾ mi inland from shore, *Wilder 06-99*. Arroyo de la Cruz, *Wilder 09-61*. SW Central Valley, 450 ft, *Felger 12389*. Haap Hill, *Felger 76-T39*. Hast Coopol, base of mountain, *Wilder 07-383*. Valley, E base of Sierra Kunkaak, 874 ft, *Wilder 06-51*. Foothills, NE portion of Sierra Kunkaak, *Wilder 06-445*. Steep, protected canyon, 720 m, N interior of Sierra Kunkaak, *Wilder 07-586*. San Miguel Peak, *Knight 996* (UNM).

Zapoteca formosa (Kunth) H.M. Hernández subsp. **rosei** (Wiggins) H.M. Hernández [*Calliandra rosei* Wiggins. *C. schottii* subsp. *rosei* (Wiggins) Felger & Lowe]

Description: Slender-stemmed shrubs 1.5+ m tall, usually taller than wide, the leafy branches mostly in the upper part of the shrub. Leaves tardily drought deciduous in canyons, and quickly drought deciduous on rocky slopes. Flowers mimosoid, rather inconspicuous, probably reddish. Pods elastically dehiscent, the valves curling back.

Local Distribution: Densely brushy areas in canyons in the Sierra Kunkaak.

Other Gulf Islands: Coronados.

Geographic Range: Subsp. *rosei* ranges from northwestern Sonora (Sierra del Viejo near Caborca) to Oaxaca and occurs in Baja California Sur. This species, with seven subspecies, ranges from southern Arizona to Argentina and the West Indies.

Tiburón: Canyon, N base of Capxölim, 395 m, not common, *Wilder 06-393*. Deep sheltered canyon, N portion of Sierra Kunkaak, *Felger 07-111*.

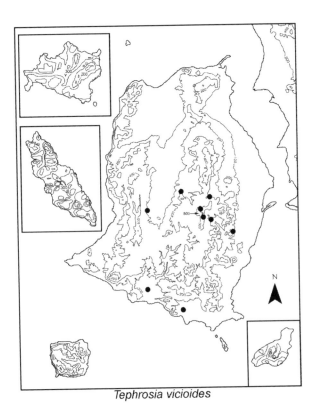

Tephrosia vicioides

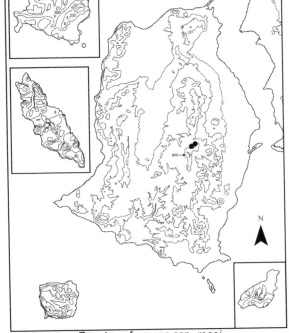

Zapoteca formosa ssp. *rosei*

Fouquieriaceae – Ocotillo Family

Large shrubs or small trees. Leaves appearing with rains and quickly drought deciduous. Flowers red and tubular, few to many in mostly near-terminal clusters. Seeds thin, papery, and winged.

1. Stems branched below and above, often crooked and branched at angles; trunk(s) and lower limbs short but thick and well developed, noticeably thicker than the upper branches. _____ **F. diguetii**

1' Stems long, straight and wand-like, ascending to erect and arising from the base, and generally few-branched above; usually appearing trunkless. _____ **F. splendens**

Figure 3.97. *Fouquieria diguetii* (left) and *F. splendens* subsp. *splendens* (right). FR (Henrickson 1972).

Figure 3.98. *Fouquieria diguetii.* Isla San Pedro Nolasco. BTW, 29 September 2008.

Figure 3.99. *Fouquieria splendens* subsp. *splendens.* Hahjöaacöl, Isla Tiburón. Someone long ago intertwined two of the stems. BTW, 6 April 2008.

Fouquieria diguetii (Van Tieghem) I.M. Johnston [*F. peninsularis* Nash]

Palo adán

Description: Shrubs or small trees to 4 m tall and often about as wide as tall, with a thick but short trunk and thick lower limbs; branching from below as well as above. Flowers in panicles about as wide as long, or reduced to only several flowers. This species somewhat resembles the common desert ocotillo (*F. splendens*) but is more shrub- or tree-like and the flowers are darker red and differ in technical features.

Local Distribution: Widespread on Nolasco but generally not on north-facing grassy slopes. Best developed at higher elevations on both sides of the island.

Reported for Tiburón and San Lorenzo, but we do not believe it occurs on those islands. The two known ocotillo plants on San Esteban might be this species or more likely *F. splendens*.

Other Gulf Islands: Ángel de la Guarda, Tortuga, San Marcos, San Ildefonso, Coronados (observation), Carmen (observation), Danzante (observation), Monserrat (observation), Santa Catalina, Santa Cruz, San Diego (observation), San José, San Francisco, Espíritu Santo, Cerralvo.

Geographic Range: Coastal Sonora from the vicinity of Tastiota southward to Puerto Yavaros, and the Baja California Peninsula and adjacent islands.

Nolasco: Middle of island, saddle at the top, 850–900 ft, small tree, *Felger 12081*. E-central side, open area on S-facing canyon slope halfway to summit, 150 m, seen at all elevations, *Felger 06-90*. Cañón de la Guacamaya, near shore, 3 May 2005, *Gallo-Reynoso* (photo). Cañón de Mellink, upper reaches of canyon, *Wilder 08-353*.

Fouquieria splendens Engelmann subsp. **splendens**

Jomjeeziz (variants: xomjeeziz, moxeeziz, xomjeezij); *ocotillo*

Description: This unique desert plant has long, wand-like, spiny branches arising from the base of the plant and with some branching above; exceptionally reaching 6.5 m tall but mostly much shorter. The bright red-orange flowers, in densely flowered panicles at the tips of the branches, appear mostly in March and attract hummingbirds.

Local Distribution: Widespread and common on Tiburón and Dátil, from low elevation to the peaks.

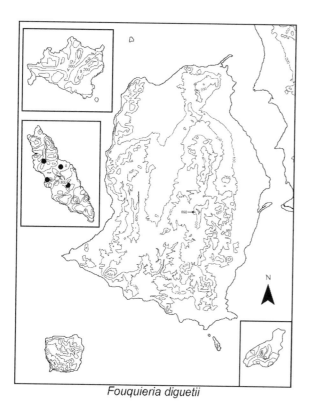

Fouquieria diguetii

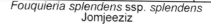

Fouquieria splendens ssp. *splendens*
Jomjeeziz

Ocotillos on Tiburón are prominent in many places in the *Frankenia* zone near the shore, and elsewhere are generally in the more exposed, arid habitats including the highest elevations of the island.

Tom Bowen photographed *Fouquieria* shrubs at the southeast peak of San Esteban; they are either this species or *F. diguetii*. There are two photos, showing two adjacent shrubs, the only ones known from the island. The larger shrub might be about 1.5 m tall. It has some rather crooked stems like those of *F. diguetii*, but we think the branching pattern looks more like *F. splendens*. Johnston (1924) infers that *F. diguetii* occurs on Islas San Esteban and San Lorenzo, and Gentry (1949) lists observations for it on those islands. Moran (1983b) shows an observation on San Lorenzo and a question mark for San Esteban. Turner et al. (1995) indicate specimens for *F. diguetii* for both San Esteban and San Lorenzo, but a search of their records shows that they did not see a specimen (Ray Turner, personnel communication, 2007). Reid Moran was the only one of these botanists who went to the top of San Esteban, but there is no indication he saw these ocotillos. Jim Henrickson (personal communication, 2007) did not see a specimen from San Esteban when he monographed the genus (Henrickson 1972). We have not located a specimen for San Esteban or San Lorenzo. Tom Bowen (personal communication, 2007) has scoured both islands and has seen only two *Fouquieria* plants on San Esteban and none on San Lorenzo.

Other Gulf Islands: Ángel de la Guarda.

Geographic Range: Sonora south to the Sierra el Aguaje north of Guaymas and northern Baja California Sur to southeastern California to Trans-Pecos Texas and southward to Nuevo León, Durango, and Zacatecas. Subsp. *splendens* is the widest ranging of the three subspecies, nearly encompassing the range of the species through most of the Sonoran and Chihuahuan deserts.

Tiburón: Ensenada Blanca (Vaporeta), *Felger 7120*. Sauzal landing strip, hills and mesas, 6–6.5 m tall, *Felger 6445*. El Monumento, *Felger 08-73*. Exposed highest ridge of Sierra Kunkaak, 26 Oct 2007, *Wilder*, observation. Palo Fierro, *Felger 6325*. Hee Inoohcö, 30 Sep 2007, *Felger & Wilder*, observation. Tecomate, *Whiting 9025*. Pooj Iime, 29 Sep 2007, *Felger & Wilder*, observation.

San Esteban: SE peak, summit, highest point on the island, Apr 1980, *Thomas Bowen* (photos).

Dátil: NE side of island, common, rocky canyon, *Felger 2554*.

Frankeniaceae – Frankenia Family

Frankenia palmeri S. Watson

Seepol

Description: Dwarf woody shrubs, 30–50 cm tall; apparently slow growing and long-lived. Leaves mostly 2.5–3.5 mm long, essentially evergreen, broadly oblong, nearly terete and bead-like with revolute margins, and grayish with sac-like hairs. Flowers white, small, and star-like. Flowering at least January–May, mostly April.

Local Distribution: Occurring along the perimeter of Tiburón on sites facing the sea including lower bajadas, arroyos, dunes, and rocky hills and mountain slopes. *Frankenia* is predictable in areas between the coastal halophytic salt scrub and mangroves on the shore and the lower margin of mixed desertscrub or creosotebush scrub. On the eastern side of the island *Frankenia* forms a zone of low, gray, and monotonous coastal vegetation in a zone parallel with the shore, where it is often the only common perennial species and predominately occurs on pavement-like areas of transported felcite. The flowers are visited by flies and small wasps, including ichneumonids.

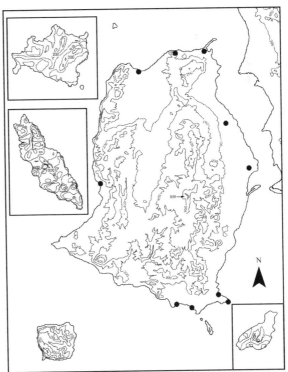

Frankenia palmeri
Seepol

Other Gulf Islands: Ángel de la Guarda.

Geographic Range: Coastal margins of southern California, both coasts of the Baja California Peninsula, and southward in Sonora to Punta Baja (between Bahía Kino and Tastiota). The leaves excrete salt and on nights with dew, water condenses on the leaves and drips down the stems onto the ground. When dry the leaves are gray and the stems brownish, but when wet the leaves become green and the stems blackish. The occurrence of coastal maritime dew seems to limit its distribution. In most places *F. palmeri* ranges no more than 1–3 km inland from the shore. However, in the central part of the Baja California Peninsula it extends far inland following the Vizcaíno maritime fog.

> **Tiburón:** Ensenada Blanca (Vaporeta), common on bench at coast and not seen farther inland, 11 Aug 1964, *Felger 10254* (specimen not located; also see Felger 1966:170–72). Ensenada de la Cruz, *Felger 12795.* Coralitos, *Wilder 06-71.* El Monumento, *Felger 15534.* Ensenada del Perro, *Wilder 07-143.* Palo Fierro, *Felger 12556.* Valle de Águila, 4 May 2007, *Wilder*, observation. Punta Perla, *Felger 74-12.* Hee Inoohcö, NE part of island, *Felger 07-71.* Tecomate, *Felger 6858.*

Hydrophyllaceae, *see* Boraginaceae

Koeberliniaceae – Allthorn Family

Koeberlinia spinosa Zuccarini

Xooml; *corona de cristo*; crucifixion thorn, allthorn

Description: Shrubs or small trees to 4 m tall with a thick hardwood trunk and rigid, thorn-tipped twigs; essentially leafless, the leaves scale-like, quickly falling with drought. Flowers on slender pedicels in small racemes. Petals and filaments white, the anthers pale yellow. Fruit a rounded berry, drying capsule-like, 3–3.5 mm diameter. Flowering at least March and visited by hoards of flies.

"The wood is exceedingly hard and produces copious oily black smoke when burned. . . . The wood was burned in a hut to drive away disease during epidemics, probably measles. During the nineteenth century, after excursions to Hermosillo to sell and trade, the Seris came home and burned *xooml* wood to disinfect their homes against diseases they had encountered in the city" (Felger & Moser 1985:318).

Local Distribution: Common on the Agua Dulce Valley floor, where there are impressive numbers of exceptionally large, often tree-sized plants. Also as shrubs scattered in lowlands in other parts of the island.

Other Gulf Islands: None.

Geographic Range: Southeastern California to Texas, and southward to Baja California Sur and Hidalgo, and scattered in Sonora southward nearly to the Sinaloa border, and disjunct in the Chaco of Bolivia and Paraguay. Two weakly defined varieties are described, both in North America. The family has a single species.

> **Tiburón:** Arroyo Sauzal, rare, near shore, small alluvial fan, *Felger 7023.* SW part of Central Valley, arroyo, 1 Feb 1965, *Felger*, observation (Felger 1966:239). NE side of Central Valley, *Wilder 07-266.* Tecomate, rare, *Whiting 9020.*

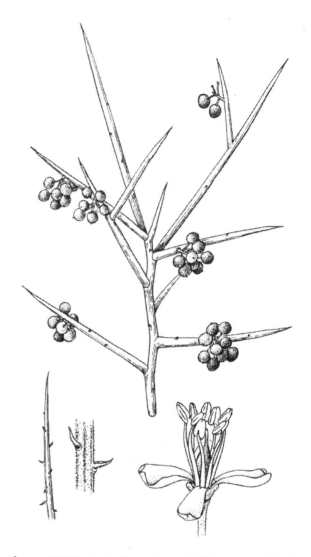

Figure 3.100. *Koeberlinia spinosa.* FB (© James Henrickson).

Cactus scrub, Isla Cholludo. Cardón (*Pachycereus pringlei*) and a few organ pipes
(*Stenocereus thurberi*). Isla Tiburón in the background to the north.
BTW, 2 September 2008.

Southern foothills of the Sierra Kunkaak, Isla Tiburón. PRG, October 2001.

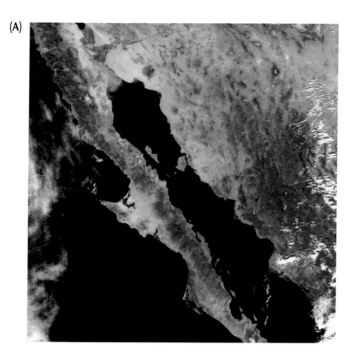

Figure 1.1. Midriff Region, Gulf of California.
(A) Satellite image of the Gulf of California and Baja
California Peninsula, with the Midriff Islands at center.
NASA Rapid Fire image, *Aqua* satellite. (B) Looking
from the northwest over the Baja California Peninsula
(foreground), the Midriff Islands (middle), and Sonora
to the southeast (background). NASA STS-004 shuttle
mission, 4 July 1982.

Figure 1.5. Mangrove vegetation at Estero Santa Rosa on the Sonora mainland opposite Isla Tiburón. NSB, 20 March 2008.

Figure 1.24. Interior canyon of Sierra Kunkaak. BTW, 25 November 2006.

Figure 1.28. Pooj lime, west side of Isla Tiburón. *Bursera microphylla,*
Pachycereus pringlei, and *Stenocereus thurberi* (foreground) and western flank of
the Sierra Menor (background). BTW, 29 September 2007.

Figure 1.36. Isla San Pedro Mártir, cardón forest at summit. BTW, 5 December 2007.

Figure 1.39. Isla San Pedro Mártir, cardons with a ground cover of *Vaseyanthus insularis* after Hurricane John of September 2006. J. A. Soriano/GECI archive.

Figure 1.41. Ridge crest of Isla San Pedro Nolasco, looking southward from above Cañón de Mellink to the summit. BTW, 29 September 2008.

Figure 3.7. *Justicia candicans*. Interior canyons of Sierra Kunkaak, Isla Tiburón. BTW, 24 November 2006.

Figure 3.21. *Dasylirion gentryi*. Near the summit of Sierra Kunkaak, Isla Tiburón. BTW, 26 October 2007.

Figure 3.31. *Bajacalia crassifolia*. Coralitos, Isla Tiburón. BTW, 2 September 2008.

Figure 3.37. *Cordia parvifolia*. Vicinity of Zozni Cmiipla, Isla Tiburón. BTW, 24 November 2006.

Figure 3.43. *Bursera fagaroides* var. *elongata*. Capxölim, Isla Tiburón. BTW, 24 October 2007.

Figure 3.44. *Bursera microphylla.* Plant in the foothills of Sierra Kunkaak, Isla Tiburón, 24 November 2006, and flower at Bahía de Kino, 25 July 2007. BTW.

Figure 3.53. *Echinocereus scopulorum*. Sierra Kunkaak, Isla Tiburón. BTW, 25 November 2006.

Figure 3.54. *Echinocereus webserianus.* Vicinity of Cañón el Farito, Isla San Pedro Nolasco. May 1952, GEL (courtesy San diego Natural History Museum).

(A)

(B)

Figure 3.56. *Ferocactus tiburonensis*. Isla Tiburón. (A) Foothills of Sierra Kunkaak, half-grown plant. HRM, January 2009. (B) Flower, Ensenada del Perro. BTW, 12 April 2007.

Figure 3.60. *Mammillaria tayloriorum*. Isla San Pedro Nolasco. BTW, 3 February 2008.

Figure 3.58. *Mammillaria estebanensis*. Arroyo Limantour, Isla San Esteban. BTW, 3 September 2008.

Figure 3.61. *Mammillaria* sp. Isla Cholludo. BTW, 2 September 2008.

Figure 3.64. *Pachycereus pringlei.* A venerable cardón with new and decaying osprey nests, Valle de Águila, Isla Tiburón. BTW, 8 November 2008.

Figure 3.65. *Stenocereus gummosus,* Isla Tiburón. Fruit from the vicinity of Zozni Cmiipla, 2 September 2006. BTW.

Figure 3.67. Imám, the Seri term for ripe fruits of columnar cacti. White and pink forms of *Pachycereus pringlei* on left, *Carnegiea gigantea* on upper right, and *Stenocereus thurberi* on bottom right and center. Seri basket made from *Jatropha cuneata* with *Krameria bicolor* root-dye. JRT, July 2008.

Figure 3.80. *Euphorbia lomelii.* Isla San Pedro Nolasco. Large colony in Cañón de Mellink, 29 September 2008, and developing fruits and withered staminate flowers, 2 February 2008. BTW.

Figure 3.73. *Jacquemontia abutiloides.* Foothills of Sierra Kunkaak, Isla Tiburón. BTW, 25 November 2006.

Figure 3.82. *Euphorbia xanti*. Near Punta Chueca, Sonora. BTW, 7 March 2007.

Figure 3.83. *Jatropha cinerea*. Staminate flowers, vicinity of Punta Tormenta, Isla Tiburón, BTW, 5 September 2008.

Figure 3.90. *Acacia willardiana*. Foothills of Sierra Kunkaak, Isla Tiburón. PRG, October 2001.

Figure 3.91. *Calliandra californica*. Isla Dátil. BTW, 10 March 2007.

Figure 3.92. *Calliandra eriophylla* var. *eriophylla*. Foothills of Sierra Kunkaak, Isla Tiburón. JRT, January 2008.

Figure 3.96. *Psorothamnus emoryi* var. *emoryi*. Zozni Cacösxaj, Isla Tiburón. BTW, 6 April 2008.

Figure 3.102. *Krameria erecta*. Freshly opening flower. Foothills of Sierra Kunkaak, Isla Tiburón. BTW, 25 November 2006.

Figure 3.103. *Krameria bicolor*. Hahjöaacöl, Isla Tiburón. BTW, 6 April 2008.

Figure 3.105. *Salvia similis*. Cañón de Mellink, Isla San Pedro Nolasco. BTW, 29 September 2008.

Figure 3.109. *Callaeum macropterum*. Foothills of Sierra Kunkaak, Isla Tiburón. Plant, 7 April 2008, and fruit, 24 May 2006. BTW.

Figure 3.110. *Echinopterys eglandulosa*. Foothills of Sierra Kunkaak, Isla Tiburón. BTW, 24 November 2006.

Figure 3.114. *Horsfordia alata*. Vicinity of Xapij waterhole, Arroyo Sauzal, Isla Tiburón. BTW, 17 March 2006.

Figure 3.115. *Sphaeralcea ambigua* var. *ambigua*. Isla Dátil. BTW, 27 February 2009.

Figure 3.124. *Passiflora arida*. Arroyo Limantour, Isla San Esteban. BTW, 8 March 2007.

Figure 3.125. *Passiflora palmeri*. Sauzal, Isla Tiburón, 17 March 2006. Plant, CDB; flower, BTW.

Figure 3.127. *Gambelia juncea*. Isla San Pedro Nolasco. BTW, 11 November 2009.

Figure 3.135. *Cenchrus palmeri*. Isla San Pedro Nolasco. BTW, 11 November 2009.

Figure 3.137. *Phragmites australis* subsp. *berlandieri*. Xapij (Sauzal waterhole), Isla Tiburón. Dense stands with Richard. BTW, 29 January 2008.

Figure 3.138. *Setaria macrostachya*. Cañón de Mellink, Isla San Pedro Nolasco. BTW, 29 September 2009.

Figure 3.145. *Rhizophora mangle*. Estero Santa Rosa, Sonora. BTW, 1 April 2006.

Figure 3.151. *Castela polyandra.* Branches with male flowers and female flower inset. Vicinity of Zozni Cmiipla, Isla Tiburón. BTW, 25 May 2006.

Figure 3.153. *Physalis crassifolia* var. *infundibularis.* Isla San Esteban. BTW, 8 March 2007.

Figure 3.154. *Physalis crassifolia* var. *versicolor.* Arroyo Sauzal, Isla Tiburón. Flower, 29 January 2008, and fruit with portion of calyx removed, 17 March 2006. BTW.

Figure 3.155. *Solanum hindsianum.* Isla Tiburón. Blister beetle feeding on an anther, Coralitos, 3 September 2008. BTW.

Figure 3.159. *Bonellia macrocarpa* subsp. *pungens.*
Flowers, Sierra Kunkaak, Isla Tiburón, 5 September 2008. BTW.

Figure 3.164. *Fagonia palmeri.* Ensenada de la Cruz, Isla
Tiburón. BTW, 28 February 2009.

Figure 3.165. *Guaiacum coulteri.* Flowers, cultivated at the
Desert Laboratory, Tumamoc Hill, Tucson. BTW, 3 August 2005.

Figure 4.24. Isla San Pedro Mártir, upper portion of small canyon at base of
southeast-facing slope that leads to the summit. BTW, 6 December 2007.

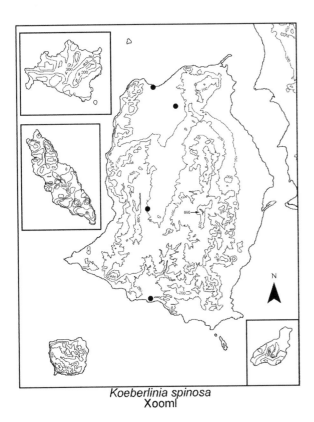

Koeberlinia spinosa
Xoomi

Krameriaceae – Ratany Family

Krameria

The family has a single genus. The pollinators are primarily *Centris* bees, which collect saturated fatty acids from floral glands. These bees also collect oil from flowers of the relatively closely related Malpighiaceae (Malpighiales) and Polygalaceae (Fabales), which are in the same taxanomic neighborhood as the order Zygophyllales in which Karameriaceae is currently placed.

Shrubs with high tannin contents, reported as root parasites on other shrubs (MacDougal & Cannon 1910). Leaves alternate, simple, and sessile; stipules none. Flowers bilateral, the sepals 5 and showy, petal-like and larger than the petals; petals 5, the 2 lower ones highly modified as fleshy oil-secreting glandular structures (elaiophores). The 3 upper petals much smaller than the sepals. Anthers opening from terminal pores, the pollen extruded in a mass when stimulated by a female *Centris* bee straddling the flower. Fruits bur-like with slender spines, resembling a miniature sea urchin.

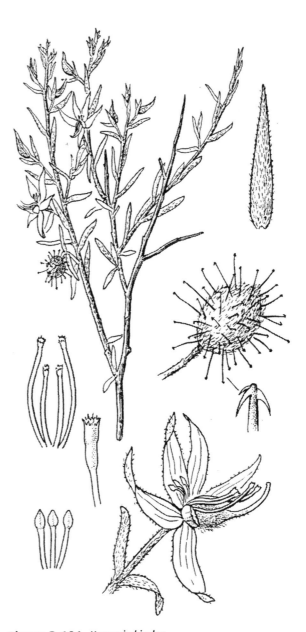

Figure 3.101. *Krameria bicolor.*
BA. (© James Henrickson).

Figure 3.102. *Krameria erecta.* Freshly opening flower. Foothills of Sierra Kunkaak, Isla Tiburón. BTW, 25 November 2006.

Figure 3.103. *Krameria bicolor.* Hahjöaacöl, Isla Tiburón. BTW, 6 April 2008.

1. Branches mostly straight and without knotty spur branches; the 3 upper petals separate, the blades nearly orbicular; spines of fruit with barbs in a terminal cluster. _____ **K. bicolor**
1' Branches tough and knotty with many very short spur branches; claws of the 3 upper petals fused basally, the blades lanceolate; spines of fruit with barbs along the upper part of shaft. _____ **K. erecta**

Krameria bicolor S. Watson [*K. grayi* Rose & Painter]

Heepol; *cósahui*; white ratany

Description: Low sprawling shrubs wider than tall, with radiating thick, lateral roots, the source of the reddish brown dye used in Seri basketry (Felger & Moser 1985). Herbage densely pubescent with short, gray hairs, the stems glabrate with age. Stems mostly branched at right angles, straight, slender, and sparsely leaved or often leafless, the stem tips often spinescent. Leaves 5–10 (14) mm long, soon drought deciduous, linear to oblong, often grayish. Flowers showy, 1.5 cm wide, solitary or in

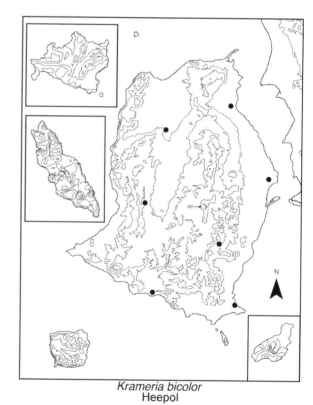

Krameria bicolor
Heepol

short racemes with leafy bracts. Sepals bright magenta-purple inside, white hairy outside. Upper petals spear shaped (narrowly spatulate), distinct, bright chartreuse with purplish tips; oil-gland petals thick, slab-like, and dark purple. Fruits more or less globose, 1 cm wide, the spines with a cluster of small terminal barbs. Flowering various seasons.

Local Distribution: Widespread across the lowlands as well as mountains on Tiburón but not documented for the arid western margin of the island. Bajadas, valley plains, and sometimes on rocky slopes.

Other Gulf Islands: San Marcos, Coronados, Espíritu Santo.

Geographic Range: Sonora southward to Bahía Kino and Hermosillo, where it is abruptly replaced southward by *K. sonorae* Britton. Also southeastern California to western Texas, Nevada, southwestern Utah, Baja California Sur, Chihuahua, and Coahuila.

Tiburón: El Sauz, *Moran 8771* (SD). Ensenada del Perro, *Felger 17712*. SW Central Valley, 1200–1400 ft, upper rocky slope of mountain [Sierra Menor] bordering valley, *Felger 12411*. E foothills of Sierra Kunkaak, *Felger 6994*. Palo Fierro, lower bajada, *Felger 10014*. Hahjöaacöl, NE coast, *Wilder 08-210*. Hast Iif (Ojo de Puma), Central Valley, broad arroyo, *Wilder 07-253*.

Krameria erecta Willdenow ex J.A. Schultes [*K. parvifolia* Bentham]

Haxz iztim "dog's hip bone"; *cósahui*; range ratany

Description: Small shrubs often 0.3–0.5 m tall. Stems tough and woody, the upper branches beset with small, knotty, short shoots. Herbage grayish with short white hairs. Leaves linear. Flowers bright magenta-purple, solitary or in short racemes with leafy bracts. Fruits more or less globose and moderately compressed, 6 mm wide, the spines 3.5 mm long with small barbs more or less evenly distributed along the upper part of the shaft. Flowering at various seasons.

Local Distribution: Widespread across most of Tiburón from near beaches to the Sierra Kunkaak, and not encountered in the arid western margin of the island; arroyos, bajadas, hills, and mountains.

Other Gulf Islands: San Marcos, Carmen.

Geographic Range: Sinaloa, Zacatecas, San Luis Potosí, Durango, and Baja California Norte to southwestern United States.

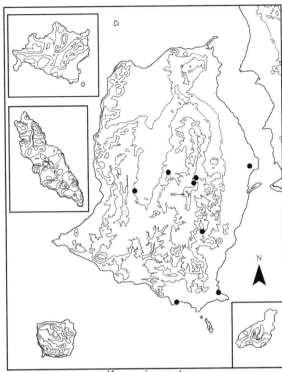

Krameria erecta
Haxz iztim

Tiburón: Ensenada de la Cruz, *Felger 9215*. Ensenada del Perro, *Tenorio 9497*. E foothills of Sierra Kunkaak, *Felger 6993*. SW Central Valley, *Felger 12381*. Haap Hill, *Felger 74-T48*. 3 km W [of] Punta Tormenta, *Scott P12* (UNM). Foothills of NE portion of Sierra Kunkaak, *Wilder 06-375*. Base of Capxölim, head of arroyo, *Wilder 05-48*. Palo Fierro, *Felger 8925*.

Lamiaceae (Labiatae) – Mint Family

Shrubs (also many herbaceous species on the mainland). Leaves opposite. Flowers bilaterally symmetrical, small, and blue (those in the flora area), attracting hummingbirds.

1. Pubescence of branched hairs; leaves mostly 1–2.5 cm long, the blades generally broadest at about middle; fertile stamens 4; Tiburón. _____ **Hyptis**
1' Pubescence of simple hairs; leaves often 2–8 cm long, the blades generally broadest below the middle; fertile stamens 2; Nolasco. _____ **Salvia**

Figure 3.105. *Salvia similis.* Cañón de Mellink, Isla San Pedro Nolasco. BTW, 29 September 2008.

Figure 3.104. *Hyptis albida.* Foothills of Sierra Kunkaak, Isla Tiburón. BTW, 1 January 2006.

Hyptis albida Kunth [*H. emoryi* Torrey. *H. emoryi* var. *amplifolia* I.M. Johnston. *H. emoryi* var. *palmeri* (S. Watson) I.M. Johnston]

Xeescl; *salvia*; desert lavender

Description: Shrubs often 1.5–2.5 (3) m tall, with many straight, slender woody stems arising from the base; densely covered with branched, whitish hairs. Leaves often nearly evergreen but tardily drought deciduous. Leaves often 1.5–3 cm long, ovate to oval, grayish or whitish to olive green depending on moisture conditions, gradually and tardily drought deciduous; the margins crenulate to broadly toothed. Flowers small and highly fragrant, produced in profusion in dense axillary and terminal clusters, small and fragrant. Calyx 5–6 mm long, the lobes slender and purplish. Corollas longer than the calyx, dark lavender-blue and attractive against the white foliage and calyces. Flowering for lengthy time periods, sometimes almost all year, mass flowering often in spring; flowers visited by many insects including honeybees and by hummingbirds. Very often at least a few flowers can be located during drought when it is one of the only plants flowering.

Local Distribution: Widespread at all elevations on Tiburón, San Esteban, and Dátil; often abundant along arroyos and canyons and also rocky slopes to the peaks. On Tiburón it is one of the more widespread and common major shrubs, and often a dominant species along arroyo bottoms.

Other Gulf Islands: Ángel de la Guarda, San Lorenzo, Tortuga, San Marcos, Carmen, Santa Catalina, San Francisco, Espíritu Santo.

Geographic Range: Tom Van Devender challenged Richard to distinguish *H. emoryi* from *H. albida*. Covering up the labels of dozens of herbarium specimens from many localities in Mexico and the United States,

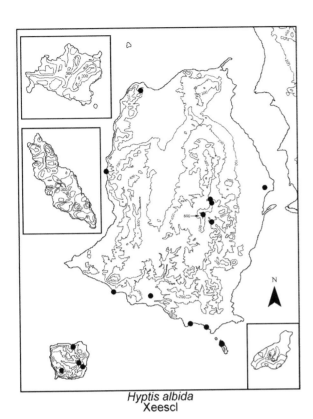

Hyptis albida
Xeescl

Examination of a wide range of specimens indicates that these features do not serve to separate them and their flowers are very similar if not identical (Jim Henrickson, personal communication, 2011; *also* see Martin et al. 1998; Turner et al. 1995).

Tiburón: Tordillitos, *Felger 15488*. Sauzal landing field, *Felger 6441*. Ensenada de la Cruz, *Felger 2573*. Coralitos, *Wilder 06-61*. E base of Sierra Kunkaak, head of arroyo, *Wilder 06-30*. N base of Sierra Kunkaak, large wash, *Felger 07-99*. Capxölim, solitary peak NE of Sierra Kunkaak Mayor, *Wilder 07-475*. Summit of island, not common, *Wilder 07-565*. Palo Fierro, middle bajada, *Felger 10018*. W of Tecomate, *Felger 6239*. Pooj Iime, *Wilder 07-449*.

San Esteban: Arroyo Limantour, *Van Devender 92-478*. E-central side of island, *Felger 12754*. Steep slopes, S-central peak, *Felger 17548B*. N side, canyon behind cove, *Felger 17582*.

Dátil: NE side of island, rocky outcrop on NE-facing slope, *Felger 9101*. E side, slightly N of center, *Wilder 07-135*. NW side of island, *Felger 15335*.

Salvia similis Brandegee

Description: Shrubs often 1–1.6 m tall with slender stems, grayish-white herbage, and thin leaves. Corollas dark blue, the calyx tinged with blue, recorded flowering

Richard failed to separate them and accepted Tom's insistence that they probably are not different species. *Hyptis emoryi* ranges from southwestern and central Arizona, southern Nevada, and southeastern California to Baja California Sur and lowland arid regions of Sonora to San Luis Potosí, Michoacán, and Guerrero. *Hyptis albida* sensu stricto ranges from southern Sonora to Chiapas. Several varieties of *H. emoryi* have been described across the large geographic range.

We are taking the broad view of the species complex as it extends from tropical to desert habitats (Felger et al. 2007b). The Sonoran Desert populations and plants from drier habitats tend to be more densely white-pubescent than those from less arid, non-desert regions farther south in Mexico, but the variation is continuous. The leaves of well-watered desert plants, especially ones growing in shaded niches and following favorable rains, can be much larger, greener (sparser pubescence), and thinner than dry-season leaves and can resemble leaves of *H. albida* sensu stricto. Differences in pedicel length (reportedly longer in *H. emoryi*) and nutlets (smooth in *H. albida* sensu stricto vs. "minutely granular-roughened" in *H. emoryi*) are key characters cited to distinguish between them (Wiggins 1964:1292–1293; also see Epling 1949).

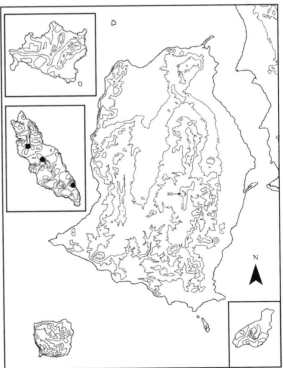

Salvia similis

in March, September, and November. The flowers are frequented by Costa's hummingbirds, vigorously zooming into the corollas.

Local Distribution: Only on Nolasco, where it is found on north- and northeast-facing slopes near the summit on the east side of the island and upper elevations of canyons on the west side. Often in aggregations of several plants.

Other Gulf Islands: None.

Geographic Range: Apparently rare on mainland Sonora, where it is known from a single record. The mainland and Nolasco populations are both on granodiorite slopes with strong maritime influences, especially on north- and east-facing slopes; there are very few mainland habitats where these conditions are duplicated. Otherwise known from Baja California Sur, where it rather widespread.

Nolasco: E side, near summit, 800 ft, localized, not common, mostly on N and NE-facing slopes, *Felger 9636*. SE side, above the cove, *Sherbrooke* [*Felger 11235*]. Cañón de Mellink, upper reaches of canyon, *Wilder 08-351*.

Sonora: Peninsula on S side of Algodones Bay, 2 km W of Cerro Teta de Cabra summit, N-facing granite slope, elev. 5–20 m, desertscrub with *Stenocereus thurberi, Jatropha cuneata, Bursera microphylla, Fouquieria diguetii*, uncommon shrub, about 1 m tall, 17 Mar 1983, *Burgess 6361*.

Loasaceae – Stickleaf or Loasa Family

Annuals or perennials with silicified or calcified and mostly barbed (barbs usually in whorls) or gland tipped hairs—the barbed hairs resulting in the leaves and fruits sticking like Velcro.

1. Shrubs or subshrub perennials; flowers white.
   2. Leaves prominently petioled, the leaf blades 2–10 cm wide, as long or slightly longer than wide, shallowly lobed; stamens many. _____ **Eucnide cordata**
   2' Leaves sessile or petioles very short, the leaf blades less than 2 cm wide, more than twice as long as wide, the margins entire; stamens 5. _____ **Petalonyx**
1' Annuals; flowers green and yellow, yellow, or orange.
   3. Stems and petioles thickened and semi-succulent; leaf blades as wide as or wider than long, the upper surface glistening; petals united below, the corollas yellow with green lobes. _____ **Eucnide rupestris**
   3' Stems and petioles not succulent; leaf blades longer than wide, dull green; petals separate, orange, yellowish, or silvery. _____ **Mentzelia**

Figure 3.106. *Eucnide cordata*. Arroyo Limantour, Isla San Esteban. BTW, 8 March 2007.

Figure 3.107. *Eucnide rupestris.* Cañón de Mellink, Isla San Pedro Nolasco. BTW, 29 September 2008.

Eucnide

1. Shrubby perennials; stems not succulent; roots well developed; flowers white. _____ **E. cordata**
1' Annuals; stems semi-succulent; roots weakly developed; flowers yellow and green. _____ **E. rupestris**

Eucnide cordata Kellogg ex Curran

Description: Bushy or shrubby perennials. Stems and leaves with hairs mostly topped with a whorl of recurved barbs. Leaf blades 2–10 cm wide, as long as or slightly longer than wide, cordate at base, shallowly 5–9 lobed, with dull surfaces. Flowers in terminal cymes or racemes; flowers white, tubular, often 2.5–3 cm long including the exerted stamens.

Local Distribution: San Esteban in sheltered nooks on arroyo and canyon walls; highly localized and not common.

Other Gulf Islands: Ángel de la Guarda, San Lorenzo, Tortuga, Coronados (observation), Carmen, Monserrat, Santa Cruz (observation), San José, San Francisco, Espíritu Santo, Cerralvo.

Geographic Range: West-central Sonora and both states of Baja California.

San Esteban: Arroyo Limantour: *Felger 17661*; 2.25 mi inland, S side of arroyo, sheltered wall of arroyo, *Wilder 07-48*.

Figure 3.108. *Petalonyx linearis.* Arroyo Limantour, Isla San Esteban. BTW, 8 March 2007.

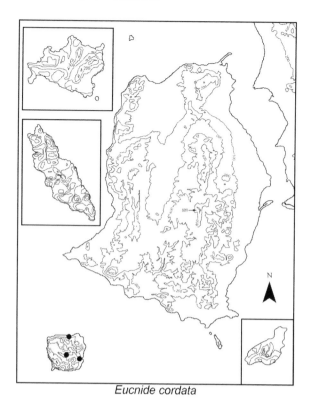

Eucnide cordata

N side of island, narrow, rocky canyon behind large cove, *Felger 17589*.

Eucnide rupestris (Baillon) H.J. Thompson & W.R. Ernst

Zaaj iti cocaai "cliff-hanger"; Velcro plant

Description: Non-seasonal annuals, the roots unusually small for the size of the plants; stems and petioles semi-succulent. Stems, leaves, and fruits tenaciously stick like Velcro due to the minutely barbed hairs. Leaf blades 3–14 cm long and about as wide, shallowly lobed, relatively thin and bright yellow-green, the upper surfaces glistening. Flowers to 2 cm long, with a yellow tube and bright green lobes, the stamens shorter than the corolla.

Local Distribution: Tiburón, San Esteban, Dátil, and Nolasco. Usually growing from crevices in rocks, cliffs, and arroyo or canyon walls and mountain slopes, from the shore to higher elevations, rarely in sandy washes.

Other Gulf Islands: Ángel de la Guarda, Salsipuedes, Partida Norte (*Johnston 3277*, CAS), San Marcos, Espíritu Santo.

Geographic Range: Gulf side of Baja California Sur to southeastern California, islands in the Gulf, and northwestern Sinaloa and western Sonora to southern Arizona.

Tiburón: Ensenada Blanca (Vaporeta), *Felger 12241*. Canyon just N of Willard's Point, N-facing cliff, *Moran 8738* (SD). Tordillitos, *Felger 15471*. Arroyo Sauzal: Sandy soil next to rock, *Felger 9993*; 2¾ mi inland, *Wilder 06-102*. El Monumento, *Felger 15540*. S-facing slope of Sierra Kunkaak, 1115 ft, rocky soil below a rock arch, *Wilder 06-34*. Cerro San Miguel, *Quijada-Mascareñas 91T004*. Agua Dulce Bay, N-facing cliff, *Moran 12996* (SD).

San Esteban: N-facing rock outcrop cut into side of arroyo wall on S side of Arroyo Limantour, 3 mi inland, *Wilder 07-83*. SE corner of island, eroded calcareous rocky hilltops and slopes, *Felger 7060*. N side of island, *Felger 15446*. SE corner, slope of high peak, on cliff, *Moran 8858* (CAS, SD).

Dátil: E side, slightly N of center, narrow ridge crest on peak, *Wilder 07-112*. NE side, very common, sea cliffs and steep rocky N-facing slopes, *Felger 9107*. NW side, cliffs, *Felger 15354*. SE side, steep, narrow canyon, *Felger 17519*.

Nolasco: Very common in sheltered rock crevices on cliffs near the sea, *Johnston 3143* (CAS, US). [Above SE cove], just below crest, 900 ft, in rocks, *Pulliam & Rosenzweig 6*

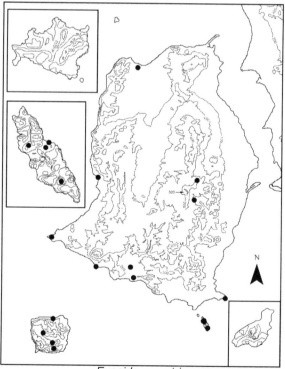

Eucnide rupestris
Zaaj iti cocaai

Oct 1974. E-central side, N exposure below crest of island, 270 m, from mid- to high elevations, *Felger 06-105*. Cañón de Mellink, *Wilder 08-346*.

Mentzelia – Blazing star

Annuals. Hairs barbed (glochidiate), especially the leaves and capsules adhesive (fresh leaf pieces stick on clothing like Velcro). Leaves usually alternate.

1. Petals orange, 10–15 mm long; flowers and fruits stalked. _____ **M. adhaerens**
1' Petals pale yellow, 12–28 mm long; flowers and fruits sessile._____ **M. hirsutissima**

Mentzelia adhaerens Bentham

Hehe cotopl "clinging plant," hehe czatx "stickery plant"; *liga, pega pega*; stickleaf

Description: Non-season annuals; often best developed after summer-fall rains. Flowers pedicelled; petals orange, opening early in the morning.

Local Distribution: Tiburón, San Esteban, Alcatraz, Dátil, and Mártir. Widespread and sometimes locally abundant, including rocky beaches, arroyos, canyons, bajadas, mesas, and rocky slopes. It is so abundant in one area of Tiburón that Seri women's skirts would get covered with the leaves (Felger & Moser 1985).

Other Gulf Islands: Ángel de la Guarda, San Lorenzo, Tortuga, San Marcos, Coronados, Carmen, Danzante, Monserrat, Santa Catalina, Santa Cruz, San José (observation), San Francisco, Espíritu Santo, Cerralvo.

Geographic Range: Western Sonora from the Pinacate region southward to the Guaymas region, and Baja California Norte from 50 km north of San Felipe southward through Baja California Sur; not known from the United States.

Tiburón: Ensenada Blanca (Vaporeta), *Felger 12219*. Canyon just N of Willard's Point, *Moran 8721* (UC). Tordillitos, *Felger 15495*. Xapij (Sauzal) waterhole, *Wilder 06-117*. Ensenada de la Cruz, *Felger 9067*. El Monumento, *Felger 15554*. SW Central Valley, arroyo of valley floor, *Felger 12328*. Estero San Miguel, *Wilder 06-278*. Hant Hax, *Felger 9348*. Arroyo Agua Dulce, *Felger 6815*. Pooj Iime, *Wilder 07-445*.

San Esteban: El Monumento, arroyo, *Felger 7087*. Arroyo Limantour, *Felger 12728*. N side of island, arroyo, *Felger 15401*.

Dátil: E side, slightly N of center, canyon bottom, *Wilder 07-97*. NE side, *Felger 9103*. NW side, *Felger 15311*. SE side, *Felger 17492*.

Alcatraz: E side of flat, *Felger 08-02*.

Mártir: Summit, among cactus and rocks, 1887, *Palmer 402* (GH, image). Frequent over higher part of island, 18 Apr 1921, *Johnston 3156* (CAS, image). High elevation plateau, NW of highest point of island, 275 m, abundant, *Wilder 07-602*.

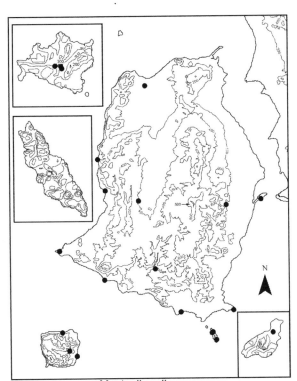

Mentzelia adhaerens
Hehe cotopl, hehe czatx

Mentzelia hirsutissima S. Watson

Hehe cotopl "clinging plant," hehe czatx "stickery plant"; hairy stickleaf

Description: Cool-season annuals. Flowers sessile; petals pale yellow (12) 14–24 (28) mm long.

Local Distribution: Western and northern coasts and mountains of Tiburón; arroyos and canyons.

Other Gulf Islands: Ángel de la Guarda (type locality).

Geographic Range: Sonoran Desert in southeastern California and Baja California Norte including Isla Cedros.

Tiburón: Canyon just N of Willard's Point, *Moran 8722* (SD). Arroyo Agua Dulce, very common, rocky-gravelly soil, *Felger 6815B*. Tecomate, floodplain, *Felger 12531*. Sierra Menor central, 13 Mar 2001, *Romero-Morales* (specimen not curated).

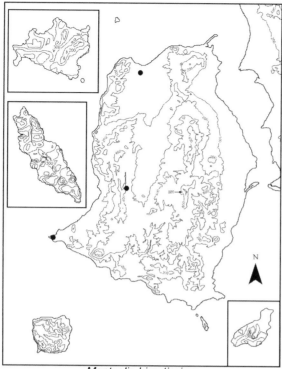

Mentzelia hirsutissima
Hehe cotopl, hehe czatx

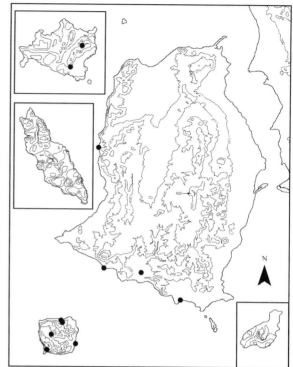

Petalonyx linearis

Petalonyx linearis Greene

Narrow-leaf sandpaper plant

Description: Perennials and often flowering during the first season, forming a dense, woody-based, rounded bush 30–50 cm tall. Leaves tardily drought deciduous, 1.2–3.5 cm long, linear-oblong to lanceolate, more than twice as long as wide, the margins entire. Flowers white; petals ca. 3 mm long. Flowering at least in spring and fall.

Local Distribution: An arid-adapted species; widespread on San Esteban in arroyos and on rocky beaches, and similar habitats on the southern and northwestern margins of Tiburón. The plants seen on Tiburón are often larger than those on San Esteban. Known from Mártir by Palmer's 1887 collection and Johnston's 1921 collection. Johnston found a single one and collected the whole plant, and there is no record for it on the island since then.

Other Gulf Islands: Ángel de la Guarda, San Lorenzo, Tortuga, San Marcos.

Geographic Range: Southeastern California to northern Baja California Sur, Isla Cedros (type locality), Gulf Islands, southwestern Arizona, and the Pinacate region in northwestern Sonora (Felger 2000a).

Tiburón: Tordillitos, *Felger 15501*. Arroyo Sauzal, 1 mi from shore, locally common, not seen closer to shore, *Wilder 06-85*. Arroyo de la Cruz, *Wilder 09-52*. Pooj Iime, arroyo at beach, base of S-facing slope, 29 Sep 2007, *Felger*, observation.

San Esteban: Canyon, SW corner of island, *Felger 16646*. N side of island, *Felger 15463*. Narrow canyon, NW corner, *Felger 17606*. El Monumento, rocky beach among sea-worn boulders, one of the closest plants to the ocean, *Felger 7082*. Arroyo Limantour, 3.25 mi inland, *Wilder 07-54*.

Mártir: In shade of *Cereus* and rocks, Oct 1887, *Palmer 411* (CAS, UC; GH, image). A single plant at foot of sea cliff on E side of island, 18 Apr 1921, *Johnston 3164* (CAS, image; young plant, annual or first season).

Malpighiaceae – Malpighia Family

Perennial vines and herbaceous perennials or subshrubs to shrubs with Malpighian hairs (each hair is attached in the middle, with two opposite-pointing arms), or the plants glabrate or glabrous. Leaves opposite (except *Echinopterys*). The Sonoran Desert malpighs have yellow flowers and dry fruits. Sepals 5; petals 5, clawed (narrowed basally), 4 of equal size, the 5th or odd petal slightly enlarged, serves as a flag to visiting bees. Most

New World species have sepals with oil glands, attracting oil-gathering anthrophorid bees, such as *Centris*, which are the principal pollinators (see *Krameria*).

The family is largely tropical and subtropical. Sixteen species occur in the subtropical Río Mayo region of southeastern Sonora, 6 in the Guaymas region and 5 of them make it as far north as Tiburón. But only *Cottsia gracilis* ranges into the harsh desert of the Pinacate region in northwestern Sonora and southern Arizona and none reach California. This pattern is similar to that of other tropically inclined plants such as the Acanthaceae and *Bursera* (Burseraceae).

Figure 3.109. *Callaeum macropterum.* Foothills of Sierra Kunkaak, Isla Tiburón. Plant, 7 April 2008, and fruit, 24 May 2006. BTW.

Figure 3.110. *Echinopterys eglandulosa.* Foothills of Sierra Kunkaak, Isla Tiburón. BTW, 24 November 2006.

1. Stems vining to trailing; sepals bearing conspicuous oil glands; fruits winged.
    2. Fruits with 4 large papery wings, each wing 2 or more cm long and wider than long. ___ **Callaeum**
    2' Fruits with 2 or 3 wings, each less than 1.5 cm long and longer than wide. _____**Cottsia**
1' Stems erect to spreading, not vining or trailing; sepals without oil glands; fruits not winged.
    3. Shrubs, usually more than 1 m tall; leaves alternate. _____ **Echinopterys**
    3' Subshrub usually less than 0.5 m tall; leaves opposite. _____**Galphimia**

Callaeum macropterum (Moçiño & Sessé ex de Candolle) D.M. Johnson [*Mascagnia macroptera* (Moçiño & Sessé ex de Candolle) Niedenzu]

Haxz oocmoj "dog's waist cord"; *gallinita*; yellow orchid-vine

Description: Bushy perennials, becoming large vines in better watered habitats with denser vegetation. Leaves 2.5–11 cm long, gradually drought deciduous but plants often retaining at least some smaller leaves. Flowers bright yellow and attractive, appearing at any season except during extreme drought; petals fringed. Fruits unique with 4 large, papery wings 2–3.5 cm long and wider than long.

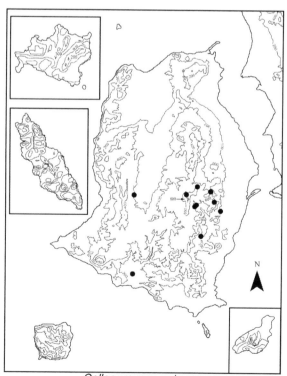

Callaeum macropterum
Haxz oocmoj

Local Distribution: Widespread on Tiburón, especially in arroyos and canyons, and extending to peak elevations on Sierra Kunkaak; not on the dry western margin or the northern part of the island.

Other Gulf Islands: Carmen, Danzante, Santa Catalina (observation).

Geographic Range: Sonora to Guerrero and Baja California Sur. Sonora in Sonoran desertscrub (except the Lower Colorado Desert), thornscrub, and tropical deciduous forest; northward to within about 50 km of the Arizona border.

Tiburón: Arroyo Sauzal, *Wilder 06-95*. SW Central Valley, arroyo, *Felger 12369*. Foothills of Sierra Kunkaak, ca. 6 km W of Punta Narragansett, not common, arroyo and canyon bottoms in lower mountains, *Felger 6963*. E base of Sierra Kunkaak, *Romero-Morales 07-9*. Arroyo below Pazj Hax, *Wilder 06-166*. Sopc Hax, canyon, *Felger 9293*. Hast Cacöla, E flank of Sierra Kunkaak, foothills, *Wilder 08-227*. S-facing slope of Sierra Kunkaak, 355 m, *Wilder 06-1*. Steep, protected canyon, 720 m, N interior of Sierra Kunkaak, *Wilder 07-513*. Capxölim, solitary peak NE of Sierra Kunkaak Mayor, *Wilder 07-480*.

Cottsia

Perennial vines with slender stems, often twining in shrubs. Foliage and flowering responding to soil moisture and warm weather at any time of the year, with flowering peaks often in spring and late summer; maximum-size leaves appear with the summer-fall rains; the larger leaves quickly fall with dry weather. Flowers often 10–15 mm wide, the petals bright lemon yellow. Fruits with 2 and sometimes 3 papery wings (samaras), usually green tinged with red, each wing often 10–15 mm long.

The two species in the flora are similar appearing and most readily distinguished by their leaf shapes; small-leaved drought-stressed plants may be difficult to distinguish. There are also subtle floral differences. Both species sometimes occur together. Anderson and Davis (2007a) report possible intermediate specimens from Sonora.

Cottsia, having three species, is one of the few common, widespread genera of vining plants in the Sonoran Desert and also occurs in thornscrub habitats. The third species is endemic to Sonora. "All three species grow in true desert. It is not rare for Neotropical Malpighiaceae to grow in open more or less xeric places, but very few can survive in a place as hot and dry as the Sonoran desert, that is surely a derived adaptation of this genus" (Anderson & Davis 2007a:159).

1. Leaves ovate, widest at middle. _____ **C. californica**
1' Leaves lanceolate, widest below middle. __ **C. gracilis**

Cottsia californica (A. Gray) W.R. Anderson & C. Davis [*Janusia californica* A. Gray]

Hehe quiijam "entwining plant"; janusia

Description: Leaves ovate and generally larger than those of *C. gracilis*. Chromosome number: *n* = 10 (Anderson 1993).

Local Distribution: Widespread across Tiburón but not on the arid southwestern margin of the island where one finds *C. gracilis*, although the two variously occur together. Also on Dátil. *Cottsia californica* reaches maximum development in sites more mesic (less harsh) than where *C. gracilis* is more numerous—as would be expected by the larger, more mesophytic leaves of *C. californica*.

Other Gulf Islands: Tortuga, Coronados, Carmen, Danzante, Monserrat, Santa Catalina, Santa Cruz, San José, Espíritu Santo, Cerralvo.

Geographic Range: Northern Sonora (Sierra del Viejo near Caborca, El Oasis, and Sinoquipe) to northern Sinaloa, southern Baja California Norte, and Baja California Sur and islands off its coast.

Tiburón: Ensenada de la Cruz, rocky hill, *Felger 9238*. El Monumento, *Felger 15525*. SW Central Valley, *Felger 12380*. Haap Hill, *Felger T74-18*. Zozni Cmiipla, *Felger 08-133*. Sopc Hax, rocky canyon, *Felger 9297*. Base of Capxölim, head of arroyo, *Wilder 05-54*. Palo Fierro, *Felger 11077*.

Dátil: NE side, arroyo bottom and steep rocky slopes, with *C. gracilis* (*9089a*), *Felger 9080b*. NW side, *Felger 15303*.

Cottsia gracilis (A. Gray) W.R. Anderson & C. Davis [*Janusia gracilis* A. Gray]

Hehe quiijam "entwining plant"; janusia

Description: Distinguished from *C. californica* by its usually smaller, narrower, and lanceolate leaves. Chromosome number: *n* = 20 (Anderson 1993).

This species is somewhat intermediate between the extremely narrow-leaved *C. linearis* and *C. californica*. "Anderson (1993) hypothesized that the intermediate species, *C. gracilis*, is an allotetraploid derived, possibly repeatedly, from hybridization between the diploids, *C. californica* and *C. linearis*, followed by doubling of the chromosome number" (Anderson & Davis 2007a:160).

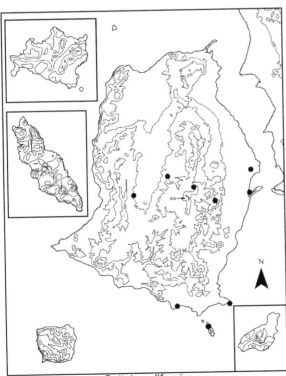

Cottsia californica
Hehe quiijam

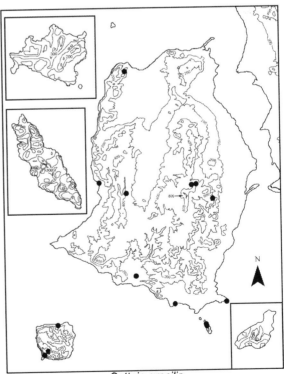

Cottsia gracilis
Hehe quiijam

Local Distribution: Widespread nearly throughout Tiburón, including arroyos, canyons, and rocky slopes. San Esteban in canyons, north-facing slopes on the north side of the island, and at higher elevations. Variously occurring with *C. californica* on Tiburón and Dátil but generally extending into more arid sites.

Other Gulf Islands: San Marcos, Carmen.

Geographic Range: Sonora south to the Guaymas region, southern Arizona to western Texas, and southward to Chihuahua, Coahuila, Durango, and both states of Baja California. "This species, one of only six Malpighiaceae native to the United States, occurs farther north than any other New World species" (Anderson & Davis 2007a:163).

Tiburón: Ensenada Blanca (Vaporeta), *Felger 14947*. Sauzal landing field, *Felger 6467*. Ensenada de la Cruz, arroyo bottom, *Felger 9150*. El Monumento, *Felger 08-49*. SW Central Valley, upper rocky slope of mountain [Sierra Menor] bordering valley, 1200–1400 ft, *Felger 12416*. Sopc Hax, *Felger 9296*. Cerro San Miguel, *Quijada-Mascareñas 90T001*. Base of Capxölim, head arroyo, *Wilder 05-47*. Rocky hill W of Tecomate, *Felger 6259*.

San Esteban: Canyon, SW corner of island, *Felger 16625*. N side of island, *Felger 15442*. Steep slopes, S-central peak, *Felger 17540*.

Dátil: NE side of island, arroyo bottom, steep narrow rocky canyon, *Felger 2555*. NE side, with *C. californica* (*9080b*), *Felger 9089a*. E-central side, E-W trending canyon, *Wilder 09-33*.

Echinopterys eglandulosa (A. Jussieu) Small

Hap oaacajam "what mule deer flay antlers on"

Description: Drought-deciduous shrubs to 2 m tall. Leaves alternate (an unusual feature among the family). Flowers showy, bright yellow in elongated inflorescences. Sepals without oil glands. Flowering at almost any time of year following rains.

Local Distribution: We encountered a single shrub (conspicuous because of its bright yellow flowers) in a densely vegetated canyon bottom in the Sierra Kunkaak, but it is probably more widely distributed in the Sierra since it is conspicuous only when in flower. Fruits of capsules, about 5 mm long, pubescent and bristly.

Other Gulf Islands: None.

Geographic Range: Sonora northward to the vicinity of El Desemboque San Ignacio and Altar, and southward to Oaxaca. Shrubs in the desert but in tropical deciduous forest it often has scandent or vining stems 4–5 m long.

Tiburón: NE base of Sierra Kunkaak, S-facing side of arroyo, in rocky talus, *Wilder 06-377*.

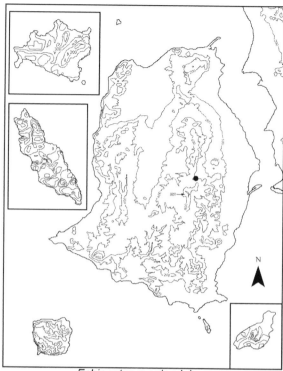

Echinopterys eglandulosa
Hap oaacajam

Galphimia angustifolia Bentham [*Thryallis angustifolia* (Bentham) Kuntze]

Hap oaacajam "what mule deer flay antlers on"

Description: Low, herbaceous or suffrutescent perennials or subshrubs. Flowers attractive, yellow (corollas yellow in bud and rarely orange after opening), often becoming reddish with age; flowering various seasons with sufficient rainfall. Sepals without oil glands.

Local Distribution: Tiburón, at the southern, eastern, and central part of the island; bajadas and valley plains, hills and mountains, mostly in arroyos and canyons. Known from Nolasco by a single collection.

Other Gulf Islands: Danzante, Monserrat, Santa Catalina, Santa Cruz, San José, Espíritu Santo.

Geographic Range: Western Sonora (from the mountains opposite Tiburón southward), northwestern Sinaloa, Baja California Sur, northeastern Mexico, and southern Texas (Anderson & Davis 2007b).

Tiburón: Tordillitos, *Felger 15487*. SW Central Valley, arroyo, *Felger 12344*. Haap Hill, *Felger T74-17*. Foothills, E side of Sierra Kunkaak, canyon bottom, *Felger 6969*. Bajada between Hant Hax camp and Zozni Cmiipla, *Felger 9356*. 0.7 mi downstream from Sopc Hax, upper bajada, *Wilder 07-207*. Hast Cacöla, E flank of Sierra Kunkaak seen only in the foothills, *Wilder 08-226*.

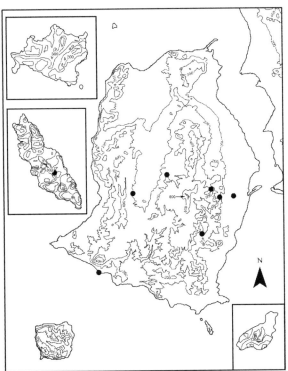

Galphimia angustifolia
Hap oaacajam

Nolasco: E-central part of island, 700 ft, bottom of steep rocky canyon, rare, only one plant encountered, perennial, bushy with dense foliage, ca. 1 m high and 1.3 m across, base scarcely woody, stems herbaceous, in full flower, flowers yellow, 26 Nov 1963, *Felger 9633*.

Malvaceae – Mallow Family
(includes Sterculiaceae)

Herbs and shrubs, generally with stellate hairs, and some also with forked or simple hairs. Leaves alternate and simple but sometimes deeply cut. Sepals united below, 5-toothed to parted; petals 5, separate. Stamens numerous, the filaments united into a tube surrounding the ovary and much of the style (Malvaceae sensu stricto) or stamens 5 and fused at least basally (*Ayenia, Melochia, Waltheria*). Fruits with few to many carpels or mericarps (segments).

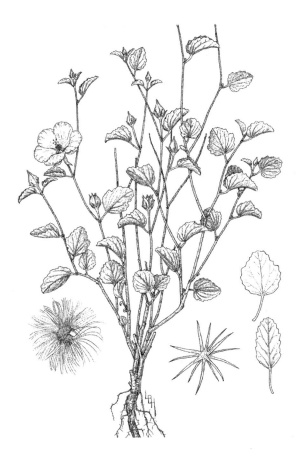

Figure 3.111. *Hibiscus denudatus* var. *denudatus*. LBH.

Figure 3.112. *Sphaeralcea ambigua* var. *ambigua*. LBH.

Figure 3.113. *Hibiscus biseptus.*
Dry season aspect with bracts,
sepals, and capsules. Foothills of
Sierra Kunkaak, Isla Tiburón.
BTW, 2 January 2006.

Figure 3.114. *Horsfordia alata.*
Vicinity of Xapij waterhole,
Arroyo Sauzal, Isla Tiburón.
BTW, 17 March 2006.

Figure 3.115. *Sphaeralcea ambigua* var. *ambigua.*
Isla Dátil. BTW, 27 February 2009.

Figure 3.116. *Sphaeralcea hainesii.* Isla San Pedro Mártir.
BTW, 11 April 2007.

1. Fruiting carpels (segments) persistent and not falling away at maturity.
 2. Capsules glabrous on outer surface; Nolasco _____ **Gossypium**
 2' Capsules hairy; not on Nolasco. _____ **Hibiscus**
1' Mericarps (segments) of fruits separating and falling at maturity.
 3. Stems very slender and weak; leaves of lower, vegetative stems petioled, leaves of flowering stems sessile and smaller; flowering and fruiting stalks thread-like and bent at a conspicuous joint; fruits resembling a tiny paper lantern and breaking apart at maturity. _____ **Herissantia**
 3' Stems usually not slender and weak (except *Sida abutifolia*); pedicels not thread-like and not bent at a joint; fruits not as above.
  4. Shrubs often 1+ m tall; flowers rose-lavender; fruits 5-winged. _____ **Melochia**
  4' Annuals to shrubs; fruits not 5-winged.
   5. Upper and lower halves of mericarps similar, not reticulate on sides, not winged.
    6. Fruits 1-seeded; flowers 5–6 mm long and bright yellow. _____ **Waltheria**
    6' Fruits multiple-seeded, flowers 5 mm or more in length, pale yellow, orange or maroon.
     7. Flowers minute and maroon; stamens 5; fruits globose, ca. 5 mm diameter, studded with short blunt spines. _____ **Ayenia**
     7' Flowers not minute, the petals pale yellow or orange; stamens numerous; fruits not globose, 8 mm or more long and wide (except *Sida*), not studded with spines.
      8. Shrubs or sometimes flowering in the first season; stems upright and not weak and decumbent; larger leaf blades often more than 3 cm long and usually less than 1.5 times longer than wide; flowers orange; fruits about 1 cm long. _____ **Abutilon**
      8' Herbaceous perennials or facultative annuals; stems rather weak and decumbent; leaf blades usually less than 3 cm long and more than 1.5 times longer than wide; flowers pale yellow-orange or perhaps whitish; fruits 2–3 mm long. _____ **Sida**

5' Upper and lower halves of mericarps dissimilar; mericarps reticulate on sides, with flared membranaceous wings above.

9. Plants often more than 1 m tall and not on Mártir or Alcatraz; floral bractlets none or minute. _____

_____ **Horsfordia**

9' Plants seldom more than 1 m tall, except on Mártir and Alcatraz; floral bractlets present. _____

_____ **Sphaeralcea**

Abutilon

Shrubs with orange flowers. Stamens numerous. Fruit a schizocarp (sometimes appearing capsule-like), 8–10 mm long (those in the flora area).

1. Petals pale orange with a maroon spot at base; capsule segments (mericarps) 5, the tips blunt and rounded; fruiting calyx shorter than the mericarps. **A. incanum**
1' Petals uniformly orange; capsule segments 7 or more, the tips slender and pointed; fruiting calyx about as long as the mericarps.
   2. Leaf blades not especially velvety; stems and mericarps with hairs less than 0.5 mm long. _____

   _____**A. californicum**
   2' Leaf blades soft and velvety; stems and mericarps with hairs 0.8–2.5 mm long. _____ **A. palmeri**

Abutilon californicum Bentham

Hant ipásaquim "what the ground is swept with"

Description: Shrubs often 1.5–2.5 m tall, gradually drought deciduous. Flowers orange. Carpel walls with the larger stellate hairs stalked, and the smaller ones sessile. Flowering response non-seasonal.

Local Distribution: Widespread across Tiburón, especially along arroyos and canyons, and to high elevation in the Sierra Kunkaak.

Other Gulf Islands: Coronados, Carmen, San José, Cerralvo.

Geographic Range: Gulf Coast of Sonora from the Pinacate region to the Río Mayo region, both states of Baja California, and Islas Revillagigedos.

Tiburón: Ensenada Blanca (Vaporeta), *Felger 17292*. Arroyo Sauzal, near shore, *Felger 7026*. Sauzal (Xapij), *Wilder 08-105*. Ensenada de la Cruz, *Felger 9136*. La Viga, *Wilder 08-145*. SW Central Valley, arroyo, *Felger 12319*. Haap Hill, *Felger 76-T1*. Zozni Cmiipla, *Wilder 06-358*. Punta Tortuga, ¾ mi inland, *Felger 12554*. Camino de Caracol, in broad arroyo just entering the foothills of the Sierra Kunkaak, *Wilder 05-11*. Steep, protected canyon, 720 m, N interior of Sierra Kunkaak, *Wilder 07-585*. Bahía Agua Dulce, *Felger 8900*.

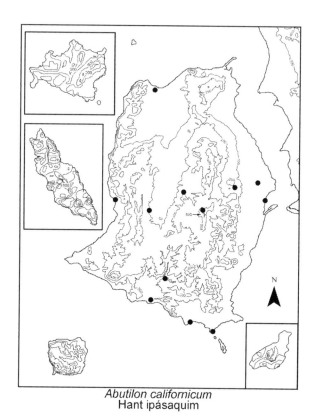

Abutilon californicum
Hant ipásaquim

Abutilon incanum (Link) Sweet

Hasla an iihom "what the ear is in," caatc ipapl "what grasshoppers are strung with"; *rama escoba, malva*; Indian mallow

Description: Slender-stem shrubs, often 0.7–1.5 m tall. Leaves bi-colored and usually gray-green, the stellate hairs minute. Pedicels usually thickened above the pedicel joint, even in bud. Petals and stamens yellow-orange; corollas with large maroon spots in center. Fruits with 5 mericarps, the tips blunt. Flowering and fruiting at various seasons.

Local Distribution: Widespread in the southern, central, and eastern part of Tiburón, including the Sierra Kunkaak; bajadas, arroyos, and canyons, and brushy hill and mountain slopes; also on Dátil.

Other Gulf Islands: Tortuga, Coronados, Carmen, Danzante, San José, Cerralvo.

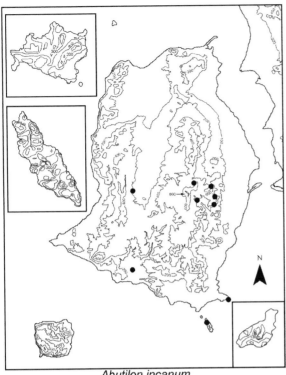

Abutilon incanum
Hasla an iihom, caatc ipapl

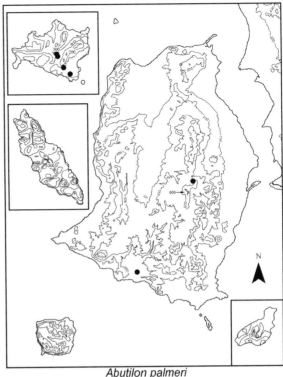

Abutilon palmeri
Caatc ipapl

Geographic Range: Western Arizona to Texas and the northern states of Mexico.

Tiburón: Arroyo Sauzal, *Felger 10110.* El Monumento, *Felger 15558.* SW Central Valley, *Felger 12378.* Pazj Hax, *Wilder 06-14.* 0.8 km below [E of] Sopc Hax, *Felger 9325.* Hast Cacöla, E flank of Sierra Kunkaak, 380 m, *Wilder 08-246* (USON). Head of arroyo, E base of Sierra Kunkaak, *Wilder 06-26.* Canyon bottom, N base of Sierra Kunkaak, 395 m, *Wilder 06-410.*

Dátil: NW side of island, *Felger 15336.*

Abutilon palmeri A. Gray

Caatc ipapl "what grasshoppers are strung with"; *pelotaso*; Indian mallow

Description: Sparsely branched, open and scarcely woody shrubs or subshrubs, often to 1.2+ m tall; also often flowering in the first season as a facultative annual. Densely stellate-hairy, the hairs on the stems and fruits long and slender. Leaf blades often 4.5–12 cm long and markedly velvety. Flowers in large, slender-stemmed, openly branched terminal panicles. Petals pale orange. Fruits 1.3–1.5 cm wide, the mericarps about 10. Flowering non-seasonally.

Local Distribution: Tiburón, documented from the south side of the island and the Sierra Kunkaak; mostly along arroyos and canyons. On Mártir, one of the most abundant and widespread plants on the island but generally not at lower elevations.

Other Gulf Islands: Ángel de la Guarda, Tortuga, Coronados, Carmen, Santa Cruz, San José, Espíritu Santo.

Geographic Range: Southwestern Arizona to Sinaloa, and southeastern California to the Cape Region of Baja California Sur, Gulf Islands, and disjunct in Tamaulipas.

Tiburón: Arroyo Sauzal, 1 mi from shore, *Wilder 06-84.* Base of Capxölim, head of arroyo, *Wilder 05-30.*

Mártir: 1887, *Palmer 401* (UC). Very common on upper part of island, 2½–4 ft high, 18 Apr 1921, *Johnston 3158* (CAS, UC). Summit, 300 m, *Moran 8806* (SD, UC). Above ruins of guano workers' village, widespread and locally abundant above ⅓ way up island, 100 m, most of the plants are flowering in their first season, *Felger 07-11.*

Ayenia filiformis S. Watson [*A. compacta* Rose]

Description: Non-seasonal annuals to short-lived perennials, slender and few-branched or with rigid, stubby branches when grazed. Leaves 1–3.8 cm long, linear to narrowly lanceolate, the dry-season leaves often reduced. Leaf margins with gland-tipped teeth. Flowers minute, maroon, and intricately sculptured; petals 5, the tips united with the stamen tube; stamens 5. Fruits of

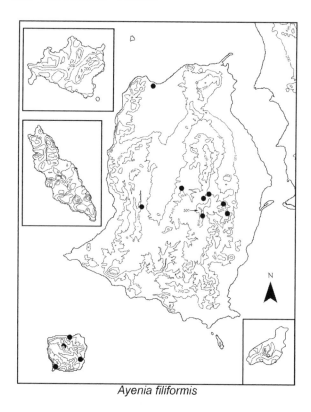

Ayenia filiformis

San Esteban: Arroyo Limantour, locally common in cracks in rocky arroyo bottom, *Van Devender 90-534*. Canyon, SW corner of island, *Felger 16622*. N side of island, near shore, *Felger 15438*. Sub-peak of Central Mountain, *Wilder 07-62*.

Gossypium davidsonii Kellogg [*G. klotzchianum* var. *davidsonii* (Kellogg) J.B. Hutchinson]

Description: Broad, low-branching shrubs often 1–1.8 m tall. Leaf blades often 4–8 cm long, ovate, and cordate at the base. Flowers in lateral inflorescences; floral bracts 1.5–2.8 cm long, with several large, often ragged teeth, and persistent even on the fruits. Petals 2.5–3 cm long, bright yellow, often with a dark maroon spot at the base. Stamens numerous. Flowering at least October and November.

Local Distribution: Nolasco, locally on steep rock slopes above Cala Güina at the southeast side of the island.

Other Gulf Islands: Coronados, Carmen, Monserrat, Cerralvo.

Geographic Range: Sonora on rocky slopes in the Guaymas region and the Baja California Peninsula south of 26°N. *Gossypium davidsonii* is not especially a desert plant but rather a member of a subtropical group that extends into the desert margin.

rounded capsules, 4.6–5 mm wide, with blunt "spines" and stellate hairs, separating into five 1-seeded segments (carpels). Seeds dark red-brown to nearly black, 2.5–3 mm long, resembling an insect pupa.

Local Distribution: Widespread on Tiburón and San Esteban, where it occurs in canyons and arroyos and on rocky slopes to higher elevations. On San Esteban the plants are often reduced to compact, fist-sized collections of stem grazed to nubbins by the endemic chuckwallas (*Sauromalus varius*).

Other Gulf Islands: Ángel de la Guarda, San Marcos, Coronados, Carmen, Danzante, Monserrat, San José, Cerralvo.

Geographic Range: Sinaloa to Coahuila northward to Arizona, New Mexico, and Texas to southeastern California. *Ayenia compacta* seems to be an arid-inhabiting conspecific form in the Sonoran Desert.

Tiburón: Upper rocky slope of mountain bordering SW Central Valley, 1200–1400 ft, *Felger 12409*. Haap Hill, *Felger T74-4*. Sopc Hax, *Wilder 07-219*. Hast Cacöla, E flank of Sierra Kunkaak, 450 m, *Wilder 08-265*. Foothills, NE portion of Sierra Kunkaak, *Wilder 06-441*. Siimen Hax, ca. ½ km to E, *Wilder 06-456*. In the "bowl" on S side of the Sierra Kunkaak, *Knight 1008*. Bahía Agua Dulce, among rocks, *Felger 6271*.

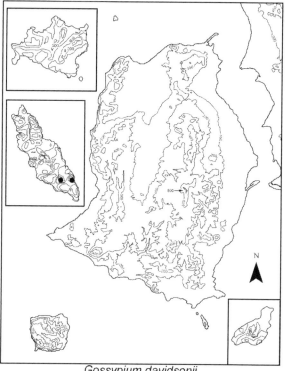

Gossypium davidsonii

Nolasco: Above SE cove, very steep rocky slope, 50 ft, low spreading bush in crevices of steep rock slopes, not common, *Felger 9659.* Above SE cove, 100 ft, *Cooper* [*Felger 10400*].

Herissantia crispa (Linnaeus) Brizicky [*Abutilon crispum* Linnaeus]

Description: Mostly short-lived perennials, also facultative annuals. Stems to 1 m long, slender and delicate, trailing, arching, or semi-vining and often growing in shrubs. Herbage and pedicels with long simple hairs in addition to shorter stellate hairs, the hairs velvety. Leaves long petioled on vegetative, non-flowering stems, and sessile and usually much smaller on flower-bearing stems; leaf blades broadly ovate, relatively thin, 3–7 cm long. Flowers pale yellow-orange, solitary in leaf axils, the pedicels jointed with a prominent elbow-bend. Stamens numerous. Fruits globose, inflated, resembling a miniature paper lantern, at maturity quickly falling apart into separate segments (mericarps). Flowering and fruiting non-seasonal, especially during warm, moist weather.

Local Distribution: Along arroyos and canyons, valley floor and bajadas, and rocky slopes; documented from the central and eastern part of Tiburón and on Dátil.

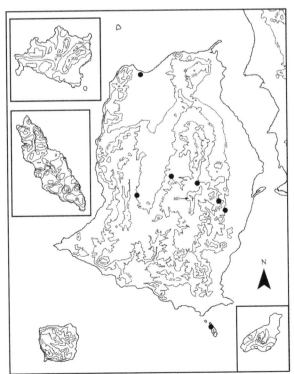

Herissantia crispa

Other Gulf Islands: Tortuga, Coronados, Danzante, San José, Cerralvo.

Geographic Range: Both Baja California states, Arizona to Texas and Florida southward through tropical America and in tropical Asia.

Tiburón: SW Central Valley, along arroyo margin, *Felger 12651.* Haap Hill, *Felger 74-T44.* Upper bajada E of Pazj Hax, *Romero-Morales 07-6.* Downstream of Sopc Hax, arroyo, *Wilder 07-225.* NE base of Sierra Kunkaak, 'bench' above arroyo bottom, *Wilder 06-379.* Agua Dulce Bay, rather scarce, *Moran 13002* (SD, not seen).

Dátil: NW side of island, *Felger 15333.*

Hibiscus

Subshrubs or coarse herbs with showy flowers and silky-haired seeds. Stamens numerous. Fruit a capsule of 5 persistent carpels spreading open and star shaped at maturity, the carpels not falling separately.

1. Leaves deeply 3-lobed to parted (except on juvenile growth); floral bracts conspicuous, more than 1 cm long, longer than the carpels; flowers yellow. __ **H. biseptus**
1' Leaves broadly ovate to obovate, not deeply lobed or parted; floral bractlets inconspicuous, less than 0.5 cm long, shorter than the carpels; flowers white or pink.
————————————————— **H. denudatus**

Hibiscus biseptus S. Watson

Description: Slender perennials, often few-branched to 1+ m tall, and also flowering in the first season. Stems with simple or 2- or 3-rayed hairs, these not appressed, and stems also with two vertical lines of small, recurved hairs extending down from the leaf bases (stipules). Leaves green to reddish green. Flowers yellow with a maroon center, the petals mostly 2.5–3 cm long; flowering during warmer months.

Local Distribution: Canyons, arroyos, and rocky slopes in the Sierra Kunkaak.

Other Gulf Islands: None.

Geographic Range: Widespread in Sonora; Jalisco and Nayarit to Chihuahua and Arizona, and both Baja California states.

Tiburón: E base of Sierra Kunkaak, flat above arroyo dominated by *Ruellia californica, Wilder 06-25.* Sopc Hax, N-facing slope, rock crevice, *Felger 9302.* Foothills at NE base of Sierra Kunkaak, *Felger 07-124.* Siimen Hax, *Wilder 06-446.* San Miguel Peak, *Knight 924, 970, 997* (UNM).

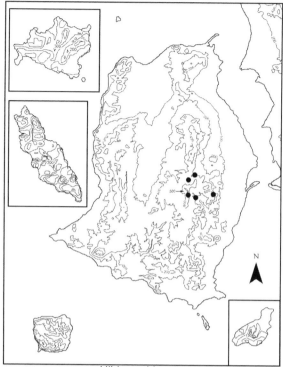

Hibiscus biseptus

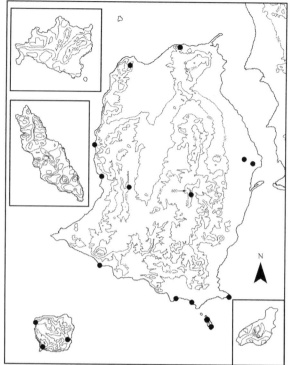

Hibiscus denudatus var. *denudatus*
Hepem ijcóa

Hibiscus denudatus Bentham var. **denudatus**

Hepem ijcoa "white-tailed deer's globe-mallow";
rock hibiscus

Description: Multiple-stem subshrubs; often grazed by animals. Herbage pale yellow-green and densely stellate-pubescent; rainy season leaves greener. Leaves tardily drought deciduous. Corollas white or pale pink with a maroon center, the petals often 2.5+ cm long. Flowering at various seasons, especially spring.

Local Distribution: Widespread and common across Tiburón, San Esteban, and Dátil. On San Esteban the plants are often severely grazed by chuckwallas (*Sauromalus varius*).

Other Gulf Islands: Ángel de la Guarda, Tortuga, San Marcos (observation), Coronados, Carmen, Danzante, Monserrat, Santa Catalina, Santa Cruz, San Diego, San José, San Francisco, Espíritu Santo, Cerralvo.

Geographic Range: Guaymas region northward to southern Arizona, and the Cape Region of Baja California Sur to southeastern California. Another variety ranges from near Tucson and northeastern Sonora to the Chihuahuan Desert.

Tiburón: Ensenada Blanca (Vaporeta), *Felger 6299.* Tordillitos, *Felger 15519.* Ensenada de la Cruz, *Felger 9218.* Coralitos, *Wilder 06-58.* El Monumento, *Felger 15570B.* SW Central

Valley, upper rocky slope of mountain [Sierra Menor] bordering valley, *Felger 12632.* Palo Fierro, *Felger 11072.* 3 km W [of] Punta Tormenta, *Scott P10* (UNM). San Miguel Peak, *Knight 1014* (UNM). Hee Inoohcö, mountains at NE coast, *Felger 07-69.* Hill W of Tecomate, *Felger 6277.* Pooj Iime, *Wilder 07-433.*
San Esteban: Canyon, SW corner of island, *Felger 17637.* Limantour, *Felger 17654.* San Pedro, *Felger 16663.*
Dátil: E side, slightly N of center, halfway up N-facing slope of peak, *Wilder 07-128.* NE side, *Felger 9121.* NW side, *Felger 15318.* SE side, *Felger 17517.*

Horsfordia

Slender and often tall spindly shrubs, densely stellate-pubescent. Stamens numerous. Fruit a schizocarp, the mericarps separating at maturity, the lower chamber indehiscent with firm, wrinkled walls and a single tightly held seed, the upper chamber dehiscent, splitting into expanded, smooth membranous wings.

1. Larger leaves mostly broadly ovate and often cordate at base; petals pink or sometimes nearly white, drying bluish to pale lavender; mericarps 1-seeded. _____
H. alata
1' Larger leaves mostly lanceolate and not cordate at base; petals orange when fresh; mericarps 2- to 3-seeded. _
H. newberryi

Horsfordia alata (S. Watson) A. Gray

Hehe coozlil "slippery plant," caatc ipapl "what grasshoppers are strung with"; pink velvet-mallow

Description: Sparsely branched slender shrubs often 2–3 m tall; often grayish pubescent. Leaves velvety, larger ones 6–10+ cm long, usually broadly ovate, cordate at the base, often thickish, but new-growth summer-fall leaves often thin and scarcely or not at all cordate (uppermost leaves of flowering branches often lanceolate). Petals white or pale pink or pale lavender, drying bluish to pale lavender. Mericarps 1-seeded, the upper chamber forming ovules but these aborting. Flowering at various seasons.

Local Distribution: Widely scattered across Tiburón from low-elevation washes, bajadas, and hills to mountains at middle elevations including Sierra Menor and Sierra Kunkaak. Not known from the westernmost margin of the island.

Other Gulf Islands: Coronados, San José, Espíritu Santo, Cerralvo.

Geographic Range: Sonora south to the vicinity of Tiburón, southeastern California, southwestern Arizona, and both Baja California states.

Tiburón: Tordillitos, *Felger 15474*. Arroyo Sauzal, 2 mi inland, *Felger 12285*. Xapij (Sauzal) waterhole, *Wilder 06-121*.

SW Central Valley, upper rocky slope of mountain [Sierra Menor] bordering valley, 1200–1400 ft, S-facing slope, *Felger 12424*. Zozni Cmiipla, *Wilder 06-357*. Punta Tortuga, bajada 2 mi inland, *Felger 12548*. Top of Sierra Caracol, *Knight 1045, 1049* (UNM). Hahjöaacöl, NE coast, *Wilder 08-216*. Tecomate, *Whiting 9060*.

Horsfordia newberryi (S. Watson) A. Gray

Mariola; orange velvet-mallow

Description: Slender, erect shrubs, usually few branched, 1–2+ m tall. Petals bright orange-yellow; flowering and fruiting at least March–May and October–December. Mericarps 2-chambered, the lower chamber with 1 tightly held seed, the upper chamber with 1 or 2 loose seeds that invariably fall out soon after maturing, thus giving two strategies for seed dispersal.

Local Distribution: Widely distributed on Tiburón, San Esteban, and Dátil; most common and nearly restricted to steep rock slopes, canyon walls, and outcrops. Not documented for the western margin of Tiburón. At higher elevations on San Esteban it often occurs with *Desmanthus fruticosus*, forming clusters of tall, slender shrubs.

Other Gulf Islands: Ángel de la Guarda, San Marcos.

Geographic Range: Widespread in the Sonoran Desert; western Sonora southward to the Guaymas region,

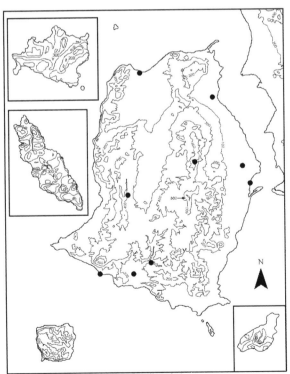

Horsfordia alata
Hehe coozlil, caatc ipapl

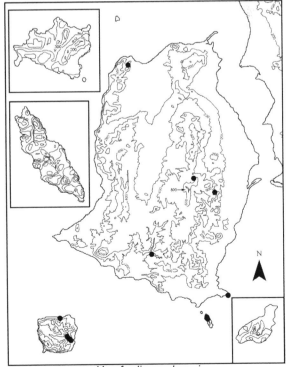

Horsfordia newberryi

both states of Baja California, Arizona, and southeastern California.

> **Tiburón:** Xapij (Sauzal) waterhole, *Wilder 06-122.* El Monumento, *Felger 08-54.* Sopc Hax, not common, *Felger 9320.* Base of Capxölim, head of arroyo, *Wilder 05-31.* Rocky hill, W of Tecomate, *Felger 6278.*
>
> **San Esteban:** Arroyo Limantour, *Montaño-Herrera 08-3.* Hilltops, not common, *Felger 7092.* E-central side of island, canyon bottom, *Felger 12755.* N side of island, rocky slopes, *Felger 15452.* Sandy wash, 4–5 ft high, flowers orange, *Johnston 3177* (CAS).
>
> **Dátil:** Steep rocky slopes, rare at ridge crest summit, *Felger 13448.* NW side of island, *Felger 15343.* E side slightly N of center, S-facing slope, *Wilder 07-134.*

Melochia tomentosa Linnaeus

Hehe coyoco "dove plant"; dove plant

Description: Shrubs to 1.5+ m tall. Leaves pubescent with deeply incised veins. Flowers showy, rose-lavender, appearing at various times of the year including moderately dry seasons. Stamens 5. Fruits of 5-winged capsules.

Local Distribution: Widespread and common on Tiburón, where it is most abundant in the Sierra Kunkaak

but not recorded from the western and northern parts of the island; also on Dátil.

> **Other Gulf Islands:** Tortuga (observation), San Marcos, Coronados, Carmen, Danzante, Santa Cruz, San Diego, San José, San Francisco, Espíritu Santo, Cerralvo.
>
> **Geographic Range:** New World tropics and subtropics.
>
> **Tiburón:** Xapij (Sauzal), *Wilder 08-127.* Ensenada de la Cruz, rocky slopes, *Felger 9216.* Coralitos, small hill above beach, *Wilder 06-64.* El Monumento, *Felger 15568.* Ensenada del Perro, *Felger 17726.* Haap Hill, *Felger T74-21.* Pazj Hax, *Wilder 06-13.* Sopc Hax, from waterhole to upper bajada, *Felger 9326.* Camino de Caracol, broad arroyo entering foothills of Sierra Kunkaak, *Wilder 05-9.* Cerro San Miguel, en el pie de monte y arroyos, *Quijada-Mascareñas 90T006.* Canyon bottom, N base of Sierra Kunkaak, *Wilder 06-419.* Cójocam Iti Yaii, saddle, *Wilder 07-534.*
>
> **Dátil:** NW side of island, *Felger 15322.*

Sida abutifolia Miller [*S. procumbens* Swartz]

Description: Inconspicuous, weakly perennial herbs or facultative annuals, the stems often procumbent; herbage and calyces densely pubescent with short, stellate hairs, and also with simple, spreading hairs except on the leaf blades. Leaf blades 6–30 mm long, generally

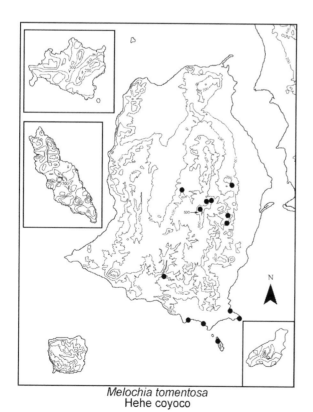

Melochia tomentosa
Hehe coyoco

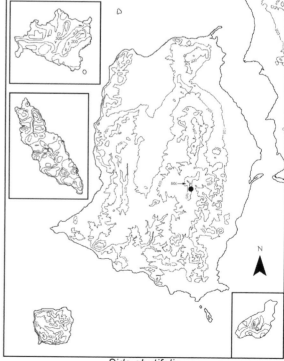

Sida abutifolia

more than twice as long as wide (lanceolate-oblong), cordate at base, the margins crenate; petioles one-half to as long or longer than the blades. Flowers solitary in leaf axils and rather small; calyx lobes ovate, acuminate at the tip; petals pale yellow-orange or perhaps whitish; probably flowering with warm weather and sufficient soil moisture. Stamens numerous. Carpels 5, 2–3 mm long.

Local Distribution: Known from a single collection on Tiburón but expected in canyons elsewhere at higher elevations on the island.

Other Gulf Islands: None.

Geographic Range: Widespread in Sonora except the dryer parts of the desert; not known along the coast north of the Guaymas region mountains. United States from California to Florida and through Mexico to northern South America and the West Indies.

Tiburón: A small canyon below "the bowl" on top of Sierra Kunkaak, 20 Oct 1979, *Knight 985.*

Sphaeralcea – Globemallow

Annual or perennial herbs or shrubs densely pubescent with stellate hairs. Stamens numerous. Fruit a schizocarp, the mericarps 1- to 3-seeded, with a ventral notch; each mericarp differentiated into (1) an upper, dehiscent section, or wings (the carpel walls smooth and spreading apart at maturity, the seed(s), if present, falling away early), and (2) a lower, indehiscent section (the body of the mericarp, reticulate and retaining its seeds); thus there are two strategies for seed dispersal. Seeds kidney shaped.

As with most mallows, leaf measurements and descriptions are for the larger leaves rather than the smaller ones in the upper stems or near the inflorescences. Flowering at various seasons depending on moisture, especially in spring. Taxonomy of this group of mallows is in need of serious study.

1. Mericarps 1-seeded, the dehiscent section (wings) short and stubby, less than half as large as the body. _____
_____ **S. coulteri**
1' Mericarps 2- or 3-seeded, the dehiscent section more than half as long as the body.
　2. Annuals to shrubs to 1.8+ m tall; Mártir._ **S. hainesii**
　2' Perennial subshrubs, less than 1 m tall; various islands but not on Mártir. _____ **S. ambigua**
　　3. Herbage often with whitish or straw-colored hairs. _____ **S. ambigua** var. **ambigua**
　　3' Herbage with yellowish hairs. _____
_____ **S. ambigua** var. **versicolor**

Sphaeralcea ambigua A. Gray var. **ambigua**

Jcoa; *mal de ojo*; desert globemallow

Description: Bushy perennials, often globose. Leaf blades mostly 1.8–5 cm long, more or less ovate, about as wide as long, often more or less 3-lobed, thinner and larger during more favorable conditions and thicker, smaller, and often wrinkled during drier conditions, the margins variously toothed to wavy. Petals orange, 2–3 cm long; anthers white. Mericarps 2-seeded, the upper, dehiscent section larger than the body.

Local Distribution: Documented from the western and northern sides of Tiburón and on Dátil.

Other Gulf Islands: San José.

Geographic Range: This highly variable species is the most widespread perennial globemallow in the Sonoran Desert, ranging from southwestern Utah and southern Nevada to southeastern California, Baja California Sur, and northwestern Sonora southward to the vicinity of Punta Baja (between Bahía Kino and Tastiota). Five often indistinct varieties or subspecies are generally recognized; var. *ambigua* is the most widespread one.

Tiburón: Ensenada Blanca (Vaporeta), ½ mi inland, S-facing rocky hillside, *Felger 12225.* Just N of Punta Willard, *Wiggins 17118* (CAS). Hee Inoohcö, mountains at

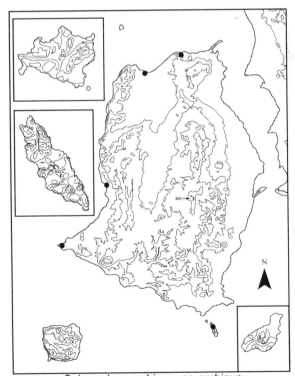

Sphaeralcea ambigua var. *ambigua*
Jcoa

NE coast, *Felger 07-72*. Tecomate, one plant on edge of dune, *Whiting 9045*.

Dátil: Steep slopes of E-central side of island, *Wilder 07-109*.

Sphaeralcea ambigua var. **versicolor** (Kearney) Kearney

Jcoa; *mal de ojo*; globemallow

Description: Differing from var. *ambigua* in having woodier stems, often smaller leaf blades characterized by being thick, wrinkled, yellow pubescent, and bi-colored.

Local Distribution: Widely distributed on San Esteban, especially along arroyos and canyons.

Other Gulf Islands: Ángel de la Guarda (type locality of var. *versicolor*).

Geographic Range: This variety also occurs on Baja California Norte and portions of the coast of northwestern Sonora.

San Esteban: Narrow canyon, behind cove at N side of island, *Felger 17588*. Large arroyo, SW [SE] corner of island, 50 m, *Hastings 71-53*. Arroyo Limantour, 1 mi inland, *Wilder 07-41*. Near S end, in large arroyo running inland from E side [Arroyo Limantour], *Wiggins 17223* (CAS).

Sphaeralcea coulteri (S. Watson) A. Gray [*S. orcuttii* Rose]

Mal de ojo; annual globemallow

Description: On Alcatraz as cool season annuals and probably also growing with summer rains, with a well-developed taproot, to short-lived perennials to ca. 1.6 m tall. Leaf blades greenish to gray. Flowers opening in early morning, the petals orange. Mericarps about as long as wide, 2–2.7 mm long, the dehiscent section much smaller than the body; each mericarp has a single seed.

The mericarps of *S. coulteri* and *S. orcuttii* are indistinguishable, the only distinguishing feature being the much larger-sized plants in the sand-adapted *S. orcuttii*.

Local Distribution: Seasonally one of the most abundant plants across the sandy flats and lowland areas of Alcatraz.

Other Gulf Islands: None.

Geographic Range: The annual form of *S. coulteri* is one of the most common spring wildflowers in the Sonoran Desert north of the Seri region. It is often found on sandy soils, as well as on fine-textured, poorly drained

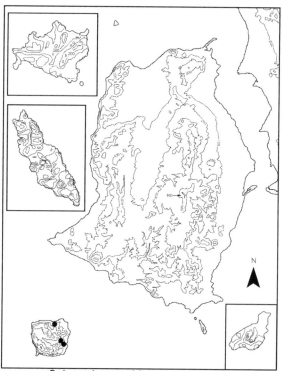

Sphaeralcea ambigua var. *versicolor*
Jcoa

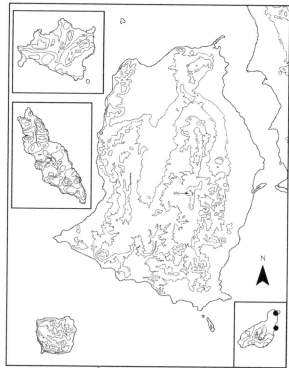

Sphaeralcea coulteri

soils. Southeastern California to Baja California Sur, and southern Arizona to Mazatlán, Sinaloa. Populations on Isla Alcatraz and from the vicinity of Bahía Kino to Tastiota are shrubby and resemble *S. orcuttii* in size.

   Alcatraz: E side, common on sandy flat, *Felger 14926.* S-central side, *Felger 07-161.*

Sphaeralcea hainesii Brandegee

   Description: Generally robust, perennial shrubs to 1.8+ m tall, probably not long-lived; also flowering in the first season as an ephemeral, these with a well-developed taproot. Flowers orange; non-seasonal. Mericarps 2- or 3-seeded, the wings larger than the carpel body.

   Local Distribution: One of the most widespread and abundant plants on Mártir, including low elevations in guano-white soils, where it is one of the pioneer plants. The plants are unusual in the genus due to their often large, shrub-sized stature and often relatively large and somewhat thick leaves. Its relationship to other Sonoran Desert globemallows warrants investigation.

   Other Gulf Islands: Ángel de la Guarda, San Lorenzo, Tortuga, San Marcos, Carmen, Danzante, Monserrat, San José.

   Geographic Range: Gulf of California islands and both Baja California states.

   Mártir: Stony ravines, Oct 1887, *Palmer 404, 405* (GH, images). Most abundant perennial on island, 18 Apr 1921, *Johnston 3145* (CAS, UC). S end, small landing below guano terraces, 100 ft, *Kipping 20 Jun 1975* (CAS). Summit, 300 m, one seen with white flowers, *Moran 8813* (CAS). Summit, 290 m, abundant, *Felger 07-198.*

Waltheria indica Linnaeus [*W. americana* Linnaeus]

   Description: Openly branched subshrub perennials, with stellate hairs. Stems slender, straight and few-branched. Leaves drought deciduous. Flowers in axillary clusters, probably 5–6 mm long, bright yellow. Stamens 5. Capsules 2.5–3 mm long, 1-seeded.

   Local Distribution: Known from Tiburón by two records from the east side of the front range of Sierra Kunkaak.

   Other Gulf Islands: Cerralvo.

   Geographic Range: Tropical and subtropical regions of the world, probably native to the Americas.

   Tiburón: Vicinity of Sopc Hax, *Romero-Morales 07-18.* Small shrub, just below the bowl, 26 Oct 1979, *Knight 999.*

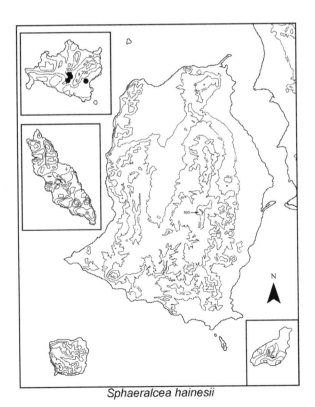

Sphaeralcea hainesii

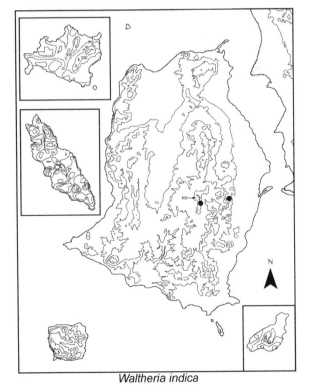

Waltheria indica

Martyniaceae – Devil's Claw Family

Proboscidea altheifolia (Bentham) Decaisne

Xonj; *cuernitos, torito*; desert unicorn plant, devil's claw

Description: Perennials from a large, thick, deeply set tuberous root, the shoots emerging about when the summer rains begin or with warm weather in spring and wither with post-summer drought. Leaves opposite below, alternate above, often 6–15 cm long, the blades broadly ovate to orbicular or kidney shaped, and shallowly lobed. Flowers 4 cm long, the corollas bright yellow inside the tube and on lobes, with purple-brown speckles and dark yellow-orange nectar guides, the tube often bronze colored. Fruits unique, persisting woody capsules with an intricately sculptured body 4–6 cm long, and two woody, curved claws 9–14 cm long (the claws hook onto an animal's leg, effecting dispersal). Seeds 6–9 mm long, black, and warty.

Local Distribution: Known from widely scattered localities on Tiburón—from the eastern bajada, foothills of the Sierra Kunkaak, and the northern part of the island. René Montaño-Herrera (personal communication, 2007) said this plant occurs on Hast Cacöla, the high front range of the Sierra Kunkaak. It is common on the opposite Sonoran mainland.

Other Gulf Islands: San José, San Francisco, Cerralvo.

Geographic Range: Southeastern California to Baja California Sur, to western Texas and through much of Sonora to Sinaloa, and disjunct in Peru.

Figure 3.117. *Proboscidea altheifolia.* Caracol, Isla Tiburón. HRM, 5 September 2008.

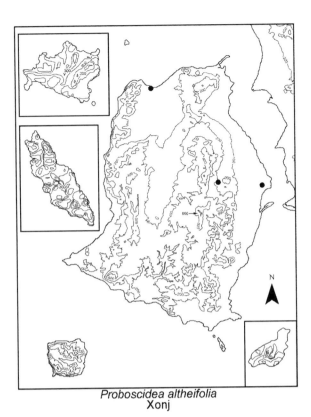

Proboscidea altheifolia
Xonj

Tiburón: Palo Fierro, 2 km W of shore, not common, small arroyo in *Frankenia* zone, *Felger 10321*. Caracol, *Romero-Morales 08-15*. Tecomate, 10 Aug 1964, *Felger 10204* (specimen not located).

Molluginaceae – Carpetweed Family

Mollugo

Delicate summer-fall annuals; glabrous and with thread-like stems. First leaves in a basal rosette; these plants are unusual among hot-weather annuals (ephemerals) in having a basal rosette of leaves. Sepals 5; petals none. Fruits of small capsules.

1. Leaves linear, stems erect-ascending; seeds with a net-like pattern, and of a single, dull color. __ **M. cerviana**
1' Leaves spatulate (broader above the middle), the stems erect-ascending to mostly prostrate-spreading; seeds ridged on back and sides, the ridges blackish or very dark brown, the rest of the seed shiny and brown. __
M. verticillata

*Mollugo cerviana (Linnaeus) Séringe

Hant iit "land's lice"; thread-stem carpetweed, Indian chickweed

Description: Stems 4–12 cm long, ascending to spreading, thread-like, often orange. First leaves mostly 3–12 mm long, the upper leaves often opposite; leaves somewhat glaucous, linear to linear-spatulate. Flowers inconspicuous, sepals 1.2–1.7 mm long, green with membranous margins. Seeds 0.3–0.4 mm wide, minutely reticulate with faint striae (low, thin lines) on the dorsal side, and uniformly dull, red-brown.

Local Distribution: Tiburón, gravelly desert flats and sandy soils at widely separated lowland localities.

Other Gulf Islands: Santa Cruz, Espíritu Santo.

Geographic Range: Widespread in the Sonoran Desert; southward in Sonora to the Guaymas region. Widely distributed in the New World; reported as introduced from the Old World.

Tiburón: Coralitos, shore, *Wilkinson 203*. SW part of Central Valley, 1 Feb 1965, *Felger 12310* (specimen not located). Zozni Cmiipla, *Felger 08-117*. 3 km W of Punta Tormenta, *Smartt 13 Oct 1976* (UNM). Bahía Agua Dulce, locally common, *Felger 8879*.

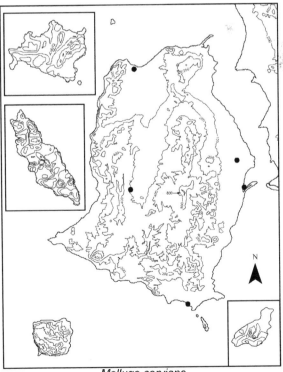

Mollugo cerviana
Hant iit

Mollugo verticillata Linnaeus

Carpetweed

Description: Leaves small and spatulate. Flowers very small, green and white. Seeds 0.54–0.63 mm wide, kidney shaped, with conspicuous snail-shaped dark ridges (striae) on their flattish sides, the ridges dark brown and areas in between (sulci) lighter brown. The plants resemble those of *M. cerviana* but are usually somewhat larger.

Local Distribution: Nolasco, seasonally common on the east side of the island.

Other Gulf Islands: San Diego, San José, Espíritu Santo.

Geographic Range: Western Sonora, from near Tastiota and the Guaymas region southward; common in subtropical scrub regions southward and eastward from the desert. Nearly worldwide; probably native in the New World.

Nolasco: E side, common on rocky slopes, *Felger 9652.*

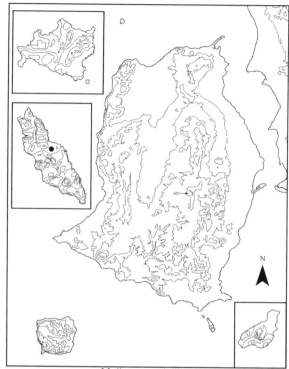

Mollugo verticillata

Figure 3.118. *Ficus petiolaris* subsp. *palmeri*. Tree in foothills of Sierra Kunkaak, Isla Tiburón, 24 May 2006. Branch with leaves and figs, Isla San Pedro Nolasco, 11 November 2009. BTW.

Moraceae – Mulberry Family

Ficus petiolaris Kunth subsp. **palmeri** (S. Watson) Felger & Lowe [*Ficus palmeri* S. Watson, Proccedings of the American Academy of Arts and Sciences 24:77, 1898]

Xpaasni; *tescalama*; cliff fig

Description: Shrubs or small trees, and occasionally large trees on Tiburón. Root and stem bark whitish. Twigs, leaves, and figs densely to sparsely pubescent, or sometimes glabrate. Leaves tardily drought deciduous (often leafless during extended drought), often 6–28 cm long, the blades often relatively firm, broadly ovate or less often oval, the base moderately cordate or not; stipules often 15–27 × 6–8 mm. The developing leaf and growth bud is enclosed by a pair of these relatively large, bract-like broad stipules that fall away as the new leaf expands, the fallen stipules leaving a ring-like scar on the twig; one stipule in the pair often partially to fully encloses the other.

This is the largest tree on the Gulf Islands, and the leaves are notably large in comparison to most other Sonoran Desert trees and shrubs. Occasional trees in the interior of Tiburón reach 10 m, and one was estimated to be more than 15 m tall. The larger trees produce a deep litter of dry leaves.

Figs 1.6–1.8 cm diameter, globose-obovoid, paired or single if one fails to develop or falls off, subtended by 2 scales or sometimes appearing as 3 due to splitting of 1 scale as the fig develops. Short, broad, thick, overlapping, and ascending reddish scales usually obscure a terminal pore—the ostiole. The figs are eaten by the Comcaac, fresh or usually cooked (Felger & Moser 1985:348).

This unique tree grows on sea cliffs, often high above the sea or sometimes in the reach of sea spray, sheer canyon walls, and mountain rock. The seedling begins life usually in a rock crevice, and soon forms numerous adventitious roots that grasp the rock and cascade over the surface, these roots fuse (coalesce), and if they reach the canyon floor or moist soil the plant may develop into a tree; otherwise, it remains dwarfed as a shrub. The fused adventitious roots may also form part of the trunk; these fused roots appear as if they had been poured molten over the rock surfaces. The fact that it can become established on exposed rock faces in the desert indicates adaptations that set it apart from typical woody trees.

Local Distribution: Xpaasni is on most Gulf Islands, including Tiburón, San Esteban, Dátil, Alcatraz, Cholludo, Nolasco, and Mártir; on these islands it extends from near sea level to peak elevations. Tiburón is the only island where *F. petiolaris* occasionally reaches large sizes, especially in the canyons of Sierra Kunkaak. On San Esteban *F. petiolaris* is widely scattered and generally restricted to steep cliffs, where it usually remains a facultatively dwarfed shrub and only rarely becomes a small tree. On Dátil it is scattered on cliff faces in the interior and coast, and a few are thick-trunked small trees—on 7 April 1958, Richard went around the island by boat and counted 132 fig plants on sea cliffs and estimated that the total island population might be 200 plants. *Ficus* on Cholludo hangs on a sea cliff inhabited by hoards of noisy seabirds. On Nolasco it is especially common near the crest on the east-facing side of the island but also is widely scattered on cliffs elsewhere. On Mártir one finds robust large shrubs scattered on high sea cliffs and high escarpments at the edge of the island. A single *Ficus* shrub was seen on Alcatraz.

Other Gulf Islands: Ángel de la Guarda, San Lorenzo, Tortuga, San Marcos, San Ildefonso, Carmen, Danzante, Monserrat, Santa Catalina, Santa Cruz, San Diego, San José, San Francisco, Espíritu Santo, Cerralvo.

Geographic Range: The *F. petiolaris* species complex ranges from Sonora and Baja California Sur to Oaxaca (Piedra-Malagón et al. 2011). Within this complex, subsp. *palmeri* is endemic to the Sonoran Desert: western Sonora from the Guaymas region northward to the Sierra Seri and Cerros Anacoretas southwest of Caborca, the islands, and the Baja California Peninsula.

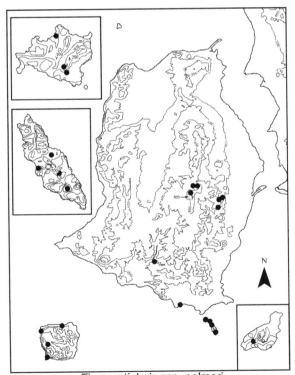

Ficus petiolaris ssp. palmeri
Xpaasni

Ficus petiolaris subsp. *palmeri* is the only fig that truly ranges into the Sonoran Desert. This subspecies and subsp. *petiolaris* are the most widespread figs in Sonora, distinguished by their habit and unusual root system, large, broadly ovate or less often oval leaves, and large and broad stipules (Felger et al. 2001). Subsp. *palmeri* tends to be more densely pubescent, have tougher, more oval leaves with yellowish green veins, while subsp. *petiolaris* tends to more often be glabrate or glabrous (except a tuft of hairs at the base of the leaf blade), and have thinner, cordate-ovate leaves with pink veins (especially in Sonora). Subsp. *petiolaris* is widespread in Mexico beyond the desert. Intermediate forms occur at the desert edge in the Sierra el Aguaje north of the San Carlos-Guaymas region and the southern portion of the Sierra Seri northeast of Bahía Kino: in the more arid, exposed habitats the plants have the aspect of the *palmeri* taxon, while in more favorable habitats, such as riparian canyons, they resemble the *petiolaris* taxon. Sometimes the more exposed branches have characteristics of *palmeri* while shaded branches bear leaves like those of *petiolaris*.

Tiburón: Xapij (Sauzal), *Wilder 08-125*. Ensenada de la Cruz, rock hill at beach, *Felger 9203*. Foothills of Sierra Kunkaak, SW side of mountain, 8 km W of shore, tree about 15 m tall (largest tree seen on island), on rock outcropping in middle of large canyon bottom, *Felger 6951*. [Tinaja Anita, 23 Dec 1895, see "Botanical Explorations" and "Gazetteer"], McGee expedition, *McGee 11* (UC). Near Pazj Hax, *Romero-Morales 07-1*. Carrizo Canyon, *Knight 1102* (UNM). Base of Capxölim, head of arroyo, *Wilder 05-23*. Siimen Hax, *Ezcurra 07-1*. Deep canyons in upper reaches of Sierra Kunkaak, 750 m, 26 Oct 2007, *Wilder*, observation.

San Esteban: NE corner of island, near beach on sea cliffs, 5 m tall, *Felger 9192*. NE shore of island, *Felger 479*. SW part of island, E-facing side canyon with dense vegetation, rocks on side of canyon, *Wilder 07-89*. La Freidera, 21 Dec 1966, *Felger*, observation.

Dátil: NE side: Arroyo bottom on N side of large rocks, *Felger 9083*; Scattered, steep rocky slopes, *Felger 13458*. Rescue Beach, NE side, small tree 3 m tall, 4+ m wide, base of cliff, 5 m elev., *Felger 07-199*. SE side, steep rocky sea cliff, *Felger 2560*.

Alcatraz: A single shrub ca. 1.5 m wide, wider than tall, clinging to a red rock cliff in a vertical crack in the rock face, N side of highest peak at the N side of island nearest the sea, 28 Jan 2008, *Felger & Wilder*, observation.

Cholludo: NW side, not common, clinging to rock cliff at a small cave, *Felger 13413*. N-central slope, *Wilder 08-317*.

Mártir: 1887, *Palmer 413* (type collection, US 796147, image; isotype, UC 11624). In a rock-hewn draw in mid-part of island, *Johnston 3153* (CAS, image). Shrub or small tree, 3–4 m high,

on cliffs and large rocks along ridge at N side of island near the top, 200+ m, population estimated at 100 mature plants, *Felger 6354* (2 sheets). Small canyon at base of SE-facing slope that leads to island summit, large shrub, 2.5 m tall and twice as wide, widely scattered, growing on rock, *Felger 07-17*.

Nolasco: NE side, small tree, on rocks in canyon bottom, *Felger 11439*. Growing from rock on a N slope near crest, *Johnston 3139* (CAS). [Above SE cove] E exposure, 50 ft below crest, 800 ft, *Pulliam & Rosenzweig 21 Mar 1974*. E-central side, N exposure below crest of island, 270 m, tree 5 m tall × 6 m wide, scattered trees of similar size on cliff faces below ridge, *Felger 06-101*. Cañón de la Guacamaya, 20 m elev., 14/15 Apr 2003, *Gallo-Reynoso* (photo).

Myrtaceae – Myrtle Family

Eucalyptus camaldulensis Dehnhardt

Hehe hataap iic cöihiipe "plant that cures a cold"; *eucalipto*; Murray red gum, eucalyptus

Description: Trees with relatively large, simple, somewhat tough, and falcate leaves. Flowers white.

Local Distribution: Biologists from Fauna Silvestre planted a small grove of tamarisk and eucalyptus trees at the military station at Tecomate in the late 1970s. Some of the eucalyptus were still surviving in 2007 but not as

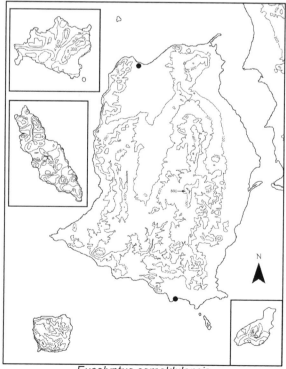

Eucalyptus camaldulensis
Hehe atap iic cöihipe

vigorously as the tamarisk trees (*Tamarix aphylla*). A few eucalyptus trees were also planted at the marine station at Ensenada de la Cruz on the south part of the island, presumably also in the late 1970s. In 2009 two of the trees were still present but not healthy and not producing much shade.

Other Gulf Islands: None.

Geographic Range: Native to Australia. The Murray red gum is the most commonly grown eucalyptus in the Sonoran Desert region and the most wide-ranging *Eucalyptus* species in Australia. *Eucalyptus* has not naturalized in Sonora or Arizona.

Tiburón: Ensenada de la Cruz, *Wilder 09-46*. Tecomate, about 6 trees, 5 m tall, *Wilder 07-264*.

Nolinaceae, *see* **Asparagaceae**

Nyctaginaceae – Four O'Clock Family

Annuals to perennials. Leaves simple, opposite (sometimes subopposite in *Boerhavia*); stipules none. Calyx corolla-like; petals none. Fruits 1-seeded and indehiscent, resembling an achene, enclosed in the base of the calyx tube, the collective structure called an anthocarp and referred to here as the fruit.

Figure 3.119. *Boerhavia triquetra*, racemose inflorescence, and umbellate cluster of *B. xanti* at right center. RSF.

Figure 3.120. *Abronia maritima* subsp. *maritima*. Punta Perla, Isla Tiburón, 9 November 2008; flower from Isla Alcatraz, 16 September 2007. BTW.

1. Involucral bracts united into a persistent tube with 5 teeth; fruits rounded or nearly so, hard and nearly smooth. ___ **Mirabilis**

1' Involucral bracts separate, 1–5, sometimes reduced and/or soon deciduous; fruits variously winged, angled, grooved, or ornamented, not rounded and smooth.

  2. Coastal plants, markedly succulent; fruits 5-winged, each wing as wide as or wider than the fruit body; stigmas longer than wide (fusiform). ___ **Abronia**

  2' Plants not succulent; fruits not winged, or the wings narrower than the fruit body, or the wings 2 and inrolled (*Allionia*); stigmas as wide as long (capitate or peltate).

    3. Stems weak, prostrate or trailing; flowers in clusters of 3, the cluster resembling a single flower; fruits oval-ellipsoid and strongly convex as view from the "back" and with a single deep cavity or groove formed by a pair of inrolled and toothed wings. ___ **Allionia**

    3' Stems erect to spreading, sometimes decumbent but not trailing; flowers often clustered but each flower clearly separate; fruits more or less club shaped, not grooved or with 3–5 grooves.

      4. Ephemeral annuals, growing mostly with summer-fall rains; stems and leaves glandular-sticky; perianth pink or white; fruits 2–3 mm long, not glandular. ___ **Boerhavia**

      4' Perennials, stems, and leaves essentially glabrous and not sticky; perianth yellow-green; fruits 8–10 mm long, with large sticky glands. ___ **Commicarpus**

Abronia maritima Nuttall ex S. Watson subsp. **maritima**

Spitj cmajiic "female coastal saltbush"; *alfombrilla playera*; coastal sand verbena

Description: Dense, mat-forming perennials; stems and leaves thick and highly succulent; stems trailing and often buried with sand sticking to glandular hairs. Flowers small (7–8 mm wide), bright purple-magenta, in densely flowered clusters. Fruits sculptured and winged, in rounded clusters. Flowering non-seasonally.

Local Distribution: Upper beaches and beach dunes on Tiburón, widely scattered and often localized; documented on the southwestern, northeastern, and northern shores but absent from many seemingly suitable beach dunes. Also on beach sand on Alcatraz. Common on the opposite Sonoran coast.

Other Gulf Islands: Ángel de la Guarda, Salsipuedes, San Lorenzo (observation), Carmen (observation), Monserrat, Santa Catalina, San Diego (observation), San José, San Francisco, Espíritu Santo, Cerralvo.

Geographic Range: This species occurs on sandy seashores. Subsp. *maritima* ranges from southern California to Nayarit, the Islas Tres Marías, and the Gulf of California. Johnson (1978) distinguished two populations of subsp. *maritima*. The "Gulf" population occurs along the Gulf of California and southward and has a relatively shorter, bright purple-pink to purple-magenta perianth with non-reflexed lobes, and smaller, crenately lobed leaves. The "Pacific" population occurs along the Pacific Coast of the Baja California Peninsula and southern California. Subsp. *capensis* A.F. Johnson occurs in the Cape Region of Baja California Sur.

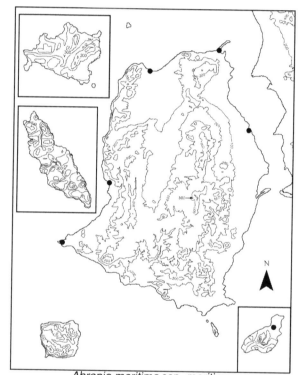

Abronia maritima ssp. *maritima*
Spitj cmajiic

Tiburón: Ensenada Blanca (Vaporeta), *Felger 14953*. Willard's Point, canyon just N, dunes near beach, *Moran 8748* (SD). Vicinity of Valle de Águila, sandy beach, *Wilder 07-272*. Punta Perla, estero, *Wilder 08-372*. Bahía Agua Dulce, high beach dunes (*Felger 1966:253*).

Alcatraz: E side of island, sandy beach, *Felger 14909*.

Allionia incarnata Linnaeus var. **incarnata** [*A. incarnata* var. *nuda* (Standley) Munz]

Hamíp cmaam "female spiderling (*Boerhavia*)"; trailing four o'clock, windmills

Description: Ephemerals, annuals, or short-lived perennial herbs with a stout taproot, dying back to the roots during drought. Plants glandular hairy and sticky viscid, often with sand sticking to the herbage. Stems slender, forked, weak and trailing, sometimes reaching more than 1 m long, although usually much shorter. Leaves petioled, variable, the larger ones (1.7) 2.5–7.5 cm long, the blades usually ovate; opposite leaves of a pair unequal in size. Flowers lavender-pink, opening in the early morning and collapsing with daytime heat; individual flowers bilateral, in 3's, each forming a wedge-shaped one-third of the cluster, the trios axillary on slender peduncles—the trio has the appearance of a single flower and apparently highly variable in size depending on moisture and temperature. Fruits 3–4.8 mm long, tan colored, oblong, one side smooth, the other side with a deep cavity between toothed wings; fruits exude copious mucilage when wet and adhere tenaciously after drying. The fruits are unique among Sonoran Desert nyctages in being bilaterally rather than radially symmetrical. With magnification the fruit structure is spectacular—the glands inside the cavity are often relatively large, whitish, and translucent. Flowering response non-seasonal; germinating mostly with summer-fall rains.

Local Distribution: Common and widely distributed on Dátil, and on Tiburón from low elevations to the interior of the Sierra Kunkaak. Sandy arroyos, canyons, bajadas and valley plains, and rocky slopes; often extending onto harsh, xeric habitats including nearly barren south-facing slopes or low ridges.

Other Gulf Islands: Ángel de la Guarda, San Lorenzo, Tortuga, San Marcos, San Ildefonso, Carmen, Danzante, Monserrat, Santa Catalina, Santa Cruz, San Diego, San José, San Francisco, Espíritu Santo, Cerralvo.

Geographic Range: Southwestern United States to South America and the West Indies. Var. *villosa* (Standley) Munz differs only in having larger flowers and supposedly slightly larger fruits, and is not known from the lowland regions of the Gulf of California.

Tiburón: Ensenada Blanca (Vaporeta), *Felger 12224*. Tordillitos, *Felger 15467*. Coralitos, *Felger 15392*. Vicinity El Monumento, *Felger 15551*. Haap Hill, *Felger T74-27* (specimen discarded). Canyon bottom, N base of Sierra Kunkaak, *Wilder 06-398*. Siimen Hax, *Wilder 06-450*.

Dátil: NE side of island, steep rocky slopes, *Felger 9120*. NW side, *Felger 15314*. E side, slightly N of center, halfway up N-facing peak, in dense vegetation, *Wilder 07-122*.

Boerhavia

Hot-weather ephemerals, sometimes persisting until or even through winter; stems slender and glandular pubescent. Leaves opposite or subopposite, the opposite leaves of a pair often unequal in size; leaves usually glandular sticky. Inflorescences of slender and usually branched stems, the flowers subtended by small bracts. Perianth white or pink. Fruits obovoid or obpyramidal, usually angled and/or grooved (furrows or sulci), fruits glabrous, not sticky when dry (those in the flora area), the angles forming short wings, these wings and fruit surfaces melt into copious slime when wet and tenaciously adhere upon drying.

Flowers opening at sunrise and collapsing with morning/daytime heat. At such time the stamens collapse onto the stigma, apparently self-fertilizing the flower if it has not been cross-pollinated. At least those with larger flowers produce nectar and are insect pollinated. The boerhavias as a whole, especially those with smaller flowers, are probably capable of selfing and the smaller-flowered ones may be entirely selfing (Spellenberg 2000, 2003).

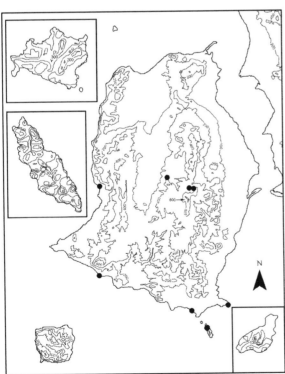

Allionia incarnata var. *incarnata*
Hamíp cmaam

1. Flowers in umbellate or sub-umbellate clusters, or some-
    times reduced to a single flower. _____**B. triquetra**
1' Flowers on elongated racemose branches. __ **B. xanti**

Boerhavia triquetra S. Watson [*B. intermedia* M.E.
Jones. *B. triquetra* var. *intermedia* (M.E. Jones) Spellenberg. *B. maculata* Standley]

Hamíp caacöl "large spiderling"; *juanamipili, juantilipín*; spiderling

Description: Leaves oblong to narrowly lanceolate,
1–3 cm long, paler beneath, often glandular-punctate.
Inflorescence branches slender. Flowers 2 to several or
more in umbellate or sub-umbellate clusters or the flowers sometimes solitary at the branch tips especially in
drought. Fruits 2–2.8 mm long, 3–5 angled, the angles or
ridges acute, broad, and smooth, the intervening grooves
(sulci) open, coarsely and deeply rugose, the fruit tip
blunt (truncated).

Local Distribution: Seasonally abundant and widespread in many habitats on Tiburón, San Esteban, Dátil,
and Nolasco. In many locations it is the most common
summer-fall ephemeral.

Among the Nolasco population the fruits are often
solitary at branch tips, 2.4–2.8 m long and with 5 angles.

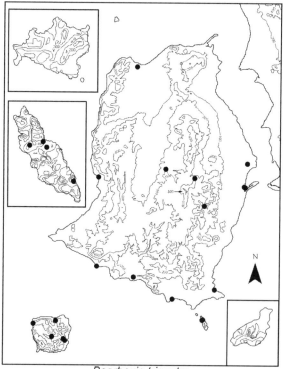

Boerhavia triquetra
Hamíp caacöl

It is seasonally one of the most common and widespread
annuals on the island. On steep slopes at Cañon el Farito,
February 2008, we found abundant gray mats of old fruits
(anthocarps) cemented together, due to dried mucilage,
covering substantial areas between rocks—this being a
significant factor holding the soil together.

Other Gulf Islands: Ángel de la Guarda, Tortuga
(*Flores 497*), San Ildefonso, San Marcos, Carmen, Mon-
serrat, Santa Catalina, Santa Cruz, San Jose, Epíritu
Santo, Cerralvo, Carmen.

Geographic Range: Widespread in the Mojave,
Chihuahuan, and Sonoran Deserts including regions
around the Gulf of California and Gulf Islands; southwestern United States and northwestern Mexico. Two
varieties are described but are only weakly segregated
and of doubtful significance (Richard Spellenberg, personal communication, 2 Jun 2008; also see Spellenberg
2007).

Tiburón: Ensenada Blanca, *Felger 14973*. Arroyo Sauzal, *Felger 2753*. Ensenada de la Cruz, *Felger 9229*. Tordillitos, *Felger
15492*. Ensenada del Perro, *Wilder 08-306*. Haap Hill, *Felger
T74-5*. Zozni Cmiipla, *Wilder 06-280, 06-359*. Palo Fierro, *Felger 8916*. Large arroyo, E base of the Sierra Kunkaak, *Wilder 06-
19*. Top of Sierra Kunkaak Segundo, E peak of Sierra Kunkaak,
490 m, *Wilder 06-485*. Bahía Agua Dulce, *Felger 8886*.

San Esteban: Arroyo Limantour, *Felger 12730*. Arroyo Limantour, 2.25 mi inland, *Wilder 07-69*. [Arroyo Limantour],
13 Oct 1977, *Wilkinson 155*. N side of island, arroyo, *Felger
15400*. W side, rocky slopes, 25 m, *Felger 16657*.

Dátil: NE side of island, *Felger 9108* (ARIZ; SD, determined
by R. Spellenberg). NW side, *Felger 15323* (ARIZ; SD, determined by R. Spellenberg).

Nolasco: SE side, *Maya 1 Nov 1963*. E-central side, N exposure below crest of island, 270 m, abundant in open areas
with little or no *Vaseyanthus*, seen at all elevations and exposures, *Felger 06-107*. N-facing slope above Cañón el Farito,
dense mats of dried anthocarps adhered together in areas
between rocks, no living plants seen, 2 Feb 2008, *Wilder
08-172a*. Cañón de Mellink, *Felger 08-146*.

Boerhavia xanti S. Watson, 1889

Hamíp; *juanamipili, juantilipín*; spiderling

Description: Flowers in loose, slender, interrupted,
raceme-like spikes. Flowers (perianth) white or pink,
opening wide and reaching at least 5 mm in width shortly
after sunrise and smaller when drought-stressed. Fruits
narrowly obovoid, 2.5–3 mm long, 5-angled, ridges

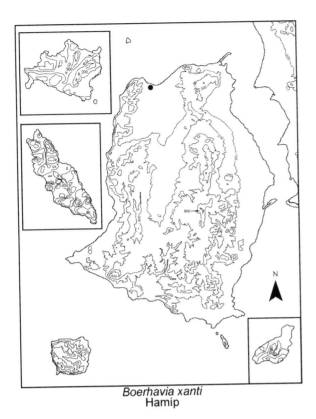

Boerhavia xanti
Hamíp

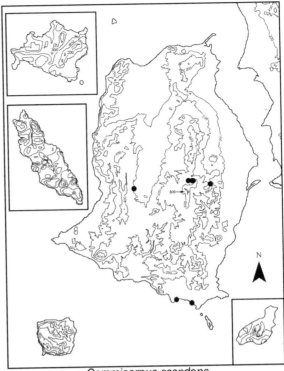

Commicarpus scandens

broad, smooth, obtuse, the grooves (sulci) closed or nearly closed, and rugulose.

Local Distribution: Tiburón, at least along the eastern and northern parts of the island.

Other Gulf Islands: San José, Carmen, Santa Catalina.

Geographic Range: Sonora, Sinaloa, both states of Baja California, and Gulf Islands. *Boerhavia xanti* is indistinguishable from *B. spicata* except for flower size—*B. xanti* has a substantially larger perianth than *B. spicata* Choisy (described in 1849). Throughout its range *B. xanti* is found at low elevations and is replaced at higher elevations by *B. spicata*. Since fresh or freshly pressed flowers are not always available, identification can be problematic.

Tiburón: Bahía Agua Dulce, abundant in dunes, *Felger 8867*.

Commicarpus scandens (Linnaeus) Standley

Description: Perennials, commonly growing through shrubs, often 1–2.3 m tall; glabrous or glabrate. Stems long, slender, and brittle, with unusually long internodes. Leaves often 3–7.5 cm long, semi-succulent. Flowers 7–8 mm wide, pale yellow-green, in umbellate clusters. Fruits 1 cm long, club shaped, 10-ribbed, terete or nearly so, with large, viscid, knobby glands that cause the fruit

to stick to clothing, feathers, or fur. The fruits produce relatively small amounts of mucilage when wet. Flowering non-seasonally but especially with summer rains.

Local Distribution: Widespread on Tiburón including valleys, arroyos, canyons, and mountain slopes; especially prevalent in canyons of the Sierra Kunkaak.

Other Gulf Islands: San José, Cerralvo.

Geographic Range: Arizona to Texas and Mexico to northern South America and the West Indies.

Tiburón: Ensenada de la Cruz, low hillslope, 23 Oct 1963, *Felger 9200* (specimen not located). Coralitos, 21 Dec 1966, *Felger 15370* (specimen not located). SW Central Valley, arroyo, 1 Feb 1965, *Felger 12358* (specimen not located; also see Felger 1966:239). Hast Cacöla, E flank of Sierra Kunkaak, 380 m, *Wilder 08-256*. Canyon bottom, N base of Sierra Kunkaak, 395 m, *Wilder 06-400, 06-436*. Siimen Hax, *Wilder 06-500*.

Mirabilis – Four o'clock

Herbaceous perennials, glandular pubescent; the stems forking. Leaves entire, the lower ones petioled, the upper ones sessile. Involucral bracts united, calyx-like, 5-lobed; 1 to many flowers per involucre; involucres often clustered at stem tips. Calyx petal-like, longer than the involucre. Flowers collapsing with daytime heat. "Fruits" seed-like, not winged.

1. Involucres 6–9 mm long, the lobes 2–3 (4) mm long.
    _____ **M. laevis**
1' Involucres 10–15 mm long, the lobes 5–9 mm long.
    _____**M. tenuiloba**

Mirabilis laevis (Bentham) Curran var. **crassifolia** (Choisy) Spellenberg

Desert four o'clock

Description: Stems erect to spreading, slender, dying back to rootstock during drought. Leaves semisucculent, larger ones 2–3+ cm long including petioles, the blades deltate to ovate. Involucres green, enlarging moderately as fruits develop, reaching 6–9 mm, the lobes 2–3 (4) mm long. Perianth white to pale pink; flowering almost any time of the year. Fruits ovoid, faintly reticulate to mottled dark brown and gray, sometimes faintly glaucous. (Description based on Baja California Norte specimens.)

Local Distribution: Known from high elevation on San Esteban.

Other Gulf Islands: None.

Geographic Range: *M. laevis*, with four varieties, occurs in western Arizona, northwestern Sonora south to the Sierra Seri, Baja California Norte and Sur, and southern California to Oregon, Nevada, and southwestern Utah. Var. *crassifolia* ranges from Baja California Sur to California.

San Esteban: Arroyo on steep N slope of NE peak, 28°42'N, 112°35'W, 450 m, 26 Apr 1966, *Moran 13051* (SD; also see *Cuscuta corymbosa* var. *grandiflora, Moran 13052*).

Mirabilis tenuiloba S. Watson

Long-lobed four o'clock

Description: Sometimes flowering in the first year, often dying back to rootstock during drought; roots stout, somewhat thickened. Herbage, especially the new growth, and inflorescences densely pubescent with sticky, glandular hairs. Leaves often 3–6+ cm long, tardily drought deciduous, the blades thick and semi-succulent, mostly deltate. Involucres green, enlarging moderately to 10–15 mm long as fruits develop, the lobes 5–9 mm long. Flowering at various seasons except during cold weather. Fruits rounded, smooth, and blackish.

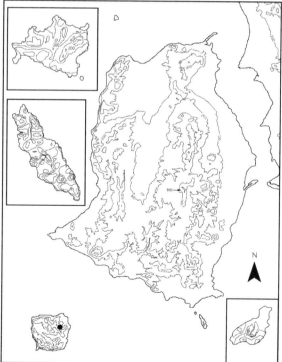

Mirabilis laevis var. crassifolia

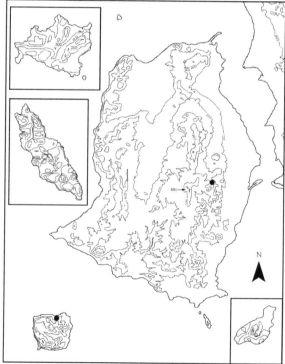

Mirabilis tenuiloba

Local Distribution: Known from two localities in the flora area: interior of Sierra Kunkaak on Tiburón and on the north side of San Esteban.

Other Gulf Islands: Ángel de la Guarda, San Lorenzo, San Marcos, Carmen.

Geographic Range: Western part of the Colorado Desert in southern California, southwestern Arizona, northwestern Sonora in the Sierra del Rosario, Gulf side of Baja California Norte and Sur, western Gulf Islands and San Esteban and Tiburón.

Tiburón: Hast Cacöla, E flank of Sierra Kunkaak, 380 m, *Wilder 08-250.*

San Esteban: Narrow, rocky canyon behind cove at N side of island, *Felger 17573.*

Oleaceae – Olive Family

Trees and shrubs or subshrubs. Leaves opposite; stipules none. Flowers radial, bisexual or unisexual. Stamens 2 (4), borne on the petals. Fruits highly variable, dry or fleshy.

1. Shrubs or small trees 1.5–4 m tall; leaves pinnate. ___ _____ **Fraxinus**
1' Subshrubs or shrubs to ca. 2 m tall; leaves simple.
    2. Woody shrubs mostly more than 1 m tall; flowers inconspicuous, without a corolla; fruits fleshy (drupes). _____ **Forestiera**
    2' Subshrubs to 1 m tall; flowers with bright yellow corollas; fruits dry (capsules). _____**Menodora**

Figure 3.121. *Fraxinus gooddingii.* Interior canyon of Sierra Kunkaak, Isla Tiburón. BTW, 25 October 2007.

Figure 3.122. *Menodora scabra.* Community of *M. scabra* (small shrubs) and *Sideroxylon occidentale* (small trees). Summit of Isla Tiburón. BTW, 26 October 2007.

Forestiera phillyreoides (Bentham) Torrey [*F. shrevei* Standley]

Desert olive

Description: Rigidly branched hardwood shrubs to about 2 m tall, the branches opposite and decussate, with numerous short spur-branches. Leaves drought deciduous, simple, the spur-branch leaves often 7–18 mm long, oblong-elliptic to narrowly oblanceolate, glandular punctate, at first densely pubescent with soft, spreading hairs, becoming glabrous with age. Flowers inconspicuous, probably bisexual, perianth absent (or perhaps with a greatly reduced calyx), the anthers dark purple; probably flowering with new leaves on branchlets of previous growing season in late winter or early spring. Fruits of drupes less than 1 cm long, moderately fleshy, longer than wide and moderately curved, and 1-seeded. (Flower and fruit descriptions based on northern Sonora and southern Arizona populations.)

Local Distribution: Known from the highest elevations of the Sierra Kunkaak on sheltered north-facing slopes and exposed ridges.

The nearest known populations are from mountains north of Caborca in Sonora, the Ajo Mountains in southwestern Arizona, and the Volcán las Tres Vírgenes in Baja California Sur. A population of *Forestiera* in the Sierra el Aguaje (e.g., Cañón Nacapule), probably *F. angustifolia* Torrey, has larger leaves and fruits than *F. phillyreoides*.

Other Gulf Islands: Carmen, Espíritu Santo.

Geographic Range: Southern California to New Mexico and Mexico in northern Sonora, the central plateau to Oaxaca, and Baja California Norte and Sur.

Tiburón: Hast Cacöla, E flank of Sierra Kunkaak, top of E-facing slope at base of cliff, ca. 540 m, ca. 1.75 m tall, *Wilder 08-268*. Exposed upper ridge of island, between peak of Cójocam Iti Yaii and summit, occasional shrub on ridge, 1–1.5 m tall, *Wilder 07-551*. Summit of island, not common, *Wilder 07-560*.

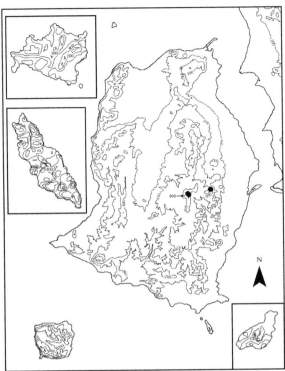

Forestiera phillyreoides

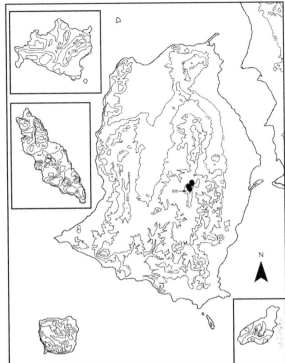

Fraxinus gooddingii

Sonora: Pass between Cerro San Luis and Sierra Santa Rosa, ca. 16 mi W of Trincheras, 950 m, *Bowers 2195A*. Sierra la Gloria, N of Caborca, 30°56'45"N, 112°11'45"W, 1220 m, *Turner 85-1*.

Fraxinus gooddingii Little

Fresnillo; Goodding ash, littleleaf ash

Description: Shrubs or small trees (1) 2–4 m tall with one to several slender trunks to 4.5 cm diameter; branches with conspicuous lenticels; herbage glabrate with age. Leaves and twigs opposite and decussate. Leaves potentially evergreen, odd pinnate with well-developed petioles, less than 10 cm long, the larger lateral leaflets reaching 3.5 cm long, the margins entire or serrated. Inflorescences about as long as the leaves or shorter, the inflorescence stalks persistent. Flowers bisexual or unisexual, small, and wind-pollinated. Fruits of single, wedge-shaped samaras probably 12–15 mm long, with an elongated, flat, firm, terminal wing.

Local Distribution: Sheltered canyon bottoms in the interior of Sierra Kunkaak and extending nearly to the peak in sheltered, north-facing canyons.

Other Gulf Islands: None.

Geographic Range: The nearest populations are in widely scattered sites in north-central and northeastern Sonora (Felger et al. 2001; Nesom 2010). Also southeast Arizona to west Texas, Chihuahua, Coahuila, and Nuevo León. This species occurs in the uppermost elevations in Sonoran desertscrub and mostly in grassland and oak woodland. Unlike most other ash trees in the Sonoran Desert region, it is not a riparian tree.

Tiburón: Deep sheltered canyon bottom, N portion of Sierra Kunkaak, shrub 1 m tall, *Felger 07-121*; Shrub 1.5 m tall, *Felger 07-116*. Steep, protected canyon, 720 m, N interior of Sierra Kunkaak, tree 3.5 m tall, several trees in one small area at N base of large rock outcrop, *Wilder 07-504*. N base of summit of island, N-facing sheltered cliff community, shrub 2 m tall, not common, *Wilder 07-553*.

Menodora scabra A. Gray [*M. scoparia* Engelmann ex A. Gray]

Twinberry

Description: Subshrubs nearly to 1 m tall with many slender, erect branches. Stems striate (with longitudinal grooves). Herbage scabrous (short, rough sandpaper-like hairs). Leaves alternate or often opposite below, gradually

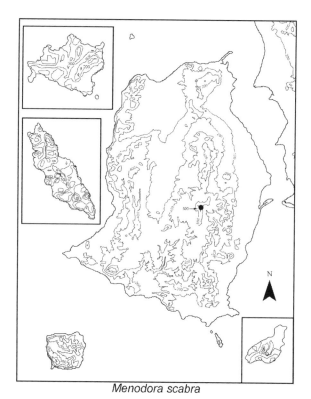

Menodora scabra

Onagraceae – Evening Primrose Family

The several species in the flora area are reproductive and grow only during the cooler seasons. The flowers are nocturnal or crepuscular and wither soon after dawn or with daytime heat.

1. Petals more than 2 cm long, yellow without spots or other markings; stigma of 4 linear lobes.   **Oenothera**
1' Petals less than 1.5 cm long, colors various, often with spots or other markings; stigma hemispherical or capitate.
   2. Leaf blades ovate, about as wide as long; flowers and capsules pedicelled. _____ **Chylismia**
   2' Leaf blades more than 3 times longer than wide, pinnately lobed to divided; flowers and capsules sessile.
      3. Leaf blades usually red spotted or reddish; flowers minute, petals 2–2.5 mm long, white. ____
      _____ **Eremothera**
        (see "Doubtful and Excluded Reports")
      3' Leaf blades not red spotted; petals 5–7 mm long, bright yellow or orange. _____ *Eulobus*

drought deciduous, reduced upward, probably 1–4 cm long, sessile or subsessile, narrowly elliptic to narrowly oblong; margins entire. Flowers showy, bright yellow, 1–1.5 cm wide; calyx with 7–10+ slender lobes. Fruits inflated, papery, testiculate capsules of 2 hemispheres opening around the middle. Seeds 4 per chamber, with a narrow (water-absorbing?) wing.

Local Distribution: Locally abundant on the summit of Tiburón, and Humberto has seen it on top of Capxölim, a solitary peak, 590 m, also in the Sierra Kunkaak range. These plants are exceptionally robust and shrubby, and ephedra-like in appearance.

Other Gulf Islands: None.

Geographic Range: Western Sonora in granitic mountains southward nearly to Puerto Libertad and the Sierra del Viejo (southwest of Caborca), and eastward across the northern part of the state. Also Colorado and Utah to Arizona to west Texas, both Baja California states, Chihuahua, Durango, and Nuevo Leon. Within the genus *M. scabra* is a generalist; its geographic range is the largest of the North American species.

Tiburón: Summit of island, dominant subshrub, dense population, the principal component of the summit community with *Sideroxylon occidentale, Wilder 07-556.*

Figure 3.123. *Chylismia cardiophylla* subsp. *cedrosensis.* El Monumento, Isla Tiburón. BTW, 30 January 2008.

Chylismia cardiophylla (Torrey) Small subsp. **cardio-phylla** [*Camissonia cardiophylla* (Torrey) P.H. Raven subsp. *cardiophylla*]

Description: Annuals to short-lived perennials; pubescence generally not glandular. Stems leafy, the leaves broadly ovate to nearly orbicular, about as wide as long, the margins coarsely toothed. Flowers vespertine, pungently fragrant, 3.8–4.5 cm long including the pedicel; petals 6–10+ mm long, bright yellow, often drying pink.

Local Distribution: San Esteban, including higher elevations. Mártir, known from several collections from different parts of the island.

Other Gulf Islands: Ángel de la Guarda, San Lorenzo, Tortuga, San Marcos.

Geographic Range: Northern Baja California Sur, Baja California Norte, southeastern California, southwestern Arizona, and northwestern Sonora nearly to Puerto Libertad. Members of this species are reported to be self-compatible but are also often out-crossing.

San Esteban: SE side, *Lindsay 3 Apr 1947* (CAS). N side of island, arroyo margin, and rare on slopes, *Felger 15427*. N side of island, narrow rocky canyon behind large cove, *Felger 17552*. San Pedro, upper beach and steep slopes of gully-like arroyo, *Felger 16674*. Steep slopes of S-central peak, *Felger 17547*. Steep arroyo on N slope of NE peak, 450 m, *Moran 13047* (CAS).

Mártir: 1887, *Palmer 403* (UC). Locally abundant on a ledge on a high sea cliff, clumps dense, 3 ft high, 18 Apr 1921, *Johnston 3147* (CAS). Canyon on N side [100 m, in field notes], occasional, 21 Mar 1962, *Moran 8820* (SD). 21 Mar 1962, *Wiggins 17186* (CAS).

Chylismia cardiophylla subsp. **cedrosensis** (Greene) W.L. Wagner & Hoch [*Camissonia cardiophylla* subsp. *cedrosensis* (Greene) P.H. Raven]

Description: This subspecies differs from subsp. *cardiophylla* by having leaves that are ovate-acuminate, longer than broad, and pubescence that is predominately glandular.

Local Distribution: Documented on Tiburón from widely scattered localities near the shore; washes and rocky slopes.

Other Gulf Islands: Ángel de la Guarda.

Geographic Range: Isla Cedros and the adjacent mainland of Baja California Norte east to Bahía de Los Ángeles and south to the Vizcaíno Desert in Baja California Sur, and the Gulf Coast of Sonora in the Sierra Bacha and Sierra Seri.

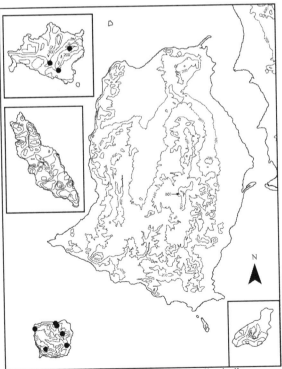
Chylismia cardiophylla ssp. *cardiophylla*

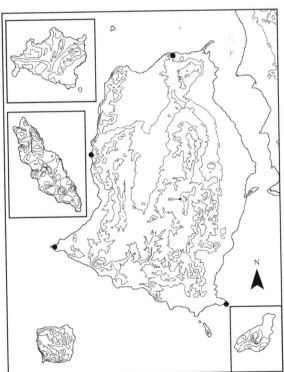
Chylismia cardiophylla ssp. *cedrosensis*

Tiburón: Canyon just N of Willard's Point, sand of arroyo, *Moran 8741* (SD). El Monumento, *Felger 08-41*. Hee Inoohcö, mountains at NE coast, *Felger 07-63*. Pooj Iime, large canyon, NW shore, *Wilder 07-436*.

Eulobus californicus Nuttall ex Torrey & A. Gray [*Camissonia californica* (Nuttall ex Torrey & A. Gray) P.H. Raven]

Hast ipenim "with which the rock is splashed"; California evening primrose

Description: Slender winter-spring ephemerals; much taller than wide, to 70 cm tall, with an erect main axis, unbranched or usually with a few ascending straight branches. First leaves in a basal rosette, or often not forming a basal rosette. Leaves linear to narrowly elliptic. Plants leafy when young, leafless or nearly so at flowering time. Flowers open near dawn, bright yellow, sometimes with red flecks, becoming orange with age, and drying pink. Capsules slender, straight to slightly curved, turning downward at maturity. This species is "mostly self-pollinating, and apparently always self-compatible" and "rarely visited by insects" (Raven 1969:198). As the flower matures the anthers collapse on top of the stigma.

Local Distribution: Tiburón, documented from the Central Valley and at Sauzal, growing in sandy-gravelly soil of watercourses.

Other Gulf Islands: Ángel de la Guarda.

Geographic Range: Southern California to Baja California Sur, Arizona, and Sonora south to the vicinity of Tiburón and Hermosillo.

Tiburón: Sauzal, *Wilder 08-102*. SW Central Valley, scattered in arroyo bed, *Felger 17305*.

Oenothera primiveris A. Gray

Yellow evening primrose

Description: Winter-spring annuals with a relatively deep taproot. Leaves in basal rosettes, nearly stemless. Pubescence dense, of bristly spreading papilla-based white hairs. Leaves mostly pinnatifid, narrowed into a long, winged petiole. Flowers fairly large and showy, opening in the evening and closing with daytime heat. Petals and stigma bright lemon yellow, the petals notched at apex (flower, leaf, and plant size correlated with soil moisture). Fruits with large spreading white hairs, each from a large conical fleshy papilla (gland). (Description based on mainland plants.) Capsules thick and woody, upright, straight, 4-angled, tapering to a conspicuously narrowed tip. The

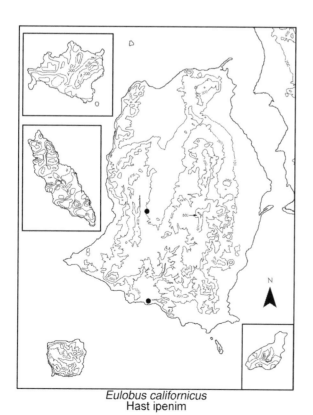

Eulobus californicus
Hast ipenim

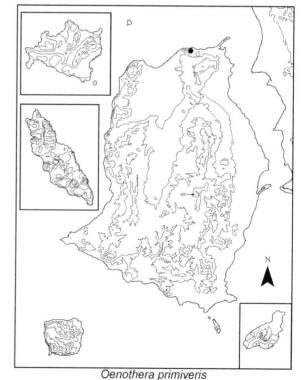

Oenothera primiveris

dry skeletons, consisting of part of the taproot and a cluster of sessile woody capsules, may persist for several years.

Local Distribution: Known from north-facing mountain slopes at the northeastern shore of Tiburón. The nearest known populations are on the coastal plain west of Hermosillo and the vicinity of El Desemboque del Río Concepción (the northern Desemboque).

Other Gulf Islands: None.

Geographic Range: Southern Arizona to western Texas, southwestern Utah to southeastern California, Baja California Norte, and northern Sonora. There are two subspecies, distinguished in part by flower size and reproductive strategy.

Tiburón: Hee Inoohcö, mountains at NE coast, several small dry plants with dried capsules, *Felger 07-95*.

Orobanchaceae – Broomrape Family

Orobanche cooperi (A. Gray) A. Heller

Matar; *flor de tierra*; desert broomrape

Description: Root parasites on various shrubs, without chlorophyll, the roots short, stubby and coral-like, and without root hairs. Apparently annuals, appearing in spring. Stems thick and semi-succulent; leaves reduced to scales; corollas purple and white, the throat often marked with yellow.

Local Distribution: Known from Tiburón by a single collection. The nearest record is from the vicinity of El Desemboque, 60 km to the north (Felger & Moser 1985), which is the southernmost record for mainland Sonora.

Other Gulf Islands: Ángel de la Guarda (*Diane E. Boyer 8 Jan 2008*, photo specimen, ARIZ).

Geographic Range: Deserts and semi-arid regions in southwestern United States and northwestern Mexico in Baja California Norte and northern Sonora.

Tiburón: Cerro Kunkaak, 700 m, *Scott 11 Apr 1978* (UNM).

Papaveraceae – Poppy Family

Argemone subintegrifolia Ownbey

Xazacöz caacöl "large prickly poppy"; *cardo*; San Esteban prickly poppy

Description: Woody shrubs (a rare feature in the genus) beset with spines, the stems exuding bright yellow latex when cut. Larger leaves often 5–9 cm long, reduced above, simple, sessile, broadly oblong to oval, bluish glaucous, and densely spiny with coarse, spine-tipped teeth. Petals white, about 5–6 cm long. Stamens numerous. Flowering at least in spring.

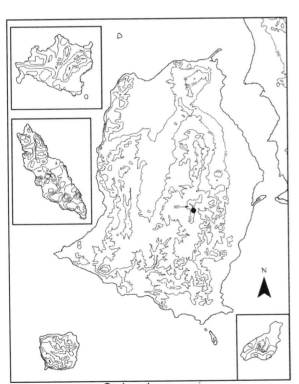

Orobanche cooperi
Matar

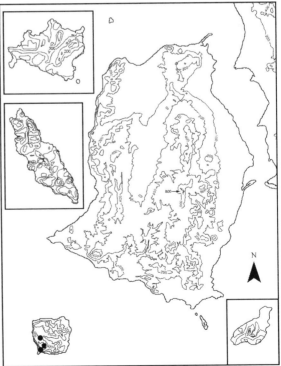

Argemone subintegrifolia
Xazacöz caacöl

Local Distribution: San Esteban, scattered in canyons and on rock slopes at the southwestern part of the island and extending to the south-central peak.

Other Gulf Islands: Ángel de la Guarda.

Geographic Range: Endemic to the Gulf Coast of Baja California Norte as far north as the Sierra Cucapa and the two islands. This species represents a sister clade to the rest of the genus *Argemone*, and due to its woodiness and rare nature it is considered a relict (Schwarzbach & Kadereit 1999). It is unlike any other member of the genus.

San Esteban: Canyon, SW corner of island, *Felger 17618*. Arroyo, SW corner, *Moran 10478* (SD). Steep slopes, S-central peak, *Felger 17538*. N slope of high peak near SE corner of island, 280 m, on cliff, only one seen, *Moran 8860* (CAS).

Passifloraceae – Passion Vine Family

Passiflora – Passion vine

Perennial vines or semi-vines with simple (unbranched) and often stout tendrils and alternate leaves. Flowers unique and showy with complex morphology. Fruits fleshy with a gelatinous edible pulp.

1. Herbage densely pubescent but not sticky, without stalked glands (or few stalked glands on sepals). ____ _____ **P. arida**
1' Herbage densely and sticky glandular-pubescent, the glands conspicuously stalked. _____ **P. palmeri**

Figure 3.124. *Passiflora arida.* Arroyo Limantour, Isla San Esteban. BTW, 8 March 2007.

Figure 3.125. *Passiflora palmeri*. Sauzal, Isla Tiburón, 17 March 2006. Plant, CDB; flower, BTW.

Passiflora arida (Masters & Rose) Killip [*P. foetida* Linnaeus var. *arida* Masters & Rose]

Oot ijoeene "coyote's passion-vine"; desert passion vine

Description: Stems semi-vining to vining, often 1–2 m long growing over shrubs, the herbage densely white-woolly (not sticky glandular), sometimes drying brownish. Leaves 3-lobed; stipules divided into slender segments. Flowers 4.5–5 cm wide, the petals white, the corona blue, curling outwards. Fruits nearly globose, 3 cm wide, green and pubescent. Flowering and fruiting at various seasons.

The fruit is eaten but not enthusiastically. Humberto said that it is ok to eat 5 or 10 fruits but not 20: "*Cuando come mucho, hace vomitar*."

Local Distribution: Tiburón, in the Sierra Kunkaak, from the foothills to high elevations and also found in the north at Tecomate. On San Esteban, common and widespread from near the shore to peak elevations.

Other Gulf Islands: San Lorenzo, Tortuga, San Marcos, Coronados, Monserrat, San Diego, San José, Espíritu Santo.

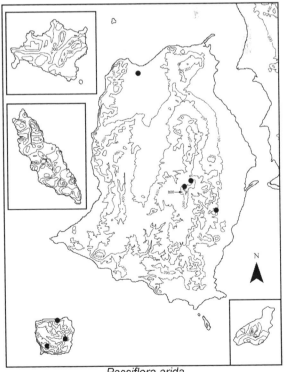

Passiflora arida
Oot ijoeene

Geographic Range: Western Sonora from near the Sinaloa boarder northward to southern Arizona, Baja California Sur and southeastern Baja California Norte.

> **Tiburón:** Hast Coopol, *Wilder 07-385*. Base of Capxölim, head of arroyo, *Wilder 05-24*. Deep canyon in the northern interior of the Sierra Kunkaak, 685 m, *Wilder 07-493*. 1 km inland, Tecomate, floodplain, *Felger 12506*.
>
> **San Esteban:** Arroyo Limantour, 1 mi inland, *Wilder 07-40*. N side of island, *Felger 15448*. Steep slopes, S-central peak, *Felger 17541*.

Passiflora palmeri Rose

Joeene (variants: xoeene, cjoeene); stinking passion-vine

Description: Stems self-supporting and forming bushes to semi-vining stems, often less than 1.5 m long, the herbage brownish and sticky with conspicuous stalked glands especially on the leaf blade margins, petioles, stipules, and sepals. Leaves 3-lobed; stipules divided into slender gland-tipped segments. Flowers 6–7+ cm wide; petals white; corona blue, bending inward, "hugging" the base of the style (not extended outward as in *P. arida*). Fruits nearly globose, 3 cm wide, green and softly pubescent.

The fruit is greatly appreciated by the Comcaac for its subtle and pleasing taste.

Local Distribution: Widespread and locally common on Tiburón and Dátil, and along Arroyo Limantour on San Esteban.

Other Gulf Islands: Ángel de la Guarda, San Lorenzo, Tortuga, San Marcos, Coronados, Carmen (type locality), Danzante, Cerralvo (*Johnston 4043*, GH not seen, cited by Goldman 2003).

Geographic Range: Gulf Coast of both Baja California states, Gulf Islands; not known from the Sonora mainland.

> **Tiburón:** Ensenada Blanca (Vaporeta), *Felger 15734*. Arroyo Sauzal, 100 m from shore, rocky outcrop, *Wilder 06-86*. Ensenada de la Cruz, 0.5 km inland, infrequent, *Felger 2588*. Tordillitos, *Felger 15485*. Coralitos, *Wilkinson 216*. La Viga, *Wilder 08-147*. Ensenada del Perro, *Felger 17710*. Hast Coopol, *Wilder 07-379*. Large arroyo heading N to E base of Sierra Kunkaak, *Wilder 06-17*. Bahía Agua Dulce, 11 Aug 1985, *Tenorio 9535* (US, not seen, cited by Goldman 2003). Pooj Iime, *Wilder 07-450A*.
>
> **San Esteban:** 0.5 km inland, Arroyo Limantour, *Felger 75-120*. Arroyo Limantour, 25 m elev., common on shrubs, *Moran 28172* (SD).
>
> **Dátil:** NW side of island, *Felger 15315*. NE side, steep rocky N-facing slope, uncommon, *Felger 9099*. E side, slightly N of center, narrow ridge crest of peak, *Wilder 07111*.

Phyllanthaceae – Phyllanthus Family

Andrachne microphylla (Lamarck) Baillon [*A. ciliato-glandulosa* (Millspaugh) Croizat]

Hasoj an hehe "plant in the river"

Description: Delicate winter-spring annuals, with tack-shaped, stalked glandular hairs, the stems very slender. Leaves alternate, light green, thin, and orbicular to broadly ovate. Flowers small and inconspicuous, unisexual with male and female flowers on the same plant. Capsules 6-seeded.

Local Distribution: Documented on Tiburón from the western and central part of the island, along larger arroyos and canyon bottoms.

Other Gulf Islands: Ángel de la Guarda, Carmen, Santa Catalina, Espíritu Santo, Cerralvo.

Geographic Range: Gulf of California region from coastal Sonora north of San Carlos to the Sierra Seri, and both states of Baja California. Disjunct in northwestern South America.

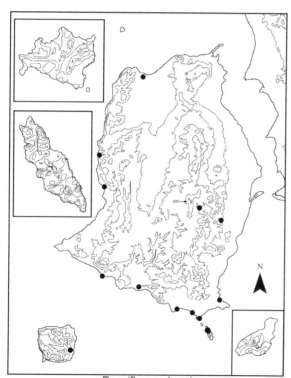

Passiflora palmeri
Joeene

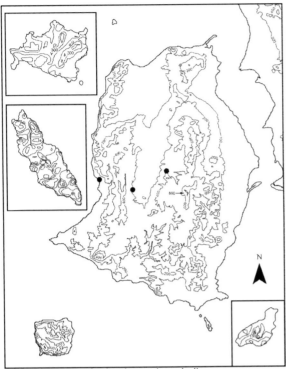

Andrachne microphylla
Hasoj an hehe

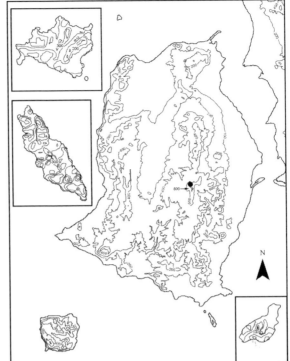

Rivinia humilis

Tiburón: Ensenada Blanca (Vaporeta), *Felger 17285.* SW Central Valley, arroyo bed, *Felger 17302.* Haap Hill, *Felger 76-T5.*

Phytolaccaceae – Pokeweed Family

Rivinia humilis Linnaeus

Chile de coyote; pigeon berry

Description: Bushy perennials to probably about 1 m tall, scarcely woody at the base; sparsely puberulent or glabrous. Stems slender and brittle. Leaves alternate, simple, thin, and entire, probably 4–10+ cm long, ovate, and quickly wilting. Herbage reddish green. Inflorescences of short, slender, terminal racemes. Flowers 4–5 mm wide, white or pink; sepals 4, petals none. Stamens 4. Fruits 3–4 mm diameter, fleshy and red, 1-seeded.

Local Distribution: Tiburón in the Sierra Kunkaak, where it is known from a single collection from a high elevation, interior canyon.

Other Gulf Islands: None.

Geographic Range: Riparian and semi-riparian habitats in the Sonoran Desert region; Baja California Sur, Arizona to Florida to northern South America.

Tiburón: Steep, protected canyon, 720 m, N interior of Sierra Kunkaak, common shrub in canyon bottom, *Wilder 07-502.*

Stegnosperma, see **Stegnospermataceae**

Plantaginaceae – Plantago Family
(includes Scrophulariaceae in part)

Annuals or perennial herbs (the species in the flora area). Leaves opposite or alternate, sometimes in basal rosettes, without stipules. Flowers bilateral or radial in *Plantago.* Fruit a capsule, with 2–numerous small seeds.

1.   Bushy perennials or subshrubs; corollas bright red. _____ **Gambelia**
1' Annuals (or short-lived herbaceous perennials in *Stemodia*); corollas not red.
   2.   Plants glabrous; corollas with a prominent, slender spur. _____ **Nuttallanthus**
   2' Plants sparsely to densely pubescent; corollas with or without a spur.
      3.   Leaves incised or toothed. _____ **Stemodia**
      3' Leaves entire.

4. Herbage viscid-sticky; leaves ovate to broadly lanceolate, prominently petioled; pedicels shorter than the flowers or fruits. _____ **Pseudorontium**

4' Herbage not viscid sticky; leaves linear, sessile or nearly so.

    5. Vegetative stems short or not developed; flowers 4-merous, the corollas whitish to brown; flowers and fruits sessile; seeds smooth, 2 per capsule. _____ **Plantago**

    5' Vegetative stems well developed, often with twining branchlets; flowers 5-merous, corollas lavender-blue; pedicels longer than the flowers or fruits; capsules multiple-seeded, the seeds corky-tuberculate. ___
_____ **Sairocarpus**

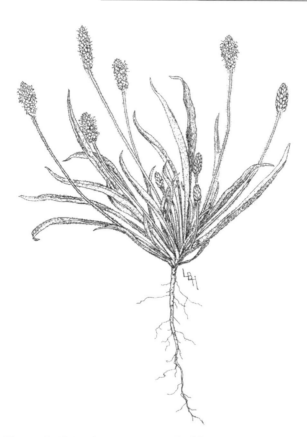

Figure 3.127. *Gambelia juncea.* Isla San Pedro Nolasco. BTW, 11 November 2009.

Figure 3.126. *Plantago ovata* var. *fastigiata.* LBH.

Figure 3.128. *Plantago ovata* var. *fastigiata.* Freshly picked plants, near Pozo Coyote, Sonora. RSF, April 1963.

Gambelia juncea (Bentham) D.A. Sutton [*Galvezia juncea* (Bentham) Ball. *G. juncea* var. *pubescens* (Brandegee) I.M. Johnston. *G. juncea* var. *foliosa* I.M. Johnston]

Nojoopis caacöl "large *chuparrosa* (*Justicia* sp.)"

Description: Bushy perennials, sometimes to about 1 m tall. Herbage moderately glaucous, glabrous or glandular pubescent even on the same plant. Leaves opposite or more often 3 per node, mostly elliptic to orbicular, often (5)10–25 mm long, drought deciduous, the foliage often sparse. Flowers 2.5–3 cm long, tubular and bright red, attracting hummingbirds.

Local Distribution: Tiburón, Sierra Kunkaak, rocky habitats in canyon bottoms, cliffs, and slopes to the summit. San Esteban, locally in sheltered canyons and mostly north-facing cliffs and steep rock slopes. Nolasco, along the island crest and below the crest on the east side of island.

Other Gulf Islands: Ángel de la Guarda, San Lorenzo, San Marcos, Carmen, Monserrat, Espíritu Santo.

Geographic Range: Sonora from the Sierra del Viejo near Caborca and the vicinity of Libertad southward to the Guaymas region, the Baja California Peninsula, and Gulf Islands.

Tiburón: Cerro San Miguel, 400 m, not common, restricted to the peak, *Quijada-Mascareñas 91T017*. Deep canyon, SW and up canyon from Siimen Hax, *Wilder 06-474*. Top of

Sierra Kunkaak Segundo, *Wilder 06-490*. Steep, protected, high canyon, 720 m, N interior of the Sierra Kunkaak, *Wilder 07-499*. Summit of island, not common, *Wilder 07-569*.

San Esteban: N side of island, N-facing cliff, *Felger 15436*. NE part of island, rare, N-facing steep rocky slope, near ridge of mountain, *Felger 7058*. SW part of island, canyon, *Wilder 07-84*.

Nolasco: [Above SE cove], E exposure, just below crest at 800 ft, *Pulliam & Rosenzweig 21 Mar 1974*. E-central side below crest of island, higher elevations on N-facing slopes in aggregations of several plants, *Felger 06-100*. Cañón de Mellink, ridgetop of island, *Wilder 08-357*.

Nuttallanthus texanus (Scheele) D.A. Sutton [*Linaria texana* Scheele. *L. canadensis* (Linnaeus) Dumont de Courset var. *texanus* (Scheele) Pennell]

Texas toadflax

Description: Winter-spring annuals with a basal rosette and one to several slender, erect nearly leafless flowering stalks. What appears to be a basal rosette of pinnate leaves is a rosette of short, leafy stems that usually wither by flowering time (elsewhere these stems may develop into flowering stems). Lower stem leaves opposite and upper leaves alternate, leaves linear, glandular punctate. Flowers blue and attractive, bilaterally symmetrical, the

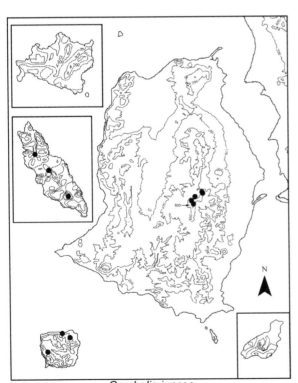

Gambelia juncea
Nojoopis caacöl

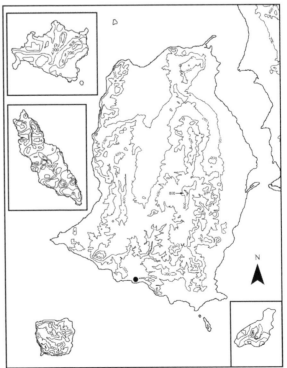

Nuttallanthus texanus

corollas with a conspicuous spur; corollas about 8 mm long plus a nectar-filled spur that attracts insects, the corolla lip serving as a landing pad; stamens 4.

Local Distribution: Documented on Tiburón at the south end of the island but probably more widespread.

Other Gulf Islands: Cerralvo.

Geographic Range: Widespread in the Sonoran Desert and across the United States to South America.

Tiburón: El Sauz, few, seen only in one place, *Moran 8792* (SD).

Plantago ovata Forsskål var. **fastigiata** (Morris) S.C. Meyers & Liston [*P. insularis* Eastwood. Not *P. insularis* Nyman ex Briquet. *P. fastigiata* E. Morris. *P. insularis* var. *fastigiata* (E. Morris) Jepson]

Hatajeen; *pastora;* woolly plantain, Indian-wheat

Description: Winter-spring annuals, highly variable in size. Leaves simple, elongated, and silky-silvery pubescent. Flowers on slender spikes, 4-merous, the petals papery, straw colored, and persistent. Seeds 2 per capsule, yellowish brown or red-brown, 2–2.5 mm long. As with other plantagos, when water contacts the seed coat it immediately forms a jacket of slime (mucilage) that on drying tenaciously glues the seed to any available substrate.

The seeds were an important food for the Comcaac on the mainland and were much appreciated. The seeds were separated and then mixed with water and allowed to sit for about a half hour and then eaten. Sugar was often added and the mixture consumed as a beverage; the seeds were not cooked (Felger & Moser 1985).

Local Distribution: Tiburón, known only from the southern part of the Central Valley. Moderate amounts of winter rain had fallen in this area of the island in late 2008 and early 2009. Humberto was in this area in February 2009 and found three plants. It is widespread and common on the opposite mainland.

Other Gulf Islands: Ángel de la Guarda, Las Ánimas (*Wiggins 17266*, CAS), San José.

Geographic Range: Var. *fastigiata* is the inland North American variety and one of the more common and widespread winter-spring annuals of the Sonoran Desert, reaching its southern limits in Sonora in the Guaymas region. This species occurs in southwestern United States and northern Mexico and the Old World from the Canary Islands, southern Spain, and North Africa to Pakistan and India.

Tiburón: Sur de Satoocj, 12 Feb 2009, *Romero-Morales 09-2.*

Pseudorontium cyathiferum (Bentham) Rothman [*Antirrhinum cyathiferum* Bentham]

Hehemonlc "curled (or crinkled) plant"; desert snapdragon

Description: Non-seasonal annuals but found mostly during cooler times of the year, prominently viscid-glandular and foul smelling. Corollas 1 cm long, purplish blue with darker veins, the lip with two yellow spots at entrance to the snapdragon throat, which is hairy inside. Seeds whitish, becoming dark brown with age, intricately sculptured with irregular corky and tuberculate ridges, 2–2.5 mm long with a wide cup-shaped wing.

Local Distribution: Widespread and often common on Tiburón, San Esteban, and Dátil from near the shore (upper beaches) to mountains at higher elevations. Also on Cholludo and Nolasco.

Other Gulf Islands: Ángel de la Guarda, Tortuga, San Marcos, San Ildefonso, Coronados, Danzante, Monserrat (observation), San José, San Francisco, Espíritu Santo, Cerralvo.

Geographic Range: Endemic to the Sonoran Desert: western Sonora south to the Guaymas region, both Baja California states, southwestern Arizona, and rare in southeastern California.

Tiburón: Ensenada Blanca (Vaporeta), *Felger 15726.* Tordillitos, *Felger 15468.* Ensenada de la Cruz, near beach, *Felger 9143.* Coralitos, *Wilkinson 220.* El Monumento, *Felger 15546.*

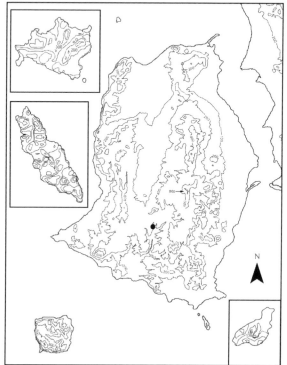

Plantago ovata var. *fastigiata*
Hatajeen

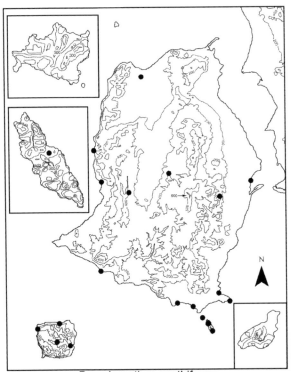

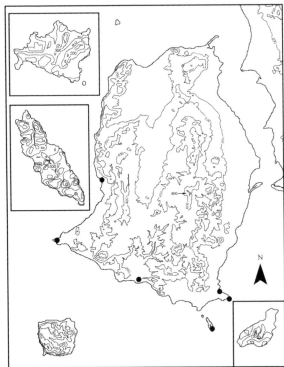

Pseudorontium cyathiferum
Hehemonlc

Sairocarpus watsonii
Comaacöl

Ensenada del Perro, *Wilder 07-146*. Upper rocky slope of mountain bordering SW Central Valley, 1200–1400 ft, *Felger 12398*. Haap Hill, *Felger 76-T6*. Between Sopc Hax and Hant Hax camp, *Felger 9345B*. Zozni Cmiipla, *Wilder 06-352*. Arroyo Agua Dulce, *Felger 6837*. Pooj Iime, *Wilder 07-450B*. **San Esteban:** Arroyo Limantour: Canyon bottom, *Felger 12753*; 2¼ mi inland, *Wilder 07-79*. N side of island, steep N- facing slope near shore, *Felger 15435*. San Pedro, rocky slopes, *Felger 16661*.

Dátil: NE side of island, local areas in arroyos and hillslopes, *Felger 9128*. NW side of island, *Felger 15344*. SE side, *Felger 17500*. E side, *Wilder 07-131*.

Cholludo: Rocky ledge near ocean, not common, *Felger 9154*.

Nolasco: E-central side of island, N-facing slope, 300 ft, 18 Jan 1965, *Felger*, observation (Felger 1966:98).

Sairocarpus watsonii (Vasey & Rose) D.A. Sutton [*Antirrhinum watsonii* Vasey & Rose. *A. kingii* S. Watson var. *watsonii* (Vasey & Rose) Munz]

Comaacöl; Sonoran Desert snapdragon

Description: Cool-season annuals. Stems erect, often slender and delicate, simple to much branched, the larger plants often with slender, twining branchlets arising from lower nodes; often sparsely pubescent with glandular hairs. Leaves alternate or the lower ones opposite or whorled.

Corollas 6–7 mm long, lavender-blue; seeds intricately sculptured with irregular corky and tuberculate ridges.

Local Distribution: Western and southern parts of Tiburón, and also on Dátil; arroyos, canyons, and rocky slopes.

Other Gulf Islands: Ángel de la Guarda, San Lorenzo, San Marcos, San Ildefonso, Carmen, Danzante, San José, Espíritu Santo.

Geographic Range: Sonora from the Guaymas region northward along the coast to Puerto Lobos and inland to Arizona in Organ Pipe Cactus National Monument, both states of Baja California, and Gulf Islands.

Tiburón: Ensenada Blanca (Vaporeta), *Felger 7100*. Canyon just N of Punta Willard, scarce, *Moran 8735* (SD). El Sauz, 200 m, *Moran 8792.5* (SD). El Monumento, *Felger 08-64*. Ensenada del Perro, *Felger 17719*.

Dátil: SE side of island, steep and narrow canyon, common except on S exposures, *Felger 17490*.

Stemodia durantifolia (Linnaeus) Swartz

Description: Annuals to short-lived perennials, to 50 cm tall; glandular pubescent. Leaves opposite, the first leaves in a basal rosette, often 8–15 mm long, and sessile; leaf margins serrated. Flowers about 1 cm long; corollas dark blue.

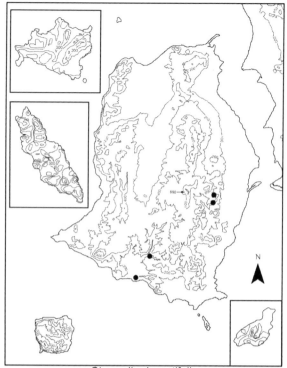

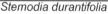

Stemodia durantifolia

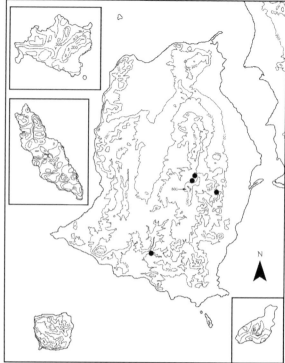

Plumbago zeylanica

Local Distribution: Tiburón, at major waterholes on wet soils and emergent from shallow water.

Other Gulf Islands: None.

Geographic Range: Southwestern United States and tropical America; characteristic of waterholes and streams in the Sonoran Desert region.

Tiburón: Arroyo Sauzal, *Felger 10004*. Sauzal waterhole (Xapij): Shallow water and wet sand at edge of waterhole, *Felger 10079*; *Wilder 06-125*. Sopc Hax, *Wilder 07-209*. Pazj Hax, *Wilder 06-16*.

Plumbaginaceae – Leadwort Family

Plumbago zeylanica Linnaeus [*P. scandens* Linnaeus]

Estrenina; doctorbush

Description: Herbaceous perennials or subshrubs with slender stems; glabrous except for stalked glands. Leaves alternate, to 16 cm long, mostly lanceolate, thin and tardily drought deciduous. Calyx with stalked glands along the ribs that stick to almost anything, and similar glands on the rachis and bracts; these glands begin exuding at about the time of anthesis, perhaps protecting the flowers and fruits from predation. Flowers 1–2 cm wide, 5-merous. Corollas, filaments, style

and stigma satiny white, the anthers violet-purple before opening longitudinally to shed white pollen, the anthers then dark blue. Calyx enclosing the fruit, these 1-seeded capsules circumscissle near the base; fruits readily detaching and adhering by means of the sticky calyx glands.

Local Distribution: Tiburón at the Sauzal waterhole and shaded places in densely vegetated canyons and waterholes in the Sierra Kunkaak.

Other Gulf Islands: None.

Geographic Range: Baja California Sur and much of Sonora but not near the coast north of the Guaymas region. Southern Arizona, Texas, and Florida, Mexico to South America, Africa, Asia, and Pacific Islands.

Tiburón: Arroyo Sauzal, rare, in shade, N side of steep rock wall in dried pool of cracked mud, *Felger 2739*. Sopc Hax, rare, seen only near waterhole, 26 Oct 1963, *Felger 9311* (specimen not located). Canyon bottom, N base of Sierra Kunkaak, *Wilder 06-439*. Deep canyon, N slope of Sierra Kunkaak, SW and up canyon from Siimen Hax, *Wilder 06-473*.

Poaceae (Graminae) – Grass Family

There are 24 genera with 35 species and 2 varieties of grasses in the flora area, representing 9 percent of the total flora of the Sonoran Islands. All are known from Tiburón except *Aristida divericata*, *Bothriochloa barbinodis*, *Enteropogon chlorideus*, and *Eragrostis pectinacea*. There are three non-native grasses, *Arundo donax*, *Cenchrus ciliaris*, and *C. echinatus*. The diversity of the grasses in the flora area is comparable to that of the composites (Asteraceae) and legumes (Fabaceae), these being the three largest families in the flora, as well as in the entire Sonoran Desert.

All grasses in the flora area are in the PACMCAD clade, a large monophyletic group that includes seven subfamilies and all C4 grasses but also includes many C3 grasses (Clark & Kellogg 2007). No non-PACMCAD grasses occur in the flora area, although there are many such species in northern and eastern Sonora and northern regions of Baja California Norte. Four of the five PACMCAD subfamilies occur in the flora area and are distributed in six tribes. The Cynodonteae and Paniceae (panicoid grasses) are the most diverse grass tribes in the flora, perhaps in part explained by their C4 photosynthetic pathway, which is associated with plants of warm climates.

The 4 subfamilies, 6 tribes, and genera of the grasses in the flora area are listed here, with the number of species present in each genus.

ARUNDINOIDEAE

Arundineae
Arundo 1
Phragmites 1

ARISTIDOIDEAE

Aristideae
Aristida 4

CHLORIDOIDEAE

Cynodonteae
Bouteloua 3
Chloris 1
Dasyochloa 1
Distichlis 2
Enteropogon 1
Eragrostis 1
Leptochloa 2
Muhlenbergia 1
Sporobolus 3
Trichloris 1
Tridens 1

Pappophoreae
Enneapogon 1

PANICOIDEAE

Andropogoneae
Bothriochloa 1
Heteropogon 1

Paniceae
Cenchrus 3
Digitaria 1
Lasiacis 1
Panicum 1
Setaria 2
Urochloa 2

Figure 3.129. *Aristida adscensionis.*
LBH (Humphrey et al. 1958).

Figure 3.130. *Arundo donax.* From Sopc Hax, Isla Tiburón. FR.

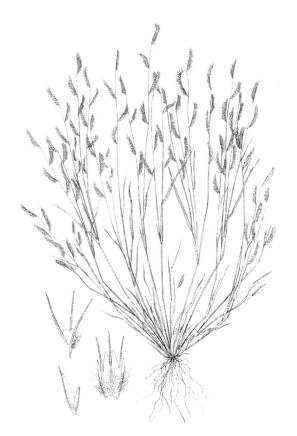

Figure 3.131. *Bouteloua barbata* var. *barbata*. LBH.

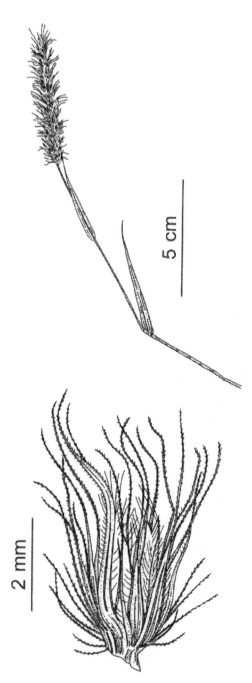

Figure 3.132. *Cenchrus ciliaris*. PB.

2 cm

Figure 3.133. *Cenchrus palmeri*. FR (Felger 1980).

Figure 3.134. *Distichlis littoralis.* From Hitchcock (1951).

Figure 3.135. *Cenchrus palmeri.* Isla San Pedro Nolasco. BTW, 11 November 2009.

Figure 3.136. *Chloris virgata.* Localized near the shore in the vicinity of Zozni Cmiipla, Isla Tiburón. BTW, 24 November 2006.

(A)

(B)

Figure 3.137. *Phragmites australis* subsp. *berlandieri*. Xapij (Sauzal waterhole), Isla Tiburón. (A) Dense stands with Richard. (B) Flowering panicle. BTW, 29 January 2008.

Figure 3.138. *Setaria macrostachya.* Cañón de Mellink, Isla San Pedro Nolasco.
BTW, 29 September 2009.

Figure 3.139. *Sporobolus cryptandrus.* Vicinity of Punta San Miguel,
Isla Tiburón. BTW, 15 September 2007.

Figure 3.140. *Sporobolus virginicus.* Vicinity of Zozni Cmiipla, Isla Tiburón. BTW, 24 November 2006.

1. Plants bamboo-like or reed-like, 1–3 m tall; associated with waterholes.
    2. Plants about 1–1.5 m tall; stems less than 1 cm thick; inflorescences openly branched, panicles not at all plume-like; spikelets globose, essentially glabrous; canyon bottom near a mountain waterhole. _____**Lasiacis**
    2' Plants 2 or more m tall; larger stems more than 1 cm thick; inflorescences densely flowered and plume-like, spikelets slender and elongated with long hairs.
        3. Rachillas glabrous, the lemmas with long hairs. _____ **Arundo**
        3' Rachillas with long hairs, the lemmas glabrous. _____ **Phragmites**
1' Plants not bamboo-like or reed-like.
    4. Coarse grasses; spikelets in pairs (terminal spikelets of *Bothriochloa* in 3's, 1 fertile and 2 sterile): 1 sessile (bisexual or pistillate, awned or not), and 1 pedicellate (smaller, staminate or sterile, not awned, deciduous or not, or reduced to a rudiment); spikelets with awns 2–7 cm long. (If you are not certain about this choice, then go to 4'—these genera key out both places.)
        5. Inflorescences cottony and whitish; awns to about 2 cm long. _____ **Bothriochloa**
        5' Inflorescences not cottony, not whitish; awns 4.5–7 cm long. _____ **Heteropogon**
    4' Coarse or delicate grasses; spikelets not in pairs (or 3's) in combination with characters given in 4; variously awned or not.
        6. Plants with stolons and/or rhizomes; perennials.
            7. Plants low and tufted, with stolons but not rhizomes, spikelets with short awns; dry desert habitats.
                8. Leaves soft and flexile, not firm; flowering stems more than twice as tall as the leaves. _____
                _____**Bouteloua diversispicula**
                8' Leaves firm; flowering stems less than twice as tall as the leaves. _____ **Dasyochloa**

7' Plants creeping and/or forming dense colonies or mats; spikelets awnless; tidal marshes, beaches, or wetland saline or alkaline soils.

 9. Leaf blades less than 1 (1.5) cm long; inflorescences inconspicuous, hidden among leaves. _____ **Distichlis littoralis**

 9' Leaf blades more than 2 cm long; inflorescences conspicuous.

  10. Male and female flowers on separate plants; spikelets several-flowered, more than 5 mm long. _____ **Distichlis spicata**

  10' Flowers bisexual; spikelets 1-flowered, about 2.5 mm long. _____ **Sporobolus virginicus**

6' Plants without conspicuous stolons or rhizomes; annuals or perennials.

 11. Spikelets in burs or fascicles, these falling as a unit with the attached bristles or spines. _____ **Cenchrus**

 11' Spikelets not in burs or fascicles.

  12. Spikelets with slender bristles below the spikelets; spikelets breaking off above the bristles. ____**Setaria**

  12' Spikelets not subtended by bristles.

   13. At least some spikelets with awns.

    14. Spikelets clearly 1-flowered and with 3 prominent terminal awns from a short to long central column. _____**Aristida** (except *A. ternipes*)

    14' Spikelets 1- to several-flowered with 1 to several or more awns, not both 1-flowered and 3-awned (if 3-awned, then spikelets with rudiments [reduced florets] above).

     15. Perennials, almost always obviously so.

      16. Lemmas with 9 feathery awns; panicles spike-like, lead-gray or green-gray. _____ **Enneapogon**

      16' Lemmas or other spikelet parts 1- to several-awned, the awns not feathery.

       17. Plants less than 15 cm tall; leaves fascicled, the leaf blades 7 cm or less in length; flowering stems (peduncles) shorter than the leaves, inflorescences compact, less than twice as long as the leaves. _____ **Dasyochloa**

       17' Plants usually more than 20 cm tall; leaves not conspicuously fascicled, the blades more than 7 cm long; flowering stems and inflorescences raised well above the basal leaves.

        18. Spikelets with 2 dissimilar florets each with 3 awns 1.5–2 cm long.__ **Trichloris**

        18' Spikelets various but florets (lemmas) not 3-awned.

         19. Awns at least 2 cm long.

          20. Inflorescences cottony and whitish; awns to about 2 cm long. _____ **Bothriochloa**

          20' Inflorescences not cottony, not whitish; awns 4.5–7 cm long. _____ **Heteropogon**

         19' Awns less than 1.5 (2) cm long.

          21. Spikelets 1-flowered; awns more than 10 mm long. _____ **Aristida ternipes** var. **ternipes**

          21' Spikelets with 2 to several florets; awns inconspicuous, to 2 mm long.

           22. Spikelets with 1 fertile and 1 sterile floret, both with lemma awns 0.7–2 mm long, the lemmas not notched or lobed. **Enteropogon**

           22' Spikelets several-flowered; lemmas notched (2 terminal lobes) with a minute bristle between the lobes. ___ **Leptochloa dubia**

     15' Annuals (mostly ephemerals).

      23. Lower leaf axils with cleistogenes; spikelet 1-flowered, with a single awn 15–28 mm long. _____ **Muhlenbergia**

      23' Plants not forming cleistogenes; spikelets with 2 to several awns and florets.

       24. Inflorescence branches or spikes racemose, not digitate; spikelets with 1 basal fertile floret and 1 or more reduced sterile florets (rudiments) above; one or more florets with 3 awns to 6 mm; spikelets without long white hairs. _____**Bouteloua**

       24' Inflorescence branches digitately arranged at top of main axis (stunted plants occasionally with a single spike); spikelets with 2 florets each with a single awn 5–20 mm long; lower floret with a tuft of long white hairs. _____ **Chloris**

13' Spikelets not awned.
    25. Spikelets with 2 or more bisexual florets.
        26. Panicles less than 1 cm wide and densely flowered with short, appressed branches. _____ **Tridens**
        26' Panicles more than 5 cm wide, openly branched with ascending to spreading branches often 5–10 cm long. _____**Leptochloa**
    25' Spikelets with 1 bisexual floret, sometimes also with a reduced or vestigial sterile floret and appearing 1-flowered unless dissected.
        27. Perennials, the stems tough with knotty bases.
            28. Spikelets 3–3.7 mm long (excluding the hairs), cottony with silky hairs. _____**Digitaria**
            28' Spikelets 2–2.5 mm long, glabrous. _____ **Sporobolus cryptandrus**
        27' Annuals, the stems not tough with knotty bases.
            29. Spikelets 1.5–2.5 mm long, the grain readily separating and falling; spikelets without a sterile lemma; lemma, palea, and grain distinct from each other. _____**Sporobolus pyramidatus**
            29' Spikelets 2.3–4 mm long, falling as a unit (breaking off below the glumes); spikelets with a sterile lemma like the upper glume in texture, the fertile lemma hard and shiny and firmly enclosing the palea and grain.
                30. Spikelets glabrous; prominent veins of spikelets longitudinal only. _____**Panicum**
                30' Spikelets hairy; prominent veins of spikelets longitudinal and transverse forming a net-like pattern on upper part of spikelet. _____**Urochloa**

Aristida – Three-awn; *zacate tres barbas*

The five species in the region can grow and become reproductive at various seasons depending on soil moisture, although the perennial species tend to become inactive in winter. The perennials sometimes become reproductive in their first season.

1. Spikelets 1-awned, or lateral awns very short and stubby; perennials. _____ **A. ternipes** var. **ternipes**
1' Spikelets 3-awned, the 3 awns usually well developed (if lateral awns reduced, then the plants clearly are stunted annuals); annuals or perennials.
    2. Annuals; awns flattened with minutely serrated margins (seen with 10× magnification), reaching 1.5 (1.7) cm long (awns occasionally very reduced, especially the lateral ones). _____**A. adscensionis**
    2' Perennials (sometimes flowering in first season); awns terete (1.6) 2 cm long or more.
        3. Awn column of mature spikelets jointed—breaking apart at a line of separation (seen as a horizontal line across the column) just above the lemma body; awns 3–5 cm long. _____ **A. californica**
        3' Awn column not jointed; awns 1.2–3 cm long.
            4. Awn column (neck) often 1–2 mm long; awns (1.6) 2–3 cm long; Tiburón. _ **A. purpurea** (see "Doubtful and Excluded Reports")
            4' Awn column (neck) 8–11 mm long and twisted; awns 0.8–1.5 cm long; Nolasco. _____ **A. divaricata**

Aristida adscensionis Linnaeus

Impós; *zacate tres barbas*; six-weeks three-awn

Description: Non-seasonal annuals, the roots often weakly developed. Awn column short; awns flattened (a useful diagnostic character and one of striking beauty under magnification, easily seen with 10× magnification; each side of the green midrib has a translucent, thin-winged, serrulate margin). Drought-stressed plants sometimes have spikelets with reduced or aborted lateral awns, or all awns reduced or rarely awnless. This is the only annual *Aristida* in the Sonoran Desert.

Local Distribution: One of most common and ubiquitous grasses on Tiburón, from low to high elevations, and also common on San Esteban, Dátil, Mártir, and Nolasco.

Other Gulf Islands: Ángel de la Guarda, San Lorenzo, Tortuga, San Marcos, Coronados, Carmen, Monserrat, Santa Catalina, San José, San Francisco, Espíritu Santo, Cerralvo.

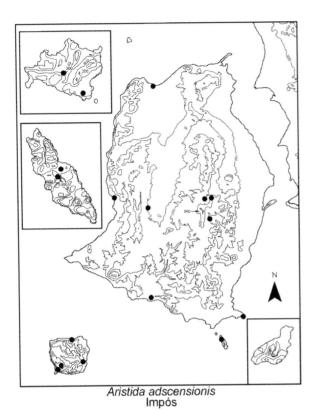

Aristida adscensionis
Impós

Geographic Range: This species occurs on all vegetated continents on earth and is widespread and common in the Sonoran Desert. Although some authors have claimed it is not native to the New World, it has been in the Sonoran Desert for at least 32,000 years (Van Devender et al. 1990b).

Tiburón: Ensenada Blanca (Vaporeta), *Felger 14977*. Arroyo Sauzal, *Wilder 06-94*. El Monumento, *Felger 15541*. SW Central Valley, *Felger 12304*. S-facing slope of Sierra Kunkaak, 354 m, *Wilder 06-3*. Top of Sierra Kunkaak Segundo, 490 m, *Wilder 06-488*. Vicinity of Siimen Hax, *Wilder 06-457*. Tecomate, *Felger 6263*.

San Esteban: Arroyo Limantour, *Van Devender 90-538*. Canyon at SE corner of island, *Felger 16633*. Steep slopes of S-central peak, *Felger 17528*. N side of island, steep slope near shore, *Felger 15433*.

Dátil: NW side of island, *Felger 15339*.

Mártir: Common, *Felger 6364*. Above ruins of the guano worker's village, about ⅓ way up island, 150 m, very common in some places between cardons, especially where the ground is not covered by *Vaseyanthus*, *Felger 07-15*.

Nolasco: NE side, very common, *Felger 11453*. E-central side, N exposure below crest of island, exposed ridgetop, locally common, *Felger 06-95*.

Aristida californica Thurber ex S. Watson

Two varieties are described.

1. Stems (at least the lower internodes) with fine to coarse white hairs. _____ var. **californica**
1' Stems glabrous. _____ var. **glabrata**

Aristida californica var. **californica**

Conee csaai "grass hairbrush"; *tres barbas de california*; California three-awn

Description: Non-seasonal tufted perennials, forming small, dense clumps. Awn column long, loosely twisted, and with a conspicuous joint (articulation), seen as a horizontal line toward the lower part of the awn column on mature spikelets; at maturity the column breaks apart at the joint; awns relatively long and slender. The Comcaac fashioned hairbrushes from bundles of the wiry roots (Felger & Moser 1985).

Local Distribution: Locally on dunes and sand soils along the eastern shore of Tiburón, and rare along large arroyos near the south shore of the island.

Other Gulf Islands: Ángel de la Guarda, San Lorenzo, San Marcos, Carmen.

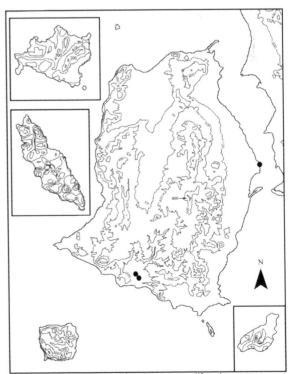

Aristida californica var. *californica*
Conee csaai

Geographic Range: Deserts in southeastern California, both Baja California states, southwestern Arizona, and coastal Sonora to northwestern Sinaloa.

Tiburón: Arroyo Sauzal: sandy floodplain, ca. 1 km inland, localized, *Felger 08-35*; 11 Sep 2007, *Romero-Morales 07-23*. Palo Fierro, locally abundant on dunes at beach, *Felger 12537*.

Aristida californica var. glabrata Vasey

Description: Similar to var. *californica* but with glabrous stems and the awns tend to be shorter.

Local Distribution: Interior of Tiburón on sandy-gravelly soils at the upper (southern) end of the Central Valley. This population is notable for occurring well within the desert.

Other Gulf Islands: Coronados (*R. Domínguez 1940*, CAS), Espíritu Santo.

Geographic Range: Arizona Upland across much of the southern half of Arizona, north-central and northeast Sonora, and both states of Baja California at the northern and southern margins of the desert. Also in grassland, oak grassland, and Chihuahuan Desert in southern Arizona and northern Sonora. This variety occurs on firmer ground than does var. *californica*, generally at higher elevation and/or areas of higher precipitation, and for the most part geographically peripheral to the desert variety (Reeder & Felger 1989).

Tiburón: Central Valley, 13 mi S of Tecomate, *Felger 17351*.

Aristida divaricata Humboldt & Bonpland ex Willdenow

Poverty grass

Description: Tufted perennials to about 60 cm tall, the inflorescence branches spreading often at 90°. Awn column often relatively long (8–11 mm long) and twisted, the lateral awns often shorter than the central awn (0.8–1.5 cm long). Reproductive during warmer months.

Local Distribution: Only on Nolasco, where it occurs in soil pockets and crevices on large rock outcrops and other open areas at high elevations. It seems to require relatively open microhabitats free from dense vegetation. The nearest known populations are in the northern part of Sierra el Aguaje (north of San Carlos) and at higher elevations in the Sierra Libre (northeast of San Carlos). These places and Nolasco are the closest this species comes to the desert.

Other Gulf Islands: None.

Geographic Range: Guatemala to southwestern United States (Kansas to Texas and southern California), including mountains above the desert in Baja California Norte, Sonora, and Arizona.

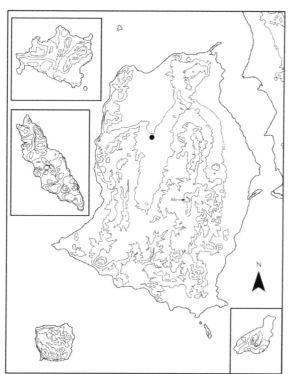

Aristida californica var. *glabrata*

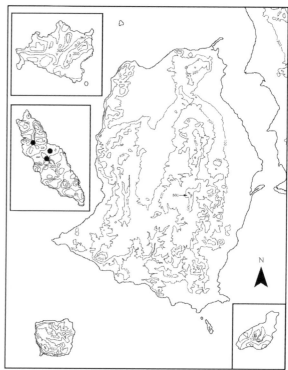

Aristida divaricata

Nolasco: E-central side, *Felger 9673b* (determined by J. R. Reeder, 1996). E-central side, N exposure about 15 m below crest of island, 250 m, open, exposed ridgetop with *Mammillaria multidigitata, Fouquieria diguetii*, etc., *Felger 06-96* (determined by J. R. Reeder, 2006). Cañón de Mellink, ridgetop of island, *Wilder 08-359*.

Aristida ternipes Cavanilles var. **ternipes**

Zacate araña; spidergrass

Description: Coarse, tufted perennials to ca. 1 m. Panicles with an open pattern of branching, the branches usually large and spreading at approximately 90°. Awns 1, well developed, straight or curved, 11–14 mm long, the two lateral awns absent or greatly reduced.

Local Distribution: Tiburón, widespread, mostly in rocky habitats, including higher elevations, except the northern and western margin of the island. Nolasco, common on both sides of the island, especially along the major canyons and grassy slopes.

Other Gulf Islands: Cerralvo.

Geographic Range: Often abundant and prominent in moderately arid and semi-arid places in the Sonoran Desert region. Arizona to Texas and Mexico to South America and the West Indies.

Tiburón: Ensenada de la Cruz, *Felger 9219*. SW Central Valley, *Felger 12427*. Haap Hill, *Felger T74-3*. Foothills of Sierra Kunkaak, SE part of island, *Felger 6984*. Pazj Hax, *Wilder 06-15*. S-facing slope of Sierra Kunkaak, 354 m, *Wilder 06-43*. Exposed upper ridge of island, between peak of Cójocam Iti Yaii and summit, rare, found only in protected localities inaccessible to bighorn grazing, *Wilder 07-575*. Base of Capxölim, N-facing rock slope, *Felger 07-106*.

Nolasco: SE side of island, *Sherbrooke* [*Felger 11242*]. NE side, 150 m, open exposed ridgetop with a S-facing exposure in a local area mostly free of *Vaseyanthus*, also scattered on grassy slopes at various exposures, *Felger 06-103*. Cañón de la Guacamaya, 14/15 Apr 2003, *Gallo-Reynoso* (photo). Cañón de Mellink, *Wilder 08-345*.

*Arundo donax Linnaeus

Xapij, xapijaacöl "large reedgrass," xapijaas; *carrizo*; giant reed

Description: Robust, leafy-stemmed bamboo-like grass to ca. 4 m tall. One might confuse it with *Phragmites australis*. The sure way to distinguish them is by the spikelets: *Phragmites* has densely hairy rachillas and glabrous lemmas, whereas *Arundo* has glabrous rachillas and conspicuously hairy lemmas. *Arundo* generally

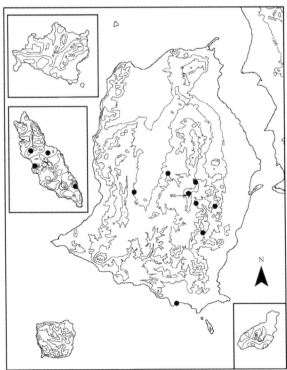

Aristida ternipes var. *ternipes*

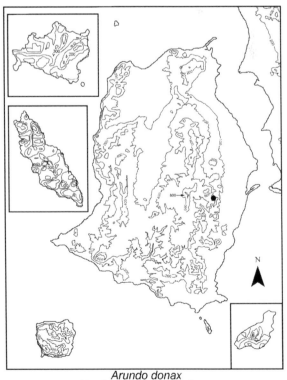

Arundo donax
Xapij, xapijaacöl, xapijaas

is more robust and generally produces only large stems; *Phragmites* usually has both larger and smaller, slender stems. The two reedgrasses can also be distinguished by microscopic examination of the phytoliths (Ollendorf et al. 1988).

Local Distribution: *Arundo* is known from Tiburón only at Sopc Hax. Given the close proximity to the Pazj Hax, where *Phragmites australis* grows, the occurrence of *Arundo* is a striking feature of the flora. *Arundo* was documented at Sopc Hax in 1963 and was still there in 2009. It grows at the rock tinaja, where it fills the waterhole, forming an impenetrable thicket about 18 × 10 m. Felger and Moser (1985:83, 305) mistakenly placed *A. donax* at Tinaja Anita (Pazj Hax) rather than at Sopc Hax.

The Comcaac used the stems to make their balsa boats or rafts, for musical instruments, and for many other purposes (Felger & Moser 1985; see *Phragmites*).

Other Gulf Islands: None.

Geographic Range: Native to the Old World and widely naturalized and cultivated in the New World including the Sonoran Desert. It long ago reached various isolated Sonoran Desert wetland sites.

Tiburón: Sopc Hax waterhole: *Felger 9284, 83-200*; *Wilder 07-223*; Carrizo Canyon [Sopc Hax], Cerro Kunkook [sic], *Knight 1085* (UNM).

Bothriochloa barbinodis (Lagasca) Herter [*Andropogon barbinodis* Lagasca]

Zacate popotillo; cane bluestem

Description: Perennials to 1 m tall. Usually villous with dense tufts of long, white hairs at nodes, ligules, and on inflorescences. Leaves drying red-brown, at least the bases semi-persistent. Panicles white and cottony, often 7–10 cm long with numerous branches clustered at the top of tall, nearly naked stems. Spikelets mostly paired and dimorphic, the sessile one with an awn 14–22 mm long.

Local Distribution: On San Esteban, one small population was discovered in a sheltered northwest-facing side canyon on the south side of the island. Nolasco, seasonally common on the relatively equable north-facing grassy slopes on the east side of the island. A record for it on Isla Espíritu Santo is the only other known Gulf Island locality. The nearest populations in Sonora are in the coastal mountains north of San Carlos and the closest peninsular location is in the Sierra San Francisco, Baja California Sur.

Other Gulf Islands: Espíritu Santo (*Wiggins 15239*, from US database, specimen not seen).

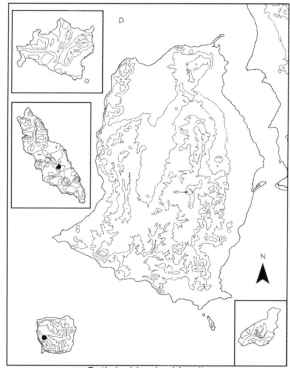
Bothriochloa barbinodis

Geographic Range: Widespread in the Americas, mostly in non-desert regions, Southwestern United States to southern Mexico including Baja California Norte and Sur, and in South America. Widely scattered within the Sonoran Desert at places of higher soil moisture and not in the open desert; often on north-facing slopes, canyons and waterholes, especially in mountains.

San Esteban: SW corner of island, sheltered canyon branched off from main drainage, rare, *Wilder 07-91*.

Nolasco: E-central side, steep N-facing slope, mid-elevation, common, *Felger 9670*.

Bouteloua – *Navajita*; grama grass

Annuals and low-growing perennials. Panicles with spicate branches.

1. Perennials with slender, arching stolons. _____
_____ **B. diversispicula**
1' Short-lived annuals, not forming stolons.
   2. Panicle branches readily deciduous, dart shaped (not pectinate), very slender, especially at base, and few-flowered. _____ **B. aristidoides**
   2' Panicle branches relatively persistent (although individual spikelets may be deciduous), conspicuously pectinate (comb shaped) and many-flowered. ___
_____ **B. barbata**

Bouteloua aristidoides (Kunth) Grisebach

Isnaap iic is "whose fruit is on one side"; *aceitilla*; six-weeks needle grama

Description: Summer-fall annuals sometimes reaching 45 (75–90) cm tall but usually much smaller; rarely occurring as spring annuals and then the plants stunted. Roots often weakly developed. Spicate branches 16–24 mm long, slender and dart-shaped, 6–17 per inflorescence stem, readily falling at maturity; individual spikelets appressed, several per spicate branch. The small, needle-like spicate branches and spikelets can be bothersome, sticking in socks and shoes. People may be supplanting animals as dispersal agents.

Local Distribution: Seasonally widespread and abundant from near the shore to higher elevations on all islands except Patos and Mártir. On Tiburón, during favorable years *B. aristidoides* and *B. barbata* may form extensive stands on the desert floor, such as in the Agua Dulce Valley. *Bouteloua aristidoides* is seasonally one of the most common grasses on the island.

Other Gulf Islands: Tortuga, Coronados, Carmen, Danzante, Santa Catalina, Santa Cruz, San Diego, San José, Espíritu Santo.

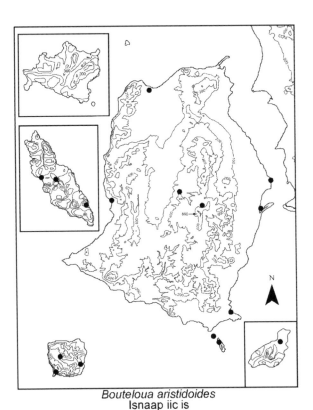

Bouteloua aristidoides
Isnaap iic is

Geographic Range: One of the most abundant hot-weather annuals in the Sonoran Desert. Southwestern North America to Central America and in Argentina.

Tiburón: Ensenada Blanca (Vaporeta), *Felger 14972*. Ensenada del Perro, *Wilder 08-301*. Zozni Cmiipla, *Felger 08-128*. Punta Tormenta, 1.5 km N of military (marine) station, *Wilder 06-348*. Haap Hill, *Felger T74-6*. Deep canyon on N slope of Sierra Kunkaak SW and up canyon from Siimen Hax, *Wilder 06-463*. Bahía Agua Dulce, *Felger 8884*.

San Esteban: SE corner of island, S fork of canyon, occasional, ca. 200 m, *Moran 8834* (SD). Arroyo Limantour: 3 mi inland, *Wilder 07-82*; Broad wash, *Van Devender 90-541*. Canyon, SW corner of island, *Felger 16632A*.

Dátil: NW side of island, steep rocky slopes, *Felger 2780*.

Alcatraz: On flat, near cardons back of beach dune, *Lowe 10 Nov 1969*. Flats, E side of island, dense colonies in open sandy areas, *Felger 14917*.

Cholludo: Abundant near top of island, *Felger 9157* (SD).

Nolasco: SE side of island, *Sherbrooke [Felger 11246]*. NE side, 15 m below crest of island, open exposed ridgetop, seen at low to high elevations on exposed sites without *Vaseyanthus*, *Felger 06-97*. Cañón de Mellink, *Wilder 08-344*.

Bouteloua barbata Lagasca var. **barbata**

Isnaap iic is "whose fruit is on one side"; *navajita barbada, navajita annual, zacate liebrero*; six-weeks grama

Description: Summer-fall annuals, sometimes persisting or growing with spring rains as stunted plants. Usually branched from near the base, the stems geniculate-spreading (abruptly bent upward like a knee-joint from near the base of the plant), 5–50+ cm tall, or sometimes reaching 80+ cm with straight, erect stems when crowded and vigorous. Spicate branches (1) 4–12 per inflorescence stem, (6) 10–25 (30) mm long, pectinate (comb shaped), arched or curved to straight, persistent, the individual spikelets spreading, many, and often falling separately from the spicate branch.

Local Distribution: Common throughout the lowlands of Tiburón; bajadas, valley plains, arroyos, and rocky mesas, hills and slopes. Also common on San Esteban and Alcatraz, and documented on Patos in 1921 where "several large colonies of this species were found with *Atriplex [barclayana]* on the guano-covered flats of Patos Island" (Johnston 1924:983).

Other Gulf Islands: Ángel de la Guarda, Partida Norte, Coronados, Carmen, Monserrat, San José, San Francisco, Espíritu Santo.

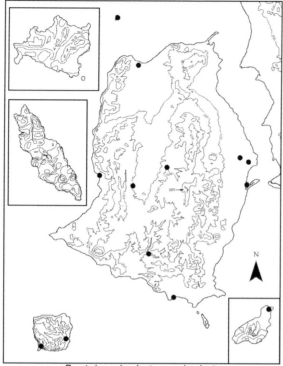

Bouteloua barbata var. *barbata*
Isnaap iic is

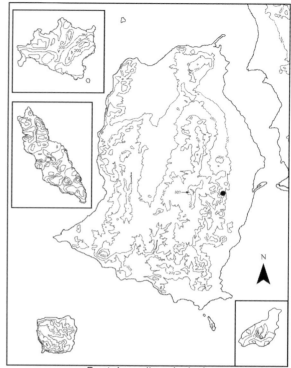

Bouteloua diversispicula

Geographic Range: Southwestern United States to southern Mexico and in Argentina. Together with *B. aristidoides* it is one of the most abundant and widespread summer annuals in the Sonoran Desert.

Tiburón: Ensenada Blanca (Vaporeta), *Felger 14971*. Xapij (Sauzal) waterhole, *Wilder 06-114*. Ensenada de la Cruz, *Felger 9233*. SW Central Valley, *Felger 17324*. Haap Hill, *Felger 76-T8*. Zozni Cmiipla, *Felger 08-127*. Palo Fierro, *Felger 8936*. 3 km W [of] Palo Fierro, *Smartt 13 Oct 1976* (UNM). Bahía Agua Dulce, *Felger 8880*.

San Esteban: Limantour, floodplain, *Felger 16607*. Canyon, SW corner of island, *Felger 16632*.

Alcatraz: E side of island, sandy flat, *Felger 14918*.

Patos: In dense masses W of old guano-gatherers' cabin on flat *Atriplex*-covered guano flats, 23 Apr 1921, *Johnston 3245* (CAS).

Bouteloua diversispicula Columbus [*Cathestecum brevifolium* Swallen]

Grama china

Description: Low, spreading perennials, forming small clumps and long, slender, arching stolons. Leaf blades short. Spicate branches probably 3–10 per inflorescence, laterally compressed and readily deciduous at maturity; male and female flowers on separate plants or spikelets variously with bisexual and/or unisexual flowers. Flowering most vigorously during the warmer times of the year.

Local Distribution: Documented on rocky hills along the east flank of Sierra Kunkaak.

Other Gulf Islands: None.

Geographic Range: Arizona and Sonora to Central America.

Tiburón: Lower foothills, between Sopc Hax and Hant Hax camp, arroyo bottom on rocky outcropping, *Felger 9327*.

Cenchrus

Annuals or perennials; spikelets enclosed in fascicles with feathery bristles or burs with hard, sharp spines, the fascicles and burs fall as a unit.

1 Spikelets enclosed in fascicles with flexible bristles; bristles scarcely united at the very base. __ **C. ciliaris**
1' Spikelets enclosed in burs with sharp, stiff spines; spines conspicuously united at least in lower ¼ of the bur.
  2. Inflorescence with more than 10 burs; larger spines of bur 3.5–6 (7) mm long. _____ **C. echinatus**
  2' Inflorescence with 1–3 burs; spines of bur (6) 9–15 mm long. _____**C. palmeri**

*Cenchrus ciliaris Linnaeus [*Pennisetum ciliare* (Linnaeus) Link]

Oot iconee "coyote's grass"; *zacate buffel, buffel;* buffelgrass

In May 2007, Humberto asked Steve and Cathy Marlett for suggestions for a Seri name for *buffel*—so that the people could recognize it by name. The three of them decided to call it oot iconee, because unusual or odd items are said to belong to the coyote.

Description: Robust perennials and often flowering in the first season; growing and flowering during the warmer months. Panicles 4.5–12 cm long, densely flowered and feathery, the spikelets in crowded fascicles with feathery bristles. The small fascicles are dispersed by wind, animals, and people.

Local Distribution: It has been in the vicinity of Caracol on Tiburón at least since 1998 (West & Nabhan 2002). This has been the most actively used and disturbed site in the interior of the island, and where the Seri guides often stay with bighorn hunters during prolonged hunting activities on the island. The Caracol buffel population was the largest infestation known on the island, and some plants were 1.5 m tall. Several additional populations were found in 2007 along the lower

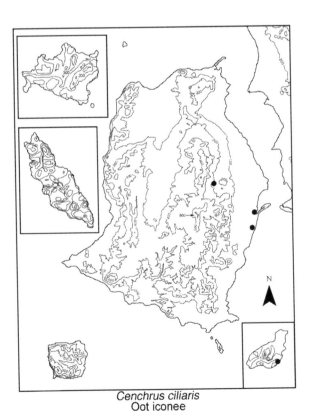

Cenchrus ciliaris
Oot iconee

bajada bordering the Infiernillo shore. In May 2007, Ben, Seth Turner, and Humberto removed the known buffelgrass plants on Tiburón, although a seed bank remained. Later in the year and in 2008 Humberto and a crew of Seri workers removed additional *oot iconee* from the Caracol area. In 2008 Humberto led "the plant team," a group of ten youthful Seris who searched all the islands in the area for new populations of buffel, including the areas on Tiburón, where buffel was previously known. No new populations were encountered, and established populations controlled in 2007 on Tiburón were revisited with new plants removed, reducing the seed bank substantially. Monitoring invasive species on the islands is an indefinite task, yet catching problem species while populations are small will likely lead to successful results.

Buffelgrass was reported for Alcatraz (West & Nabhan 2002; West et al. 2002). In 1997, Tad Pfister (personal communication, 23 May 2006) and Patty West found a patch of buffelgrass just below the main cave on east side of Alcatraz. Tad pulled it up and bagged it, and remembers seeing no new plants. We did not see it when we surveyed the island in 2007 and 2008.

On the Sonoran mainland buffelgrass is extensively planted for cattle grazing after removing the desert vegetation. For example, a large plot was observed in 2007 at the western base of Pico Johnson of the Sierra Seri. Buffelgrass is beginning to colonize slopes of the Sierra Seri and adjacent peaks. Buffelgrass lines Highway 10 linking Hermosillo and Bahía de Kino and is especially thick at the entrance to the coastal town. It is also along the length of the road between Kino Nuevo and Punta Chueca, and from Punta Chueca to El Desemboque, and is expanding in this area.

The Midriff Islands are free of significant populations of buffelgrass. Unlike most of the Sonoran Desert, a unique opportunity exists to control this invasive species on the Gulf Islands before it becomes nearly impossible to thwart its invasion and the subsequent ecosystem transformation associated with this noxious weed (see Búrquez et al. 2002; Franklin et al. 2006). This species is one of the most serious conservation threats to Tiburón and other Gulf Islands (West & Nabhan 2002; Wilder et al. 2007b).

Other Gulf Islands: None.

Geographic Range: Native to the Old World and widely introduced for forage and fodder in arid and semi-arid regions worldwide, often becoming seriously invasive.

Tiburón: Caracol, small canyon arroyo 100 m W of station: Scattered plants along a 100 m section of the arroyo, the total [observed] population around 50 plants, 24 May 2006, *Wilder 06-151*; Population expanded since May 2006, buffel lines arroyo for 200 m, all plants removed, 3 May 2007, *Wilder 07-231*. Vicinity of Estero San Miguel, 0.75 km S of Zozni Cmiipla, 400 m inland, scattered clusters of plants, 28°57.824'N, 112°13.241'W, total ca. 200 plants, many seedlings, all removed, 4 May 2007, *Wilder 07-273*; revisited on 26 Sep 2008, ca. 100 new seedlings, all removed, and adjacent area searched with no other plants seen. Lower E bajada, 3.75 km S of Zozni Cmiipla, 500 m inland, along road to Pazj Hax, 28°56.171'N, 112°13.503'W, single plant, removed, 2 May 2007, *Wilder*, observation.

Alcatraz: Reported for the island (West & Nabhan 2002).

*Cenchrus echinatus Linnaeus [*C. insularis* Scribner]

Guachapori, zacate toboso; southern sandbur

Description: Warm weather annuals. Inflorescences spike-like to 7.5 cm long, with 18–40 burs. Burs often 4–5 mm diameter (not including spines), with a basal ring of often 30–50 small, slender bristles; larger spines flattened, 3.5–6 (7) mm long.

Local Distribution: Known from the islands by a single record. This unpleasant weed is common at Bahía

Kino and Kino Nuevo and may have reached the island from there. Its occurrence at a fishing camp is evidence of the role of fishermen, who often spend one to several nights on the islands, or even tourists being vectors.

West and Nabhan (2002) report *C. brownii* Roemer & Schultes for Tiburón, but it is in fact *C. echinatus*, and their report is based on the specimen cited here. The taxonomic distinction between the two species is subtle.

Other Gulf Islands: None.

Geographic Range: Widespread, unwelcome and often weedy in disturbed habitats, mostly sandy or fine-textured soils including upper beaches. Southern United States to South America and adventive in the Old World.

Tiburón: Fishing camp at Ensenada del Perro, just above high tide, *Bourillón 31 Oct 1993*.

Cenchrus palmeri Vasey

Cözazni "tangled" (variant: czazni); *guachapori*; giant sandbur

Description: Annuals, found at least from August to May, depending upon soil moisture and winter cold, highly variable in size, stems often (5) 8–30 cm long. Roots unusually small and weakly developed. Burs 1–3

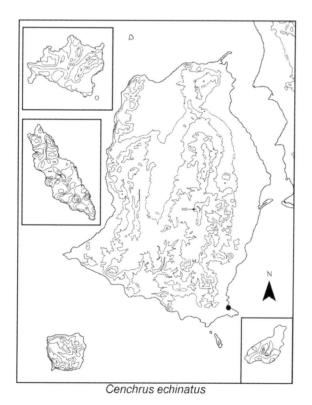

Cenchrus echinatus

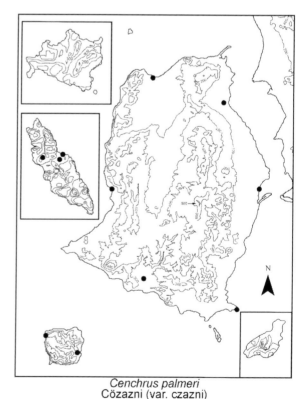

Cenchrus palmeri
Cözazni (var. czazni)

per inflorescence, 20–31 mm diameter, with sharp stiff spines (6) 9–15 mm long, dark purple-brown most of the year, green during hot, humid weather. The obnoxious burs persist long after the plant dies, often hidden in the sand.

No other *Cenchrus* species has so small a geographic range, such large burs, or so few burs per inflorescence (DeLisle 1963). As with other *Cenchrus*, seedlings develop within the bur, which may remain attached to the root of the mature plant. Johnston (1924:984) reported, "When present the plant was common, for the vicious burs were ubiquitous, and heedless kneeling on the ground nearly always produced specimens." The burs not only puncture sleeping bags, tents, and bare feet at the beach, but they cling together tenaciously in nasty clusters, often attaching to pant legs and shoes.

Local Distribution: On Tiburón, widespread in lowlands, especially on sandy soils and near the shore but also on rocky slopes. Sometimes it becomes incredibly abundant, such as on the low dunes and sandy expanses at Tecomate. On San Esteban, common in arroyos, especially near the shore. On Nolasco, common across the island and sometimes seasonally abundant, the inflorescences sometimes blackened with a smut.

Other Gulf Islands: Ángel de la Guarda (observation), San Lorenzo, San Marcos (observation), Coronados, Carmen, Danzante (observation), Monserrat, San José, San Francisco, Espíritu Santo, Cerralvo.

Geographic Range: Both states of Baja California, most islands in the Gulf of California, western Sonora northward to the Pinacate region near the United States border, and northwestern Sinaloa. This is one of the few grasses nearly endemic to the Sonoran Desert, only ranging south of the desert along the arid coast to northwestern Sinaloa.

Tiburón: Ensenada Blanca (Vaporeta), rocky hillside, 29 Jan 1965, *Felger*, observation (Felger 1966:172). Arroyo Sauzal, 2¾ mi inland, arroyo, *Wilder 06-93*. El Monumento, *Felger 15555*. Zozni Cmiipla, *Wilder 06-356*. Hahjöaacöl, NE coast, *Wilder 08-221*. Tecomate, abundant on dunes, and seen over wide areas of desert floor, *Whiting 9069*.

San Esteban: Arroyo Limantour, *Van Devender 90-539*. Rocky cove, W side of island, *Felger 16656*.

Nolasco: NE side, N-facing grassy slopes, *Felger 11447*. E side, above landfall, 25 m, *Turner 79-253*. Cañón el Faro, 60 m, ridgetop with sparse vegetation, on various exposures where *Vaseyanthus* is not dense or absent, not seen at higher elevations, *Felger 06-84*. Cañón de Mellink, S-facing canyon slope, 110 m, *Felger 08-137*.

Chloris – Fingergrass

Enteropogon, treated here as a separate genus, was formerly included in *Chloris*. These grasses have spikelets with 2 florets and each is awned. The lower floret is fertile and larger, and the upper one is sterile.

1. Perennials; florets without a tuft of long hairs, the fertile and sterile florets similar except in size. _____ _____ **Enteropogon chlorideus**
1' Annuals; fertile and sterile florets very different from each other, the fertile lemma humpbacked on keel with a conspicuous tuft of hair. _____ **C. virgata**

Chloris virgata Swartz

Zacate lagunero; feather fingergrass

Description: Warm-weather annuals, highly variable in size, 10–80 cm tall. Spikes 4–17 in number, digitate at stem apex, appearing feathery, densely flowered with whitish to tawny silky hairs. Spikelets with 2 florets, each with a prominent awn. Fertile and sterile florets differently shaped. Fertile lemma humpbacked on keel, bearing a conspicuous tuft of hair at apex and a stout awn 5–7.5 mm long.

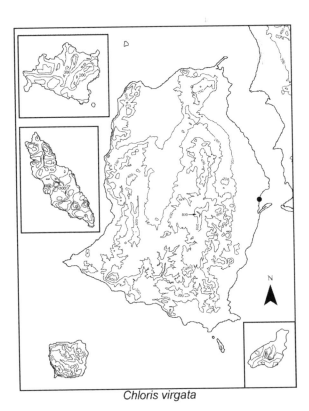

Chloris virgata

Local Distribution: We found it on Tiburón after the strong warm-weather rainy season of 2006. It was abundant on the lower portion of the eastern bajada, extending southward from Zozni Cmiipla near the shore for several kilometers. It is common and weedy on the Sonoran mainland in the Bahía Kino region. The closest peninsular location is in the Sierra San Francisco.

Other Gulf Islands: Tortuga, Espíritu Santo.

Geographic Range: Southwest and Midwest United States through tropical America, and introduced in the Old World.

> **Tiburón:** Zozni Cmiipla, abundant in water channels and decreasing in density closer to the *Frankenia* zone, *Wilder 06-353.*

Dasyochloa pulchella (Kunth) Willdenow ex Rydberg [*Erioneuron pulchellum* (Kunth) Tateoka. *Tridens pulchellum* (Kunth) Hitchcock]

Conee ccosyat "spiny grass," conee ccapxl "sour grass"; *zacate borreguero*; fluff-grass

Description: Dwarf, tufted perennials less than 14 cm tall, with stolons of one long internode, the longer ones 7–11 cm, bearing at the top a tightly fascicled cluster of short leaves and inflorescences, the stolons often arching and bending to the ground but seldom if ever taking root. Leaf blades 1.7–7 cm long, less than 0.5 mm wide, the margins firm and often white. Panicles compact and dense, on peduncles shorter than the longer leaves. Spikelets 6–10 mm long, with (4) 6–8 florets; lemmas deeply 2-lobed at the tip with an awn 1–2 mm long from between the lobes. Glumes and lemmas papery. Reproductive during warm weather and sufficient soil moisture; recruitment occurs with summer rains. During the driest times the herbage, or even the entire plant, may be drought killed but persist, acting as a nurse plant for seedlings or shelter for new emerging shoots.

Local Distribution: Tiburón on the arid western and southern parts of the island. Known from one record on San Esteban and one on Dátil.

Other Gulf Islands: Ángel de la Guarda, San Lorenzo.

Geographic Range: Arid and semi-arid southwestern United States and northern Mexico from Baja California Norte to Coahuila, Zacatecas, and Aguascalientes. This is one of the most arid-inhabiting perennial grasses in the Sonoran Desert, including Sonora southward to the Sierra Seri and the vicinity of Hermosillo, Baja California Norte southward to Sierra San Borja, and the larger northern islands in the Gulf of California.

> **Tiburón:** Ensenada Blanca (Vaporeta), *Felger 17282.* Willard's Point, canyon just N, *Moran 8751* (SD). El Sauz, beach, *Moran 8798* (SD).
>
> **San Esteban:** Canyon, SW corner of island, *Felger 16639.*
>
> **Dátil:** Seen only in one place, *Moran 13034* (SD).

Digitaria californica (Bentham) Henrard var. **californica** [*Trichachne californica* (Bentham) Chase]

Cpooj; *zacate punta blanca*; Arizona cottontop

Description: Tufted perennials with hard, knotty bases; winter or dry-season dormant. Stems about 0.5+ m tall, firm, with felt-like hairs. Panicles white to purplish due to silvery or purple-tinged silky hairs on the spikelets, giving a cottony appearance. Spikelets in pairs, one long-pedicelled, the other one short-pedicelled, or the second spikelet on portions of a branch sometimes not developing. Spikelets 3–3.7 mm long (excluding the hairs). Lower glumes much reduced.

Local Distribution: Widely scattered from low to high elevations on Tiburón and San Esteban, and also on Dátil and Cholludo; canyons and arroyos, and rocky slopes. On Mártir, localized at the top of the island. On Nolasco,

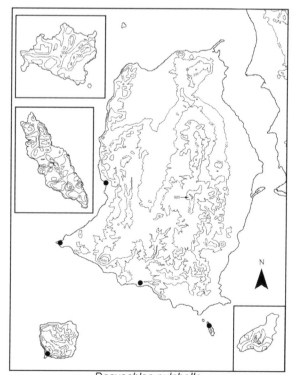

Dasyochloa pulchella
Conee ccosyat, conee ccapxl

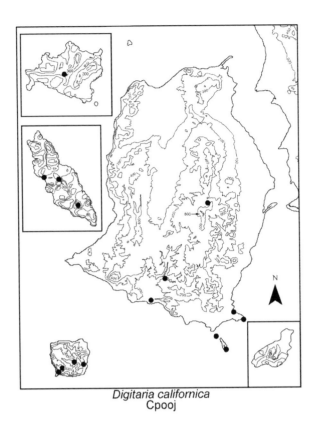

Digitaria californica
Cpooj

common on the east side, especially north-facing grassy slopes, and canyons at lower elevations on the west side.

Other Gulf Islands: San Marcos, Cerralvo.

Geographic Range: Southwestern United States (rare in California) and Mexico, the Caribbean and in South America. An additional variety occurs in South America and the Caribbean.

Tiburón: El Sauz, occasional on N slope, *Moran 8788* (SD). Xapij (Sauzal) waterhole, *Wilder 06-115*. El Monumento, *Felger 15549*. Ensenada del Perro, *Tenorio 9504*. Canyon, N base of Sierra Kunkaak, 340 m, *Wilder 06-415*.

San Esteban: Canyon at SW corner of island, *Felger 16642*. Arroyo Limantour, uncommon perennial in rocky canyon, *Van Devender 90-536*. Steep slopes, S-central peak, *Felger 17536*. Rock crevice, main arroyo [Limantour], 130 m elev., *Moran 28173* (SD).

Dátil: SE side, canyon, *Felger 17507*.

Cholludo: Near top of island, not common, *Felger 9167B*.

Mártir: High elevation plateau, to the NW of highest point of island, 275 m, seen only at this locality, mostly at bases of *Pachycereus pringlei*, 28.38072°N, 112.30832°W, 6 Dec 2007, *Wilder 07-604*.

Nolasco: [S]E side, 250 ft, *Pulliam & Rosenzweig 20 Mar 1974*. E-central side, N exposure below crest of island, 270 m,

perennials and some are reproductive in first year or season, scattered at all elevations, *Felger 06-102*. Cañón de Mellink, *Wilder 08-347*.

Distichlis

Perennial saltgrasses, mat-forming with rhizomes and stolons. Leaves 2-ranked (distichous), firm and sharp-pointed, with salt-excreting bicellular hairs sunken in leaf tissue. Male and female flowers on separate plants; spikelets few- to many-flowered and awnless.

We have looked for *D. palmeri* on Tiburón but have not found it there. It grows in tidal marshes, the southern limit in Sonora being Estero Santa Cruz at Bahía Kino, where it is locally abundant. Sterile specimens are often not reliably distinguishable from sterile *D. spicata* or *Sporobolus virginicus*.

1. Leaf blades less than 1 (1.5) cm long; inflorescences inconspicuous, hidden among leaves. _____**D. littoralis**
1' Leaf blades more than 2 cm long; inflorescences conspicuous, not hidden.
   2. Lemmas 7–16 mm long; anthers 3.8–4.9 mm long. _____**D. palmeri** (not known from the flora area)
   2' Lemmas 3–6 mm long; anthers 1.8–2.6 mm long. _____**D. spicata**

Distichlis littoralis (Engelmann) H. Bell & Columbus
[*Monanthochloe littoralis* Engelmann]

Cötep; *zacate playero*; shore grass

Description: Perennials forming dense mats of wiry stems, upright, creeping, or spreading from numerous rhizomes and stolons. Leaves firm and sharp-pointed, short (smallest leaves of any grass in region), the blades (4) 5–9 mm long (one specimen, *Felger 9069*, has some long-shoot leaf blades 15 mm long). Inflorescences reduced to inconspicuous, leaf-like single spikelets nearly hidden among leaves at ends of short shoots but becoming conspicuous on staminate plants when the anthers are exerted. Glumes none. Simultaneous mass flowering in March.

According to a humorous Seri legend, one who feels pain when walking barefoot on this sharp-leaved grass was said to be stingy. It was used as roofing for traditional brush houses, which provided protection against light rain. This grass was considered one of the best plants for making smoke signals (Felger & Moser 1985).

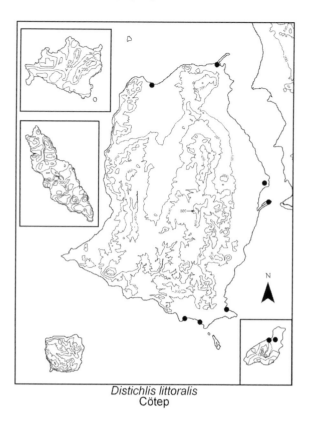

Distichlis littoralis
Cötep

Local Distribution: Dense and extensive colonies occur on Tiburón in muddy soils of esteros and tidal flats inundated by higher tides, at the inland margins of mangroves with other maritime halophytes, and other low-lying areas of saline soils near the shore. Also locally common on Alcatraz near the shore and in the saline flat.

Other Gulf Islands: Ángel de la Guarda, Rasa, Coronados, Santa Catalina, San José, San Francisco, Espíritu Santo, Cerralvo.

Geographic Range: Coastal marshes in southern California, the Pacific and Atlantic sides of the northern half of Mexico including both coasts of the Gulf of California and the Baja California Peninsula, and Texas to Florida, and Cuba. Also inland in southern Texas and Cuatro Ciénegas, Coahuila.

> **Tiburón:** Ensenada de la Cruz, mud flats near coast, *Felger 9069*. Coralitos, *Felger 15378*. Ensenada del Perro, *Wilder 07-159*. Punta San Miguel, *Wilder 07-199*. Palo Fierro, edge of mangroves, *Felger 12562*. Punta Perla, *Felger 21312*. Tecomate, arroyo bottom at waterhole, *Felger 8891* (SD).
>
> **Alcatraz:** E side of island, locally common in *Allenrolfea*-dominated saline flat, *Felger 12717*. NW side, cobble/shell rock berm along shore, locally common with *Cressa*, *Atriplex barclayana*, *Allenrolfea occidentalis*, *Sesuvium portulacastrum*, *Felger 07-179*.

Distichlis spicata (Linnaeus) Greene [*D. stricta* A. Gray. *D. spicata* (Linnaeus) Greene var. *stricta* (A. Gray) Beetle]

Zacate salado; saltgrass

Description: Perennials with creeping rhizomes, forming extensive colonies. Leaf blades often 4–8 cm long. Panicles slender, often 3–7 cm long, the spikelets 5.5–14 mm long, laterally compressed, with few to many florets, soon becoming straw colored. Male panicles usually overtopping the leaves, the spikelets sometimes longer than the female spikelets. Female panicles more or less as tall as the leaves. Growing and reproductive during the warmer months and winter dormant.

Local Distribution: At opposite ends of Tiburón. Sauzal at the south, where it is locally common at the waterhole in alkaline, wet soil and also in saline soil among coastal halophytes at the shore. Also at the estero at Punta Perla on the northeast side, where it is common at the edge of mangroves. The nearest population is in the vicinity of Bahía Kino.

Other Gulf Islands: None.

Geographic Range: Coastal regions and interior basins, Canada to Mexico, the West Indies, and South America. Both Baja California states and Sonora in widely scattered localities near the coast, and especially abundant and widespread in the Río Colorado delta region.

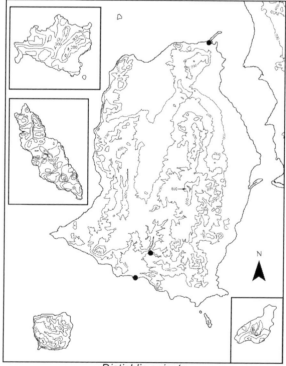

Distichlis spicata

Tiburón: Sauzal waterhole, *Felger 2749.* Arroyo Sauzal, with other halophytes near the shore, *Felger 08-34.* Punta Perla, estero, *Wilder 08-378.*

Enneapogon desvauxii P. Beauvois

Zacate lobero; spike pappusgrass

Description: Small, tufted perennials mostly 20–35 cm tall, with hard, knotty bases containing cleistogamous, awnless spikelets. Blades wiry, less than 10 cm long and 0.5 mm wide, rolled up during drier times, or flat and reaching 1.5 mm wide during moist times (e.g., with summer-fall rains). Panicles slender. Spikelets 4.5–7 mm long including awns, lead-gray or gray-green, mostly 3-flowered, only the lower one bisexual. Lemmas rounded on the back, with 9 prominent veins extending into 9 feathery awns, making this grass unmistakable (*Enneapogon* is derived from the Greek words for "nine beards").

Local Distribution: Known from the Central Valley and Sierra Kunkaak on Tiburón. The nearest known populations are in western Sonora in the vicinity of Hermosillo, but it can be expected closer in coastal mountains. The closest peninsular population is in Baja California Norte just north of Calmallí, about 140 km from Tiburón.

Other Gulf Islands: San José.

Geographic Range: Southwestern United States and southward in Mexico to Oaxaca, South America, and widespread in the Old World.

Tiburón: Haap Hill, *Felger 76-T17.* Deep canyon, N slope of Sierra Kunkaak, SW and up canyon from Siimen Hax, *Wilder 06-464.*

Enteropogon chlorideus (J. Presl) Clayton [*Chloris chlorridea* (J. Presl) Hitchcock. *C. brandegeei* (Vasey) Swallen. *Enteropogon brandegeei* (Vasey) Clayton. *Gouinia brandegei* (Vasey) Hitchcock]

Verdillo cacahuatoide; buryseed umbrella-grass

Description: Coarse tufted perennials; glabrous. This grass is unusual in forming large cleistogamous spikelets at the rhizome tips (resembling little scaly tubers). Panicles with 3–15 racemosely arranged, slender and often spreading spicate branches to about 10 cm long, the branches usually more than one at a node. Spikelets slender and not crowded, with 1 bisexual floret and 1 sterile floret, each with a 1-awned lemma. Fertile floret below and much larger than the sterile floret; fertile lemma 5.5–7 mm long (not including the awn).

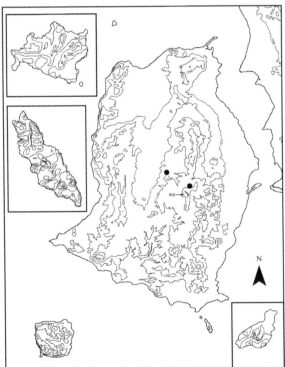

Enneapogon desvauxii

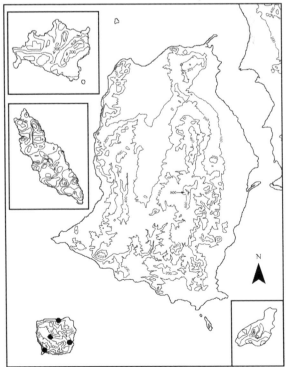

Enteropogon chlorideus

Gould and Moran (1981) indicated that *E. brandegeei* might be conspecific with *E. chlorideus*, from which it differs most notably by the supposed lack of buried cleistogenes and in having shorter awns on the spikelets. The reported absence of cleistogenes among *E. brandegeei* is not convincing because most herbarium specimens lack extensive root system, and collectors seldom inspect plants for presence of cleistogenes. Furthermore, we found a number of *E. brandegeei* specimens with rhizomes and cleistogenes. The shortest awns (0.7–1.9 mm) are seen on plants from San Esteban. Plants from Baja California Sur have awns 2.7–9.4 mm long; the longer awns seem correlated with increased soil moisture. The variation from short awned to longer awned is continuous, but none of the peninsular specimens have awns as long as some non-desert mainland populations. Thus, apart from their usually shorter awns, *E. brandegeei* appears indistinguishable from the more widespread *E. chlorideus*.

Local Distribution: This is the most common perennial grass on San Esteban, where it is most numerous on rocky benches along arroyos.

Other Gulf Islands: Coronados, Carmen, Monserrat, Santa Catalina, Santa Cruz, San José, Cerralvo.

Geographic Range: *Enteropogon chlorideus* occurs in Baja California Sur, and Arizona and southern Texas to Central America. *Enteropogon brandegeei* is nearly endemic to Baja California Sur, from the vicinity of San Ignacio to the Cape Region and adjacent islands, and on San Esteban.

San Esteban: E bay [Limantour], *Knobloch 2394* (SD, pencil annotation, 1970, probably by Frank Gould, points out two rhizomes). Arroyo Limantour, at shore, broad plain of arroyo, *Wilder 07-72* (cleistogenes present among roots). Arroyo Limantour, uncommon perennial in broad wash, *Van Devender 90-542* (awns 1.6 mm long); *Van Devender 92-487* (awns 0.7–1.9 mm long). N side of island, *Felger 15407*. Canyon at SW corner of island, *Felger 16634*. In large tufts along border of washes, usually in small colonies, 20 Apr 1921, *Johnston 4399* (CAS).

Eragrostis pectinacea (Michaux) Nees var. **pectinacea** [*E. diffusa* Buckley. *E. pectinacea* var. *miserrima* (E. Fournier) J. Reeder. *E. arida* Hitchcock. *E. tephrosanthos* Schultes]

Carolina lovegrass

Description: Summer-fall annuals, sometimes growing with winter-spring rains or germinating in fall and persisting through the winter; highly variable in size, to 75 cm tall, the panicles with few to many branches.

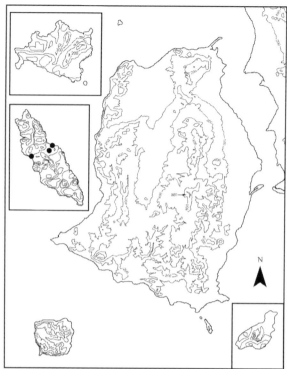
Eragrostis pectinacea var. *pectinacea*

Local Distribution: On Nolasco, where it is widespread and often common on both sides of the island, canyon bottoms and slopes at all elevations, especially on north-facing slopes and larger canyons to ridgetops. It is strange that it is not known from any other Gulf Island.

Other Gulf Islands: None.

Geographic Range: Across much of the Sonoran Desert except the driest areas of the Lower Colorado Valley, and one of the most widespread hot-weather annual grasses in Sonora, Baja California Sur, and southern Arizona. Canada to Argentina. An additional variety occurs in Florida. This is the most widespread and abundant New World species of *Eragrostis*, vying only with the now cosmopolitan *E. cilianensis* (native to the Old World) for geographic expanse.

Nolasco: E side of island, above landfall, 25 m, growing near a moist tinaja, 30 Sep 1979, *Turner 79-248*. Cañón el Faro, 60 m, abundant, highly variable in size, canyon bottom and also on slopes, seen at all elevations, *Felger 06-88*. Cañón de Mellink, at base of canyon, *Felger 08-140*.

Heteropogon contortus (Linnaeus) P. Beauvois ex Roemer & Schultes

Zacate colorado; tanglehead

Description: Robust tufted perennials; dry leaves rust colored and persistent. Reproductive during warmer

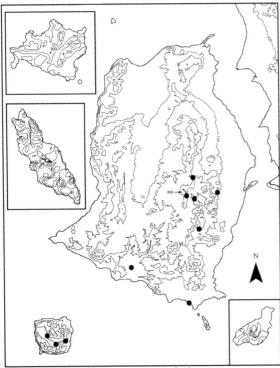

Heteropogon contortus

months. Inflorescence a solitary spike-like 1-sided raceme (the awns are all on one side). The reproductive structure of this andropogonid grass is complex. Spikelets in pairs, one very different from the other, one larger and pedicelled and one sessile, except a few spikelet pairs at lower nodes of the inflorescence are staminate and sessile. Pedicelled spikelet awnless and staminate or sterile, with a slender pedicel-like callus, the actual pedicel reduced to a little stump. Sessile spikelet bisexual or pistillate with a large, stout, hairy awn 4.5–7+ cm long, twisted and twice bent when mature, tawny brown with a needle-sharp base that "readily penetrates clothing and is a ferociously efficient dispersal mechanism" (Clayton & Renvoize 1986:359). Each spikelet of the pair is 1-flowered or appearing 1-flowered, the lower floret greatly reduced or absent.

Local Distribution: Tiburón, arroyos, canyons, hills and mountains in scattered localities in the Sierra Kunkaak and at the south side of the island. San Esteban, rocky canyon sites and along Arroyo Limantour. Nolasco, occasional at the ridge crest of the island.

Other Gulf Islands: Ángel de la Guarda, San Lorenzo, San Marcos, Carmen, Coronados, Danzante, Monserrat, Santa Catalina, San Diego, San José, San Francisco, Espíritu Santo, Cerralvo.

Geographic Range: Southwestern United States to South America and warm regions of the Old World.

Although some authors have claimed it is "adventive in America since the time of Columbus" (Correll & Johnston 1970:201; also see Barkworth et al. 2003), it has been in southwestern North America for at least 7900 years (Van Devender et al. 1990b; Felger 2000a).

Tiburón: El Sauz, only one place, on N slope, ca. 200 m, *Moran 8791* (CAS). Coralitos, *Wilkinson 12 Oct 1977.* Foothills of [E side] Sierra Kunkaak, locally abundant in arroyo bottom at edge of mountains, *Felger 6959.* Hant Hax camp, not common, upper edge of bajada, *Felger 9353.* Hill above valley, E base of Sierra Kunkaak, 312 m, *Wilder 06-5.* San Miguel Peak, *Knight 974* (UNM). Capxölim, *Wilder 07-484.* **San Esteban:** Canyon, SW corner of island, *Felger 16650.* Arroyo Limantour, 3 mi inland, bench of arroyo, *Wilder 07-81.* Near S end, large arroyo running inland from E [Arroyo Limantour], *Wiggins 17226* (CAS). Rocky canyon, *Johnston 3208* (Johnston 1924:985, specimen not located). **Nolasco:** Summit of island, *Wilder 09-141.*

Lasiacis ruscifolia (Kunth) Hitchcock var. ruscifolia

Carrizito

Description: Bamboo-like perennials, 1.2–1.6+ m tall with the major stems woody. Leaf blades broadly ovate to elliptic, 7.4–8.4 cm long, 2.1–2.6 cm wide. Panicles openly branched, few- to many-flowered. Spikelets firm

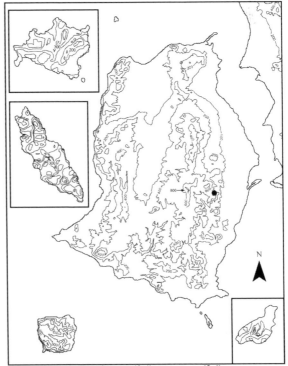

Lasiacis ruscifolia var. *ruscifolia*

and globose, 2.3 mm wide, the immature fruits green, the ripe fruits blackish and somewhat fleshy (glumes blackening and accumulating oil when mature); fertile lemma smooth and blunt with a tuft of hair at the tip.

Apart from the reed grasses (*Arundo* and *Phragmites*), *L. ruscifolia* is the only broad-leaved grass in the flora area. The genus is decidedly mesic-inhabiting and of tropical affinity. Fruit dispersal is by birds, and the grain is viable after passing through a bird gut. Adaptations for bird dispersal include the berry-like fruit, glumes modified for oil production, spikelets shiny black when ripe, and thick, tough lemmas and paleas to protect the grain. This adaptation is unique among the grasses (Davidse & Morton 1973).

Local Distribution: The only record in the region is from Sopc Hax in the Sierra Kunkaak, where a small population grows next to the waterhole in the shaded shelter of a north-facing canyon wall and beneath trees.

Other Gulf Islands: None.

Geographic Range: This tropical species extends northward in western Mexico to the southern margin of the Sonoran Desert, where it is localized in densely vegetated riparian canyons, in Sonora in mountains north of San Carlos, Baja California Sur mountains including the Sierra de la Giganta, and on Tiburón. Florida, Sonora, and Baja California Sur to South America. Another variety occurs in Central America and Venezuela.

Tiburón: Carrizo Canyon [Sopc Hax], *Knight 1086* (UNM). Sopc Hax, shade of *Lysiloma divaricatum*, about 4 ft high, *Felger 9312*.

Leptochloa

1. Perennials with hard, knotty bases often bearing cleistogenes (sometimes flowering in the first season); spikelets with 5–12 florets; lemma truncate with 2 broad lobes, the midrib often extending into a short bristle between the lobes. _____**L. dubia**
1' Annuals, without cleistogenes; spikelets (1) 2- or 3-flowered; lemmas not lobed and without a bristle. _____ **L. panicea**

Leptochloa dubia (Kunth) Nees

Green sprangletop

Tufted perennials to 70+ cm tall, with a tough, knotty base and well-developed roots, or sometimes flowering in the first season. Cleistogamous spikelets often

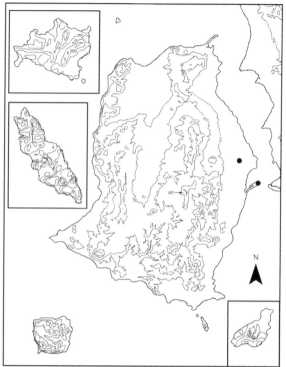

Leptochloa dubia

at stem bases. Inflorescences often 10–15+ cm long, often with 10 or more branches. Florets of young spikelets are closely spaced, but at maturity they are well separated from each other and the rachis is readily visible. Spikelets with 5–12 florets. Glumes persistent, the spikelets breaking apart above the glumes and between the florets, each floret falling with its segment of the rachis. Lemmas 2.1–5 mm long, broadly truncate (blunt-ended), the tip with 2 broad lobes and minutely fringed, the midrib extending into a bristle within the notch.

Local Distribution: Known from two lowland records from Tiburón, but we would expect its occurrence elsewhere such as in the Sierra Kunkaak.

Other Gulf Islands: San José, Espíritu Santo.

Geographic Range: Arizona to Colorado and Texas, Florida, Mexico including Baja California Sur to Central America, and in South America.

There are few records for this species at lower elevations within the Sonoran Desert. In north-central and northeastern Sonora, both states of Baja California, and Arizona it commonly occurs at elevations above the desert and sometimes extends into the upper elevation limits of the desert. Cattle eagerly seek this highly palatable grass, and it can be assumed that it has been eliminated from some moist sites in the desert. There are seven specimens from Guaymas collected by Edward Palmer in the nineteenth century but no other records from that region. Most likely the habitat has been lost.

Tiburón: 3 km W [of] Punta Tormenta, *Smartt 13 Oct 1976* (UNM 84145). Base and N side of Punta San Miguel, *Wilder 06-371.*

Leptochloa panicea (Retzius) Ohwi subsp. **brachiata** (Steudel) N.W. Snow [*L. filiformis* (Persoon) P. Beauvois]

Desparramo rojo; sprangletop

Description: Summer-fall annuals, sometimes persisting through the winter, green or reddish, highly variable in size, mostly less than 70 cm tall. Glabrous or the leaf sheaths sometimes sparsely hispid with papilla-based hairs. Panicles of several or more racemosely arranged spicate branches. Spikelets often reddish purple, 2.2–3.2 mm long (1) 2- or 3-flowered (mature spikelets may appear 1-flowered). Lemmas of first (larger) floret 1–1.5 mm long, the tip obtuse to acute, not lobed.

Local Distribution: On Tiburón, seasonally widespread but not documented from the western margin or south side of the island; often in arroyos and canyons, and also on rocky slopes. On San Esteban, common along Arroyo Limantour. On Nolasco, seasonally abundant on the east side of the island and canyons on the west side, at all elevations, and sometimes persisting and growing with cool-season rains.

Other Gulf Islands: Coronados, Carmen, Danzante, Santa Catalina, San Diego.

Geographic Range: Southern half of the United States to Argentina and Peru, and introduced and spreading worldwide, often weedy.

Leptochloa panicea is a nearly worldwide complex of 3 subspecies. Subsp. *brachiata* is the more widespread of the two New World subspecies. This subspecies has previously been known as *L. mucronata* (Michaux) Kunth, *L. panicea* subsp. *mucronata* (Michaux) Nowack (in part), and *L. filiformis*. Subsp. *mucronata* is restricted to southeastern United States to Iowa and Texas; subsp. *panicea* is native to the Old World.

Tiburón: Haap Hill, *Felger 76-T26.* Canyon bottom, N base of Sierra Kunkaak, *Wilder 06-430.* Siimen Hax, *Wilder 06-511.* 1 km inland from Zozni Cmiipla, *Wilder 06-371.* Bahía Agua Dulce, *Felger 8891.*

San Esteban: [Arroyo Limantour], 13 Oct 1977, *Wilkinson 173.*

Nolasco: Cañón el Farito, common, 2 Feb 2008, *Wilder 08-170.* E-central side, halfway to summit, seen at all elevations, highly variable in size, 28 Nov 2006, *Felger 06-92.* Cañón de Mellink, ca. 10 m, 29 Sep 2008, *Felger 08-138.*

Muhlenbergia microsperma (de Candolle) Trinius

Impós, ziizil; *liendrilla chica*; littleseed muhly

Description: Non-seasonal annuals, often best developed during winter-spring seasons. Plants soft and delicate, the stems often weak and frequently growing through other plants. Roots usually weakly developed. Leaves often reduced when stressed by cold or drought. Panicles terminal, longer than wide, open and loosely flowered, the spikelets (lemmas) tapering to a slender awn 14–28 mm long. Lower leaf axils bearing single or clustered cleistogamous spikelets. The cleistogenes are narrowly conical, awnless, 4–10 (12) mm long, sometimes 2- or 3-flowered, and the caryopsis (grain) is larger than those of the terminal panicles.

The cleistogene, much more variable than the normal spikelet, is a contracted inflorescence; its grain is difficult to extract, is not shed, and germinates within its hard protective sheaths. The cleistogenes, because they are heavier than normal grains, lack awns, and are situated at the base of the plant, have a much reduced range of

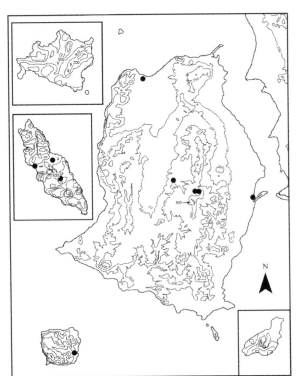
Leptochloa panicea ssp. *brachiata*

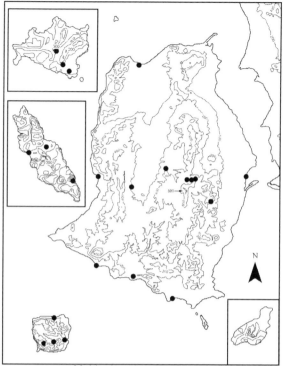

Muhlenbergia microsperma
Impós, ziizil

dispersal with a greater probability of remaining where their parent was biologically successful. The terminal panicles are often eaten by animals, giving additional advantage to the dual reproductive strategy.

Local Distribution: Widespread on Tiburón and San Esteban to higher elevations, generally most common in better watered niches such as washes and canyon bottoms. Reported from Cholludo (see "Doubtful and Excluded Reports"). Seasonally abundant and widespread on Mártir and Nolasco. On Mártir, during times of favorable rains, it can form 100 percent cover in areas above the guano-covered lower elevations.

Other Gulf Islands: Ángel de la Guarda, San Lorenzo, Tortuga, San Marcos, Carmen, Danzante, Monserrat, Santa Catalina, Santa Cruz, San José, Espíritu Santo, Cerralvo.

Geographic Range: Southwestern United States to South America.

Tiburón: Ensenada Blanca (Vaporeta), *Felger 17277.* Tordillitos, *Felger 15510.* El Sauz, *Moran 8784* (SD). Ensenada de la Cruz, 100 m inland, *Felger 9135.* SW Central Valley, *Felger 17322.* Haap Hill, *Felger 76-T28.* Zozni Cmiipla, *Wilder 06-354.* Pazj Hax, *Wilder 06-42.* Cerro San Miguel, *Quijada-Mascareñas 91T022.* Siimen Hax waterhole, *Wilder 06-466.* S side of Sierra Kunkaak, rocky slopes, *Knight 1035.* Tecomate, *Felger 12515.*

San Esteban: Arroyo Limantour, *Wilkinson 13 Oct 1977.* N side of island, arroyo, *Felger 15402.* S-central peak, steep

slopes, *Felger 17532.* SE corner of island, E ridge of high peak, NE talus slope, few, *Moran 8843* (SD).

Mártir: 1887, *Palmer 416* (UC, image). Abundant on all parts of island, the only grass, 18 Apr 1921, *Johnston 4398* (CAS, image). Above ruins of the guano workers' village, abundant from above ca. ¼ way up island, forming 100 percent coverage in some places between cardons, especially where the ground is not covered by *Vaseyanthus,* 11 Apr 2007, *Felger 07-14.*

Nolasco: SE side of island, *Sherbrooke [Felger 11238].* NE side, 150 m, very common, mostly in open areas on canyon slope, local area mostly free of *Vaseyanthus, Felger 06111.* Cañón de Mellink, *Wilder 08-350.*

Panicum hirticaule J. Presl var. **hirticaule**

Zacate brujo, panizo capiro; Mexican panicgrass

Description: Summer-fall annuals, the roots often weakly developed. Stems and leaves with spreading, usually papilla-based hispid hairs, or glabrate; flowering branches and spikelets glabrous. Spikelets 2.7–3.1 mm long, lower glume ½ to ¾ as long as the spikelet. Upper (fertile) lemma smooth and shiny, cream-white, becoming dark brown with age, and tightly enclosing the palea and fruit.

Local Distribution: There are surprisingly few records on the islands for this usually common Sonoran Desert summer grass. Its occurrence in so arid an area as the west

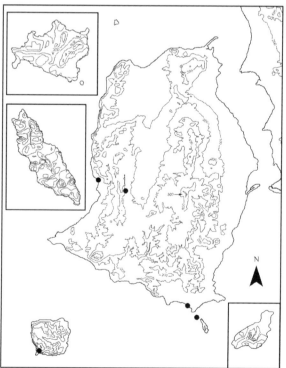

Panicum hirticaule var. *hirticaule*

side of Tiburón, as well as on San Esteban, indicates that it is undoubtedly more widespread in favorable seasons.

Other Gulf Islands: San José.

Geographic Range: Widespread in the Sonoran Desert; southwestern United States to South America and the West Indies.

Tiburón: Ensenada Blanca (Vaporeta), *Felger 14952A*. Coralitos, *Barnett-Morales 08-6*. SW Central Valley, upper rocky slope of mountain bordering valley, *Felger 12401*.

San Esteban: Canyon, SW corner of island, *Felger 16639A* (SD).

Cholludo: N-central slope, seedlings with attached grain, 2 Sep 2008, *Wilder*, observation.

Phragmites australis (Cavanilles) Trinius ex Steudel subsp. **berlandieri** (E. Fournier) Saltonstall & Hauber

Xapij; *carrizo nativo*; reedgrass, cane, common reed

Description: Bamboo-like reeds 2–3+ m tall, emergent from shallow water or alkaline wet soil, with strong rhizomes and tough roots. Stems stout, reaching 1–1.5 cm diameter, and also often producing numerous smaller stems as slender as 1.7 mm diameter. Leaves large, 2-ranked and evenly spaced, the blades (7) 30–40 cm long. Panicles terminal and plume-like, reaching 30–45 cm long. Spikelets several-flowered. Rachillas with long silky white hairs, the lemmas glabrous (distinguishing it

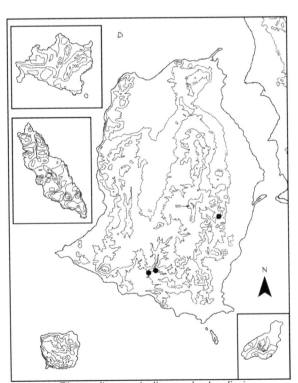

Phragmites australis ssp. *berlandieri*
Xapij

from *Arundo donax*, which has glabrous rachillas and conspicuously hairy lemmas). The fruits usually break off below the long-bearded rachillas, which aids in wind dispersal. In arid regions such as the Sonoran Desert, birds flying between waterholes are the likely dispersal agents.

Local Distribution: Reedgrass forms dense thickets at the Xapij and Pazj Hax waterholes on Tiburón. It was planted at Tecomate at the north end of the island by personnel of the Mexican federal wildlife department but did not survive. The most extensive stand in the Seri region is at Xapij (Sauzal) at the south end of the island. This was the locality where the San Esteban people would gather xapij. (Also see *Arundo*.)

In 2008 the Xapij (Sauzal) stand occurred intermittently along about 1.2 km of the watercourse, apparently getting water from a spring at the rock edge of the arroyo. The densest patch occupies a roughly linear-ellipsoid area ca. 25+ m long and 4–5 m wide. The main areas of carrizo are impenetrable with 100 percent cover. The taller parts of the colony are about 3 m tall, and one area of the colony forms a wall along the edge of the arroyo waterhole and is more than 4 m tall but is so dense that it is not possible to find the base. On 29 January 2008 most of the culms were mature with terminal inflorescences and showed no sign of seasonal die-back. The leaves were green and healthy, and a minority of leaves glistened from sticky exudate, probably from aphids.

Carrizo was one of the most important resources to the Comcaac (Felger & Moser 1985). The cane was cut and carried to shore, each man carrying two large bundles suspended from a carrying yoke (*palanca*). The stems were intricately interwoven into three large bundles tied together with mesquite-root cord to construct ocean-going reed boats (hascám, *balsas*). The balsa was used for transport to the islands and coastal camps and for hunting and fishing. The smaller balsas carried one man, while the larger ones could carry three families (six adults and nine or ten children) or six to seven large sea turtles. The balsa rode low in the water, and all aboard would be wet. Crossing open water could be harrowing. María Antonia Colosio described a childhood trip from Tiburon Island to San Esteban Island in the first decade of the twentieth century. This account was recorded in El Desemboque in 1978 by Mary Beck Moser:

> We were at *Cyajoj*. We were on two balsas that were tied together, side by side. We were going to . . . what's it called, the mountain out in the sea? I didn't want to go. We were going to *Coftécöl*. They had put blankets and water on, jugs as big as this [gesturing, indicating

a large water vessel]. Since the jugs were full of water, they were tied in place. Plants were stuck in the mouth of the jugs, and the jugs were in carrying nets. I didn't want to get on, but my father caught me and put me on. After he caught me and put me on, he tied me behind a blind man who went along to paddle. Then I cried a lot, but he didn't pay any attention to me. That's how we went to *Coftécöl*. It was so dangerous when we entered the area called *Ixötáacoj* [Big Whirlpool]. The sea just swirled and churned. The wind wasn't blowing but the water was choppy. It just churned, it was dangerous. Everything just roared. The children and old women all cried. The old man Pozoli just said, "We'll land really soon." As we were going to land, he sang to the shore. And it seemed we landed right away. The men paddled with all their strength, and we landed near the rocks. (Felger & Moser 1985:131–132)

The use of balsas was discontinued by the 1920s in favor of wooden boats, and these more recently replaced by fiberglass *pangas* (open Mexican fishing boats used throughout the Gulf of California). The Comcaac also made use of xapij stems for containers, fishing, food gathering, games, hunting, musical instruments including flutes and whistles, shelter, arrow shafts, and tobacco pipes (Felger & Moser 1985).

Other Gulf Islands: Cerralvo.

Geographic Range: This species occurs on every vegetated continent on earth and is perhaps the world's most widespread species of flowering plant. New World populations show difference in their genome from Old World populations, with three subspecies, one introduced and invasive, especially in the eastern United States, of Old World origin and two native to the New World (Saltonstall & Hauber 2007). Subsp. *berlandieri*, known as the Gulf Coast lineage, ranges across southernmost United States and through Mexico to South America.

Phragmites was probably once at numerous Sonoran Desert waterholes and regional river deltas, but in many places *Arundo* has replaced it or the wetland habitat has disappeared. *Phragmites* remains abundant in the delta region of the Río Colorado and the lower Colorado River.

Tiburón: Xapij (Sauzal) waterhole: *Moser & Moser* cataloged as *Felger 9986*; *Wilder 06-91, 06-123, 08-96*. Pazj Hax, wet soil, *Wilder 06-46*.

Setaria – Bristlegrass

Annuals or perennials growing with warm weather. Panicles contracted, often spike-like. Spikelets subtended by one (the two species in the flora area) or more conspicuous, persistent scabrous bristles (the bristles represent reduced panicle branches or branchlets); individual spikelets detach below the glumes and above the bristles. Spikelets awnless, panicoid with the fertile lemma firm and tightly grasping the fertile palea and grain.

1. Summer-fall annuals, roots weak._____ **S. liebmannii**
1' Perennials, roots stout and coarse. _ **S. macrostachya**

Setaria liebmannii E. Fournier

Ziizil; *cola de zorra*; summer bristlegrass

Description: Summer annuals, highly variable in size, unbranched or with several or occasionally many, slender stems, usually branched from or near the base. Panicles cylindrical, loosely flowered, usually held well above the few leaves on a relatively long, slender peduncle. Bristles 1 below each spikelet.

Local Distribution: Tiburón, known from two records, the Central Valley and the Sierra Kunkaak, and it is expected to be more widespread on the island. On Nolasco, common on the east side of the island.

Other Gulf Islands: Carmen, San Diego, Espíritu Santo, Cerralvo.

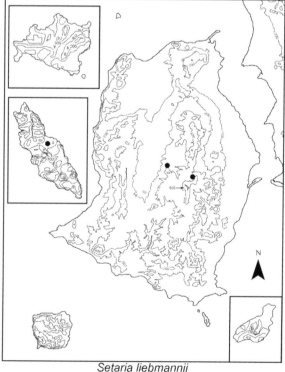

Setaria liebmannii
Ziizil

Geographic Range: Arizona to Central America. This is the only annual bristlegrass in the lowland desert and thornscrub of western Sonora and northwestern Sinaloa. In Arizona and Sonora it occurs mostly at elevations slightly above the desert, and it reaches greater size and density in the thornscrub and tropical deciduous forest to the east and south of the Sonoran Desert.

Tiburón: Haap Hill, *Felger T74-1*. Canyon bottom, N base of Sierra Kunkaak, ca. 340 m, *Wilder 06-421*.

Nolasco: Cañón el Faro, 60 m, common, canyon bottom and also on rock slopes, seen at all elevations, *Felger 06-89*.

Setaria macrostachya Kunth [including *S. leucopila* (Scribner & Merrill) K. Schumann]

Hasac, xiica quiix "globular things"; white-haired bristlegrass

Description: Densely tufted perennials, often 0.5–1 m tall. Panicles often slender, cylindrical, and densely flowered. Bristles 1 below each spikelet. Growing and reproductive during the warmer months, especially following rains.

The *S. macrostachya* complex represents a polyploid complex split into five species (Rominger 1962, 2003) that are sometimes difficult to distinguish. *Setaria leucopila* is distinguished by narrower leaves, the spikelets are not as

fat (gibbous), the fertile palea flat or not as convex, and the fertile palea shorter than in *S. macrostachya* sensu stricto. McVaugh (1983:361) reported that leaf width is a useful character but that the other features are subjective, although "after a little practice one can see the difference," but he concludes that the characters are "not very useful in intermediate forms." In *S. macrostachya* sensu stricto the fertile lemma is inflated at the base (gibbous) and in *S. leucopila* it is not. Rominger (1962) also distinguished *S. leucopila* partly on the basis of leaf width; *S. leucopila* is supposed to have a narrower leaf. But this distinction does not hold up; leaf width seems strongly influenced by rainfall (soil moisture) and temperature.

The Comcaac harvested the seed as a minor grain (Felger & Moser 1985).

Local Distribution: Common on Tiburón in the Sierra Kunkaak and the southern part of the island but apparently rare or absent elsewhere. On San Esteban, it has been found along the Arroyo Limantour. Also on Dátil and Cholludo. Common on Nolasco, especially along canyons and often dominating north- and east-facing grassy slopes.

Other Gulf Islands: San Marcos, San Ildefonso, Carmen, Santa Catalina.

Geographic Range: *Setaria leucopila* occurs in Central and northwestern Mexico including Baja California Norte and Sur and southwestern United States, while *S. macrostachya* ranges from the United States to South America.

Tiburón: La Pescadita, S shore, *Wilkinson 11 Oct 1977*. Coralitos, just inland from beach, *Wilder 06-55*. El Monumento, *Felger 08-75*. Canyon, N base of Sierra Kunkaak, 340 m, *Wilder 06-416* (determined as *S. macrsostachya* by J.R. Reeder, 2007). Rock outcrop above steep, protected canyon, 720 m, N interior of Sierra Kunkaak, *Wilder 07-523*. Deep sheltered canyon, N portion of Sierra Kunkaak, *Felger 07-114* (determined by J.R. Reeder, 2007 as "*S. leucopila*, but some spikelets look like *S. macrostachya*").

San Esteban: Arroyo Limantour: Solitary perennial along wash, *Van Devender 92-482*; ½ mile inland, *Wilder 07-75*. "A few small colonies were seen on San Esteban," *Johnston 4396* (Johnston 1924:987, specimen not located).

Dátil: NE side of island, not common, canyon bottom, steep rocky slopes, *Felger 13473*. NW side, *Felger 15310*. Arroyo near beach and N-facing steep rocky slopes, *Felger 9115*.

Cholludo: NE slope, population of ca. 75 plants, ca. 20 m elev., in an area 15 × 15 m, *Wilder 08-320*.

Nolasco: Abundant on N facing slopes, which look like hay fields because of the plant, 17 Apr 1921, *Johnston 4397* (CAS). SE side, *Sherbrooke [Felger 11247]*. Steep granitic mountain slope with scattered trees and giant cactus, shady slope with

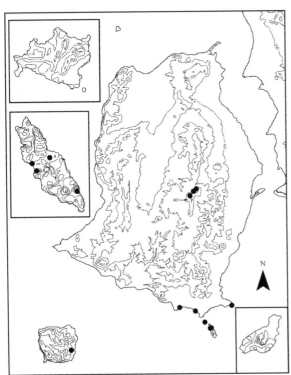

Setaria macrostachya
Hasac, xiica quiix

soil, *Gentry 11357*. N-facing slope above Cañón el Farito, abundant, mostly dry, some with reddish/purple leaf blades, with nearly 100 percent coverage in some areas, 2 Feb 2008, *Wilder 08-160*. Cañón de la Guacamaya, adjacent to cave near shore, Jul 2003, *Gallo-Reynoso* (photo). Cañón de Mellink, 10 m elev., *Felger 08-139*.

Sporobolus – Dropseed

Annuals or perennials of diverse sizes and growth habits. Ligules a line of hairs, sometimes also with large hairs at the collar and leaf sheath "behind the collar." Spikelets small, 1-flowered, glabrous, and awnless. Lemmas thin and 1-veined. The grain readily falls from the spikelets at maturity (hence the name "dropseed"); the pericarp is usually thin and closely enclosing the seed but free from it and readily soaks up water to become mucilaginous when wet and forcibly ejects the seed, the "naked" seed often clinging to the tip of the palea and lemma. Generally growing and reproductive during the warmer months.

The young panicles are strongly contracted in *S. cryptandrus* and *S. pyramidatus*. As the panicle matures the branches usually spread—the contrast is striking. "Those species of *Sporobolus* in which some or all of the inflorescences remain enclosed in the subtending leaf sheaths tend to reproduce cleistogamously, that is, pollen cannot be dispersed and obligately pollinates the stigma of the same floret. The percentage of seed-set in these inbred plants is usually quite high" (Yatskievych 1999:727).

1. Perennials with long stolons, not tufted; beaches. ___
   _____ **S. virginicus**
1' Annuals or tufted perennials, not stoloniferous; not restricted to beaches.
   2. Perennials; branches at lower nodes mostly solitary, very rarely with 3 branches (lower nodes of young inflorescences often concealed by a leaf sheath); spikelets 1.8–2.4 mm long. _____ **S. cryptandrus**
  2' Annuals; branches at lowest node of panicle whorled with (4) 5 or more branches (concealed by a leaf sheath in younger inflorescences); spikelets 1.5–1.8 mm long. _____ **S. pyramidatus**

Sporobolus cryptandrus (Torrey) A. Gray

Cpooj; sand dropseed

Description: Tufted perennials. Leaf sheaths usually pilose with relatively dense, long, white and often tangled hairs at the summit. Panicle base, or sometimes the entire

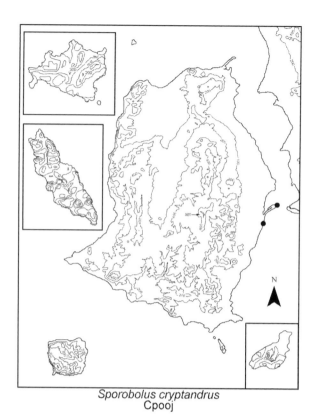

Sporobolus cryptandrus
Cpooj

panicle, enclosed by an enlarged, inflated leaf sheath, the upper part of the panicle often but not always becoming free and the branchlets spreading.

Local Distribution: Known in the flora area by two records from Tiburón, both in sandy soil near Punta San Miguel. It seems unusual that there are not more records since it is widespread in nearby western Sonora.

Other Gulf Islands: None.

Geographic Range: Widespread in Sonora including coastal areas, especially on sandy soils of strands and beach dunes. Temperate and subtropical North America south to central Mexico.

  Tiburón: 1 km S of Punta San Miguel, *Wilder 07-386*. Punta San Miguel, low dunes 5–50 m inland, *Wilder 07-196*.

Sporobolus pyramidatus (Lamarck) Hitchcock [*S. patens* Swallen. *S. pulvinatus* Swallen]

Zacate pirámide, zacate de arena; whorled dropseed, sand dropseed

Description: Hot-weather annuals in the Sonoran Desert, highly variable in size. Upper margin of leaf sheaths sparsely pubescent with long white hairs but not in a dense tuft. Panicles at first contracted and narrow,

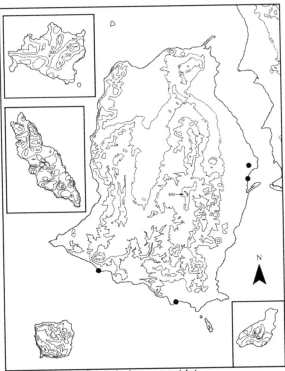

Sporobolus pyramidatus

5–10 cm long; spikelets 2 mm long. Grain (0.7) 0.9–12 mm long, pale orange-brown; seed production is ample.

Local Distribution: Infiernillo shore of Tiburón, documented at Punta San Miguel and Punta Perla; upper beaches and margins of tidal esteros including the high tide zone. Also along the shore on Alcatraz.

Other Gulf Islands: Coronados (*Moran 9119*, SD, originally labeled as *Distichlis palmeri*, which does not occur this far south), Carmen, Danzante, Monserrat, Santa Catalina, San José, Espiritu Santo (*Moran 9626*, SD).

Geographic Range: Tropical and subtropical shores worldwide. Sonora from near Bahía Kino southward, and the Baja California Peninsula from Bahía San Francisquito southward.

The Gulf of California plants are notably robust, and plants from the Pacific side of the Baja California Peninsula are smaller, with leaves 2–7 cm long and much smaller inflorescences and more slender stems.

Peterson et al. (2003:121) state, "No fruits of this species have been found despite examination of several natural populations and over 200 herbarium specimens." Yet the fruits are described in the literature (e.g., Laegaard & Peterson 2001; Pohl 1980; Pohl et al. 1994). A number of specimens at ARIZ from distant parts of the world have

often partly covered by the leaf sheath, the branchlets ultimately spreading and the panicles somewhat pyramidal; lower branches whorled, (4) 5–11 branches per node; upper branches single at the nodes.

Local Distribution: Documented for southern and eastern coastal areas of Tiburón, most often on sandy-gravelly soils.

Other Gulf Islands: Coronados, Carmen, San José.

Geographic Range: Central United States to South America, including southern Arizona, Baja California Sur, and through much of Sonora.

Tiburón: Tordillitos, *Felger 15477*. Ensenada de la Cruz, arroyo bottom and low flat terrain, *Felger 9139*. Zozni Cmiipla, *Wilder 06-351*. Palo Fierro, *Felger 8917*.

Sporobolus virginicus (Linnaeus) Kunth

Xojasjc; *zacate salado de la playa*; seashore dropseed

Description: Perennial saltgrass. Erect leafy stems emerge from long, yellow, scaly rhizomes running across sandy beaches. Leaves inrolled, firm and sharp-pointed, larger leaves 8–20 cm long. Inflorescences of dense, many-flowered, spike-like panicles mostly (3.7)

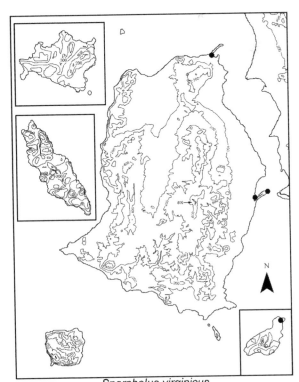

Sporobolus virginicus
Xojasjc

fruits and indicate copious and synchronous flowering, with fruiting February–April in the Southern Hemisphere and June–October in the Northern Hemisphere.

Sterile specimens from Gulf of California shores might be indistinguishable from the saltgrass *Distichlis palmeri* (see *Distichlis*).

Tiburón: Zozni Cmiipla at Estero San Miguel, *Wilder 06-361.* Punta San Miguel, *Wilder 07-184.* Punta Perla, estero margin, 9 Nov 2008, *Wilder,* observation.

Alcatraz: E side, sand beach, *Felger 14928.* NE end of island, near lighthouse, low flat, sand soil just above high tide line, *Felger 07-191.*

Trichloris crinita (Lagasca) Parodi [*Chloris crinita* Lagasca] *Zacate escoba*; feather fingergrass

Description: Large, tufted perennials; growing and reproductive during the warmer months; often growing through desert shrubs with stems reaching 1.5–2 m tall on Tiburón and much shorter on Nolasco. Panicles of 6–20 white, bristly spikes clustered at the top of the stem; spikelets with 1 fertile and 1 sterile floret. This species is unusual in having 3 long awns on each floret.

Local Distribution: On Tiburón, known from the lower portion of the extensive eastern bajada in a swale among dense desertscrub. Here it was growing among a large population of *Chloris virgata.* Nolasco, grassy, north-facing slopes on the east side of the island and near the shore at the northwest side of the island. The nearest known populations are on the opposite mainland at Bahía San Pedro (*Felger 11620*).

Other Gulf Islands: None.

Geographic Range: Arizona to western Texas and south to Durango, Coahuila, Sonora, and Baja California Norte and Sur, and disjunct in South America.

Tiburón: Zozni Cmiipla, 1.5 m tall, *Wilder 06-512.*

Nolasco: NE side of island, N-facing slope, mid-elev., common, *Felger 9671.* Cañón de Mellink, ca. 10 m elev., *Felger 08-135.*

Tridens muticus (Torrey) Nash var. **muticus**

Slim tridens

Description: Tightly clumping small perennials with slender, erect stems; often bluish glaucous; leaf blades usually tightly rolled. Panicles slender (less than 1 cm wide), spike-like, with closely appressed short branches, each branch with 1 to several spikelets. Spikelets 6–13 mm long, and several flowered. Glumes persistent, shorter than the spikelets. Florets strongly overlapping,

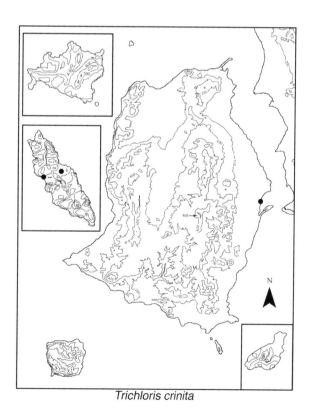

Trichloris crinita

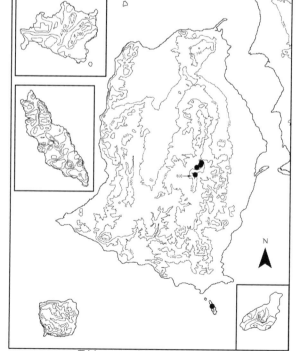

Tridens muticus var. *muticus*

the lemmas 3.5–5 mm long, the lower half markedly hairy (pilose). Growth and flowering response apparently non-seasonal.

Local Distribution: Tiburón in deep canyons in the interior of Sierra Kunkaak and rocky slopes to the peak, and rare on Dátil. The nearest known population is 150 km to the north in the Sierra del Viejo southwest of Caborca, which is the southernmost documentation for this grass for mainland Sonora (Felger 2000a).

Other Gulf Islands: None.

Geographic Range: Southwestern United States and Baja California Norte, northern Sonora to Coahuila and Nuevo León; deserts, grasslands, and oak woodlands. One of the most drought-tolerant perennial grasses of the Sonoran Desert, often in surprisingly harsh, xeric habitats such as the Pinacate region. A second variety ranges into higher elevation and occurs farther eastward and northward in the United States than does var. *muticus*.

Tiburón: Base of Capxölim, N-facing rock slope, *Felger 07-103*. Canyon bottom, N base of Sierra Kunkaak, between Sierra Kunkaak Mayor and Sierra Kunkaak Segundo, 405 m, *Wilder 06-401*. Deep canyon, N slope of Sierra Kunkaak SW and up canyon from Siimen Hax, *Wilder 06-479*. Summit of island, rare, *Wilder 07-570*.

Dátil: E-central side, E-W trending canyon, three plants just above canyon bottom in stabilized soil, *Wilder 09-35*.

Urochloa – Signalgrass

Summer ephemerals (and perennials elsewhere); plant size and leaf size and width varying with soil moisture and warmth. Spikelets panicoid–awnless, the fertile lemma firm and tightly grasping the fertile palea and grain. Plant size and leaf size and width varying with soil moisture and warmth.

1. Spikelets hairy or sometimes glabrous, generally elliptic, widest at middle; cross venation (on sterile lemma and upper glume) near tip of spikelet, not extending below middle of spikelet (cross veins sometimes absent but generally evident on at least some spikelets). _____ _____ **U. arizonica**
1' Spikelets glabrous, generally obovoid, widest at or above the middle; cross venation (on sterile lemma and upper glume) near tip of spikelet and extending below middle of spikelet (cross veins sometimes absent but generally evident on at least some spikelets)._____ **U. fusca**

Urochloa arizonica (Scribner & Merrill) R.D. Webster [*Panicum arizonicum* (Scribner & Merrill) S.T. Blake. *Brachiaria arizonica* (Scribner & Merrill) S.T. Blake]

Conee ccapxl "sour grass"; *piojillo de arizona*; Arizona signalgrass

Description: Plants mostly slender and pale green, variously soft pubescent, the hairs often papilla based, or sometimes glabrous or essentially so. Pedicels and panicle branchlets usually conspicuously hairy. Spikelets generally elliptic, 3.0–3.7 mm long, mostly hairy or sometimes glabrous, often pale green or reddish, the upper glume and sterile lemma with green longitudinal and cross veins forming a net-like pattern toward the tip, or cross veins sometimes absent. Lower glume often to about half as long as the spikelet.

Local Distribution: Widespread across Tiburón in many habitats, mostly localized; drainageways and rocky slopes, from near the shore to higher elevations in the Sierra Kunkaak and locally on three other islands.

Other Gulf Islands: Coronados, Carmen, Santa Cruz, San José, San Francisco, Espíritu Santo, Cerralvo.

Geographic Range: Arizona to Texas and Mexico southward to Oaxaca.

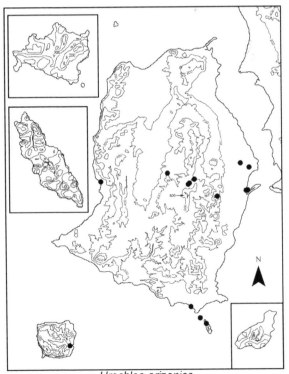

Urochloa arizonica
Conee ccapxl

Tiburón: Ensenada Blanca (Vaporeta), *Felger 14952*. Coralitos, *Wilkinson 226*. Haap Hill, *Felger T74-7*. Zozni Cmiipla: 100–300 m inland, *Wilder 06-287*; *Felger 08-130*. Palo Fierro, *Felger 8917*. 3 km W [of] Punta Tormenta, *Smartt 13 Apr 1976* (UNM). Lower foothills, between Sopc Hax and Hant Hax camp, arroyo bottom, *Felger 9347*. Arroyo bottom, NE base of Sierra Kunkaak, *Wilder 06-382*. Siimen Hax, *Wilder 06-465*. Capxölim, upper N-facing slope, *Wilder 06-498*.

San Esteban: Arroyo Limantour, not common, ½ km inland, floodplain, *Felger 12750*.

Dátil: NE side of island, steep rocky slopes, *Felger 9111*.

Cholludo: Steep rocky slopes, very common, *Felger 9167*.

Urochloa fusca (Swartz) B.F. Hansen & Wunderlin [*Panicum fasciculatum* Swartz. *Brachiaria* fasciculata (Swartz) Parodi]

Browntop signalgrass

Description: Pedicels and panicle branchlets hairy. Spikelets usually obovoid, 2.8–3.1 mm long, glabrous, pale green at first, generally bronze or brownish at maturity, the upper glume and sterile lemma with longitudinal veins and cross veins in a net-like pattern from near the tip to below the middle, or cross veins sometimes absent. Lower glume often about one-third as long as the spikelet.

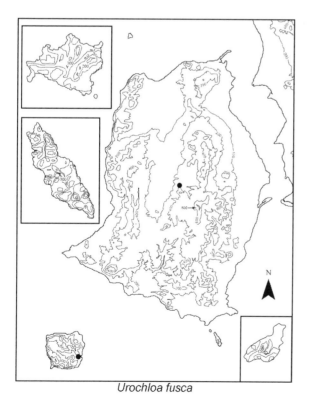

Urochloa fusca

Differences between *U. arizonica* and *U. fusca* are often subtle and not entirely satisfactory. The pedicels of *U. arizonica* tend to be densely pubescent with papilla-based hairs while those of *U. fusca* tend to have fewer or no papilla-based hairs. Spikelets of *U. arizonica* tend to be rather densely hairy but are sometimes glabrous; all *U. fusca* specimens examined from the Sonoran Desert region have glabrous spikelets; *U. arizonica* spikelets tend to be more elegant and narrower, and widest near the middle, while those of *U. fusca* are relatively fat and widest above the middle. The cross veins on the spikelets (upper glume and sterile lemma) of *U. arizonica* are usually near the tip of the spikelet, well above the middle, while *U. fusca* commonly has cross veins extending well below the middle of the spikelet. The net-like, cross veins, however, may be absent from many spikelets and some specimens of both species. The fertile lemma ("grain") of *U. arizonica* is reportedly not as rough (rugose) as that of *U. fusca*.

Local Distribution: Known from two records, one on Tiburón and one on San Esteban; presumably this species is seasonally more widely distributed. The nearest record is from Bahía de Kino (*Van Devender 90-519*).

Other Gulf Islands: Tortuga, Carmen, Espíritu Santo.

Geographic Range: Widely scattered in the Sonoran Desert but not known from California or Baja California Norte; generally not in the most arid regions, and more common toward the margins of the desert. It is especially common in the southern part of Baja California Sur and widespread in non-desert regions of Sonora. Within most of the desert it is encountered far less often than *U. arizonica*. Southwestern United States to South America and the West Indies.

Tiburón: Haap Hill, *Felger 76-T10*.

San Esteban: Arroyo Limantour, uncommon, *Van Devender 90-537*.

Polygonaceae – Buckwheat Family

Eriogonum inflatum Torrey & Frémont

Tee; desert trumpet, bladder stem

Description: Herbaceous perennials, also flowering in the first season. Leaves mostly basal, green to reddish, the dry leaves semi-persistent, the blades oblong to orbicular or kidney shaped with wavy margins. Flowering stems with the first internode often 10–30 cm long, the upper part inflated (swollen) or not. Flowers about 5 mm

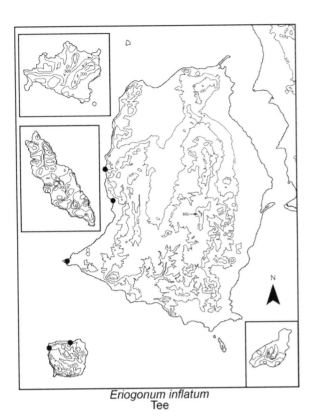

Eriogonum inflatum
Tee

Felger 7112. 1 km N of Punta Willard, wind-blown sand piled against rock hills, *Felger 17755.*

San Esteban: N side of island, N-facing cliff, *Felger 15416.* San Pedro, steep rock slopes, *Felger 16668.*

Portulacaceae – Portulaca Family

Succulent annuals or perennial herbs, growing with summer-fall rains. Flowers small; fruits of capsules. Many genera in the classical interpretation of the Portulaca family have been transferred to other families, including *Talinum*, which was segregated as a separate family, the Talinaceae, even as we were completing this book (Stevens 2008).

1. Annuals; leaves mostly less than 2.5 (3) cm long; flowers sessile; capsules opening horizontally to shed the top like a cap or lid (circumscissle). _____ **Portulaca**
1' Perennials; leaves 2.5–11+ cm long; flowers stalked (pedicelled) in panicles 30+ cm long; capsules opening longitudinally (valvate). _____ **Talinum**

wide, yellow, and on slender, upper branches; flowering in spring and summer.

Local Distribution: Coastal areas on the west side of Tiburón and the north side of San Esteban; mostly among rocks of arroyos, cliffs, small mesas, and north-facing slopes; sometimes on cobble rock beaches just inland from the high tide zone.

Other Gulf Islands: Ángel de la Guarda, San Lorenzo, Tortuga, San Marcos, Coronados, Carmen.

Geographic Range: Baja California Sur to eastern California to Colorado and much of Arizona, and northwestern Sonora southward to the vicinity of El Desemboque.

In contrast to the nine taxa of *Eriogonum* found on Isla Ángel de la Guarda (Moran 1983a), the presence of only one species on the eastern Midriff Islands is indicative of the differences encountered between the two sides of the Gulf. Most significant in this case is the relative paucity of winter rainfall in the flora area in comparison to the western and northern portions of the Sonoran Desert. The Pinacate region of northwestern Sonora also supports nine taxa of *Eriogonum*, while none occur in western Sonora south of the vicinity of Bahía Kino.

Tiburón: Pooj Iime, *Wilder 07-458.* Ensenada Blanca (Vaporeta), localized, small "bench" area near mouth of arroyo,

Figure 3.141. *Talinum paniculatum.* Vicinity of Zozni Cmiipla. BTW, 26 September 2008.

Portulaca – Purslane

Small summer annuals, highly succulent including the leaves. Flowers brightly colored, opening in sunlight, the petal-like perianth usually lasting only a few hours and then deliquescent (dissolving, or melting away).

1. Stems scarcely succulent; leaves generally terete (rounded in cross section; appearing flat when dry); leaf axils and flower clusters densely white-hairy. ___
_____ **P. halimoides**
1' Stems succulent; leaves thick but flattened; plants glabrous (except a few inconspicuous hairs).
  2. Fruits not collar-winged, the capsules opening at about the middle, the lid often taller than the capsule body; seeds iridescent blackish, the projections only scarcely peg-like. _____**P. oleracea**
  2' Fruits with a distinctive collar-like wing 1–2 mm wide surrounding the capsule rim, the capsules opening above the middle, the capsule lid shallow and saucer-like; seeds dull gray with prominent peg-like projections._____ **P. umbraticola**

Portulaca halimoides Linnaeus [*P. parvula* A. Gray]

Dwarf purslane

Description: Plants often diminutive but potentially fairly robust when well watered, the stems slender.

Flowers 3–7 mm wide; sepals red-pink and relatively persistent; petals, anthers, and stigma golden yellow, but these are only evident while the flowers are open—generally from a few hours after sunrise until late morning or midday. Capsules to 2 mm wide; seeds without conspicuous tubercles.

Local Distribution: Known from the northern and eastern margins of Tiburón.

Other Gulf Islands: None.

Geographic Range: Arid and tropical Americas and southwestern North America including Arizona and Sonora southward to the vicinity of Hermosillo and the Baja California Peninsula.

Tiburón: Zozni Cmiipla, *Wilder 06-350.* Bahía Agua Dulce, very common, *Felger 10223.* Tecomate, localized, not common, *Felger 8881.*

Portulaca oleracea Linnaeus

Verdolaga; purslane

Description: Plants succulent throughout, erect to spreading or prostrate with age and size; glabrous except for a few inconspicuous axillary hairs (visible with magnification). Flowers yellow. Capsules separating at about the middle, the capsule lid conical and somewhat constricted at about the middle, with at least 1 seed usually

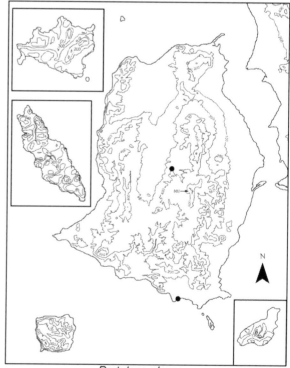

Portulaca halimoides

Portulaca oleracea

remaining with the lid, the others falling quickly. Seeds with low, star-shaped tubercles.

Local Distribution: Known from widely separated localities on Tiburón, indicating widespread distribution on the island.

Other Gulf Islands: Coronados.

Geographic Range: Worldwide in tropical to warm-temperate climates, including Baja California Norte and Sur and Sonora, often weedy. It is often difficult to determine which populations might be native and which are non-native. Native to the Old World and present in the New World in pre-Columbian times. The plants, used as a pot herb, are sold in markets in Sonora and southern Arizona.

Tiburón: Ensenada de la Cruz, *Felger 12778*. Haap Hill, *Felger T74-24*.

Portulaca umbraticola Kunth subsp. **lanceolata** J.F. Matthews & Ketron

Description: Distinguished from *P. oleracea* by a conspicuous collar-like wing 1–2 mm wide surrounding the capsule rim; capsules opening above the middle to shed a shallow saucer-like lid and dull-gray seeds with prominent peg-like projections.

Local Distribution: Known from three widely separated localities on Tiburón, indicating widespread distribution on the island.

Other Gulf Islands: Cerralvo.

Geographic Range: Widespread in the Americas with three subspecies.

Tiburón: Haap Hill, *Felger 76-T35*. Palo Fierro, *Felger 8940*. Bahía Agua Dulce, *Felger 10222*.

Talinum paniculatum (Jacquin) Gaertner

Pnaacoj quistj "leafy mangrove"; *rama de sapo*; pink baby-breath

Description: Perennial herbs from thick, fleshy, tuberous roots, and also flowering in the first season; appearing during the summer rainy season and virtually no trace of the plant is evident when it is dormant. Stems and leaves succulent, the leaves broadly elliptic to obovate, 2.5–11+ cm long, falling quickly as the soil dries. Inflorescences of loose, open panicles 30–100 cm long. Flowers small, pink to dark red-purple, open for about three hours in the late afternoon.

Local Distribution: Tiburón at the eastern shore near Punta Tormenta and Estero San Miguel, where we found

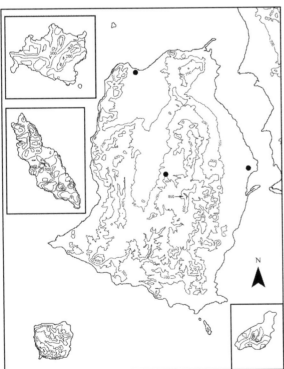

Portulaca umbraticola ssp. *lanceolata*

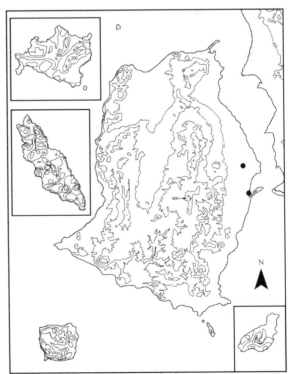

Talinum paniculatum
Pnaacoj quistj

it at the bases of larger shrubs in swales that create pockets of diverse desertscrub species within the *Frankenia* zone.

Other Gulf Islands: Cerralvo.

Geographic Range: Arizona to Florida and through Mexico to South America and the West Indies, and introduced in Africa and Asia.

> **Tiburón:** Near Punta Tormenta, *Knight 910* (UNM). Zozni Cmiipla, 100–300 m inland, *Wilder 06-285.*

Resedaceae – Mignonette Family

Oligomeris linifolia (Vahl) J.F. Macbride

Xamaasa (variants: tomaasa, xomaasa); desert cambess

Description: Annuals, October–May; plants generally upright, glabrous, and with a taproot. Stems leafy; leaves mostly 1.5–4 cm long, narrowly linear, often

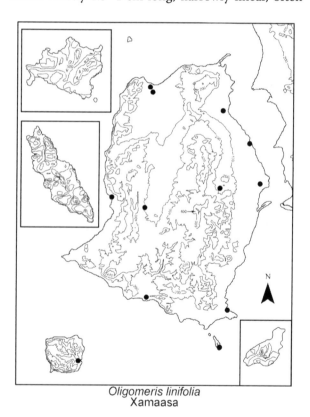

Oligomeris linifolia
Xamaasa

appearing fascicled because of dense clusters of axillary short shoots. Flowers 1–2 mm long, green, on slender, branched spikes; ovaries and fruits gaping open at apex. Fruits (capsules) 2.5–3 mm wide, globose, moderately persistent. Seeds 0.5 mm long, numerous, shiny black.

The Comcaac gathered the seeds in substantial quantities during favorable years, toasted and ground them, mixed the flour in water, and consumed it as *atole* (Felger & Moser 1985).

Local Distribution: Widespread through the lowlands on Tiburón, and on San Esteban and Dátil.

Other Gulf Islands: Ángel de la Guarda, Salsipuedes, San Lorenzo, Tortuga, San Marcos, Coronados, Espíritu Santo.

Geographic Range: Mostly in desert regions; northwestern Mexico and southwestern United States, and arid regions of the Old World from northern Africa to southwestern Asia and southern China. While no other member of the family is native to the New World, Martín-Bravo et al. (2009) provide evidence for a long-distance dispersal for *O. linifolia* from the Old World during the Quaternary.

> **Tiburón:** Ensenada Blanca (Vaporeta), arroyos and mesas, *Felger 17250.* Arroyo Sauzal, near beach and inland on hills, *Felger 7033.* Ensenada del Perro, *Felger 17722.* SW Central Valley, *Felger 12373.* Palo Fierro, *Felger 12561.* Zoojapa, NE foothills of Sierra Kunkaak, 180 m, *Wilder 08-224.* Vicinity of Valle de Águila, within 50 m of shore, *Wilder 07-270.* Hahjöaacöl, NE coast, *Wilder 08-209.* Tecomate, *Felger 12498.* Bahía Agua Dulce, one of the most abundant spring ephemerals, *Felger 6840.*
>
> **San Esteban:** Limantour, *Felger 17656.*
>
> **Dátil:** SE side of island, steep, narrow canyon, locally common, N- and E-facing slopes, *Felger 17502.*

Rhamnaceae – Buckthorn Family

Hardwood shrubs to small trees, the twigs often spinescent tipped. Leaves alternate (opposite or subopposite in *Sageretia*), simple. Inflorescences of small axillary clusters shorter than the leaves. Flowers small and radially symmetrical; calyx lobes or sepals 5; petals 5 or none.

1. Leaves 3.5–12 (20) mm long, widest well above middle; petals none; fruits 3–4.5 mm long. _____**Condalia**

1' Leaves 5–30 mm long, widest at or below middle (except *Sageretia* leaves that may be widest above the middle); petals present but deciduous; fruits 5 mm or more in length or diameter.

  2. Fruits dry, rounded capsules 5–6 mm wide. _____**Colubrina**

  2' Fruits fleshy, 8–10+ mm long and longer than wide.

    3. Shrubs leafy but drought deciduous, the leaves thin and green, soon becoming shiny; flowers in slender terminal spike-like inflorescences, the flowers 1.2–1.5 mm wide, sessile; fruits 3-seeded. _____**Sageretia**

    3' Shrubs often leafless or with sparse foliage, the leaves dull gray-green and relatively thick; flowers in umbellate clusters on short shoots, the flowers 2.5–3 mm wide, on short, persistent pedicels; fruits 1-seeded. **Ziziphus**

Colubrina viridis (M.E. Jones) M.C. Johnston [*C. glabra* S. Watson]

Ptaact; *granadita, palo colorado*

Description: Shrubs often 2–3 m tall, with rigid branches and very hard wood; bark often with conspicuous smooth, reddish patches; shorter long-shoot twigs often ending in a thorn. Leaves drought deciduous, 5–30 mm long including slender petioles 3–10 mm long, glabrous or sparsely pubescent, the blades obovate to orbicular, bright green, and thin, the margins entire; leaves promptly produced non-seasonally on short shoots following rains. Flowers in small axillary clusters or solitary, on slender pedicels 3–10 mm long; flowers 5–6 mm wide, yellow-green, the floral disk awash in nectar at anthesis; massive flowering during the summer-fall rainy season. Fruits of globose capsules 5–6 mm wide, breaking apart into 3 one-seeded segments.

The Comcaac primarily used this shrub for its wood, for firewood, as a pry bar for harvesting agaves, and to make the detachable foreshaft of the sea turtle harpoon (Felger & Moser 1985).

Local Distribution: Tiburón, widely distributed and common, including bajadas, valley plains, arroyos, canyon bottoms, and rocky slopes. On San Esteban, the dominant shrub in arroyos and canyons. Also on Dátil and Cholludo. Nolasco, widespread, canyons and slopes but generally not on north-facing slopes and reaching maximum development near the summit.

Other Gulf Islands: Ángel de la Guarda, Tortuga, San Marcos, Coronados, Carmen, Danzante, Monserrat,

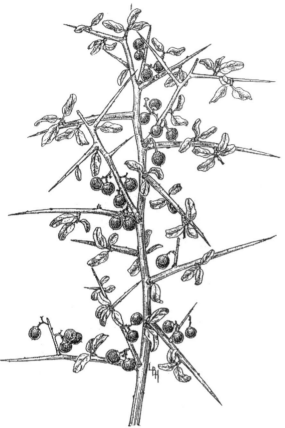

Figure 3.142. *Ziziphus obtusifolia* var. *canescens*. LBH.

Figure 3.143. *Colubrina viridis.* Pedro Comito next to a mature shrub, near Pozo Peña, southeast of Punta Chueca, Sonora, RSF, April 1983.

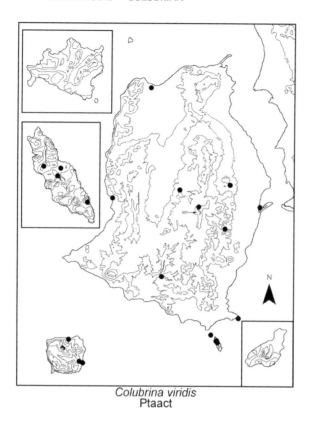

Colubrina viridis
Ptaact

Santa Catalina, Santa Cruz, San José, Espíritu Santo, Cerralvo.

Geographic Range: Western Sonora from the vicinity of El Desemboque and Ures southward, Sinaloa, both states of Baja California, and disjunct in Durango and Coahuila.

Tiburón: Ensenada Blanca (Vaporeta), *Felger 14968*. Xapij (Sauzal), *Wilder 08-140*. El Monumento, *Felger 08-59*. Haap Hill, *Felger T74-35*. Deep canyon in the northern interior of the Sierra Kunkaak, *Wilder 07-486*. Arroyo 1 km SW from Pazj Hax, *Wilder 06-41*. Zozni Cmiipla, *Felger 08-114*. Camino de Caracol, broad arroyo entering foothills of Sierra Kunkaak, *Wilder 05-2*. Freshwater Bay, shrub 8–9 ft high, frequent in a sandy draw, *Johnston 3273* (CAS).

San Esteban: Arroyo Limantour, *Van Devender 90-531*. Main arroyo [Limantour] at El Monumento, *Felger 9179*. N side of island, *Felger 15451*. S-facing canyon, central mountain, *Wilder 07-67*. Common shrub in sandy wash, *Johnston 3197* (CAS).

Dátil: NE side of island, arroyo bottom, *Felger 9092*. NW end of island, steep rocky slopes, *Felger 2778*. E-central side, E-W trending canyon, *Wilder 09-32*.

Cholludo: Near top of island, not common, *Felger 9155*.

Nolasco: Infrequent in draws about upper part of island, 17 Apr 1921, *Johnston 3136* (CAS). SE side, *Sherbrooke [Felger 11245]*. NE side, about 5 m below crest, fairly common near the ridge crest, *Felger 06-109*. Cañón de Mellink, 110 m, canyon bottom, *Felger 08-142*.

Condalia globosa I.M. Johnston var. **pubescens** I.M. Johnston

Cozi; *crucillo*; bitter condalia

Description: Medium size to large shrubs, and occasionally small trees, with very hard wood and thorn-tipped twigs. A few trees in the Central Valley of Tiburón reach 6 m and have massive trunks. [Many of the larger individuals on the mainland desert have re-grown from large, axe-cut stumps. The largest plants are in remote areas where there has been little or no woodcutting. The hard wood is highly desirable as firewood (Felger et al. 2001).] Leaves gradually drought deciduous, alternate and petioled on long shoots, and crowded, fascicled, and subsessile on short shoots (spur branches), mostly 3.5–12 (20) mm long. Flowers in small clusters on spur branches, yellow-green, 3 mm wide, the disk at anthesis awash with sticky, glistening nectar; petals none. Fruits dark brown, globose drupes, 3–5 mm wide, the style persistent. Flowering various seasons.

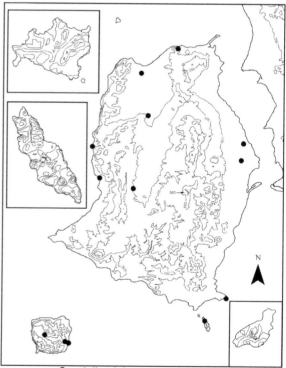

Condalia globosa var. *pubescens*
Cozi

The Comcaac said the wood is as good as ironwood (*Olneya tesota*) for fires, and the sharp, thorn-tipped twigs and mashed leaves were used in tattooing (Felger & Moser 1985).

Local Distribution: Widespread and common on Tiburón across a broad span of habitats: bajadas and valley plains, arroyos, canyons, and rocky slopes. Also on San Esteban, especially in arroyos and similar habitats on Dátil.

Other Gulf Islands: Ángel de la Guarda, San José, Espíritu Santo.

Geographic Range: Southeastern California and southwestern Arizona, western Sonora to northwestern Sinaloa, and the Baja California Peninsula. Var. *pubescens* is distinguished only by having pubescent leaves, while var. *globosa* is glabrous or essentially so. Only var. *pubescens* occurs in the northernmost part of the range and only var. *globosa* in the southernmost part (e.g., south of Guaymas). In between, the majority of the range, are plants of both varieties.

Tiburón: Ensenada Blanca (Vaporeta), 29 Jan 1965, *Felger*, observation (Felger 1966:181). El Monumento, *Felger 08-69*. Central Valley, valley plain, *Felger 17359G* (Felger 1966:239). 3 km W [of] Punta Tormenta, *Scott P4* (UNM). Valle de Águila, middle bajada, *Wilder 07-232*. Hee Inoohcö, mountains at NE coast, *Felger 07-66B*. Tecomate, 1 km inland, floodplain, *Felger*, observation (Felger 1966:261). Hast Iif (Ojo de Puma), Central Valley, broad arroyo, *Wilder 07-255*. Pooj Iime, rare, *Wilder 07-442*.

San Esteban: El Monumento, arroyo, *Felger 9185*. SW corner of island, large arroyo, *Hastings 71-54A*. Arroyo Limantour, 3.25 mi inland, *Wilder 07-55*.

Dátil: NE side of island, arroyo bottom near shore, *Felger 9125*.

Sageretia wrightii S. Watson

Wright's mock buckthorn

Description: Shrubs at least 1 m tall, the branchlets may be spinose tipped. Leaves opposite or subopposite, about 7–10+ mm long, elliptic to obovate, thin, at first sparsely pubescent, soon becoming glabrous, shiny, bright green (margins may be entire or serrated), and short petioled. Flowers on slender terminal spike-like inflorescences from short shoots; flowers sessile or nearly so, about 1.2–1.5 mm wide with 5 sepals and 5 petals, the petals white. Fruits probably less than 1 cm long, fleshy, smooth, and purple-black, and reported to be sweet and edible, with 3 seeds.

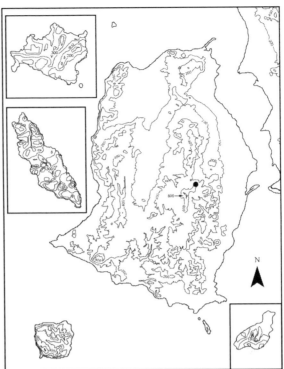

Sageretia wrightii

Local Distribution: Higher elevations in the Sierra Kunkaak, where it is known from a single collection. The nearest population is probably in northeastern Sonora in the vicinity of Rancho La Brisca about 225 km northeast of Tiburón.

Other Gulf Islands: None.

Geographic Range: Arizona and Sonora at elevations above the desert (known from northeastern Sonora but not widespread), southwestern New Mexico, Trans-Pecos Texas, Baja California Sur, Chihuahua, Coahuila, and Durango. Reports of the species in Jalisco are in error (Jim Henrickson, personal communication, 2007).

Tiburón: Capxölim, rare, shrub 1 m tall, *Wilder 07-482*.

Ziziphus obtusifolia (Hooker ex Torrey & A. Gray) A. Gray var. **canescens** (A. Gray) M.C. Johnston [*Condalia lycioides* (A. Gray) Weberbauer var. *canescens* (A. Gray) Trelease. *Condaliopsis lycioides* (A. Gray) Suessenguth var. *canescens* (A. Gray) Suessenguth]

Haaca, xiica imám coopol "black-fruited things"; *abrojo, barchata*; graythorn, lotebush

Description: Sprawling shrubs with rigid branches, forming briar-like thorny tangles; often leafless or the foliage sparse. Stems gray-green, the twigs thorn tipped

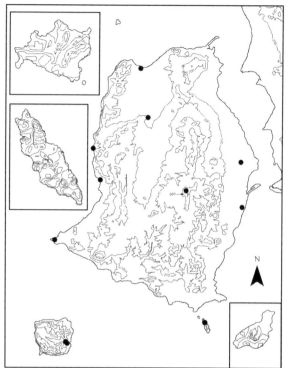

Ziziphus obtusifolia var. *canescens*
Haaca, xiica imám coopol

Tiburón: Ensenada Blanca (Vaporeta), arroyo bottom, *Felger 7122*. N of Punta Willard, *Perrill 5115*. Between shore and Sierra Kunkaak, bajada flats, *Wilder 06-40*. Palo Fierro, ca. 2.5 km inland, 14 Feb 1965, *Felger*, observation (Felger 1966:200). E-facing slope just below highest ridge on island, occasional, *Wilder 07-529*. Tecomate, Arroyo Agua Dulce, *Felger 6832*. Hast Iif (Ojo de Puma), Central Valley, broad arroyo, *Wilder 07-254*. Pooj Iime, *Wilder 07-438*.

San Esteban: Main E draining canyon [Arroyo Limantour], *Felger 7041*. SW corner of island, large arroyo, *Hastings 71-56*.

Dátil: NE side of island, arroyo bottom, near beach, *Felger 9126*. E shore, N of island center, canyon bottom, *Wilder 07-181*.

Rhizophoraceae – Red Mangrove Family

Rhizophora mangle Linnaeus

Pnaacoj xnaazolcam (variants: xnaazolcam, pnaazolcam) "criss-crossed mangrove," mojépe pisj—the fruit (embryo) "saguaro pisj"; *mangle rojo*; red mangrove

Description: Shrubs or small trees 2–4+ m tall. Distinguished from other Gulf of California mangroves (*Avicennia* and *Laguncularia*) by the prominent stilt roots, shiny green and semi-succulent leaves, and the unique fruit. Leaves opposite, 7–15 cm long, leathery, semi-succulent, ovate to obovate or broadly elliptic, and entire. The opposite leaves are not quite decussate since successive "pairs of leaves diverge from each other at an angle of less than 90°. This arrangement has important architectural consequences, since mutual shading is reduced and branches diverge at diverse angles" (Tomlinson 1986:319). Stipules sheathing the terminal bud and falling away as the new leaf unfolds. Inflorescences axillary, much shorter than the leaves, usually 2- to 5-flowered. Flowers white and yellow, soon turning downward (pendulous), and mostly or entirely wind-pollinated. Calyx 4-lobed, 10–12 mm long, yellow to green, thick and leathery. Petals 4, cream colored, densely white-hairy along the margins of the inside surface, and soon deciduous. Stamens 8, the anthers 4–6 mm long, sessile, thick and honeycomb-like. Flowering during warmer months.

The anthers open in the bud before the flower opens, so that the pollen is trapped among the numerous hairs of the petals. The pollen is drawn out with the petals and disperses as the petals expand and recurve. The petals fall within a day. The wind-pollinated flowers of *Rhizophora*

and mostly spreading at right angles. Leaves mostly 8–22 mm long, and quickly drought deciduous; leaf edges often chewed by insects. Flowers in short sub-umbellate clusters usually less than 1 cm long, the old inflorescences with the basal disk of the fruit semi-persistent. Flowers 2.5–3 mm wide, pedicelled, with white petals loosely enfolding the stamens and quickly falling after the anthers mature. Fruits 8–10 mm long, glaucous, blackish blue or purple-brown, the pericarp thin, fleshy, and edible but not very sweet. Flowering at least May–September, the flowers visited by honeybees, native bees, large orange-winged spider wasps (*Pepsis* or *Hemipepsis*), and other insects.

Local Distribution: Widespread on Tiburón and San Esteban, often forming dense thickets along arroyos, and sometimes on bajada flats, and occasionally on rocky slopes. Occasional in the Sierra Kunkaak to near peak elevation.

Other Gulf Islands: Ángel de la Guarda, San Marcos, San José, Cerralvo.

Geographic Range: Var. *canescens* is mostly a Sonoran Desert element: southeastern California, southern Nevada, and northern Arizona to southern Sonora and Baja California Sur; nearly ubiquitous in Sonora except at higher elevations. It is replaced by var. *obtusifolia* in the Chihuahuan Desert region.

Figure 3.144. *Rhizophora mangle.* Estero Sargento, Sonora. LBH.

Figure 3.145. *Rhizophora mangle.* Estero Santa Rosa, Sonora. BTW, 1 April 2006.

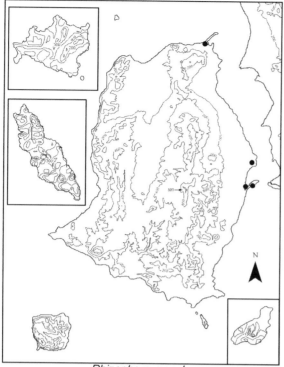

Rhizophora mangle
Pnaacoj xnaazolcam

have evolved from an insect-pollinated ancestor. Nectar secretion is low (Tomlinson 1986). The seedling develops while still attached to the parent tree. This club-shaped, photosynthetic fruit-like seedling, often 20–30 cm long, can drift far away from its parent. On 29 September 2008 at Punta Chueca we saw numerous "fruits" floating in shallow, warm water riding vertically with the root-end down and the shoot-end bobbing at the sea surface.

Oysters were gathered from the stilt roots. The "fruit" or enlarged embryo, mojepe pisj, was eaten when green. It was cooked in ashes, the outside "shell" broken with a stone, and the inside eaten with sea turtle fat. Another method of preparation was to scrape the fruit with a knife to remove part of the outside "shell," wrap the fruit in a cloth, and cook it in water with sea turtle fat (Felger & Moser 1985).

Local Distribution: Locally abundant in the three mangrove esteros along the east shore of Tiburón. Red mangroves grow on the seaward (deeper water) side of the other two mangroves.

Other Gulf Islands: Rasa (one small plant, *Velarde 20 Apr 2008*, SD), Carmen, Danzante (observation), San Diego, Espíritu Santo.

Geographic Range: Both coasts of the Gulf of California and Pacific and Atlantic tropical shores of the New World and in West Africa. The northernmost red mangroves in Sonora are at Estero Sargento at the north end of the Canal del Infiernillo. The occasionally winter freezing weather is the obvious factor limiting further northward distribution. The Sargento mangroves periodically suffer from extensive frost damage.

Tiburón: Estero San Miguel, *Wilder 06-274*. Zozni Cmiipla, seedlings in mud, *Wilder 06-367*. Punta Tormenta estero, 5 Sep 2008, *Wilder*, observation. Cyazim It, Punta Perla, estero, *Wilder 08-376*.

Rubiaceae – Madder Family

Leaves simple, entire (sometimes notched at tip in *Randia*), opposite and decussate (each pair at right angles to the one above and below) or appearing whorled (the stipules transformed into leaves in *Galium*). Flowers 4- or 5-merous.

1. Annuals or small shrubs; leaves appearing whorled with 4 per node; fruits bristly or hairy. _____ **Galium**
1' Woody shrubs; leaves 2 per node; fruits smooth.
    2. Plants unarmed. _____**Chiococca**
    2' Twigs armed with sharp spines. _____**Randia**

Chiococca petrina Wiggins

Description: Unarmed woody shrubs often 1–1.5 m tall. Leaves opposite, dull olive green, often 1–4.3 cm long; evergreen to tardily drought deciduous. Flowers small and white, 5-merous, in short axillary clusters shorter than the leaves. Fruits fleshy and whitish, rounded to moderately compressed, 5–8 mm wide.

Local Distribution: Scattered canyons, north-facing rock slopes, and higher elevations of Sierra Kunkaak. The nearest known population is the Sierra el Aguaje north of Guaymas.

Other Gulf Islands: None.

Geographic Range: Sonora in scattered canyons and mountains from the north-central (southeast of Magdalena de Kino) to southeastern part, and canyons on the north side of the Sierra el Aguaje, southwestern Chihuahua, and Sinaloa.

Tiburón: [Locality on the island not provided but based on sequence of collection numbers it is from near Sopc Hax waterhole in the Sierra Kunkaak], Oct 1979, *Knight 1106* (UNM; photocopy SD). Near Caracol Research station, shrub in upper canyon near tinaja, *Knight 28 Oct 1979*.

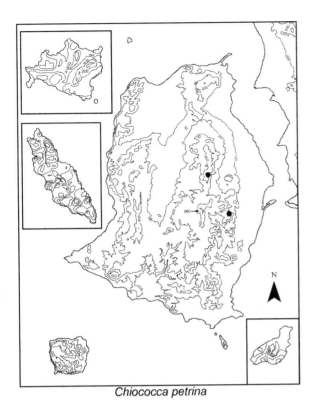

Chiococca petrina

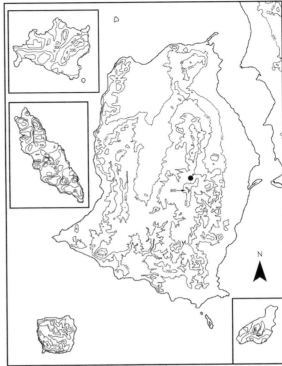

Galium proliferum

Galium – Bedstraw

Annuals or small shrubs. Stems 4-angled. Leaves appearing whorled (2 leaves and 2 leaf-like stipular appendages). Flowers small, 4-merous. Calyx minute or absent; corollas deeply 4-parted. Fruits bristly or hairy.

1. Small herbaceous annuals. _____**G. proliferum**
1' Small shrubs or subshrubs, perennials. _ **G. stellatum**

Galium proliferum A. Gray

Desert bedstraw

Description: Delicate, small and often minute winter-spring annuals. Leaves 3–7 mm long, oblong to ovate, the leaf tips blunt or rounded; fruit body less than 3 mm wide and densely covered with hooked hairs.

Local Distribution: A population was found scattered along a shaded canyon floor in the Sierra Kunkaak. This is the only record for a Gulf Island. The nearest known locality is in the Sierra el Aguaje (La Balandrona, *Felger 01-650*).

Other Gulf Islands: None.

Geographic Range: Southeastern California to western Texas and northern Mexico. In Sonora known from canyons in the northern part of Sierra el Aguaje, the Sierra Libre, and in the eastern part of the state including tropical deciduous forest in the Alamos region and northward mostly at higher elevations.

Tiburón: 2 km E of Siimen Hax, 360 m, sand-gravel soil in shaded margin of arroyo bed among leaf litter, *Wilder 06-458*.

Galium stellatum Kellogg [*G. stellatum* var. *eremicum* Hilend & J.T. Howell]

Desert bedstraw

Description: Untidy small shrubs or subshrubs. Stems slender, square with white-margined corners. Leaves and young stems scabrous with short stiff white hairs. The old dry leaves whitish and often persisting, the leaves probably less than 10 mm long, narrowed to a spinescent tip. Male and female flowers on separate plants; flowers small, whitish to pale yellow with purplish net-like veins, in small leafy panicles among the short shoots. Ovaries and fruits densely covered with spreading straight silky white hairs, the hairs longer than the fruit. Flowering in spring. *Galium stellatum* is the shrubbiest of all New World galiums.

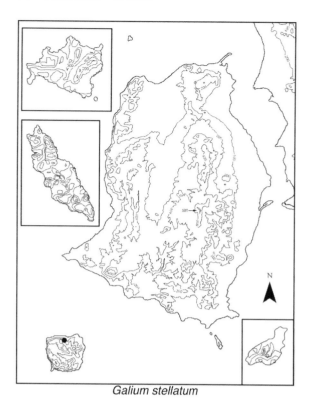

Galium stellatum

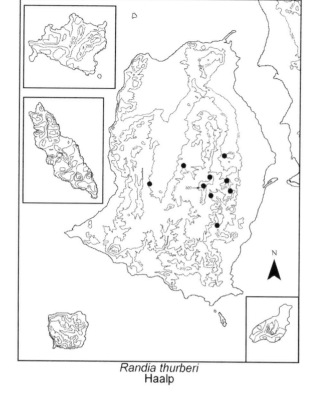

Randia thurberi
Haalp

Local Distribution: Known from San Esteban by a single collection, where it is presumably rare or highly localized. It might be found in mountains on the northeast side of Tiburón since it occurs on the nearby mainland at Cerro Tepopa.

Other Gulf Islands: Ángel de la Guarda, San Lorenzo.

Geographic Range: Mostly in desert mountains; southwestern Utah, southern Nevada, western and southern Arizona, southeastern California to northern Baja California Sur, and northern Sonora southward in coastal mountains to Cerro Tepopa (29°21'N).

San Esteban: Inland on NE quadrant of island, UTM 12 3178413 N, 12 345752 E, NAD 83, 30 Mar 2005, *Wegscheider 101* (SD).

Randia thurberi S. Watson

Haalp; *papache borracho*

Description: Shrubs with rigid woody branches and sharp, paired spines. Leaves drought deciduous. Flowers 10–12 mm long, white and fragrant; flowering with summer rains. Fruits globose, hard shelled, mottled green and white, 1.5–2.5 cm diameter, ripening at least in spring, the mesocarp (pulp) black, sweet and edible (Felger & Moser 1985).

Local Distribution: Tiburón in the Central Valley and upper bajadas, foothills, canyons, and slopes of the Sierra Kunkaak.

Other Gulf Islands: None.

Geographic Range: Central Sonora and Tiburón to northwestern Sinaloa.

Tiburón: SW part of Agua Dulce Valley (Central Valley), *Felger 12375.* Haap Hill, *Felger 74-T58.* Foothills, E side Sierra Kunkaak, arroyo, *Felger 6949.* Sopc Hax, *Felger 9316.* Hast Cacöla, E flank of Sierra Kunkaak, foothills, *Wilder 08-229.* S-facing slope of Sierra Kunkaak, 355 m, *Wilder 06-27.* Base of Capxölim, cliff face of canyon, 415 m, *Wilder 06-403.* Sierra Kunkaak, highest ridge of island, 820 m, *Wilder 07-548.* SE of Caracol, *Wilder 08-325.*

Ruppiaceae – Ditch-Grass Family

Ruppia maritima Linnaeus

Zimjötaa; ditch-grass

Description: Delicate submerged aquatics, annuals or perhaps perennials. Leaves alternate, expanded at the base into a nearly transparent sheath, the blades thread-like, less than 10 cm long, 1 mm wide, the tip acute (pointed)

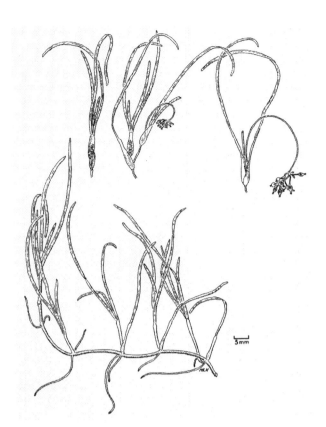

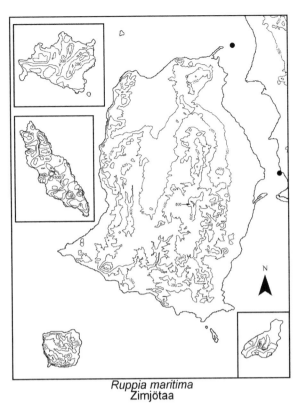

Ruppia maritima
Zimjötaa

Figure 3.146. *Ruppia maritima.* Beach drift at Bahía Kino, Sonora, April 1974. NLN.

to obtuse (rather blunt) with microscopic teeth. Inflorescences of 2 flowers, one above the other on opposite sides of the peduncle (flowering stalk), at first enclosed in a membranous spathe inside the leaf sheath, the peduncles pushing through the sheath and greatly elongating as the fruits develop; flowers bisexual and without a perianth. Fruits of 4 single-seeded nutlets 2 × 1 mm.

Rooted in shallow subtidal waters in protected bays and esteros; most luxuriant during hot weather when fragments of the plants wash ashore in the beach drift. Sometimes found year round in esteros, and also sometimes in quiet, freshwater pools or tanks in Sonora and elsewhere.

Ruppia is readily distinguished from eelgrass (*Zostera*) by its much narrower and shorter leaves, more slender stems and distinctive inflorescence and small seeds, and generally different season of growth. It is distinguished from *Halodule* by its elongated stems, expanded leaf sheaths, acute to obtuse leaf tips with microscopic teeth, and small but visible flowers and fruits.

Local Distribution: During the 1970s and 1980s, Richard found detached pieces of the stems common during summer and early fall in the warm water of the Infiernillo along the Tiburón and mainland shores. According to Alf Meling-López (personal communication, 2007), the source of this *Ruppia* is from a population in Estero Santa Cruz at Bahía Kino. Some *Ruppia*, however, was found rooted in the bay at Punta Chueca.

Other Gulf Islands: Ángel de la Guarda, Espíritu Santo.

Geographic Range: Widespread in the Gulf of California as a seagrass, and worldwide in fresh to brackish water.

Canal del Infiernillo: Punta Chueca, growing in one fathom of seawater, inside bay, and common in beach drift, 28 Jun 1973, *Felger 20943b*. Near mouth of Estero Sargento, N end of Infiernillo, common in beach drift, 28 Jun 1973, *Felger 20929*.

Ruscaceae, *see* Asparagaceae

Salicaceae – Willow Family

Salix exigua Nuttall

Paaij; *sauce*; narrowleaf willow

Description: Shrubs 2–5+ m tall. Stems erect or ascending, often reddish or yellow-brown especially

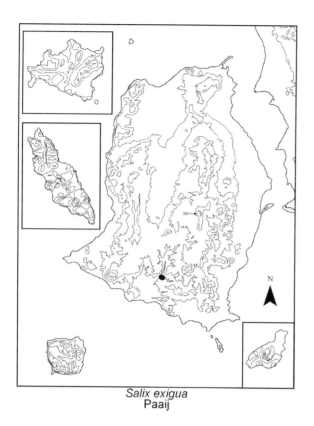

Salix exigua
Paaij

Santalaceae – Sandalwood Family
(includes Viscaceae)

Phoradendron – *Toji*; mistletoe

Parasitic on various trees and shrubs; male and female flowers on separate plants, the flowers small, the fruits fleshy, disseminated by birds.

1. Stems appearing leafless, the leaves scale-like. _____
_____ **P. californicum**
1' Stems leafy, the leaves well developed. **P. diguetianum**

Phoradendron californicum Nuttall

Eaxt, preceded by the host's tree, e.g., comitin eaxt, ironwood's (*Olneya*) mistletoe; *toji*; desert mistletoe

Description: Stems essentially leafless, with small scale leaves; often festooning common desert trees. Flowers yellow and fragrant, appearing at various seasons but especially in mid-winter. Berries white to pale reddish.

Local Distribution: Widespread and common on Tiburón, mostly on ironwood (*Olneya*) and sometimes on other legume trees.

Other Gulf Islands: Ángel de la Guarda, Santa Catalina, Cerralvo.

when young. Leaves silvery to grayish green, short petioled, linear, 3.5–17 cm long; winter deciduous with new leaves and inflorescences emerging in late January. Peak flowering in early spring and continuing sporadically until fall.

Local Distribution: In 2008 it was locally common at the Sauzal waterhole, forming impenetrable thickets, the largest ones 5+ m tall growing from permanently wet slickrock and gravelly soil. Sauzal is a place of large or many willows, *sauces*.

No other willow population is known for a Gulf of California island. The nearest known population of narrowleaf willow is at the La Salina pozos near the Río Colorado delta (Felger 2000a).

Other Gulf Islands: None.

Geographic Range: Western North America from Canada to northwestern Mexico. The closely related sandbar willow, *S. interior* Rowlee [*S. exigua* subsp. *interior* (Rowlee) Cronquist] replaces *S. exigua* eastward and northward in North America.

Tiburón: Sauzal waterhole (Xapij): Localized in shallow water and wet soil at upper carrizo patch, *Felger 10076*; Xapij, *Wilder 08-97*.

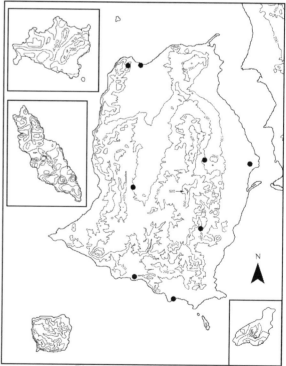

Phoradendron californicum
Eaxt

Geographic Range: Northwestern Sinaloa and the Cape Region of Baja California Sur to southwestern United States.

Tiburón: Arroyo Sauzal, near shore, common, *Felger 7032*. Ensenada de la Cruz, *Felger 12803*. SW Central Valley, arroyo, *Felger 12340*. Foothills of Sierra Kunkaak, bajada, *Felger 6995*. Palo Fierro, *Felger 10013*. Caracol station, *Wilder 06-153*. Tecomate, *Whiting 9013*. Rocky hill W of Tecomate airstrip, common on legume trees, *Felger 6258*.

Phoradendron diguetianum Van Tieghem

Eaxt, preceded by the host tree, e.g., cof eaxt, san juanico's (*Bonellia*) mistletoe; *toji*

Description: Leaves 1.5–3 cm long, broad, semi-succulent, yellow-green, and dimorphic, some branches with elongated and narrower leaves, and others with shorter and broader leaves, the two leaf forms on the same or different plants. Leaf shape apparently varies with growth rate, seasonal weather conditions, and flowering conditions. Recorded on *Bonellia, Castela, Maytenus*, and *Ziziphus*.

Local Distribution: Widespread across the southern and eastern side of Tiburón and the Sierra Kunkaak nearly to the summit.

Other Gulf Islands: Carmen, Danzante, Monserrat, Santa Catalina, Santa Cruz, San Diego, San José, Espíritu Santo, Cerralvo.

Geographic Range: Coastal Sonora from the southern part of the state northward to El Desemboque, Gulf Islands on the Baja California side, and the southern half of the Baja California Peninsula.

Tiburón: Sauzal waterhole, not common, *Felger 10073*. Ensenada de la Cruz, 1 km inland, bottom of large arroyo, *Felger 12790*. Bajada between Hant Hax camp and Zozni Cmiipla, *Felger 9360*. San Miguel Peak, 1000 ft, on *Jacquinia, Knight 986* (UNM). Capxölim, *Wilder 07-477*. Steep, protected canyon, 720 m, N interior of Sierra Kunkaak, *Wilder 07-518*. E-facing slope above canyon and just below highest ridge on island, *Wilder 07-527*. Deep sheltered canyon, N portion of Sierra Kunkaak, *Felger 07-119*. Punta Perla, mangrove estero, *Felger 21317*.

Sapindaceae – Soapberry Family

1. Vines or scandent plants; leave pinnately compound; inflorescences bearing tendrils; flowers bilateral; fruits globose, inflated, and balloon-like. _ **Cardiospermum**
1' Shrubs; leaves simple; without tendrils; flowers radial; fruits 3-winged, not inflated. _____ **Dodonaea**

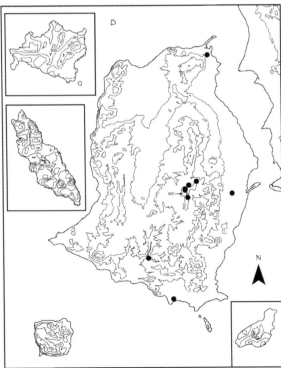

Phoradendron diguetianum
Eaxt

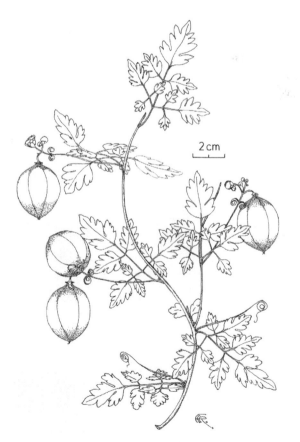

Figure 3.147. *Cardiospermum corindum*. FR.

Figure 3.148. *Dodonaea viscosa*. Male flower and branches above, female flower above right, fruit and fruiting branch below. LBH.

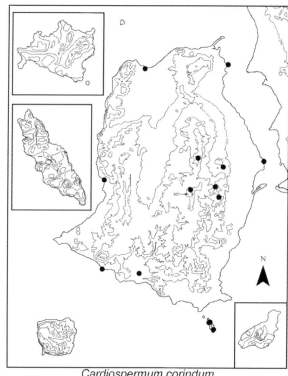

Cardiospermum corindum
Hax quipoiin

Cardiospermum corindum Linnaeus

Hax quipoiin *maybe inferring* "covering water (as in covering a water-filled *olla*)"; *tronador, farolito*; balloon vine

Description: Herbaceous to somewhat scandent climbers and vines with tendrils at the base of each flowering cluster. Leaves once or twice pinnately compound, the segments lobed to crenulate. Flowers small and white; sepals and petals each 4. Fruits 3-lobed and 3-celled, inflated like a miniature balloon or paper lantern, often 2–3.5 cm wide. Seeds 3, smooth and rounded.

Local Distribution: Widespread across Tiburón, including arroyos, bajadas, canyons, and slopes to higher elevations in the Sierra Kunkaak. Common on Dátil.

Other Gulf Islands: San Marcos, Coronados, Carmen, Danzante, Monserrat, Santa Catalina, Santa Cruz, San José, Espíritu Santo, Cerralvo.

Geographic Range: Tropical and subtropical regions worldwide. Widespread in the Sonoran Desert, the northern limit in the Coyote Mountains of southern Arizona.

Tiburón: Ensenada Blanca (Vaporeta), *Felger 14931*. Tordillitos, *Felger 15503*. Arroyo Sauzal, ¾ mi from shore, *Wilder 06-81*. Sopc Hax, *Wilder 07-216*. Near marine base and airstrip, *Knight 904* (UNM). Hast Cacöla, E flank of Sierra Kunkaak, 380 m, *Wilder 08-241* (USON). Foothills of Sierra Kunkaak, *Wilder 05-14*. Deep canyon in N interior of Sierra Kunkaak, *Wilder 07-583*. Top of Sierra Caracol, *Knight 1066,1108* (UNM). Zozni Cacösxaj, NE coast, *Wilder 08-202*. Tecomate, *Whiting 9058*.

Dátil: NE side of island, steep rocky slopes, widespread, most common vine on island, *Felger 9096*. NW side, *Felger 15337*. E side, slightly N of center of island, halfway up N-facing slope of peak, *Wilder 07-107*. SE side, *Felger 17521*.

Dodonaea viscosa (Linnaeus) Jacquin

Caasol caacöl "large caasol"; *tarachiqui, tarachico*; hop bush

Description: Slender-stemmed shrubs, mostly 1–2 m tall; bark shredding in narrow, braided ridges. Herbage, especially the young shoots, resinous-sticky. Leaves evergreen except in extreme drought, 3–11 cm long, linear-oblanceolate, essentially sessile. Male and female flowers mostly on separate plants but often with some bisexual flowers. Flowers small and yellow-green, without

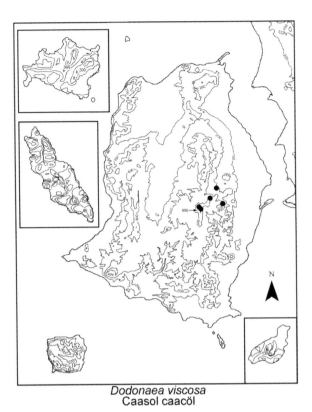

Dodonaea viscosa
Caasol caacöl

petals, in small clusters among the leaves. Fruits papery, 3-winged capsules 1–2 cm wide.

Local Distribution: Sierra Kunkaak, from the foothills to the summit, canyons and rocky, often north-facing slopes especially at higher elevations.

Other Gulf Islands: Cerralvo.

Geographic Range: Widespread in Arizona, both states of Baja California, and Sonora, mostly at elevations above the desert. Tropical and subtropical regions worldwide.

Tiburón: Hast Cacöla, E flank of Sierra Kunkaak, top of E-facing slope at base of cliff, ca. 540 m, *Wilder 08-269*. Foothills, Sierra Kunkaak, *Wilder 05-15*. Cerro San Miguel, 300 m, not common, *Quijada-Mascareñas 91T015*. Cójocam Iti Yaii, saddle, *Wilder 07-540*. Summit of island, not common, *Wilder 07-567*.

Sapotaceae – Sapote Family

Sideroxylon

Hardwood trees and shrubs, with latex in trunks, limbs and fruits; hairs 2-branched (T-shaped), with one long and one short branch. Leaves alternate in a spiral arrangement, or fascicled on axillary short shoots, simple, and petioled;

Figure 3.149. *Sideroxylon leucophyllum.* Capxölim, Isla Tiburón. BTW, 24 October 2007.

Figure 3.150. *Sideroxylon occidentale.* With Humberto, Sierra Kunkaak, Isla Tiburón. BTW, 25 November 2006.

stipules none. Fruits fleshy, usually with an edible pericarp, 1-seeded. Mostly tropical and subtropical; Americas, Africa, Madagascar, and Mascarene Islands.

1. Plants unarmed; leaves 4–16 cm long, densely white-pubescent. _____**S. leucophyllum**
1' Twigs often thorn tipped; leaves 1–3 cm long, dull gray-green. _____ **S. occidentale**

Sideroxylon leucophyllum S. Watson

Hehe pnaacoj "mangrove plant"

Description: Shrubs or trees 3–5 (7) m tall, often with a well-defined trunk to 30+ cm diameter. Leaves on long shoots (without short shoots). Leaves oblong and densely tomentose, 4–9 (16) cm long. Flowers small, cream colored, in crowded fascicles in leaf axils. Fruits 1.5–2 cm long, ovoid to globose, at first tomentose, eventually glabrate.

Pnaacoj is the general term for mangrove, especially the black mangrove (*Avicennia germinans*), a shrub or tree with leaves of similar color, size, and shape to those of *S. leucophyllum.* The leaves were mixed with cinnamon and held in the mouths to bring good luck in card games (Felger & Moser 1985).

Local Distribution: Tiburón in the Sierra Kunkaak, mostly on north-facing rocky slopes, above approximately 380 m elevation, often on steep talus of large boulders. Here it is locally common, developing substantial trunks. San Esteban in a few, scattered canyons and ridge crests. The peaks of San Esteban and Tiburón are often shrouded in fog clouds—these mountains make their own microclimate. The densely pubescent leaves of *S. leucophyllum* might be efficient in capturing moisture.

Other Gulf Islands: Ángel de la Guarda.

Geographic Range: Primarily Baja California Norte and Isla Ángel de la Guarda, and disjunct on Tiburón

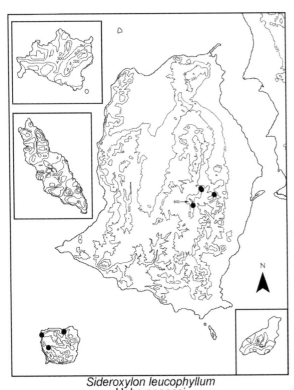

Sideroxylon leucophyllum
Hehe pnaacoj

and San Esteban, and perhaps in the Sierra Seri (Wilder et al. 2007a).

Tiburón: Hast Cacöla, E flank of Sierra Kunkaak, 380–540 m, *Wilder 08-244.* Cerro Kunkaak, 700 m, *Scott 11 Apr 1978* (UNM 53277). Cerro San Miguel, árbol de baja estatura, común en laderas rocosas, principalmente en la ladera norte, 4–5 m, con tronco definido, *Quijada-Mascareñas 91T009.* Capxölim, solitary peak NE of Sierra Kunkaak Mayor, 400 to 590 m, trees to 7 m tall with deeply twisted and old branches, about 20 trees on this slope, *Wilder 07-471.*

San Esteban: SW corner of island, occasional in canyon bottom, *Wilder 07-85.* N side of island, narrow rocky canyon behind large cove, *Felger 17568.* San Pedro, common on steep slopes and rocky arroyo bottom, *Felger 16664.*

Sideroxylon occidentale (Hemsley) T.D. Pennington [*Bumelia occidentalis* Hemsley]

Hehe hateen ccaptax "plant that punctures the mouth," paaza; *bebelama*

Description: Large shrubs to heavy-trunked trees sometimes 6–10+ m tall with checkered, light gray bark; hairs mostly appressed. Branches rigid, the twigs often thorn tipped. Leaves alternate on long shoots and crowded on axillary short shoots, 1–3 cm long, obovate to spatulate, grayish with dense pubescence. Flowers dirty white and relatively inconspicuous. Fruits oblong, 1.5 cm long, with white latex, ripening dark purple and prune-like, the pericarp sweet and edible. Flowering in July and also in other seasons. The female flowers have stamens modified into large staminodes, which is an unusual character and known elsewhere in the genus only among Old World species (Pennington 1990:126).

Local Distribution: Tiburón, from the south-central part of the island to the Central Valley, and most numerous in the Sierra Kunkaak ranging from lower elevations to the summit. Along washes and canyon bottoms, sometimes in small groves, and often exceeding 10 m in height but dwarfed as shrubs on steep, high slopes.

Other Gulf Islands: None.

Geographic Range: Northern Sonora near Saric, Magdalena, and Arizpe, through Sonora to Sinaloa, Baja California Sur, and Tiburón.

Tiburón: Arroyo Sauzal, 1 mi from shore, *Wilder 06-73.* Rock outcrop at Xapij, tree 6 m tall, 29 Jan 2008, *Felger & Wilder*, observation. Arroyo de la Cruz, *Wilder 09-59.* SW Central Valley, widely scattered, tree about 35 ft tall, *Felger 12653.* Near Haap Hill, *Felger 74-T61* (specimen discarded). Arroyo 1 km SW from Pazj Hax, *Wilder 06-24.* Foothills of

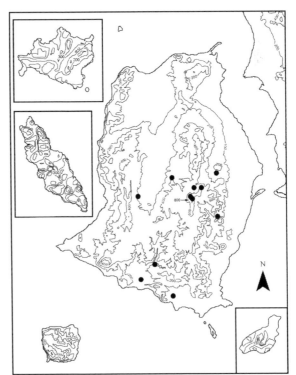

Sideroxylon occidentale
Hehe hateen ccaptax, paaza

Sierra Kunkaak, fork in road leading to Caracol station or N flank of Sierra Kunkaak, *Wilder 06-158*. 0.5 km E of Siimen Hax, *Wilder 06-469*. Capxölim, *Wilder 07-478*. E-facing slope above canyon and just below highest ridge on island, *Wilder 07-526*. Exposed upper ridge of island, between peak of Cójocam Iti Yaii and summit, occasional large shrubs, 1.5–3 m tall, *Wilder 07-547*. Summit of island, common to dominant, ca. 2 m tall, a principal component of the community at the top with Menodora scabra, *Wilder 07-555*.

Scrophulariaceae, *see* Plantaginaceae

Simaroubaceae – Quassia Family

Castela polyandra Moran & Felger

Snaazx, zazjc

Description: Low, spreading shrubs sometimes to 1 m tall. Stems rigid, laterally compressed, moderately glaucous, thorn tipped and at first finely pubescent, becoming glabrous with age. Leaves alternate, few and very quickly deciduous, simple, often 0.5–2.5 cm long, the plants leafless much of the year. Male and female flowers on separate plants. Flowers in compact axillary clusters, sessile or in panicles to 1 cm long. Flowers 4-merous, 8–15 mm wide; sepals and petals red. Petals 3–7 mm long. Stigma orange; anthers yellow. Flowering May–July, apparently with maximum flowering in July. Fruits 1–1.5 cm long, very bitter. *Pseudomymex* ants were seen visiting extrafloral nectaries (observed on *Wilder 07-387*, 15 Sep 2007, by Alexander Swanson).

This species is morphologically intermediate between *Holacantha* and *Castela*, having the branching pattern and petal dimorphism of *Holacantha* and the 4-parted flowers and deciduous fruits of *Castela*, which was the deciding factor in uniting *Holacantha* and *Castela* (Grieco et al. 1999; Moran & Felger 1968).

Figure 3.151. *Castela polyandra.* Branches with male flowers and female flower inset. Vicinity of Zozni Cmiipla, Isla Tiburón. BTW, 25 May 2006.

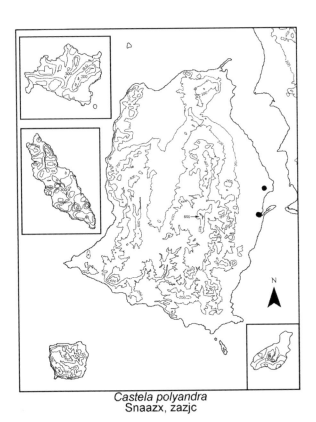

Castela polyandra
Snaazx, zazjc

Simmondsiaceae – Jojoba Family

Simmondsia chinensis (Link) Schneider

Pnaacöl; jojoba

Description: Shrubs, evergreen or leaves markedly reduced in number during dry seasons, or most of the leaves may be shed during prolonged drought. Leaves opposite, 2–5 cm long, simple, leathery, entire. Male and female flowers on separate plants, flowers axillary, petals none, flowering in mid-winter. Male flowers many in clusters shorter than the leaves, yellow due to the anthers. Female flowers usually single in leaf axils, green and much larger than the male flowers, the sepals about 1 cm long at anthesis, enlarging to about 1.5 cm in fruit. Ovary with three ovules, but only one seed develops. Seeds 1.5–2 cm long (one of the largest seeds of any Sonoran Desert plant). Cotyledons thick and fleshy, containing simmondsin, a cyanogenic glucoside, and a high percentage of a unique liquid wax, the basis of the jojoba industry.

Local Distribution: Tiburón on the eastern bajada in a narrow band parallel to the shore, just inland from the *Frankenia* zone and extending into the lower mixed desertscrub vegetation. Often growing with *Bursera microphylla, Cordia parvifolia, Ferocactus tiburonensis, Fouquieria splendens, Frankenia palmeri, Hibiscus denudatus, Horsfordia alata, Jatropha cinerea,* and *Stenocereus thurberi,* and sometimes a host for *Phoradendron diguetianum.*

Other Gulf Islands: San Marcos.

Geographic Range: Baja California Norte in the vicinity of Bahía de Los Ángeles, Baja California Sur in the vicinity of Santa Rosalía and the nearby Isla San Marcos, and Isla Tiburón. Throughout its range, *C. polyandra* grows exclusively on east-sloping, gravelly bajadas within a few kilometers of the sea.

Tiburón: 1 mi inland from shore at Santa Rosa [Punta San Miguel], locally common, *Felger 9359.* Palo Fierro, just inland of *Frankenia* zone, *Felger 10135.* Between the shore and Sierra Kunkaak, *Frankenia* dominated flats and desertscrub, 5 m elev., *Wilder 06-49.* E bajada, N of Estero San Miguel, 4 m elev., *Wilder 07-387.*

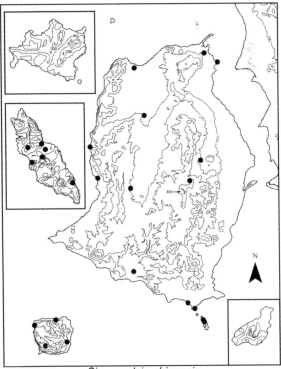

Simmondsia chinensis
Pnaacöl

Local Distribution: Common across most of Tiburón, San Esteban, Dátil, and Nolasco; bajadas, arroyos, canyons, and rocky slopes.

Other Gulf Islands: Ángel de la Guarda, San Marcos (observation), Coronados, Carmen, Danzante, Monserrat (observation), Santa Catalina, Santa Cruz, San José, San Francisco, Espíritu Santo.

Geographic Range: Both states of Baja California, southern California, Arizona, and Sonora southward to the Guaymas region. The family has a single species.

Tiburón: Ensenada Blanca (Vaporeta), *Felger 12234*. Arroyo Sauzal, ¾ mi inland, *Wilder 06-79*. Coralitos, *Wilder 06-56*. La Viga, *Wilder 08-152*. SW Central Valley, mostly along arroyo, *Felger 17321*. Canyon at base of Capxölim, *Wilder 06-425*. Caracol station, *Wilder 06-154*. Zozni Cacösxaj, NE coast, *Wilder 08-204*. Punta Perla, *Felger 74-10*. Bahía Agua Dulce, *Felger 10206*. Hast Iif (Ojo de Puma), Central Valley, broad arroyo, *Wilder 07-256*. Pooj Iime, *Wilder 07-434*.

San Esteban: Arroyo Limantour, *Felger 12739*. N side of island, *Felger 15415*. San Pedro, *Felger 16665*. Steep slopes of S-central peak, *Felger 17545*.

Dátil: NE side of island, *Felger 9079*. E-central side, canyon bottom, *Wilder 09-21*.

Nolasco: Common shrubs all over island, 17 Apr 1921, *Johnston 3129* (CAS, UC). Steep slopes very scattered, some plants showing drought die-back, 16 Dec 1951, *Gentry 11358*. Above SE cove, *Cooper [Felger 10403]*. E-central side, N exposure below crest of island, 270 m, staminate, *Felger 06-93*; pistillate, *Felger 06-94*. Cañón de la Guacamaya, 14/15 Apr 2003, *Gallo-Reynoso* (photo). Cañón de Mellink, 110 m, canyon bottom, *Felger 08-148a* (pistillate), *08-148b* (staminate).

Solanaceae – Nightshade or Potato Family

Herbs to shrubs. Leaves alternate and simple; stipules none. Flowers mostly 5-merous, sympetalous.

(A)

(B)

Figure 3.152. *Lycium brevipes* var. *brevipes*. (A) Flower, Isla Alcatraz, 17 September 2007. (B) Fruit, vicinity of Estero Santa Rosa, Sonora, 25 July 2007. BTW.

Figure 3.153. *Physalis crassifolia* var. *infundibularis.* Isla San Esteban.
BTW, 8 March 2007.

Figure 3.154. *Physalis crassifolia* var. *versicolor.* Arroyo
Sauzal, Isla Tiburón. Flower, 29 January 2008, and fruit with
portion of calyx removed, 17 March 2006. BTW.

Figure 3.155. *Solanum hindsianum.* Isla Tiburón. (A) Flower, Sierra Kunkaak, 24 November 2006. (B) Blister beetle feeding on an anther, Coralitos, 3 September 2008 BTW.

1. Shrubs, often 1 m or more in height.
   2. Glabrous or pubescent but hairs not stellate; twigs often thorn tipped; leaves and stems not spiny; flowers less than 1 cm wide. _____ **Lycium**
   2' Plants densely pubescent with stellate hairs; stems not thorn tipped; usually with slender spines on stems and leaves; flowers more than 2.5 cm wide. _____ **Solanum**
1' Plants herbaceous.
   3. Flowers more than 10 cm long; fruits spiny. _____ **Datura**
   3' Flowers less than 4 cm long; fruits not spiny.
      4. Corollas tubular; fruit a capsule (dry), the calyx not like a paper bag. _____ **Nicotiana**
      4' Corollas as broad as, or broader than long or deep (not tubular); fruit a berry (fleshy), the calyx growing around the berry like a paper bag. _____ **Physalis**

Datura discolor Bernhardi

Hehe camostim "plant that causes grimacing"; *toloache*; desert thorn-apple

Description: Foul-smelling non-seasonal annuals but responding poorly to colder weather. Larger leaves often 7–20 cm long. Flowers opening at dusk and remaining open for a short time after sunrise, or later on cooler days; corollas 8–17 cm long, white with a purple flush in the throat. This is the largest-flowered plant in the Sonoran Desert. Fruits of globose capsules mostly 4–6 cm wide, covered with spines, and turning down at maturity. Seeds with a conspicuous white aril (relatively large in comparison with most other daturas).

Local Distribution: Widespread at lower elevations on Tiburón; also on Dátil and common on Alcatraz.

Other Gulf Islands: Ángel de la Guarda, Partida Norte, San Lorenzo, San Marcos, Carmen, Monserrat, Santa Catalina (observation), Santa Cruz, San José (observation), Espíritu Santo, Cerralvo.

Geographic Range: Southern Mexico and Baja California Sur to southeastern California and southern Arizona.

Tiburón: Ensenada de la Cruz, arroyo margin, *Felger 9141.* SW Central Valley, *Felger 12322.* Tecomate, near beach and on dunes at beach, *Whiting 9063.*

Dátil: NE side of island, *Felger 9130.*

Alcatraz: Sandy flats on E side of island, *Felger 14914.* S-central coast, upper cobble-rock-shell beach, *Felger 07-166.*

Lycium – *Salicieso*; wolfberry, desert thorn

Sonoran Desert lyciums are densely branched shrubs, often 1–2.5 m tall on the Sonoran Islands, have hard wood and rigid branches, and twigs ending in sharp thorns. They are drought deciduous and often flower in midwinter to early-spring depending on soil moisture, or with sufficient rain some may flower at any time of year, especially *L. brevipes.* Leaves entire, often succulent or semi-succulent, clustered (fascicled) in short shoots, as well as widely spaced on long shoots (primary growth). Fruits (those in the flora area) many-seeded orange to red berries.

Identification can be risky or hopeless without flowers or fruits. However, even in dry season you can often find at least a few dried flowers or fruits, allowing identification.

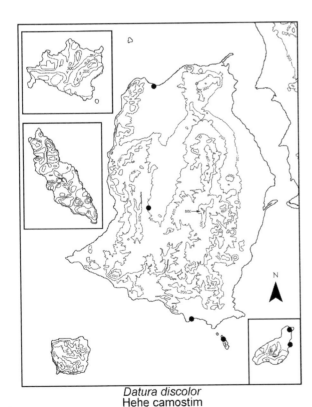

Datura discolor
Hehe camostim

1. Calyx lobes about as long as, or longer than the calyx tube (lobes sometimes shorter on some flowers on a shrub, but some longer-lobed calyces are usually present). _____ **L. brevipes**
1' Calyx lobes shorter than the tube (sometimes as long as the tube in *L. berlandieri*).
   2. Plants glandular-hairy (leaves, pedicels, and calyces); pedicels 1–16 mm long.
      3. Pedicels to 7 mm long; flowers bisexual. _____ **L. andersonii** var. **pubescens**
      3' Pedicels 6–16 mm long; male and female flowers on separate plants. _____**L. fremontii**
   2' Plants glabrous or essentially so (at least the mature herbage); pedicels mostly 1–7 mm long.
      4. Flowers as wide as, or wider than long—corollas campanulate, the tube conspicuously expanded (widened) above, the corollas (including lobes) white. _____ **L. berlandieri**
      4' Flowers longer than wide, the corollas usually with some lavender color.
         5. Flowers bisexual, notably slender, the corolla tube whitish, narrowly cylindrical or nearly cylindrical, the lobes lavender. _____ **L. andersonii** var. **andersonii**
         5' Flowers apparently unisexual, male and female flowers apparently on separate plants; flowers not notably slender, somewhat flared at tip (apex), the corolla tube greenish purple, the lobes whitish or pale lavender. _____ **L. megacarpum** (see "Doubtful and Excluded Reports")

Lycium andersonii A. Gray

Hahjöenej "empty *Lycium*," hahjö inaail coopol "black-barked *Lycium*"; *salicieso*; desert wolfberry

Description: Generally distinguished by its relatively narrow leaves, narrow and nearly tubular flowers, and non-exerted or moderately exerted stamens. Stems usually tan. Corolla lobes 4 or 5 (sometimes even on the same plant), lavender, 1.4–2.5 mm long. Fruits bright orange.

Local Distribution: Tiburón, Dátil, and San Esteban.

Other Gulf Islands: Ángel de la Guarda, San Lorenzo (Chiang-Cabrera 1981:234), Tortuga, San Marcos, Carmen, Danzante, Monserrat, Espíritu Santo, Cerralvo.

Geographic Range: Northern Sinaloa and Baja California Sur to New Mexico, Utah, Nevada, and California. Four varieties are recognized, but probably only two are worthy of recognition (Felger 2000a), both of which occur in the flora area.

Lycium andersonii var. andersonii [*L. andersonii* var. *deserticola* (C.L. Hitchcock) Jepson]

Description: Plants glabrous or essentially so, at least the mature herbage. *Lycium andersonii* var. *andersonii* might be confused with *L. berlandieri* and is reliably distinguished by lighter-colored stems and the narrow flowers and calyx.

Local Distribution: Widespread across Tiburón, including washes, canyons, and rocky slopes. Apparently not on the arid west side and southern parts of the island, where it is replaced by var. *pubescens*.

Geographic Range: Var. *andersonii* occurs through most of the species range.

Tiburón: Bahía Agua Dulce, common, *Felger 10191*. SW Central Valley, *Felger 17350*. Head of arroyo at base of the

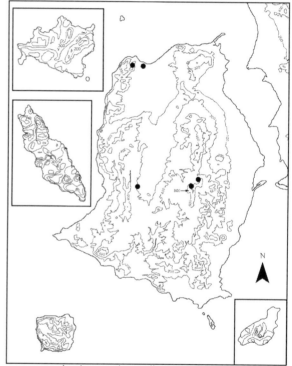

Lycium andersonii var. *andersonii*
Hahjöenej, hahjö inaail coopol

Sierra Kunkaak, between Sierra Kunkaak Mayor and Sierra Kunkaak Segundo, *Wilder 05-49*. Steep, protected, high canyon, 720 m, N interior of Sierra Kunkaak, *Wilder 07-501*. Rocky hill W of Tecomate, *Felger 6274*.

Lycium andersonii var. pubescens S. Watson

Description: This variety is distinguished by its pubescent leaves, pedicels, and calyces.

Local Distribution: Western and southern margin of Tiburón and on Dátil and San Esteban; often along

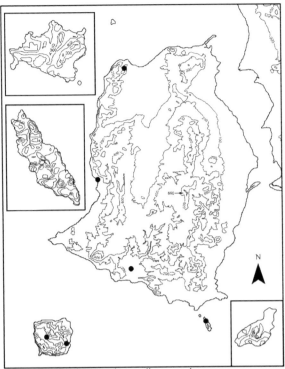

Lycium andersonii var. pubescens
Hahjöenej, hahjö inaail coopol

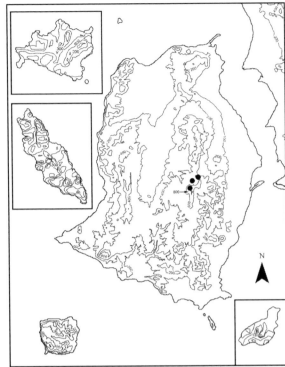

Lycium berlandieri var. longistylum

washes and canyons. The large chuckwallas (*Sauromalus varius*) on San Esteban feed on the leaves.

Geographic Range: Gulf Coast of southern Baja California Norte and Baja California Sur and Gulf Islands.

Tiburón: Ensenada Blanca (Vaporeta), common, *Felger 17246.* Arroyo Sauzal, 2 mi inland from shore, trail to Sauzal waterhole, *Felger 6476.* Sauzal, *Felger 6476.* W of Tecomate, *Felger 6268.*

San Esteban: Arroyo Limantour: Main canyon, common, *Felger 17652.* Main arroyo, 150 m elevation, *Moran 28174* (SD). **Dátil:** *Felger 17508.*

Lycium berlandieri Dunal var. **longistylum** C.L. Hitchcock

Description: Plants glabrous. Leaves narrow. Shrubs resembling *L. andersonii* and distinguished in part by its smooth, dark red-brown bark, white or pale yellow-white flowers, and flaring or wide (campanulate) 5-lobed corollas. (In the literature these two species are often distinguished by differences in hairiness of the basal free portion of the filaments. These are relative characters and reliable if you are comparing both at the same time or are familiar with one or the other. Var. *longistylum* is identified by having filaments densely pubescent below their free portion all the way down to the base of the corolla tube.)

Local Distribution: Sierra Kunkaak on Tiburón, along canyon bottoms and mostly north-facing rocky slopes to higher elevations. Nearest population is in the Sierra Seri (*Felger 08-229*).

Other Gulf Islands: Var. *longistylum*: none. Var. *peninsulare*: Ángel de la Guarda, San Lorenzo, Monserrat, San José.

Geographic Range: This species occurs in Baja California Sur, and Arizona to Texas and southward to Sinaloa and Zacatecas. Chiang-Cabrera (1981) recognized four geographically segregated varieties. Var. *longistylum* ranges from central and southwestern Arizona southward in western Sonora to the Guaymas region.

Tiburón: Siimen Hax, *Wilder 06-454.* Base of Capxölim, N-facing base of cliff, *Felger 07-102.* N slope of San Miguel Peak, 1000 ft, *Knight 915* (UNM).

Lycium brevipes Bentham var. **brevipes** [*L. richii* A. Gray]

Hahjö an quinelca "empty *Lycium*"; *salicieso*; desert wolfberry

Description: Densely branched shrubs beset with glandular hairs, or perhaps sometimes glabrate or glabrous (see *L. megacarpum* in "Doubtful and Excluded Reports"), the bark often tan. Flowers with 4 or 5 corolla

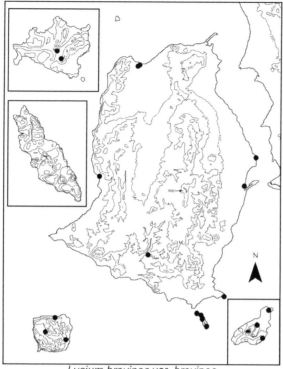

Lycium brevipes var. *brevipes*
Hahjö an quinelca

and calyx lobes, sometimes even on the same branch. Calyx tube often 4- or 5-keeled, especially when dried, the lobes 1.5–4.5 mm long, the sinuses (gaps) between lobes rounded. Corollas pale violet with a white throat, the lobes mostly 3–4 mm long and spreading. Stamens well exerted, about equal in length to corolla lobes, the free base of the filaments hairy. Flowering branches, especially stem tips, often with very crowded flowers and fruits. Fruits orange to red. Apparently flowering during the warmer times of the year except during extreme drought.

Local Distribution: This is the most common and widespread *Lycium* on the Sonoran Islands. Tiburón, lowlands across the island, especially near the coast, from just inland from the mangroves to the lower foothills, often along washes. Also on San Esteban, Dátil, Alcatraz, and Cholludo. On Mártir locally common at scattered sites at higher elevations.

Other Gulf Islands: Ángel de la Guarda, Rasa, Salsipuedes, San Lorenzo (*Wiggins 17293*, CAS, cited by Chiang-Cabrera 1981:217), Tortuga, San Marcos, San Ildefonso, Carmen, Danzante, San José, Partida Sur (*Johnston 3233*, CAS, cited by Chiang-Cabrera 1981:218), Espíritu Santo, Cerralvo.

Geographic Range: Abundant in coastal habitats of the Gulf of California. Also inland in the Sonoran and Mojave Deserts, often in xeroriparian or alkaline, semi-saline habitats. Sinaloa, Sonora, south-central Arizona, Baja California Sur to southern California, and Gulf Islands. Var. *hassei* (Greene) C.L. Hitchcock occurs on the Channel Islands and nearby coastal southern California.

Tiburón: Ensenada Blanca (Vaporeta), not common, *Felger 17247*. Xapij (Sauzal), *Wilder 08-126*. El Monumento, *Felger 08-67*. Zozni Cmiipla, *Wilder 06-281*. Palo Fierro, *Felger 6392* (specimen not located). Tecomate, *Felger 07-53*. Agua Verde Bay, *Rempel 118*.

San Esteban: Arroyo Limantour, scattered along large wash, *Van Devender 90-532*. Lower portion of S-facing canyon on central mountain, *Wilder 07-68*. N side, *Felger 15454* (TEX, cited by Chiang-Cabrera 1981:220).

Dátil: NE side of island, Rescue Beach, common along upper cobble beach, *Wilder 07-608*. NW side of island, *Felger 15348*. SE side, steep and narrow canyon, *Felger 17505*. E side, S-facing slope, *Wilder 07-141*.

Alcatraz: E side, not common, sandy soil, *Felger 14913*. E-central side, guano flats, intermixed with *L. fremontii*, *Felger 07-159*. SE side of sand flat, *Felger 08-06*. E slopes of mountain near base, *Felger 12720*.

Cholludo: Common shrub, especially near sea at N shore, *Felger 2724*. N-central slope, common, *Wilder 08-314*.

Mártir: Frequent at middle altitude along S side of island, *Johnston 3154* (CAS, image). Summit, 290 m, locally common at a few places at higher elevations, *Felger 07-195*.

Lycium fremontii A. Gray var. **fremontii**

Hahjö cacat "bitter *Lycium*"; *salicieso*; desert wolfberry

Description: Glandular-pubescent shrubs. Flowers with pedicels often about as long as the flower. Male and female flowers on separate plants. Male flowers have slightly exerted stamens, included (not exerted) styles, and do not produce fruits. Female flowers are noticeably smaller, have exerted styles, produce fruits, and have included stamens with sterile anthers. Fruits fleshy, orange, juicy, and edible with a tart flavor. Flowering mostly in winter and early spring, fruits ripening in the same season.

Local Distribution: Tiburón, along washes and bajada plains of the Agua Dulce Valley (Central Valley). Alcatraz locally at the edge of the guano flats and sand flats, intermixed with *L. brevipes*, where we found both species flowering at the same time.

Other Gulf Islands: San José.

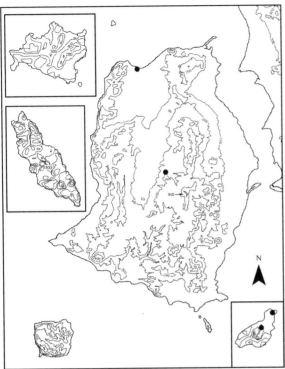

Lycium fremontii var. *fremontii*
Hahjö cacat

Geographic Range: Western Sonora from the Río Mayo region northward to Arizona and Baja California Sur to California.

Tiburón: Haap Hill, *Felger 76-T27*. N side, *Mallery & Turnage 3 May 1937* (CAS, pistillate, with fruit).

Alcatraz: E-central side, guano flats, intermixed with *L. brevipes* but less common, staminate, *Felger 07-173*; Pistillate, *Felger 07-174*. E side of sand flat, *Felger 08-07*.

Nicotiana – *Tabaco*; tobacco

Annuals and perennials. Herbage notably sticky (viscid) glandular-hairy (those in the flora area). Calyx 5-lobed and persistent. Corolla 5-lobed, tubular, and white. Fruit a capsule; seeds numerous, minute. The genus includes commercial tobacco.

1. Winter-spring annuals; leaves sessile to short petioled, not clasping; flowers nocturnal or open on cooler days or hours. _____**N. clevelandii**
1' Perennials, sometimes flowering in the first season; leaves all sessile, the stem leaves clasping (the leaf base wraps around the stem); flowers diurnal. _____
_____**N. obtusifolia**

Nicotiana clevelandii A. Gray

Xeezej islitx "badger's inner ear"; *tabaco del coyote*; coyote tobacco, desert tobacco

Description: Winter-spring annuals, with a single main stem or the more robust plants with several major stems. Leaves lanceolate to elliptic or ovate, the larger ones 5–13 cm long, the tip acute to sometimes obtuse; first leaves in a basal rosette, the lower stem leaves with winged petioles, not clasping, upper leaves sessile. Calyx lobes slender, 1 lobe wider than the others and longer than the capsule, the other lobes about equal to, or shorter than the capsule. Corollas 12–20 mm long, white, nocturnal, closing with warmer daytime temperatures. Capsules 5.8–8 mm long.

Local Distribution: Widespread across the lowlands of Tiburón; mostly in sandy-gravelly soils along washes, valley plains and bajadas, and sometimes on dunes.

Other Gulf Islands: None.

Geographic Range: Southern California to southeastern Arizona and southward through Baja California Sur, and Sonora southward to the Guaymas region.

Tiburón: 1 km N of Punta Willard, wind blown sand piled against rock, *Felger 17747*. Sauzal near shore, *Felger 7031*.

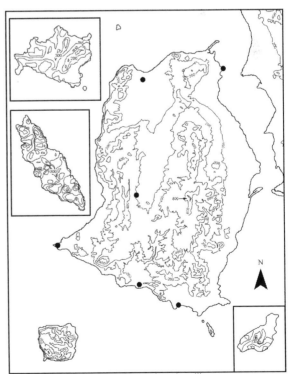

Nicotiana clevelandii
Xeezej islitx

Arroyo de la Cruz, *Wilder 09-57*. SW Central Valley, *Felger 17314*. Zozni Cacösxaj, NE coast, *Wilder 08-197*. Tecomate, *Felger 12500*.

Nicotiana obtusifolia M. Martens & Galeotti [*N. trigonophylla* Dunal]

Haapis casa "putrid tobacco"; *tabaco del coyote*; coyote tobacco, desert tobacco

Description: Herbaceous perennials and flowering in the first season. Herbage sticky glandular-pubescent. Leaves 6–12 cm long, sessile, the larger ones with the bases clasping (wrapping around) the stem. Calyx and its lobes as long as, to much longer than the capsules. Corollas 15–22 mm long, cream white, remaining open all day. Flowering in winter and spring, the plants usually dying back severely during drought and often flowering again with summer rains, or almost any time of year with sufficient soil moisture. Apparently germinating during the winter-spring season.

Local Distribution: On all islands in the flora area except Patos. Mostly in arroyos and canyons, and rocky slopes to higher elevations, and also sometimes just above the reach of high tides. Widespread across Tiburón including the Sierra Kunkaak to higher elevations but not found between the mountains and the Infiernillo shore. Widely scattered at all elevations on San Esteban, Dátil, Cholludo, Mártir, and Nolasco. Seasonally common on Alcatraz on the east side of mountain, reaching maximum density and development near the top of mountain on north- and northeast-facing slopes, and also on the cobble and rocky shores. On Alcatraz, Mártir, and Nolasco this tobacco often extends into otherwise barren guano-covered rocky slopes.

Other Gulf Islands: Ángel de la Guarda, Partida Norte, Salsipuedes, San Lorenzo, Tortuga, San Ildefonso, Carmen, Santa Catalina.

Geographic Range: California deserts to Nevada and Texas, and southward to Nayarit.

Tiburón: 1 km N of Punta Willard, *Felger 17756*. Sauzal, *Wilder 06-129*. La Cruz waterhole, W side of island, *Parsons 32-38* (UC). Ensenada de la Cruz, *Felger 9237*. Coralitos, *Wilder 06-59*. SW Central Valley, *Felger 17355*. Hast Cacöla, E flank of Sierra Kunkaak, 450 m, *Wilder 08-267*. San Miguel Peak, *Knight 989* (UNM). Hee Inoohcö, mountains at NE coast, *Felger 07-87*. Bahía Agua Dulce, *Felger 6263*. St. Maria waterhole, 12 mi N of S end of island, *Parsons 32-37* (UC; we are not able to locate "St. Maria waterhole").

San Esteban: Canyon, SW corner of island, *Felger 17612*. NE part of island, hilltop, *Felger 7075*. Arroyo Limantour, 1 mi inland, *Wilder 07-43*.

Dátil: NE side of island, common, edge of beach and in arroyo, *Felger 9133*.

Alcatraz: Common perennials on rocky hillsides, *Johnston 4282* (UC). E and S slopes, common among rocks, *Felger 13403*. Base of mountain, SW from *Allenrolfea* flats, sea ledges just above the shore, *Felger 07-171*.

Cholludo: Near top of island, fairly common, *Felger 9156*. N-central slope, *Wilder 08-315*.

Mártir: 1887, *Palmer 410* (2 sheets, UC 26175 & 173577). Common in rocky ground in all parts of island, *Johnston 3150* (UC). Summit, 300 m, occasional, *Moran 8811* (SD). Cardonal de *Pachycereus pringlei* con abundante *Sphaeralcea*, roca ignea gris, hierba annual 50 cm, flores amarillo pálidos, 4 May 1985, *Lott 2430 & Atkinson* (CAS). N side of island, ca. 8 m elev., steep rock slopes white with bird guano, among ruins of guano works rock-wall platform, in areas devoid of *Vaseyanthus*, *Felger 07-09*.

Nolasco: [S]E side, N exposure, in gullies, 20 ft elev., *Pulliam & Rosenzweig 22 Mar 1974*. On ledges, *Stinson 29 Apr 1974* (SD). Base of Cañón el Faro, *Felger 06-74*.

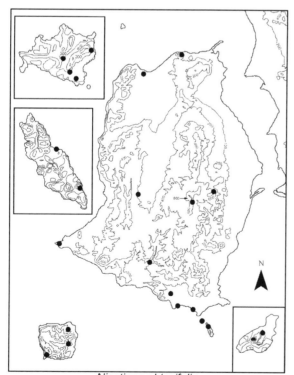

Nicotiana obtusifolia
Haapis casa

Physalis – *Tomatillo*; ground cherry, husk tomato

Annual or perennial herbs. Leaves petioled. Flowers pendent, mostly solitary in leaf axils. Calyx enlarging and growing over the fruit to form a bladder-like structure (like a miniature paper lantern or bag). Fruit a globose, many-seeded berry.

1. Perennials and also flowering in the first season; herbage not slimy; corollas pale yellow; anthers yellow. _____
   _____ **P. crassifolia**
1' Annuals, herbage clammy or "slimy" with glandular hairs; corollas pale yellow with maroon spots in the center; anthers dark colored. _____**P. pubescens**

Physalis crassifolia Bentham

Perennials and sometimes flowering in first season or year. Corollas and stamens (including anthers) pale yellow. Three varieties are recorded for the Baja California Peninsula and Gulf Islands and var. *crassifolia* also occurs on the mainland.

1. Leaves often thickish; corollas funnelform (longer than wide). _____**P. crassifolia** var. **infundibularis**
1' Leaves relatively thin; corollas rotate (about as wide as long). _____ **P. crassifolia** var. **versicolor**

Physalis crassifolia var. infundibularis I.M. Johnston

Xtoozp insaacaj

Description: Leaves thicker than those of var. *crassifolia* and the leaf surfaces tend to be viscid. The plants are distinctive and perhaps should be treated as a distinct species.

Local Distribution: Common on San Esteban, primarily in canyons, arroyos, and north-facing slopes.

Other Gulf Islands: Ángel de la Guarda (type locality), San Lorenzo, San Marcos, Santa Cruz, San Francisco.

Geographic Range: Gulf coast of Baja California Norte and adjacent islands, and extreme northern Baja California Sur.

San Esteban: Arroyo Limantour: *Felger 16606*; along wash, *Van Devender 90-530*; ca. 2.25 mi inland, *Wilder 07-49*. NW corner of island, narrow canyon, *Felger 17602*. SW corner of island, canyon, *Felger 16647*. SE side, *Lindsay 3 Apr 1947* (CAS). Common on a W-facing shade slope in canyon, corolla clear yellow, *Johnston 3174* (CAS, 2 sheets).

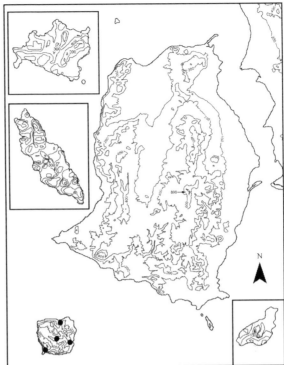

Physalis crassifolia var. *infundibularis*
Xtoozp insaacaj

Physalis crassifolia var. versicolor (Rydberg) Waterfall

Xtoozp; *tomatillo*; desert ground-cherry

Description: Plants often globose, drought deciduous and dying back severely in drought. Leaves highly variable, 1.5–9 cm long, the petioles as long as to much longer than the blades. Corollas 15–20 mm wide. Fruiting calyx 2.5–3.5 cm long. Fruits edible, resembling the domesticated tomatillo (*P. philadelphica*) but much smaller and the flavor inferior. Flowering and fruiting at various times of the year.

Local Distribution: Widely scattered across Tiburón; often along arroyos and canyons, and on rocky slopes. Dátil in canyons and rocky slopes to the peaks.

Other Gulf Islands: San Diego (*Moran 9591*, SD), San José, San Francisco, Espíritu Santo.

Geographic Range: Northwestern Sinaloa, western Sonora, both Baja California states, western Arizona, southeastern California, and Nevada.

Tiburón: Ensenada Blanca (Vaporeta), *Felger 15736*. Tordillitos, *Felger 15500*. Arroyo Sauzal, 3 mi inland, *Wilder 06-107*. Coralitos, *Barnett-Morales 08-5*. Ensenada del Perro, *Felger 17718*. SW Central Valley, arroyo bed, *Felger 17311*.

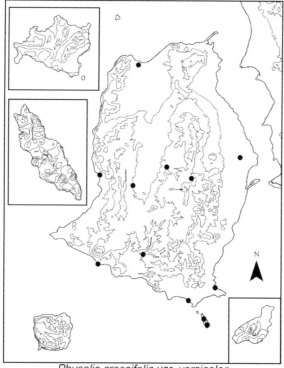

Physalis crassifolia var. *versicolor*
Xtoozp

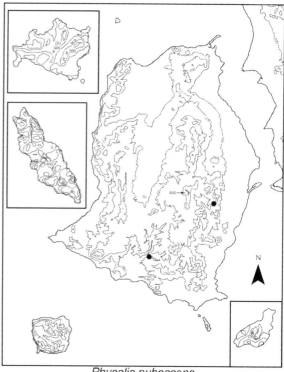

Physalis pubescens
Insaacaj

Haap Hill, *Felger 76-T34*. Canyon, N base of Sierra Kunkaak, 340 m, *Wilder 06-404*. Palo Fierro, bajada 2 mi inland, *Felger 12457*. Bahía Agua Dulce, *Felger 10212*.

Dátil: E side, ridge crest on peak, *Wilder 07-126*. SE side, steep, narrow canyon, *Felger 17518*.

*Physalis pubescens Linnaeus

Insaacaj; *tomatillo*; hairy ground-cherry

Description: Annuals, apparently non-seasonal, densely glandular-pubescent and slimy, or wet, to the touch. Stems semi-succulent, the nodes swollen, the stems and leaves pale green. Corollas pale yellow with 5 dark-maroon spots in the center, the filaments swollen and maroon-brown, the anthers dark colored (greenish to purplish, blue-green, or dark gray).

Local Distribution: Known from two waterholes on Tiburón. Although supposedly not native, its distribution is indicative of a native.

Other Gulf Islands: None.

Geographic Range: Probably native to eastern United States, now weedy in warm regions worldwide.

Tiburón: Pazj Hax, wet soil, *Wilder 06-12*. Sauzal waterhole, rare, dense shade in rocky wet soil at edge of water, *Felger 10092*.

Solanum hindsianum Bentham

Hap itapxeen "inner corner of mule deer's eye"; *mariola, mala mujer, tomatillo espinoso*

Description: Sparsely branched spiny shrubs often 1–2+ m tall; densely pubescent with stellate hairs and mostly with slender prickles to stout spines on stems, leaves, inflorescence branches and calyces. Flowers showy, largest during periods of high soil moisture and hot weather, the corollas 3–5+ cm wide, lavender, the anthers and stigma large and yellow; anthers opening by terminal pores. Berries 2 cm diameter, mottled dark and light green. Flowering for extended periods at various times of the year depending upon soil moisture and temperature.

Local Distribution: A common and widespread shrub on Tiburón, San Esteban, and Dátil, occurring across a wide range of habitats. It extends to the summit on Tiburón.

Other Gulf Islands: Ángel de la Guarda, San Lorenzo, Tortuga, San Marcos (observation), Carmen, Danzante, Monserrat, Santa Catalina, Santa Cruz, San José, San Francisco, Espíritu Santo, Cerralvo.

Geographic Range: Endemic to the Gulf of California region: western Sonora from the Guaymas region northward, most of the Baja California Peninsula, Gulf Islands, and Arizona in Organ Pipe Cactus National Monument.

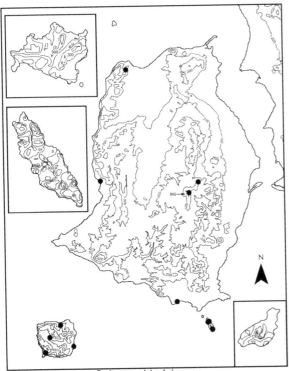

Solanum hindsianum
Hap itapxeen

Figure 3.156. *Stegnosperma halimifolium*. Foothills of Sierra Kunkaak, Isla Tiburón. BTW, 24 November 2006.

Tiburón: Ensenada Blanca (Vaporeta), *Felger 6304*. Ensenada de la Cruz, rocky hill near beach, *Felger 9242*. Cerro San Miguel [probably ca. 500 m], the highest local elevation, *Quijada-Mascareñas 90T008*. Summit of island, not common, *Wilder 07-561*. Hill, W of Tecomate, from base of hill to peak, *Felger 6245*.

San Esteban: El Monumento, arroyo, rocky slopes and floodplain, *Felger 9184*. Arroyo Limantour, 2.25 mi inland, arroyo, *Wilder 07-52*. N side of island, narrow rocky canyon behind large cove, *Felger 17567*. Canyon, SW corner of island, *Felger 16628*.

Dátil: NE end of island, rocky hills, just above high tide zone, *Felger 2547*. E side, slightly N of center, canyon bottom, *Wilder 07-101*. NW end of island, common, steep rocky slope, *Felger 2770*. SE side, steep, narrow canyon, *Felger 17509*.

Stegnospermataceae

Stegnosperma halimifolium Bentham [*S. watsonii* D.J. Rogers]

Xnejamsiictoj "red xneejam"; *chapacolor, ojo de zanate*

Description: Sprawling shrubs often more than 1.5 m tall, sometimes to 3+ m tall, often forming dense mounds with nearly evergreen or very tardily drought-deciduous

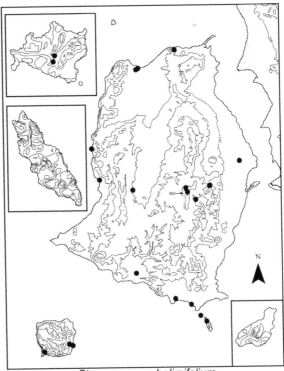

Stegnosperma halimifolium
Xnejamsiictoj

foliage; glabrous. Leaves alternate, simple, and entire, 1.5–4.5+ cm long, and often semi-succulent. Flowers 1 cm wide, fragrant, star shaped, and whitish, in terminal or axillary racemes. Sepals 5, semi-succulent; petals 5, white. Stamens 10 or more. Fruits ovoid, 6–7 mm diameter, at first fleshy, maturing as reddish capsules, with 1–5 smooth black seeds embedded in a red aril. Flowering and fruiting response non-seasonal.

It was said by the Comcaac that if bitten by a rattlesnake the leaves could be chewed and made into a paste and applied to the wound and one would not die. For this reason women carried the plant with them while gathering firewood (Felger & Moser 1985).

Local Distribution: Widespread across Tiburón, often along arroyos and canyons, and sometimes on rocky slopes, and extending to the summit. Common on San Esteban in canyons and along the canyon floor and floodplain of Arroyo Limantour. Also on Dátil and Cholludo. Known from Mártir by two records; in 1962 Moran found it to be rare on the island, and it has not been recorded there since.

Other Gulf Islands: Ángel de la Guarda, San Lorenzo, Tortuga, San Marcos, Coronados (observation), Carmen, Danzante, Monserrat (observation), Santa Catalina, Santa Cruz (observation), San Diego, San José, Espíritu Santo, Cerralvo.

Geographic Range: Widespread along the shores and coastal desert on both sides of the Gulf of California and sometimes ranging 100–150 km inland.

Tiburón: Ensenada Blanca (Vaporeta), *Felger 14963*. Arroyo Sauzal, *Wilder 06-74*. Arroyo de la Cruz, 50 m, shrub 3 m high, occasional, *Moran 13011* (CAS). Coralitos, *Felger 15374*. SW Central Valley, *Felger 12318*. Palo Fierro, bajada 2 mi inland, *Felger 12541*. Hast Cacöla, E flank of Sierra Kunkaak, 380 m, *Wilder 08-243*. Head of arroyo, E base of Sierra Kunkaak, *Wilder 06-18*. Deep canyon, N interior of Sierra Kunkaak, *Wilder 07-487*. Summit of island, 885 m, *Wilder 07-574*. Hee Inoohcö, mountains at NE coast, *Felger 07-77*. N Coast, *Mallery & Turnage 3 May 1937*. Tecomate, *Whiting 9055*. Pooj lime, *Wilder 07-452*.

San Esteban: El Monumento, floodplain, *Felger 9180*. Arroyo Limantour, *Van Devender 92-485*. Frequent shrub in sandy wash, *Johnston 3166* (UC). Canyon, SW corner of island, *Felger 16652*.

Dátil: NE side of island, scattered, steep rocky slopes and arroyo, *Felger 13439*.

Cholludo: Large shrub, scattered, not common, *Felger 9158* (CAS). Scattered, not common, *Felger 13417*.

Mártir: Center [of island], Oct 1887, *Palmer 400* (GH, image). 300 m, shrub 4 m tall, only two seen, 21 Mar 1962, *Moran 8816* (SD).

Sterculiaceae, *see* Malvaceae

Tamaricaceae – Tamarisk Family

*Tamarix

Trees and shrubs; leaves scale-like, with salt-excreting glands. Flowers small, subtended by bracts, 5-merous (those in the flora area) and in densely flowered, branched spikes or racemes (panicles). Fruits of small capsules; seeds many, minute, with feathery hairs. Native to arid and semi-arid regions in the Old World.

1. Trees 5+ m tall with a thick trunk and limbs; leaves completely encircling the stem, the blade reduced to a cusp less than 0.5 mm long; flowers white, sessile. ___
_____ **T. aphylla**

1' Shrubs to 3+m; leaves not completely encircling the stem, the blade scale-like but evident, 0.7–3 mm long, triangular-ovate; flowers pink or pinkish white, sometimes white becoming pink with age, on pedicels 0.6–0.8 mm long. _____ **T. chinensis**

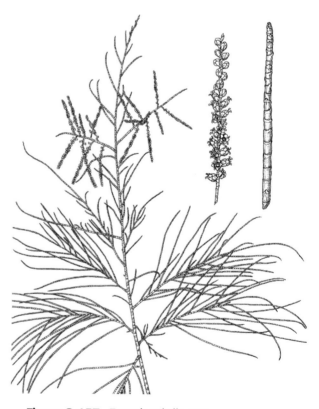

Figure 3.157. *Tamarix aphylla*. LBH.

Figure 3.158. *Tamarix chinensis.* LBH (Parker 1972).

*Tamarix aphylla (Linnaeus) H. Karsten

Hocö hapéc (translated as any planted tree from outside of the area); *pino*; athel tree, salt cedar

Description: Evergreen trees to 8+ m tall with thick and sometimes massive trunks and limbs. Readily propagated from cuttings of almost any size, either stuck in the ground or in water. Once established the trees usually persist, especially on sandy soils near the coast. During the cooler months dew condenses on the salt-encrusted twigs and drips onto the soil.

Local Distribution: Tamarisk trees were planted on Tiburón to a significant extent at the now abandoned small military station at Tecomate and also several trees at the active Mexican marine station at Punta Tormenta. These trees were planted in the late 1970s and 1980s and have developed into large, healthy shade trees but are not reproducing. The dense litter of dead twigs beneath the trees prevents other plants from growing.

Other Gulf Islands: Cerralvo.

Geographic Range: Native to North Africa and the eastern Mediterranean. Extensively planted in the

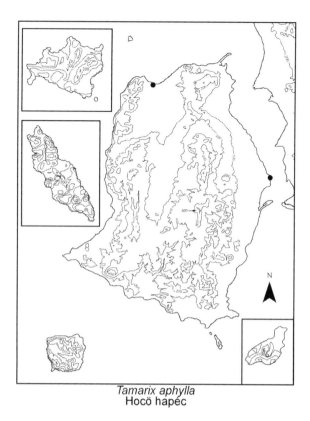

Tamarix aphylla
Hocö hapéc

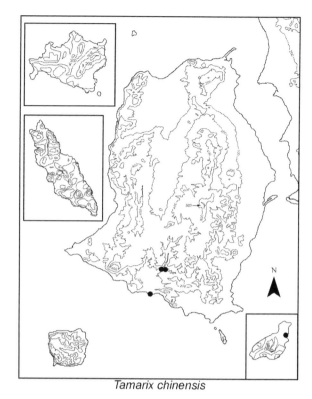

Tamarix chinensis

Sonoran Desert including the Sonoran mainland opposite Tiburón (Felger & Moser 1985).

> **Tiburón:** Palo Fierro, just inland from beach, *Wilder 06-143*. Tecomate, ca. 30 trees in rows, ca. 6 m tall forming dense cover, a few also near two decaying buildings where *Eucalyptus* are also planted, *Wilder 07-251*.

*Tamarix chinensis Loureiro [*T. ramosissima* Ledebour]

Pino salado, salado; salt cedar, tamarisk, shrub tamarisk

Description: Shrubs 2.5–3 m tall. Branchlets (short shoots) or ultimate twigs winter deciduous, their internodes shorter than, and obscured by the overlapping scale leaves. Long shoots perennial. Leaves and young stems with regularly spaced, alkali- or salt-excreting glands (seen with 10× magnification); these glands glistening gold when fresh and first developing, soon becoming white with the buildup of alkali or salt. Scale leaves of ultimate branchlets (short shoots) 0.7–1.5 mm long, those of the long shoots mostly 2.3–3 mm long. Flowers 5-merous, pinkish white to pink, or sometimes white and becoming pink with age; floral disk glands or nectary maroon, producing glistening nectar droplets; pedicels 0.5–0.8 mm long. Seeds 0.5 mm long, yellow-brown.

Local Distribution: At the Sauzal waterhole on Tiburón it was first documented in 1958 when it was already common and well established with adult shrubs. In 2006 and 2008 we found it locally abundant with numerous seedlings and small plants. This is a major invasive species of concern for the island, since it often dominates and transforms riparian settings, which on Tiburón are rare and host unique plant assemblages. In reality, however, it seems to have reached equilibrium at Sauzal with the other wetland plants. Attempts to remove it would undoubtedly be futile and the process would likely inflict serious damage to the native wetland community. We recommend monitoring the site but not disturbing it. In 2008 we found well-established shrubs at the mouth of Arroyo Sauzal just inland from the beach. These shrubs were among coastal halophytes at the edge of stagnant, saline pools; we recommend removing the coastal tamarisks.

In 1997, Tad Pfister (personal communication, 23 May 2006) found a tamarisk shrub on Alcatraz. It was about 1 m tall, on the low, flat area on the east side of the island by the shore, and he and Patricia Ann West removed it.

In 2007, Ben and Humberto found a few small tamarisk shrubs in about the same place in the *Allenrolfea* flat and these were promptly removed.

This tamarisk is also on San Lorenzo and Ángel de la Guarda. A large stand on Isla Ángel de la Guarda was observed in the early 1970s by Bahía de los Ángeles resident Pablo Murillo and documented in 1983 by Ray Turner (Turner et al. 1995:384). This stand, which is directly behind the beach, covers about 7 hectares and appears to be expanding (Tom Bowen, personal communications, 2007 and 2009). A second unrelated stand, reported by Peter Garcia in 2010, consists of 15 plants in a dry arroyo 500 m from shore (Tom Bowen, personal communication 2011). Lawrence A. Johnson discovered *T. chinensis* on San Lorenzo in 2004, when he found two large shrubs along the shore (Tom Bowen, personal communication, 2007; Ben Wilder, observation, 2010). Johnson and Bowen observed some seedlings on San Lorenzo in 2004 and 2005, but in June 2009 and June 2010 Bowen found no seedlings, and the two largest plants had been eradicated.

The presence of this invasive species along the shores of Midriff Islands is of concern and eradication efforts are needed while the populations remain relatively small and control is feasible.

Other Gulf Islands: San Lorenzo, Ángel de la Guarda.

Geographic Range: Native to the Old World, now widespread, weedy, and invasive in many warm, dry parts of the world, especially in disturbed desert riparian habitats.

Tiburón: Sauzal waterhole (Xapij): Locally common, *Felger 10086*; waterhole, *Wilder 06-112*; Southern waterhole, pocket of above-ground water, ¼ mi S of main waterhole, *Wilder 06-113*. Mouth of Arroyo Sauzal near shore, *Wilder 08-141*.

Alcatraz: Wet area of flat, *Wilder 07-407*.

Theophrastaceae – Theophrasta Family

Bonellia macrocarpa (Cavanilles) B. Ståhl & Källersjö subsp. **pungens** (A. Gray) B. Ståhl & Källersjö [*Jacquinia macrocarpa* Cavanilles subsp. *pungens* (A. Gray) B. Ståhl. *J. pungens* A. Gray]

Cof; san juanico

Description: Large shrubs or more often small trees with a thick trunk and smooth gray bark. Foliage dense and essentially evergreen. Leaves alternate, simple, 2.5–6 cm long, elliptic to lanceolate, dark green above, firm and spine tipped. Leaves of juvenile plants linear and conspicuously narrower than those of adult trees. Corollas 10–14 mm wide, bright orange, readily detaching and falling as a unit with the attached stamens. Fruits 2–2.5 cm long, ovoid, and hard shelled, the seeds embedded in fleshy-gelatinous mesocarp (pulp). Flowering and fruiting at various seasons with peak flowering often in late spring and early summer.

The Comcaac ate the sweet fruit pulp, but "if one ate too much it was said to cause dizziness and possible intestinal disorder" (Felger & Moser 1985: 373). Girls and women string the fresh flowers into attractive necklaces called cof yapxöt (Moser & Marlett 2005). The dried flowers reopen when put in water, taking on a fresh and newly picked appearance, and may be revived repeatedly even after more than one hundred years of storage (Felger & Moser 1985; Felger 2000b).

Local Distribution: Widespread and common across the lowlands of Tiburón except the northern and arid western margins of the island. Most often along washes and canyon bottoms.

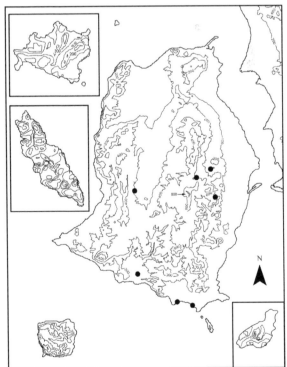

Bonellia macrocarpa ssp. *pungens*
Cof

(A)

(B)

Figure 3.159. *Bonellia macrocarpa*
subsp. *pungens*. (A) Tree, with Ed Gilbert,
carrying plant specimens in plastic bags,
canyon of Sierra Kunkaak, Isla Tiburón,
29 December 2005. (B) Flowers, Sierra
Kunkaak, Isla Tiburón, 5 September 2008.
BTW.

Other Gulf Islands: None.

Geographic Range: This species extends from Panama to Sonora; subsp. *pungens* occurs along the west coast of Mexico southward to Guerrero, Jalisco, and Michoacán. The northernmost localities are northeast of El Desemboque.

Tiburón: Arroyo Sauzal, 2 km inland, arroyo margin, *Felger 10099*. Ensenada de la Cruz, arroyo bottom, and some dwarfed shrubs on beach dunes, *Felger 12788*. Coralitos, *Felger 15367*. SW Central Valley, *Felger 12348*. Sopc Hax, *Wilder 07-221*. E foothills of Sierra Kunkaak, *Wilder 05-12*. Capxölim, wash at base of mountain, *Wilder 07-465*.

Typhaceae – Cattail Family

Typha domingensis Persoon

Pat; *tule*; cattail

Description: Robust, perennial herbs with thick, starchy rhizomes, emergent from freshwater. Leaves 2–2.5 m long, erect, slender, strap shaped, spongy, and flat on one side and convex on the other side. Spikes with a gap between the male and female parts, individual flowers minute.

Local Distribution: In the flora area known only from the Sauzal waterhole, where it is well established,

growing with *Phragmites* and *Salix*. It is locally common at wetland sites in the Kino region and elsewhere along the coast.

Other Gulf Islands: None.

Geographic Range: Southern two-thirds of the United States and most of the Americas in tropical and subtropical regions.

Tiburón: Brought from southern part of island by a turtle hunting expedition [undoubtedly from the Sauzal waterhole], 14–22 Jun 1951, *Whiting 9037*. Sauzal waterhole, locally common in shallow water, *Felger 10075*. Southern Sauzal waterhole, small pocket of above-ground water, ¼ mi S of main waterhole, *Wilder 06-103*.

Ulmaceae (in part, **Celtis**), *see* Cannabaceae

Urticaceae – Nettle Family

Parietaria hespera B.D. Hinton var. **hespera** [*P. floridana* Nuttall, of authors, in part]

Description: Delicate winter-spring annuals, the stems often 10–30 cm long and trailing or decumbent. Root systems small relative to the size of the plant. Stems semisucculent and brittle, the leaves thin and quickly wilting. Flowers inconspicuous.

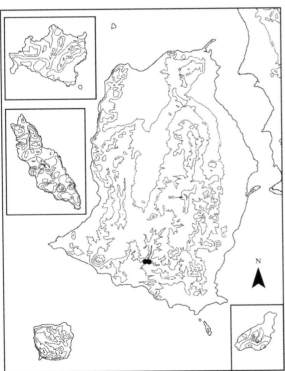

Typha domingensis
Pat

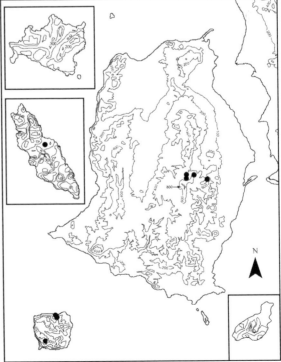

Parietaria hespera var. *hespera*

Local Distribution: Widespread in the Sierra Kunkaak on Tiburón. Also widespread on San Esteban and recorded from Nolasco. Often in shaded niches, among large rocks, in washes, and at the base of shrubs.

Other Gulf Islands: Ángel de la Guarda, San Lorenzo.

Geographic Range: Northern Mexico and southern United States.

Tiburón: Hast Cacöla, E flank of Sierra Kunkaak, 380 m, *Wilder 08-255*. Cerro San Miguel, 400 m, *Quijada-Mascareñas 91T023*. Base of N portion of Sierra Kunkaak, 2 km E of Siimen Hax, 360 m, sand-gravel soil in shaded arroyo bed among leaf litter, *Wilder 06-459*. Deep canyon, N slope of the Sierra Kunkaak, SW and up canyon from Siimen Hax, *Wilder 06-483*.

San Esteban: N side of island, *Felger 15450*. N side of island, narrow rocky canyon behind large cove, *Felger 17581*. Steep slopes of S-central peak, *Felger 17533*. Rather common on N side of E ridge of high peak near SE corner of Island, 380 m, *Moran 8840* (CAS).

Nolasco: NE side, on grassy N-facing slope with *Amaranthus fimbriatus, Perityle californica, Muhlenbergia microsperma,* and *Setaria leucopila, Felger 12075*.

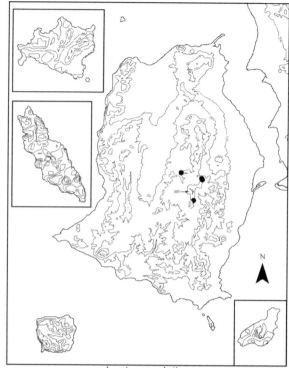

Lantana velutina

Verbenaceae – Verbena Family

Two shrubs in the flora area, the stems slender and 4-angled (square in cross section). Leaves opposite and simple; stipules none.

1. Leaves not noticeably aromatic; flowers in a head-like cluster about as wide or wider than long, the corollas white to pink; fruits fleshy. _____**Lantana**
1' Leaves noticeably aromatic (fresh or dried); flowers in cone-like structures often longer than wide, the corollas cream color to bright yellow; fruits dry._____ **Lippia**

Lantana velutina M. Martens & Galeotti

Confiturilla blanca

Description: Unarmed shrubs often to 1 m tall with slender stems. Flowers attractive, in compact heads ca. 10 mm wide, subtended by conspicuous broad, overlapping bracts. Corollas 4 mm wide, white with a pale yellow center, often becoming pink to pale lavender with age. Fruits pinkish, sweet and edible.

Local Distribution: Sierra Kunkaak in canyons and steep north-facing slopes to higher elevations. The nearest known locality in Sonora is in the Sierra el Aguaje

(La Balandrona, *Gutiérrez 00-07*, USON) and the closest peninsular population is in the Sierra San Francisco.

Other Gulf Islands: Cerralvo.

Geographic Range: The only other Gulf Island record is for Cerralvo, the most southern Gulf Island. Baja California Sur and Sonora to Panama.

Tiburón: Cerro San Miguel, flores de color blanco a crema, de común a abundante en la pediente del cerro, *Quijada-Mascareñas 90T011*. Top of Sierra Kunkaak Segundo, E peak of Sierra Kunkaak, 425–490 m, common on slope, becoming less so closer to the peak, *Wilder 06-484*. Sierra Kunkaak mayor, parte sur, 22 Jan 2008, *Romero-Morales 08-8*. Sierra Kunkaak, NW foothills, Jan 2009, *Romero-Morales* (photo).

Lippia palmeri S. Watson

Xomcahiift; *orégano, mariola*; Sonoran oregano

Description: Shrubs with slender, brittle stems. Leaves highly aromatic, gradually drought deciduous and appearing after each soaking rain, 1–3 cm long, the blades with deeply incised veins. Flowers in cone-like spikes often 4–10 mm long; flowers 3 mm wide, cream or pale yellow, becoming bright yellow with age or when dry.

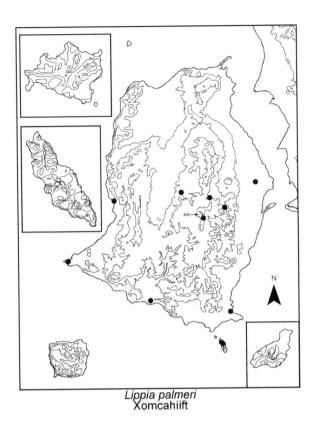

Lippia palmeri
Xomcahiift

Flowering response apparently non-seasonal depending on soil moisture.

The dried crushed leaves are used extensively locally as *orégano*. Elsewhere in Sonora it is commercially harvested and sold in Sonoran markets, and also exported for the gourmet market. This species was featured in the gourmet food periodical *Saveur* (Brennan 2005; it was misidentified as *L. graveolens* in the article). It is also taken as a medicinal tea.

Local Distribution: Widespread across Tiburón but not known from the northern part of the island; bajadas, arroyos, canyons, and rock slopes. Also common on Dátil.

Other Gulf Islands: San Marcos, Coronados, Carmen, Monserrat, San José, Espíritu Santo.

Geographic Range: The adjacent Sonoran coast and the vicinity of Hermosillo to Sinaloa and in Baja California Sur.

> **Tiburón:** Ensenada Blanca (Vaporeta), *Felger 15001.* Just N of Punta Willard, *Wiggins 17149* (CAS). Arroyo Sauzal, *Felger 08-34.* Ensenada del Perro, arroyo *Wilder 07-160.* Haap Hill, *Felger 74-T50.* 3 km W [of] Punta Tormenta, *Scott 9 Apr 1978* (UNM). Hast Cacöla, E flank of Sierra Kunkaak, 450 m, *Wilder 08-263.* Foothills of NE portion of the Sierra Kunkaak, *Wilder 06-374.* Cerro Kunkaak, *Scott 9 Apr 1978* (UNM).

> **Dátil:** NE side of island, usually in arroyo bottoms, *Felger 13444.* E-central side, N-facing slope above beach and base of canyon, *Wilder 09-12.*

Violaceae – Violet Family

Hybanthus fruticulosus (Bentham) I.M. Johnston

Description: Tufted, herbaceous perennials from a semi-woody base. Stems yellow-green. Leaves alternate, quickly drought deciduous, best developed during summer-fall wet season, lanceolate-elliptic with toothed margins, and smaller and narrower and sometimes entire during drier conditions. Flowering in spring and summer with rains, the flowers small and white, often hidden among the uppermost leaves; petals white, to 3 mm long. Capsules 3 mm wide and elastically dehiscent.

Local Distribution: Tiburón in canyons in the Sierra Kunkaak to higher elevations and along its eastern flanks.

Other Gulf Islands: San José, Cerralvo.

Geographic Range: Sinaloa to western Sonora at least as far north as the Sierra el Aguaje, Tiburón, and Baja California Sur.

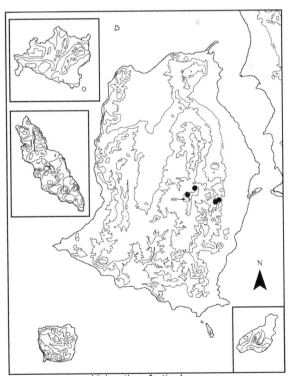

Hybanthus fruticulosus

Tiburón: Sopc Hax, rare, canyon bottom in shade, among rocks, *Felger 9301*. Between Sopc Hax and Hant Hax camp, not common, in shade of trees, arroyo bottom, *Felger 9344*. Capxölim, canyon bottom, *Wilder 06-396*. Steep, protected canyon, 720 m, N interior of Sierra Kunkaak, *Wilder 07-587*.

Viscaceae, *see* Santalaceae

Zosteraceae – Eelgrass Family

Zostera marina Linnaeus var. **atam** T.W.H. Backman

Eaz (eelgrass dislodged or washed ashore, mostly in spring), hatám (when growing on the sea floor), xepe an yaail (a form of eelgrass), xepenipóosj (described as eelgrass with long transparent leaves–perhaps senescent leaves without chlorophyll), xnoois (eelgrass grain/seed); *zacate de mar, trigo de mar*; eelgrass.

Description: Seagrasses with slender rhizomes and long, slender, and leafy stems. Leaves alternate, ribbon-like, 0.5–1.5 m long in deeper water and much shorter in shallow water, and 3.2–5 mm wide. Flowers unisexual, both occurring on the same plant, in two rows enclosed in spathe-like sheaths, inconspicuous and without perianth. Fruits brown when ripe, flask shaped, 3 mm long with a single seed.

In the Gulf of California it grows about 0.5–3 m below the low tide level and is not exposed to the desert air. Eelgrass in the Gulf is a winter annual. The plants begin growth in October and early November as the water temperatures begin to drop. In late October and November vigorous and rapidly growing rhizomes and shoots begin to develop. Seedlings are abundant as late as early December in the beach drift on Alcatraz and the Infiernillo coast, especially following storms. Rapid growth continues through the mild winter. As water temperatures rise in late February and March, 100 percent of the shoots become reproductive. Massive and essentially simultaneous fruit ripening occurs in mid- to late April. Reproductive shoots, with ripe or rapidly ripening fruits break off and often form extensive rafts floating on the sea surface. Much of this floating eelgrass washes ashore in large quantities, accumulating in the beach drift.

Zostera in the Seri region is probably the most distinctive of all of the major populations of *Z. marina*.

Among the unique features of the Gulf of California eelgrass, recognized as var. *atam* (Backman 1991), is the annual habit. In contrast, eelgrass along the Pacific Coast and elsewhere is a perennial plant and in these cooler regions it is often exposed at low tide. The high light intensity and seasonal fluctuations in water temperatures apparently produces optimum conditions for

Figure 3.160. *Zostera marina*. Single mature shoot, from shallow water in the vicinity of Campo Viboras, Infiernillo coast, Sonora. The four lower leaf-like spathes are pistillate and the two upper ones are staminate. NLN.

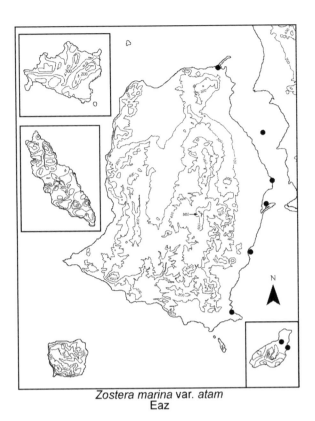

Zostera marina var. *atam*
Eaz

flowering and seed-set for the eelgrass population in the Seri region. One hundred percent of eelgrass shoots in the Infiernillo produce seeds, while on the Pacific Coast the percentage of shoots becoming reproductive each year declines northward to Alaska, where only 2 percent of the shoots are reproductive. The Infiernillo eelgrass population is the largest, densest, and most productive anywhere in the Gulf of California. Eelgrass requires a sandy-muddy, non-rocky substrate, as well as protected waters.

Local Distribution: Eelgrass forms undersea meadows with the roots and rhizomes in sandy mud of the seafloor in protected waters of the Canal del Infiernillo. It is seasonally common in the beach drift on Tiburón, especially along the east shore, and is also found in beach drift on Alcatraz. Jorge Torre-Cosío (2002) provides extensive information on *Zostera* in the Seri region.

Eelgrass provides a diverse habitat for many organisms and is a favorite food of sea turtles, especially the green turtle (*Chelonia mydas*). Seri men often hunted sea turtles (*Chelonia*) in the eelgrass beds in the Canal del Infiernillo. They located specific named eelgrass areas at sea by day or night. For example, if not evident from the surface, the men would place an ear to the end of an oar pushed into the water blade first and listen for the

distinctive swish of the eelgrass moving in the current. At night they located eelgrass beds by aligning the boat with mountains and other features on the mainland and Tiburón, visible even by starlight. Jorge Torre mapped these eelgrass beds with the aid of a GPS.

Seri knowledge of eelgrass is extensive, and they harvested the grain in late spring as one of their staples (Felger & Moser 1973, 1985). Nowhere else in the world have people harvested a grain from the sea. (Cocopah people harvested *nipa*, *Distichlis palmeri*, at the delta of the Río Colorado, a marshland grass that is often tidally flooded with seawater.) After gathering eelgrass that washes ashore in great quantities in spring, it was essential to be able to dry it so that the grain did not rot and to be able to separate the grain from the chaff. This was possible because at this time of year the prevailing winds shift from a sea breeze to a land breeze. Winds from the desert allowed for the drying of eelgrass. Women gathered salty, wet masses of eelgrass and spread them out on deer hides (later they used canvas and plastic tarps) to dry. These multiple factors allowed for a bountiful and unique harvest.

The dry eelgrass was threshed with stout sticks, such as ocotillo poles, and the grain was then separated from the chaff by wind winnowing with shallow baskets. The grain (fruit) was toasted and then ground on a metate to separate the seeds from the fruit husks, further winnowed, and the pure seeds were then ground into flour on a metate. The flour was variously consumed, for example, as *atole* (flour mixed with water) or combined with oil-rich foods. Eelgrass is rich in protein but has very little oil (an oil seed would not sink in the sea), and the flour was often prepared with sea turtle oil or combined with ground seeds of cardón (*Pachycereus pringlei*). In 1975, in Tucson, Hazel Fontana baked the first loaf of bread from a grain from the sea using eelgrass flour that Mary Beck Moser obtained from Seri women. The bread was green inside and unfortunately gritty with sand, an unavoidable consequence of grinding foodstuff on a metate resting on sand. We had learned the value of a consuming flour as *atole*, which allows the sand and sediment to settle to the bottom and be avoided.

The moon, or month, corresponding to April is called xnoois ihaat iizax "eelgrass-seed time-of-ripening moon" or "moon of the eelgrass seed" (Felger & Moser 1985:58).

Sea turtles (*Chelonia*) feeding on eelgrass were said to have "sweet, good-tasting meat, whereas those from west coast of Tiburón—which ate xpanaams, or marine algae—were cheemt 'stinky.' Eelgrass was often found in the stomach of the green turtle" (Felger & Moser 1985:378). The large sea hare, *Aplysia californica* (a gastropod mollusk), also seasonally feeds extensively on eelgrass in the Canal del Infiernillo, and these were consumed by sea turtles (Felger and Moser 1985:23). Seventeen large *Aplysia*, their guts filled with eelgrass, were found in the gut of a butchered sea turtle from the Infiernillo.

Other Gulf Islands: None reported but undoubtedly to be found as beach drift or growing in waters adjacent to some islands along the Baja California coast.

Geographic Range: This seagrass occurs along the Atlantic and Pacific Coasts of North America and Eurasia and is found on both coasts of the Baja California Peninsula. On the mainland it extends from the vicinity of Tiburón southward to the vicinity of Altata, Sinaloa.

Tiburón: Ensenada del Perro, beach drift, *Felger 17734*. 50 m off SE shore of Tiburón, 12 Apr 2007, *Wilder 07-165*. Bahía San Miguel, water 1 m deep, 30°C, *Zostera* almost entirely seasonal dead and disintegrated; *Halodule wrightii* abundant and vigorous, *Torre 17 June 1999*. Palo Fierro, on beach, was washed up on shore, 24 May 06, *Wilder 06-145*. Punta Tormenta, beach drift, *Felger 07-97*. Canal del Infiernillo, in Seri panga, hauled in as fishing bycatch, 24 Nov 2006, *Wilder 06-505*. Canal del Infiernillo, midway between Isla Tiburón and mainland, W of Campo Víboras, 19 Dec 1973, *Norris 4730*. Punta Perla, beach drift, 4 Apr 1974, *Felger 74-13*.

Alcatraz: Shore at S-central side, in beach drift among large quantities of a red alga, *Felger 07-158*. Beach drift, E side of flat, *Felger 08-04*.

Zygophyllaceae – Caltrop Family

Hot-weather ephemerals, perennials, and hardwood shrubs and trees. *Fagonia*, *Larrea*, and *Viscainoa* occur in some of the most arid or extreme habitats in the region.

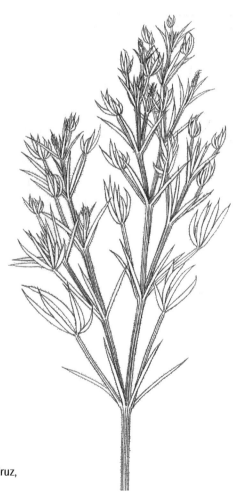

Figure 3.161. *Fagonia palmeri*. Ensenada de la Cruz, Isla Tiburón. LBH.

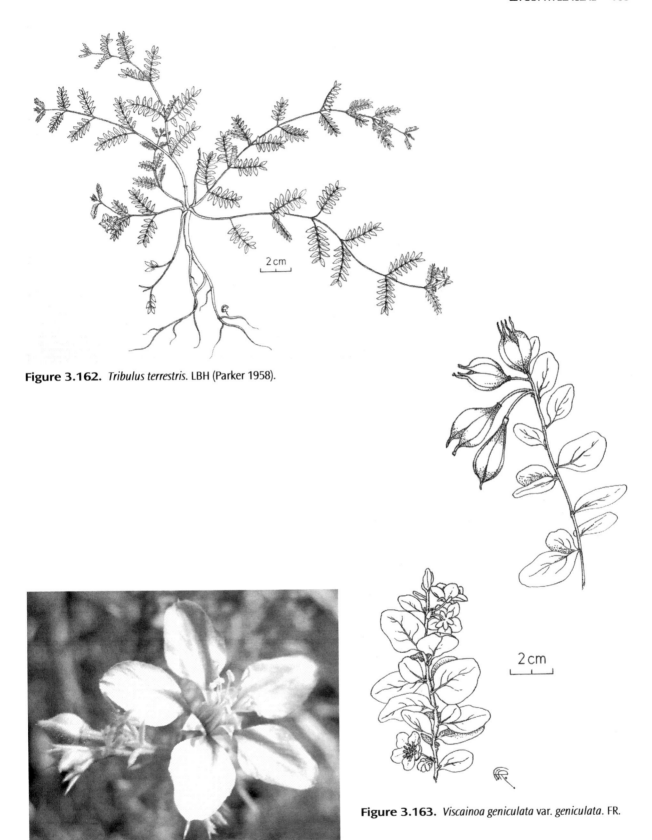

Figure 3.162. *Tribulus terrestris*. LBH (Parker 1958).

Figure 3.163. *Viscainoa geniculata* var. *geniculata*. FR.

Figure 3.164. *Fagonia palmeri*. Ensenada de la Cruz, Isla Tiburón. BTW, 28 February 2009.

Figure 3.165. *Guaiacum coulteri*, cultivated at the Desert Laboratory, Tumamoc Hill, Tucson. BTW, 3 August 2005.

Figure 3.166. *Viscainoa geniculata* var. *geniculata*, clusters of old capsules and seeds. Note the thin, sticky arils surrounding the seeds. Capxölim, Isla Tiburón. BB, 24 October 2007.

1. Woody shrubs or small trees, and unarmed (without spines).
    2. Leaves pinnate; flower dark blue; large shrubs or small trees. _____**Guaiacum**
    2' Leaves simple or bifoliolate (2 fused leaflets); flowers white or yellow; shrubs.
        3. Leaves opposite; herbage viscid/sticky; flowers yellow. _____**Larrea**
        3' Leaves alternate; herbage softly pubescent, not sticky; flowers white. _____**Viscainoa**
1' Herbaceous annuals or perennials, or if partially woody then spiny and less than 1 m tall.
    4. Perennials (also flowering in first season) or small shrubs; stems spiny; leaves digitately compound with 3 leaflets (5–7 in *F. palmeri*); flowers rose-pink. _____**Fagonia**
    4' Annuals; stems unarmed (not spiny); leaves pinnately compound with 3 or more pairs of leaflets; flowers yellow or orange.
        5. Opposite leaflets in each pair of different sizes; fruits not spiny. _____**Kallstroemia**
        5' Opposite leaflets in each pair equal in size; fruits spiny. _____**Tribulus**

Fagonia

Low-growing perennial herbs to small shrubs, and also flowering in the first season. Stems brittle, leaves opposite, the paired stipules modified as sharp spines. Flowers rose-pink; flowering with warm weather at various seasons except in drought. No other genus is so wide-ranging yet so closely restricted to the hot, arid deserts of the world. The seeds, which become mucilaginous when wet and adhere tenaciously upon drying, are probably a major factor in the unique and widely disjunct distribution.

1. Small woody shrubs; leaves with 5–7 leaflets. _____ _____**F. palmeri**
1' Not shrubby and not woody; leaves with 3 leaflets.
    2. Plants essentially glabrous (stems scabrous). ____ _____**F. californica**
    2' Plants conspicuously glandular-pubescent. _____ _____**F. pachyacantha**

Fagonia californica Bentham [*F. laevis* Standley]

Xtamosnoohit "what desert tortoises eat"

Description: Perennials or sometimes flowering in first season, much branched with very slender, brittle stems; glabrous or essentially so, the stems rough (scabrous). Leaves with 3 slender leaflets, the middle one often slightly wider; stipular spines mostly 1.5–3 mm long, sharp and at least some are moderately recurved.

Local Distribution: Widespread on Tiburón and Dátil, especially in the more arid habitats.

Other Gulf Islands: San Marcos, Coronados, San José.

Geographic Range: Arizona, southern California, both Baja California states, and western Sonora south to the Sierra Seri.

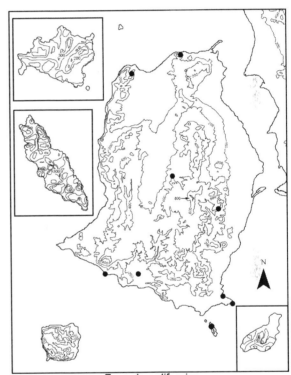

Fagonia californica
Xtamosnoohit

Tiburón: Near Pazj Hax waterhole, *Romero-Morales 07-7.* Arroyo Sauzal, 2 mi from shore, *Felger 12289.* Tordillitos, *Felger 15475.* El Monumento, *Felger 15533.* Ensenada del Perro, *Felger 17714.* Haap Hill, *Felger 76-T21.* Pazj Hax, *Wilder 06-48.* Hee Inoohcö, mountains at NE coast, *Felger 07-73.* Tecomate, 1 mi W on rocky hill, *Felger 12521.*

Dátil: NE side of island, not common, N and S slopes, *Felger 9112.* NW side of island, common, *Felger 15321.* E side of island, halfway up N-facing slope of peak, *Wilder 07-127.*

Fagonia pachyacantha Rydberg

Xtamosnoohit "what desert tortoises eat"

Description: Forming spreading or semi-prostrate mats to 1+ m across during favorable times. New growth with golden yellow glands (rarely glabrous or glabrate during times of high rainfall); leaves bright green, the leaflets often becoming semi-succulent during favorable seasons. Stipular spines straight, at least some on each plant 5–12 mm long, often stout, with age bending back toward the stem but not curved.

Local Distribution: Common on the arid southern and western sides of Tiburón; often on exposed ridges, mesas, and hills and in arroyos.

Other Gulf Islands: Ángel de la Guarda.

Geographic Range: Southward in western Sonora to Cerro Tepopa (vicinity 29°22'N), extreme southwestern Arizona, southeastern California, and Baja California Norte. Its apparent closest relative, *F. barclayana* (Bentham) Rydberg, abruptly replaces it to the south in Sonora and on the Baja California Peninsula. These two species are allopatric with no indication of intermediates.

Tiburón: Ensenada Blanca (Vaporeta): Very common, to 3½ ft across, arroyo, rocky hills, and ridges, *Felger 7141* (SD); Sparsely vegetated flat ridge or mesa, ¼–½ mi inland, *Felger 12221* (SD). Willard's Point, just to the N, occasional, *Moran 8736* (SD). Arroyo de la Cruz, *Wilder 09-49*.

Fagonia palmeri Vasey & Rose

Description: Densely branched dwarf woody shrubs, with numerous closely set upright branches. Herbage yellow-green, conspicuously glandular pubescent. Stipular spines straight and sharp, resembling the leaflets or petioles. Leaflets 5–7, linear, 1 mm or less in width, soon deciduous from the more prominent petiole.

Local Distribution: In the flora area known only from the vicinity of Ensenada de la Cruz on the south shore of the Tiburón, where it is locally abundant.

Other Gulf Islands: San Marcos.

Geographic Range: Gulf side of Baja California Sur.

Tiburón: Arroyo de la Cruz, colony on N slope, 50 m elev., ½ m high, *Moran 13012* (SD). Ensenada de la Cruz, gravelly arroyo bottom at base of E-facing rock wall, rare and localized, woody bush 2½ ft tall, bright yellow-green and dense foliage, *Felger 12787* (SD).

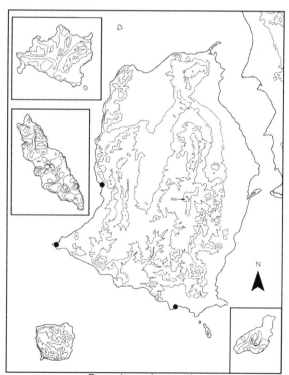

Fagonia pachyacantha
Xtamosnoohit

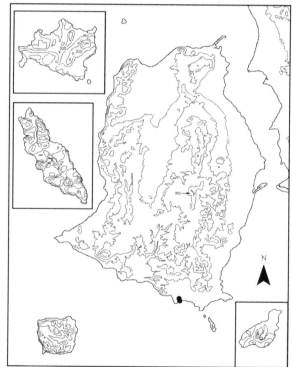

Fagonia palmeri

Guaiacum coulteri A. Gray

Mocni; *guayacán*; guaiacum

Description: Rigidly branched large shrubs to small trees with very hard wood. Evergreen or tardily deciduous during extreme drought. Leaves opposite, 0.5–2 cm long, dark green, pinnate. Petals and filaments dark indigo blue, the anthers lemon yellow before anthesis. Seeds large, enveloped in a thin, bright red aril, and conspicuously protruding from the orange, semi-fleshy fruit capsule. Spectacular masses of dark blue flowers produced from mid-May through June, the hottest, driest time of year.

Guaiacum officinale and *G. sanctum*, closely related species from the American tropics, are the source of lignum vitae, among the hardest commercial woods.

Local Distribution: Common in the foothills, upper bajadas, and margins of major arroyos on the east side of the Sierra Kunkaak on Tiburón.

Other Gulf Islands: None.

Geographic Range: Widespread and common across much of Sonora, and southward to Puebla and Oaxaca.

Tiburón: Foothills of [E side of] Sierra Kunkaak, canyon bottoms and adjacent benches in mountains, arroyos and bajada near base of mountain, *Felger 6953.* Between shore and Sierra Kunkaak, on bajada flats, *Wilder 06-37.* Cerca de primera loma, Pazj Hax, lado este, *Romero-Morales 07-5.* Sopc Hax, 26 Oct 1963, *Felger 9322* (specimen not located).

Kallstroemia

Summer annuals; stems spreading to prostrate, the leaves pinnately compound.

1. Petals yellow to yellow-orange, 4–6 mm long; sepals usually deciduous; beak of fruit less than 5 mm long; fruiting pedicels 1–2.3 cm long. _____**K. californica**
1' Petals bright orange with a darker base (15) 20–35 mm long; sepals persistent; beak of fruit (5) 8–12 mm long; fruiting pedicels (2) 3–7 cm long.____ **K. grandiflora**

Kallstroemia californica (S. Watson) Vail

California caltrop

Description: Leaves 2–4.5 (6) cm long, with 3–6 (7) pairs of leaflets. Flowers small and yellow to yellow-orange. Fruiting pedicels 1–2.3 cm long. Body of fruit 4–5 mm long; beak of fruit usually shorter than to about as long as the body.

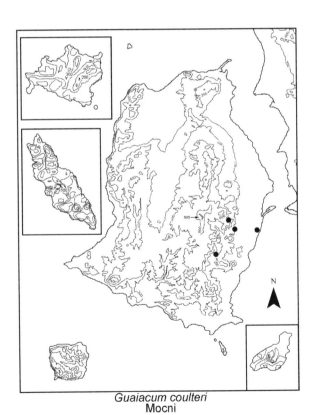

Guaiacum coulteri
Mocni

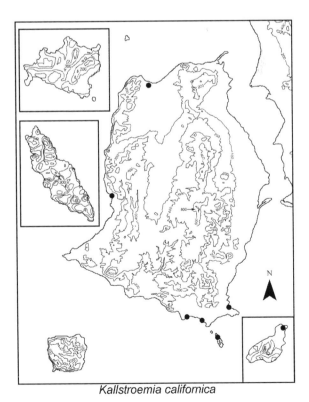

Kallstroemia californica

Local Distribution: Widespread and common across the lower elevations on Tiburón and also common on Dátil and the flats on Alcatraz.

Other Gulf Islands: Coronados, Carmen, Danzante, San Diego, San José, Cerralvo.

Geographic Range: Western Mexico including Baja California Sur and southwestern United States.

Tiburón: Ensenada Blanca (Vaporeta), *Felger 14950*. Ensenada de la Cruz, abundant, hills near coast, *Felger 9202*. Coralitos, common, gravelly soil near wash, *Wilkinson 209*. Ensenada del Perro, *Wilder 08-305*. Bahía Agua Dulce, edge of dune near beach, *Felger 8909*.

Dátil: NE side of island, S-facing slopes near arroyo bottom and infrequent elsewhere, *Felger 9080*.

Alcatraz: 28°49'22"N, 111°56'27"W, sandy substrate, *Wickenheiser 17* (Prescott College Field Station Herbarium, Bahía Kino).

Kallstroemia grandiflora Torrey ex A. Gray

Hast ipenim "with which the rock is splashed"; *baiburín, mal de ojo*; summer poppy, orange caltrop

Description: Leaves 4.5–12 cm long, with 5–9 pairs of leaflets. Flowers large and showy; corollas orange and often reddish in the center. Flowers opening about an hour after dawn and brightest in the morning and withering in the afternoon heat. Ants and a variety of flying insects eagerly feed at green nectaries between the sepals and petals. Fruiting pedicels (2) 3–7 cm. Body of fruits 4–5+ mm long and knobby; beaks 8–12 mm long (as short as 5 mm when drought stressed).

Local Distribution: Central, western, and southern regions of Tiburón; apparently not at higher elevations.

Other Gulf Islands: None.

Geographic Range: Western Mexico and border states of the United States; not known from the Baja California Peninsula.

Tiburón: Ensenada Blanca (Vaporeta), 29 Jan 1965, *Felger*, observation (Felger 1966:173). Arroyo Sauzal, 2 km inland, rocky low ridgetop, *Felger 12279*. Coralitos, *Wilkinson 191*. SW Central Valley, upper bajada at base of mountain, *Felger 12663*. Haap Hill, *Felger T74-20*.

Larrea divaricata Cavanilles subsp. **tridentata** (Sessé & Moçiño ex de Candolle) Felger & Lowe [*L. tridentata* (Sessé & Moçiño ex de Candolle) Coville]

Haaxat "with smoke"; *gobernadora, hediondilla*; creosotebush

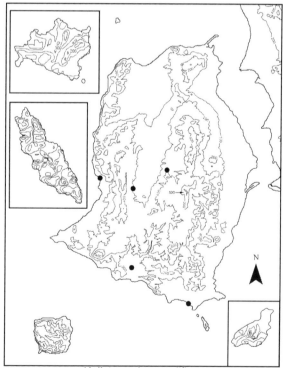

Kallstroemia grandiflora
Hast ipenim

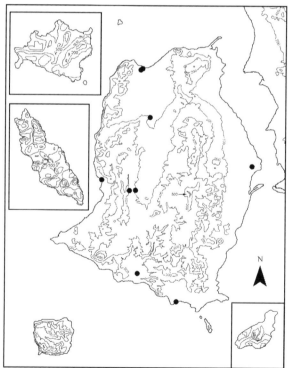

Larrea divaricata ssp. *tridentata*
Haaxat

Description: Multiple-stem shrubs, 1–2 (2.5) m tall, with very hard, brittle wood. Herbage gummy-resinous, the leaves shriveling during extreme drought. Leaves of two leaflets fused at their bases. Stipules broadly ovate-triangular, at first green and clasping the new growth bud, soon becoming red-brown, and extremely gummy-resinous. Flowers bright yellow, the filaments each with a well-developed yellow scale or appendage, these serving to cup nectar produced at the base of the style. Fruits like a small fuzzy white ball, separating into 5, single-seeded segments. After a rain the aroma of terpines from the wet foliage fills the desert air.

The stems are sometimes encrusted with ant-tended, reddish colored lac produced by the scale insect *Tachardiella*. This lac was used as an allpurpose adhesive by the Comcaac (Felger & Moser 1985).

Local Distribution: One of the most widespread and common shrubs across most of lowland desertscrub vegetation on Tiburón. It extends to the ridge crests of Sierra Menor but has not been found at higher elevations of the Sierra Kunkaak. It is the dominant landscape element across most of the expansive Agua Dulce/Central Valley. Much of the valley plain away from streamways at the western edge of the valley contains nearly pure stands of *Larrea* plus seasonal annuals, with coverage values for *Larrea* of 6–12 (average 9) percent (Felger 1966:232). *Larrea* and *Cordia parvifolia* are dominant landscape features in a narrow zone inland from the *Frankenia palmeri* zone along the lower bajada bordering the Infiernillo shore.

Other Gulf Islands: Ángel de la Guarda, San Marcos, Carmen, Santa Catalina, San Francisco.

Geographic Range: *Larrea* reaches its southern limits in the Guaymas region for the west coast of mainland Mexico. Subsp. *tridentata* is widespread in all North American deserts except the Great Basin Desert and is the primary element in mapping and defining these deserts. Subsp. *divaricata* occupies deserts in South America.

Tiburón: Ensenada Blanca (Vaporeta), 30 Jan 1965, *Felger*, observation (Felger 1966:176). Sauzal landing strip, *Felger 6444*. Ensenada de la Cruz, low hills, *Felger 2591*. Sierra Menor, ridge crest, 2 Feb 1965, *Felger*, observation (Felger 1966:245). SW Central Valley, *Felger 12394*. Palo Fierro, *Felger 11081*. Hast Iif (Ojo de Puma), Central Valley, broad arroyo, *Wilder 07-262*. Bahía Agua Dulce, *Felger 10196*. Tecomate, abundant below mountains W of village, growth stunted compared to Tucson, rocky flats, *Whiting 9021*.

*Tribulus terrestris Linnaeus

Cosi cahoota "horn that causes one to urinate," cözazni caacöl "large *Cenchrus palmeri*," hee inoosj "jackrabbit's claw," heen ilít "cow's head," hehe ccosyat "spiny plant"; *torito, toboso*; goathead, puncture vine

Description: Hot weather annuals with spreading to prostrate stems. Leaves pinnate, 1–4.5 cm long, with 4–7 leaflet pairs, the leaflets of the lower pair unequal in size. Flowers yellow, the petals 5 mm long or less. Fruits 1.5–1.8 cm wide, intricately sculptured and spiny, at maturity breaking into 5 sharp, tack-like nutlets that land with the largest spine upward. The nutlets can be transported on tires, shoes, and sandals.

Local Distribution: It is abundant along the nearby Sonora mainland. It was documented on Alcatraz in 1966. The population has not expanded from the northeast sandy flat but is very large with a massive seed bank, and potential expansion should be closely monitored. On Tiburón it has been found only near the boat landing at the marine station, where there is almost daily traffic from the mainland.

Other Gulf Islands: None.

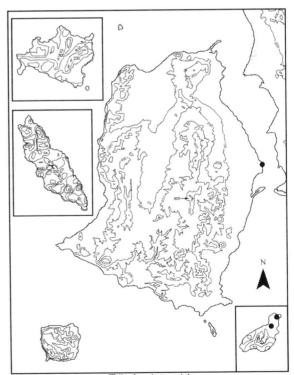

Tribulus terrestris
Cosi cahoota

Geographic Range: Native to the Old World, this obnoxious weed is now widespread in the warmer regions of the world. The scientific name translates as "tribulation of the earth."

Tiburón: Punta Tormenta, 23 Nov 2006, *Wilder 06-345.*

Alcatraz: Sand flat, [N]E side of island, not common, 8 Oct 1966, *Felger 14923.* Low flat near shore, abundant in sand soil at NE tip of island near lighthouse, 4 Dec 2007, *Felger 07-188.*

Viscainoa geniculata (Kellogg) Greene var. **geniculata**

Xneeejam is hayaa "xneejam whose seeds are owned"

Description: Shrubs often 1–2 m tall, with soft hairs. Leaves tardily drought deciduous, alternate, simple, mostly 2.5–4.5 cm long, obovate to oval. Flowers 2 cm wide; cream-white; flowering at various seasons. Fruits velvety green capsules with 5 low ridges or wings. Seeds ovoid, black, 5 mm long with a sticky aril that probably assists in dispersal, especially via birds.

Local Distribution: Widespread and common on Tiburón, San Esteban, Dátil, Alcatraz, and present on Mártir and Cholludo. Sometimes extending into habitats of guano-rich soils. On Mártir it is widely scattered, mostly at higher elevations, often in the shelter of large cardons. Ivan Johnston (1924:1054) wrote, "One of the most characteristic and widely distributed shrubs in

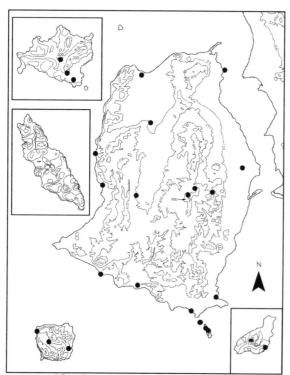

Viscainoa geniculata var. *geniculata*
Xneeejam is hayaa

the gulf area" and he specifically noted the islands from which it is absent including San Pedro Mártir. Neither he nor Palmer collected or reported this species on the island at the end of the nineteenth and early twentieth centuries. Reid Moran saw only one on his hike to the top of the island in 1962, and the next year Felger also found it was at the top and noted it was rare. In 2007 it was common at the top of the island and scattered at middle elevations. In addition, this species was a new arrival to Isla Rasa in 2006.

Other Gulf Islands: Ángel de la Guarda, Partida Norte, Rasa (one plant, first seen in 2006 when it was 1 m tall in 2008; *Velarde 20 Apr 2008,* SD), San Lorenzo, Tortuga, San Ildefonso, Coronados (observation), San Diego, San José, Espíritu Santo.

Geographic Range: Sonora from the mainland coast opposite Tiburón to Guaymas, both states of Baja California, and most of the Gulf Islands. Another variety in Baja California Sur has pinnate leaves.

Tiburón: Ensenada Blanca (Vaporeta), *Felger 17292a.* Tordillitos, *Felger 15493.* Arroyo Sauzal, *Felger 08-33.* Coralitos, *Wilder 06-57.* Ensenada del Perro, *Wilder 07-151.* SW Central Valley, arroyo, *Felger 17318.* Palo Fierro, *Felger,* 14 Feb 1965, observation (Felger 1966:204). Hast Cacöla, E flank of Sierra Kunkaak, 380 m, *Wilder 08-261* (USON). Base of Capxölim, head of arroyo, *Wilder 05-53.* Deep canyon in N interior of Sierra Kunkaak, 720 m, *Wilder 07-579.* Zozni Cacösxaj, NE coast, *Wilder 08-198.* N side, *Mallery & Turnage 3 May 1937.* Bahía Agua Dulce, *Felger 8902.* Hast Iif (Ojo de Puma), Central Valley, *Wilder 07-258.* Pooj Iime, *Wilder 07-440.*

San Esteban: Arroyo Limantour, *Barnett-Diaz 08-4.* Near center of island, rocky basaltic hills and broad rocky floodplain, *Felger 473.* San Pedro, steep rocky slopes, *Felger 16666.* Cañones en el SE de la isla, *Lott & Atkinson 2464* (CAS).

Dátil: NW side of island, *Felger 15338.* E side of island in canyon, *R. Russell 6 Aug 1962.*

Alcatraz: Fairly common throughout E side of island, *Felger 13401.* Base of SE side of mountain, *Felger 12823.*

Cholludo: Scattered, not common, *Felger 13412.* E-facing, steep rocky slope just above sea cliff, rare, ca. 1.3 m tall, *Wilder 09-79.*

Mártir: Near summit, 300 m, only one seen, shrub 3 m tall, 21 Mar 1962, *Moran 8812* (SD). Only seen near top of island, rare, shrub 3 m tall, 25 Jan 1963, *Felger 6355* [probably the same individual as *Moran 8812*]. 125 m, 17 Mar 1971, *Hastings 71-58.* Above ruins of guano workers' village, 150 m, to ca. 1.5 m tall, several scattered shrubs, mostly partially shaded or sheltered by large cardons, 11 Apr 2007, *Felger*

07-12. Summit of island, common shrub 2–2.5 m tall, corollas white, 5 Dec 2007, *Felger 07-196.*

DOUBTFUL AND EXCLUDED REPORTS AND ONES IN NEED OF VERIFICATION

A number of reports for island plants for which we have not located specimens actually may be correct, but we are accepting only specimen-based records or, in some cases, reliable observervations.

AIZOACEAE

Sesuvium verrucosum Rafinesque

Listed for Tiburón by Felger and Lowe (1976) and Rebman et al. (2002) instead of *S. portulacastrum.*

APOCYNACEAE

Metastelma pringlei A. Gray

The report of *M. pringlei* on Nolasco (Felger & Lowe 1976; Rebman et al. 2002) was based on misidentified specimens of *M. californicum.*

ASTERACEAE

Ambrosia ambrosioides (Cavanilles) W.W. Payne

Although Felger and Moser (1985) indicated it was on Tiburón, it apparently does not occur on the island. It is common on the nearby Sonora mainland.

Bahiopsis chenopodina

Reported for San Pedro Mártir (Rebman et al. 2002), but we have not seen a specimen and do not believe it occurs on the island.

Coreocarpus sonoranus var. sonoranus

Reported for San Esteban (Felger & Lowe 1976; Rebman et al. 2002), but we have not located any specimen of *Coreocarpus* on this island.

Pelucha trifida

Rebman et al. (2002) report an observation for San Esteban. We are not able to verify the report.

Perityle emoryi

Two Nolasco collections previously identified as *P. emoryi* are probably *P. californica*; *P. emoryi* apparently does not occur on Nolasco. *Perityle emoryi* has white rays and *P. californica* has yellow rays; both have a yellow disk and are winter-spring annuals. The two specimens in question have ambiguously colored rays and other distinguishing features are not evident.

Perityle leptoglossa Harvey & A. Gray

Reported for mainland mountains and the Sierra Kunkaak on Tiburón by Felger and Moser (1985), but we have not found it on the island.

Peucephyllum schottii

A specimen seen at ARIZ has the following information: "Tiburón, con *Larrea divaricata, Fouquieria, Carnegiea gigantea, Stenocereus,* suelo pedregoso, arbusto de 1 m, en los riscos pegado a la playa, aquenios cafes, solo, se observo en los riscos, 14 May 1985, *Valiente 569.*" At the MEXU herbarium in Mexico City, where the primary set of vouchers from the mid-1980s Gulf Island collecting trips led by the Instituto de Biologia at UNAM are curated, is another sheet of *Valiente 569.* It has the same label as described above from Tiburón but just above is an additional label indicating the same specimen, number 569, but collected on a different date, 1 August 1985, from Puerto Refugio, Isla Ángel de la Guarda. The full label reads: "Matorral xerófolio con *Lycium berladieri* var. *pubescens, Bursera microphylla, Pachycereus pringlei, Jatropha* sp. etc., arbusto de 1 m, en los riscos cercanos a la playa, aquenios de color café, solo se observó en los riscos, 1° de agosto de 1985, *Valiente-Banuet 569* y F. Chiang Cabrera. Det. J. L. Villaseñor. MEXU # 579935."

Based on our knowledge and personal experiences in the field and the evidence that the above specimen in question is not from Tiburón but in fact is from Ángel de la Guarda, where it is known to occur, we report *Peucephyllum schottii* as occurring only on San Esteban among the Sonoran Islands.

Xanthisma incisifolium

Moran (1983b) and Rebman et al. (2002) list it for Tiburón, but we have not seen a specimen from this island. It occurs on San Esteban.

BORAGINACEAE

Phacelia crenulata

Reported for San Esteban by Rebman et al. (2002) as *P. crenulata* and *P. ambigua*. We have not seen a specimen from the island.

BRASSICACEAE

Draba cuneifolia

Listed by Felger and Lowe (1976) for Dátil, but no specimen has been located.

CACTACEAE

Cylindropuntia alcahes var. alcahes

In his dissertation on the *Cylindropuntia* of the Baja California Peninsula, Rebman (1995) reports two specimens of *C. alcahes* var. *alcahes* from San Pedro Nolasco: 3 May 1952, *Lindsay 2226* (ASU) and 29 May 1937, *Rempel 304a* (RSA, first annotated by Dawson 5/10/1948 as "near *O. fulgida*"; second annotation by Rebman 1995, as "*O. alcahes* var. *alcahes*"). Both collections are from Cañón el Farito, the place where we and others have conducted extensive fieldwork and have not encountered this cholla. In our opinion both specimens are *C. fulgida*, the only cholla that has been found on the island. We do not believe that *C. alcahes* occurs on Nolasco.

Cylindropuntia cholla and C. fulgida

Rebman (1995:161 & 163) cites two cholla specimens for San Esteban, based on "13 April 1911, *Rose 16824*" (US, not seen by us)—one as *C. cholla* and one as *C. fulgida* var. *mamillata*. There are no other records for these cacti on San Esteban. Rose's field book at US shows #16824 only as *Mammillaria* from Isla San Esteban, but in the database at US there are two cacti listed for *Rose 16824*: *Cylindropuntia fulgida* and *Mammillaria*. Rose collected living cactus specimens, which were often made into herbarium specimens at a later date (see Joseph Rose in "Botanical Explorations"). Furthermore, Rose's field book shows that on the same day, 13 April 1911, he collected a *Cylindropuntia* on Isla Cholludo (Seal Island), and *C. fulgida* is abundant on that island. We conclude that *Rose 16824* as *Cylindropuntia cholla* and *C. fulgida* is not from Isla San Esteban.

CAMPANULACEAE

Nemacladus orientalis

Listed by Rebman et al. (2002) for Tiburón and Dátil, but we have not located a specimen.

CANNABACEAE

*Cannabis sativa Linnaeus

Haapis cooil "green tobacco"; marijuana

Annuals growing with warm weather. Rarely encountered as clandestinely grown plants at waterholes and as packages abandoned, stashed, or washed ashore after storms. The region is infamous for illicit trafficking. This Old World species is widely grown and traded in Mexico but is not commercially grown in the flora area.

Tiburón: Undisclosed waterhole, 1990, *Cosme Becerra*, observation (personal communication, 30 Jan 2008).

Alcatraz: Several packages ("kilos") near the shore, 4 Dec 2007, *Felger & Wilder*, observation.

CAPPARACEAE

Atamisquea emarginata

Rebman et al. (2002) list this shrub for Nolasco in their Gulf checklist, which is undoubtedly based on Gentry's (1949) indication of it on the island. We have not found this species on the island and have not located a specimen.

Forchhammeria watsonii Rose

Reported for Tiburón by Turner et al. (1995) based on a sighting by David E. Brown. There are no records for this tree north of the Bahía Kino region, and the sighting is probably mistaken for *Bonellia macrocarpa* (Theophrastaceae).

CONVOLVULACEAE

Cuscuta leptantha

Listed by Rebman et al. (2002) for Dátil. A previously unidentified specimen of *Cuscuta* from Dátil is *C. desmouliniana* and this is the likely source for the listing of *C. leptantha*.

CUCURBITACEAE

Ibervillea sonorae (S. Watson) Greene var. **sonorae** [*Maximowiczia sonorae* S. Watson var. *sonorae*]

Hant yax; *guarequi*; cow-pie plant

The report of it on Tiburón (Felger & Lowe 1976; Rebman et al. 2002) was based on misidentification of a sterile specimen of *Tumamoca macdougalii* from Punta Tormenta (*Felger 8935*).

EBENACEAE – PERSIMMON FAMILY

Diospyros intricata (A. Gray) Standley [*Maba intricata* (A. Gray) Hiern]

Baja California persimmon

The report of this species for Tiburón (Provance et al. 2008) was based on our initial and mistaken identification of sterile specimens of *Forestiera phillyreoides*.

FABACEAE

Acacia occidentalis Rose [*Senegalia occidentalis* (Rose) Britton & Rose]

Sonoran catclaw acacia

Turner et al. (1995) report *A. occidentalis* on Tiburón, but we have not seen specimens of this species from the island and believe the report is based on misidentified *A. greggii*. They are closely related: *A. greggii* has cylindrical, spike-like inflorescences and generally smaller leaves, while *A. occidentalis* has rounded, head-like (capitate) inflorescences and generally larger leaves. They are not known to occur together (Felger et al. 2001). *Acacia occidentalis* occurs between Guaymas and Hermosillo and elsewhere in Sonora.

Astragalus magdalenae Greene var. **magdalenae**

Reported from beach dunes on Tiburón by Felger and Moser (1985), but no specimens are known from the island. The nearest documented specimens are from the vicinity of El Desemboque.

Caesalpinia palmeri S. Watson

Reported from Tiburón by Felger and Moser (1985), but no specimens are known from the island. The nearest documented specimens are from the Sierra Seri.

Senna covesii

This species is listed for San Esteban by Moran (1983b) and Rebman et al. (2002). We have not found it on the island and these reports are probably based on *S. confinis*, which does occur on San Esteban. We have not found these similar-appearing species growing together.

FOUQUIERIACEAE

Fouquieria diguetii

Rebman et al. (2002) report this species on Tiburón and list it as questionable on San Esteban. We have no record of this species from either island and doubt that *F. diguetii* occurs on Tiburón. The two known ocotillo plants on San Esteban might be this species or *F. splendens*. There are various reports of *F. diguetii* from Isla San Lorenzo (Rebman et al. 2002 show it as an observation), but no specimen has been located and it does seem to occur on that island (see "Species Accounts").

LOASACEAE

Mentzelia adhaerens

This species is reported for Nolasco (Felger & Lowe 1976; Rebman et al. 2002), but we have not located a specimen.

Mentzelia involucrata S. Watson

Reported for Tiburón by Felger & Moser (1985), but no specimen is known from the island.

MALVACEAE

Ayenia jaliscana S. Watson

Reported as occurring in the Sierra Kunkaak (Felger 1966:208), but this apparently was based on misidentification of the common *A. filiformis*.

Horsfordia rotundifolia S. Watson

Reported from Tiburón: 12 Apr 1911, *Rose 16809* (US). Rose collected this specimen at the southeastern part of the island along with the rest of his collection for the same day, recording it in his field notebook as *Horsfordia rotundifolia* (see Joseph Rose in "Botanical Explorations"). Fryxell (1988) also cited this specimen as *H. rotundifolia*. The specimen, however, seen by Felger in 2009, is *Hibiscus denudatus*.

Horsfordia rotundifolia is distinctive and abundantly documented from the southern half of the Baja California Peninsula and the adjacent Islas Coronados, San José, San Francisco, and Espíritu Santo. The type collection is described as "annual, erect, slender, 2 feet high or more. . . . On an island in Guaymas harbor," *Palmer 351 in 1887* (Watson 1889; NY, three isotypes, images at http://sciweb.nybg.org/Science2/hcol/vasc/index.asp, accessed 8 August 2009). The type collection is the only mainland report for this species. In spite of more than 120 years of extensive collecting in the Guaymas region and elsewhere in Sonora, *H. rotundifolia* has never again been reported for the mainland. Palmer probably collected it on the Baja California Peninsula—he made a very large collection from the Peninsula, as well as Sonora, during his several months of collecting in the region, and there are other instances of similar locality mix-ups on his labels (see Felger 2000a; McVaugh 1956).

Melochia tomentosa

Although reported for Nolasco (Felger & Lowe 1976; Rebman et al. 2002), we have not found a specimen or seen it on this island. Gentry (1949) shows it for Nolasco as an observation.

ONAGRACEAE

Oenothera arizonica (Munz) W.L. Wagner [*O. californica* (S. Watson) S. Watson subsp. *arizonica* (Munz) W. Klein. *O. avita* (W. Klein) W. Klein subsp. *arizonica* (Munz) W. Klein]

Hantoosinaj cooxp; Arizona evening primrose

Reported for Tiburón by Felger and Lowe (1976), but no specimen for the island has been located. It is present on coastal dunes in Sonora from the vicinity of Tastiota northward and sandy soils in southern Arizona. If it is on Tiburón it most likely is on dunes in the vicinity of Tecomate or the northeast portion of the island (e.g., Zozni Cacösxaj).

PAPAVERACEAE

Eschscholzia minutiflora S. Watson

Reported for Tiburón by Felger and Lowe (1976), but no specimen has been located.

POACEAE

Aristida purpurea Nuttall var. **wrightii** (Nash) Allred

Tres barbas; Wright's three-awn, purple three-awn

Reported for Tiburón by Rebman et al. (2002) as *A. wrightii* Nash, but no specimen has been located. Most Sonoran Desert populations of *A. purpurea* are var. *nealleyi* (Vasey) Allred (Felger 2000a). The nearest known populations are from Ángel de la Guarda (*Moran 12452* [SD], identification not conclusive) and San Lorenzo (not seen, cited by Gould & Moran 1981).

*Cenchrus brownii Roemer & Schultes

West and Nabhan (2002) report it for Tiburón, but it is in fact *C. echinatus* (see "Species Accounts"). The taxonomic distinctions between the two species are subtle (John Reeder, personnel communication, 2007).

Muhlenbergia microsperma

Reported by Felger and Lowe (1976) for Cholludo. We have not located a specimen.

SELAGINELLACEAE – Spike-Moss Family

Selaginella arizonica Maxon. Arizona spike-moss

Reported for Tiburón by Felger and Lowe (1976:33, 53) based on "presence verified by Seri Indians." We would expect it on north-facing rock slopes at higher elevations in the Sierra Kunkaak, but we have not found it there. In retrospect, Felger suspects that there may have been confusion with desert ferns. This species and specimens keying to *S. eremophila* Maxon grow intermixed in the Sierra Seri on the mainland opposite Tiburón (Felger 2000a).

SOLANACEAE

Lycium megacarpum Wiggins

This enigmatic taxon is reported to differ from *L. exsertum* A. Gray primarily by being glabrous and having larger fruits. Chiang-Cabrera (1981) suggests *L. megacarpum* is probably different only at the subspecific level. Furthermore, "the distinction between *Lycium exsertum* and *L. fremontii* is usually not difficult in living plants but often so with dried specimens as habitat and flower color are important" (Chiang & Landrum 2009:21). Like *L. exsertum, L. megacarpum* is presumed to have unisexual and dimorphic flowers with male and female flowers on different plants.

A number of specimens from San Esteban (and one from Dátil) are cited or identified in herbaria as *L. megacarpum* (e.g., Chiang-Cabrera 1981), but we have not been able to verify occurrence in the flora area, in the field or with the various specimens seen, as either *L. megacarpum* or *L. exsertum*. Part of the problem is the lack of sufficient diagnostic specimens. Some specimens labeled as *L. megacarpum* could be glabrous or glabrate examples of the widespread and variable *L. brevipes*. The situation calls for further study.

Lycium megacarpum is reported from both states of Baja California, Isla Cedros, and several Gulf Islands including Ángel de la Guarda, Monserrat, and San José. Enigmatic specimens identified or labeled as *L. megacarpum* include the following:

San Esteban: Arroyo on N slope of NE peak, shrub on N slope of NE peak, 450 m, 26 Apr 1966, *Moran 13050*

(CAS [large elongated flowers, herbage short-pubescent, apparently becoming glabrous]). NE peak, few on talus on N slope, 200 m, shrub 1.5 m high × 2 m, with arching flexuous branches, fruit orange, ca. 5 mm, *Moran 13053* (CAS [without fruit, the herbage glabrous]; SD). Matorral diminutiva de *Atriplex* con unos pocos individuos de *Pachycereus pringlei, Lophocereus schottii, Stenocereus gummosus*, arbusto redondeado 1 m, hojas suculentas, flores rosada-lilas, frutos rojos, 8 May 1985, *Lott 2487* (CAS).

Dátil: *Felger 9100* (TEX, not seen, reported by Chiang-Cabrera 1981).

URTICACEAE

Parietaria hespera var. **hespera** [*P. floridana*]

Report for Dátil by Felger and Lowe (1976) and Rebman et al. (2002). We have no record for it on this island.

Addendum

After completing this book, our continued explorations in the region and study of specimens resulted in the discovery of additional populations and species of interest. These collections and additional herbarium discoveries add five species to the flora of Tiburón, increasing the total flora of the island to 345 taxa and the total for the Sonoran Islands to 393. We also report several distribution extensions for Tiburón and two exceptional new records for the Sierra Seri on the mainland opposite Tiburón.

Ephedraceae

Ephedra aspera. Haaxt; *tepopote*; boundary ephedra

Humberto discovered a population at high elevation in the Sierra Kunkaak. *Ephedra* in the flora area was previously known only from San Esteban.

> **Tiburón:** Exposed upper ridge of Sierra Kunkaak, rare, 18 Dec 2011, *Romero-Morales 11-1*.

Apiaceae

Daucus pusillus. *Zanahoria silvestre*; wild carrot

An additional specimen was located that expands the distribution on Tiburón to the west coast.

> **Tiburón:** Ensenada Blanca (Vaporeta), *Felger 17297*.

Asteraceae

Pluchea salicifolia (Miller) S.F. Blake. Mexican camphorweed

Description Robust herbaceous perennials, pungently aromatic and sticky with glandular hairs. Stems winged from decurrent leaf bases; the leaves, probably 3–8 cm long, are sessile and narrowly elliptic to oblanceolate, with toothed margins. Inflorescences panicle-like, with many flower heads of numerous pink disk florets; flowering in warmer months. Achenes about 1 mm long, the pappus of slender, minutely barbed bristles.

Local Distribution: A new record for a Sonoran Island, discovered at a remote waterhole in the Sierra Kunkaak, where several plants grow rooted in shallow water.

Other Gulf Islands: Cerralvo.

Geographic Range: The nearest known populations occur in riparian canyons with permanent water in the Sierra El Aguaje northwest of Guaymas (see Felger 1999). Baja California Sur and southern Sonora to Guatemala. *Pluchea*, with about 40 species in warm regions worldwide, often occurs in brackish marshes and other saline or alkaline habitats.

> **Tiburón:** Unnamed waterhole in the interior of the Sierra Kunkaak, ca. 3.5 km W of Hast Coopol, 30 Mar 2012, *Wilder 12-8*.

Porophyllum pausodynum

Local Distribution: A new record was found on Tiburón. This species was previously thought to be endemic to the Guaymas–San Pedro Nolasco region.

> **Tiburón:** Interior canyon of the Sierra Kunkaak, 2 km W of Hast Coopol, 30 Mar 2012, *Wilder 12-1*.

Boraginaceae

Nama hispidum. Hohroohit; sand bells, bristly nama

Local Distribution: An additional specimen was located that expands the distribution on Tiburón to the west coast.

> **Tiburón:** Ensenada Blanca (Vaporeta), *Felger 17298*.

Pectocarya recurvata I.M. Johnson. Mixed-nut combbur

Description: Small, cool-season annuals. Leaves narrow. Flowers very small and white. Fruits of 4 nutlets in 2 pairs, these spreading and recurved, the margins at least partially bristly.

Local Distribution: A specimen from the west coast of Tiburón provides a new species record for the Gulf of California islands.

Other Gulf Islands: None.

Geographic Range: Southwestern United States to northern Baja California Sur and Sonora southward to the Sierra El Aguaje.

> **Tiburón:** Ensenada Blanca (Vaporeta), *Felger 17299A*.

Brassicaceae

Draba cuneifolia. Cocool cmaam; wedge-leaf draba

Local Distribution: Two additional specimens were located that expand the distribution on Tiburón.

> **Tiburón:** Ensenada Blanca (Vaporeta), *Felger 17296*. Central Valley, 28°57′59″N, 112°25′43″W, valley plain with *Larrea*, *Felger 17338*.

Celastraceae

Canotia holacantha. Xooml icös caacöl; canotia, crucifixion thorn

Local Distribution: Ben and Humberto found canotia on Pico Johnson, the tallest portion of the Sierra Seri, giving another regional disjunct population in addition to the one on Tiburón in the Sierra Kunkaak.

> **Sonora:** N-facing slope of Pico Johnson, common above 600 m, some to 3 m tall, mostly 1.5–2 m in height, 30 Sep 2011, *Wilder 11-238*.

Convolvulaceae

Cuscuta desmouliniana. Hamt itoozj; *fideo*; dodder

Previously known from Tiburón and Dátil by a single record from each island. An additional specimen was located on Tiburón, indicating a wider distribution.

Tiburón: SE side of Agua Dulce Valley, 28°57'20"N, 112°24.5'W, 280 m, 8 Sep 1974, *Felger 76-T14.*

Onagraceae

Eremothera chamaenerioides (A. Gray) W.L. Wagner & Hoch [*Camissonia chamaenerioides* (A. Gray) P.H. Raven. *Oenothera chamaenerioides* A. Gray] Willow-herb evening primrose

Description: Delicate annuals growing and reproductive during the cooler months. Stems slender. Leaf blades often red-spotted or reddish. Flowers 2–2.5 mm long, whitish and pink, opening near sunset and closing the next morning. This is the smallest-flowered evening primrose in the Sonoran Desert—the floral structure and modifications are characteristic of self-pollinated flowers.

Local Distribution: Reported for Tiburón by Felger and Lowe (1976) and Rebman et al. (2002) but no specimens were located until the ultimate stages of the publication of this book.

Other Gulf Islands: Ángel de la Guarda.

Geographic Range: Southeastern United States and northwestern Sonora.

Tiburón: Ensenada Blanca (Vaporeta), *Felger 17299B.* Central Valley; 28°57'59"N, 112°25'43"W, valley plain, *Felger 17306.*

Rubiaceae

Randia capitata de Candolle [*R. megacarpa* Brandegee]

Possibly the plant referred to by the Comcaac as "ma-hyan" (Felger & Moser 1985:388).

Description: Shrubs often 3–6 m tall, here 2 m tall, usually forming 4 axillary or terminal spines, but the Sierra Seri specimens were spineless. Leaves opposite, to about 8 cm long. Flowers white, to 5 cm wide. Fruits globose, 3–5 cm wide, with a soft exterior and seeds embedded in a black pulp, which is sweet and with a musty but pleasing smell.

Local Distribution: A population found in the Sierra Seri is the first mainland record within the Sonoran Desert.

Other Gulf Islands: Espiritu Santo, Cerralvo.

Geographic Range: Baja California Sur and southeastern Sonora to Chiapas.

Sonora: Canyon that leads from the N-facing slope of Pico Johnson, 460 m, rare, ca. 2 m tall, 30 Sep 2011, *Wilder 11-229.*

PART IV

Gazetteer

Place names listed here are those mentioned in the text and localities for specimens cited. The coordinates are from our GPS readings and from Google Earth (2009), both based on WGS 84. The most prominent place names are shown in fig. 4.1.

We are fortunate to have satellite-based global mapping systems that provide accurate and user-friendly mapping. Earlier maps were often inaccurate, and determining coordinates was often a difficult task. Detailed, accurate topographic maps of the islands were not available before the 1980s. However, many coordinates and elevations still represent only our best estimate.

Coordinates for larger features, such as a bay, peninsula, or mountainside, are for the approximate middle of the area. Unless otherwise noted, coordinates for linear features (canyons, arroyos, etc.) are located at their mouths. Coordinates for mountains, unless otherwise noted, are for the peak. In preparing this listing we have relied heavily on *Unknown Island* (Bowen 2000) and information from Tom Bowen (personal communication, 2007–2010), and Seri place-names are largely from Moser and Marlett (2005, 2010), Stephen Marlett (personal communication, 2008–2011), Luque-Agraz and Robles-Torres (2006), and Humberto's knowledge of Isla Tiburón. Other place-names are from "Place-Names of Seriland" in McGee (1898) and from Felger and Moser (1985). In addition, many names and much information result from our fieldwork with local people in the region.

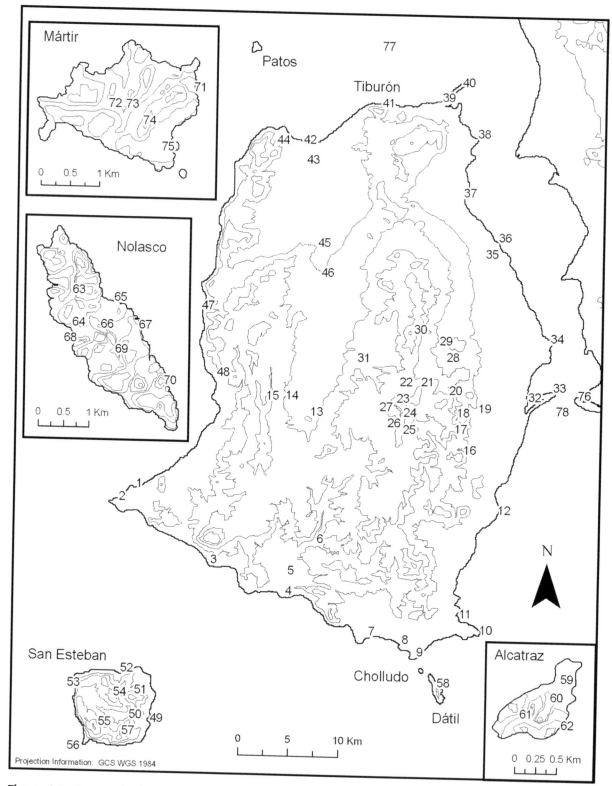

Figure 4.1. Sonoran Islands, with major place-names, KH.

Tiburón

1 Ensenada Blanca
2 Punta Willard
3 Tordillitos
4 Arroyo Sauzal, at the coast
5 Arroyo Sauzal, airfield, ca. 2 km inland
6 Xapij (Sauzal waterhole)
7 Ensenada de la Cruz
8 Coralitos
9 La Viga
10 El Monumento
11 Ensenada del Perro
12 Punta Narragansett, Hast Caacoj Quih Iyat
13 Satoocj
14 Central Valley, southwest part
15 Sierra Menor, above southwest part of Central Valley
16 Hast Coopol
17 Pazj Hax
18 Sopc Hax
19 Hant Hax
20 Hast Cacöla
21 Capxölim
22 Siimen Hax
23 Cójocam Iti Yaii, main peak
24 Sierra Kunkaak, summit of island
25 The "bowl"
26 San Miguel Peak
27 Sierra Kunkaak, upper ridge between Cójocam Iti Yaii and summit
28 Zoojapa
29 Caracol research station
30 Sierra Caracol
31 Haap Hill
32 Zozni Cmiipla
33 Punta San Miguel, Haanc
34 Punta Tormenta, Cösecöla Iyat
35 Valle de Águila
36 Iifa Hamoiij Quih Iti Ihiij
37 Hahjöaacöl
38 Zozni Cacösxaj
39 Cyazim It
40 Punta Perla, Hant Cmaa Coocp
41 Hee Inoohcö
42 Tecomate, Hajháx
43 Arroyo Agua Dulce
44 Tecomate, rocky hill west of
45 Hast Iif
46 Central Valley, ca. 13 mi south of Tecomate
47 Pooj Iime
48 Ensenada Blanca (Vaporeta)

San Esteban

49 El Monumento, Pajii
50 Arroyo Limantour, terminus at coast
51 Pico Rojo
52 La Freidera, Hant Haaizj
53 San Pedro, Insóc Yaahit
54 Central Peak, Icámajoj Zaaj
55 Pico San Esteban
56 Cascajal, Cofteecöl Iifa
57 Pico del Sur

Dátil

58 Central mountain

Alcatraz

59 East side of island, flat
60 Guano flats, base of mountain
61 Peak of island, north portion of mountain
62 Southeast part of island, base of mountain

Nolasco

63 Cañón de Mellink, ridgetop
64 Cañón de Mellink, base
65 Cañón el Farito, rock landing at cove
66 Cañón el Farito, ridge crest above canyon
67 Agua Amarga
68 Cañón de la Guacamaya
69 Summit
70 Cala Güina

San Pedro Mártir

71 Northeast landing
72 High-elevation plateau
73 Summit
74 Small canyon at base of southeast-facing slope that leads to island summit
75 Guano miners' village

Sonora

76 Punta Santa Rosa

Canal del Infiernillo, Xepe Coosot

77 North end
78 South end

ISLA TIBURÓN; TAHEJÖC

Tiburón is the Spanish name for shark. Etymology of the Seri name is unknown.

Agua Dulce Valley. *See* Central Valley.

Agua Verde Bay. Botanist Peter J. Rempel used this name when he was on the island on 10 March 1937; it is probably Bahía Agua Dulce. *See* Bahía Agua Dulce.

Arroyo Agua Dulce. The broad arroyo that drains the Central Valley and reaches the Gulf at Tecomate (*see* "Bahía Agua Dulce"). We are applying this name to the arroyo in the vicinity of Tecomate, with collections from the arroyo farther south being attributed to the Central Valley. 29°10'30"N, 112°24'53"W.

Arroyo de la Cruz. *See* Ensenada de la Cruz.

Arroyo Sauzal (El Sauz, sometimes incorrectly spelled Sausal). See fig. 1.30. The major drainage of the south part of the island. At the coast, 28°47'53"N, 112°25'28"W. Ike Russell had a landing place on a little mesa about 2 km inland, 28°48'57"N, 112°25'28"W. It was a miserable landing place because it was rather short and the ground was riddled with rodent burrows that caused the wheels to sink into the loose soil, slowing takeoff speed. Jean Russell would go there only once, and Richard not after the second time. *Also see* XAPIJ.

Bahía Agua Dulce (Freshwater Bay). The bay at north end of the island, named "Freshwater Bay" in 1826 by Lt. Hardy (1829:293) for the waterhole (at Tecomate). 29°11'03"N, 112°25'12"W.

Bahía San Miguel. The sheltered cove on the north side of Punta San Miguel, at the southwest part of the Canal del Infiernillo. 28°58'40"N, 112°12'25"W.

The "bowl." A Paul Knight place-name high in the Sierra Kunkaak below his locality San Miguel Peak (see Paul Knight in part II, "Botanical Explorations"). The exact locality of the "bowl" is not clear, but based on his field notes and our experience, it is probably in the vicinity of 28°57'00"N, 112°19'28"W.

Canal del Infiernillo; XEPE COOSOT "narrow sea," XEPE HEEQUE "small sea." The shallow channel separating Tiburón from the mainland. The north end of the Canal is between Punta Perla on Tiburón and Punta Sargento on the mainland. The south end is defined by the narrow passage between Punta San Miguel of Tiburón and Punta Santa Rosa of the mainland. These four small peninsulas or points have mangrove esteros and dune habitats, which are more extensive on the mainland sides. The Canal del Infiernillo is a rich maritime ecosystem, recognized as a RAMSAR site (the Convention on Wetlands, signed in Ramsar, Iran, in 1971, an intergovernmental treaty for wetlands and their resources). The fisheries are discussed by Basurto (2005, 2006, 2008) and Bourillón-Moreno (2002); the ethnobiology by Felger and Moser (1973, 1985), Felger et al. (1980), and Torre-Cosío (2002); and sea turtles by Felger et al. (1976) and Felger & Moser (1987). North end at 29°16'05"N, 112°20'54"W, south end at 28°57'58"N, 112°10'57"W.

CAPXÖLIM "shattered" (for the extensive talus field on the north side of the mountain); (Cerro San Miguel, Sierra Kunkaak Segundo). Fig. 4.2. Large solitary peak northeast of Sierra Kunkaak Mayor. 28°58'52.27"N, 112°18'30.96"W, ca. 600+ m.

Caracol research station. Fig. 4.3. Northeast side and base of Sierra Kunkaak. A facility built in the 1970s by the Mexican Department of Wildlife as a research center (Ezcurra et al. 2002). In 1979 there was a resident biologist stationed there (see part II, "Botanical Explorations," Richard Felger, 24 May 2006, and Paul Knight). Later it was used primarily for bighorn hunting operations led by the Comcaac. 29°00'57"N, 112°17'37"W, 190 m.

Carrizo Canyon. This is a place-name given by Paul Knight, where he collected on 27 October 1979. We are certain this site is SOPC HAX because he collected *Arundo donax* (*Knight 1085*) there, and this is the only locality for it on the island. *See* SOPC HAX.

Central Valley (Agua Dulce Valley). Expansive valley between Sierra Kunkaak on the east and Sierra Menor on the west, running the majority of the length of the island. The valley has a principal arroyo with a network of smaller arroyos that drain northward to the shore, where it terminates at Tecomate. A prominent collecting site used by Richard in his earlier fieldwork is called "Central Valley, about 13 miles south of Tecomate (shore)." He

Figure 4.2. Capxölim, solitary peak in the northeast portion of the Sierra Kunkaak. BTW, 25 November 2006.

Figure 4.3. Caracol research station, Isla Tiburón. The main building is to the left, and the sleeping quarters are to the right. BTW, 24 May 2006.

would reach this remote locality with backcountry pilot Ike Russell (Felger 2002; for a collection of stories of Ike and his flying adventures, see Bowen 2002). Two of Richard's Central Valley collection sites are about 13 miles south of Tecomate, 29°04'21"N, 112°23'54"W; and southwest part of Central Valley, 28°57'59"N, 112°25'43"W.

Cerro Kunkaak. This place-name was used by Norman Scott, who climbed to the top of the island on 11 April 1978. We are unsure of his exact route to reach the higher elevations of the Sierra Kunkaak and are using the same location for Paul Knight's collections from the top of the Sierra Kunkaak. *See* San Miguel Peak.

Cerro San Miguel. This is a place-name used by Adrian Quijada-Mascareñas for collections he made from a mountain in the Sierra Kunkaak on 9 November 1990 and 9 March 1991: *Quijada 90T001* to *90T013* and *91T001* to *91T024*. Based on discussions and photos shown to us by Gil Gillenwater, Troy Gillenwater, and Adrian, we determined that Cerro San Miguel is Capxölim, the northeastern peak of the Sierra Kunkaak. *See* Capxölim.

Cójocam Iti Yaii "where the flee-ers were" (Sierra Kunkaak Mayor). Fig. 4.4. This peak is the second- or third-highest point on the island; the summit is about 0.5 km to the north of the island's summit, to which it is connected by an extensive ridge that links the majority of the highest elevations. 28°57'56"N, 112°19'50"W, 855 m.

Coralitos. Sheltered cove on the southeast side of the island, frequently used as a campsite by fishermen. 28°45'30"N, 112°19'20"W.

Cyazim It "dorsal fin base" (cyazim "have a dorsal fin"; it "its base"). Sand spit with a mangrove estero (estuary, hypersaline tidal marshes) at Punta Perla. Humberto said that some Comcaac used these mangroves to hide from the military. Seris often took refuge among mangroves when eluding the military (Felger & Moser 1985:24). 29°13'30"N, 112°17'29"W.

Ensenada Blanca (Tacata Inoohcö "Tacata Bay (Ensenada)," named for a Seri encampment or a group of Seris who lived there; Moser & Marlett 2005). An area

Figure 4.4. Cójocam Iti Yaii, Isla Tiburón. View is looking from the summit of the island to the north. BB, 26 October 2007.

with coastal dunes, stretching several kilometers northeastward from Punta Willard at the southwest corner of the island. Vicinity of 28°53'16"N, 112°33'36"W probably marks the southern area of Ensenada Blanca. Ivan Johnston's collection site "3 miles north of Punta Willard" is probably at the northern end of Ensenada Blanca (see part III, "Species Accounts," for *Bahiopsis chenopodina, Helianthus niveus,* and *Justicia californica*).

Ensenada Blanca (Vaporeta). See fig. 1.27. West side of island. Richard used the name "Ensenada Blanca" for a locality on the west side of Tiburón, which was reached with Ike Russell in his airplane. Numerous trips and collections were made at this locality. Ike's landing place was a short, narrow airfield that had a small gully near one end that had to be avoided on landing and takeoff. Ike and Richard used Ensenada Blanca as the name for this place based on local knowledge of the area, although Ike knew the place was also called Vaporeta, because years earlier a small steamboat was wrecked in the vicinity. In 2008, Ben and Richard learned that Ensenada Blanca is actually a site just north of Punta Willard (see above) and that the site frequented by Richard and Ike is Vaporeta. We retain Ensenada Blanca (Vaporeta) to accommodate Richard's collection labels. 28°59'06"N, 112°29'27"W.

Ensenada de la Cruz (La Pescadita, Arroyo de la Cruz). Fig. 4.5. Protected bay and associated arroyo at the southeastern shore, frequently used as an overnight fishing camp. Ruins of the abandoned Mexican Marine outpost persisted in 2009. 28°45'54"N, 112°21'05"W.

Ensenada del Perro (Ensenada de la Perra, Ensenada de los Perros, Ensenada de Perros). Fig. 4.6. Small bay on the

Figure 4.5. Ensenada de la Cruz, Isla Tiburón. SMT, 28 February 2009.

Figure 4.6. Ensenada del Perro, Isla Tiburón. BTW, 2 September 2009.

southeast shore, just to the north of El Monumento, often used as a temporary fishing camp. 28°46'54"N, 112°16'20"W.

Estero San Miguel. See fig. 1.21. Tidal wetland with mangroves at Punta San Miguel. 28°58'14"N, 112°12'09"W.

Freshwater Bay. *See* Bahía Agua Dulce.

Haap Hill. Southeast side of Central Valley, about 19 km southward from Tecomate. This collection site includes the north side and base of a large basaltic hill in the vicinity of the Seri camp Haap Caaizi Quih Yaii "tepary bean users' place," named for gathering of wild tepary beans (Felger & Moser 1985; Moser & Marlett 2005). Felger collections are 8 September 1974 with Cayetano Montaño, Hank Gunn, and Alexander Russell; and 11 December 1976 with Rosa Flores, Cathy Moser, and Alexander Russell. See *Phaseolus acutifolius* in "Species Accounts" (part III) for further discussion. 29°00'00"N, 112°21'54"W. Labels for specimens collected here give the locality as the vicinity of 28°57'20"N, 112°24.5'W, about 280 m, which is the locality of Satoocj, not the hill at the Haap camp site.

Hahjöaacöl "hahjö (*Lycium*, here probably *L. brevipes*) large." Northeast part of the island. The beach has ancient dunes and inland there is an extensive and gently sloping bajada. There were reports from several members in the Seri community that a prickly pear or *nopal* (*Opuntia* sp.) occurs here. No prickly pear is known from Tiburón, although *Opuntia engelmannii* is on Isla Alcatraz. In April 2008, despite substantial searching effort by Ben, Humberto, and Ben's friend Brigitte Marazzi, no *nopal* was encountered, which led Humberto to believe that its report here was incorrect. 29°08'37"N, 112°16'28"W.

Hant Hax "land water." Former Seri camp at the east base of Sierra Kunkaak, between the Sopc Hax waterhole and Zozni Cmiipla. Vicinity of 28°57'30"N, 112°15'30"W.

Hast Cacöla "high mountains." Fig. 4.7. Name for the large front range of the east side of the Sierra Kunkaak. From the mainland, such as at Punta Chueca, these peaks make up the prominent visible portion of the Sierra Kunkaak. 28°58'24"N, 112°17'04"W.

Figure 4.7. Hast Cacöla, Isla Tiburón, with Humberto. BTW, 7 April 2007.

HAST COOPOL "black hill." Fig. 4.8. A large, dark volcanic hill on the east side of Sierra Kunkaak, about 1 km south of the PAZJ HAX waterhole. 28°55'20"N, 112°16'16"W.

HAST IIF "the hill's nose" (Ojo de Puma). Hill on the northwest side of the Central Valley, bordering the main central arroyo, and one of the few rocky outcrops in the valley. There are two places with this name, one on Tiburón and one near Pozo Coyote, northeast of El Desemboque. Ojo de Puma and other Spanish names are neo-toponyms, ones that workers for the bighorn hunters are giving to various places—the Seris don't use those names (Stephen Marlett, personal communication, 2007). 29°05'53"N, 112°24'07"W.

HEE INOOHCÖ "bay of the antelope jackrabbit." Mountains at northeast coast of the island. Site of collections by Felger, Wilder, & Romero, 29 September 2007 (labels for these specimens have the place-name incorrectly given as HEE IMOCÖ). 29°13'10"N, 112°20'51"W.

IIFA HAMOIIJ QUIH ITI IHIIJ "point in which there is a circle." A flat point jutting into the Canal del Infiernillo. This was a gathering place for the Comcaac before and after battles waged on the mainland against Spaniards and Mexicans—where the traditional victory dance was held (Moser & Marlett 2005:404). A large circle for the dance was drawn into the soft ground of the flat. 29°06'20"N, 112°14'34"W.

El Monumento. Fig. 4.9. The name denotes the tall rock pillar, separated by a low and short land bridge from the rest of the island, at the exact southeast corner of the island. 28°46'05"N, 112°15'11"W.

Figure 4.8. Hast Coopol, Isla Tiburón. BTW, 15 September 2007.

Figure 4.9. El Monumento, Isla Tiburón. BTW, 30 January 2008.

Palo Fierro. Place-name used on many of Richard's and some of Ben's collections for the airfield and adjacent areas at Punta Tormenta. *See* Punta Tormenta.

Pazj Hax "Pazj (etymology unknown) water" (Tinaja Anita). See fig. 1.12. Waterhole at the east base of Sierra Kunkaak and one of the principal year-round water sources on the island. This locality is associated with one of the most tragic events in Comcaac history. According to oral history, in the late 1800s the San Esteban people were rounded up and brought to Pazj Hax by soldiers dispatched by the governor of Sonora in retribution for raids made by other Seris. A stone corral, still present, was constructed by the soldiers, and it is said this is where many of the San Esteban people were killed, and the rest taken away by ship and killed later or relocated inland (Bowen 2000). 28°56'25"N, 112°16'46"W.

La Pescadita. *See* Ensenada de la Cruz.

Pooj Iime, etymology of Pooj is unknown, iime "home." See fig. 1.28. Large canyon at the northwest side of the island. This is the locality given for the elusive *Brahea* palm (see José Juan Moreno, in part II, "Botanical Explorations"). 29°02'31"N, 112°30'16"W.

Punta Mala. *See* Punta Perla.

Punta Narragansett; Hast Caacoj Quih Iyat "large mountain point." The name "Narragansett Point" was bestowed in 1873 by the U.S. Navy Hydrographic Survey for a place on the east coast of the island south of Punta San Miguel. The name derives from the USS *Narragansett* of the U.S. Navy, a 186-foot wooden-hull steam-powered sloop (Bowen 2000). (Narragansett is a place in Rhode Island and a New England tribe of the Algonquian language group.) From 1873 to 1875 the ship ran surveys in the Gulf of California under the command of George E. Dewey (Bowen 2000). In March 1875, while at Punta Monumento at the southeast tip of Tiburón, friendly relations were established with some Seris, and while several of them were onboard the *Narragansett* the first known photos of Seris were taken—one of the two surviving photos was restored and published by Tom Bowen (2000:215). 28°52'17"N, 112°14'23"W.

Punta Perla; Hant Cmaa Coocp "recently formed land" (Punta Mala). Fig. 4.10. Sand spit and associated mangrove estero at the northeast corner of island. Punta Perla marks the northwestern boundary of the Canal del Infiernillo. Sand spit, 29°14'19"N, 112°16'38"W; mangrove estero, 29°13'20"N, 112°17'36"W; rocky hill just inland, 29°12'39"N, 112°17'14"W.

Figure 4.10. Punta Perla, Isla Tiburón, looking southeast, with mainland in background. BTW, 9 November 2008.

Punta San Miguel; HAANC "the name is derived historically from 'HAAN' 'smooth Pacific Venus clam'" [*Chionista fluctifraga*] (Moser & Marlett 2005:306). The sandy and sea-cobble point that shelters Estero San Miguel and along with Punta Santa Rosa of the Sonoran mainland marks the southern end of the Canal del Infiernillo. 28°58'38"N, 112°11'33"W.

Punta Tormenta; CÖSECÖLA IYAT, from ICÖS CACÖLA "large spines" (Palo Fierro landing field, Punta Tortuga). Fig. 4.11. The principal entrance to Tiburón, across the Canal del Infiernillo from Punta Chueca. A small Mexican marine station there is the only active base on the island. The primary role is to discourage narcotics traffic and restrict access of non-Seris to the island. Punta Tormenta also serves as the parking area for the Seris' sport-utility vehicles used in bighorn hunts, often with four or more vehicles present at a time in the 2000s. The cars are brought to the island on two pangas tied together and covered with planks—the cars are driven onto the pangas at low tide and, after the tide rises, carefully rafted across the Canal and disembarked at Punta Tormenta. The long and excellent landing field, called Palo Fierro, is on one of the sparsely vegetated, low mesas in the *Frankenia palmeri* zone. Ike Russell

and Richard often landed and camped here (before the marine station was installed). 29°01'11"N, 112°11'43"W.

Punta Tortuga. *See* Punta Tormenta.

Punta Willard. Southwest corner of island, named for Alexander Willard, U.S. Consul in Guaymas from 1868 to 1891 (Bowen 2000). 28°52'35"N, 112°34'35"W. *Also see* Ensenada Blanca.

San Miguel Peak (Cerro Kunkaak). A Paul Knight collecting locality, the highest point he reached on the island on 26 October 1979 (*see* Cerro Kunkaak, and Paul Knight in part II, "Botanical Explorations"). 28°57'10"N, 112°19'30"W.

SATOOCJ, from HÁSATOJ COOCJ "two hills." Large hill at the southern end of the Central Valley. 28°57'09"N, 112°24'23"W.

El Sauz (Sauzal). Collecting locality used by Reid Moran. *See* Arroyo Sauzal.

Figure 4.11. The marine station at Punta Tormenta, Isla Tiburón. BTW, 24 May 2006.

Sauzal. *See* XAPIJ, and Arroyo Sauzal.

Sauzal waterhole. *See* XAPIJ. *Sauzal* denotes a place of willows or *sauces*, here the narrowleaf willow, *Salix exigua*.

Sierra Caracol. A locality name used by Paul Knight and likely the mountains in the vicinity of the Caracol research station. This place-name is not clearly described in his field notes, and thus the location is based in part on assumption. Vicinity of 29°01'31"N, 112°18'58"W.

Sierra Kunkaak. See fig. 1.24. This name is attributed to the entire eastern mountain range of the island—an extensive series of mountains including connected and isolated peaks. The Comcaac do not have an equivalent name for the range and instead name individual peaks. Classifying an entire mountain range does not make much sense in the Comcaac worldview (Stephen Marlett, personal communication, 2007). The name Sierra Kunkaak is on McGee's (1898) map and is based on pre-conventionalized spelling but is accepted by the Seris and is the term used on all maps and in Spanish to identify the eastern mountain range. The highest point is an unnamed peak, one peak to the south of CÓJOCAM ITI YAII. Upper ridge, between CÓJOCAM ITI YAII and

the summit, 28°57'49"N, 112°19'40"W, 825 m; summit, 28°57'43"N, 112°19'34"W, 885 m (see fig. 1.14).

Sierra Kunkaak Mayor. *See* CÓJOCAM ITI YAII.

Sierra Kunkaak Segundo. This is a name used by Ben on labels for collections from his first trip to the island before learning the name used by the Comcaac for this mountain. *See* CAPXÖLIM.

Sierra Menor. Fig. 4.12. The western mountain range of the island, with the highest points in the southern part of the range. Far fewer collections have been made on this remote and arid western range in comparison to all other locations on the island. Richard's collecting locality on east-facing slopes, reached from Ike Russell's landing place in the Central Valley. Vicinity of 28°57'56"N, 112°26'24"W.

SIIMEN HAX "Kestrel's water." Fig. 4.13. A water source in a narrow stretch of a main canyon in the northern portion of the Sierra Kunkaak. Archeological evidence here is scarce compared to XAPIJ and PAZJ HAX. The SIIMEN HAX tinaja was likely remote from main Comcaac camps on the island and dries up in times of severe drought, as seen in October 2007. 28°58'47"N, 112°19'31"W.

Figure 4.12. Sierra Menor, Isla Tiburón, looking westward from summit of Sierra Kunkaak, with Isla Ángel de la Guarda in background. BTW, 26 October 2007.

Figure 4.13. Siimen Hax, Isla Tiburón. BTW, 24 November 2006.

SOPC HAX, etymology of SOPC unknown, HAX "water" (Carrizo Canyon). Fig. 4.14. Waterhole at the eastern base of the Sierra Kunkaak. There is a dense stand of carrizo, notable as the only *Arundo donax* on the island (as opposed to the native reedgrass or carrizo, *Phragmites australis*). When Paul Knight visited this waterhole in 1979, which he called Carrizo Canyon, he made the following observation, "About ⅓ mile up the canyon I noticed a large stand of Carrizo covering the canyon bottom. As we approached Chapo [Francisco "Chapo" Barnett, his Seri guide] led us along a well-trodden trail that had been cut through the reeds. Finally after crossing through this maze the view opened up into a large pool of water perhaps 30 feet across and over 6 feet deep

Figure 4.14. Sopc Hax, Isla Tiburón. The waterhole is in the canyon bottom in foreground. RSF, April 1963.

in the center." However, neither Richard in April 1963 nor Ben in May 2007 saw such a large pool. 28°57'15"N, 112°16'39"W.

Tecomate; Hajháx "any water." See fig. 1.29. Historical Comcaac camp or village on the north shore and more recently the site of a now abandoned small Mexican marine station. The name is based on the freshwater seep just inland from the coast. This water supported the largest and most heavily used Comcaac settlement on the island. Historical camp of the Comcaac, 29°11'12"N, 112°24'58"W; rocky hill to the west of the camp, 29°11'09"N, 112°26'23"W.

I had read about the famous waterhole at Tecomate at the north end of Tiburón. The whole bay is called Agua Dulce, as is the wide central valley, which is perhaps twenty kilometers long. The waterhole is shown on the maps, even the oldest ones. The water was so dependable, and right at the shore, that the Seri camp there was essentially permanent. When the Seris told us about plants on the north end of island, it was often in terms of the distance or direction from Tecomate. So when Ike said it's easy to land there I was really looking forward to seeing this famous waterhole. We landed at the airfield, one of the better ones, and walked over to the waterhole. We took our canteens to fill, came to a broad depression, and walked down to the bottom. I don't know what I was expecting, but here was a little mud puddle totally covered by a long-dead pelican. (Felger 2000b:527–528)

Tinaja Anita. *See* Pazj Hax. Name given by WJ Mc-Gee for this site, where he obtained much-needed water shortly after dusk on 19 December 1895. "On we went, and at last reached a little patch of canes (a carrisalito) with water slipping over the rocks and lodged in nooks and potholes. How we drank!" (Fontana & Fontana 2000:85). "A specific tribute to Anita Newcomb," WJ's wife (McGee 1898:19).

Tordillitos. Point on the south coast of the island between Punta Willard and Arroyo Sauzal. This was a place used by Richard when going to and from San Esteban with Mexican fishermen from Bahía Kino, who called the place Tordillitos. A point called La Tordilla occurs several km to the west, just below Punta Willard. 28°49'26"N, 112°29'36"W.

Valle de Águila "valley of the osprey." A neo-toponym for an area of the extensive east-facing bajada on the northeast side of the island. Vicinity of 29°05'28"N, 112°15'15"W.

Vaporeta. *See* Ensenada Blanca (Vaporeta).

La Viga. Southeast part of island, on the coast between El Monumento and Coralitos. This is the part of Tiburón that is closest to Cholludo and Dátil. 28°45'10"N, 112°18'18"W.

Willard's Point. *See* Punta Willard.

Xapij "reedgrass"; Sauzal waterhole. See fig. 1.31. The largest freshwater spring on the island, about 7 km inland from where Arroyo Sauzal terminates at the coast in the southern part of the island. Reedgrass, *Phragmites australis*, is locally abundant at this waterhole. Sauzal connotes a place of *sauces*, or willows (here the narrowleaf willow, *Salix exigua*). 28°50'36"N, 112°23'54"W.

Xepe Coosot. *See* Canal del Infiernillo.

Xepe Heeque. *See* Canal del Infiernillo.

Zoojapa (meaning unknown, name provided by Humberto, April 2008). Former Seri camp in the northeastern foothills of the Sierra Kunkaak, below a prominent peak in the foothills called Hast Yeen. During the late 1990s and 2000s this place was occasionally used as a base camp for hunting bighorn sheep into the central part of the Sierra Kunkaak. 29°00'07"N, 112°17'14"W.

Zozni Cacösxaj "zozni (etymology unknown) large." Fig. 4.15. Northeast part of island, current and historical Seri beach camp at the north tip of the Canal del Infiernillo. In April 2008, this beach was used by Seri families as a camp and work area for harvesting *callo de hacha* (*Atrina tuberculosa*, *A. maura*, and/or *Pinna rugosa*; Basurto 2005) from the northern part of the Canal. 29°11'40"N, 112°15'45"W.

Zozni Cmiipla "zozni bad." Fig. 4.16. Historical Seri camp near the base of Punta San Miguel on the east side of the island. 28°58'08"N, 112°12'52"W.

Figure 4.15. Zozni Cacösxaj, Isla Tiburón, with Sierra Seri on mainland in background. BM, 6 April 2008.

Figure 4.16. Zozni Cmiipla, Isla Tiburón: ocotillo frames of traditional Seri shelters. BTW, 1 January 2006.

ISLA SAN ESTEBAN; Cofteecöl

The indigenous name is evidently based on coft (plural for cof) for the *Bonellia macrocarpa (san juanico)* tree. Coof is the piebald (San Esteban) chuckwalla (*Sauromalus varius*). However, the cof tree does not occur on San Esteban today, while the piebald chuckwalla occurs in great numbers on the island. "Seri consultants have always and consistently volunteered the etymology based on the *san juanico* tree and have claimed that is what they heard from their ancestors. They are undeterred by the lack of a present-day existence of the tree on the island" (Stephen Marlett, personal communication, 2011).

Much of the information for the island is from "Informal Gazetteer of San Esteban Island" in Tom Bowen's *Unknown Island* (2000:457–463).

Arroyo Limantour (incorrectly sometimes as Limansur). Fig. 4.17. The main drainage of San Esteban, extending from the interior mountain drainages in the northwestern part of the island, where the arroyo begins, to the eastern shore, where it terminates. The lower elevations of this arroyo system form a substantial valley. "The arroyo was likely named around the turn of the century for José Yves Limantour the younger (1854–1935), who served as Secretary of the Treasury under President Porfirio Díaz from 1893 to 1911" (Bowen 2000:462). Most visitors and researchers going to the island went to Arroyo Limantour, which offers the easiest landfall and access to the island. Terminus at coast, 28°41'15"N, 112°32'54"W; ca. 1.5 km inland, 28°41'46"N, 112°33'34"W; ca. 5 km inland, 28°42'04"N, 112°35'21"W.

(A)

(B)

Figure 4.17. Arroyo Limantour, Isla San Esteban. (A) The terminus of the arroyo at the southeast coast. RSF, 1960s. (B) Interior, ca. 3 km inland, looking to the southeast. BTW, 8 March 2007.

Canyon at southwest corner of island. A collecting locality that reflects explorations in the canyons on the steep seaward side of the mountains south of the upper reaches of Arroyo Limantour and is reached from the base of Cascajal. The locality used here is an approximation for the area visited by various botanists. 28°40'32"N, 112°35'53"W.

Cascajal "pebble beach"; COFTEECÖL IIFA "peninsula of Isla San Esteban." A low shingle-rock spit about 0.5 km long on the southwest corner of the island. The point has been built by opposing currents that run along the southern and western sides of the island. It is a prominent area where sea lions gather. 28°40'19"N, 112°36'03"W.

Central peak; ICÁMAJOJ ZAAJ "cave where agaves are cooked." Fig. 4.18. "This high and massive mountain in the center of San Esteban dominates the interior of the island" (Bowen 2000:461). 28°42'29"N, 112°34'32"W.

East-central side of island. *See* Arroyo Limantour.

La Freidera "frying place"; HANT HAAIZJ "mashed land, i.e., mud." The large shingle beach on the north coast that constitutes the only major break in the sea cliffs on this side of the island. The name "Freidera" originates from Bahía Kino fishermen because it was a place where sea lion hunters brought their catch to render the blubber into oil. 28°43'45"N, 112°34'21"W.

Limantour. *See* Arroyo Limantour.

El Monumento; PAJII "flint." Prominent landmark at the southeast corner of island, at the south end of Playa Limantour. It is a tall outcrop of rock separated from the rest of the island by a low land bridge. 28°41'11"N, 112°32'44"W.

North side of island, narrow rocky canyon behind large cove. One of Richard's collecting areas, inland from La Freidera. 28°43'30.73"N, 112°34'11.41"W.

Peak near southeast corner of island. *See* Pico del Sur.

Pico del Sur. Peak and highest point at the southeastern part of the island. Reid Moran made collections from this part of the island. 28°40'31"N, 112°34'13"W.

Pico Rojo. "This steep-sided mountain dominates the northeastern part of the island. Its talus slopes lend it a distinctive ruddy hue which sometimes turns a brilliant deep red in the fading rays of sunset" (Bowen 2000:461). Botanically explored by Reid Moran and also by Richard. 28°42'37"N, 112°33'36"W.

Pico San Esteban. The south-central peak of the island, the tall mountain in the southwestern part of the island, the highest peak on the island. Reid Moran and Richard collected plants on this peak. 28°40'57"N, 112°35'30"W, 550 m.

Figure 4.18. Central Peak, Isla San Esteban. View is looking to the southeast. BTW, 8 March 2007.

San Pedro; Insóc Yaahit "where the black skipjack is fished." Rocky cove at the northwest side of island. This "is the summer camp of Mexican and Seri fishermen and is the only sheltered beach on the west side of the island" (Bowen 2000:460). 28°43'10"N, 112°36'42"W.

South-Central Peak. *See* Pico San Esteban.

Southeast corner of island. Arroyo that drains the Pico del Sur area of the southeast corner of the island into Arroyo Limantour. 28°41'07.36"N, 112°34'23.04"W.

ISLA DÁTIL; Hastaacoj; TURNERS ISLAND

"Large hill" is the literal translation of the Seri name. Dátil is the date and the palm (*Phoenix dactylifera*) and is also applied to fleshy-fruited *Yucca* species. Here it refers to the small century plant (*Agave subsimplex*). According to McGee (1898:17) the name of the island is "probably applied in honor of Rear Admiral Thomas Turner, U.S.N. by the Hydrographic Office, U.S.N."

There are few named localities on Isla Dátil, and none that we are aware of that are used by Seris, Mexicans, or Americans. Thus, we identify the location of collections by using geographic descriptions (e.g., northeast side of island). Additionally, there are a few place-names we made up for this small yet diverse island.

Central Mountain. One of the highest, or perhaps the highest, mountain on the island, at about the middle of the island. 28°43'18"N, 112°17'27"W.

East-central side, east-west trending canyon. Fig. 4.19. Canyon between the "northeast side of island" and the central mountain. Ben, Humberto, and others made collections here in late February 2009. 28°43'27"N, 112°17'32"W.

East side of island. Collection locality on the north slope of the central mountain. 28°43'23"N, 112°17'23"W.

Northeast side of island. The first large canyon on the north side of the island when heading from north to south. The canyon lies on the north side of a tall mountain ridge. A cobble-rock beach on the east side affords easy access from small boats. This is the most commonly visited locality on the island. Canyon, 28°43'40"N,

Figure 4.19. Isla Dátil, east-central side, east-west trending canyon. BTW, 27 February 2009.

112°17'36"W; summit above the canyon, 28°43'32"N, 112°17'34"W.

Northwest side of island. The west end of the canyon at the northwest side of the island. 28°43'37"N, 112°17'41"W.

Rattlesnake Canyon. Named for the place where Charles Lowe found small diamondback rattlesnakes, *Crotalus atrox*, which at first were thought to be dwarfed. *See* Southeast side of island.

Rescue Beach. First accessible beach on the northeast side of the island. This locality was named by us after a tumultuous night at sea on 7 December 2007 when returning from Isla San Pedro Mártir. 28°44'04"N, 112°17'50"W.

Southeast side of island (Rattlesnake Canyon). Steep and narrow canyon at the southeast corner of the island. 28°42'52"N, 112°17'11"W.

ISLA CHOLLUDO; Hastisil

The Seri name derives from HAST "hill or mountain" and ISIL "shoulder joint" or "pectoral fin." On some older maps the island is called Isla Roca la Foca, Isla Roca Foca, or Seal Island (Isla Lobos). This is the smallest island in the flora area and does not have place-names. Only the north-facing side is accessible and vegetated (with an extremely dense cactus forest). Steep sea cliffs at the south side of the isle are occupied by blue-footed boobies and other seabirds. We identify two sites on the island.

Default location. Fig. 4.20. We have assigned 28°44'17"N, 112°18'18"W, on the north-facing slope, for all historical collections that do not specify precise place of collection.

Peak of island. The highest point on the island is 61 m and is located toward the south end of the island. 28°44'14.42"N, 112°18'18.76"W.

Figure 4.20. Isla Cholludo. Jean Russell in middle. RSF, April 1968.

ISLA ALCATRAZ; SOOSNI

The etymology of SOOSNI is unknown. As with Dátil and Cholludo, there are no formal place-names on this small island.

East base of mountain. A dense patch of cardón (*Pachycereus pringlei*), cholla (*Cylindropuntia fulgida*), and other species occur on the lower portions of the east side of the mountain. 28°48'39"N, 111°58'10"W.

East side of island, flat. Fig. 4.21. Half of the island is composed of a flat area of sandy soils surrounding a low central area dominated by *Allenrolfea occidentalis*. Collections from the low, flat eastern part of the island are in the vicinity of 28°48'49"N, 111°57'57"W.

Guano flats, base of mountain. At the base of the mountain, stretching between both shores below the cactus patch of the "east base of the mountain," is a whitewashed area where the guano runoff from the mountain has created a nearly barren zone with relatively few perennial plants. 28°48'45"N, 111°58'03"W.

Northeast end of island. On the corner of the island closest to the town of Bahía Kino is a sandy beach area with the abandoned lighthouse (*faro*). This is the location of a well-established population of puncture vine (*Tribulus terrestris*). 28°48'57"N, 111°57'53"W.

Northwest side. Area where the flat begins to grade into the lower portions of the mountain. A halophytic community on saline-wet soil occurs here just inland from the shore. 28°48'49"N, 111°58'03"W.

Peak of island. Fig. 4.22. Highest point on the island, at the southwest part of the mountain. 28°48'35.86"N, 111°58'13.87"W, ca. 165 m.

South-central part of flat. A rocky pebble beach that extends inland. There is a small shrine in this area. 28°48'45"N, 111°57'53"W.

Southeast part of island, base of mountain. The eastern base of the mountain at the coast. 28°48'34"N, 111°57'58"W.

Figure 4.21. Isla Alcatraz, flat on the east side, looking east to village of Bahía Kino on mainland. RSF, 1960s.

Figure 4.22. Isla Alcatraz, looking westward from *Allenrolfea* flat, the highest peak on the right. BTW, 16 September 2007.

ISLA PATOS; HAST OTIIPA

Etymology of OTÍIPA is unknown. This is a small, mostly flat guano island ringed by sea bluffs roughly 3–5 m high. The major topographic feature is a hill on the north side of the island. The vegetation is scattered across the island, with the most prolific plant being *Atriplex barclayana*. We recognize several places on the island.

Guano-mining shelters. Near the center of the island are the ruins of structures from guano mining probably in the 1940s (see Osorio-Tafall 1944, 1946). 29°16'13"N, 112°27'33"W.

North side. 29°16'26"N, 112°27'41"W.

Peak of hill. Fig. 4.23. The highest point on the island, a cone-shaped hill where there is an abandoned lighthouse. The major population of cardons (*Pachycereus pringlei*) occurs on the slopes near the summit. 29°16'19"N, 112°27'36"W, ca. 70 m.

South base of hill. The flat area at the southern base of the hill. In 2007 we found a few sinitas (*Lophocereus schottii*) here. 29°16'15"N, 112°27'33"W.

Southeast side. 29°16'07"N, 112°27'23"W.

Figure 4.23. Isla Patos, looking northward with lighthouse. Note the remains of guano mining in the foreground. BTW, 30 September 2007.

South edge. 29°16'06"N, 112°27'33"W.

Southwest side. 29°16'03"N, 112°27'41"W.

ISLA SAN PEDRO MÁRTIR; Iicj Icooz

The indigenous name may mean "stolen sand," perhaps because there is one main point where one can enter the island, and the rest is sea cliffs—the one easy point of ingress is the IICJ ICOOZ.

From Steve Marlett, 18 September 2008: "I'm interested to hear about the explanation for the name of San Pedro Mártir. I can believe it has something to do with 'sand' (as that is clear) and something to do with 'steal,' but the actual translation must be something different than 'stolen sand' because the wrong verb form is used—and that's why the etymology gets tricky. If it were 'stolen,' it would be HACOOZ—whereas the form ICOOZ means 'where she/he stole it' (or something along those lines)—it's not even passive. Some of the older names have words that mangle the true forms, and so they become unclear."

Cliffs on north side. The lighthouse, or faro, is located on this part of the island. 28°23'15"N, 112°18'27"W.

Guano miners' village. See fig. 1.38. The ruins of the village used by the guano miners are at the southeast part of the island. Most of the white guano-washed rock walls are still standing. The ruins are primarily used by the nesting blue-footed and brown boobies. This is also the area where the occasional visiting researchers camp. Next to the village site an area has been cleared as a helicopter landing site and a large water-storage tank installed to supply emergency water in case of fishermen being stranded in a storm. The sea cliffs are not as abrupt just south of the village, and this is the easiest point of ingress to the island. 28°22'26"N, 112°18'08"W.

High elevation plateau. At about the middle of the island is a large, gently sloping plateau slightly below the summit of the island. This is the location of a permanent quadrat site established by Richard and Ben in 2007. 28°22'49"N, 112°18'29"W, 275 m.

Northeast landing. Small inlet on the northeast side of the island that affords an ingress to the island; however, it is not the easiest access point. There are substantial rock walls built by the guano miners, including walls for a substantial platform. 28°22'58"N, 112°17'50"W.

Small canyon at base of southeast-facing slope leading to the summit. Fig. 4.24. This canyon runs southwest down to the northeast along the majority of the upper part of the island. Top of canyon, 28°22'40"N, 112°18'18"W.

Steep north-facing slope above landing at northeast side. Locality used by botanists when collecting on the northeast side of the island after entering the island from

Figure 4.24. Isla San Pedro Mártir, upper portion of small canyon at base of southeast-facing slope that leads to the summit. BTW, 6 December 2007.

the northeast landing. Bottom of "Small canyon at base of southeast-facing slope," 28°22'58"N, 112°18'02"W.

Summit. The highest point on the island, at about the middle of the island. The ridge and slopes below rise relatively gradually in a dome-like fashion to this high point. 28°22'49"N, 112°18'23"W, 300 m.

ISLA SAN PEDRO NOLASCO; HAST HEEPNI IT IIHOM

The Seri name translates as "mountain where there are iguanas."

Agua Amarga. Fig. 4.25. This tiny freshwater seep emerges from near the base of high cliffs at about 5 m above high tide line on the east-central side of the island. This is the only freshwater source on the island and has sustained fishermen and others in times of misfortune (Juan Pablo Gallo-Reynoso, personal communication, 2008). The water is highly alkaline, as indicated by the name *amargo* "bitter." It supports the only wetland plant, *Cyperus elegans*, on the island. 27°58'06.71"N, 111°22'22.87"W.

Cala Güina. *Cala* is a Spanish term for a place to set an anchor, and *güina* is a chigger, and applied to this cove because it is very small and narrow like a güina. The shorthand name Güina is often used by local fishermen. Cala Güina has a short and narrow cobble-rock "beach." This small cove, at the southeast side of the island, is one of only two landfalls along the east side of the island. It is not suitable for camping, as there is no dry place at high tide, but during nights of calm seas fishermen often set anchor here and sleep in their pangas. Cliffs rise directly above the high tide line. It is possible to ascend the cliffs, and a number of biologists have done so, although it is dangerous and not recommended. See part III, "Botanical Explorations," Felger, 26 November 1963. 27°57'33"N, 111°22'07"W.

Cañón de la Guacamaya (Arroyo de las Guacamayas). Canyon at the west-central side of the island. Juan Pablo Gallo-Reynoso named the canyon for the guacamayas (military macaw, *Ara militaris*) that arrived on the

Figure 4.25. Agua Amarga, Isla San Pedro Nolasco. *Cyperus elegans* at seep in center. BTW, 3 February 2008.

island in 2000 (Gallo-Reynoso et al. 2012). 27°58'00"N, 111°23'01"W in the lower part of the canyon.

Cañón de Mellink. Fig. 4.26. Canyon at the west-central side of the island. It is the next canyon north of Cañón de la Guacamaya and named for Dr. Eric Mellink, noted expert on the vertebrate fauna of northwestern Mexico. Like Cañón el Farito, access is gained by jumping from a small boat onto slippery rocks and scrambling up guano-covered rocks. From there the ascent up the canyon is not difficult. Base, near the shore, 27°58'09"N, 111°23'07"W, 10 m; mid-elevation, 27°58'19"N, 111°23'06"W, 110 m; ridgetop, 27°58'27"N, 111°23'05"W, 263 m.

Cañón el Farito. Fig. 4.27. Canyon at the northeast side above the most accessible landfall on the island, which is a small, rock-walled cove or inlet. To gain access to the

landfall, a small boat is brought up close to rocks and you jump onto rocks, being careful of the slippery seaweeds and sea lion excrement. The canyon is steep and rocky but easily traversed and leads to the top of the island. This is the most frequented collecting locality over the last century. Fishermen and researchers have camped here. Nowadays fishermen and tourists occasionally come ashore, sometimes illegally camping and leaving trash, feces, and toilet paper.

The name derives from a position light. In Spanish these are technically called *balizas*, but people refer to them as *faritos*. The original farito was a lighthouse beacon on a scaffold at the southeast corner of the inlet, or Ensenada el Farito. The beacon was installed in the 1980s to serve the tankers plying the sea between Guaymas and Puerto Libertad and the ferry between Guaymas and Santa Rosalía. The farito was destroyed

Figure 4.26. Isla San Pedro Nolasco, Cañón de Mellink. BTW, 29 September 2008.

Figure 4.27. Isla San Pedro Nolasco, Cañón el Farito. BTW, 3 February 2008.

by Hurricane Juliette in late September 2001 and was replaced with a little provisional one that works with solar cells. However, it was destroyed by Tropical Storm Jimena in early September 2009. A large metal conservation sign was installed on the rock landing site by CONANP, but it had rusted and was sunk by Hurricane Juliette with waves about 5 m high. The sign was recovered from the sea, and a new one is scheduled to be installed by CONANP (Juan Pablo Gallo, personal communication, 2012).

The locality is called Cañón el Faro on some herbarium labels. Rock landing at the cove, 27°58'23.46"N, 111°22'38.91"W, several meters above sea level; mid-elevation, 27°58'16.86"N, 111°22'43.08"W, about 215 m; ridge crest above the canyon, 27°58'07"N, 111°22'47"W, about 280 m.

East-central side of island. Collections were made here by Richard when he traversed the island from Cala Güina to Cañón el Farito on 26 January 1963. Vicinity of 27°57'53"N, 111°22'33"W.

Southeast side, talus area. Collection locality of Mike Rosenzweig and Ron Pulliam in 1974, above Cala Güina. Vicinity of 27°57'32"N, 111°22'21"W.

Summit of island. See fig. 1.41. The highest point on the island near the center and is accessible by following the precipitous ridge crest. Active erosion is evident. 111°22'38"W, 111°22'38"W, ca. 315 m.

SONORA MAINLAND

Bahía de Kino, Bahía Kino. Coastal fishing and vacation town, the largest population center in the vicinity of the Midriff Islands. It was named for the intrepid Jesuit priest-explorer Padre Eusebio Kino. Kino Viejo, 28°49'23"N, 111°56'23"W; Kino Nuevo, 28°51'29"N, 112°00'47"W.

Bahía San Pedro. The closest safe anchorage place from Isla San Pedro Nolasco, about 20 km northwest of San

Carlos and 15 km northeast of Isla San Pedro Nolasco. Local fishermen and sailors called it Ensenada Grande. 28°03'17"N, 111°14'39"W.

La Balandrona (Cañón la Balandrona). Deep riparian canyon with a very rich flora on the north side of the Sierra el Aguaje. Canyon mouth in the vicinity of 28°06'N, 111°04'17"W, about 210 m.

Caborca; XAPIJ ICAPCOJ "carrizo stems." City in northwestern Sonora. 30°43'N, 112°09'W.

Canal del Infiernillo. *See* Tiburón, Canal del Infiernillo.

Cañón del Nacapule. This desert-bounded canyon supports a rich tropical-derived flora with a history of botanical collections spanning more than a century (Felger 1999). The canyon is 6 km north of the beach resorts of San Carlos and about 20 km northwest of Guaymas. About 1.2 km long, the canyon slices into the southeastern flank of the Sierra el Aguaje. *Nacapul* is one of the native figs, *Ficus pertusa*. Canyon entrance at 28°00'56"N, 111°02'58"W, ca. 150 m.

Cerro Pelón; HAST YAXAXOJ "fat-pouch mountain." YAXAXOJ refers to the fat pouch on each side of the abdomen (Moser & Marlett, 2010); *pelón* means bald. Granitic mountain about 9 km northeast of El Desemboque with a partially barren west-facing slope. 29°32'39"N, 112°20'24"W.

Cerros Anacoretas. A series of hills or low mountains in northwestern Sonora, southwest of Caborca.

Cerro Tepopa; PAJQUEEME, variant is XPAJQUEEME, etymology unknown. Isolated mountain at Punta Tepopa that juts into the Gulf south of El Desemboque and north of Estero Sargento. The peak at 29°21'40"N, 112°23'57"W, ca. 590 m.

El Desemboque del Río de la Concepción. This coastal town is north of Puerto Libertad and Puerto Lobos, north of the present-day Seri region. It is distinguished in the text from the Seri village by its full name. 30°34'06"N, 113°00'31"W.

El Desemboque del Río San Ignacio (El Desemboque); HAXÖL IIHOM "place of the clams" (HAXÖL is a Venus clam *Leukoma grata* [*Protothaca grata*]). See fig. 1.16. Northern and oldest of the two Seri villages. References to El Desemboque in the text refer to this Seri village, at the mouth of the usually dry Río San Ignacio. 29°30'15"N, 112°23'46"W.

Estero Sargento; XTAASI HANT IIJÖC TAZO ITI COOM "one side of Estero Sargento" (literally, the dry beach of the estero). The largest estero in the Seri region, located south of Cerro Tepopa at the north end of the Canal del Infiernillo. Sargent's Point was named by Lt. Hardy in 1826 (Hardy 1829:293). 29°19'43"N, 112°19'08"W.

Guaymas; HASOJ IYAT "river edge." City at the southern limit of the Sonoran Desert, established in Spanish colonial times on a well-protected harbor surrounded by desert mountains. The Guaymas people were the southern group of Seris who settled in the multi-ethnic town of Belem in the Yaqui region (Moser 1963; Felger & Moser 1985; Pérez de Ribas 1645). 27°56'N, 110°54'W.

Hermosillo; HEZITMISOJ "real town." The capital of Sonora, about 100 km inland from Bahía Kino. 29°04'N, 110°58'W.

Islas Mellizas. Small cardón-forested islands in Guaymas Bay (see Turner et al. 2003). Westernmost island: 27°53'57"N, 110°54'03"W. Easternmost island: 27°54'02"N, 110°53'55"W.

Pico Johnson; HASTAACOJ CACÖSXAJ "tall large mountain." Named for WJ McGee's topographer, Willard T. Johnson, who ascended the peak as a survey station on 7 and 8 December 1895 (Bowen 2000; McGee 1898). It is the highest peak in the Sierra Seri. Summit 29°12'40"N, 112°07'24"W, about 1000 m.

Prescott College Field Station, Bahía Kino. The Prescott College Kino Bay Center for Cultural and Ecological Studies is one of a few permanent academic field stations on the coast of Sonora. The station has accommodated researchers and has generously hosted and assisted us in our studies in the region. 28°51'32"N, 112°01'15"W.

Puerto Libertad; XPANOHÁX "freshwater in the sea" (from XEPE ANO HAX "freshwater in sea water"). XPANOHÁX at Puerto Libertad is a spring where freshwater would flow out of the sand below the high tide line. Coastal town about 50 km north of El Desemboque and 100 km southwest of Caborca. A large power plant receives shipments of fuel from tankers that dock on a 1-km-long pier in the bay, and there is concern for a possible oil spill. 29°54′21″N, 112°40′57″W.

Puerto Lobos. "Lobos" refers to sea lions. Coastal fishing town north of Puerto Libertad. 30°16′20″N, 112°50′52″W.

Punta Baja. A broad, sandy point between Kino and Tastiota. Vicinity of 28°27′36″N, 111°42′38″W.

Punta Chueca; SOCAAIX "possibly related to the word CÖCAAIX 'carry to a place' where they stored things on the mainland" (Moser & Marlett 2005:560). See fig. 1.17. Chueca is the term for crooked, bent, or twisted, referring to the hook-like sandy point extending into the sea at the west side of the village. Southern and more recent of the two Seri villages. 29°00′54″N, 112°09′37″W.

Punta Santa Rosa; IT IYAT "base of the point." The sandy peninsula that shelters the large estero Santa Rosa just south of Punta Chueca. Punta Santa Rosa and Punta San Miguel on Tiburón mark the southern boundary of the Canal del Infiernillo. Vicinity of the western tip, 28°58′16″N, 112°10′14″W.

San Carlos, Bahía San Carlos. This extensive resort, established in the middle of the twentieth century, has a sheltered, mountain-enclosed bay at the south end of the Sierra el Aguaje. Extensive mangroves as well as expansive shell middens once bordered the bay. The imposing Cerro Tetas de Cabra marked the acclaimed north-south demarcation between the Seris to the north and Yaquis to the south (Spicer 1980). 27°58′N, 111°04′W.

Sierra Bacha (Sierra Cirio). Granitic mountains south of Puerto Libertad and north of El Desemboque del San Ignacio. This is the only mainland locality for the boojum, or *cirio* (*Fouquieria columnaris*). North end, 29°50′40″N, 112°38′18″W; south end, 29°35′16″N, 112°24′19″W.

Sierra del Viejo. Large limestone mountain mass about 45 km southwest of Caborca. 30°20′35″N, 112°19′54″W.

Sierra el Aguaje. Fig. 4.28. Extensive coastal mountain mass to the north of San Carlos (north of Guaymas). The red and yellow rhyolite slopes of this range rise steep and ragged from the surrounding bajada plains and contain deep, narrow riparian canyons, including La Balandrona, Barajitas, and Cañón del Nacapule. Howard Scott Gentry (1949) referred to this mountain mass as the Guaymas Monadnoc. Vicinity of 28°02′50″N, 111°09′50″W, about 860 m, at the peak.

Sierra Seri. See fig. 1.9. Extensive range on the mainland opposite the Sierra Kunkaak (*see* Pico Johnson). North end, 29°28′10″N, 112°17′19″W; south end, 29°04′13″N, 112°02′28″W. Peak at ca. 1000 m.

Figure 4.28. Southwest portion of Sierra el Aguaje in the vicinity of La Manga, Sonora. RSF, 29 September 2008.

Tinaja Picu. Collection site of Ira Wiggins, in the Sierra Picu, along the road from Pitiquito to Puerto Libertad, about 20 miles from the coast. Vicinity of 30°04'N, 112°27'W.

XNAPOFC, from COCPOFC "to throw, strong waves hitting the beach." Maybe the first part has to do with the wind XNAAI; the second part has to do with that verb of hitting (Stephen Marlett, personal communication, 2009). Fig. 4.29. Northwest-facing beach and dunes about 4.5 km north of El Desemboque. This area is an old Seri campsite. 29°32'00"N, 112°25'51"W.

BAJA CALIFORNIA PENINSULA

We generally refer to the state of Baja California as Baja California Norte, in order to avoid confusion with the southern state, Baja California Sur. We identify the entire peninsula, or unspecified localities thereon, as the Baja California Peninsula, or use the term "Peninsula."

Bahía de San Francisquito, Baja California Norte. Southeast part of the state. 28°26'24"N, 112°52'14"W.

Calmallí, Baja California Norte. Village in the southern part of the state, between Las Palomas and El Arco, in the vicinity of 28°06' N, 113°27'W.

Sierra San Francisco, Baja California Sur. Large volcanic mountain range, about 1600 m elevation. The range provides a link between tropical components of the southern Peninsula and temperate influences of the north (Jon Rebman, personal communication, 2007). Peak, 27°39'20"N, 112°54'58"W.

Figure 4.29. Xnapofc, north of El Desemboque, Sonora. BTW, 5 April 2008.

Appendix A

Checklist of the Flora of the Sonoran Islands

Non-native plants are marked with an asterisk (*).

Family and species	Seri name(s)	Tiburón	Esteban	Dátil	Alcatraz	Patos	Cholludo	Mártir	Nolasco
Pteridaceae									
Astrolepis sinuata subsp. sinuata		TIB							
Cheilanthes wrightii		TIB							
Notholaena californica subsp. californica	Hehe quina	TIB	EST						
Notholaena lemmonii var. lemmonii		TIB							NOL
Notholaena standleyi	Hehe quina ·	TIB							
Ephedraceae									
Ephedra aspera	Haaxt		EST						
Acanthaceae									
Avicennia germinans	Pnacojiscl	TIB			ALC				
Carlowrightia arizonica		TIB	EST	DAT					
Dicliptera resupinata		TIB							
Elytraria imbricata		TIB							
Holographis virgata subsp. virgata	Hooinalca	TIB		DAT					
Justicia californica	Nojoopis	TIB	EST		ALC		CHO		
Justicia candicans	Nojoopis caacöl	TIB							
Justicia longii		TIB							
Ruellia californica	Satooml	TIB							
Tetramerium fruticosum		TIB							
Achatocarpaceae									
Phaulothamnus spinescens	Aniicös	TIB		DAT	ALC		CHO		NOL
Aizoaceae									
*Mesembryanthemum crystallinum	Hax ixapz iti yacp			DAT					
Sesuvium portulacastrum	Spitj caacöl, spitj ctamcö	TIB			ALC				
Trianthema portulacastrum	Comaacöl	TIB			ALC	PAT			
Amaranthaceae									
Allenrolfea occidentalis	Tacs	TIB			ALC				
Amaranthus fimbriatus	Ziim caaitic	TIB		DAT	ALC	PAT			NOL

Appendix A Continued

Family and species	Seri name(s)	Island							
		Tiburón	Esteban	Dátil	Alcatraz	Patos	Cholludo	Mártir	Nolasco
Amaranthus watsonii	Ziim quicös	TIB	EST	DAT	ALC				
Atriplex barclayana	Spitj	TIB	EST	DAT	ALC	PAT	CHO		
Atriplex linearis	Hatajixp, hatajísijc	TIB							
Atriplex polycarpa	Hatajixp, hatajísijc	TIB	EST	DAT					
*Chenopodium murale	Ziim xat	TIB			ALC	PAT	CHO		
Salicornia bigelovii	Xnaa caaa	TIB							
Salicornia subterminalis		TIB			ALC				
Suaeda esteroa	Sipjö yaneaax	TIB							
Suaeda nigra	Hatajipol	TIB	EST		ALC	PAT			
Tidestromia lanuginosa subsp. eliassoniana	Halít an caascl	TIB							
Apiaceae									
Daucus pusillus		TIB							
Apocynaceae									
Asclepias albicans	Najcaazjc	TIB	EST	DAT					
Asclepias subulata	Najcaazjc	TIB							
Funastrum hartwegii	Hexe	TIB							
Haplophyton cimicidum		TIB							
Marsdenia edulis	Xomee	TIB							
Matelea cordifolia	Comot	TIB							
Matelea pringlei	Nas, ziix is quicös	TIB		DAT					
Metastelma arizonicum		TIB							
Metastelma californicum									NOL
Vallesia glabra	Tanoopa	TIB							
Arecaceae									
Brahea armata	Zamij cmaam	TIB							
Aristolochiaceae									
Aristolochia watsonii	Hataast an ihiih	TIB							
Asparagaceae									
Agave cerulata subsp. dentiens	Heme, xiica istj caaitic		EST						
Agave chrysoglossa	Hasot	TIB							NOL
Agave subsimplex	Haamjö	TIB		DAT			CHO		
Agave sp.	Haamjö caacöl	TIB							
Dasylirion gentryi	Istj ano caap	TIB							
Triteleiopsis palmeri	Caal oohit	TIB							
Asteraceae									
Ambrosia camphorata var. leptophylla		TIB							
Ambrosia carduacea	Tincl caacöl	TIB							
Ambrosia divaricata	An icoqueetc	TIB	EST	DAT					
Ambrosia dumosa	Xcoctz	TIB							
Ambrosia ilicifolia	Tincl, xcoctz	TIB	EST						
Ambrosia salsola var. pentalepis	Caasol cacat	TIB							
Baccharis salicifolia	Caajö	TIB							
Baccharis sarothroides	Caasol caacöl							MAR	
Bahiopsis chenopodina	Hehe imoz coopol	TIB		DAT					

Family and species	Seri name(s)	Island							
		Tiburón	Esteban	Dátil	Alcatraz	Patos	Cholludo	Mártir	Nolasco
Bahiopsis triangularis			EST						NOL
Bajacalia crassifolia		TIB	EST	DAT					
Bebbia juncea var. *aspera*	Sapatx	TIB	EST	DAT					NOL
Brickellia coulteri var. *coulteri*	Comima	TIB							
Chromolaena sagittata	Comima	TIB							
Coreocarpus sanpedroensis									NOL
Coreocarpus sonoranus var. *sonoranus*		TIB		DAT					
Encelia farinosa var. *farinosa*	Cotx	TIB		DAT	ALC	PAT			
Gymnosperma glutinosum		TIB							
Helianthus niveus var. *niveus*		TIB							
Heliopsis anomala		TIB							
Hofmeisteria crassifolia									NOL
Hofmeisteria fasciculata var. *fasciculata*	Taca imas	TIB	EST	DAT			CHO		
Palafoxia arida var. *arida*	Moosni iiha, mosnoohit	TIB			ALC				
Pectis papposa var. *papposa*	Caasol heecto	TIB			ALC				
Pectis rusbyi	Caasol heecto caacöl	TIB					CHO		
Pelucha trifida		TIB						MAR	
Perityle aurea		TIB	EST						
Perityle californica									NOL
Perityle emoryi	Hee imcát, hehe cotopl, zaah coocta	TIB	EST	DAT	ALC	PAT	CHO	MAR	
Peucephyllum schottii			EST						
Pleurocoronis laphamioides	Hamt inoosj, hast yapxöt	TIB	EST	DAT	ALC			MAR	NOL
Porophyllum gracile	Xtisil (var. xtesel, xtisel)	TIB	EST	DAT					
Porophyllum pausodynum									NOL
Stephanomeria pauciflora	Hehe imixaa, posapatx camoz	TIB							
Thymophylla concinna	Cahaahazxot	TIB							
Trixis californica var. *californica*	Cocaznootizx	TIB	EST	DAT		PAT		MAR	NOL
Verbesina palmeri	Cocaznootizx caacöl	TIB							
Xanthisma incisifolium			EST						
Xanthisma spinulosum var. *scabrella*		TIB							
Xylothamia diffusa	Caasol cacat	TIB			ALC				
Bataceae									
Batis maritima	Pajoocsim, xpajoocsim, xpacoocsim	TIB			ALC				
Boraginaceae									
Cordia parvifolia	Naz (var. nooz), hehet inaail coopl	TIB							
Cryptantha angelica		TIB							
Cryptantha angustifolia	Hehe cotopl, hehe czatx	TIB	EST						
Cryptantha fastigiata		TIB	EST	DAT					
Cryptantha maritima	Hehe cotopl, hehe czatx	TIB	EST						
Eucrypta micrantha		TIB							
Heliotropium curassavicum	Hant otopl, potacs camoz	TIB			ALC				
Nama hispidum	Hohroohit	TIB							
Phacelia affinis		TIB							

Appendix A Continued

Family and species	Seri name(s)	Island							
		Tiburón	Esteban	Dátil	Alcatraz	Patos	Cholludo	Mártir	Nolasco
Phacelia crenulata	Cahaahazxot ctam, najmís	TIB							
Phacelia pedicellata		TIB	EST	DAT					
Tiquilia canescens		TIB		DAT					
Tiquilia palmeri	Hee ijoját	TIB							
Tournefortia hartwegiana		TIB							
Brassicaceae									
Descurainia pinnata	Cocool	TIB		DAT					
Draba cuneifolia	Cocool cmaam	TIB	EST						
Lyrocarpa coulteri	Ponás camoz	TIB		DAT					
Lyrocarpa linearifolia			EST						
Burseraceae									
Bursera fagaroides var. *elongata*	Xoop isoj	TIB							
Bursera hindsiana	Xopinl	TIB	EST	DAT			CHO		
Bursera laxiflora	Xoop caacöl	TIB							
Bursera microphylla	Xoop	TIB	EST	DAT	ALC				NOL
Cactaceae									
Carnegiea gigantea	Mojepe	TIB			ALC	PAT	CHO		
Cylindropuntia alcahes var. *alcahes*	Heem icös cmasl		EST					MAR	
Cylindropuntia bigelovii	Coote, sea	TIB	EST	DAT	ALC				
Cylindropuntia cholla								MAR	
Cylindropuntia fulgida (both vars.)	Sea cotopl, sea icös cooxp	TIB			ALC	PAT	CHO		NOL
Cylindropuntia leptocaulis	Iipxö	TIB		DAT					
Cylindropuntia versicolor	Heem icös cmáxlilca	TIB		DAT					
Echinocereus grandis	Hant ipzx íteja caacöl		EST						
Echinocereus scopulorum	Hant ipzx íteja caacöl	TIB							
Echinocereus websterianus									NOL
Ferocactus emoryi	Siml caacöl, siml yapxöt cheel				ALC				
Ferocactus tiburónensis	Siml, siml áa	TIB							
Lophocereus schottii var. *schottii*	Hasahcapjö, hehe is quizil	TIB		DAT	ALC	PAT	CHO		
Mammilaria estebanensis	Hant ipzx íteja caacöl		EST						
Mammilaria grahamii subsp. *sheldonii*	Hant ipzx íteja	TIB			ALC				
Mammillaria multidigitata									NOL
Mammillaria tayloriorum									NOL
Mammillaria sp.				DAT					
Opuntia bravoana									NOL
Opuntia engelmannii var. *engelmannii*	Heel hayeen ipaii				ALC				
Pachycereus pringlei	Xaasj, xcocni	TIB	EST	DAT	ALC	PAT	CHO	MAR	NOL
Peniocereus striatus	Xtooxt	TIB		DAT					
Stenocereus gummosus	Ziix is ccapxl	TIB	EST	DAT		PAT	CHO		
Stenocereus thurberi	Ool	TIB	EST	DAT	ALC		CHO		NOL
Campanulaceae									
Nemacladus orientalis	Xtamaaija oohit		EST						

Family and species	Seri name(s)	Tiburón	Esteban	Dátil	Alcatraz	Patos	Cholludo	Mártir	Nolasco
					Island				
Cannabaceae									
Celtis pallida	Ptaacal	TIB							
Celtis reticulata		TIB							
Capparaceae									
Atamisquea emarginata	Cöset	TIB	EST				CHO		
Caryophyllaceae									
Achyronychia cooperi	Hant yapxöt	TIB							
Drymaria holosteoides var. *holosteoides*		TIB							
Celastraceae									
Canotia holacantha	Xooml icös caacöl	TIB							
Maytenus phyllanthoides	Cos	TIB			ALC				
Cleomaceae									
Cleome tenuis subsp. *tenuis*	Cocool	TIB							
Cochlospermaceae									
Amoreuxia palmatifida	Joját (var. xoját)	TIB							
Combretaceae									
Laguncularia racemosa	Pnaacoj hacaaiz	TIB							
Convolvulaceae									
Cressa truxillensis	Ziix casa insii	TIB			ALC				
Cuscuta americana	Hamt itoozj	TIB							
Cuscuta corymbosa var. *grandiflora*	Hamt itoozj		EST						NOL
Cuscuta desmouliana	Hamt itoozj	TIB		DAT					
Cuscuta leptantha	Hamt itoozj	TIB							
Evolvulus alsinoides		TIB							
Ipomoea hederacea	Hehe quiijam, hataaij	TIB							
Jacquemontia abutiloides		TIB							
Jacquemontia agrestis		TIB							
Crossosomataceae									
Crossosoma bigelovii		TIB							
Cucurbitaceae									
Tumamoca macdougalii	Hatoj caaihjö	TIB							
Vaseyanthus insularis	Hant caitoj	TIB	EST					MAR	NOL
Cymodaceae									
Halodule wrightii	Zimjötaa	TIB							
Cyperaceae									
Cyperus elegans		TIB							NOL
Cyperus squarrosus								MAR	NOL
Eleocharis geniculata		TIB							
Euphorbiaceae									
Acalypha californica	Queeejam iti hacniiix	TIB	EST	DAT					
Bernardia viridis									NOL
Cnidosculus palmeri	Coaap	TIB		DAT					

Appendix A Continued

Family and species	Seri name(s)	Island							
		Tiburón	Esteban	Dátil	Alcatraz	Patos	Cholludo	Mártir	Nolasco
Croton californicus	Hacaain cooscl, moosni iti hateepx	TIB							
Croton magdalenae	Hehe ziix capete	TIB							
Croton sonorae	Hooinalca	TIB							
Ditaxis lanceolata	Hehe czatx caacöl	TIB	EST	DAT					
Ditaxis neomexicana	Hehe czatx caacöl	TIB		DAT					
Ditaxis serrata			EST						
Euphorbia abramsiana		TIB							
Euphorbia arizonica		TIB							
Euphorbia eriantha	Pteept	TIB		DAT					
Euphorbia florida		TIB							
Euphorbia leucophylla subsp. *comcaacorum*		TIB							
Euphorbia lomelii									NOL
Euphorbia magdalenae		TIB							NOL
Euphorbia misera	Hamacj	TIB	EST	DAT					
Euphorbia pediculifera var. *pediculifera*		TIB	EST	DAT					
Euphorbia petrina	Tomitom hant cocpeetij	TIB	EST		ALC			MAR	
Euphorbia polycarpa	Tomitom hant cocpeetij	TIB	EST	DAT					
Euphorbia prostrata		TIB							
Euphorbia setiloba	Tomitom hant cocpeetij	TIB	EST						
Euphorbia tomentulosa	Tomitom hant cocpeetij caacöl	TIB							
Euphorbia xanti	Hehe ix cooxp	TIB							
Jatropha cinerea	Hamisj, oot iquejöc	TIB							
Jatropha cuneata	Haat	TIB	EST	DAT					NOL
Sebastiania bilocularis	Hehe coaanj	TIB							
Tragia jonesii	Hehe cocozxlim	TIB		DAT					
Fabaceae									
Acacia greggii	Tis	TIB							
Acacia willardiana	Cap	TIB							NOL
Acmispon maritimus var. *brevivexillus*	Hee inoosj	TIB							
Acmispon strigosus	Hee inoosj	TIB	EST						
Calliandra californica	Hast iti heepol cheel	TIB		DAT					
Calliandra eriophylla var. *eriophylla*	Haxz iztim	TIB							
Coursetia glandulosa	Cmapxöquij, hehe ctoozi	TIB							
Dalea bicolor var. *orcuttiana*		TIB							
Dalea mollis	Hanaj itaamt	TIB		DAT	ALC				
Desmanthus covillei	Pohaas camoz	TIB							
Desmanthus fruticosus	Hehe casa	TIB	EST	DAT					
Desmodium procumbens		TIB							
Ebenopsis confinis	Heejac	TIB		DAT					
Errazurizia megacarpa	Cooxi iheet	TIB							
Hoffmannseggia intricata	Haxz iztim	TIB	EST	DAT					
Lupinus arizonicus	Zaah coocta	TIB	EST						
Lysiloma divaricatum	Hehe ctoozi	TIB							
Marina evanescens			EST						

Family and species	Seri name(s)	Tiburón	Esteban	Dátil	Alcatraz	Patos	Cholludo	Mártir	Nolasco
							Island		
Marina parryi	Hanaj iit ixac	TIB	EST	DAT					
Marina vetula		TIB							
Mimosa distachya var. *laxiflora*	Hehe cotázita	TIB							
Olneya tesota	Comitin, hesen	TIB	EST	DAT	ALC		CHO		
Parkinsonia florida	Ziij, iiz	TIB							
Parkinsonia microphylla	Ziipxöl	TIB	EST						
Phaseolus acutifolius var. *latifolius*	Haap	TIB							
Phaseolus filiformis	Haamoja ihaap	TIB	EST	DAT					
*Pithecellobium dulce	Camootzila	TIB							
Prosopis glandulosa var. *torreyana*	Haas	TIB	EST	DAT	ALC				
Psorothamnus emoryi var. *emoryi*	Xomeete	TIB							
Rhynchosia precatoria		TIB							
Senna confinis		TIB	EST						
Senna covesii	Hehe quiinla	TIB		DAT					
Tephrosia palmeri	Cozi hax ihapooin	TIB							
Tephrosia vicioides		TIB							
Zapoteca formosa subsp. *rosei*		TIB							
Fouquieriaceae									
Fouquieria diguetii									NOL
Fouquieria splendens subsp. *splendens*	Jomjeeziz	TIB	EST	DAT					
Frankeniaceae									
Frankenia palmeri	Seepol	TIB							
Koeberlineaceae									
Koeberlinia spinosa	Xooml	TIB							
Krameriaceae									
Krameria bicolor	Heepol	TIB							
Krameria erecta	Haxz iztim	TIB							
Lamiaceae									
Hyptis albida	Xeescl	TIB	EST	DAT					
Salvia similis									NOL
Loasaceae									
Eucnide cordata			EST						
Eucnide rupestris	Zaaj iti cocaai	TIB	EST	DAT					NOL
Mentzelia adhaerans	Hehe cotopl, hehe czatx	TIB	EST	DAT	ALC			MAR	
Mentzelia hirsutissima	Hehe cotopl, hehe czatx	TIB							
Petalonyx linearis		TIB	EST					MAR	
Malpighiaceae									
Callaeum macropterum	Haxz oocmoj	TIB							
Cottsia californica	Hehe quiijam	TIB		DAT					
Cottsia gracilis	Hehe quiijam	TIB	EST	DAT					
Echinopterys eglandulosa	Hap oaacajam	TIB							
Galphimia angustifolia	Hap oaacajam	TIB							NOL
Malvaceae									
Abutilon californicum	Hant ipásaquim	TIB							

Appendix A Continued

Family and species	Seri name(s)	Island							
		Tiburón	Esteban	Dátil	Alcatraz	Patos	Cholludo	Mártir	Nolasco
Abutilon incanum	Hasla an iihom, caatc ipapl	TIB		DAT					
Abutilon palmeri	Caatc ipapl	TIB						MAR	
Ayenia filiformis		TIB	EST						
Gossypium davidsonii									NOL
Herissantia crispa		TIB		DAT					
Hibiscus biseptus		TIB							
Hibiscus denudatus var. *denduatus*	Hepem ijcoa	TIB	EST	DAT					
Horsfordia alata	Hehe coozlil, caatc ipapl	TIB							
Horsfordia newberryi		TIB	EST	DAT					
Melochia tomentosa	Hehe coyoco	TIB		DAT					
Sida abutifolia		TIB							
Sphaeralcea ambigua var. *ambigua*	Jcoa	TIB		DAT					
Sphaeralcea ambigua var. *versicolor*	Jcoa		EST						
Sphaeralcea coulteri					ALC				
Sphaeralcea hainesii								MAR	
Waltheria indica		TIB							
Martyniaceae									
Proboscidea altheifolia	Xonj	TIB							
Molluginaceae									
*Mollugo cerviana	Hant iit	TIB							
Mollugo verticillata									NOL
Moraceae									
Ficus petiolaris subsp. *palmeri*	Xpaasni	TIB	EST	DAT	ALC		CHO	MAR	NOL
Myrtaceae									
*Eucalyptus camaldulensis	Hehe hataap iic cöihiipe	TIB							
Nyctaginaceae									
Abronia maritima subsp. *maritima*	Spitj cmajiic	TIB			ALC				
Allionia incarnata var. *incarnata*	Hamíp cmaam	TIB		DAT					
Boerhavia triquetra	Hamíp caacöl	TIB	EST	DAT					NOL
Boerhavia xanti	Hamíp	TIB							
Commicarpus scandens		TIB							
Mirabilis laevis var. *crassifolia*			EST						
Mirabilis tenuiloba		TIB	EST						
Oleaceae									
Forestiera phillyreoides		TIB							
Fraxinus gooddingii		TIB							
Menodora scabra		TIB							
Onagraceae									
Chylismia cardiophylla subsp. *cardiophylla*			EST					MAR	
Chylismia cardiophylla subsp. *cedrosensis*		TIB							
Eulobus californicus	Hast ipenim	TIB							
Oenothera primiveris		TIB							

Family and species	Seri name(s)	Island							
		Tiburón	Esteban	Dátil	Alcatraz	Patos	Cholludo	Mártir	Nolasco
Orobanchaceae									
Orobanche cooperi	Matar	TIB							
Papaveraceae									
Argemone subintegrifolia	Xazacöz caacöl		EST						
Passifloraceae									
Passiflora arida	Oot ijoeene	TIB	EST						
Passiflora palmeri	Joeene	TIB	EST	DAT					
Phyllanthaceae									
Andrachne microphylla	Hasoj an hehe	TIB							
Phytolaccaceae									
Rivinia humilis		TIB							
Plantaginaceae									
Gambelia juncea	Nojoopis caacöl	TIB	EST						NOL
Nuttallanthus texanus		TIB							
Plantago ovata var. fastigiata	Hatajeen	TIB							
Pseudorontium cyathiferum	Hehemonlc	TIB	EST	DAT			CHO		NOL
Sairocarpus watsonii	Comaacöl	TIB		DAT					
Stemodia durantifolia		TIB							
Plumbaginaceae									
Plumbago zeylanica		TIB							
Poaceae									
Aristida adscensionis	Impós	TIB	EST	DAT				MAR	NOL
Aristida californica var. californica	Conee csaai	TIB							
Aristida californica var. glabrata		TIB							
Aristida divaricata									NOL
Aristida ternipes var. ternipes		TIB							NOL
*Arundo donax	Xapijaacöl, xapijaas	TIB							
Bothriochloa barbinodis			EST						NOL
Bouteloua aristidoides	Isnaap iic is	TIB	EST	DAT	ALC		CHO		NOL
Bouteloua barbata var. barbata	Isnaap iic is	TIB	EST		ALC	PAT			
Bouteloua diversispicula		TIB							
*Cenchrus ciliaris	Oot iconee	TIB			ALC				
*Cenchrus echinatus		TIB							
Cenchrus palmeri	Cözazni (var. czazni)	TIB	EST						NOL
Chloris virgata		TIB							
Dasyochloa pulchella	Conee ccosyat, conee ccapxl	TIB	EST	DAT					
Digitaria californica	Cpooj	TIB	EST	DAT			CHO	MAR	NOL
Distichlis littoralis	Cötep	TIB			ALC				
Distichlis spicata		TIB							
Enneapogon desvauxii		TIB							
Enteropogon chlorideus			EST						
Eragrostis pectinacea var. pectinacea									NOL
Heteropogon contortus		TIB	EST						NOL

Appendix A Continued

Family and species	Seri name(s)	Tiburón	Esteban	Dátil	Alcatraz	Patos	Cholludo	Mártir	Nolasco
					Island				
Lasiacis ruscifolia var. *ruscifolia*		TIB							
Leptochloa dubia		TIB							
Leptochloa panicea subsp. *brachiata*		TIB	EST						NOL
Muhlenbergia microsperma	Impós, ziizil	TIB	EST					MAR	NOL
Panicum hirticaule var. *hirticaule*		TIB	EST				CHO		
Phragmites australis subsp. *berlandieri*	Xapij	TIB							
Setaria liebmannii	Ziizil	TIB							NOL
Setaria macrostachya	Hasac, xiica quiix	TIB	EST	DAT			CHO		NOL
Sporobolus cryptandrus	Cpooj	TIB							
Sporobolus pyramidatus		TIB							
Sporobolus virginicus	Xojasjc	TIB			ALC				
Trichloris crinita		TIB							NOL
Tridens muticus var. *muticus*		TIB		DAT					
Urochloa arizonica	Conee ccapxl	TIB	EST	DAT			CHO		
Urochloa fusca		TIB	EST						
Polygonaceae									
Eriogonum inflatum	Tee	TIB	EST						
Portulacaceae									
Portulaca halimoides		TIB							
Portulaca oleracea		TIB							
Portulaca umbraticola subsp. *lanceolata*		TIB							
Talinum paniculatum	Pnaacoj quistj	TIB							
Resedaceae									
Oligomeris linifolia	Xamaasa	TIB	EST	DAT					
Rhamnaceae									
Colubrina viridis	Ptaact	TIB	EST	DAT			CHO		NOL
Condalia globosa var. *pubescens*	Cozi	TIB	EST	DAT					
Sageretia wrightii		TIB							
Ziziphus obtusifolia var. *canescens*	Haaca, xiica imám coopol	TIB	EST	DAT					
Rhizophoraceae									
Rhizophora mangle	Pnaacoj xnaazolcam	TIB							
Rubiaceae									
Chiococca petrina		TIB							
Galium proliferum		TIB							
Galium stellatum			EST						
Randia thurberi	Haalp	TIB							
Ruppiaceae									
Ruppia maritima	Zimjötaa	TIB							
Salicaceae									
Salix exigua	Paaij	TIB							

Family and species	Seri name(s)	Island							
		Tiburón	Esteban	Dátil	Alcatraz	Patos	Cholludo	Mártir	Nolasco
Santalaceae									
Phoradendron californicum	Eaxt (+ host plant)	TIB							
Phoradendron diguetianum	Eaxt (+ host plant)	TIB							
Sapindaceae									
Cardiospermum corindum	Hax quipoiin	TIB		DAT					
Dodonaea viscosa	Caasol caacöl	TIB							
Sapotaceae									
Sideroxylon leucophyllum	Hehe pnaacoj	TIB	EST						
Sideroxylon occidentale	Hehe hateen ccaptax, paaza	TIB							
Simaroubaceae									
Castela polyandra	Snaazx, zazjc	TIB							
Simmondsiaceae									
Simmondsia chinensis	Pnaacöl	TIB	EST	DAT					NOL
Solanaceae									
Datura discolor	Hehe camostim	TIB		DAT	ALC				
Lycium andersonii var. andersonii	Hahjöenej, hahjö inaail coopol	TIB							
Lycium andersonii var. pubescens	Hahjöenej, hahjö inaail coopol	TIB	EST	DAT					
Lycium berlandieri var. longistylum		TIB							
Lycium brevipes var. brevipes	Hahjö an quinelca	TIB	EST	DAT	ALC		CHO	MAR	
Lycium fremontii var. fremontii	Hahjö cacat	TIB			ALC				
Nicotiana clevelandii	Xeezej islitx	TIB							
Nicotiana obtusifolia	Haapis casa	TIB	EST	DAT	ALC		CHO	MAR	NOL
Physalis crassifolia var. infundibularis	Xtoozp insaacaj		EST						
Physalis crassifolia var. versicolor	Xtoozp	TIB		DAT					
*Physalis pubescens	Insaacaj	TIB							
Solanum hindsianum	Hap itapxeen	TIB	EST	DAT					
Stegnospermataceae									
Stegnosperma halimifolium	Xnejamsiictoj	TIB	EST	DAT			CHO	MAR	
Tamaricaceae									
*Tamarix aphylla	Hocö hapéc	TIB							
*Tamarix chinensis		TIB			ALC				
Theophrastaceae									
Bonellia macrocarpa subsp. pungens	Cof	TIB							
Typhaceae									
Typha domingensis	Pat	TIB							
Urticaceae									
Parietaria hespera var. hespera		TIB	EST						NOL
Verbenaceae									
Lantana velutina		TIB							
Lippia palmeri	Xomcahiift	TIB		DAT					

Appendix A Continued

Family and species	Seri name(s)	Island							
		Tiburón	Esteban	Dátil	Alcatraz	Patos	Cholludo	Mártir	Nolasco
Violaceae									
Hybanthus fruticulosus		TIB							
Zosteraceae									
Zostera marina var. *atam*	Eaz	TIB			ALC				
Zygophyllaceae									
Fagonia californica	Xtamosnoohit	TIB		DAT					
Fagonia pachyacantha	Xtamosnoohit	TIB							
Fagonia palmeri		TIB							
Guaiacum coulteri	Mocni	TIB							
Kallstroemia californica		TIB		DAT	ALC				
Kallstroemia grandiflora	Hast ipenim	TIB							
Larrea divaricata subsp. *tridentata*	Haaxat	TIB							
Tribulus terrestris	Cosi cahoota, cözazni caacöl, hee inoosj, heen ilít, hehe ccosyat	TIB			ALC				
Viscainoa geniculata var. *geniculata*	Xneeejam is hayaa	TIB	EST	DAT	ALC		CHO	MAR	

Appendix B

Species Mutually Absent from
Isla Tiburón and Mainland Sonora

Non-native plants are marked with an asterisk (*).

PLANTS ON TIBURÓN NOT KNOWN FROM MAINLAND SONORA

ASTERACEAE
Bajacalia crassifolia
Pelucha trifida

BORAGINACEAE
Cryptantha angelica

CARYOPHYLLACEAE
Drymaria holosteoides *var.* holosteoides

CONVOLVULACEAE
Jacquemontia abutiloides

EUPHORBIACEAE
Euphorbia magdalenae

FABACEAE
Senna covesii
Tephrosia palmeri

PASSIFLORACEAE
Passiflora palmeri

SAPOTACEAE
Sideroxylon leucophyllum

SIMAROUBACEAE
Castela polyandra

ZYGOPHYLLACEAE
Fagonia palmeri

PLANTS ON THE SONORA MAINLAND IN THE SERI REGION (Vicinity of Bahía Kino to the vicinity of El Desemboque San Ignacio, and Inland to Pozo Coyote, Sierra Seri, and Playa San Bartolo)
NOT KNOWN FROM ISLA TIBURÓN

AIZOACEAE

*Mesembryanthemum crystallinum

A single specimen is known from Dátil. It could be a serious invasive species if it establishes on Tiburón (see part III, "Species Accounts").
El Desemboque, *Wilder 08-185*.

Sesuvium verrucosum Rafinesque. Western sea purslane

Succulent annuals localized on alkaline or saline soils. Suitable habitat does not occur on the Sonoran Islands.
Playa San Bartolo, *Felger 08-222*.

AMARANTHACEAE

Atriplex canescens (Pursh) Nuttall. Hatajísijc; *chamizo cenizo*; four-wing saltbush

Gray-leaved shrubs with four-winged fruits. Coastal dunes, occasional along washes, and low-lying silty flats.
0.5 mi S of El Desemboque, dunes at beach, *Felger 6210.*

*Salsola tragus Linnaeus [*S. kali* of authors, not *S. kali* Linnaeus]. Ziim caacöl; *chamizo volador*; Russian thistle, tumbleweed

Annuals. "This common agricultural weed has invaded disturbed sites along the Infiernillo coast" (Felger & Moser 1985:277).
3.2 mi by rd W of Rancho La Yuta, on road to El Desemboque, 29° 06'N, 111°47', *Felger 16962.*

APODANTHACEAE

Pilostyles thurberi A. Gray

Minute parasitic plants on *Psorothamnus emoryi*; flowers globose and maroon-colored. Highly localized on dunes at the north side of Cerro Tepopa (Felger & Moser 1985). The host plant is common on Tiburón, and we have searched in vain for *Pilostyles* on the island.
N side of Cerro Tepopa, *Felger 17946.*

ASPARAGACEAE

Dichelostemma capitatum (Bentham) Alph. Wood subsp. **pauciflorum** (Torrey) Keator [*D. pulchellum* (Salisbury) A. Heller var. *pauciflorum* (Torrey) Hoover. *Brodiaea capitata* Bentham var. *pauciflora* Torrey]. Caal oohit caacöl; *coveria cobena, ajo blanco*; desert hyacinth

Lavender-blue flowers in spring from a single, small bulb. Granitic bajadas near Cerro Pelón southeast of El Desemboque.
Vicinity of Cerro Pelón, 5 mi SE of El Desemboque, *Felger 17916.*

ASTERACEAE

Ambrosia ambrosioides (Cavanilles) W.W. Payne. Tincl; *chicura*; canyon ragweed

Slender-stem shrubs with large, rough-surfaced leaves, and cocklebur-like burs. Arroyos and canyons along the Seri coast.
1 km E of El Desemboque, *Felger 14202.*

Ambrosia confertiflora de Candolle. Paxaaza; ragweed

Herbaceous perennials. Playas and weedy places (Felger & Moser 1985). Suitable habitat probably does not occur on Tiburón.
Roadside, 5 mi E of town of Bahía Kino, *Felger 9049.*

Ambrosia deltoidea (A. Gray) W.W. Payne. Xcotz; *chamizo*; triangle-leaf bursage

Small shrubs common along the Sonora coast. Reported for Tiburón by Felger and Moser (1985), but no specimens are known from the island. Suitable habitat would seem to occur on Tiburón and its absence there is baffling.
10 km S of El Desemboque, edge of playa, *Felger 18164.*

Ambrosia monogyra (Torrey & A. Gray ex A. Gray) Strother & B.G. Baldwin [*Hymenoclea monogyra* Torrey & A. Gray ex A. Gray]. Caasol cacat; *jécota*; burro brush

"Common wispy shrub 2 to 3 m tall found along arroyos and floodplains of major drainageways, and as a weed on farmland" (Felger & Moser 1985:284).
3.5 mi NE of El Desemboque, *Felger 17386.*

Arida coulteri (A. Gray) D.R. Morgan & R.L. Hartman [*Machaeranthera coulteri* (A. Gray) B.L. Turner & D.B. Horne. Listed as *M. parvifolia* A. Gray by Felger & Moser 1985]. Zaah coocta, hehe cacaatajc; arid tansy-aster

Annuals or short-lived perennials with lavender rays and yellow disk florets. Common in the Bahía Kino region.
Rancho Carrizo, NE of Kino, *Felger 07-136.*

Baccharis sarothroides A. Gray. Caasol caacöl; *romerillo*; desert broom

Sparsely leaved shrubs. Large washes, margins of dry lakebeds, and weedy in the vicinity of Bahía Kino. Its absence from the island seems strange. Also on Isla San Pedro Mártir.

Vicinity of El Desemboque, *Moser 30 Oct 1961.*

Bahiopsis laciniata (A. Gray) E.E. Schiling & Panero [*Viguiera laciniata* A. Gray]

Small shrubs; leaf margins serrated. A species of Baja California and disjunct in the Sierra Bacha and higher elevations on Cerro Tepopa. It should be sought at higher elevations on Tiburón.

Summit of Cerro Tepopa, *Robert Hopewell 2 Feb 1970.*

Baileya multiradiata Torrey. Cahaahazxot; desert marigold

Cool-season annuals with large, yellow flower heads. Common at Puerto Libertad and infrequent southward (Felger & Moser 1985).

Chaenactis carphoclinia A. Gray. Xtamosnoohit, hehe cacaatajc

Cool-season annuals with white flowers. Common in the northern part of the Seri region (Felger & Moser 1985).

Las Cuevitas, ca. 26.5 km NNW (by air) from El Desemboque, *Felger 14273.*

Chaenactis stevioides Hooker & Arnott var. **stevioides.** Desert pincushion

Cool-season annuals with white flowers. In the northern part of the Seri region.

Floodplain of Río San Ignacio, 3.5 mi NE of El Desemboque, *Felger 17399.*

Coreocarpus arizonicus (A. Gray) S.F. Blake

Herbaceous perennials or facultative annuals. Flowers bright yellow. Canyons and slopes on east side of the Sierra Seri. Comparable habitats occur in the Sierra Kunkaak.

East side and base of Sierra Nochebuena, mountains W of Playa San Bartolo, *Felger 08-224.*

Encelia halimifolia Cavanilles

Shrubs with yellow flower heads. Common on sandy or silty soils in the Bahía Kino region.

Bahía Kino Nuevo, beach dunes, *Felger 74-31.*

Gnaphalium palustre Nuttall

Small cool-season annuals with inconspicuous flowers.

1.5 km S of El Desemboque San Ignacio, *Felger 17166.*

Isocoma tenuisecta Greene. Burroweed

Small shrubs with small yellow flowers. It thrives on sandy-clayish soils.

Rancho La Cabaña, W side of SW part of Playa San Bartolo, *Felger 08-218.*

Logfia arizonica (A. Gray) Holub [*Filago arizonica* A. Gray]

Diminutive cool-season annuals with minute flowers. Northern part of the Seri region.

Floodplain of Río San Ignacio, 3.5 mi NE of El Desemboque, *Felger 17430.*

Logfia depressa (A. Gray) Holub [*Filago depressa* A. Gray]

Diminutive cool-season annuals with minute flowers. Northern part of the Seri region.

Floodplain of Río San Ignacio, 3.5 mi NE of El Desemboque, *Felger 17376.*

Logfia filaginoides (Hooker & Arnott) Morefield [*L. californica* (Nuttall) Holub. *Filago californica* Nuttall]

Diminutive cool-season annuals with minute flowers. Northern part of the Seri region.

Floodplain of Río San Ignacio, 3.5 mi NE of El Desemboque, *Felger 17430B.*

Malperia tenuis S. Watson

Delicate cool-season annuals. Common in the northern part of the Seri region.

Pozo Coyote, 6.5 mi by road NE of El Desemboque, *Felger 17766.*

Perityle leptoglossa Harvey & A. Gray. Hee imcát caacöl; giant rock-daisy

Annuals or small perennials with bright yellow flower heads. Usually on rock faces in the Sierra Seri.

Vicinity of Cerro Pelón, 5 mi SE of El Desemboque, *Felger 17922* (USON).

Stylocline micropoides A. Gray. Woollyhead neststraw

Diminutive cool-season annuals with minute flowers.

> 1 mi E of 19 mi S of El Desemboque, granitic low mountain slopes and upper bajada, *Felger 17201.*

Thymophylla pentachaeta (de Candolle) Small var. **belenidium** (de Candolle) Strother [*Dyssodia pentachaeta* (de Candolle) B.L. Robinson var. *belenidium* (de Candolle) Strother]

Small, short-lived perennials with bright yellow flowers. Uncommon in the northern part of the Seri region.

> Arroyo San Ignacio, 4 mi NE of El Desemboque, *Felger 17817.*

*Xanthium strumarium Linnaeus. Cozazni caacöl; *cardillo*; cocklebur

Robust annuals producing burs with hooked spines. Weedy places, near the north end of the Seri region and near Bahía Kino (Felger & Moser 1985).

> 2 mi NW of El Desemboque, *Felger 17784.*

Zinnia acerosa (de Candolle) A. Gray. Saapom ipemt, cmajiic ihásaquim; desert zinnia

Small woody-stemmed perennials with white rays and yellow disk flowers.

> 19 mi S of El Desemboque, *Felger 14072.*

BORAGINACEAE

Eucrypta chrysanthemifolia (Bentham) Greene var. **bipinnatifida** (Greene) Constance

Delicate, cool-season annuals, the flowers small and pale lavender.

> 19 mi S of El Desemboque, granitic slopes, mountaintops and flats, *Felger 17183.*

Pectocarya heterocarpa (I.M. Johnston) I.M. Johnston. Mixed-nut comb-bur

Small cool-season annuals with minute white flower and comb-like nutlets.

> Floodplain of Río San Ignacio, 3.5 mi NE of El Desemboque, *Felger 17407.*

Phacelia cryptantha Greene

Cool-season annuals, the flowers small and pale violet.

> 19 mi S of El Desemboque, granitic low mountain slopes and upper bajada, *Felger 17186.* Cerro Pelón, *Felger 17936.*

BRASSICACEAE

*Brassica tournefortii Gouan. Sahara mustard

Cool-season annuals, highly variable in size, with small pale-yellow flowers. A serious invasive species in northwestern Sonora and occurring sporadically southward in the state (Felger 2000a). It can be expected to spread to the Bahía Kino and Seri region and is a potentially serious threat to the island ecosystems. Native to the Old World.

> Hermosillo, Carr. Hermosillo a Bahía de Kino km. 24, a orilla del camino, *Reina 95-25* (USON). Vicinity of Rancho Picu, moist arroyos and about spring, *Turner 88-14.* 7.2 km (by air) SE of Puerto Libertad, *Van Devender 91-29.*

Caulanthus lasiophyllus (Hooker & Arnott) Payson. Jewell flower

Slender cool-season annuals with small white flowers.

> 4 mi E of El Desemboque, Arroyo San Ignacio floodplain, *Felger 14291.*

Dithyrea californica Harvey. Hehe imixaa; spectacle pod

Winter-spring annuals with fragrant white flowers. Sandy soils and dunes in the northern part of the Seri region.

> El Desemboque, S of town, *Wilder 08-193.*

Lepidium lasiocarpum Nuttall. Coquee; pepper-weed

Cool-season annuals with a spicy, chili-like flavor. Arroyo San Ignacio near El Desemboque and mesquite bosques near Bahía Kino. It seems unusual that it has not been found on Tiburón.

> 3.5 mi NE of El Desemboque, *Felger 17408.*

Sibara laxa (S. Watson) Greene

Cool-season annuals. The plants are easily overlooked.

> 1.5 mi S of El Desemboque, *Felger 17158.*

*Sisymbrium irio Linnaeus. Cocool; London rocket

Cool-season annuals with small yellow flowers. Arroyo San Ignacio near El Desemboque and expected in weedy places.

Rancho Miramar, 2 mi NW of El Desemboque, abandoned farmland, *Felger 17796.*

CACTACEAE

Cylindropuntia arbuscula (Engelmann) F.M. Knuth [*Opuntia arbuscula* Engelmann]. Heem, heem áa; *siviri, tasajo*; pencil cholla

Cholla shrubs with green stems and fruits. Locally on upper bajadas along the Infiernillo coast (Felger & Moser 1985).

0.7 mi W of paved rd to Libertad on Las Cuevitas rd, E side of Sierra Bacha, 2 m tree form, *Van Devender 88-852.*

Echinocereus engelmannii (Parry ex Engelmann) Rumpler. Hant ipzx íteja caacöl; strawberry hedgehog cactus

Spines mostly gray. Flowers large and bright purple-magenta.

1 km E of El Desemboque, *Felger 17765.*

Echinocereus nicholii (L.D. Benson) B. Parfitt. Hant ipzx íteja caacöl; golden hedgehog cactus

Robust multiple-stem cacti with yellow spines and pale pink flowers. Rocky slopes in the Sierra Seri (Felger & Moser 1985).

Hasteemla ["Hast Empla"], Sierra Seri, *Felger 18125.*

Ferocactus cylindraceus (Engelmann) Orcutt [*F. acanthodes* (Lemaire) Britton & Rose]. Mojepe siml; *biznaga*; mountain barrel cactus

Tall, slender barrel cacti in the vicinity of El Desemboque (Felger & Moser 1985).

0.4 mi E of Punta Cirio, Sierra Bacha, *Van Devender 91-75.*

Ferocactus emoryi (Engelmann) Orcutt [*F. covillei* Britton & Rose]. Siml caacöl; *biznaga*; Emory barrel cactus

Robust barrel cacti with stout spines and red flowers. Rocky slopes along the Sierra Seri (Felger & Moser 1985) and desert plains in the vicinity of Bahía Kino. Also on Isla Alcatraz.

Grusonia marenae (W.E. Parsons) E.F. Anderson [*Opuntia marenae* W.E. Parsons]. Xomcahoij

Inconspicuous, scrawny, cholla-like cacti with tuberous roots, grayish spines, and white flowers. Locally common among *Frankenia palmeri* at a few places along the Infiernillo coast and near Bahía Kino. We searched for this cactus in the *Frankenia* zone along the east side of Tiburón but did not find it.

Campo Almon, N of Punta Chueca, within 50 m of shore, *Felger 20572.*

Opuntia engelmannii var. **engelmannii**. Desert prickly pear

Low, usually spreading or sprawling prickly pears. Also on Alcatraz.

Sand flat desert near sea level, ca. 1 mi inland from north edge of Estero Sargento, *Felger 18160.*

Opuntia gosseliniana F.A.C. Weber. Heel, saapom; *duraznilla*; purple prickly pear

The "pads" become purplish during winter and dry season. Rocky slopes in the Sierra Seri. "One of several species planted near Punta Sargento by people of the Sargento Region" (Felger & Moser 1985:273).

Sierra Seri, 4 mi by road NW of Rancho Nochebuena, *Felger 14048a.*

CAPPARACEAE

Wislizenia palmeri A. Gray. Jackass clover

Perennials with bright yellow flowers. Semi-saline or alkaline soils in low-lying places.

3.1 mi NW of Kino, *Felger 15277.*

CONVOLVULACEAE

*Ipomoea pes-caprae (Linnaeus) R. Brown

Cultivated at Punta Chueca, and at Bahía Kino where it grows on fences and also untended and extending onto the upper beach. Tropical coastal regions worldwide.

Bahía Kino Nuevo, dunes, *Lindsay 12 Jun 1989.*

CRASSULACEAE

Crassula connata (Ruiz & Pavón) A. Berger [*C. erecta* (Hooker & Arnott) A. Berger. *Tillaea erecta* Hooker & Arnott]. Hant yapxöt; sand pygmyweed

Diminutive and succulent cool-season annuals. Common in the northern part of the Seri region.

1 km S of El Desemboque, *Felger 12460.*

Dudleya arizonica Rose. Hast yapxöt; Arizona liveforever

Herbaceous rosette succulents with thick leaves. Rock crevices on Cerro Tepopa.

NW side of Cerro Tepopa, *Felger 1423a.*

CUCURBITACEAE

Brandegea bigelovii (S. Watson) Cogniaux. Hehe iti scahjiit, hant caitoj; desert star-vine

Cool-season annual vines. Common in Arroyo San Ignacio near El Desemboque. These vines resemble *Vaseyanthus insularis*, which replace *Brandegea* to the south and on the Sonoran Islands.

Arroyo San Ignacio, 2 mi N of El Desemboque, *Felger 21270.*

Cucurbita digitata A. Gray. Ziix is cmasol; *chichi coyote*; coyote gourd

Sprawling warm-season vines from a thick, tuberous root, and producing rounded yellowish gourds. Sandy soils including those of abandoned farms.

3.5 mi NE of El Desemboque, floodplain of Río San Ignacio, *Felger 17384.*

Ibervillea sonorae (S. Watson) Greene var. **sonorae** [*Maximowiczia sonorae* S. Watson]. Hant yax; *guarequi*; cow-pie plant

Summer-fall vines from a swollen above-ground caudex. "North of Pozo Coyote by the big ranches" (Felger & Moser 1985:290).

EUPHORBIACEAE

Euphorbia incerta Brandegee [*Chamaesyce incerta* (Brandegee) Millspaugh] Seaside spurge

Low-growing annuals. Beaches and seaward margins of dunes.

2 mi S of Bahía Kino village, dunes and low hills, *Felger 17478.*

*Ricinus communis Linnaeus. Hehe caacoj; *higuerilla*; castor bean

Shrubs with large, deeply lobed leaves. Disturbed habitats at the margin of the Seri region including farmland at Rancho Miramar north of El Desemboque and occasionally along the Río San Ignacio riverbed. Native to the Old World.

Arroyo San Ignacio, 2 mi NW of El Desemboque, *Felger 21269.*

FABACEAE

Acacia farnesiana (Linnaeus) Willdenow [*Vachellia farnesiana* (Linnaeus) Wright & Arnott]. Oenoraama; *vinorama*; sweet acacia

Large woody shrubs with fragrant, yellow flowers. Scattered in the mainland Seri region, mostly in agricultural areas, probably spread by cattle.

Pozo Coyote, *Felger 17783.*

Astragalus magdalenae Greene var. **magdalenae**. Iix casa insii; satiny milkvetch

Robust cool-season annuals with silver foliage and inflated pods; coastal dunes.

Vicinity of El Desemboque, *Felger 6211.*

Caesalpinia palmeri S. Watson. Hap oaacajam; *piojito*

Spineless shrubs with bright yellow flowers. Recorded from the Sierra Seri.

Hasteemla, Sierra Seri, *Moser & Moser 2 Feb 1969.*

*Leucaena leucocephala (Lamark) de Wit subsp. **glabrata** (Rose) S. Zárate. *Guaje*; white lead-tree

Large shrubs or small trees in weedy washes or arroyos in Bahía Kino.

Bahía Kino Nuevo, arroyo, *Wilder 07-462.*

*Parkinsonia aculeata Linnaeus. Snapxöl; *bagota*; Mexican palo verde

Trees with long, slender leaves and bright yellow flowers in spring. Widely cultivated in the region and weedy in the Bahía Kino region.

Kino Nuevo, roadside, 28 Sep 2007, *Felger & Wilder*, observation.

FOUQUIERIACEAE

Fouquieria columnaris (Kellogg) Curran [*Idria columnaris* Kellogg]. Cótotaj; *cirio*; boojum tree

Slender, spiny trees with white flowers. Common in the Sierra Bacha from about 5 miles north of El Desemboque to Punta Cirio about 9 miles south of Puerto Libertad. The only island population occurs near the summit of Ángel de la Guarda (Moran 1983b).

5 mi N of El Desemboque San Ignacio along coast, near sea, some inland on steep slopes and canyons, *Felger 5249.*

FRANKENIACEAE

Frankenia salina (Molina) I.M. Johnston. Alkali heath

Herbaceous perennials in salt scrub at esteros along the Infiernillo.

Estero Sargento, *Felger 84–86.*

GERANIACEAE

Erodium texanum A. Gray

Cool-season annuals with small pink flowers.

Cerro Pelón, 5 mi SE of El Desemboque; *Felger 17920.*

JUNCACEAE

Juncus acutus Linnaeus subsp. **leopoldii** (Parlatore) Snogerup. Caail oocmoj; *junco espinoso*; spiny rush

Large perennial rush in low, seasonally wet mud flats in the Bahía Kino region.

3.6 mi NE of village of Bahía Kino, *Felger 15262.*

LAMIACEAE

Salvia columbariae Bentham. Hehe yapxöt; chia

Cool-season annuals with small blue flowers and edible seeds. Northern part of the Seri region, often on bajadas and sandy-gravelly soils.

Vicinity of Cerro Pelón, 5 mi SE of El Desemboque, sandy arroyo, *Felger 17896.*

Teucrium cubense Jacquin. Hehe itac coozalc

Herbs in low, wet places such as edges of dry lakebeds.

Río San Ignacio, 2 mi N of El Desemboque, riverbed, *Felger 17021.*

Teucrium glandulosum Kellogg. Hehe itac coozalc

Upright-growing perennial herbs with attractive blue and white flowers. Occasional along arroyos and margins of dry lakes.

3.8 mi S of El Desemboque, dry lakebed, *Felger 14091.*

LOASACEAE

Mentzelia involucrata S. Watson. Hehe cotopl

Cool-season annuals, the flowers silvery yellow.

Floodplain of Río San Ignacio, 3.5 mi NE of El Desemboque, *Felger 17425.*

MALVACEAE

Hermannia pauciflora S. Watson

Small herbaceous perennials with yellow flowers. Rock slopes in the Sierra Seri.

East side and base of Sierra Nochebuena, mountains W of Playa San Bartolo, *Felger 08-233.*

*Malva parviflora Linnaeus. *Malva*; cheeseweed

Cool-season annuals. Common weed in disturbed habitats. Native to the Old World. "Apparently a relatively recent invader in El Desemboque and Punta Chueca" (Felger & Moser 1985:346).

El Desemboque, *Wilder 08-186.*

Sphaeralcea coulteri (S. Watson) A. Gray [*S. orcuttii* Rose]. *Mal de ojo*; annual globemallow

Annuals, probably non-seasonal, the flowers bright orange. Highly variable in size; exceptionally large, shrubby plants grow on dunes near Bahía Kino and also occur on Isla Alcatraz.

Vicinity of El Desemboque, *M. B. Moser 10 Feb 1962.*

MENNISPERMACEAE

Cocculus diversifolius de Candolle

Woody, perennial vines. Canyons and arroyos north and eastward from the vicinity of Pozo Coyote.

Pozo Coyote, *Felger 6168.*

NYCTAGINACEAE

Abronia villosa S. Watson. Hantoosinaj; sand verbena

Attractive spring wildflowers with pink-purple flowers. Coastal dunes and sandy soils from Cerro Tepopa northward.

NW side of Cerro Tepopa, *Felger 14986.*

*Boerhavia coccinea Miller. Scarlet spiderling

Herbaceous perennials or annuals with sticky foliage and small, bright magenta flowers. Weedy places and probably not native to the Seri region.

Bahía Kino Nuevo, weedy arroyo, 1 Oct 2007, *Felger & Wilder,* observation.

Boerhavia wrightii A. Gray. Large-bract spiderling

Hot-weather annuals with small pink flowers withering by mid-morning.

Canyon del Coyote, 4 mi NE of El Desemboque de San Ignacio, *M. B. Moser 28 Sep 1961.*

Mirabilis laevis (Bentham) Curran var. **villosa** (Kellogg) Spellenberg [*M. bigelovii* A. Gray]. Hepem isla; desert four o'clock

Herbaceous perennials with sticky foliage and small white flowers. In the Sierra Seri and other mainland mountains.

Hasteemla ["Hast Empla"], Sierra Seri, *Felger 18146.*

ONAGRACEAE

Chylismia claviformis (Torrey & Frémont) A. Heller subsp. **yumae** (P.H. Raven) W.L. Wagner & Hoch [*Camissonia claviformis* (Torrey & Frémont) P.H. Raven subsp. *yumae* P.H. Raven]. Hantoosinaj ctam; browneyes

Cool-season annuals with pale yellowish flowers. In the northern part of the Seri region.

6.5 mi by rd (to Pozo Coyote) NE of El Desemboque, *Felger 17776-J.*

Oenothera arizonica (Munz) W.L. Wagner [*O. californica* (S. Watson) S. Watson subsp. *arizonica* (Munz) Klein]. Hantoosinaj; Arizona evening primrose

Cool-season annuals, flowers pink to white. Dunes near El Desemboque. Reported for Tiburón by Felger and Moser (1985), but no specimens are known from the island.

1 mi NW of El Desemboque, coastal dunes, *Van Devender 11 Mar 1977.*

PAPAVERACEAE

Argemone gracilenta Greene. Xazacöz; *cardo;* spiny poppy

Robust herbaceous perennials, spiny throughout, the flowers large with white petals and numerous yellow stamens. Weedy habitats and perhaps not native to the Seri region.

Rancho Miramar, 2 mi NW of El Desemboque, common in farmland, *Felger 17792.*

Eschscholzia minutiflora S. Watson. Little gold poppy

Cool-season annuals with small orange flowers. Common in the northern part of the Seri region.

3.5 mi NE of El Desemboque, *Felger 17391.*

PLANTAGINACEAE

Maurandya antirrhiniflora Humboldt & Bonpland ex Willdenow. Snapdragon vine

Herbaceous vines, the corollas blue with a yellow lip. Major drainageways in the northern part of the Seri region.

Rancho Miramar, 2 mi NW of El Desemboque, *Felger 17777.*

Penstemon parryi (A. Gray) A. Gray. Nojoopis caacöl; Parry penstemon

Cool-season annuals with attractive pink flowers. Mountains including the Sierra Seri.

Cerro Pelón, ca. 5 mi SE of El Desemboque, *Felger 17093.*

POACEAE

*Cenchrus setaceus (Forsskål) Morrone [*Pennisetum setaceum* (Forsskål) Chiovenda] Fountain grass

Perennial clumping grass. Native to the Old World. Often planted as an ornamental across the Sonoran Desert and a potentially serious invasive species.

*Cenchrus spinifex Cavanilles [*C. incertus* M.A. Curtis. *C. pauciflorus* Bentham]. *Huizapori, guachapori, toboso;* field sandbur, common sandbur

Aggressive weed and on dunes in the Bahía Kino region; not native to the Seri region. A potentially invasive species on Tiburón.

Bahía Kino Nuevo, dunes and roadside, *Van Devender 83-86.*

*Chloris barbata Swartz. Swollen windmill-grass

Annuals, often robust. Common in sandy soil in weedy, uncultivated places in the Bahía Kino region, where it was first documented in 1998. Native in eastern Mexico and adventive in western Mexico.

Bahía de Kino Nuevo, locally common in wet sand in yard, *Van Devender 98-793.*

*Cynodon dactylon (Linnaeus) Persoon var. **dactylon.** *Zacate bermuda, zacate inglés*; Bermuda grass

Common and widespread weed in the Bahía Kino region and expected to spread into other disturbed and wetland areas. Native to the Old World and very widespread in Sonora. A potentially serious invasive species in wetland habitats on Tiburón.

Kino Nuevo, vicinity of Prescott College Field Station, 23 May 2006, *Felger & Wilder,* observation.

*Dactyloctenium aegyptium (Linnaeus) Willdenow. *Zacate pata de cuervo*; crowfoot grass

Warm-weather annuals. Mostly in disturbed habitats. Native to the Old World.

W of Hermosillo between the Seven Sisters [Siete Cerros] and the coast, *Beetle M-5267.*

Distichlis palmeri (Vasey) Fassett. *Zacate salado*; nipa

Perennial saltgrass, male and female flowers on different plants. Common in parts of Estero Santa Cruz at Bahía Kino, this being the southernmost record for this species on mainland Mexico. We have searched potential habitats on Tiburón and have not found it there.

Estero de la Cruz [Santa Cruz], *Felger 07-150.*

*Eragrostis cilianensis (Allioni) Vignolo Lutati ex Janchen. *Zacate apestoso*; stinking lovegrass, stink grass

Warm-weather annuals. Widespread in western Sonora, especially on summer-wet sandy soils. Native to the Old World.

Arroyo San Ignacio, 6.3 mi N-NE of El Desemboque, localized, sandy riverbank, *Felger 20499.*

Festuca octoflora Walter [*Vulpia octoflora* (Walter) Rydberg]. Six-weeks fescue, eight-flowered fescue

Diminutive cool-season annuals, ranging southward to the vicinity of El Desemboque.

3.5 mi by rd NE of El Desemboque, *Felger 17432.*

Hilaria rigida (Thurber) Bentham ex Scribner [*Pleuraphis rigida* Thurber]. *Tobosa*; big galleta

Large perennial clumping grass. Reaching its southern limits in the Seri region.

Punta Santa Rosa, sandy soil, *Felger 20096.*

Jouvea pilosa (J. Presl) Scribner. Cocasjc, xojasjc

Mound-forming saltgrass; male and female flowers on different plants. The northernmost limits of this tropical species are three small and isolated populations along the mainland shore of the Infiernillo Channel.

6.5 mi N of Punta Chueca, *Felger 83-116a & 116b.* N of Kino Bay, 8 Dec 1931, *Shantz & McGinnies M.T.-9.*

Leptochloa viscida (Scribner) Beal [*Diplachne viscida* Scribner]. Sticky sprangletop

Warm-weather annuals. Widespread in western Sonora, mostly on seasonally damp soils.

10.5 mi SE of Bahía Kino, edge of drying mud flat, *Helmkamp 1-3.*

Panicum alatum Zuloaga & Morrone var. **minus** (Andersson) Zuloaga & Morrone [*P. hirticaule* var. *minus* Andersson]. Winged panicgrass

Hot-weather annuals, resembling *P. hirticaule* but with ear-like fleshy expansions at the base of the lemma.

1 mi inland from Punta Sargento, *Felger 20516.*

Sporobolus wrightii Munro ex Scribner [*S. airoides* (Torrey) Torrey var. *wrightii* (Munro ex Scribner) Gould]. Big sacaton

Large perennial grasses. Margins of inland salt flats and dry lakebeds.

3.6 mi NE of Bahía Kino, low saline flats, *Felger 15260.*

POLEMONIACEAE

Linanthus bigelovii (A. Gray) Greene

Cool-season annuals. Corollas white, opening in the evening.

19 mi S of El Desemboque, on flats, *Felger 17178.*

POLYGONACEAE

Chorizanthe brevicornu Torrey subsp. **brevicornu**. Brittle spineflower, short-horn spineflower

Winter-spring annuals with minute white flowers, the plants breaking apart at maturity. Northern part of the Seri region.

6.5 mi by rd NE of El Desemboque, *Felger 17776-B*.

Chorizanthe corrugata (Torrey) Torrey & A. Gray

Winter-spring annuals with minute flowers, the plants breaking apart at maturity. Northern part of the Seri region.

Sierra Bacha, Las Cuevitas, *Felger 14272*.

Eriogonum trichopes Torrey. Little desert-trumpet

Winter-spring annuals. Northern part of the Seri region.

23 mi by rd S of El Desemboque, along Infiernillo, 200 m inland, *Felger 14058*.

Rumex inconspicuus Rechinger f. *Cañaigre, hierba colorado*; dock

Winter-spring annuals with a stout taproot. Wet mud in roadside ditches and other temporarily wet habitats.

Roadside, 10.5 mi E of the shore at Bahía Kino, *Felger 07-1*.

SANTALACEAE

Struthanthus palmeri Kuijt [*S. haenkei* (Presl) Engler, in part]. *Toji*

Parasitic on various trees including mesquite (*Prosopis glandulosa*), the stems pendent, leafy, and often twisted in loose spirals.

Rancho Carrizo, NE of Bahía Kino, *Felger 07-134*.

SELAGINELLACEAE

Selaginella arizonica Maxon. Hehe quina caacöl; *flora de piedra*; desert spike-moss

Low, moss-like plants among rocks. Higher elevations on Sierra Seri on north- and east-facing slopes. Despite our searches, we have not found any Selagninella on Isla Tiburón in places that appear to be suitable habitat.

Hasteemla ["Hast Empla"], ca. 500 m, NE-facing slope near summit, *Felger 74-9*.

Selaginella eremophila Maxon. Hehe quina caacöl; *flora de piedra*; desert spike-moss

Low, moss-like plants among rocks. Higher elevations on Sierra Seri on north- and east-facing slopes. Resembling *S. arizonica*; they are distinguished by technical features.

Hasteemla ["Hast Empla"], ca. 500 m, NE-facing slope near summit, *Felger 18135*.

SOLANACEAE

Lycium californicum A. Gray. Hahjöizij; California desert-thorn

Low shrubs with rigid branches and thorn-tipped twigs, and small white flowers. Locally common in scattered localities near the shore.

1 km S of El Desemboque, semi-saline flat immediately inland from coastal dunes, *Felger 17238A*.

Lycium macrodon A. Gray var. **macrodon**. Hehe ix cooil; desert wolfberry

Thorny shrubs with glaucous leaves and greenish flowers. Large arroyos in the northern part of the Seri region.

Arroyo San Ignacio, 1.6 mi SW of Pozo Coyote, *Felger 17059*.

*Nicotiana glauca Graham. Nojoopis caacöl; *juan loco*; tree tobacco

Sparsely branched shrubs with yellow, tubular flowers. Weedy places such as ranches and around Bahía Kino (Felger & Moser 1985).

VERBENACEAE

Aloysia gratissima (Gillies & Hooker) Troncoso [*A. lycioides* Chamisso]. Hapsx iti icoocax; whitebrush

Slender-stemmed shrubs with small fragrant white flowers. Arroyo San Ignacio northeast of El Desemboque (Felger & Moser 1985).

Pozo Coyote, 10 km N from El Desemboque along Arroyo San Ignacio, *Felger 83-105* (USON).

Verbena menthifolia Bentham. Mint vervain

Herbaceous perennials with small blue flowers. Playas (dry lakebeds) and low-lying mesquite thickets.

3.8 mi by rd S of El Desemboque, dry lakebed, corollas blue, *Felger 14094*.

Appendix C

Botanical Name Changes

We are following the APG III (Angiosperm Phylogeny Group) designations of plant families as of June 2011, reflecting current knowledge of phylogenetic relationships. This dynamic system reflects an amalgamation of information with the goal of having a natural classification system (one that reflects evolutionary relationships). In the last decades a multitude of revisions of species, genera, and families across the tree of life have been produced, significantly altering the placement of taxa set forth in previous treatments. While many of these changes may be inconvenient, newer nomenclature can enhance understanding of the relationships and histories of biological diversity. The production of this flora falls in the midst of what may be viewed in the future as the "era of re-classification," and one can expect revisions of some scientific names that are used in this book—science marches on. Furthermore, differences of opinion and disagreements are to be expected. The ever-changing science of biological classification can be frustrating and risks alienating those who have become familiar with the "old names." When Ben was explaining these scientific name changes to Humberto, he responded, "Entonces, vamos a cambiar nuestros nombres cada año y vamos a ver cómo ustedes pueden hacer." (Well then, we are going to change our [Seri] names each year and then see how you are able to do.)

The following listing reflects changes in botanical nomenclature since the publication of *People of the Desert and Sea: Ethnobotany of the Seri Indians* (Felger & Moser 1985). We also include other recent changes in nomenclature for the flora area, not just those that differ from the ethnobotany book. Name changes are arranged in alphabetical order of the previously used name ("old"). The newly accepted or currently used name ("new") is in **boldface**. The Cmiique Iitom (language of the Comcaac) name is included when known.

533

AGAVACEAE

Old: all species in Agavaceae

New: all species in **Asparagaceae**

AIZOACEAE

Old: *Mollugo* in Aizoaceae

New: *Mollugo* in **Molluginaceae**

Old: *Sesuvium verrucosum* (a valid name but incorrectly applied to plants on the Sonoran Islands)

New: *Sesuvium portulacastrum* (the species on Alcatraz and Tiburón)

ASCLEPIADACEAE

Old: Asclepiadaceae

New: all species in **Apocynaceae**

Old: *Sarcostemma cynanchoides* subsp. *hartwegii*

New: *Funastrum hartwegii*
     Hexe

ASTERACEAE

Old: *Dyssodia concinna*

New: *Thymophylla concinna*
     Cahaahazxot

Old: *Eupatorium sagittatum*

New: *Chromolaena sagittata*
     Comima

Old: *Hofmeisteria laphamioides*

New: *Pleurocoronis laphamioides*
     Hamt inoosj, hast yapxöt

Old: *Hymenoclea salsola*

New: *Ambrosia salsola*
     Caasol cacat, caasol coozlil, caasol ziix iic cöihiipe

Old: *Machaeranthera pinnatifida* var. *incisifolia*

New: *Xanthisma incisifolium*

Old: *Machaeranthera pinnatifida* var. *scabrella*

New: *Xanthisma spinulosum* var. *scabrella*

Old: *Porophyllum crassifolium*

New: *Bajacalia crassifolia*

Old: *Viguiera deltoidea* var. *chenopodina*

New: *Bahiopsis chenopodina*
     Hehe imoz coopol

Old: *Viguiera triangularis*

New: *Bahiopsis triangularis*

AVICENNIACEAE

Old: *Avicennia germinans* in Avicenniaceae or Verbenaceae

New: *Avicennia germinans* in **Acanthaceae**
     Pnacojiscl

CACTACEAE

Old: *Opuntia bigelovii*

New: *Cylindropuntia bigelovii*
     Coote, sea

Old: *Opuntia* cf. *burrageana*

New: *Cylindropuntia alcahes* var. *alcahes*
     Heem icös cmasl

Old: *Opuntia cholla*

New: *Cylindropuntia cholla*

Old: *Opuntia fulgida*

New: *Cylindropuntia fulgida*
     Sea cotopl, sea icös cooxp

Old: *Opuntia leptocaulis*

New: *Cylindropuntia leptocaulis*
     Iipxö

Old: *Opuntia phaeacantha* var. *discata*

New: *Opuntia engelmannii* var. *engelmannii*
     Heel hayeen ipaii

Old: *Opuntia versicolor*

New: *Cylindropuntia versicolor*
     Heem icös cmáxlilca

Old: *Neoevansia striata*

New: *Peniocereus striatus*
     Xtooxt

CAMPANULACEAE

Old: *Nemacladus glanduliferus* var. *orientalis*

New: **Nemacladus orientalis**
Xtamaaija oohit

CAPPARACEAE

Old: *Cleome tenuis* in Capparaceae

New: *Cleome tenuis* in **Cleomaceae**
Cocool

CHENOPODIACEAE

Old: Chenopodiaceae

New: all species in **Amaranthaceae**

COCHLOSPERMACEAE

Old: *Amoreuxia palmatifida* in Cochlospermaceae

New: *Amoreuxia palmatifida* in **Bixaceae**
Joját

EUPHORBIACEAE

Old: *Andrachne microphylla* in Euphorbiaceae

New: *Andrachne microphylla* in **Phyllanthaceae**
Hasoj an hehe

Old: *Pedilanthus macrocarpus*

New: **Euphorbia lomelii**

FABACEAE

Old: *Cercidium floridum*

New: **Parkinsonia florida**
Ziij, iiz

Old: *Cercidium microphyllum*

New: **Parkinsonia microphylla**
Ziipxöl

Old: *Lotus salsuginosus, L. salsuginosus* var. *brevivexillus*

New: **Acmispon maritimus** var. **brevivexillus**
Hee inoosj

Old: *Lotus tomentellus, L. strigosus* var. *tomentellus*

New: **Acmispon strigosus**
Hee inoosj

Old: *Pithecellobium confine*

New: **Ebenopsis confinis**
Heejac

HYDROPHYLLACEAE

Old: all species in Hydrophyllaceae

New: all species in **Boraginaceae**

Old: *Phacelia ambigua*

New: **Phacelia crenulata**
Caháahazxot ctam, najmís

KRAMERIACEAE

Old: *Krameria grayi*

New: **Krameria bicolor**

LILIACEAE

Old: *Triteleiopsis palmeri* in Liliaceae

New: *Triteleiopsis palmeri* in **Asparagaceae**
Caal oohit

MALPIGHIACEAE

Old: *Janusia californica*

New: **Cottsia californica**
Hehe quiijam

Old: *Janusia gracilis*

New: **Cottsia gracilis**
Hehe quiijam

Old: *Mascagnia macropterum*

New: **Callaeum macroptera**
Haxz oocmoj

NOLINACEAE

Old: *Dasylirion gentryi* in Nolinaceae

New: *Dasylirion gentryi* in **Asparagaceae**

NYCTAGINACEAE

Old: *Boerhavia intermedia*, etc. *B. coulteri* (misapplied)

New: **Boerhavia triquetra**
Hamíp caacöl

ONAGRACEAE

Old: *Camissonia californica*

New: **Eulobus californicus**

Old: *Camissonia cardiophylla* subsp. *cardiophylla* and subsp. *cedrosensis*

New: **Chylismia cardiophylla** subsp. **cardiophylla** and subsp. **cedrosensis**

PHYTOLACCACEAE

Old: *Phaulothamnus spinescens* in Phytolaccaceae

New: *Phaulothamnus spinescens* in **Achatocarpaceae**
      Aniicös

Old: *Stegnosperma halimifolium* in Phytolaccaceae

New: *Stegnosperma halimifolium* in **Stegnospermataceae**
      Xnejamsiictoj

PLANTAGINACEAE

Old: *Plantago insularis* var. *fastigiata*

New: **Plantago ovata** var. **fastigiata**
      Hatajeen

PLUMBAGINACEAE

Old: *Plumbago scandens*

New: **Plumbago zeylanica**
      Hatajeen

POACEAE

Old: *Brachiaria arizonica*

New: **Urochloa arizonica**
      Conee ccapxl

Old: *Brachiaria fasciculata*

New: **Urochloa fusca**

Old: *Cathestecum brevifolium*

New: **Bouteloua diversispicula**

Old: *Erioneuron pulchellum*

New: **Dasyochloa pulchella**
      Conee ccosyat, conee ccapxl

Old: *Chloris brandegeei, C. chloridea, Enteropogon brandegeei*

New: **Enteropogon chlorideus**

Old: *Leptochloa filiformis*

New: **Leptochloa panicea** subsp. **brachiata**

Old: *Monanthochloe littoralis*

New: **Distichlis littoralis**
      Cötep

PTERIDACEAE

Old: *Notholaena sinuata*

New: **Astrolepis sinuata**

SAPOTACEAE

Old: *Bumelia occidentalis*

New: **Sideroxylon occidentale**
      Hehe hateen ccaptax, paaza

SCROPHULARIACEAE

Old: all species in the flora area in Scrophulariaceae

New: all species in the flora area in **Plantaginaceae**

Old: *Antirrhinum cyathiferum*

New: **Pseudorontium cyathiferum** in **Plantaginaceae**
      Hehemonlc

Old: *Antirrhinum kingii, A. kingii* var. *watsonii*

New: **Sairocarpus watsonii** in **Plantaginaceae**
      Comaacöl

Old: *Galvezia juncea*

New: **Gambelia juncea** in **Plantaginaceae**
      Nojoopis caacöl

Old: *Linaria texana*

New: **Nuttallanthus texanus** in **Plantaginaceae**

STERCULIACEAE

Old: Sterculiaceae

New: all species in **Malvaceae**

TAMARICACEAE

Old: *Tamarix ramosissima*

New: **Tamarix chinensis**

THEOPHRASTACEAE

Old: *Jacquinia macrocarpa* subsp. *pungens*

New: **Bonellia macrocarpa** subsp. *pungens*
    Cof

ULMACEAE

Old: *Celtis pallida* in Ulmaceae

New: *Celtis pallida* in **Cannabaceae**
    Ptaacal

Old: *Celtis reticulata* in Ulmaceae

New: *Celtis reticulata* in **Cannabaceae**

VISCACEAE

Old: *Phoradendron californicum* in Viscaceae

New: *Phoradendron californicum* in **Santalaceae**
    Eaxt

Old: *Phoradendron diguetianum* in Viscaceae

Old: *Phoradendron diguetianum* in **Santalaceae**
    Eaxt

Literature Cited

Anderson, C. 2007. Revision of *Galphimia* (Mapighiaceae). *Contributions from the University of Michigan Herbarium* 25:1–82.

Anderson, W. R. 1993. Note on neotropical Malpighiaceae—IV. *Contributions from the University of Michigan Herbarium* 19:355–392.

Anderson, W. R., & W. C. Davis. 2007a. Generic adjustments in neotropical Mapighiaceae. *Contributions from the University of Michigan Herbarium* 25:137–166.

———. 2007b. Revision of *Galphimia* (Mapighiaceae). *Contributions from the University of Michigan Herbarium* 25:1–82.

Aragón-Arreola, M., & A. Martín-Barajas. 2007. Westward migration of extension in the northern Gulf of California, Mexico. *Geology* 35:571–574.

Aragón-Arreola, M., M. Morandi, A. Martín-Barajas, L. Delgado-Argote, & A. González-Fernández. 2005. Structure of the rift basins in the central Gulf of California: kinematic implications for oblique rifting. *Tectonophysics* 409:19–38.

Austin, D. F. 2008. *Evolvulus alsinoides* (Convolvulaceae): an American herb in the Old World. *Journal of Ethnopharmacology* 117:185–198.

Backman, T. W. H. 1991. Genotypic and phenotypic variability of *Zostera marina* on the West Coast of North America. *Canadian Journal of Botany* 69:1361–1371.

Baldwin, B. G., & B. L. Wessa. 2000. Phylogenetic Placement of *Pelucha* and new subtribes in Helenieae sensu stricto (Compositae). *Systematic Botany* 25:522–528.

Barkworth, M. E., K. M. Capels, S. Long, & M. B. Piep, editors. 2003. Flora of North America Editorial Committee, editors, *Flora of North America North of Mexico*, vol. 25. Oxford University Press, New York.

Basurto, X. 2005. How locally designed access and use controls can prevent the Tragedy of the Commons in a Mexican small-scale fishing community. *Society and Natural Resources* 18:643–659.

———. 2006. Commercial diving and the callo de hacha fishery in Seri Territory. *Journal of the Southwest* 48:189–209.

———. 2008. Biological and ecological mechanisms supporting marine self-governance: the Seri callo de hacha fishery in Mexico. *Ecology and Society* 13(2):20.

Beaty, J. J. 1964. *Plants in His Pack: A Life of Edward Palmer, Adventurous Botanist and Collector*. Pantheon Books, New York.

Becerra, J. X., & D. L. Venable. 1999. Nuclear ribosomal DNA phylogeny and its implications for evolutionary trends in Mexican *Bursera* (Burseraceae). *American Journal of Botany* 86:1047–1057.

Betancourt, J. L., T. R. Van Devender, & P. S. Martin, editors. 1990. *Packrat Middens: The Last 40,000 Years of Biotic Change*. University of Arizona Press, Tucson.

Bostic, D. L. 1971. Herpetofauna of the Pacific coast of north central Baja California, Mexico, with a description of a new subspecies of *Phyllodactylus xanti*. *Transactions of the San Diego Society of Natural History* 16(10):237–263.

———. 1975. *A Natural History Guide to the Pacific Coast of North Central Baja California and Adjacent Islands*. Biological Educational Expeditions, Vista, CA.

Bourillón-Moreno, L. 2002. *Exclusive Fishing Zone as a Strategy for Managing Fishery Resources by the Seri Indians, Gulf of California, Mexico*. PhD diss., University of Arizona, Tucson.

Bowen, T. 1976. *Seri Prehistory: The Archaeology of the Central Coast of Sonora, Mexico*. Anthropological Papers of the University of Arizona 27. University of Arizona, Tucson.

———. 1983. Seri. Pages 230–249 *in* A. Ortiz, editor, *Handbook of North American Indians*, vol. 10. Smithsonian Institution, Washington, DC.

————. 2000. *Unknown Island: Seri Indians, Europeans, and San Esteban Island in the Gulf of California*. University of New Mexico Press, Albuquerque.

————. editor. 2002. *Backcountry Pilot: Flying Adventures with Ike Russell*. University of Arizona Press, Tucson.

————. 2003. Hunting the elusive organ pipe cactus on San Esteban Island in the Gulf of California. *Desert Plants* 19:15–28.

————. 2008. Bad day at Black Rock. *New Mexico Historical Review* 83:451–474.

————. 2009. *The Record of Native People on Gulf of California Islands*. Arizona State Museum Archaeological Series, no. 201. University of Arizona, Tucson.

Bowen, T., R. S. Felger, & R. J. Hills. 2004. Chollas, circles and Seris: did Seri Indians plant cactus at Circle 6? *Desert Plants* 20(2):26–35.

Bowen, T., D. W. Bench, & L. A. Johnson. 2006. Recent colonization of Midriff Islands, Gulf of California, Mexico, by feral honeybees, *Apis mellifera*. *Southwestern Naturalist* 51:542–551.

Bowers, J. E. 1988. *A Sense of Place: The Life and Work of Forrest Shreve*. University of Arizona Press, Tucson.

Brennan, K. 2005. Heady harvest: from Mexico's Seri Indians comes an intense, aromatic, very flavorful oregano. *Saveur* 88:46.

Britton, N. L., & J. N. Rose. 1919–1923. *The Cactaceae*, 4 vols. Carnegie Institution of Washington Publication no. 248. Carnegie Institution, Washington, DC.

Brown, D. E. 1982. Biotic communities of the American Southwest—United States and Mexico. *Desert Plants* 4:3–341.

Brummitt, R. K., & C. E. Powell, editors. 1992. *Authors of Plant Names*. Royal Botanic Gardens, Kew.

Búrquez, A., A. Martínez-Yrízar, R. S. Felger, & D. Yetman. 1999. Vegetation and habitat diversity at the southern edge of the Sonoran Desert. Pages 36–67 *in* R. H. Robichaux, editor, *Ecology of Sonoran Desert Plants and Plant Communities*. University of Arizona Press, Tucson.

Búrquez-Montijo, A., M. E. Miller, & A. Martínez-Yrízar. 2002. Mexican grasslands, thornscrub, and the transformation of the Sonoran Desert by invasive exotic buffelgrass (*Pennisetum ciliare*). Pages 126–146 *in* B. Tellman, editor, *Invasive Exotic Species in the Sonoran Region*. University of Arizona Press, Tucson.

California Invasive Plant Inventory. 2009. California Invasive Plant Council. http://www.cal-ipc.org/ip/inventory/index.php (accessed 5 July 2010).

Carabillas Lillo, J., J. de la Maza Elvira, D. Gutiérrez Carbonell, M. Gómez Cruz, G. Anaya Reina, A. Zavala González, A. L. Figuroa, & B. Bermúdez Almada. 2000. *Programa de Manejo Área de Protección de Flora y Fauna Islas del Golfo de California, México*. Comisión Nacional de Áreas Naturales Protegidas, Secretará de Medio Ambiente, Recursos Naturales y Pesca, México, DF.

Carlquist, S. J. 1965. *Island Life: A Natural History of the Islands of the World*. Natural History Press, New York.

Carpenter, A. 2003. Element Stewardship Abstract for *Tamaix ramosissima* Ledebour. The Nature Conservancy Wildland Weed Management and Research Program. http://tncweeds.ucdavis.edu/esadocs/documnts/tamaram.pdf (accessed 8 April 2008).

Carreño, A. L., & J. Helenes. 2002. Geology and ages of the islands. Pages 14–40 *in* T. J. Case, M. L. Cody, & E. Ezcurra, editors, *A New Island Biogeography of the Sea of Cortés*. Oxford University Press, New York.

Case, T. J., & M. L. Cody, editors. 1983. *Island Biogeography of the Sea of Cortéz*. University of California Press, Berkeley.

Case, T. J., M. L. Cody, & E. Ezcurra, editors. 2002. *A New Island Biogeography of the Sea of Cortés*. Oxford University Press, New York.

Chiang, F., & L. R. Landrum. 2009. Vascular plants of Arizona: Solanaceae part three: *Lycium* L., wolf berry, desert thorn. *Canotia* 5(1):17–26.

Chiang-Cabrera, F. 1981. *A Taxonomic Study of the North American Species of Lycium (Solanaceae)*. PhD diss., University of Texas, Austin.

Clark, L. G., & E. A. Kellogg. 2007. Poaceae Barnhart. Grass family. Pages 3–10 *in* Flora of North America Editorial Committee, editors, *Flora of North America North of Mexico*, vol. 24. Oxford University Press, New York.

Clark, P. U., & A. C. Mix. 2002. Ice sheets and sea level of the Last Glacial Maximum. *Quaternary Science Reviews* 21:1–7.

Clarke, P. J., P K. Latz, & D. E. Albrecht. 2005. Long-term changes in semi-arid vegetation: invasion of an exotic perennial grass has larger effects than rainfall variablity. *Journal of Vegetation Science* 16:237–248.

Clark-Tapia, R., & F. Molina-Freaner. 2003. The genetic structure of a columnar cactus with a disjunct distribution: *Stenocereus gummosus* in the Sonoran Desert. *Heredity* 90:443–450.

Clayton, W. D., & S. A. Renvoize. 1986. *Genera Graminum, Grasses of the World*. Kew Bulletin Additional Series no. 8. Her Majesty's Stationery Office, London.

Cliffton, K., D. O. Cornejo, & R. S. Felger. 1982. Sea turtles of the Pacific coast of Mexico. Pages 199–209 *in* K. Bjorndal, editor, *Biological Conservation of Sea Turtles*. Smithsonian Institution Press, Washington, DC.

Cody, M. L., R. Moran, & H. Thompson. 1983. The plants. Pages 49–97 *in* T. J. Case & M. L. Cody, editors, *Island Biogeography of the Sea of Cortéz*. University of California Press, Berkeley.

Colchero, F., R. A. Medellin, J. S. Clark, R. Lee, & G. G. Katul. 2009. Predicting population survival under future climate change: density dependence, drought and extraction in an insular bighorn sheep. *Journal of Animal Ecology* 78:666–673.

CONANP (Comisión Nacional de Áreas Naturales Protegidas). 2007. Programa de Conservación y Manejo Reserva de la Biosfera Isla San Pedro Mártir, Mexico. SEMARNAT (Secretaría de Medio Ambiente y Recursos Naturales, México, DF). http://www2.ine.gob.mx/publicaciones/consultaPublicacion.html?id_pub=569&id_tema=12&dir=Consultas (accessed 10 January 2010).

Correll, D. S., & M. C. Johnston. 1970. *Manual of the Vascular Plants of Texas.* Texas Research Foundation, Renner.

Costea, M. 2009. Digital Atlas of *Cuscuta* (Convolvulaceae). http://www.wlu.ca/page.php?grp_id=2147&p=8968 (accessed 6 May 2009).

Cronquist, A. 1981. *An Integrated System of Classification of the Flowering Plants.* Columbia University Press, New York.

Daniel, T. F. 1984. The Acanthaceae of the southwestern United States. *Desert Plants* 5:162–179.

———. 1997. The Acanthaceae of California and the peninsula of Baja California. *Proceedings of the California Academy of Sciences* 49:309–403.

———. 2004. Acanthaceae of Sonora: taxonomy and phytogeography. *Proceedings of the California Academy of Sciences* 55:690–805.

Davidse, G., & E. Morton. 1973. Bird-mediated fruit dispersal in the tropical grass genus *Lasiacis* (Gramineae: Paniceae). *Biotropica* 5:162–167.

Davis, E., & E. Y. Dawson. 1945. The savage Seris of Sonora—I. *Scientific Monthly* 60(3):193–202; II. *Scientific Monthly* 60(4):261–268.

Davis, L. G. 2006. Baja California's paleoenvironmental context. Pages 14–23 *in* D. Laylander & J. D. Moore, editors, *The Prehistory of Baja California.* University Press of Florida, Gainesville.

Dawson, E. Y. 1944. The marine algae of the Gulf of California. *Allan Hancock Pacific Expeditions* 3: i–v, 189–454, plates 31–77.

DeLisle, D. G. 1963. Taxonomy and distribution of the genus *Cenchrus. Iowa State Journal of Science* 37:259–351.

Desonie, D. L. 1992. Geologic and geochemical reconnaissance of Isla San Esteban: post-subduction orogenic volcanism in the Gulf of California. *Journal of Volcanology and Geothermal Research* 52:123–140.

Diario Oficial. 2002. Derecto por el que se declara área natural protegida con la categoría de reserva de la biosfera, la región denominada Isla San Pedro Mártir, ubicada en el Golfo de California, frente a las Hermosillo, Estado de Sonora, con una superficie total de 30,165-23-76.165 hectáreas. Secretaría de Gobernación, Mexico, DF. http://www.conanp.gob.mx/sig/decretos/reservas/Sanpedromartirisla.pdf (accessed 3 June 2011).

Diario Oficial. 2011. Acuerdo por el que se da a conocer el resumen del Programa de Manejo de la Reserva de la Biosfera Isla San Pedro Mártir. Secretaría de Gobernación, Mexico, DF. http://www.dof.gob.mx/nota_detalle.php?codigo=5176266&fecha=01/02/2011 (accessed 12 June 2011).

Díaz, J. S., A. E. Meling-López, & D. C. Escobedo-Urías. 2008. Vegetación y flora de las islas de las lagunas Navachiste y Macapule, norte de Sinaloa. Pages 51–62 *in* L. M. Flores-Campaña, editor, *Estudios de las Islas del Golfo de California.* Universidad Autónoma de Sinaloa, Culiacán, Sinaloa.

Duffy, D. E. 1990. On continuity-corrected residuals in logistic regression. *Biometrika* 77:287–293.

Emmelin, N., & W. Feldberg. 1949. Distribution of acetylcholine and histamine in nettle plants. *New Phytologist* 48:143–148.

Epling, C. 1949. Revisión del género *Hyptis* (Labiatae). *Revisrta Museo La Plata, New Series (Sección Botanica)* 7(30):153–497.

Ezcurra, E., L. Bourillón, A. Cantú, M. Elena Martínez, & A. Robles. 2002. Ecological Conservation. Pages 417–444 *in* T. J. Case, M. L. Cody, & E. Ezcurra. *A New Island Biogeography of the Sea of Cortés.* Oxford University Press, New York.

Fehlberg, S. D., & T. A. Ranker. 2009. Evolutionary history and phylogeography of *Encelia farinosa* (Asteraceae) from the Sonoran, Mojave, and Peninsular Deserts. *Molecular Phylogenetics and Evolution* 50:326–335.

Felger, R. S. 1966. *Ecology of the Islands and Gulf Coast of Sonora, Mexico.* PhD diss., University of Arizona, Tucson.

———. 1980. Vegetation and flora of the Gran Desierto, Sonora, Mexico. *Desert Plants* 2:87–114.

———. 1999. The flora of Cañón del Nacapule: a desert-bounded tropical canyon near Guaymas, Sonora, Mexico. *Proceedings of the San Diego Society of Natural History* 35:1–42.

———. 2000a. *Flora of the Gran Desierto and Río Colorado of Northwestern Mexico.* University of Arizona Press, Tucson.

———. 2000b. The Seris and the guy who cuts the tops off plants. *Journal of the Southwest* 42:521–543.

———. 2002. Sinaloa shootout. Pages 2–14 *in* T. Bowen, editor, *Backcountry Pilot: Flying Adventures with Ike Russell.* University of Arizona Press, Tucson.

———. 2004. Seed plants. Pages 147–163 *in* R. C. Brusca, E. Kimrey, & W. Moore, editors, *A Seashore Guide to the Northern Gulf of California.* Arizona-Sonora Desert Museum, Tucson.

———. 2007. Living resources at the center of the Sonoran Desert: Native American plant and animal utilization. Pages 147–192 *in* R. S. Felger & B. Broyles, editors, *Dry Borders: Great Natural Reserves of the Sonoran Desert.* University of Utah Press, Salt Lake City.

Felger, R. S., & J. Henrickson. 1997. Convergent adaptive morphology of a Sonoran Desert cactus (*Peniocereus striatus*) and an African spurge (*Euphorbia cryptospinosa*). *Haseltonia* 5:77–85.

Felger, R. S., & E. Joyal. 1999. The palms (Arecaceae) of Sonora, Mexico. *Aliso* 18:1–18.

Felger, R. S., & C. H. Lowe. 1967. Clinal variation in the surface-volume relationships of the columnar cactus *Lophocereus schottii* in northwestern Mexico. *Ecology* 48:530–536.

———. 1976. The island and coastal vegetation and flora of the northern part of Gulf of California. *Natural History Museum of Los Angeles County, Contributions in Science* 285:1–59.

Felger, R. S., & M. B. Moser. 1970. Seri use of *Agave. Kiva* 35:159–167.

———. 1973. Eelgrass (*Zostera marina* L.) in the Gulf of California: discovery of its nutritional value by the Seri Indians. *Science* 181:355–356.

———. 1974. Columnar cacti in Seri culture. *Kiva* 39:257–275.

———. 1985. *People of the Desert and Sea: Ethnobotany of the Seri Indians.* University of Arizona Press, Tucson. Reprinted 1991.

———. 1987. Sea turtles in Seri Indian culture. *Environment Southwest* 519:18–21.

Felger, R. S., and B. T. Wilder, in collaboration with H. Romero-Morales. 2012. *Plant Life of a Desert Archipelago: Flora of the Sonoran Islands in the Gulf of California.* University of Arizona Press, Tucson.

Felger, R. S., & A. Zimmerman. 2000. Cactaceae. Pages 194–226 *in* R. S. Felger, *Flora of the Gran Desierto and Río Colorado of Northwestern Mexico.* University of Arizona Press, Tucson.

Felger, R. S., K. Cliffton, & P. J. Regal. 1976. Winter dormancy in sea turtles: independent discovery and exploitation in the Gulf of California by two local cultures. *Science* 191:283–285.

Felger, R. S., M. B. Moser, & E. W. Moser. 1980. Seagrasses in Seri Indian culture. Pages 260–276 *in* R. C. Phillips & C. P. McRoy, editors, *Handbook of Seagrass Biology: An Ecosystem Perspective.* Garland STPM Press, New York.

Felger, R. S., M. B. Johnson, & M. F. Wilson. 2001. *Trees of Sonora, Mexico.* Oxford University Press, New York.

Felger, R. S., W. J. Nichols, & J. A. Seminoff. 2005. Sea turtle conservation, diversity and desperation in northwestern Mexico. Pages 405–424 *in* J.-L. E. Cartron, G. Ceballos, & R. S. Felger, editors, *Biodiversity, Ecosystems, and Conservation in Northern Mexico.* Oxford University Press, New York.

Felger, R. S., B. Broyles, M. F. Wilson, G. P. Nabhan, & D. S. Turner. 2007a. Six grand reserves, one grand desert. Pages 3–26 *in* R. S. Felger & B. Broyles, editors, *Dry Borders: Great Natural Reserves of the Sonoran Desert.* University of Utah Press, Salt Lake City.

Felger, R. S., M. Wilson, K. Mauz, & S. Rutman. 2007b. Botanical diversity of southwestern Arizona and northwestern Sonora. Pages 202–271 *in* Felger & Broyles, editors, *Dry Borders: Great Natural Reserves of the Sonoran Desert.* University of Utah Press, Salt Lake City.

Felger, R. S., B. T. Wilder, & J. P. Gallo-Reynoso. 2011. Floristic diversity and long-term vegetation dynamics of Isla San Pedro Nolasco, Gulf of California, Mexico. *Proceedings, San Diego Natural History Museum* 43:1–42.

Fishbein, M., D. Chuba, C. Ellison, R. J. Mason-Gamer, & S. P. Lynch. 2011. Phylogenetic relationships of *Asclepias* (Apocynaceae) inferred from non-coding chloroplast DNA sequences. *Systematic Botany* 36:1008–1023.

Fleming, T. H. 2002. Pollination biology of four species of Sonoran Desert columnar cacti. Pages 207–224 *in* T. H. Fleming & A. Valiente-Banuet, editors, *Columnar Cacti and Their Mutualists: Evolution, Ecology and Conservation.* University of Arizona Press, Tuscon.

Fleming, T. H., & A. Valiente-Banuet, editors. 2002. *Columnar Cacti and Their Mutualists: Evolution, Ecology and Conservation.* University of Arizona Press, Tuscon.

Fontana, H. M., & B. L. Fontana. 2000. *Trails to Tiburón, the 1894 and 1895 Field Diaries of WJ McGee.* University of Arizona Press, Tucson.

Franklin, K. A., K. Lyons, P. L. Nagler, D. Lampkin, E. P. Glenn, F. Molina-Freaner, T. Markow, & A. R. Huete. 2006. Buffelgrass (*Pennisetum ciliare*) land conversion and productivity in the Plains of Sonora, Mexico. *Biological Conservation* 127:62–71.

Fryxell, P. A. 1988. Malvaceae of Mexico. *Systematic Botany Monographs* 25:1–522.

Gastil, G. R., J. Neuhaus, M. Cassidy, J. T. Smith, J. C. Ingle Jr., & D. Krummenacher. 1999. Geology and paleontology of southwestern Isla Tiburón, Sonora, Mexico. *Revista Mexicana de Ciencias Geológicas* 16(1):1–34.

Gallo-Reynoso, J. P., R. S. Felger, & B. T. Wilder. 2012. Near colonization of a desert island by a tropical bird: military macaw (*Ara militaris*) at Isla San Pedro Nolasco, Sonora, Mexico. *Southwest Naturalist* 57 (4): (in press).

Gentry, H. S. 1942. *Rio Mayo Plants.* Carnegie Institution of Washington Publication no. 527. Carnegie Institution, Washington, DC.

———. 1949. *Land Plants Collected by the* Vallero III, *Allan Hancock Pacific Expeditions 1937–1951.* Allan Hancock Pacific Expeditions 13. University of Southern California Press, Los Angeles.

———. 1950. Taxonomy and evolution of *Vaseyanthus. Madroño* 10:142–155.

———. 1982. *Agaves of Continental North America.* University of Arizona Press, Tucson.

Giral, F. 1994. *Ciencia Española en el Exilio (1939–1989): El Exilio de los Científicos Españoles.* Centro de Investigación y Estudios Republicanos, Barcelona.

Glass, C., & R. Foster. 1975. *Mammillaria tayloriorum,* a new species from San Pedro Nolasco Island. *Cactus and Succulent Journal (US)* 47:173–176.

Goldman, D. H. 2003. Two species of *Passiflora* (Passifloraceae) in the Sonoran Desert and vicinity: a new taxonomic combination and an introduced species in Arizona. *Madroño* 50:243–264.

Goss, N. S. 1888. New and rare birds found breeding on San Pedro Martir Isle. *Auk* 5:240–244.

Gould, F. W., & R. Moran. 1981. *The Grasses of Baja California, Mexico*. San Diego Society of Natural History Memoir 12. San Diego Society of Natural History, San Diego, CA.

Grieco, P. A., J. Haddad, M. M. Piñeiro-Núñez, & J. C. Huffman. 1999. Non-quassinoid constituents from the twigs and thorns of *Castela polyandra*. *Phytochemistry* 51:575–578.

Grismer, L. L. 1999. Checklist of amphibians and reptiles on islands in the Gulf of California, Mexico. *Bulletin of the Southern California Academy of Sciences* 98(1):45–56.

———. 2002. *Amphibians and Reptiles of Baja California, Including Its Pacific Islands and the Islands in the Sea of Cortés*. University of California Press, Berkeley.

Hardy, R. W. H. 1829. *Travels in the Interior of Mexico in 1825, 1826, 1827, and 1828*. Colburn and Bently, London.

Hastings, J. R. 1963. *Historical Changes in the Vegetation of a Desert Region*. PhD diss., University of Arizona, Tucson.

Hastings, J. R., & R. M. Turner. 1965. *The Changing Mile: An Ecological Study of Vegetation Change with Time in the Lower Mile of an Arid and Semiarid Region*. University of Arizona Press, Tucson.

Hausback, B. P. 1984. Cenozoic volcanic and tectonic evolution of Baja California Sur, Mexico. Pages 219–236 *in* V. A. Frizzell, editor, *Geology of the Baja Peninsula: Pacific Section*. Society of Economic Paleontologists and Mineralogists Special Paper 39.

Helenes, J., & A. L. Carreño. 1999. Neogene sedimentary evolution of Baja California in relation to regional tectonics. *Journal of South American Earth Sciences* 12:589–605.

Helenes, J., A. L. Carreño, & R. M. Carrillo. 2009. Middle to late Miocene chronostratigraphy and development of the northern Gulf of California. *Marine Micropaleontology* 72(1–2):10–25.

Henderson, A., G. Galeano, & R. Bernal. 1995. *Field Guide to the Palms of the Americas*. Princeton University Press, Princeton.

Henrickson, J. 1972. A taxonomic revision of the Fouquieriaceae. *Aliso* 7:439–537.

Hitchcock, C. L. 1951. *Manual of the Grasses of the United States*. 2nd ed., revised by A. Chase. USDA Miscellaneous Publication 200. Government Printing Office, Washington, DC.

Humphrey, R. R. 1974. Fire in the desert and desert grassland of North America. Pages 365–400 *in* T. T. Kozlowski & C. E. Ahlgren, editors, *Fire and Ecosystems*. Academic Press, New York.

Humphrey, R. R., A. L. Brown, & A. C. Everson. 1958. Arizona range grasses. *University of Arizona, Agricultural Experiment Station Bulletin* 298:1–104.

Hunt, D. R. 1998. *Mammillaria dioica* subsp. *estebanensis* (G.E. Linds.) D.R. Hunt. Mammillaria Postscripts 7:3 (England).

Index Herbariorum. 2009. *Index Herbariorum: A Global Directory of Public Herbaria and Associated Staff*. http://sciweb.nybg.org/science2/IndexHerbariorum.asp (accessed 8 August 2010).

In memoriam, Irving W. Knobloch. 2000. *American Association of Plant Taxonomists Newsletter* 14(1). http://www.inhs.uiuc.edu/~kenr/ASPT/newsletter14_1.html (accessed 1 August 2009).

International Plant Names Index. 2011. http://www.ipni.org/ (accessed 15 June 2011).

Johnson, A. F. 1978. A new subspecies of *Abronia maritima* from Baja California, Mexico. *Madroño* 25:224–227.

Johnston, I. M. 1924. Expedition of the California Academy of Sciences to the Gulf of California in 1921: the botany (vascular plants). *Proceedings of the California Academy of Sciences*, IV, 12:951–1218.

———. 1940. New phanerogams from Mexico, III. *Journal of the Arnold Arboretum* 21:253–265.

Knobloch, I. W. 1983. A preliminary verified list of plant collectors in Mexico. *Phytologia Memoirs* 4 (Plainfield, NJ).

Knobloch, I. W., & D. S. Correll. 1962. *Ferns and Fern Allies of Chihuahua, Mexico*. Texas Research Foundation, Renner.

Kroeber, A. L. 1931. The Seri. *Southwest Museum Papers* 6:1–60.

Laegaard, S., & P. M. Peterson, editors. 2001. *Gramineae (Part 2): Subfam. Chloridoideae*. Flora of Ecuador, vol. 68. Botanical Institute, University of Göteborg, Göteborg, and Section for Botany, Riksmuseum, Stockholm.

Lambeck, K., & J. Chappell. 2001. Sea level change through the last glacial cycle. *Science* 292:679–686.

Lawler, T. E., D. J. Hafner, P. Stapp, B. R. Riddle, & S. T. Alvarez-Casteñeda. 2002. The mammals. Pages 326–361 *in* T. J. Case, M. L. Cody, & E. Ezcurra, editors, *A New Island Biogeography of the Sea of Cortés*. Oxford University Press, New York.

Lee, R. M., & E. E. López-Saavedra. 1994. A second helicopter survey of desert bighorn sheep in Sonora, Mexico. *Transactions of the Desert Bighorn Council* 38:12–13.

Levin, G. A., & R. Moran. 1989. *The Vascular Flora of Isla Socorro, Mexico*. San Diego Society of Natural History Memoir 16. San Diego Society of Natural History, San Diego, CA.

Lindsay, G. E. 1947. Cacti of San Pedro Nolasco Island. *Desert Plant Life* 19:71–76.

———. 1948. A cruise in the Gulf of California, part III. *Cactus and Succulent Journal (US)* 20(2):17–20.

———. 1955a. *Notes Concerning the Botanical Explorers and Expedition of Lower California, Mexico*. Belvedere Scientific Fund [Library, California Academy of Science, San Francisco].

———. 1955b. *The Taxonomy and Ecology of the Genus Ferocactus*. PhD diss., Stanford University, Stanford.

———. 1962. A short trip to Sonora. *Cactus and Succulent Journal (US)* 34(3):115–122.

———. 1966. The Gulf Islands Expedition of 1966. *Proceedings of the California Academy of Sciences, IV,* 30:309–355.

Lindsay, G., et al. 1996. *The Taxonomy and Ecology of the Genus Ferocactus, Exploration in the USA and Mexico, etc.* Tireless Termite Press, USA.

Lindsay, G. E., & I. Engstrand. 2002. History of scientific explorations in the Sea of Cortés. Pages 3–13 *in* T. J. Case, M. L. Cody, & E. Ezcurra, editors, *A New Island Biogeography of the Sea of Cortés*. Oxford University Press, New York.

Lopez-Portillo, J., & E. Ezcurra. 1989. Zonation in mangrove and salt marsh vegetation at Laguna de Mecoacan, Mexico. *Biotropica* 21:107–114.

Lowe, C. H. 1955. An evolutionary study of island faunas in the Gulf of California, Mexico, with a method for comparative analysis. *Evolution* 9:339–344.

Luque-Agraz, D., & A. Robles-Torres. 2006. Naturalezas, Saberes y Territorios Comcáac (Seri). INE-SEMARNAT, México, DF.

MacArthur, R. H., & E. O. Wilson. 1967. *Theory of Island Biogeography*. Princeton University Press, Princeton.

MacDougal, D. T., & W. A. Cannon. 1910. The Conditions of Parasitism in Plants. Carnegie Institution of Washington Publication no. 129. Carnegie Institution, Washington, D.C.

Marlett, C. M. 2000. A Desemboque childhood. *Journal of the Southwest* 42:411–426.

———. 2002. A good day for playing hooky. Pages 26–29 *in* T. Bowen, editor, *Backcountry Pilot: Flying Adventures with Ike Russell*. University of Arizona Press, Tucson.

Martin, P. S., D. Yetman, M. Fishbein, P. Jenkins, T. R. Van Devender, & R. K. Wilson. 1998. *Gentry's Río Mayo Plants: The Tropical Deciduous Forest and Environs of Northwest Mexico*. University of Arizona Press, Tucson.

Martín-Bravo, S, P. Vargas, & M. Luceño. 2009. Is *Oligomeris* (Resedaceae) indigenous to North America? Molecular evidence for a natural colonization from the Old World. *American Journal of Botany* 96:507–518.

Martínez-Yrízar, A., R. S. Felger, & A. Búrquez. 2010. Los ecosistemas terrestres de Sonora: un diverso capital natural. Pages 129–156 *in* F. Molina-Freanar & T. R. Van Devender, editors, *Diversidad Biológica del estado de Sonora*. Universidad Nacional Autónoma de México, Hermosillo.

Maya, J. A. 1968. *The Natural History of the Fish-Eating Bat, Pizonyx vivesi*. PhD diss., University of Arizona, Tucson.

McGee, W. J. 1898. *The Seri Indians*. Pages 1–344 *in Seventeenth Annual Report*, part 1. Bureau of American Ethnology, Washington, DC.

McLaughlin, S. P., & J. E. Bowers. 1982. Effects of wildfire on a Sonoran Desert plant community. *Ecology* 63:246–248.

McMillan, C., & R. Phillips. 1979. *Halodule wrightii* Aschers in the Sea of Cortez, Mexico. *Aquatic Botany* 6:393–396.

McVaugh, R. 1956. *Edward Palmer, Plant Explorer of the American West*. University of Oklahoma Press, Norman.

———. 1983. *Flora Novo-Galiciana*, vol. 14, *Gramineae*. University of Michigan Press, Ann Arbor.

Medel-Narvaez, A., J. L. León de la Luz, F. Freaner-Martínez, & F. Molina-Freaner. 2006. Patterns of abundance and population structure of *Pachycereus pringlei* (Cactaceae), a columnar cactus of the Sonoran Desert. *Vegetation* 187:1–14.

Medellín, R. A., C. Manterola, M. Valdéz, D. G. Hewitt, D. Doan-Crider, & T. E. Fulbright. 2005. History, ecology, and conservation of the pronghorn antelope, bighorn sheep, and black bear in Mexico. Pages 387–404 *in* J.-L. E. Cartron, G. Ceballos, & R. S. Felger, editors, *Biodiversity, Ecosystems, and Conservation in Northern Mexico*. Oxford University Press, New York.

Meling-López, A. E., & S. E. Ibarra-Obando. 1999. Annual life cycles of two *Zostera marina* L. populations in the Gulf of California: contrasts in seasonality and reproductive effort. *Aquatic Botany* 65:59–69.

Mellink, E. 2002. Invasive vertebrates on islands of the Sea of Cortés. Pages 112–125 *in* B. Tellman, editor, *Invasive Exotic Species in the Sonoran Region*. University of Arizona Press, Tucson.

Mitich, L. W. 1993. Cacti, shells, and music—the Charles Glass story. *Cactus and Succulent Journal (US)* 65:3–11.

———. 2000. Comments—Dudley Gold, beloved cactophile. *Cactus and Succulent Society of America Newsletter* 72:71–72.

Molina-Freaner, F. E., & T. R. Van Devender. 2009. *Diversidad Biológica del Estado de Sonora*. Universidad Nacional Autónoma de México, Hermosillo.

Montoya, B., & G. Gates. 1975. Bighorn capture and transplant in Mexico. *Transactions of the Desert Bighorn Council* 19:28–32.

Mooney, H., & W. A. Emboden. 1968. The relationship of terpine composition, morphology, and distribution of populations of *Bursera microphylla* (Burseraceae). *Brittonia* 20:44–51.

Moran, R. 1936–1993. *The Field Notes of Reid Moran, the Flora of Baja California, San Diego Natural History Museum*. http://bajaflora.org/MoranNotesSearch.aspx (accessed 8 September 2009). Also on file, Botany Department, San Diego Natural History Museum.

———. 1983a. The vascular flora of Isla Ángel de la Guarda. Pages 382–403 *in* T. J. Case & M. L. Cody, editors, *Island Biogeography of the Sea of Cortéz*. University of California Press, Berkeley.

——. 1983b. Vascular plants of the Gulf Islands. Pages 348–381 *in* T. J. Case & M. L. Cody, editors, *Island Biogeography of the Sea of Cortéz*. University of California Press, Berkeley.

——. 1996. *The Flora of Guadalupe Island, Mexico*. Memoirs of the California Academy of Science 19, San Francisco.

——. 2009. Crassulaceae. Pages 147–229 *in* Flora of North America Editorial Committee, editors, *Flora of North America North of Mexico*, vol. 8. Oxford University Press, New York.

Moran, R., & R. S. Felger. 1968. *Castela polyandra*, a new species in a new section; union of *Holacantha* with *Castela* (Simaroubaceae). *Transactions of the San Diego Society of Natural History* 15(4):31–40.

Moran, R., & J. Rebman. 2002. Plants on some small Gulf Islands. Pages 527–534 *in* T. J. Case, M. L. Cody, & E. Ezcurra, editors, *A New Island Biogeography of the Sea of Cortés*. Oxford University Press, New York.

Moser, E. 1963. Seri Bands. *Kiva* 28:14–27.

Moser, M. B. 1970. Seri: From conception through infancy. *Kiva* 35: 201–210.

——. 1973. Seri basketry. *Kiva* 38:105–140.

Moser, M. B., & S. A. Marlett, compilers. 2005. *Comcáac quih yaza quih hant ihíip hac; Diccionario seri-español-inglés*. Editorial UniSon, Hermosillo & Plaza y Váldés Editores, México, DF.

——. 2010. *Comcaac quih yaza quih hant ihíip hac; Diccionario seri-español-inglés*, 2nd ed. Editorial UniSon, Hermosillo & Plaza y Váldés Editores, México, DF. http://www.sil.org/mexico/seri/G004c-DiccionarioEd2-sei.pdf (accessed 9 June 2012).

Murphy, R. W., & G. Aguierre-Léon. 2002. Distributional checklist of nonavian reptiles and amphibians on the islands in the Sea of Cortés. Pages 580–591 *in* T. J. Case, M. L. Cody, & E. Ezcurra, editors, *A New Island Biogeography of the Sea of Cortés*. Oxford University Press, New York.

Murphy, R. W., F. Sanchez-Peñero, G. A. Pollis, & R. L. Aalbu. 2002. New measurements of area and distance for islands in the Sea of Cortés. Pages 447–464 *in* T. J. Case, M. L. Cody, & E. Ezcurra, editors, *A New Island Biogeography of the Sea of Cortés*. Oxford University Press, New York.

Nabhan, G. P. 2000. Cultural dispersal of plants and reptiles to the Midriff Islands of the Sea of Cortés: integrating indigenous human dispersal agents into island biogeography. *Journal of the Southwest* 42:545–558.

——. 2002. Cultural dispersal of plants and reptiles. Pages 407–416 *in* T. J. Case, M. L. Cody, & E. Ezcurra, editors, *A New Island Biogeography of the Sea of Cortés*. Oxford University Press, New York.

——. 2003. *Singing the Turtles to Sea: The Comcáac (Seri) Art and Science of Reptiles*. University of California Press, Berkeley.

Nabhan, G. P., & R. S. Felger. 1978. Teparies in Southwestern North America. *Economic Botany* 32:2–19.

Nason, J. D., J. L. Hamrick, & T. H. Fleming. 2002. Historical vicariance and postglacial colonization effects on the evolution of genetic structure in *Lophocereus*, a Sonoran Desert columnar cactus. *Evolution* 56:2214–2226.

Navarro, F. 1999. Unidad de Manejo y Aprovechamiento Sustenable de la Vida Silvesrtre (UMA) en Isla Tiburón: un ejemplar de desarrollo sustenable. *Insulario (Gaceta Informativa de al Reserva islas del Golfo de California)* 8:13–15.

Navarro-Quezada, A., R. Gonzalez-Chauvet, F. Molina-Freaner, & L. E. Eguiarte. 2003. Genetic differentiation in the *Agave deserti* (Agavaceae) complex of the Sonoran desert. *Heredity* 90:220–227.

Nesom, G. L. 2010. Notes on *Fraxinus cuspidata* and *F. gooddingii* (Oleaceae). *Phytoneuron* 2010-38:1–14.

Ollendorf, A. L., S. C. Mulholland, & G. Rapp Jr. 1988. Phytolith analysis as a means of plant identification: *Arundo donax* and *Phragmites communis*. *Annals of Botany* 61:209–214.

Oskin, M., & J. Stock. 2003a. Cenozoic volcanism and tectonics of the continental margins of the Upper Delfin Basin, northeastern Baja California and western Sonora. Pages 421–438 *in* S. E. Johnson, S. R. Paterson, J. M. Fletcher, G. H. Girty, D. L. Kimbrough, & A. Martin-Barajas, editors, *Tectonic Evolution of Northwestern Mexico and the Southwestern USA*. Geological Society of America Special Paper 374. Geological Society of America, Boulder, CO.

——. 2003b. Marine incursion synchronous with plate-boundary localization in the Gulf of California. *Geology* 31:23–26.

Osorio-Tafall, B. F. 1944. La expedición del *M. N. Gracioso* por aguas del extremo Noroeste Mexicano. 1: Resumen General. *Anales de la Escuela Nacional de Ciencias Biológicas* (3–4):331–360. Secretaría de Educación Pública, Insituto Politécnico Nacional, México, DF.

——. 1946. Contribución al conocimiento del Mar de Cortés. *Boletín de la Sociedad Mexicana de Geografía y Estadístico* 62(1):89–130.

Palacios, R. A. 2006. Los mezquites mexicanos: biodiversidad y distribución geográfica. Boletín de la Sociedad Argentina de Botánica 41(1–2):99–121.

Paredes Aguilar, R., T. R. Van Devender, & R. S. Felger. 2000. *Las Cactáceas de Sonora: Su Diversidad, Uso y Conservatión*. Arizona-Sonora Desert Museum Press, Tucson.

Parfitt, B. D., & D. J. Pinkava. 1988. Nomenclatural and systematic reassessment of *Opuntia engelmannii* and *O. lindheimeri* (Cactaceae). *Madroño* 35:342–349.

Parker, K. F. 1958. *Arizona Ranch, Farm, and Garden Weeds*. Agricultural Extension Service Circular 265. University of Arizona, Tucson.

———. 1972. *An illustrated Guide to Arizona Weeds.* University of Arizona Press, Tucson.

Parsons, S. H. 1937. North America's most primitive savages. *Travel* 68(6):48–51, 66–67.

Payne, W. W. 1964. A re-evaluation of the genus *Ambrosia* (Compositae). *Journal of the Arnold Arboretum* 45:401–438.

Peltier, W. R. 2002. On eustatic sea level history: Last Glacial Maximum to Holocene. *Quaternary Science Reviews* 21:377–396.

Pennington, T. D. 1990. *Sapotaceae.* Flora Neotropica Monograph 52. New York Botanical Garden Press, Bronx, NY.

Pérez de Ribas, A. 1645. *Historia de los trivmphos de nvestra santa fee entre gentes las mas barbaras, y fieras del Nuevo orbe: A. de Paredes, Madrid.* Reprinted 1999. *History of the Triumphs of Our Holy Faith Amongst the Most Barbarous and Fierce Peoples of the New World.* Edited and translated by Daniel T. Reff, Marueen Ahern, & Richard K. Danford. University of Arizona Press, Tucson.

Peterson, P. M., S. L. Hatch, & A. S. Weakley. 2003. *Sporobolus* R. Br. Pages 115–139 *in* Flora of North America Editorial Committee, editors, *Flora of North America North of Mexico,* vol. 25. Oxford University Press, New York.

Piedra-Malagón, E. M., V. Sosa, & G. Ibarra-Manríquez. 2011. Clinal variation and species boundaries in the *Ficus petiolaris* complex (Moraceae). *Systematic Botany* 36:80–87.

Pohl, R. W. 1980. Family #15, Gramineae. Flora Costaricensis. *Fieldiana Botany, New Series* 4:1–595.

Pohl, R. W., C. G. Reeder, & G. Davidse. 1994. *Sporobolus* R. Br. Pages 273–276 *in* G. Davidse, M. Sousa S., & A. O. Chater, editors, Alismataceae a Cyperaceae. *Flora Mesoamerica,* vol. 6. Universidad Nacional Autónoma de México, México, DF.

Polis, G. A., S. D. Hurd, C. T. Jackson, & F. S. Pinero. 1997. El Niño effects on the dynamics and control of an island ecosystem in the Gulf of California. *Ecology* 78:1884–1897.

Polis, G. A., M. D. Rose, F. Sánchez-Piñero, P. T. Stapp, & W. B. Anderson. 2002. Island food webs. Pages 362–380 *in* T. J. Case, M. L. Cody, & E. Ezcurra, editors, *A New Island Biogeography of the Sea of Cortés.* Oxford University Press, New York.

Powell, A. M. 1974. Taxonomy of *Perityle* section *Perityle* (Compositae—Peritylinae). *Rhodora* 76:229–306.

Provance, M. C., I. García-Ruiz, & A. C. Sanders. 2008. The *Diospyros salicifolia* complex (Ebenaceae) in Mesoamerica. *Journal of the Botanical Research Institute of Texas* 2:1009–1100.

Quammen, D. 1996. *The Song of the Dodo: Island Biogeography in an Age of Extinctions.* Touchstone, New York.

Ramírez, R. M. A. 1996. *Revisión taxonómica de* Euphorbia *subgénero* Agaloma *sección* Alectoroctonum *(Euphorbiaceae) en México.* Tesis de maestria, Universidad Nacional Autónoma de México, Facultad de Ciéncias, México, DF.

Raven, P. H. 1969. A revision of the genus *Camissonia* (Onagraceae). *Contributions from the United States National Herbarium* 37:161–396.

Readdie, M. D., M. Ranelletti, & R. M. McCord. 2006. *Common Seaweeds of the Gulf of California.* Sea Challengers, Monterrey, CA.

Rebman, J. P. 1995. *Biosystematics of* Opuntia *subgenus* Cylindropuntia *(Cactaceae): The Chollas of Lower California, Mexico.* Ph.D diss., Arizona State University, Tempe.

———. 2002. Plants endemic to the Gulf Islands. Pages 540–544 *in* T. J. Case, M. L. Cody, & E. Ezcurra, editors, *A New Island Biogeography of the Sea of Cortés.* Oxford University Press, New York.

Rebman, J. P., J. L. León de la Luz, & R. V. Moran. 2002. Vascular plants of the Gulf Islands. Pages 465–510 *in* T. J. Case, M. L. Cody, & E. Ezcurra, editors, *A New Island Biogeography of the Sea of Cortés.* Oxford University Press, New York.

Reeder, J. R., & R. S. Felger. 1989. The *Aristida californica-glabrata* complex (Gramineae). *Madroño* 36:187–197.

Reichenbacher, F. W. 1990. *Tumamoc globeberry studies in Arizona and Sonora, Mexico.* Final report prepared for the U.S. Department of the Interior, Bureau of Reclamation, Phoenix, AZ.

Robertson, K. R. 1971. *A Revision of the Genus* Jacquemontia *(Convolvulaceae) in North* and *Central America* and *the West Indies.* PhD diss., Washington University, St. Louis.

Rominger, J. M. 1962. Taxonomy of *Setaria* (Gramineae) in North America. *Illinois Biological Monographs* 29:1–132.

———. 2003. *Setaria* P. Beauv. Pages 539–558 *in* Flora of North America Editorial Committee, editors, *Flora of North America North of Mexico,* vol. 25. Oxford University Press, New York.

Rose, M. D., & G. A. Polis. 2000. On the insularity of islands. *Ecography* 23:693–701.

Russell, D. 2002. Grace. Pages 155–165 *in* T. Bowen, editor, *Backcountry Pilot: Flying Adventures with Ike Russell.* University of Arizona Press, Tucson.

Saltonstall, K., & D. Hauber. 2007. Notes on *Phragmites australis* (Poaceae: Arundinoideae) in North America. *Journal of the Botanical Research Institute of Texas* 1:385–388.

Sánchez-del Pino, I., & T. J. Motley. 2010. Evolution of *Tidestromia* (Amaranthaceae) in the deserts of the southwestern United States and Mexico. *Taxon* 59: 38–48.

Schilling, E. E. 1990. Taxonomic revision of *Viguiera* subgenus *Bahiopsis* (Asteraceae: Heliantheae). *Madroño* 37:149–170.

Schwarzbach, A. E., & J. W. Kadereit. 1999. Phylogeny of prickly poppies, *Argemone* (Papaveraceae), and the evolution of morphological and alkaloid characters based on ITS rDNA sequence variation. *Plant Systematics and Evolution* 218:257–279.

Seigler, D. S., & E. Wollenweber. 1983. Chemical variation in *Notholaena standleyi*. *American Journal of Botany* 70:790–798.

Severson, S. 2007. Alfred Frank Whiting, edited by M. L. Voelker. http://www.mnsu.edu/emuseum/information/biography/uvwxyz/whiting_alfred.html (accessed 4 August 2009).

Sheridan, T. E. 1999. *Empire of Sand: The Seri Indians and the Struggle for Spanish Sonora, 1645–1803.* The University of Arizona Press, Tucson.

Shreve, F. 1915. *The Vegetation of a Desert Mountain Range as Conditioned by Climatic Factors.* Carnegie Institution of Washington Publication no. 217. Carnegie Institution, Washington, DC.

———. 1951. *Vegetation of the Sonoran Desert.* Carnegie Institution of Washington Publication no. 591. Carnegie Institution, Washington, DC.

Shreve, F., & I. L. Wiggins. 1964. *Flora and Vegetation of the Sonoran Desert,* 2 vols. Stanford University Press, Stanford.

Slevin, J. R. 1923. Expedition of the California Academy of Sciences to the Gulf of California in 1921. *Proceedings of the California Academy of Sciences,* IV, 12:55–72.

Smith, E. B. 1985. A new species of *Coreocarpus* (Compositae: Heliantheae) from San Pedro Nolasco Island, Mexico. *American Journal of Botany* 72:626–628.

Smithsonian Institution Archives. 2011. Record Unit 7097, E. Yale Dawson Paper, 1934–1966. http://siarchives.si.edu/findingaids/FARU7097.htm (accesssed 5 January 2011).

Spellenberg, R. 2000. Blooming "behavior" in five species of *Boerhavia* (Nyctaginaceae). *Sida* 19:311–323.

———. 2003. *Boerhavia.* Pages 17–28 in Flora of North America Editorial Committee, editors, *Flora of North America North of Mexico,* vol. 4. Oxford University Press, New York.

———. 2007. *Boerhavia triquetra* S. Watson var. *intermedia* (Nyctaginaceae): a new combination and varietal status for the widespread Southwestern North American *B. intermedia. Journal of the Botanical Research Institute of Texas* 1:871–874.

Spicer, E. H. 1980. *The Yaquis: A Cultural History.* University of Arizona Press, Tucson.

Steinmann, V. W., & R. S. Felger. 1997. The Euphorbiaceae of Sonora, Mexico. *Aliso* 16:1–71.

Stevens, P. F. 2008. Angiosperm Phylogeny Website, version 9, 2001 onward. http://www.mobot.org/MOBOT/research/APweb/ (accessed 1 May 2011).

Summer Institute of Linguistics in Mexico. 2008. Edward W. Moser. Instituto Lingüístico de Verano, A.C. http://www.sil.org/mexico/bio/imosered.htm (accessed 12 May 2011).

Sylber, C. K. 1988. Feeding habits of the lizards *Sauromalus varius* and *S. hispidus* in the Gulf of California. *Journal of Herpetology* 22:413–424.

Taylor, N. P. 1984. A review of *Ferocactus* Britton & Rose. *Bradleya* 2:19–38.

———. 1985. *The Genus* Echinocereus. Timber Press, Portland, OR.

Tellman, B., editor. 2002. *Invasive Exotic Species in the Sonoran Region.* University of Arizona Press, Tucson.

Tershy, B. R., & D. Breese. 1997. The birds of San Pedro Mártir Island, Gulf of California, Mexico. *Western Birds* 28:96–107.

Tershy, B. R., D. Breese, A. Angeles-P., M. Cervantes-A., M. Mandujano-H., E. Hernández-N. & A. Córdoba-A. 1992. Natural History and Management of Isla San Pedro Mártir. Unpublished Report, Conservation International-México, 59-A Col. Miramar, Guaymas, Sonora, México 85450. 83 pages.

Tershy, B. R., D. Breese, & D. A. Croll. 1997. Human perturbations and conservation strategies for San Pedro Mártir Island, Islas de Golfo de California Reserve, México. *Environmental Conservation* 24:261–270.

Tershy, B. R., L. Bourillión, L. Metzler, & J. Barnes. 1999. A survey of ecotourism on islands in northwestern México. *Environmental Conservation* 26:212–217.

Thurston, E. L., & N. R. Lersten. 1969. The morphology and toxicology of plant stinging hairs. *Botanical Review* 35: 393–412.

Tomlinson, P. B. 1986. *The Botany of Mangroves.* Cambridge University Press, Cambridge, UK.

Torre-Cosío, J. 2002. *Inventory, Monitoring and Impact Assessment of Marine Biodiversity in the Seri Indian Territory, Gulf of California, Mexico.* PhD diss., University of Arizona, Tucson.

Townsend, C. H. 1916. Voyage of the "Albatross" to the Gulf of California in 1911. *Bulletin of the American Museum of Natural History* 35:399–476.

Tropicos 2011. Tropicos.org. Missouri Botanical Garden. http://www.tropicos.org (last accessed 1 June 2011).

Tucker, G. C. 1994. Revision of the Mexican species of *Cyperus* (Cyperaceae). *Systematic Botany Monographs* 432:1–212.

Turnage, W. V., & A. L. Hinckley. 1938. Freezing weather in relation to plant distribution in the Sonoran Desert. *Ecological Monographs* 8:529–549.

Turner, B. L., & M. I. Morris. 1975. New taxa of *Palafoxia* (Asteraceae: Helenieae). *Madroño* 23:79–80.

Turner, R. M. 2007. Confessions of a repeat photographer. Pages 50–57 in R. S. Felger & B. Broyles, editors, *Dry Borders: Great Natural Reserves of the Sonoran Desert.* University of Utah Press, Salt Lake City.

Turner, R. M., J. E. Bowers, & T. L. Burgess. 1995. *Sonoran Desert Plants: An Ecological Atlas.* University of Arizona Press, Tucson.

Turner, R. M., R. H. Webb, J. E. Bowers, & J. R. Hastings. 2003. *The Changing Mile Revisited.* University of Arizona Press, Tucson.

Uphof, J. C. T. 1968. *Dictionary of Economic Plants,* 2nd ed. J. Cramer, Lehre, Germany.

Van Devender, T. R. 1990. Late Quaternary vegetation and climate of the Sonoran Desert, United States and Mexico. Pages 134–165 *in* J. L. Betancourt, T. R. Van Devender, & P. S. Martin, editors, *Packrat Middens: The Last 40,000 Years of Biotic Change.* University of Arizona Press, Tucson.

Van Devender, T. R., T. L. Burgess, R. S. Felger, & R. M. Turner. 1990a. Holocene vegetation of the Hornaday Mountains of northwestern Sonora, Mexico. *Proceedings of the San Diego Society of Natural History* 2:1–19.

Van Devender, T. R., L. J. Toolin, & T. L. Burgess. 1990b. The ecology and paleoecology of grasses in selected Sonoran Desert plant communities. Pages 326–49 *in* J. L. Betancourt, T. R. Van Devender, & P. S. Martin, editors, *Packrat Middens: The Last 40,000 Years of Biotic Change.* University of Arizona Press, Tucson.

Van Devender, T. R., T. L. Burgess, J. C. Piper, & R. M. Turner. 1994. Paleoclimatic implications of Holocene plant remains from the Sierra Bacha, Sonora, Mexico. *Quarternary Research* 41:99–108

Van Devender, T. R., R. S. Felger, A. L. Reina-Guerrero, & J. J. Sánchez-Escalante. 2009. Sonora: non-native and invasive plants. Pages 85–124 *in* T. R. Van Devender, F. J. Espinosa-García, B. L. Harper-Lore, & T. Hubbard, editors, *Invasive Plants on the Move: Controlling Them in North America.* Arizona-Sonora Desert Museum, Tucson.

Van Devender, T. R., R. S. Felger, M. Fishbein, F. Molina-Freaner, J. J. Sánchez-Escalante, & A. L. Reina-Guerrero. 2010. Biodiversidad de las plantas vasculares. Pages 230–262 *in* F. Molina-Freaner & T. R. Van Devender, editors, *Diversidad Biológica del Estado de Sonora.* Universidad Nacional Autónoma de México, México, DF.

Vasey, G., & J. N. Rose. 1890. List of plants collected by Dr. Edward Palmer in Lower California and western Mexico in 1890. *Contributions from the United States National Herbarium* 1:63–90.

Velarde, E., & D. W. Anderson. 1994. Conservation and management of seabird islands in the Gulf of California: setbacks and successes. Pages 229–243 *in* D. N. Nettleship, J. Burger, & M. Gochfeld, editors, *Seabirds on Islands.* BirdLife Conservation Series no. 1. Birdlife International, Cambridge.

Wait, D. A., D. P. Aubrey, & W. B. Anderson. 2005. Seabird guano influences on desert islands: soil chemistry and herbaceous species richness and productivity. *Journal of Arid Environments* 60:681–695.

Watson, S. 1889. Upon a collection of plants made by Dr. E. Palmer in 1887 about Guaymas, etc. *Proceedings of the American Academy* 24:36–87.

Welsh, S. L., N. D. Atwood, S. Goodrich, & L. C. Higgins, editors. 2008. *A Utah Flora,* 4th ed. Brigham Young University Print Services, Provo, UT.

West, P., & G. P. Nabhan. 2002. Invasive plants: their occurrence and possible impact on the central Gulf Coast of Sonora and the Midriff Islands in the Sea of Cortés. Pages 91–111 *in* B. Tellman, editor, *Invasive Exotic Species in the Sonoran Region.* University of Arizona Press, Tucson.

West, P., J. P. Rebman, G. P. Polis, L. D. Humphrey, & R. S. Felger. 2002. Plants of small islands in Bahía de Los Angeles. Pages 535–39 *in* T. J. Case, M. L. Cody, & E. Ezcurra, editors, *A New Island Biogeography of the Sea of Cortés.* Oxford University Press, New York.

Whiting, A. F. 1939. *Ethnobotany of the Hopi.* Museum of Northern Arizona Bulletin no. 15. Museum of Northern Arizona, Flagstaff.

Wiggins, I. L. 1964. Flora of the Sonoran Desert. Pages 189–1740 *in* F. Shreve & I. L. Wiggins. *Vegetation and Flora of the Sonoran Desert.* Stanford University Press, Stanford, CA.

———. 1980. *Flora of Baja California.* Stanford University Press, Stanford.

Wilder, B. T., & R. S. Felger. 2010. Dwarf giants, guano, and isolation: vegetation and floristic diversity of Isla San Pedro Mártir, Gulf of California, Mexico. *Proceedings, San Diego Natural History Museum* 42:1–24.

Wilder, B. T., R. S. Felger, & H. Romero-Morales. 2007a. Succulents and bighorn of Isla Tiburón, Gulf of California. *The Plant Press (Arizona Native Plant Society)* 31(2):9–11.

Wilder, B. T., R. S. Felger, H. Romero-Morales, & A. Quijada-Mascareñas. 2007b. New plant records for the Sonoran Islands, Gulf of California, Mexico. *Journal of the Botanical Research Institute of Texas* 1:1203–1227.

Wilder, B. T., R. S. Felger, & H. Romero-Morales. 2008a. Succulent plant diversity of the Sonoran Islands, Gulf of California, Mexico. *Haseltonia* 14:127–160.

Wilder, B. T., R. S. Felger, T. R. Van Devender, & H. Romero-Morales. 2008b. *Canotia holacantha* on Isla Tiburón, Gulf of California, Mexico. *Canotia* 4(1):1–7.

Windham, M. D. 1993. *Notholaena.* Pages 143–149 *in* Flora of North America Editorial Committee, editors, *Flora of North America North of Mexico,* vol. 2. Oxford University Press, New York.

Wollenweber, E. 1984. Exudate flavonoids of Mexican ferns as chemotaxonomic markers. *Revista Latinoamericana de Quimica* 15(1):3–11.

Yatskievych, G. 1999. *Styermark's Flora of Missouri,* vol. 1. Missouri Department of Conservation, Jefferson City.

Yuncker, T. G. 1921. Revision of the North American and West Indian species of *Cuscuta. Illinois Biological Monographs* 6:91–231.

About the Authors

*R*ichard Felger received his Ph.D. at the University of Arizona in 1966. His dissertation analyzed the vegetation and flora of the islands and Gulf Coast of Sonora, Mexico. Subsequently he was on the faculty of the University of Colorado, Boulder, and then senior curator of botany at the Los Angeles County Museum of Natural History. Returning to Tucson, he continued his research and conservation activities in arid lands, concentrating on the Gulf of California and Sonoran Desert region. Working at the Arizona-Sonora Desert Museum from 1978 to 1982, he founded the research department there. He has been active in regional and international conservation, including pioneer conservation of sea turtles of the eastern Pacific primarily during the 1980s. In 1988 he founded the Drylands Institute in Tucson and was executive director until 2007. From 1980 to 2002 he was adjunct senior research scientist at the Environmental Research Laboratory, University of Arizona, and is currently research associate with Sky Island Alliance and Associated Researcher, University of Arizona Herbarium.

A leading botanical authority of the Sonoran Desert, Dr. Felger has close to a half-century of experience in this region. His classic work with Mary Beck Moser, *People of the Desert and Sea: Ethnobotany of the Seri Indians* (1985, 1991, University of Arizona Press) has preserved considerable indigenous knowledge of the Comcaac (Seris) and their homeland. The current publication marks the continuation of his contributions to the understanding of botanical diversity in this region.

Benjamin Wilder was born and raised in Tucson, Arizona, and has become a young desert rat. After testing the waters in Portland, Oregon, for two years while studying conservation biology at Lewis and Clark College, he returned to the Old Pueblo to earnestly begin discovering the desert. While completing his undergraduate degree at the University of Arizona's Department of Ecology and Evolutionary Biology, he also worked on the buffelgrass eradication and outreach project at the Desert Laboratory on Tumamoc Hill, at the University of Arizona Herbarium, and on the production of this work. In 2009 Wilder entered a PhD program in the Department of Botany and Plant Sciences at the University of California, Riverside, in the lab of Dr. Exequiel Ezcurra. His dissertation looks at the deep history of the flora of the Gulf of California island realm to interpret the unique botanical diversity documented in this book.

Humberto Romero is a descendant of the Comcaac who lived on Isla Tiburón. From an early age he learned from his mother the cultural significance and knowledge held in the Comcaac community of the desert plants of the Sonoran region of the Gulf of California. He is one of the most knowledgeable members of the community concerning Isla Tiburón and the distribution and uses of plants. He has worked with a number of scientists in the past several decades and has gained considerable knowledge from such opportunities. For the past decade he has been the head guide for the bighorn sheep hunting operation on Isla Tiburón.

Sr. Romero has collected significant herbarium specimens and is co-author on several recent botanical publications with Felger and Wilder dealing with Gulf islands. He is also in charge of efforts to eradicate non-native, invasive species on islands in the Gulf of California.

INDEX

Page numbers in **boldface** indicate the location of accepted scientific names in the species accounts (Part III). Scientific names for genera, species, and infraspecific taxa are *italicized*. Plant names in Cmiique Iitom (the language of the Comcaac, or the Seri people) are in Roman font. Spanish-language (Mexican) common name are *italicized* and English-language (American) common names are in Roman font. This index is the work of Mary Stofflet, Richard Felger, and Benjamin Wilder.

The Southwest Center Series

Joseph C. Wilder, Editor

Ignaz Pfefferkorn, *Sonora: A Description of the Province*

Carl Lumholtz, *New Trails in Mexico*

Buford Pickens, *The Missions of Northern Sonora: A 1935 Field Documentation*

Gary Paul Nabhan, editor, *Counting Sheep: Twenty Ways of Seeing Desert Bighorn*

Eileen Oktavec, *Answered Prayers: Miracles and Milagros along the Border*

Curtis M. Hinsley and David R. Wilcox, editors, *Frank Hamilton Cushing and the Hemenway Southwestern Archaeological Expedition, 1886–1889*, vol. 1: *The Southwest in the American Imagination: The Writings of Sylvester Baxter, 1881–1899*

Lawrence J. Taylor and Maeve Hickey, *The Road to Mexico*

Donna J. Guy and Thomas E. Sheridan, editors, *Contested Ground: Comparative Frontiers on the Northern and Southern Edges of the Spanish Empire*

Julian D. Hayden, *The Sierra Pinacate*

Paul S. Martin, David Yetman, Mark Fishbein, Phil Jenkins, Thomas R. Van Devender, and Rebecca K. Wilson, editors, *Gentry's Rio Mayo Plants: The Tropical Deciduous Forest and Environs of Northwest Mexico*

W J McGee, *Trails to Tiburón: The 1894 and 1895 Field Diaries of W J McGee*, transcribed by Hazel McFeely Fontana, annotated and with an introduction by Bernard L. Fontana

Richard Stephen Felger, *Flora of the Gran Desierto and Río Colorado of Northwestern Mexico*

Donald Bahr, editor, *O'odham Creation and Related Events: As Told to Ruth Benedict in 1927 in Prose, Oratory, and Song by the Pimas William Blackwater, Thomas Vanyiko, Clara Ahiel, William Stevens, Oliver Wellington, and Kisto*

Dan L. Fischer, *Early Southwest Ornithologists, 1528–1900*

Thomas Bowen, editor, *Backcountry Pilot: Flying Adventures with Ike Russell*